Statistical Analysis for Business Decisions

Statistical Analysis for Business Decisions

GARY E. MEEK

University of Akron

STEPHEN J. TURNER

Babson College

HOUGHTON MIFFLIN COMPANY BOSTON
Dallas Geneva, Illinois Hopewell, New Jersey Palo Alto London

DEDICATION

To Barbra, Courtney, Hilary, Jason, and Wyatt
Who far too often heard: "Later, please, Daddy's busy," during the course of this project. Perhaps now, later has come.

Cover design by Jill Haber.

Printed in the U.S.A.

Library of Congress Catalog Card Number: 82-80074

ISBN: 0-395-32274-X

Contents

Preface

With the ever-increasing emphasis on the scientific method, statistics has become an integral part of every discipline from music and library science to business, and on to medicine and engineering. This emphasis has led to a great deal of misuse of statistics and statistical procedures. Occasionally, this misuse has been intentional, but for the most part, it has occurred because the user was unaware of the assumptions underlying the procedures and the corresponding limitations imposed by these assumptions. The user often selected a procedure because it is the only one with which he is familiar, never realizing what assumptions he has made and never knowing that alternatives are available. We have attempted to keep these points foremost in our minds throughout the writing of this text.

GENERAL COMMENTS

This text provides a solid base for a one- or two-semester course in statistical analysis for use in a business environment. The basic material presupposes a background in introductory algebra and familiarity with set notation. For students having a higher level of mathematical sophistication, some topics and exercises that require basic calculus are included. The instructor may or may not elect to cover such material without loss of continuity. If the choice is to include this material, there is a calculus review in the Appendix. In some chapters, we have included appendices in order to provide the advanced student with more depth in a given area.

The text also can be used for a one-semester course with the omission of some topics. However, the omitted material can provide the student with an excellent reference in the future. We make no suggestions as to coverage; the individual instructor knows better than we the topics that he or she prefers to emphasize in a limited amount of time, based on the background of the students involved. At the University of Akron, only Chapters 6, 16, and 17 are omitted in the one semester MBA level course. These are omitted because the topics are covered in another course.

Our emphasis throughout the text is directed toward the student's understanding when to use a given procedure and toward knowing what its limitations are. To this end, we have tried to provide an intuitive derivation for most procedures and we encourage the student to think through each development. Occasionally, we provide a theoretical derivation to stress the application and importance of a particular concept in the overall picture, e.g., the importance of Cochran's theorem.

To reinforce the thinking aspect, we have constructed our Chapter Exercises on three levels. Exercises at the end of each section are primarily for giving the student practice in using the formulas and concepts of that section. The Level 1 exercises in the Supplementary Sets are similar in difficulty to those at the ends of the sections, but are positioned randomly to force the student to consider each problem individually. The Level 2 exercises are of a slightly higher level of difficulty with the majority involving minor variations of material in the chapter. These require a little more thought than the Level 1 problems and a few are simple derivations. The Level 3 exercises are meant to be thought provoking in nature and require a solid understanding of concepts. These last exercises either involve major variations on chapter material or more complicated derivations.

As a second method of stressing the assumptions and of aiding the student to assimilate the material, we have included a section of supplementary examples at the ends of Chapters 2 through 17. In addition to presenting real-world applications, these examples highlight caveats and often extend the material presented in the chapter itself.

Our development of the basic probability concepts uses the conventional set notation we feel is easier for the student to follow. A review of set notation is included in the Appendix even though most introductory mathematics courses cover it, and it is currently introduced at the junior high school level. We have attempted to maintain the usage of standard notation when such exists. Any deviations from standard are for purposes of clarity and easier sight recognition.

ORGANIZATION

In this text we have made some major changes which set it apart from the other introductory business statistics texts that are currently on the market. The first of these changes is the placement of the chapter on descriptive statistics. Most current texts start with this as the first chapter of substance and then drop the topic until the basic materials on probability and probability distributions have been covered. We have found that the traditional approach has two major drawbacks. First, it causes all but the exceptional student to confuse the formulas for sample data with the corresponding formulas that relate to the total populations. The second drawback is that the student not only wonders about the relevancy of the sample statistics but also forgets them by the time they are actually used. Having used this new placement for more than three years, we have found that students obtain a better understanding of the purpose of sampling and exhibit less confusion between sample and population values. This approach eliminates the need to cover sample measures twice.

The other major change in organization is the inclusion of classical procedures and their nonparametric alternatives in the same chapter. We also have included more nonparametric tests than can be found in some more traditional texts. Our reason for the expanded coverage is to provide, whenever possible, at least one alternative for each of the classical (normal-based) tests. In terms of positioning, to our knowledge, all current texts place the nonparametric material near the end of the book. Doing so fails to stress the importance of the assumptions underlying all statistical procedures. Thus, the student is not required to make a *complete* analysis of a problem or to make a decision between alternative procedures. We believe that our incorporation of classical and nonparametric tests in the same chapter will make the student more fully cognizant of the limitations imposed by the assumptions and much less likely to contribute to the misuse of statistics which is so prevalent today.

Other changes from the conventional coverage are:

1. discussing levels of measurement in Chapter 1 rather than in a nonparametric chapter only
2. including the test for adequacy (linearity) of the model in the simple linear regression chapter
3. providing more extensive coverage on analysis of variance and types of experimental designs than almost any other competing text. (Too many users of statistics are unaware of the changes required in a random effects model and of the very existence of nested designs.)

We believe that the student should be aware that all procedures require a minimum level of measurement *not just the nonparametric ones.* Therefore, we have emphasized this by stating the assumptions for *every* procedure. As for point 2, most textbooks at this level assume the linear model and go blithely on their way. This leads students to believe that all relations are linear so that they may never know that a simple procedure is available for testing that aspect of the model.

Acknowledgements

We would like to thank the reviewers who made many excellent and helpful suggestions. We have tried to incorporate as many of these suggestions as possible. We express our sincere thanks to: Guiseppi Forgionne, California State Polytechnic University at Pomona; Paul Kuzdrall, University of Western Ontario; John N. Moore, University of New Haven; George F. Rhodes, Jr., University of Colorado; Stanley R. Schultz, Cleveland State University; Justin Stolen, University of Nebraska, Omaha; and William D. Warde, Oklahoma State University.

Our special appreciation goes to: Paul N. Dixon of St. Mary's University, Halifax, N.S., who contributed time and helpful comments of his own free will; Po-Young (Jeff) Chu and Fu Tang Su who aided in some of the computer solutions to examples and exercises; to the students who made many suggestions in the developmental stage; to Debra Catanzarite, Carolyn Ford, and Julie Sweet without whose typing expertise all would have been for naught; and to Laura Jones for her continual support and patience.

Gary E. Meek
University of Akron

Stephen J. Turner
Babson College

Statistical Analysis for Business Decisions

Chapter 1
Introduction to Statistics

KEY WORDS AND TERMS

Statistics	Nominal Level of Measurement
Descriptive	Ordinal Level of Measurement
Inferential	Interval Level of Measurement
Probability	Ratio Level of Measurement

1.1 INTRODUCTION

This text is written for business students who need to know why statistics is used as part of the decision-making process in the business world today. If you are similar to most students with whom we have come in contact, you may dread the adventure that lies before you. Your trepidation is based in part upon hearsay from your peers and in part upon a premonition that "statistics" means "mathematics" and mathematics means trouble. If this description comes close to summing up your current state of mind, then the approach taken in this book is tailored for you. We hope that you can put aside some of your fear for the moment so that you can begin your study with an open mind.

As you might expect, the word *statistics* conveys many things to many people. Historically, until World War II the primary concern of statistics was the collection and description of numerical data (or the equivalent of numerical data). Its use as a decision-oriented tool up to this time was, in the main, restricted to gambling. This use was, of course, a function of the assimilation of "tendencies" over time as the cards, wheels, or dice were dealt, rotated, and rolled over and over again. The primary exception to this was the work of Sir Ronald A. Fisher, who worked toward a more structured design and analysis of agricultural experiments in England. It was not until after World War II that the business world began recognizing the value of sound, established procedures of data analysis. Business people began to realize that these procedures were a vast improvement over the "seat of the pants" decision that is possible under many circumstances that continue to prevail in business environments today. The current role of the procedure we call statistics is to compose a judgment, or inference (hence the term *inferential*), about a process or population (the universal total of all events of interest) on the basis of the examination of a small part (a sample) of the process or population. For our purposes, *statistics* refers to the overall approach, composed of

1. The planning of the process to gather reliable information efficiently
2. The execution of the plan outlined in step 1
3. A valid description of the information (data) gathered in step 2 and analysis of the results
4. Making a decision

1.2 THE PURPOSE OF STATISTICS

Most of us can recall at least one situation in which we took a gamble that resulted in an undesirable event. "Murphy's law" (if anything can go wrong, it will) and other similar platitudes reflect our dismay at the "natural" events that often result from well-intentioned plans. It may come as a surprise to you that, in fact, the purpose of statistics is to discover whether or not such undesirable events are due more to bad luck than to poor planning. Thus, the problem is to decide whether the resulting information or data could have occurred as a result of chance alone (based upon well-established assumptions), or whether the evidence indicates a variation resulting from the presence of a nonrandom element.

Statistical analysis is used in business to compare what happens with what one expects to happen under a given set of assumptions. The process then becomes a system composed of model construction, system feedback, and interpretation. That is, we make a few assumptions (formulate a model) that seem reasonable, look at what the process tells us is actually happening, then compare the results with the chance variation our model says should be present in the system. Essentially we are trying to determine the likelihood that the deviation between the actual observed events and those that were expected could be normally attributed to the chance variation indicated by our model. That is, we attempt to determine how likely the deviation is, assuming that only chance variation is present. Thus, the use of statistical analysis as a decision tool in the business world today includes the development and application of procedures for evaluating the likelihood of an event's occurring by chance *and* the identification of all necessary assumptions.

1.3 LEVELS OF MEASUREMENT

Perhaps the most fundamental assumption implicitly present in all statistical procedures is that dealing with levels of measurement. Although this assumption is summarily ignored by most textbooks, we consider it essential to the proper use of statistical analysis in business. There are four basic levels of measurement: nominal, ordinal, interval, and ratio. Since our choice of model (our assumptions) is dependent on the level of measurement present (or to be chosen), it is important to describe each of the four categories.

The first and lowest level of measurement possible requires us to be able to identify, or name, the objects in our sample from the process of interest. That is, we must be able to classify each result, or object, into one of possibly several distinct categories. The term *nominal* means that data satisfying *only* this level will imply nothing about possible relative positions or distances between any two observations. The sex of an individual, quality of an item (defective or nondefective), and voter preference are all examples of nominal-level information.

The *ordinal* level of measurement requires us not only to be able to categorize observations, but also to rank them on a list according to some attribute. Thus, the listing indicates the relative positions of the objects with respect to one another as indicated by use of the attribute in question. This will be possible in many cases even if no numbers are associated with the observations originally. Color brightness, mattress firmness, and lead hardness are examples of phenomena that could be rank-ordered and thus would qualify as being at the ordinal level of measurement—should sample values be collected. Note that if numerical values are assigned to the observations, these values can be used to identify relative position but not necessarily the distance, that is, the magnitude of the difference, between objects.

If we can assign to sample observations numerical values that identify both the relative positions of the observations and the magnitude of the difference between any two observations, then the sample values are said to be of an *interval* level of measurement. If weights are determined by scales or temperatures are determined by thermometers or sales volumes are measured in dollars of sales, each of these sets of measurements would be an example of values that are at an interval level of measurement. Notice that in each of these cases we assume an objective means of determining the values. With a "seat of the pants," guess-oriented procedure, the resultant values would necessarily be *ordinal*-level data.

Finally, you can notice whether or not an interval level of measurement utilizes a natural origin, that is, a universally accepted reference point, such as the weight of a kilogram or the length of a meter. Values determined on the basis of such standardized reference points or natural origins are classified as being of the *ratio* level of measurement.

The one point that cannot be overstressed is the fact that you must not be misled by the mere association of numerical values with the observations. In many situations the numbers do no more than identify a category. Thus it is imperative that you know exactly how the numbers were obtained. As you have probably surmised, data that satisfy one level of measurement will also satisfy any level that is less restrictive.

1.4 OBJECTIVES OF THE TEXT

It is important for you, the student, to realize how this book came into being. Basically, it is the production of two authors who had become dismayed at the lack of substance in many standard business statistics texts. Reduced to reliance on classroom notes, we recognized the need to present our time-tested techniques and examples to a greater audience for the good of all concerned. We also noted that the usual result of the "cookbook" approach was that the students passed the course mindlessly. They could not transfer their ability to do the problems worked out for them to problems that required them to think more creatively.

Thus, this text utilizes many valuable methods that have proven effective in our classes in the sense that they both encourage the students and build their confidence in their understanding of the basics of statistical applications used in business.

We have also tried to stress the assumptions underlying each procedure and to identify the ramifications of violating an assumption. We have done this because if you understand that there are assumptions, you are less likely to contribute to statistical "lies" and their attendant misconceptions. We also think this text will teach you both to be conversant in the language of statistics and to think in the logical pattern required for valid statistical analysis. The intent is not to make statisticians of you but to make you intelligent *users* of statistics.

This is an *introductory* text, and as such it surveys only the simpler and more commonly used statistical techniques. If nothing else, we hope that after using this text you will be better prepared to evaluate statistical arguments, to question results with confidence when assumptions are omitted, and to realize when the tools at your disposal are not sufficient to handle the problem at hand. Above all else, we hope you will *think* rather than use the "grab a formula and force the data to fit" approach. We all have heard the cliche that "a little knowledge is a dangerous thing." This appears to be more true in statistics than in any other field. We believe that when you have completed *this* text, the cliche will not apply to you.

1.5 RATIONALE FOR TOPIC ORDERING

For the most part our ordering of topical coverage in this text is fairly standard. There are three basic changes that we consider of major importance. The first of these is the placement of the chapter on descriptive statistics. Most textbooks start with descriptive procedures, then drop the topic for three or four chapters before using it. We have found that our students retain and understand the reason for calculating sample measures better when these measures are introduced and used at the same time. Thus we present descriptive procedures in Chapter 7, when we are ready to use them.

Since our main emphasis is on statistical inference and since all inferences are based on the likelihood of occurrence, we formally begin with probability. Probability is the measure of likelihood on which *all* statistical inferences are based. It is imperative that you have a basic understanding of what a probability is and how probabilities may be combined. Therefore, we have started with simple probabilities and tried to highlight their connection with everything else that we do.

The second major change that we have incorporated is the combining of the classical procedures and their alternatives in the same chapters. In doing so we have grouped procedures by the number of samples to which they refer. We believe that this grouping reinforces the assumptions required by a given procedure and helps to underline their importance. In this way, we feel we can help you to be more aware of the limitations and/or value of the techniques, so that you are less likely to use methods that are not applicable.

Our third change in topic ordering is the placement of analysis of variance before regression. Some textbooks do this but fail to stress the relationship between the two techniques. To recognize this relationship is important because any extensive analysis in a regression problem should utilize analysis-of-variance procedures. Thus, we believe that understanding the analysis-of-variance approach *first* helps you to understand the basic test procedures required in a regression analysis.

1.6 EXAMPLES OF BUSINESS PROBLEMS THAT MIGHT UTILIZE STATISTICS

The possible applications of statistical inference in the business environment are almost limitless. They range from rather simple sampling plans that aid in deciding whether or not to accept a shipment of goods to very sophisticated forecasting models that are far beyond the scope of this text.

In the shipment of goods case, suppose that you are in charge of quality control for incoming packaging materials for a pharmaceutical company. Imagine that you have just received a shipment of 10,000 fifty-milliliter vials used in packaging a liquid vitamin supplement. You must decide on the disposition of the shipment—either to send it on to production or to send it back to the supplier. Before making your decision, you can choose to

1. Inspect every vial (a time-consuming and costly process)
2. Inspect none of the vials (possibly increasing the chance of production delays)
3. Inspect a sample of, say, 100 vials

The costs involved would probably indicate the last choice, but after inspection of the vials in the sample, you will still have to make a decision. For example, if we assume that shipments with 2% of the items defective are acceptable, what is your decision if you find three defectives in the sample, or four, or five? The role of statistics is to determine the likelihood that each of these outcomes would occur by chance alone if the shipment is actually acceptable. We shall address problems of this type in later chapters.

The previous example was concerned with an attribute of a single population, the 10,000 vials in the shipment. To consider a second area of application, let us assume that you are in charge of marketing this vitamin supplement and have to choose among four proposed package designs. How do you choose? One possibility is to draw a number from a hat (unsophisticated, but it *does* resolve the problem). The statistical approach would be to test-market the supplement using each of the package designs in a controlled environment (this comes under the heading of experimental design). Suppose that you do this and observe sales, in thousands of units, of 2.1, 3.3, 2.7, and 1.9, respectively, for the four designs. The question you must ask yourself, and that statistical analysis helps you answer, is, "Are these sales differences due to random variation in customer buying habits alone, or are they due, in part, to differences in design?" Questions of this type are considered in Chapters 12 and 13.

In the previous two situations, we were comparing sample data with a standard or with another population. In both cases, since the *sample* information provides the basis for the decision, there is a chance of making a wrong decision. Never forget that statistics will never "prove" anything beyond a shadow of a doubt. The procedures merely tend to support or tend to contradict a contention. The role of statistics is to help try to ascertain whether a particular sample could reasonably have occurred as a result of chance variation alone, or whether the evidence indicates that some factor *in addition to chance* influenced the result. The only way to eliminate the chance aspect is to inspect every object in the population, an impossible task in most situations.

For a final example of a business application, assume that production asks you for an estimate of the demand for a product over the next six months. Production is attempting to set up a schedule that will minimize overtime, inventory costs, costs of retooling for different products, and so forth, and would appreciate fairly accurate estimates. One choice for your forecast is to use the actual demand during the previous six months (perhaps with an extra 5% added). A statistician might ask you for a set of variables that you feel should be related to the demand for this product and might use these to develop a forecasting model. Hopefully, the model would be more accurate than using last period's demand plus 5%. Problems of this type are considered in Chapters 14 and 15.

Although these examples illustrate possible business applications, there are just as many opportunities to apply statistics in the public sector. We will utilize examples of both private- and public-sector applications throughout the text.

1.7 SUMMARY

In this chapter, we introduced the basic concept of statistics as the collection, organization, and analysis of data for the purpose of making a decision about a population of objects. Basically, statistics originated as a descriptive science, but today the descriptive aspects play a minor role. The primary emphasis now is on the inferential or decision-making aspect.

We also introduced four levels of measurement, called nominal, ordinal, interval, and ratio. We did so because *all* statistical procedures assume a certain minimum level of measurement, and if the assumed level of measurement is not valid for the data being used, then any conclusions derived from the analysis are open to question. We stressed the importance of *all* assumptions and will continue to do so in the hope that you will be one of those for whom a great knowledge of statistics becomes a joyous thing.

Chapter 2
Basic Probability and Counting Techniques

KEY WORDS AND TERMS

Experiment	Conditional Probability
Outcome	Multiplication Law
Sample Space	Statistical Independence
Partition	Permutation
Event	Combination
Probability	Bayes' Rule
Addition Law	

2.1 INTRODUCTION

Most of us have used basic concepts of probability on occasion. Typically, we try to balance a subjective probability value against some cost associated with a purchase. For instance, when choosing a calculator, we might try to balance the extra cost of added capabilities against the probability that we will need to use the added functions. If this probability is small enough, then we could decide on a less expensive machine and accept the occasional inconvenience we would expect after purchase. Thus, we define the term *probability* rather loosely here, although it *is* an important foundation for mathematical methods used in business today. This chapter will introduce a more structured definition of probability than you may have encountered in the past, since we will cover both *relative frequency* probability and *objective* probability as essential concepts for proper mathematical analysis of business decision problems.

Our objective is to provide methods for making inferences concerning the general truth of a phenomenon of interest. We now proceed with some definitions and axioms that will provide the basis for the inferential procedures to be used in the analysis of management decision problems.

2.2 CALCULATING THE PROBABILITY OF AN EVENT FOR AN EXPERIMENT

In general, we use probability to help us reach a decision or an action in response to some random phenomenon we are observing. Formally, we are referring to an experiment.

DEFINITION 2.1 An *experiment* is any process that generates information (that is, generates data).

A manager who runs a simulation of a process several times under the same conditions is likely to obtain different results or measurements. The different measurements reflect the element of uncertainty prevailing in the system being modeled. Each such observation is called an *outcome.*

DEFINITION 2.2 A set whose elements represent all possible outcomes of an experiment is called the *sample space.*

The sample space is typically denoted by the letter S. Note that several different sample spaces can be used to describe the outcomes of an experiment. Not only do the elements depend on the outcomes of interest, but they must always satisfy the following conditions:

1. If one outcome occurs, then no other outcome occurs (each outcome is distinct).
2. At least one outcome in S must occur whenever the experiment is conducted.

The first condition expresses the idea that the sample space is a set of *mutually exclusive* outcomes (that is, that it is impossible for any two of the outcomes to occur simultaneously). The second condition says that the outcomes must be *collectively exhaustive* (that is, one of the outcomes must occur). When these two conditions are satisfied, S is said to *partition* the experiment.

EXAMPLE 2.1 Consider the experiment of observing defective items among the next two items produced by a manufacturing process. If we are interested in the number of defects, then an appropriate sample space would be given by

$$S = \{\text{NN}, \text{ND}, \text{DN}, \text{DD}\}$$

where N indicates a nondefective item and D indicates a defective item.

This sample space provides information on both the number of defectives and how the number of defectives occurred. Although we can use more than one sample space to describe the same experiment, we most often choose the one that gives the most information about each outcome.

Frequently, we are interested in experimental "events" rather than specific outcomes. For instance, we might be interested in whether or not more than 5% of a given shipment of items is defective. In this case, the event of interest includes all those outcomes for which the number of defectives exceeds 5% of the total number of items in the shipment. Thus, we are actually concerned about the subset of outcomes in the sample space in which we could say the event of interest has occurred.

DEFINITION 2.3 An *event* is a subset of a sample space.

EXAMPLE 2.2 The time between arrivals at a loading dock can be represented by the sample space $S = \{t > 0\}$, where t indicates the time elapsed in minutes between arrivals. In this case, the subset $A = \{t > 10\}$ represents the event that the time elapsed exceeds 10 minutes.[1]

We can also observe that outcomes are subsets of the sample space and, under the previous definition, are events. In order to distinguish between outcomes and more complex events, we provide the following definition.

DEFINITION 2.4 An outcome is called a *simple event.* A *compound event* is any event consisting of more than one outcome.

Set A in Example 2.2 is a compound event. The experiment of observing the total number of pips on the upturned faces of a pair of dice has a sample space given by

$$S = \{2, 3, 4, 5, 6, 7, 8, 9, 10, 11, 12\}$$

and some compound events for this experiment are

$$\begin{aligned} A &= \{2, 4, 6, 8, 10, 12\} \\ B &= \{2, 3, 4, 5, 6\} \\ C &= \{7, 8, 9, 10, 11, 12\} \end{aligned}$$

Here A represents the event that an even number of pips are observed, B represents the event that the number of pips does not exceed six, and

1. We will employ the usual set theory conventions and notation throughout this text.

C represents the event that the number of pips exceeds six (indicating that B did not occur). In general, we say that an event occurs whenever any one of its outcomes occurs. Since we can determine the probability of occurrence of a compound event by adding the probabilities for each of its outcomes, we are now ready to state the definitions and axioms of probability.

DEFINITION 2.5*a* Suppose that there are n possible outcomes for an experiment and that event A occurs in m of these cases. If the n outcomes are equally likely, then the *objective* probability of event A is given by m/n.

DEFINITION 2.5*b* Let a_i be the number of times A occurs in the ith experiment, and let n_i be the number of trials in the ith experiment. Then the *relative frequency* probability of event A is given by

$$\text{pr}(A) = \lim_{k \to \infty} \left(\frac{\sum_{i=1}^{k} a_i}{\sum_{i=1}^{k} n_i} \right) \quad \text{if the limit exists}$$

The definition of relative frequency probability is valuable when it is not possible to enumerate all experimental outcomes, when it is difficult to establish the number of outcomes that make up the event of interest, or when outcomes may not be equally likely. Applications of probability concepts in business problems often involve extrapolation from historical records and the manipulation of these records into relative frequency probabilities that are used to predict future patterns. Let us illustrate the meaning of Definition 2.5*b* with the following example.

EXAMPLE 2.3 The Stereo Shack, Ltd., a Crown corporation, currently is starting to produce and sell Fuzz Busters. Their main concern at this point is to determine the probability of a Fuzz Buster being defective. Since 100% sampling would be too costly and time consuming, they randomly sample a few pieces each shift and tabulate the results for the first 12 shifts. These are given below.

Shift (k)	1	2	3	4	5	6	7	8	9	10	11	12
Number of items sampled (n_i)	10	7	8	15	12	10	9	11	8	10	6	13
Number of items defective (a_i)	0	1	0	2	1	1	0	1	0	1	0	1

Using these data, construct their "running" (updated) estimates of pr(defective) = pr(D).

Reorganizing the data and using Definition 2.5b, we have

k	1	2	3	4	5	6	7	8	9	10	11	12
$\sum_1^k a_i$	0	1	1	3	4	5	5	6	6	7	7	8
$\sum_1^k n_i$	10	17	25	40	52	62	71	82	90	100	106	119
pr($D \mid k$)	0/10	1/17	1/25	3/40	4/52	5/62	5/71	6/82	6/90	7/100	7/106	8/119

These estimates are shown graphically in Fig. 2.1.

Figure 2.1 illustrates how the probabilities stabilize as the number of experiments increases.

EXAMPLE 2.4 Shoppers are as likely to choose a blue box of soap as they are a red box containing the same amount of soap. Suppose that one red and one blue box of soap are on the shelf and that a customer chooses one of the two boxes. We obtain the objective probability for the event that the blue box is chosen by noting that there are two equally likely outcomes ($n = 2$), with one ($m = 1$) corresponding to the event of interest. Thus, the probability that the customer chooses a blue box of soap is given by $m/n = 0.5$. In order to

Figure 2.1 Graphical Representation of Probabilities for the Fuzz Buster Problem

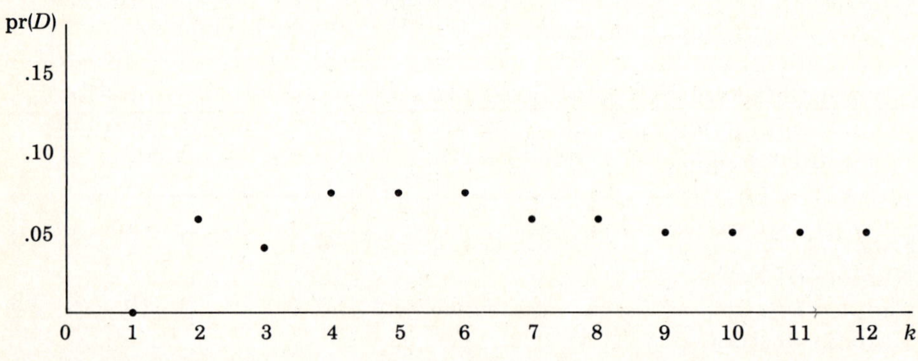

calculate a relative frequency probability, we need to look at the store records. Suppose we note that among 500 different occasions (when an equal number of blue and red boxes were on the shelf), the customer chose a red box 232 times. Since we do not know the actual probability, our best choice for the probability that a customer chooses a red box (when an equal number of red and blue boxes are available) would be given by 232/500, or 0.464.

When the number of outcomes or trials is finite and the event of interest happens in all cases, we say the event is *certain* and has a probability of 1.0. Its complementary event is said to be impossible and is called the *null event* $\emptyset$. Since $\emptyset$ is impossible, its associated probability is 0. We are now ready to state the axioms of probability.

Let the letters A and B represent events in the same sample space, and let A' denote the complement of event A, indicating that event A did not occur. (That is, A' represents the set of all outcomes in S that are not in A.) Further, let $\text{pr}(A)$ denote the probability of A.

AXIOM I

$$0 \leqslant \text{pr}(A) \leqslant 1 \tag{2.1}$$

This axiom states that the probability for an event must be a number between zero and one. These probabilities are sometimes confused with percentages, as in "There is a 25% probability of occurrence" rather than the correct statement, "There is a 25% chance of occurrence."

We will use set notation to indicate common event combinations, such as joint occurrence. If A and B occur simultaneously, this combined event is represented by

$$A \cap B \qquad \text{(read "}A \text{ and } B\text{" or "}A \text{ intersect } B\text{")}$$

which denotes all those outcomes common to both A and B.

DEFINITION 2.6 When A and B cannot occur simultaneously, they are said to be *mutually exclusive*.

We will use $A \cup B$, read "A or B" or "A union B," to denote the outcomes of event A or event B or both A and B. That is, at least one of the two events must occur, so that the symbol $\cup$ implies the inclusive interpretation of the word "or."

From this we can show that if A and B are mutually exclusive events, then

$$\text{pr}(A \cap B) = 0 \tag{2.2}$$

and

$$\text{pr}(A \cup A') = 1 \tag{2.3}$$

EXAMPLE 2.5 Consider the experiment of rolling a blue die and a red die, both of which are "fair" in the sense that any one face is as likely to occur as any other face. Let A be the event that the total number of pips on the upturned faces is less than seven, and let B be the event that at least one of the dice shows an even number of pips on its upturned face. An informative display of the outcome composition for these events is found in the following grid, in which the inscription of the event letter indicates an outcome for that event.

		Blue Die					
		1	2	3	4	5	6
	1	A	$A\ B$	A	$A\ B$	A	B
	2	$A\ B$	$A\ B$	$A\ B$	$A\ B$	B	B
Red	3	A	$A\ B$	A	B		B
Die	4	$A\ B$	$A\ B$	B	B	B	B
	5	A	B		B		B
	6	B	B	B	B	B	B

Since the outcomes are equally likely, we can use objective probability, as shown in Table 2.1.

Table 2.1 Objective Probabilities for the Die Experiment

Event	m	n	m/n
A	15	36	$\frac{15}{36}$
B	27	36	$\frac{27}{36}$
$A \cap B$	9	36	$\frac{9}{36}$
$A \cup B$	33	36	$\frac{33}{36}$

Notice that

$$\text{pr}(A \cup B) = \frac{15}{36} + \frac{27}{36} - \frac{9}{36} = \text{pr}(A) + \text{pr}(B) - \text{pr}(A \cap B)$$

This result expresses an important axiom of probability known as the *addition law of probability*, which is given as Axiom II.

AXIOM II $$\text{pr}(A \cup B) = \text{pr}(A) + \text{pr}(B) - \text{pr}(A \cap B) \qquad (2.4)$$

You can see from the previous grid that we must subtract the quantity $\text{pr}(A \cap B)$, since the outcomes of $A \cap B$ are counted twice in the calculation $\text{pr}(A) + \text{pr}(B)$.

DEFINITION 2.7 The probability that A occurs given that B has already occurred is called the *conditional probability* of A given that B occurred. This conditional probability is written $\text{pr}(A|B)$. The vertical line segment should be read "given the occurrence of."

In contrast, $\text{pr}(A)$ is sometimes called the *unconditional probability* of A or the *absolute probability* of A. The act of "conditioning," as given here, reduces the sample space so that our attention is restricted to those outcomes in B. That is, the condition defines a new sample space.

EXAMPLE 2.6 In order to calculate $\text{pr}(A|B)$ for the events given in Example 2.5, we first note that the condition restricts us to those outcomes in event B. That is, the number of equiprobable outcomes has been reduced from 36 to 27. We determine the probability by counting the number of outcomes in B that are also outcomes in A. Since there are nine of these outcomes, we have $m = 9$ and $n = 27$, giving us

$$\text{pr}(A|B) = \frac{9}{27}$$

Example 2.6 brings us to another important axiom of probability, the *multiplication law of probability*.

AXIOM III $$\text{pr}(A \cap B) = \text{pr}(A|B)\,\text{pr}(B) \qquad (2.5)$$

This example reflects the algebraically equivalent expression

$$\text{pr}(A|B) = \frac{\text{pr}(A \cap B)}{\text{pr}(B)} \qquad \text{if } B \neq \emptyset \qquad (2.6)$$

[assuming that $\text{pr}(B) \neq 0$ or $B \neq \emptyset$]. If we refer to Table 2.1, it gives us the same result:

$$\text{pr}(A|B) = \frac{9/36}{27/36} = \frac{9}{27}$$

This formula lets us make a quick calculation for the probability that at least one of the dice shows an even number, given that the total number of pips is less than seven:

$$\text{pr}(B|A) = \frac{\text{pr}(B \cap A)}{\text{pr}(A)} = \frac{\text{pr}(A \cap B)}{\text{pr}(A)} = \frac{9/36}{15/36} = \frac{9}{15}$$

We now introduce a final concept before proceeding with a presentation of methods that are helpful for determining the number of outcomes in a sample space.

DEFINITION 2.8 If the absolute probability of A is the same as the conditional probability $\text{pr}(A|B)$, then the events A and B are said to be *statistically independent*.

Note that this is a definition of *statistical* independence, which is a mathematical concept and *not* an intuitive concept. The mathematical and intuitive interpretations often conflict when we compare statistically independent events with mutually exclusive events.

EXAMPLE 2.7 Consider the following three events, defined for the die experiment of Example 2.4 (here outcomes are denoted by ordered pairs in which the left entry indicates the red die result and the right entry indicates the blue die result):

$$D = \{(1,5),\ (2,5),\ (3,5),\ (4,5),\ (5,5),\ (6,5)\}$$
$$E = \{(5,5),\ (5,6),\ (6,5),\ (6,6)\}$$
$$F = \{(2,1),\ (2,2),\ (2,3),\ (2,4),\ (2,5),\ (2,6)\}$$

Thus, D is the event that the blue die shows a 5, E is the event that the total number of pips showing is greater than or equal to 10, and F is the event that the red die shows a 2. Now notice that

$$\text{pr}(D|F) = \text{pr}(D) \quad \text{but} \quad D \cap F \text{ is not empty}$$
$$\text{pr}(D|E) \neq \text{pr}(D) \quad \text{but} \quad D \cap E \text{ is not empty}$$

Thus, D and F are independent and are not mutually exclusive, whereas D and E are not independent and are not mutually exclusive. In fact, all mutually exclusive events are dependent unless one of them is impossible.

EXERCISES

2.2.1 An experiment consists of flipping a coin until a head occurs but never flipping more than three times.
(*a*) List the elements of the sample space S.
(*b*) List the elements of S corresponding to the event A that exactly two tails occurred.

2.2.2 If A and B are disjoint events and $\text{pr}(A) = 0.2$ and $\text{pr}(B) = 0.7$, find:
(*a*) $\text{pr}(A|B)$
(*b*) $\text{pr}(A \cup B)$
(*c*) $\text{pr}(A' \cap B)$

2.2.3 A pair of dice are tossed. What is the probability of:
(*a*) Getting a total of 7 (call this event A)?
(*b*) Getting an even number on at least one of the dice (call this event B)?
(*c*) Event A given B?
(*d*) Event B given A?

2.2.4 Refer to Example 2.5. Show that:
(*a*) $\text{pr}(A) = \text{pr}(A \cup B) + \text{pr}(A \cap B) - \text{pr}(B)$
(*b*) $\text{pr}(A \cap B) = \text{pr}(A) + \text{pr}(B) - \text{pr}(A \cup B)$
(*c*) $\text{pr}(B|A) = \text{pr}(A \cap B)/\text{pr}(A)$
(*d*) $\text{pr}(B) = \text{pr}(A \cap B)/\text{pr}(A|B)$

2.3 COUNTING SAMPLE POINTS

In many cases we can simplify the calculation of an objective probability for an event by characterizing the event's sample space as one of a family of sample spaces. If we had a method of counting that would count the number of elements in any member of the family, then we could apply that method to calculate the number of sample space elements for the problem at hand. One of the most fundamental methods for this type of counting is the $m \cdot n$ rule.

2.3.1 The $m \cdot n$ Rule

If there are m possible outcomes in one experiment and n in another experiment, then the total number of outcomes in the sample space that describes the two experiments together is given by the product $m \cdot n$. This is known as *the* $m \cdot n$ *rule*.

EXAMPLE 2.8 A factory manager decides to use a two-letter code for her employees in an effort to reduce processing costs. If the factory employs 500 people, will the two-letter code be enough?

The answer is yes, because the $m \cdot n$ rule says that the total number of two-letter codes is given by $26 \times 26 = 676$.

Note that the $m \cdot n$ rule can be extended to the multiexperiment case, so that if experiment i contains n_i possible outcomes, $i = 1, 2, \ldots, k$, then the total number of outcome combinations across the k events is given by the product:

$$n_1 \cdot n_2 \cdot n_3 \cdot \cdots \cdot n_k$$

EXAMPLE 2.9 A popular buffet boasts eight appetizers, ten entrees, and seven desserts. How many different complete dinners are available?

Referring to the above, we have $k = 3$ with $n_1 = 8, n_2 = 10$, and $n_3 = 7$. Thus, there are $8 \cdot 10 \cdot 7 = 560$ different complete dinner selections.

2.3.2 Permutations

The calculation of a permutation allows us to count the number of ways in which a set of outcomes can be ordered. Frequently, this is a count of the number of ways in which r objects can be chosen from n when a different order of choice indicates a different way to choose the r objects.

EXAMPLE 2.10 How many ways can five executives be assigned to five projects?

A few possible choices for assignment sequences follow (the letters indicate the executives).

		Project		
1	2	3	4	5
A	B	C	D	E
A	B	C	E	D
A	B	D	C	E
A	B	D	E	C
.	.	.	.	.
.	.	.	.	.
.	.	.	.	.

The key to calculation in problems of this type can be found in the $m \cdot n$ rule. There are five choices for project 1, after which there are four left to choose from for project 2, and so forth. The $m \cdot n$ rule tells us that the total number of ways in which the five executives can be assigned to the five projects is $5 \cdot 4 \cdot 3 \cdot 2 \cdot 1 = 120$.

We could expand this problem to ask how many ways we could choose five executives from twelve executives to be assigned to five projects. The approach would be the same in that there would be twelve choices for project 1, leaving eleven for project 2, and so forth. The answer is this case would be $12 \cdot 11 \cdot 10 \cdot 9 \cdot 8 = 95{,}040$, leaving seven executives unassigned.

We can express this last result using factorials[1] as

$$\frac{12!}{(12-5)!}$$

in which we can observe both the number of executives to choose from (12) and the number we want to choose (5). We call this result a *permutation*, which, in the general case, is denoted by

$$P_r^n = \frac{n!}{(n-r)!} \tag{2.7}$$

This is read "permutation of n items chosen r at a time."

EXAMPLE 2.11 You are one of ten finalists in a state lottery from which three are chosen for large first, second, and third prize awards. How many ways can the prizes be awarded?

This is a permutation problem with $n = 10$ and $r = 3$. The answer is given by

$$P_3^{10} = \frac{10!}{(10-3)!} = \frac{10 \cdot 9 \cdot 8 \cdot 7 \cdot 6 \cdot 5 \cdot 4 \cdot 3 \cdot 2 \cdot 1}{7 \cdot 6 \cdot 5 \cdot 4 \cdot 3 \cdot 2 \cdot 1}$$
$$= 10 \cdot 9 \cdot 8 = 720$$

2.3.3 Combinations

When order of choice is not important, we can use a different counting formula to count the number of sample points. For instance, if our in-

1. We use an exclamation point to indicate a factorial, which is defined by

$$0! = 1 \qquad 1! = 1 \qquad n! = (n-1)(n-2)\ldots 3 \cdot 2 \cdot 1$$

for any positive integer n.

terest in the executive-project assignment problem changes so that we are interested in who is assigned without regard to project, then we would use a combination formula. As you might suspect, the number of combinations of n items chosen r at a time is smaller than the number of permutations of n items chosen r at a time. We will now proceed to justify exactly how much smaller a figure to expect.

EXAMPLE 2.12 Suppose that we want to choose three people from five in order to form a committee of three. Our only interest is how many committees of size three can be formed, not the order of choice for committee membership. Enumeration of cases tells us that there are ten different committees of size three that can be formed with five people.

If we let A, B, C, D, and E represent the people, then the different committees of size three are

$$\begin{array}{ll} ABC & ADE \\ ABD & BCD \\ ABE & BCE \\ ACD & BDE \\ ACE & CDE \end{array}$$

We can avoid the enumeration in Example 2.12 by using the permutation concept. Note that there are

$$P_3^5 = 60$$

ways to choose three people from five when order of choice is important. Within the 60 cases are the following permutations of the letters A, B, and C:

$$\begin{array}{lll} ABC & BAC & CAB \\ ACB & BCA & CBA \end{array}$$

all of which would be considered as the same committee. Thus, all we need to do to eliminate this duplication from the P_3^5 calculation is to divide P_3^5 by the number of ways to permute three items chosen three at a time. Note that $P_3^3 = 6$ is the number of replications we have for any three letters. This means that the number of combinations of five people chosen three at a time is given by

$$\frac{P_3^5}{P_3^3} = \frac{5!/(5-3)!}{3!/(3-3)!} = \frac{5}{(5-3)!3!} = 10$$

In general, the number of combinations of n items chosen r at a time is given by

$$C_r^n = \frac{P_r^n}{P_r^r} = \frac{n!}{(n-r)!r!} \tag{2.8a}$$

Another popular notation for combinations is

$$\binom{n}{r} = \frac{n!}{(n-r)!r!} \tag{2.8b}$$

which is the one we will use in the rest of this book. Occasionally, there is some confusion regarding which counting method, permutations or combinations, is appropriate for a given problem. Our suggestion is to use permutations if the problem can be characterized as an assignment problem in the sense that items are chosen and assigned in some ordered fashion. Otherwise, the combination formula might be appropriate. Sometimes permutations, combinations, and the $m \cdot n$ rule will be required concurrently in a single problem.

EXAMPLE 2.13 Fenway Fred operates a weenie stand at the ballpark where customers can request that onions, relish, mustard, ketchup, mayonnaise, chili, or nothing be placed on the roll along with the weenie. Fred wants to advertise the number of ways his stand can prepare a weenie. How many ways are there?

The solution to this problem depends on your point of view, since one could maintain that ketchup applied before mustard is just *not* the same as mustard followed by ketchup. In this case, the order of application of condiment is important, and the weenie could be prepared in

$$\sum_{k=0}^{6} P_k^6 = 1957$$

different ways. Otherwise, the count could be determined using combinations to get

$$\sum_{k=0}^{6} \binom{6}{k} = 64$$

EXAMPLE 2.14 Campus senate members want to form a representative committee of five. The 50-member senate consists of three different interest groups whose members number 10, 20, and 20, respectively. How many different representative committees can be found?

In order for the committee to be representative, we want to choose 1, 2, and 2 members, respectively, from the three interest groups. The calculation requires use of the combination formula and the $m \cdot n$ rule:

$$\binom{10}{1}\binom{20}{2}\binom{20}{2} = 361{,}000$$

EXERCISES

2.3.1 A nameless country attaches license plates to its mode of transportation, on each of which the following sequence can be found: letter, number, number, number, letter. Calculate the number of different plates that can be issued if a number is defined as being one of the integers 0, 1, 2, 3, 4, 5, 6, 7, 8, 9.

2.3.2 How many ways can four people be lined up to receive service at a grocery checkout counter?

2.3.3 From a group of eight men and three women, calculate the number of committees of five people that are possible:
(*a*) With no restrictions.
(*b*) With three men and two women.

2.3.4 Three men and two women go to the movies and want to sit in the same row. Determine the number of ways they can sit with a woman between any two men.

2.4 BAYES' RULE

A common pursuit in the business decision process is the estimation of the value of sample information. For instance, many companies buy stock in gross quantities knowing that a certain percentage of items in each package is defective. If someone were to offer to take a random sample from each package to check for defectives, we would need a method of determining how much we should pay for the inspection. Historically, if few defectives are observed, the inspection may be of little value. As the penalty for use of defective stock increases, the value of the inspection increases. The key to analysis of this type lies in Bayes' rule, which tells us how to judge the state of nature (how many defectives are in the package) given an event of interest (the number of defectives in the sample).

We will use an example to illustrate the rule. Assume that Mulligan's Golf Supply orders golf balls representing excess production from popular manufacturers. These are ordered sight unseen in lots of 500.

Some of the balls will be of good quality and can be sold for a premium price, whereas some will be defective (called "x-outs" in the profession) and must be sold at discount. Mulligan pays a fixed price per lot, but is told that he may inspect each shipment and return (in its entirety) any shipment that he judges to be unacceptable. However, total inspection is out of the question, since it would be too expensive in relation to the value of the information it would produce. On the other hand, he could inspect a small random sample of balls from each package.

For simplification, let us assume that he inspects 50 balls from each package. The idea is that the number of defects found in the sample is to be used to estimate the number of defects in the total package of 500. If the estimate is high, Mulligan will send the package back to the manufacturer. The problem can be characterized by the fact that there are 501 states of nature reflecting the possible number of defective balls in the package—0, 1, 2, . . . , 500. Further suppose that the relative frequency probabilities for these states of nature are given by $P_0, P_1, P_2, \ldots, P_{500}$, respectively. What we want to know is the probability of i defectives in the package (denoted by S_i and indicating the ith state of nature) given that there are m defectives in the sample of size 50, $m = 0, 1, 2, \ldots, 50$. If A represents the event that we observe m defectives in a sample of size 50, then we want to compute

$$\text{pr}(S_i \cap A)$$

Using the multiplication law of probability, we can write

$$\text{pr}(S_i \cap A) = \text{pr}(S_i | A)\,\text{pr}(A)$$

or,

$$\text{pr}(S_i | A) = \frac{\text{pr}(S_i \cap A)}{\text{pr}(A)} = \frac{\text{pr}(A | S_i)\,\text{pr}(S_i)}{\text{pr}(A)}$$

Now note that

$$\text{pr}(A) = \text{pr}(A \cap S_0) + \text{pr}(A \cap S_1) + \ldots + \text{pr}(A \cap S_{500})$$

because the events $(A \cap S_j)$ and $(A \cap S_k)$ are mutually exclusive whenever $j \neq k$ (if the package contains j defectives, it can't contain k defectives), and because the S_j, $j = 0, 1, 2, \ldots, 500$, account for all possible states of nature. As a result,

$$\text{pr}(S_i | A) = \frac{\text{pr}(A | S_i)\,\text{pr}(S_i)}{\sum_{j=0}^{500} \text{pr}(A \cap S_j)} = \frac{\text{pr}(A | S_i)\,\text{pr}(S_i)}{\sum_{j=0}^{500} \text{pr}(A | S_j)\,\text{pr}(S_j)}$$

This last result illustrates Bayes' rule, which is used to calculate *posterior* probabilities [that is, $\text{pr}(S_i \mid A)$] given prior probabilities [$\text{pr}(S_i)$]. Thus, using relative frequency probabilities, we can get

$$\text{pr}(S_i \mid A) = \frac{\text{pr}(A \mid S_i) P_i}{\sum_{j=0}^{500} \text{pr}(A \mid S_j) P_j}$$

We will finish the problem in Chapter 6, where the fundamental concepts used in Bayesian decision analysis are presented. However, we should note some important characteristics of this problem that can be used *outside* the sample information context.

One of the basic concepts used in this explanation is that of an *event partition* of the sample space. Just as the sample space contains outcomes that partition an experiment, we can choose a set of events that partition the sample space. The same two conditions must be met. That is, if the events $E_1, E_2, \ldots, E_k$ partition the sample space, then

1. E_j and E_k are mutually exclusive.
2. $E_1, E_2, \ldots, E_k$ are collectively exhaustive.

The two conditions together tell us that exactly one of the k events will occur each time the experiment is conducted. Second, we used a fundamental result in set theory that says that $A \cap (E_1 \cup E_2 \cup E_3 \cup \cdots \cup E_k)$ is the same as

$$(A \cap E_1) \cup (A \cap E_2) \cup (A \cap E_3) \cup \cdots \cup (A \cap E_k)$$

Third, we can use the additive law of probability to show that if $B_1, B_2, \ldots, B_k$ are mutually exclusive, then

$$\text{pr}(B_1 \cup B_2 \cup \cdots \cup B_k) = \sum_{j=1}^{k} \text{pr}(B_j)$$

Thus, letting $B_j = A \cap E_j$, we get

$$\begin{aligned} \text{pr}(A) &= \text{pr}[A \cap (E_1 \cup E_2 \cup \cdots \cup E_k)] \\ &= \text{pr}(A \cap E_1) \cup (A \cap E_2) \cup (A \cap E_3) \cup \cdots \cup (A \cap E_k) \\ &= \sum_{j=1}^{k} \text{pr}(A \cap E_j) \end{aligned}$$

Finally, using the multiplication law of probability, we can say that $\text{pr}(A \cap E_j) = \text{pr}(A \mid E_j)\,\text{pr}(E_j)$ for $j = 1, 2, \ldots, k$. Using this latter result gives the general form of Bayes' rule:

$$\text{pr}(E_i \mid A) = \frac{\text{pr}(A \mid E_i)\,\text{pr}(E_i)}{\sum_{j=1}^{k} \text{pr}(A \mid E_j)\,\text{pr}(E_j)} \tag{2.9}$$

in which we assume that $E_1, E_2, \ldots, E_k$ partition the experiment sample space.

EXAMPLE 2.15 Four members of a fraternity have been nominated for the office of treasurer. The probability that John will be elected is 0.12, the probability that Jim will be elected is 0.26, the probability that Jake will be elected is 0.38, and the probability that Jerry will be elected is 0.24. If John or Jim is elected, then the probabilities for a board and room increase are 0.2 and 0.3, respectively. If Jake or Jerry is elected, then the probabilities for an increase are 0.4 and 0.5, respectively. If a student joins the fraternity after the election and observes a board and room fee increase, what is the probability that Jim was elected?

The first step in the solution is to identify the events as follows:

- E_1: John is elected treasurer
- E_2: Jim is elected treasurer
- E_3: Jake is elected treasurer
- E_4: Jerry is elected treasurer
- A: A board and room fee increase occurred after the election

According to this notation, we want to find

$$\text{pr}(E_2 \mid A)$$

Using Bayes' rule, we get

$$\begin{aligned}\text{pr}(E_2 \mid A) &= \frac{\text{pr}(A \mid E_2)\,\text{pr}(E_2)}{\sum_{i=1}^{4} \text{pr}(A \mid E_i)\,\text{pr}(E_i)} \\ &= \frac{(0.3)(0.26)}{(0.2)(0.12) + (0.3)(0.26) + (0.4)(0.38) + (0.5)(0.24)} \\ &= 0.21\end{aligned}$$

We will use Bayes' rule in a more constructive sense when we discuss Bayesian decision analysis in Chapter 6.

EXERCISES

2.4.1 Suppose that colored balls are distributed in three identical boxes as follows:

		Box		
		1	2	3
Color	Black	2	1	3
	White	4	4	1
	Red	3	5	2
	Green	3	3	5

If a box is selected at random from which a ball is selected at random, and the ball is white, calculate the probability that box 2 was selected.

2.4.2 Refer to Exercise 2.4.1. If a box is selected at random from which a ball is selected at random, and the ball is black, calculate the probability that box 3 was selected.

2.4.3 A tax audit is known to be 95% reliable when the taxpayer is guilty of willful evasion of tax laws and 99.9% reliable otherwise (that is, the audit detects evasion when there is none on only 0.1% of the audited forms). If a form was selected from a group of tax forms of which 10% represent willful evasion of tax laws, and the audit indicates that the taxpayer is guilty of tax-law evasion, what is the probability that the taxpayer is innocent?

2.5 SUPPLEMENTARY EXAMPLES

EXAMPLE 2.16 Three boys are deciding who will pull a prank by playing "odd man out." They each flip fair coins simultaneously until one flips something the other two don't flip.

(*a*) List the elements of the sample space S.

(*b*) List the elements of S corresponding to event A, that one boy flips something the other two don't flip.

(*c*) Calculate the probability that at least two flips are required to get an "odd" man.

SOLUTION

(*a*) Letting T denote a tail and H denote a head, we get

$$S = \{\text{TTT, TTH, THT, HTT, THH, HTH, HHT, HHH}\}$$

(*b*) $A = \{\text{TTH, THT, HTT, THH, HTH, HHT}\}$

(*c*) From (*a*) and (*b*), we get $\text{pr}(A) = \frac{6}{8}$ and note that

$$\begin{aligned} &\text{pr}\,\{\text{"at least 2 flips to get an odd man"}\} \\ &= 1 - \text{pr}\,\{\text{"exactly 1 flip to get an odd man"}\} \\ &= 1 - \text{pr}(A) = 1 - \frac{6}{8} = \frac{2}{8} \end{aligned}$$

EXAMPLE 2.17 Suppose that you have three quality-control checkpoints, say A, B, and C, for an item. Each checkpoint measures the quality of an attribute different from the attributes measured by the other two. Assume that you have a box containing four items for which

- Exactly one passed only at checkpoint A
- Exactly one passed only at checkpoint B
- Exactly one passed only at checkpoint C
- Exactly one passed at all three checkpoints

Now we propose choosing one item of the four at random from the box. Consider the following three events:

- A: The item chosen at random passed at checkpoint A
- B: The item chosen at random passed at checkpoint B
- C: The item chosen at random passed at checkpoint C

Show that the three events are pairwise independent but are not mutually independent [that is, $\text{pr}(A \cap B \cap C) \neq \text{pr}(A)\,\text{pr}(B)\,\text{pr}(C)$].

SOLUTION First note that $\text{pr}(A) = \text{pr}(B) = \text{pr}(C) = 1/2$ and $\text{pr}(A \cap B) = \text{pr}(A \cap C) = \text{pr}(B \cap C) = 1/4$. Thus

$$\begin{aligned} \text{pr}(A \cap B) &= \text{pr}(A)\,\text{pr}(B) \\ \text{pr}(A \cap C) &= \text{pr}(A)\,\text{pr}(C) \\ \text{pr}(B \cap C) &= \text{pr}(B)\,\text{pr}(C) \end{aligned}$$

so that the case for pairwise independence is supported. Now note that

$$\text{pr}(A \cap B \cap C) = \frac{1}{4}$$

but

$$\text{pr}(A)\,\text{pr}(B)\,\text{pr}(C) = \frac{1}{8}$$

This example shows that pairwise independence of events does not guarantee mutual independence of the same events. That is, the condition of pairwise independence,

$$\text{pr}(E_i \cap E_j) = \text{pr}(E_i)\,\text{pr}(E_j)$$

whenever $i \neq j$ for n events $E_1, E_2, \ldots, E_n$, does not guarantee any higher-order independence (e.g., in threes or fours), and in particular does not imply

$$\text{pr}(E_1 \cap E_2 \cap \ldots \cap E_n) = \text{pr}(E_1)\,\text{pr}(E_2)\ldots\text{pr}(E_n)$$

EXAMPLE 2.18 Suppose that there are two pails containing black and white balls according to the following composition:

	Pail I	Pail II
Number of black balls	32	21
Number of white balls	68	79

Further, suppose that we conduct an experiment in which we select, at random, two balls by using two successive draws. Let

- B_1 be the event of selecting a black ball on the first draw
- B_2 be the event of selecting a black ball on the second draw
- W_1 be the event of selecting a white ball on the first draw
- W_2 be the event of selecting a white ball on the second draw

(*a*) Select two balls in two draws without replacement from pail II and calculate $\text{pr}(B_1 \cap B_2)$, $\text{pr}(B_2 \cup B_1)$, and $\text{pr}(W_2 \mid B_1)$.

(*b*) Select a ball from pail II and place it in pail I if it is black; otherwise replace it in pail II. Then select another ball from pail I. Calculate $\text{pr}(W_1 \cap W_2)$, $\text{pr}(B_2 \mid W_1)$, and $\text{pr}(W_2 \mid B_1)$.

SOLUTION

(*a*) $\text{pr}(B_1 \cap B_2) = \text{pr}(B_2 \mid B_1)\,\text{pr}(B_1)$. Since we are drawing from pail II, $\text{pr}(B_1) = 21/100$. Since the drawing is without replacement, we have $\text{pr}(B_2 \mid B_1) = 20/99$, so that

$$\text{pr}(B_1 \cap B_2) = \left(\frac{20}{99}\right)\left(\frac{21}{100}\right) = 0.0424$$

Similarly,

$$\begin{aligned}\text{pr}(B_2 \cup B_1) &= \text{pr}(B_2) + \text{pr}(B_1) - \text{pr}(B_1 \cap B_2)\\ &= \text{pr}(B_2) + \frac{21}{100} - 0.0424\end{aligned}$$

To complete this calculation, we can use the fact that the first draw must result in either a black ball or a white ball, so that

$$\begin{aligned} \text{pr}(B_2) &= \text{pr}(B_2 \cap B_1) + \text{pr}(B_2 \cap W_1) \\ &= 0.0424 + \text{pr}(B_2 \mid W_1)\,\text{pr}(W_1) \\ &= 0.0424 + \left(\frac{21}{99}\right)\left(\frac{79}{100}\right) \\ &= 0.0424 + 0.1675 = 0.2099 \end{aligned}$$

and so

$$\text{pr}(B_2 \cup B_1) = 0.2099 + 0.2100 - 0.0424 = 0.3775$$

For the third calculation, we have

$$\text{pr}(W_2 \mid B_1) = \frac{79}{99}$$

(*b*) First,

$$\text{pr}(W_1 \cap W_2) = \text{pr}(W_2 \mid W_1)\,\text{pr}(W_1) = \text{pr}(W_2 \mid W_1)\left(\frac{79}{100}\right)$$

Note that the notation $\text{pr}(W_2 \mid W_1)$ tells us that a white ball was drawn from pail II and replaced in pail II, and so we choose the second ball from pail I. Thus

$$\text{pr}(W_2 \mid W_1) = \frac{68}{100}$$

and

$$\text{pr}(W_1 \cap W_2) = \left(\frac{68}{100}\right)\left(\frac{79}{100}\right) = 0.5372$$

For the calculation of $\text{pr}(B_2 \mid W_1)$, we know that we are choosing from pail I for the second draw, so

$$\text{pr}(B_2 \mid W_1) = \frac{32}{100} = 0.32$$

Last,

$$\text{pr}(W_2 \mid B_1) = \frac{68}{101} = 0.6733$$

since we choose from pail I, which will contain the extra black ball.

EXAMPLE 2.19 Suppose that a penny, two dimes, and three nickels are in a box and each is an unfair coin in the sense that if

- PH is the event that a toss of a penny results in a head
- DH is the event that a toss of a dime results in a head
- NH is the event that a toss of a nickel results in a head

then pr(PH) = 0.2, pr(DH) = 0.3, and pr(NH) = 0.4. One of the coins is chosen at random, is tossed, and results in a tail. What is the probability that the coin selected was a dime?

SOLUTION Let P, D, and N represent the events that a penny, dime, and nickel are chosen, respectively. Then

$$\text{pr(P)} = \frac{1}{6}$$

$$\text{pr(D)} = \frac{2}{6}$$

$$\text{pr(N)} = \frac{3}{6}$$

and we want to calculate pr(D|T), where T is the event that a toss results in a tail. This is a case easily handled using Bayes' rule:

$$\begin{aligned}\text{pr(D|T)} &= \frac{\text{pr(T|D) pr(D)}}{\text{pr(T|P) pr(P)} + \text{pr(T|D) pr(D)} + \text{pr(T|N) pr(N)}} \\ &= \frac{(0.7)(2/6)}{(0.8)(1/6) + (0.7)(2/6) + (0.6)(3/6)} = 0.35\end{aligned}$$

2.6 SUMMARY

In this chapter you have been introduced to the basis for all statistical decision making: probability. It is imperative that you understand these basic concepts and that you be able to use them in future chapters, since most statistical decisions are based on the likelihood of occurrence of the sample event.

Events are subsets of the sample space and fall into two basic categories, simple and compound. A simple event contains a single experimental outcome, whereas a compound event contains more than one outcome. Probability is merely a measure of the likelihood of occurrence of an event and is restricted to values between 0 and 1, inclusive. In assigning probabilities to events, we have three primary methods available, namely, objective assessment, subjective assessment, and the relative frequency approach. These methods basically give rise to probabilities for a single event.

If we are combining events, it generally will be necessary to use the addition rule and/or the multiplication rule of probability. The addition rule is implied by the word "or"; it is necessary to check to determine whether the events are mutually exclusive. The multiplication rule applies when events are connected by the word "and"; it is necessary to check for independence of the events in question. When the events are not independent, the multiplication rule will require determination of a conditional probability. The condition in a conditional probability merely redefines the sample space in light of the information given by the condition.

In most practical problems, enumerating the sample space will be quite tedious and time consuming. Generally, we will not be interested in the actual outcomes themselves, but rather in how many of them there are. Thus, you were provided with some basic counting techniques, namely, the $m \cdot n$ rule, permutations, and combinations. If you are in doubt about whether to use a permutation or a combination, you should ask yourself, "Do I get a different outcome when the order is changed?" If the answer is yes, then permutations apply, since a permutation is an ordering, whereas a combination is simply a set and the order is irrelevant.

The final concept introduced was Bayes' rule, which will be used extensively in Chapter 6. In this case, you have available marginal (absolute) probabilities for a set of mutually exclusive and collectively exhaustive events. Also provided will be a set of conditional probabilities for some other event where the conditions are the original events. The question of interest then reverses the conditional probability, and Bayes' rule summarizes the intermediate steps necessary for finding the answer.

For ease in reference, we present the following tabular summary of these basic concepts.

Probability Axioms

1. $0 \leqslant \text{pr}(A) \leqslant 1$
2. $\text{pr}(A \cap B) = 0$ whenever A and B are mutually exclusive
3. $\text{pr}(A \cup A') = 1$, where A' denotes the complement of event A
4. $\text{pr}(A \cup B) = \text{pr}(A) + \text{pr}(B) - \text{pr}(A \cap B)$*
5. $\text{pr}(A \cap B) = \text{pr}(A|B)\ \text{pr}(B)$†

*Axiom 4 is known as the addition law of probability.
†Axiom 5 is known as the multiplication law of probability.

Counting Rules

1. Extended $m \cdot n$ rule

   If event i contains n_i possible outcomes, $i = 1, 2, \ldots, k$, and the k events are independent, then the total number of outcome combinations across the k events is given by the product:

$$n_1 \cdot n_2 \cdot n_3 \cdots n_k$$

2. Permutations

   If we want to count the number of ways in which r objects can be chosen from n objects when a different order of choice is considered to be a different way to choose the r objects, then the total number of different outcome sequences is given by

$$\frac{n!}{(n-r)!}$$

3. Combinations

   If we want to count the number of ways in which r objects can be chosen from n objects when a different order of choice is not considered to be a different way to choose the r objects, then the number of different choices of the r objects is given by

$$\frac{n!}{(n-r)!r!}$$

Other Points of Interest

1. Statistical independence

   Two events A and B are said to be statistically independent whenever $\text{pr}(A|B) = \text{pr}(A)$, that is, when information on event B leaves the probability of event A unchanged. The following statements are mathematically equivalent to this:

$$\text{pr}(B|A) = \text{pr}(B)$$
$$\text{pr}(A \cap B) = \text{pr}(A)\,\text{pr}(B)$$

2. Bayes' rule

   If $S = E_1 \cup E_2 \cup E_3 \cup E_4 \cup E_5 \cup \cdots \cup E_k$ and E_j and E_k are disjoint whenever $j \neq k$, then

$$\mathrm{pr}(E_i|A) = \frac{\mathrm{pr}(A|E_i)\,\mathrm{pr}(E_i)}{\sum_{j=1}^{k} \mathrm{pr}(A|E_j)\,\mathrm{pr}(E_j)}$$

for any event A [note that we assume that $\mathrm{pr}(A) \neq 0$; otherwise, the fraction above would indicate division by zero].

REVIEW EXERCISES

Level 1

2.1 A box of fuses contains 2 ten-amp, 3 fifteen-amp, 2 twenty-amp, and 2 twenty-five-amp fuses. During an electrical storm, 2 fifteen-amp fuses and 1 twenty-amp fuse burn out. In the dark, the owner randomly selects three fuses from the box. What is the probability that he will have:
(*a*) The correct replacements?
(*b*) Exactly two of the three correct?

2.2 A basketball recruiter observes that 50% of all good players stand taller than 6′8″, whereas only 10% of the poor players do. If 30% of all players are good, what is the probability that a player over 6′8″ is good? What is the probability that a player will exceed 6′8″?

2.3 An advertising executive has an office on Madison Avenue and a home in Connecticut. The probability that her house will be burglarized in any one year is 0.10, whereas for her office, this probability is 0.30. Find the probabilities of the following events:
(*a*) Both house and office are burglarized in the same year.
(*b*) Neither is burglarized in a given year.
(*c*) One or the other, but not both, is burglarized in a given year.

2.4 In how many ways can a company select a president, a controller, a vice president in charge of sales, and a vice president in charge of operations from among ten equally qualified candidates?

2.5 Determine the probability for each of the following events:
(*a*) An odd number appears in a single toss of a fair die.
(*b*) At least one head appears in two tosses of a fair coin.
(*c*) An ace, the ten of diamonds, or a spade appears in drawing a single card from a well-shuffled bridge deck.
(*d*) A sum divisible by 3 appears in a single roll of a pair of fair dice.

2.6 A box contains six red, four white, five blue, and two black balls. If three balls are drawn randomly one at a time, find the probability that they are drawn in the order red, white, blue if they are selected:
(*a*) With replacement.
(*b*) Without replacement.
(*c*) Suppose that we select two balls without replacement from this box. What is the probability that the balls will be black and blue?

2.7 Out of five accountants and seven salespeople, a committee consisting of two accountants and three salespeople is to be selected. In how many ways can this be done, if:
(*a*) Any of the people may be included?
(*b*) One particular salesperson must be on the committee?
(*c*) Two particular accountants cannot be on the committee?

2.8 Good Scout training teaches you to "be prepared." In the past, you were caught without your boots when you needed them 25% of the time. Only 5% of the time have you had your boots when you didn't need them. If the weather report indicates an 80% chance of rain, what is the probability of rain if you wear your boots?

2.9 The residents of Halifax were surveyed to determine readership of available newspapers within the city. It was found that 70% of the residents read the *Daily Dope*, 20% read the *Halifax Herald*, and 5% read both. If a single resident of Halifax is selected at random, what is the probability that he reads either the *Daily Dope*, the *Halifax Herald*, or both?

2.10 Allprovince Insurance Company has insured 40,000 class A drivers (good risks), 50,000 class B drivers (average risks), and 10,000 class C drivers (bad risks). The probabilities of an accident during any year are 0.01, 0.04, and 0.16, respectively, for class A, B, and C drivers. The company sold Ms. Taylor an insurance policy, and she had a crack-up within one year. What is the probability that she was a class A driver?

2.11 The personnel office for Kamikaze Aircraft, Ltd., has records that show the following breakdown of its 420 engineers.

Age	Bachelor's Degree Only	At Least a Master's Degree	Totals
Under 30	180	10	190
30–40	40	60	100
Over 40	80	50	130
Totals	300	120	420

If one engineer is selected at random from the company, what is the probability that he or she

(*a*) Has only a bachelor's degree?
(*b*) Is under 30, given that she or he has a master's degree?
(*c*) Is over 40 and has a bachelor's degree only?
(*d*) Is between 30 and 40 inclusive or has a master's degree?
(*e*) Are age and educational level independent of each other?

2.12 For their first ball game under a new manager, three players, Long Ball, Whiffer, and Line Drive, as a practical joke, exchanged numbers by putting all their shirts in a locker and randomly drawing one out to wear. Needing a pinch hitter in the ninth inning, the manager sent the player wearing Long Ball's number to bat. This player hit a home run. From their records, the players in the past have hit home runs with the following frequencies: Long Ball, 0.10; Whiffer, 0.005; Line Drive, 0.03. What is the probability that Long Ball hit the home run?

2.13 Ray's Salvage has just purchased a truckload of damaged merchandise. Included in the purchase are ten color TVs, of which four are inoperable because of internal damage. Ray is selling the sets, as is, for \$100 each, and all sales are final. The same set purchased new at Bighee's Department Store sells for \$379.00. However, Ray will not let you plug a set in before purchasing it.

(*a*) If you buy one set at random from Ray, what is the probability that it will be a good one?
(*b*) If you strongly desire a color TV in working order, what is the probability that you will get at least one if you buy two of Ray's sets at random?

2.14 Fifteen percent of college students suffer from a disease known as Turner's Trauma. A second disease, known as Meek's Malady, is contracted by 20% of all college students. Both diseases cannot be present simultaneously. The remainder of the students are free of both diseases. A test has been devised to detect Meek's Malady and does so when it is present with a probability of 0.90. If a student has Turner's Trauma, the test will indicate Meek's Malady 25% of the time; if the student is healthy, the test will indicate Meek's Malady with a probability of 0.10.

(*a*) If the test is administered to a randomly selected student, what is the probability that it will indicate Meek's Malady?
(*b*) You have just taken the test and it indicated Meek's Malady. What is the probability that you have Meek's Malady?

2.15 In a particular class, there are 15 students. The instructor has decided in advance to give 2 A's, 3 B's, 7 C's, 2 D's, and 1 F. If he randomly assigns grades to students, in how many ways can he do it?

Level 2

2.16 Prove that the multiplication rule for two events can be extended to three events, giving the following:

$$\text{pr}(A \cap B \cap C) = \text{pr}(A)\,\text{pr}(B|A)\,\text{pr}(C|A \cap B)$$

2.17 If twelve people are chosen at random, what is the probability that at least two of them will have the same birthday? (Ignore leap years.)

2.18 A piece of electronic equipment has three essential parts. In the past, part A has failed 20% of the time, part B has failed 30% of the time, and part C has failed 40% of the time. Part A operates independently of B and C, but parts B and C are interconnected, so that failure of either part affects the other. In those instances where C failed, the chances were 2 out of 3 that B would also fail. If at least two of the three parts must operate for the equipment to function, what is the probability of its functioning?

2.19 At a particular dinner party, there are seven people to be seated around a circular table. If the orientation of the table does not matter, in how many ways can this be accomplished?

2.20 If A, B, and C are not mutually exclusive events, find the expression for the probability that at least one of them will occur.

2.21 In a display case, an item is packaged in two different colors. Initially, there were an equal number of packages in each color, say red and blue, and the colors were thoroughly mixed. Each time a customer takes an item, another package of the same color is used to restock it. You are told that the first six selections included four red and two blue packages. If the items were selected at random, what is the probability that, at the end of each selection, the combined number of red items exceeded the combined number of blue items selected?

2.22 A fruit inspector accepts or rejects shipments of bananas after measuring a few sample bananas. He rejects 20% of all shipments received. Thus far, he has rejected 95% of shipments of short bananas received and 80% of long bananas received. In addition, 10% of all shipments have been long bananas and 8% have been short bananas. Find the probability that:
(*a*) a shipment has long bananas and will be rejected.
(*b*) a shipment is acceptable if he rejected it.
(*c*) a shipment is unacceptable and will be rejected.
(*d*) a shipment is acceptable and will be accepted.
(*e*) a shipment is unacceptable if he rejected it.

2.23 The key to a certain file cabinet is one of 15 in a drawer and all 15 look alike. The file cabinet in question is locked 70% of the time. If you select 2 keys at random before going to the file cabinet, what is the probability that you can open the cabinet without returning for more keys? What is the probability that you must visit the drawer a total of three times if you select 2 keys at random each time? (Assume that you keep incorrect keys in your pocket.)

2.24 An owner of a small motel having five units has television sets in two of the units. In the past, 50% of all customers have requested a room with a television set. If they don't request one, they are assigned a room with no television (if available). Suppose you stop at this motel for the night and desire to watch television. Assuming you can get a room and that occupancy rates are one at 20%, two at 35%, three at 30%, four at 10%, and five at 5%, then what is the probability that you also will be able to watch television?

2.25 Three people are working independently on the design of an information system. If the probability of being successful is .5 for the first, .4 for the second, and .2 for the third, what is the probability:

(*a*) of getting a successful design?

(*b*) that exactly two of the people will be successful?

Level 3

2.26 A couple start at a corner and stroll four blocks. Each time they come to a corner, they choose one of the directions—N, S, E, W—at random. What is the probability of their ending the walk at the same corner from which they started?

2.27 Prove that if $\text{pr}(A|B \cap C) = \text{pr}(A|B \cap C')$, then $\text{pr}(A|B) = \text{pr}(A|B \cap C)$.

2.28 Prove that two events are independent if and only if $\text{pr}(A \cap B) = \text{pr}(A)\,\text{pr}(B)$

2.29 There are 2-person tables at a party, and guests are seated in pairs as they arrive. Suppose 27 people arrive *randomly* to attend the party. Give a formula for the number of different seating plans that can be formed by this method.

2.30 Two door-to-door salespersons work independently of each other. The probability that the first one will make a sale on any particular call is .2, whereas the same probability for the second is .1.

(*a*) What is the probability, if each makes five calls, that they will make three sales in all?

(*b*) If they make calls simultaneously, what is the probability that a total of 10 calls will be required to get three sales?

Chapter 3
Describing Stochastic Phenomena Using Random Variables

KEY WORDS AND TERMS

Random Variable
Probability Distribution
Probability Function
Density Function
Parameter
Expected Value
Variance
Chebyshev's Inequality
The Empirical Law

3.1 INTRODUCTION

In the previous chapter we introduced the concept of probability and its connection with the outcomes of an experiment. When an experiment can have more than one outcome, with at least two of these having positive probability, then we say the experiment has a *stochastic* (as opposed to deterministic) nature because the next experimental outcome cannot be predicted with certainty. In performing an experiment or in observing stochastic phenomena, we are frequently interested in some function of the outcome rather than the actual outcome itself. Racetrack betting is a good example of a stochastic phenomenon for which the outcome is secondary to the payoff associated with the outcome. Mathematically, we call this association between outcome and payoff a function from the sample space (set of outcomes) to the real line (set of payoffs). Such associations are called *random variables.*

3.2 RANDOM VARIABLES

Consider a game in which two fair dice are rolled and the total number of pips on the up faces determine a payoff, according to the following table:

Sum	Payoff
2	20
3	0
4	0
5	0
6	0
7	10
8	0
9	0
10	0
11	10
12	20

The sample space containing the most information for this experiment is the set of all ordered pairs containing the first six integers, and we can construct an association between these outcomes and the payoffs as shown:

Sample Space Elements	Payoff
(1, 2), (2, 1), (3, 1), (4, 1), (5, 1), (6, 2) (1, 3), (2, 2), (3, 2), (4, 2), (5, 3), (6, 3) (1, 4), (2, 3), (3, 3), (4, 4), (5, 4), (6, 4) (1, 5), (2, 4), (3, 5), (4, 5), (5, 5) (2, 6), (3, 6), (4, 6)	0
(1, 6), (2, 5), (3, 4), (4, 3), (5, 2), (6, 1), (5, 6), (6, 5)	10
(1, 1), (6, 6)	20

What you see here is a tabular presentation of a random variable, which is defined formally as follows:

DEFINITION 3.1 A random variable is a function whose domain is a sample space and whose range is a subset of real numbers.

This definition means that each outcome of the experiment is assigned exactly one real number. In the previous example, the assigned real numbers correspond to the payoff for each type of outcome.

Almost all random variables of any interest in the business world are one of two types. The first of these types, called *discrete*, is defined here.

DEFINITION 3.2 When a random variable of interest is assigned either a finite or a countable number of possible values, then it is said to be discrete.

Since the possible values that the random variable in the example can assume are 0, 10, and 20, it can be assigned any one of a finite number of values. Thus, Definition 3.2 tells us that it must be a discrete random variable. Other examples of discrete random variables would be a count of the number of vehicles activating a highway counter per unit of time or the number of customers entering a store per unit of time.

The second type of random variable of interest is called *continuous*; it is used in business analysis whenever a continuum of possible values is assigned.

DEFINITION 3.3 When the random variable of interest assigns a continuum of possible values, then it is said to be continuous.

Continuous random variables are used in reliability studies in which we want to determine the characteristics of time until failure or time between overhauls. In fact, most measurement questions of a stochastic nature are adequately modeled using continuous random variables. Typical measurements would be time (until failure, between events, until events have occurred, and so forth), length (taphole diameters drilled by a machine), weight (ounces of fill dispensed by a coin-operated coffee machine), volume (kiloliters of river water flowing through a flood gate per unit time), or area (amount of material produced by a weaving machine). We will provide additional examples of discrete and continuous random variables in the remainder of this chapter. The next step is to construct an association between the assigned values and their probability of occurrence.

EXERCISES

3.2.1 Determine which of the following phenomena would be properly described using discrete random variables and which would be properly described using continuous random variables:

(*a*) The number of bubble defects per square meter of glass produced by a manufacturer of window glass

(*b*) The BTU level produced by a parabolic sunlight reflector on a summer day

(*c*) The average of the wind velocity off Peggy's Cove, Nova Scotia, and Nantucket, Massachusetts, on January 1, 2000

(*d*) The number of Japanese beetles I will find in my garden next July

3.2.2 Suppose that we roll a fair die and flip a fair coin in a game that awards 10 dollars whenever one pip shows on the up face of the die and 2 dollars whenever a head shows on the up side of the coin. Give a tabular representation of the random variable for this experiment.

3.2.3 Which of the following would be considered random variables?
- (*a*) The number of pips occurring on the upturned faces in rolling a pair of dice
- (*b*) The number of points that will be scored in a football game
- (*c*) The thickness of the next piece of sheet metal purchased by a body shop
- (*d*) The day of the week on which July 4 falls
- (*e*) The distance from Halifax to Toronto
- (*f*) The number of blemishes on the next piece of wood paneling to come off a production line

3.2.4 In Exercise 3.2.3, identify which of the random variables are continuous and which are discrete.

3.3 PROBABILITY DISTRIBUTIONS OF RANDOM VARIABLES

Let us return our attention to the die experiment in which we total the number of pips on the upturned faces. By counting the number of sample points associated with the payoff values of 0, 10, and 20, we can calculate the probability that the random variable will assume each value. Thus, we can add a third column to our mathematical description of the experiment. This is shown in Table 3.1.

Table 3.1 Mathematical Description for a Die-tossing Experiment

Sample Space Elements	Payoff	Probability of Payoff
(1, 2), (2, 1), (3, 1), (4, 1), (5, 1), (6, 2) (1, 3), (2, 2), (3, 2), (4, 2), (5, 3), (6, 3) (1, 4), (2, 3), (3, 3), (4, 4), (5, 4), (6, 4) (1, 5), (2, 4), (3, 5), (4, 5), (5, 5) (2, 6), (3, 6), (4, 6)	0	26/36
(1, 6), (2, 5), (3, 4), (4, 3), (5, 2), (6, 1) (5, 6), (6, 5)	10	8/36
(1, 1), (6, 6)	20	2/36

From the previous discussion, we know that the first two columns of Table 3.1 define a random variable. The second two columns define what we call a *probability distribution* for the random variable; for discrete random variables, this is said to be a *probability function.*

DEFINITION 3.4 A probability function is any function whose *domain* is the range of a discrete random variable and whose *range* is the set of probabilities associated with the domain values.

Since a random variable *is* a function, we can show that it is equivalent to a partition of the sample space and from this fact can extrapolate that the probability function range values must sum to 1. For instance, the range values in the previous example are 26/36, 8/36, and 2/36, which sum to 1.

EXAMPLE 3.1 Suppose that we flip a fair coin three times and observe the number of heads. The random variable and probability function for this experiment are given by Table 3.2.

Table 3.2 Random Variable and Probability Function for Observing the Number of Heads in Three Tosses of a Fair Coin

Sample Space Elements	Random Variable Range Values	Probability Function Range Values
T, T, T	0	1/8
T, T, H T, H, T H, T, T	1	3/8
T, H, H H, T, H H, H, T	2	3/8
H, H, H	3	1/8

Again, the random variable is defined by the first two columns, whereas the probability function is defined by the second two columns.

It is common practice in statistics to condense such tabular presentations whenever possible. The aim is to provide a compact mathematical function in equation form. The example of Table 3.2 is easily expressed as

$$\text{pr}(X = k) = \binom{3}{k}\left(\frac{1}{2}\right)^3 \qquad \text{for } k = 0, 1, 2, 3$$

in which X represents the random variable and k represents the possible number of heads. You should justify for yourself that this function in equation form (which we also call a probability function) gives the same numbers as those in the last column of Table 3.2.

EXAMPLE 3.2 A machine produces defective items 10% of the time and does so on a random basis. Give the probability function for the number of defective items in the next lot of size 25 (assume independence of quality in production).

The independence assumption allows us to use the concept of independent events to calculate the probability of exactly k defectives, where k can be any number between 0 and 25. The probability of any particular sequence of k defective and $25 - k$ nondefective items is given by

$$\left(\frac{1}{10}\right)^k \left(\frac{9}{10}\right)^{25-k}$$

Since there are $\binom{25}{k}$ ways in which the k defective items could be produced in the sequence, the addition law of probability applied to $\binom{25}{k}$ mutually exclusive events tells us that the total probability of k defective items is given by

$$\binom{25}{k}\left(\frac{1}{10}\right)^k \left(\frac{9}{10}\right)^{25-k}$$

Again, this result follows because each event has the same identical probability, $(1/10)^k\ (9/10)^{25-k}$, and there are $\binom{25}{k}$ of these that could provide k defective items. Thus, if X represents the random variable that counts the number of defective items in a lot of size 25 produced by the machine, its probability function is given by

$$\text{pr}(X = k) = \binom{25}{k}\left(\frac{1}{10}\right)^k \left(\frac{9}{10}\right)^{25-k} \qquad k = 0, 1, 2, \ldots, 25$$

Notice that this last result is a substantial improvement over the equivalent tabular representation, which would take 26 lines. The next chapter discusses the type of experiment for which this probability function is appropriate.

Probability distributions for continuous random variables are called *density functions.* However, these functions are quite different from probability functions because their range values do not represent probabilities.

DEFINITION 3.5 A density function is any nonnegative function whose *domain* is the range of a continuous random variable and whose integration over any interval is the probability that the random variable assumes a value within the interval.

The word *integration* refers to the process of calculating the area underneath the curve but above the horizontal axis. We require that this area be unity because we determine probabilities for continuous random variables by calculating the area underneath the curve corresponding to the interval of interest.

EXAMPLE 3.3 A liquid duplicating machine produces copies that fade with age, none of which last more than ten years. Calculate the probability that the next copy produced by the machine will last at least four years if the density function for the age random variable is given by

$$f(x) = \begin{cases} 0.2 - 0.02x & 0 \leqslant x \leqslant 10 \\ 0 & \text{otherwise} \end{cases}$$

This is a tall order, but the problem will definitely allow us to discuss all the aspects of working with continuous random variables. First, age is a time measurement, and so it seems reasonable that we are talking about a continuous random variable. Second, the probability that the fade is completed at any specified point in time is zero. Think about how ridiculous it is to even presuppose the idea that the fade is completed at precisely 3.75 years or precisely any number of years. The problem is that the fade can be completed at precisely so many different times that the probability that it is accomplished at exactly one of them is zero. You can rationalize the situation by thinking about simple raffles; the more people there are in the raffle, the less likely it is that you will win (assume that everyone buys one ticket). As the number of people increases, the probability that you will be the winner approaches zero, and the only intelligent questions that can be asked center around groups of individuals. That is, it makes sense to talk about 1 of 100 people being the winner. The difference between this situation and that of Example 3.3 is that there are many more real numbers between zero and ten than there are people on the earth. Thus, density function values cannot represent probabilities, and we must concentrate our attention on determining probabilities over age intervals. For instance, the example really asks us to calculate the probability that the fade is accomplished somewhere in the interval from four years to ten years. To find this probability we must calculate the area *underneath* the density function and *above* the horizontal axis that lies between the points 4.0 and 10.0.

The situation is portrayed graphically in Figure 3.1, in which the shaded area represents the probability we need to determine. This type of distribution is aptly called a *triangular distribution* because its plot depicts a triangle section on the grid. This fact also makes it quite easy to calculate areas, since we know from geometry that the area of a triangle is given by $\frac{1}{2}bh$, where b represents the measure of the base and h represents the measure of the height. Thus

$$b = 10 - 4 = 6$$

and

$$h = 0.2 - 0.02(4) = 0.12$$

so that

$$\frac{1}{2}bh = \frac{1}{2}(6)(0.12) = 0.36$$

This tells us that the probability that the next copy produced by the machine will last at least four years is 0.36 or, equivalently, that 36% of

Figure 3.1

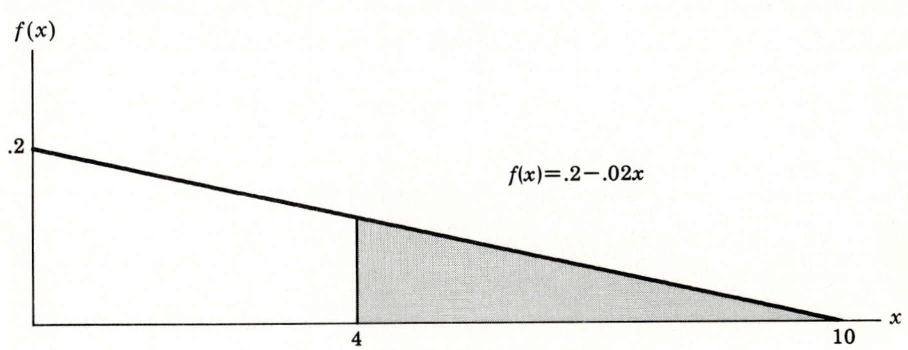

the copies produced by the machine will last at least four years before complete fade.

It should be noted that the probability and density functions we have presented are actually members of families of distributions that can be characterized using a concept presented in the next section.

EXERCISES

3.3.1 A fair coin is flipped five times. Determine the probability function for a random variable that counts the number of heads observed over the five trials.

3.3.2 Nim and Rod decide to go fishing. Nim will catch no fish with probability 0.5, one fish with probability 0.3, and two fish with probability 0.2. Rod will catch no fish with probability 0.1, one fish with probability 0.2, two fish with probability 0.6, and three fish with probability 0.1. Determine the probability distribution for the total number of fish caught by Nim and Rod, assuming that their catches are independent of each other.

3.3.3 A certain small company has ten employees, seven men and three women. From the personnel files, four cards are drawn at random. Construct the probability distribution of the random variable x, the number of women drawn.

3.3.4 A baker knows that 90% of all wedding cakes she bakes will be good. She has orders for four such cakes for next week. Calculate the probability distribution of Y, the number of good wedding cakes resulting from the next four attempts. What is the probability that Y will equal 3?

3.3.5 For the triangular distribution given by

$$f(x) = \begin{cases} \dfrac{2x}{9} & 0 \leqslant x \leqslant 3 \\ 0 & \text{elsewhere} \end{cases}$$

Determine:

(a) $\text{pr}(x \leqslant 1.5)$

(b) $\text{pr}(0.5 \leqslant x \leqslant 2.5)$

3.4 PROBABILITY DISTRIBUTION PARAMETERS

Example 3.1 shows how we can characterize a coin-tossing experiment by using a probability function that tells us the probability of observing k heads in three tosses ($k = 0$ or 1 or 2 or 3). But notice that the problem does not change much if we ask the same type of question for five tosses or six, or for n tosses where n could be 1, 2, 3, or any number of tosses we want to examine. This general case (in which we leave the number of tosses n unspecified) has the probability function (since the coin is fair)

$$\text{pr}(x = k) = \binom{n}{k}\left(\frac{1}{2}\right)^n \qquad \text{for } k = 0, 1, 2, \ldots, n \tag{3.1}$$

We say that this last probability function represents the family of probability functions for modeling the number of heads in a specified number of flips of a *fair* coin. If we want the probability function for a specific number of flips, we need only replace the letter n with the number of flips we are executing. The value of n is what we call a *parameter*, which is defined in Definition 3.6.

DEFINITION 3.6 Probability distribution parameters are those values that must be determined in order to isolate a particular probability distribution from its family.

From this definition, n represents a parameter but k does not, and we say that the distribution for modeling the number of heads in n tosses of a coin has two parameters: n, the number of tosses, and p, the probability of a head on any given toss.

Continuous distributions also have parameters. For instance, Example 3.3 actually gives a density function of the form

$$f(x) = \frac{2}{10} - \frac{2}{10^2}x \qquad 0 \leqslant x \leqslant 10$$

in which the parameter value is 10, the maximum number of years until complete fade. In general, a maximum number of β (beta) years until complete fade corresponds to a density function as given by

$$f(x) = \frac{2}{\beta} - \frac{2}{\beta^2}x \qquad 0 \leqslant x \leqslant \beta \tag{3.2}$$

In practice, we have a dual pursuit when we use probability distributions as decision tools. One problem is to select an appropriate family of distributions for the problem, and the other is to determine the parameters for the family we select. For instance, if we were to select the density function family in Equation 3.2 as an appropriate model for the fade process, our next problem would be to determine a value for β (the maximum number of years until complete fade). Determination of β may involve estimation, a process that will be outlined in detail in Chapter 8.

EXERCISES

3.4.1 A probability function describes the distribution of probabilities for the number of tosses of a biased coin until a certain number of heads are observed in a row. Determine the number of parameters for the function.

3.4.2 A probability function is given by

$$\text{pr}(X = k) = \binom{n}{k} p^k (1 - p)^{n-k} \qquad k = 0, 1, 2, \ldots, n$$

What letters represent the parameters of this distribution?

3.4.3 A probability function is given by

$$\text{pr}(X = k) = p(1 - p)^{k-1} \qquad k = 1, 2, 3, \ldots$$

How many parameters are there in this distribution?

3.4.4 Identify the values of the parameters for:
(*a*) Exercise 3.3.3.
(*b*) Exercise 3.3.4.
(*c*) Exercise 3.3.5.

3.5 EXPECTED VALUES OF RANDOM VARIABLES

Because of the stochastic nature of a random variable, it is not possible to predict an exact value in an experiment. Also, in many cases the complete information provided by a probability or density function is not required (or available) in order to make a decision. That is, we can get by with descriptors that summarize the dominant features of the range of the random variable. Usually a descriptor takes the form of a weighted average of a function of the random variable. This weighted average is called the *expected value* of the function. Expected values prove to be easier to obtain estimates for and easier to work with than the complete probability distribution. The definition of the expected value of a random variable follows.

We discuss expected values for both continuous and discrete random variables, but we will begin with the discrete case.

DEFINITION 3.7 The expected value of a discrete random variable is given by

$$E(X) = \sum_{i=1}^{n} x_i \text{ pr}(X = x_i) \tag{3.3}$$

Notice that the expected value of X, denoted μ_x, is a weighted average of the range values in which the weights are the probabilities associated with each range value. This can be visualized as a plank and fulcrum: if weights are placed at positions $x_1, x_2, \ldots, x_n$ along the plank such that the weight at x_i is $\text{pr}(X = x_i)$, $i = 1, 2, \ldots, n$, then the fulcrum placed at the balance point represents the expected value.

EXAMPLE 3.4 Suppose that we toss a fair coin three times and will award a dollar for each head observed. What is the expected value of the game (what should we charge each player if we want to break even in the long run)?

Table 3.2 tells us that this question is equivalent to asking for the expected value of the random variable whose probability function is given by

$$\text{pr}(X = k) = \binom{3}{k}\left(\frac{1}{2}\right)^3$$

By Definition 3.7, we know that the expected value can be calculated using the formula

$$E(X) = \sum_{i=1}^{n} x_i \text{ pr}(X = x_i)$$

For our problem this gives us

$$\begin{aligned} E(X) &= \sum_{i=1}^{4} x_i \text{ pr}(X = x_i) \\ &= x_1 \text{ pr}(X = x_1) + x_2 \text{ pr}(X = x_2) + x_3 \text{ pr}(X = x_3) + x_4 \text{ pr}(X = x_4) \\ &= 0\left(\frac{1}{8}\right) + 1\left(\frac{3}{8}\right) + 2\left(\frac{3}{8}\right) + 3\left(\frac{1}{8}\right) = \frac{12}{8} = 1.5 \end{aligned}$$

Therefore, we should charge each player \$1.50 to play the game.

This example emphasizes the difference between the phrases "what value do you expect?" and "what is the expected value?" That is, we do not expect a payoff of \$1.50 for *any* game; the expected value or payoff is not one of the observable values or possible payoffs of 0, 1, 2, or 3 dollars. From the definition of probability, we know that the probability values of 1/8, 3/8, 3/8, and 1/8 represent long-run relative frequencies, so that the expected value of \$1.50 is what we would expect to pay out, on the average, over the long term.

The expected value for a continuous random variable uses integration instead of summation.

DEFINITION 3.8 The expected value of a continuous random variable X with density function $f(x)$ is given by

$$E(X) = \int_{-\infty}^{\infty} xf(x)\,dx \tag{3.4}$$

The symbolism in Definition 3.8 tells us that we want to compute the area underneath the curve given by multiplying x by the density function $f(x)$.

EXAMPLE 3.5 What is the expected value for time until complete fade for the copy in Example 3.2?

Recall that the density function for this example is given by

$$f(x) = \begin{cases} 0.2 - 0.02x & 0 \leqslant x \leqslant 10 \\ 0 & \text{elsewhere} \end{cases}$$

thus

$$xf(x) = \begin{cases} 0.2x - 0.02x^2 & 0 \leqslant x \leqslant 10 \\ 0 & \text{elsewhere} \end{cases}$$

so that

$$E(X) = \int_{-\infty}^{\infty} xf(x)\,dx = \int_{0}^{10} (0.2x - 0.02x^2)\,dx$$

This situation is shown in Figure 3.2 with the integral represented as the shaded area. It is difficult to use geometrical formulas to find this area, since the curve is parabolic. The review of basic calculus in the Appendix shows how we can compute areas of this type, using a

Figure 3.2 Region Showing the Expected Value Area for the Copy Fade Problem of Example 3.2

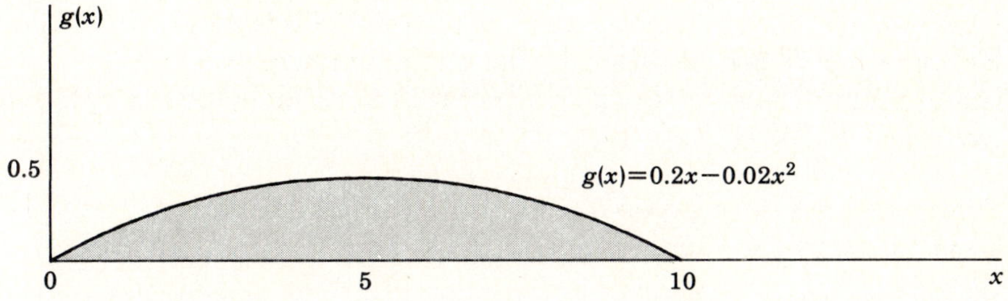

method developed independently by Sir Isaac Newton and Gottfried Wilhelm Liebnitz in 1665. Using this method, we get

$$\mu_x = \int_0^{10} (0.2x - 0.02x^2)\, dx = 0.2\frac{x^2}{2} - 0.02\frac{x^3}{3}\Bigg|_0^{10}$$

$$= 0.2\frac{10^2}{2} - 0.02\frac{10^3}{3}$$

$$= 10 - \frac{20}{3} = \frac{10}{3}$$

Thus, the area is $3\frac{1}{3}$ units, so that in the long run the average number of years until complete fade will be $3\frac{1}{3}$ years.

There are many useful rules for finding expected values that will come in handy in Chapters 4 and 5, in which we present probability distributions for discrete and continuous phenomena common to business problems. These rules apply to both discrete and continuous random variables.

RULE 3.1 The expected value of a constant is the constant. That is, if c represents a constant, then

$$E(c) = c \tag{3.5}$$

RULE 3.2 If X is a random variable and c is a constant, then

$$E(cX) = cE(X) \tag{3.6}$$

RULE 3.3 If X and Y are random variables and a and b are constants, then

$$E(aX \pm bY) = aE(X) \pm bE(Y) \tag{3.7}$$

These three rules are not difficult to show; hence we will confine ourselves to the derivation of the discrete case of Rule 3.2. Note that here

$$g(X) = cX$$

so

$$E(cX) = E[g(X)] = \sum_{i=1}^{n} (cx_i)\, \mathrm{pr}(X = x_i) = c \sum_{i=1}^{n} x_i\, \mathrm{pr}(X = x_i)$$
$$= cE(X)$$

in which we assume the range of X to be $x_1, x_2, \ldots, x_n$.

3.5.1 Law of the Unconscious Statistician

Definition 3.7 can be generalized to give expected values for functions of the random variable X, that is, for a new random variable $g(X)$. This generalization is sometimes called the law of the unconscious statistician.

DEFINITION 3.9 Let $g(X)$ be a function of a discrete random variable X that has the probability function $\mathrm{pr}(X = x_i)$. The expected value of $g(X)$ is given by

$$E[g(X)] = \sum_{i=1}^{n} g(x_i)\, \mathrm{pr}(X = x_i) \tag{3.8}$$

in which we assume that X has a range given by $x_1, x_2, \ldots, x_n$.

This general form of the expected value of a function of a random variable has great value in higher statistical analysis, but for our purposes we will be concerned only with two special functions for $g(X)$. The first of these is the identity function,

$$g(X) = X$$

which we have already considered. The second is the quadratic function,

$$g(X) = (X - \mu_x)^2$$

whose expectation is called the variance of the random variable x.

3.5.2 The Variance of a Random Variable

The second popular descriptor for probability distributions is provided by using the function

$$g(X) = [X - E(X)]^2$$

in Definition 3.9. This gives us what we call the *variance* of the random variable.

DEFINITION 3.10 The variance of a random variable X is given by

$$E[X - E(X)]^2 \tag{3.9}$$

The two E entries sometimes cause a bit of confusion. So we will use an alternative notation, substituting μ_x for $E(X)$. Thus, the variance of a random variable can now be written

$$E(X - \mu_x)^2$$

Before we show you an example, we will justify the importance of the variance as a descriptor *in addition* to the expected value. In Table 3.3 you can observe three games you could play, each with the same expected value. The decision as to which game to play is based on consideration of the variance as well as the expected value.

Table 3.3 Three Possible Payoff Sets for a Coin-tossing Game

Event	Payoff Set 1	Payoff Set 2	Payoff Set 3
Head	51.00	75.00	100.00
Tail	49.00	25.00	0

Assuming that the coin is fair, the expected value (the long-run payoff) for each payoff set is $50.00 per game. If you have to pay $50.00 to play the game and have money problems at home, you probably would choose payoff set 1 over payoff set 3. If you are the adventurous type who likes a little risk, then you might prefer payoff set 3. The element of risk is reflected in the variance, and we will now use the formula $E(X - \mu_x)^2$ to compute the variance for each of these payoff sets. The variance for payoff set 1 is given in tabular form as follows, where σ_x^2 (read "sigma-squared") denotes the variance:

x_i	$\text{pr}(X = x_i)$	$x_i \text{ pr}(X = x_i)$	$(x_i - \mu_x)^2$	$(x_i - \mu_x)^2 \text{ pr}(X = x_i)$
51	1/2	25.5	1	1/2
49	1/2	24.5	1	1/2
		$\mu_x = 50.0$		$\sigma_x^2 = 1$

For payoff set 3, we compute

y_i	$\text{pr}(Y = y_i)$	$x_i \text{ pr}(Y = y_i)$	$(y_i - \mu_y)^2$	$(y_i - \mu_y)^2 \text{ pr}(Y = y_i)$
100	1/2	50	2500	1250
0	1/2	0	2500	1250
		$\mu_y = 50.0$		$\sigma_y^2 = 2500$

Thus, the third payoff set has a much larger variance than the first payoff set, which reflects the risk relationship between the two games. Before noting some useful rules for the variance, we will use the expected value rules to convert the expression $E(X - \mu_x)^2$ into something much easier to manipulate computationally. Basic algebra tells us that

$$(X - \mu_x)^2 = X^2 - 2\mu_x X + \mu_x^2$$

so that

$$\begin{aligned} E(X - \mu_x)^2 &= E(X^2 - 2\mu_x X + \mu_x^2) \\ &= E(X^2) - E(2\mu_x X) + E(\mu_x^2) \qquad \text{(by Rule 3.3)} \\ &= E(X^2) - E(2\mu_x X) + \mu_x^2 \qquad \text{(by Rule 3.1)} \\ &= E(X^2) - 2\mu_x E(X) + \mu_x^2 \qquad \text{(by Rule 3.2)} \\ &= E(X^2) - 2\mu_x^2 + \mu_x^2 \\ &= E(X^2) - \mu_x^2 \qquad (3.10) \end{aligned}$$

Thus, the variance for payoff set 2 can be computed by the tabular routine shown here

x_i	$\text{pr}(X = x_i)$	$x_i \text{ pr}(X = x_i)$	x_i^2	$x_i^2 \text{ pr}(X = x_i)$
75	1/2	37.5	5625	2812.5
25	1/2	12.5	625	312.5
		$\mu_x = 50.0$		3125.0

So $\sigma^2 = 3125 - (50)^2 = 3125 - 2500 = 625$. Frequently, the variance is an awkward quantity to express verbally, since it is actually in squared units (for instance, we just computed a variance of 625 dollars squared). We use the positive square root to avoid this awkward situation. It is called the *standard deviation* and is denoted by σ_x (sigma). We will come back to the standard deviation in the next two sections.

Useful rules for the variance are as follows:

RULE 3.4 If X is a random variable with variance σ_x^2 and c is a constant, then the variance of $X + c$ is σ_x^2 (the variance of a random variable is unchanged by the addition of a constant).

RULE 3.5 If X is a random variable with variance σ^2 and c is a constant, then

$$\mathrm{Var}(cX) = E(cX - c\mu_x)^2 = c^2\sigma_x^2 \tag{3.11}$$

RULE 3.6 If X and Y are random variables with variance σ_x^2 and σ_y^2, respectively, and expected values μ_x and μ_y, respectively, and a and b are constants, then the variance of the linear combination $aX + bY$ is given by

$$\sigma_{ax+by}^2 = a^2\sigma_x^2 + b^2\sigma_y^2 + 2ab[E(XY) - \mu_x\mu_y] \tag{3.12}$$

The quantity

$$E(XY) - \mu_x\mu_y$$

is called the *covariance* of X and Y. Its value will be zero whenever X and Y are *independent** random variables. In this case, Equation 3.12 simplifies to

$$\sigma_{ax+by}^2 = a^2\sigma_x^2 + b^2\sigma_y^2 \tag{3.13}$$

*If X and Y are two random variables, the probability distribution for their simultaneous occurrence is called the *joint probability distribution* of X and Y, and it is given by (in the discrete case)

$$\mathrm{pr}(X = x_i,\ Y = y_i)$$

The notation denotes the probability that the range values of x_i and y_i are assumed at the same time. If

$$\mathrm{pr}(X = x_i,\ Y = y_i) = \mathrm{pr}(X = x_i)\,\mathrm{pr}(Y = y_i)$$

for all possible choices of x_i and y_i in their respective ranges, then X and Y are said to be independent.

EXERCISES

3.5.1 Find the expected value and variance for the following probability distribution X.

i	x_i	$\text{pr}(X = x_i)$
1	0	0.0778
2	1	0.2592
3	2	0.3456
4	3	0.2304
5	4	0.0768
6	5	0.0102

3.5.2 Find the variance for the following probability distribution X.

(*a*) Using the formula $\sum_{i=1}^{4} [x_i - E(X)]^2 \text{ pr}(X = x_i)$

(*b*) Using the formula $\sum_{i=1}^{4} x_i^2 \text{ pr}(X = x_i) - E(X)^2$

i	x_i	$\text{pr}(X = x_i)$
1	-2	0.4
2	0	0.3
3	4	0.2
4	10	0.1

3.5.3 Paul, a lumberjack, has daily earnings with an expected value of \$25.00 and a standard deviation of \$2.00 for his job in the north woods. His cousin Pete's daily earnings have an expected value of \$20.00 with a standard deviation of \$2.00 on his job picking produce in the Annapolis Valley. Calculate an expected value and standard deviation for the total amount of money Paul and Peter will earn on the next day they both work.

3.5.4 Suppose that a real estate investor is considering investment in two pieces of property, one in central Montana, say X, and one in northern Florida, say Y. The possible returns on investment for property X are (in thousands of dollars) -5, 0, +10 with corresponding probabilities of 0.2, 0.5, and 0.3, respectively, whereas for Y they are (in thousands of dollars) -10, 0, and 100 with respective probabilities of 0.3, 0.5, and 0.2. As can be verified, the means and variances of X and Y are $\mu_x = 2$, $\sigma_x^2 = 31$, $\mu_y = 17$, $\sigma_y^2 = 1741$. The total return on the two investments is then given by

$$V = X + Y$$

Find the expected total return and the variance of the total return.

3.5.5 Refer to Exercise 3.3.3 and calculate the expected value and variance for the number of women in the sample.

3.5.6 Refer to Exercise 3.3.4 and determine the standard deviation of the number of good cakes resulting from four attempts.

3.5.7 The Cleveland Browns have been averaging 21 points per game with a standard deviation of 7, whereas the Cleveland Cavaliers have been averaging 98 points per game with a standard deviation of 9. Both teams are playing on a particular Sunday. Calculate the mean and variance of the total number of points scored by the two teams.

3.5.8 Referring to Exercise 3.3.5, determine the mean and standard deviation for the random variable having that distribution.

3.6 CHEBYSHEV'S INEQUALITY

The variance of a random variable conveys information about the variability of the range values around the expected value. Chebyshev, a Russian mathematician, proved a relationship between the expected value and the standard deviation that allows us to compute a bound on the probability between any two values that are equidistant from the expected value.

CHEBYSHEV'S INEQUALITY The probability that a random variable X assumes a value within k units of its expected value is at least $1 - \left(\frac{\sigma_x}{k}\right)^2$, where σ_x denotes the standard deviation of X. That is,

$$\mathrm{pr}(\mu_x - k \leqslant X \leqslant \mu_x + k) \geqslant \begin{cases} 1 - \left(\dfrac{\sigma_x}{k}\right)^2 & k \geqslant \sigma_x \\ 0 & k < \sigma_x \end{cases} \tag{3.14}$$

or, equivalently,

$$\mathrm{pr}(|X - \mu_x| > k) < \begin{cases} \left(\dfrac{\sigma_x}{k}\right)^2 & k \geqslant \sigma_x \\ 1 & k < \sigma_x \end{cases} \tag{3.15}$$

The following example illustrates the use of Chebyshev's inequality and shows the improvement in the bound that results from knowing the actual type of distribution.

EXAMPLE 3.6 Suppose that we flip a fair coin ten times and want to calculate the probability that we will observe at least four heads but not more than six heads. We can obtain a bound for this probability by using Chebyshev's inequality and the fact that $\mu_x = 5$ with $\sigma_x = 1.581$:

$$\begin{aligned}\text{pr}(4 \leqslant X \leqslant 6) &= \text{pr}(3 < X < 7)\\ &= \text{pr}(5 - 2 < X < 5 + 2)\\ &\geqslant 1 - \left(\frac{\sigma_x}{2}\right)^2\\ &= 1 - \left(\frac{1.581}{2}\right)^2 = 0.375\end{aligned}$$

We can find the exact probability by using the probability function $\binom{10}{x}\left(\frac{1}{2}\right)^{10}$, evaluating for $x = 4$, 5, and 6, and adding the results. That is,

$$\begin{aligned}\text{pr}(4 \leqslant X \leqslant 6) &= \binom{10}{4}\left(\frac{1}{2}\right)^{10} + \binom{10}{5}\left(\frac{1}{2}\right)^{10} + \binom{10}{6}\left(\frac{1}{2}\right)^{10}\\ &= 0.2051 + 0.2461 + 0.2051 = 0.6563\end{aligned}$$

These two results emphasize how rough an approximation Chebyshev's inequality gives for the actual probability that we could obtain if we knew the probability function for the random variable. This inequality holds for any distribution. The cost is that we never know how much more the true probability might be beyond this lower bound. This encourages us to use probability functions whenever possible in order to obtain more accurate results in the decision-making process.

EXERCISES

3.6.1 Paul's daily earnings have an expected value and standard deviation given by 25 and 2, respectively. His cousin Peter's daily earnings have an expected value and standard deviation given by 20 and 2, respectively. Calculate:

(*a*) A lower bound for the probability that Paul earns between \$22.25 and \$27.75 on a given day.

(*b*) An upper bound for the probability that Peter does *not* earn between \$16.00 and \$24.00.

3.6.2 Refer to Example 3.6.1. Within what range will:
(*a*) Paul's daily earnings be at least 75% of the time?
(*b*) Peter's daily earnings be at least 50% of the time?

3.6.3 Refer to Example 3.5.7. What is the minimum probability that, on any Sunday, the Browns will score between 7 and 35 points? Within what point range can the Cavaliers be expected to score at least 75% of the time?

3.6.4 A particular machine produces items that have an expected value of 25 and a standard deviation of 2. What percent of the items will be between 20 and 30? Within what interval (symmetric about the mean) will 95% of the population lie?

3.7 THE EMPIRICAL LAW

Many probability distributions dealing with measurement are adequately described by the bell-shaped density function graph shown in Figure 3.3. Statisticians call this graph the *normal curve*. If we have a random sample of 25 measurements or more, then we can calculate good approximations for the expected value and variance, and the empirical law is often used to estimate probabilities.

Figure 3.3 The Normal Curve

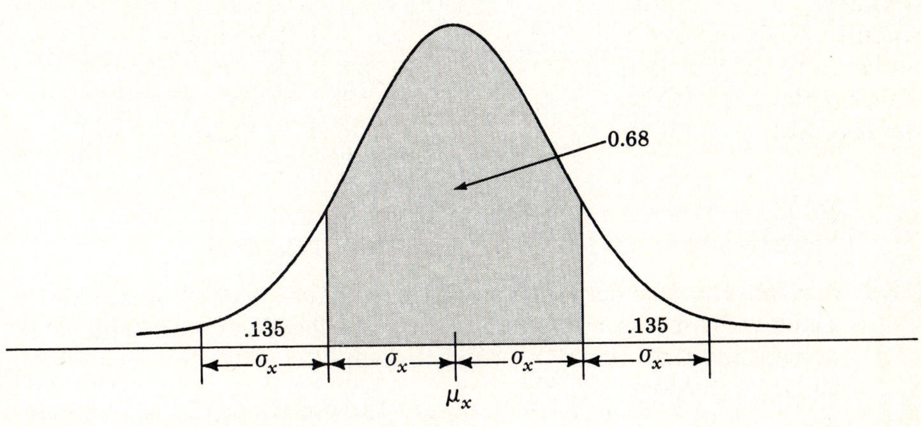

THE EMPIRICAL LAW A probability distribution whose density function graph is approximately the same as a normal curve will have approximately 68% of the range values in the interval $\mu_x \pm \sigma_x$, approximately 95% of the range values in the interval $\mu_x \pm 2\sigma_x$, and approximately 99% of the range values in the interval $\mu_x \pm 3\sigma_x$.

EXAMPLE 3.7 An accounting firm wants to make some decisions regarding the hiring of new staff and needs to know a range that represents 95% of the values for times to complete an audit. The expected value and standard deviation for these times are two hours and one-half hour, respectively. Use the empirical law to provide the required range.

SOLUTION The range begins at $\mu_x - 2\sigma_x$ and ends at $\mu_x + 2\sigma_x$, or

$$2 - 2\left(\frac{1}{2}\right) = 1 \quad \text{to} \quad 2 + 2\left(\frac{1}{2}\right) = 3 \text{ hours}$$

Note that using Chebyshev's inequality to obtain an answer for Example 3.7.1 would require going to approximately 4.48 standard deviations on each side of the mean. In this case, the bounds on the range of values would be $2 \pm 4.48\left(\frac{1}{2}\right)$ or -0.24 (actually 0) and 4.24. This interval is much wider because the assumption that a bell-shaped curve is a reasonable approximation for the actual distribution is much more restrictive than that required for use of Chebyshev's inequality.

EXERCISES

3.7.1 Moses Payday is a deacon in his church and wants to estimate how much the congregation can afford to pay the new minister. Since Moses has been around a long time, he has a good idea of how much money to expect on Sunday from the worshipers. Moses says that the expected value is about $400.00, and that he has never seen an offering of less than $155.00 nor more than $620.00. Use the empirical law to set bounds on the amount of the offering to be expected 95 percent of the time.

3.7.2 Assume that the distribution of the populations in Exercise 3.6.3 can be reasonably approximated by a bell-shaped curve and answer the first question posed therein.

3.7.3 Refer to Exercise 3.6.4 and answer the questions if a bell-shaped curve is assumed to be appropriate.

3.8 SUMMARY

Random variables help us to represent the outcomes of an experiment in a more beneficial fashion by allowing us to use the same distribution to model many experiments. Particular models are determined by specifying the parameters of the distribution. There are two types of random variables in common use, discrete and continuous, and which is used depends on the experiment. Generally speaking, if the experimental outcome is a function of one type of measurement that can assume any of a continuum of values, then a continuous random variable is appropriate.

Random variables have useful descriptors called the expected value and standard deviation, which are measures of location and dispersion of the range values, respectively. These values are usually easy to determine and are sufficient for many decision problems, so that their use eliminates the need to manipulate complex distribution functions. These descriptors can be used to get rough approximations for probabilities over restricted ranges by employing Chebyshev's inequality or the empirical law; when accuracy becomes important, however, an effort should be made to determine an appropriate random variable and its distribution function. The next two chapters show how we can proceed along these lines.

REVIEW EXERCISES

Level 1

3.1 In a grocery store, the probability that a randomly selected loaf of bread is more than one day old is 0.25. If you purchase four loaves at random, what is the probability distribution for the number of loaves purchased that are more than one day old?

3.2 Given the following probability distribution of grades in an introductory business statistics class:

Grade	A	B	C	D	F
Probability	0.1	0.2	0.4	0.2	0.1

Assume that a student is selected at random on the first day of class. Let the quality points associated with A, B, C, D, and F be 4, 3, 2, 1, and 0, respectively. Determine:

(*a*) The number of quality points the student can expect to earn.

(*b*) The standard deviation of quality points.

3.3 A certain lottery has possible payoffs of 0, 25, 100, 500, 1000, and 10,000 dollars. Its proponents have advertised that the probability distribution of payoffs to date is 0.6, 0.25, 0.1, 0.08, 0.05, 0.03, respectively. Could they be prosecuted for false advertising?

3.4 A government contractor is working on three independent projects, X, Y, and Z. Expected revenues on the projects are \$500,000, \$200,000, and \$100,000, with variances of 50,000, 30,000, and 10,000 $(\$)^2$, respectively. What is the contractor's total expected revenue from these projects?

3.5 The following table gives the probability distribution of the number of customers arriving in a bank during an arbitrary four-minute period.

X = Number of Customers	$p(x)$ = Probability
0	0.14
1	0.29
2	0.25
3	0.18
4	0.10
5	0.04

Find the expected number of customers arriving in the next four-minute period.

3.6 Among eight nominees for the board of directors of a grocery cooperative are four men and four women. Four persons are to be elected. Determine the probability distribution for the number of men elected (assuming that any four people among the eight are as likely to be elected as any other set of four people).

3.7 A particular investment has a 0.4 probability of yielding a 10 percent gain, 0.3 of no gain, 0.2 of a 10 percent loss, and 0.1 of a 20 percent loss. You are to invest \$2000 in this opportunity.
(*a*) Construct the probability distribution for X, your net change in value.
(*b*) Calculate your expected monetary return on investment.
(*c*) Calculate the standard deviation of your return.

3.8 Determine whether or not the following is a probability distribution and justify your answer.

x	-7	-2	0	5	13
$p(x)$	0.20	0.15	0.30	0.25	0.15

3.9 Assume scores on a standard exam have an expected value of 70 and a standard deviation of 10. Assume that the distribution is symmetric, but that nothing else is known about it. If the upper 10% are to receive A's, what should be the minimum score for an A?

3.10 Novelties, Inc., produces items that would generally be classified as fad items. They have found that the probability distribution that appears to describe the market lives of their products generally shows a steady increase for the first 1.5 years after introduction and a steady decrease for the next 1.5 years. No product has had a demand for more than 3 years. The actual distribution is as follows, with m representing the market life in years:

$$f(m) = \begin{cases} \frac{20}{45}m & 0 \leqslant m \leqslant 1.5 \\ \frac{4}{3} - \frac{20}{45}m & 1.5 < m \leqslant 3 \\ 0 & \text{elsewhere} \end{cases}$$

(*a*) Graph the distribution.
(*b*) Determine the probability that a product will have a market life in excess of one year.
(*c*) Determine the probability that a product will phase out in less than two years.
(*d*) Verify that the expected market life of a product is 1.5 years.

3.11 In Exercise 3.5, determine the standard deviation of the number of customers arriving during a four-minute period.

3.12 The expected number of customers arriving in the bank in Exercise 3.5 was found to be 1.93 and the standard deviation was found to be 1.3361.
(*a*) Using Chebyshev's inequality, what is the minimum percentage of all four-minute periods in which between 0 and 3.93 customers will arrive? What is this percentage using the actual distribution given in Exercise 3.5?
(*b*) Use Chebyshev's inequality to find bounds on the number of arriving customers such that at least 75% of all four-minute periods will be included. What percentage of periods will be in your interval for the actual distribution?

3.13 Rework Exercise 3.12(*a*) using the empirical law and indicate which gives a better approximation of the actual.

3.14 A particular product is known to have an average demand of ten units per week and a standard deviation of three units per week. The accounting office has determined that from a cost standpoint, it is best to produce eight weeks' supply at a time. If demand is assumed to be independent from one week to the next, what is:
(*a*) The size of a production run?
(*b*) The standard deviation of demand over the reorder period? (The reorder period is the time from one production run to the next.)

3.15 An accountant is auditing income tax returns. She has selected five returns at random to examine. Historical records show that 10% of the returns are fraudulent. Construct the probability distribution for the number of fraudulent returns in the five selected.

Level 2

3.16 The CPI Company is considering the feasibility of developing a memory system using a laser beam. The company feels that the average R&D cost will be about \$9 million, with a standard deviation of \$2 million. They also believe that, if developed, the system will have an average manufacturing cost per unit of \$400,000 with a standard deviation of \$100,000. If they plan to market the system for \$700,000 and expect to sell 40 units, what are the expected value and standard deviation of the total profit of the project? (*Note*: Total profit = total revenue - R&D cost - total variable cost.)

3.17 A certain probability distribution is given by

$$f(x) = \begin{cases} +\sqrt{r^2 - x^2} & -r \leqslant x \leqslant +r \\ 0 & \text{elsewhere} \end{cases}$$

Determine the value of r that will satisfy the conditions for a probability density function. (*Hint*: Graphically, this function will describe a semicircle.)

3.18 Suppose that there are five men in a restaurant. A careless check-room attendant loses the hat checks and returns their five hats at random. Determine the probability distribution of the number of men receiving their own hats.

3.19 Prove that, if X is a discrete random variable,

$$\text{Var}(cX) = c^2 \ \text{Var}(X)$$

3.20 In Exercise 3.4, determine the variance of the average expected revenue.

3.21 Two machines produce a certain item. The capacity per day of machine A is 1 unit and that of B is 2 units. Let (X_A, X_B) be the discrete random variable that measures the actual production on each machine each day. Each entry in the following table represents the joint probability; for example,

$$\text{pr}(0, 0) = \text{pr}(X_A = 0, X_B = 0) = \frac{1}{8}$$

X_B	X_A: 0	X_A: 1
0	1/8	0
1	1/4	1/4
2	1/8	1/4

(*a*) Find the distributions of X_A and X_B.
(*b*) Find the distribution of X_A, given $X_B = 2$.
(*c*) Are X_A and X_B independent random variables? (Justify your answer.)
(*d*) Find the probability distribution of the daily total output.

3.22 You are about to contract for the construction of a filling station. The regional gasoline supplier has informed you that gasoline will be delivered once every week. You assess the probability of the weekly volume of sales X (in thousands of gallons) to be

$$\text{pr}(X = k) = \left(\frac{1}{2}\right)^k \qquad k = 1, 2, 3, \ldots$$

and you need to decide on the minimum capacity of the tank to be constructed in order to meet the weekly demand with probability of 93.75%. What size tank should you order?

3.23 Determine the standard deviation of the market life for one of the Novelties, Inc., products from Exercise 3.10.

Level 3

3.24 The distribution of X, the life of a transistor, in hours is approximated by a triangular distribution having a probability density function given by

$$f(x) = \begin{cases} \dfrac{a}{10{,}000}\left(1 - \dfrac{x}{10{,}000}\right) & 0 \leqslant x \leqslant 10{,}000 \\ 0 & \text{elsewhere} \end{cases}$$

(*a*) What is the value of a?
(*b*) What percentage of transistors have a life exceeding 8000 hours?
(*c*) What is the average life of a transistor?
(*d*) What is the standard deviation of transistor lives?

3.25 A manufacturing process involves processing a part on two machines, A and B, in that order. The part is subject to breakage in each machine; when a part breaks, the machine is out of service temporarily for repairs. If the probability that a part will break is 0.05 in machine A and 0.08 in B, find the expected number of parts that will be completely processed before the first breakdown. (*Hint*: Use the algebraic identity

$$y + 2y^2 + 3y^3 + \ldots = \sum_{i=1}^{\infty} iy^i = \frac{y}{(1-y)^2} \qquad \text{if } -1 < y < 1)$$

3.26 The R&D department of a large company has found that the time from development of a new product to the point at which it begins to show an overall profit is distributed according to the following distribution, where x is in years.

$$f(x) = cx^2 e^{-x} \qquad x \geqslant 0$$

Use the fact that

$$\int_0^{\infty} x^k e^{-x}\, dx = k!$$

for k a positive integer and find:
(*a*) The value of c.
(*b*) The expected time before a new product shows an overall profit.

(c) The standard deviation of time from development to first overall profit.

3.27 Prove that $E(aX + bY) = aE(X) + bE(Y)$, where a and b are constants and X and Y are discrete random variables.

3.28 A particular quality control plan used in a destructive testing situation calls for sampling items at random from the process and testing them to determine whether they are defective or nondefective. The plan calls for testing the items one at a time until either two defectives are found, in which case it is decided that the process is out of control, or five items are tested with fewer than two being defective, indicating that the process is in control. Determine:

(a) The probability distribution for the total number of items tested.

(b) The expected number of items to be tested each time.

Chapter 4
Probability Distributions for Discrete Phenomena

KEY WORDS AND TERMS

Bernoulli Trial
Bernoulli Distribution
Binomial Distribution
Geometric Distribution
Negative Binomial Distribution
Hypergeometric Distribution
Multivariate
Multinomial Distribution
Poisson Distribution

4.1 INTRODUCTION

The previous chapter explained how the term *statistical experiment* is used to describe any process that can be thought of as a chance phenomenon by which several chance outcomes can be obtained. The sample space was then defined as the set of all such chance outcomes. The concept of a random variable was used to simplify the description of the experiment by assigning a number to each sample point. Discrete random variables have a sample space containing a countable number of elements as their domain and a subset of the real line as their range. For instance, the experiment "roll a pair of fair dice" could be described by the sample space and discrete random variable given in the first two columns of Table 4.1.

This random variable is a good choice if we are interested in the sum of the number of pips on the upper faces of the dice. As a result of this association, we have the economy of saying, "what is the probability of a sum of 9?" rather than, "what is the probability of $\{3, 6\}$ or $\{6, 3\}$ or $\{4, 5\}$ or $\{5, 4\}$?"

The next step in the development was to associate a probability with each numerical value of the random variable; whether in tabular or functional form, this is defined as the probability function of the random variable. Thus, the random variable in this example is defined by the first two columns of Table 4.1, whereas the probability function is

Table 4.1 An Example of a Discrete Random Variable and Its Distribution

Sample Space Elements	Associated Random Variable Value	Probability of Value
{1, 1}	2	1/36
{1, 2}, {2, 1}	3	2/36
{1, 3}, {3, 1}, {2, 2}	4	3/36
{1, 4}, {4, 1}, {2, 3}, {3, 2}	5	4/36
{1, 5}, {5, 1}, {2, 4}, {4, 2}, {3, 3}	6	5/36
{1, 6}, {6, 1}, {2, 5}, {5, 2}, {3, 4}, {4, 3}	7	6/36
{2, 6}, {6, 2}, {3, 5}, {5, 3}, {4, 4}	8	5/36
{3, 6}, {6, 3}, {4, 5}, {5, 4}	9	4/36
{4, 6}, {6, 4}, {5, 5}	10	3/36
{5, 6}, {6, 5}	11	2/36
{6, 6}	12	1/36

defined by the last two columns. The probability function is also given by

$$f(x) = \frac{\min\{x - 1, 13 - x\}}{36} \qquad x = 2, 3, 4, \ldots, 12$$

This last result suggests a way in which we might exploit the functional form of the distribution by using one model to describe the outcomes of a variety of different experiments that have similar properties. For instance, if the two objects had four sides instead of six, the distribution would be changed to give

$$f(x) = \frac{\min\{x - 1, 9 - x\}}{16} \qquad x = 2, 3, 4, \ldots, 8$$

as the probability of a sum of x for the number of pips showing (assuming that all sides are equiprobable).

This illustrates the flexibility of using random variables and their distributions to describe random phenomena. One of the best illustrations of the economy of this method can be found in what we call the *Bernoulli family*.

4.2 THE BERNOULLI DISTRIBUTION AND THE BERNOULLI FAMILY

Probably the simplest but the most useful distribution is the Bernoulli distribution. It models a wide spectrum of situations in business and provides the foundation for such a variety of other distributions that it is truly an outstanding asset for elementary analysis of business data. Its formal definition is as follows.

DEFINITION 4.1 A Bernoulli distribution is appropriate for any experiment that can be thought of as resulting in one of only two possible outcomes; this type of experiment is called a Bernoulli experiment. One of the outcomes is typically labeled a "success" and the other a "failure." The Bernoulli distribution is defined by the probability function

$$f(x|p) = \begin{cases} p & \text{if } x = 1 \text{ (success)} \\ 1 - p & \text{if } x = 0 \text{ (failure)} \end{cases} \tag{4.1}$$

where p represents the probability of a success.

We use the notation $f(x|p)$ to emphasize that the Bernoulli distribution probabilities depend on the probability of success p. That is, we *must* know the value of p before we can calculate any of the probabilities. Critical values of this type are called *parameters*, and the notation should be read, "the probability that a Bernoulli random variable assumes the value x when the probability of success is p."

Using the expressions for expected value and variance given in the previous chapter, we can compute $E(X)$ and $\mathrm{Var}(X)$ as shown.

$$E(X) = \sum_{x=0}^{1} x f(x|p) = 0(1 - p) + 1(p) = p \tag{4.2}$$

$$\begin{aligned} \mathrm{Var}(X) = E(X^2) - [E(X)]^2 &= \sum_{x=0}^{1} x^2 f(x^2|p) - p^2 \\ &= (0^2)f(0^2|p) + (1^2)f(1^2|p) - p^2 \\ &= (0)f(0|p) + (1)f(1|p) - p^2 \\ &= (0)(1 - p) + (1)(p) - p^2 \\ &= p - p^2 \end{aligned} \tag{4.3}$$

This rather simple random variable gives rise to the binomial, geometric, and negative binomial distributions, all of which are very useful in business analysis problems. We now present each of these distributions that, together with the Bernoulli distribution, constitute the Bernoulli family.

DEFINITION 4.2 A *binomial distribution* is appropriate for any experiment that consists of n mutually independent Bernoulli experiments (hereafter we will call them trials) in which the probability of success p remains the same from trial to trial.

The binomial random variable maps the number of successes observed during the n trials, and thus its distribution has two parameters: the number of trials n and the probability of success on any given trial p. The binomial distribution is given in functional form by

$$f(x|n, p) = \binom{n}{x} p^x (1 - p)^{n-x} \qquad \text{for } x = 0, 1, 2, \ldots, n \tag{4.4}$$

EXAMPLE 4.1 A supervisor of customer service uses a "plant" ten times each month to check on the service personnel. The plant walks up to the service counter with a problem, then reports to the supervisor, who makes a judgment as to whether the service rendered was acceptable or unacceptable. The times for testing the service are chosen at random. Unknown to the supervisor, there is a 0.9 probability that service, in general, will be satisfactory.

(*a*) What is the probability that nine of the ten plants for this month will receive service that is acceptable to the supervisor?

(*b*) What is the probability that at least seven out of the ten plants for this month will receive service that is acceptable to the supervisor?

SOLUTION The binomial assumptions appear to be reasonable in this situation. Thus, we may utilize Equation 4.4, using $n = 10$ and $p = 0.9$.

(*a*) We have

$$\text{pr}(x = 9) = f(9|n = 10, p = 0.9) = \binom{10}{9}(0.9)^9 (0.1)^1 = 0.3874$$

(*b*) pr(at least seven will be acceptable)

$$\text{pr}(x \geqslant 7) = \sum_{x=7}^{10} f(x|n = 10, p = 0.9)$$

$$= \sum_{x=7}^{10} \binom{10}{x}(0.9)^x (0.1)^{10-x} = 0.9872$$

We can justify Equation 4.4 if we think of the sample space as consisting of all success-failure sequences of size n. Thus, each sample point is characterized by a sequence of size n of the letters S (for success) and F (for failure). One way we could see x successes in n trials could be represented by the sequence

$$\underbrace{\text{SSS}\cdots\text{S}}_{x}\ \underbrace{\text{FFF}\cdots\text{F}}_{n-x}$$

Thus, the problem of counting the number of ways in which we could see x successes in n trials reduces to the problem of calculating the number of ways in which we can select x positions for the letter S in the sequence. Since order is not important, the x positions can be selected in $\binom{n}{x}$ different ways. Next we use two basic laws of probability given in Chapter 2. First, we use Equation 2.5, which shows us how to calculate the probability of an event that can be expressed as the intersection of n mutually independent events. Specifically, if

$$E = E_1 \cap E_2 \cap \cdots \cap E_n$$

then

$$\text{pr}(E) = \text{pr}(E_1)\,\text{pr}(E_2)\cdots\text{pr}(E_n)$$

Therefore, if we let E_i represent the outcome of the ith Bernoulli trial and let E represent any outcome containing exactly x successes, then, for instance,

$$\begin{aligned}
\text{pr}\{\underbrace{\text{SSS}\cdots\text{S}}_{x}\ \underbrace{\text{FFF}\cdots\text{F}}_{n-x}\} &= \text{pr}(E) \\
&= \text{pr}(E_1)\,\text{pr}(E_2)\cdots\text{pr}(E_n) \\
&= \underbrace{ppp\cdots p}_{x}\ \underbrace{(1-p)(1-p)\cdots(1-p)}_{n-x} \\
&= p^x(1-p)^{n-x}
\end{aligned}$$

Next, we use Equation 2.4, which shows us how to calculate the probability of an event that can be expressed as the union of k mutually exclusive events. Specifically, if

$$W = W_1 \cup W_2 \cup \cdots \cup W_k$$

and

$$W_i \cap W_j = \emptyset \qquad (\text{whenever } i \neq j)$$

then

$$\text{pr}(W) = \sum_{i=1}^{k} \text{pr}(W_i)$$

Therefore, if we let W_i represent the ith way in which we can observe exactly x successes in n Bernoulli trials and let W represent the event of observing exactly x successes in n Bernoulli trials, then

$$\begin{aligned} f(x|n,p) &= \text{pr}\ \{\text{exactly } x \text{ successes in } n \text{ Bernoulli trials}\} \\ &= \text{pr}(W) = \sum_{i=1}^{k} \text{pr}(W_i) \\ &= \sum_{i=1}^{k} p^x(1-p)^{n-x} \\ &= kp^x(1-p)^{n-x} \end{aligned}$$

But we have already shown that $k = \binom{n}{x}$. Thus,

$$f(x|n,p) = \binom{n}{x} p^x (1-p)^{n-x}$$

We have presented this derivation to illustrate how the basic laws of probability and experimental assumptions interact to provide the binomial model. Notice that *all* the assumptions come into play in the derivation. If any of the assumptions is violated, then the binomial model is not strictly correct. However, if we judge that the experimental conditions contain only minor violations of the basic assumptions, then the binomial model may be a suitable approximation. The following example gives a possible set of circumstances.

EXAMPLE 4.2 A tire manufacturer guarantees truck tires for 20,000 miles under normal driving conditions. The manufacturer makes this guarantee after consulting with the quality control staff, who maintain that the probability of a blowout before the 20,000-mile limit is about 0.05. If a trucking company purchases ten tires, what is the probability that at least two of the tires will blow out before 20,000 miles?

SOLUTION If the assumptions for the binomial model are satisfied, then the probability can be calculated as follows:

$$\begin{aligned} \text{pr}\{\text{at least 2 blowouts}\} &= \text{pr}\{\text{exactly 2 blowouts or exactly 3 blowouts or} \ldots \text{or exactly 10 blowouts}\} \\ &= \sum_{x=2}^{10} f(x|n=10, p=0.05) \\ &= \sum_{x=2}^{10} \binom{10}{x} (0.05)^x (1-0.05)^{10-x} \end{aligned}$$

We can shorten this same calculation by observing that

$$\sum_{x=j}^{n} f(x|n,p) = 1 - \sum_{x=0}^{j-1} f(x|n,p) \qquad \text{for } j = 1, 2, \ldots, n \tag{4.5}$$

In our case $j = 2$, and so

$$\begin{aligned}
&\text{pr}\{\text{at least 2 blowouts}\} = \\
&\quad \sum_{x=2}^{10} f(x|n=10, p=0.05) \\
&= 1 - \sum_{x=0}^{1} f(x|n=10, p=0.05) \\
&= 1 - f(0|n=10, p=0.05) - f(1|n=10, p=0.05) \\
&= 1 - \binom{10}{0}(0.05)^0(1-0.05)^{10} - \binom{10}{1}(0.05)^1(1-0.05)^9 \\
&= 1 - (0.95)^{10} - 10(0.05)(0.95)^9 \\
&= 1 - 0.5987 - 0.3151 = 0.0862
\end{aligned}$$

Perhaps you have noticed that calculating the binomial distribution probabilities can be tedious. Not surprisingly, many other people have noticed the same thing and, consequently, have published tables that can be used once the parameters n and p are known. Table A.1 in the back of the book gives binomial probabilities of the type

$$f(x|n,p)$$

and Table A.2 in the back of the book gives cumulative probabilities of the type

$$\sum_{x=0}^{j} f(x|n,p) \qquad j = 0, 1, 2, \ldots, n$$

Probabilities of the latter type are called *cumulative binomial probabilities* because they represent a sum of probabilities. In particular, the function $F(j|n,p)$ given by

$$\begin{aligned}
F(j|n,p) &= \text{pr}\{X \leqslant j\} \\
&= \sum_{x=0}^{j} f(x|n,p)
\end{aligned} \tag{4.6}$$

is called the distribution function for the binomial probability function $f(x|n,p)$. The introduction given with each of the tables explains how to find specific probabilities.

Next, we should clarify the conditions under which the answer of 0.0862 would be appropriate.

CONDITION 1 We must be conducting a sequence of ten Bernoulli trials.

This condition is satisfied, since each of the ten tires either experiences a blowout (corresponding to a success) or does not (corresponding to a failure).

CONDITION 2 The probability of success is constant from trial to trial.

This condition is satisfied only if the manufacturing process was uniform at a 0.05 blowout probability level for each of the ten tires purchased by the trucking company. We must also assume that the trucks on which the tires are to be placed are comparably equipped, will travel roads of a comparable composition, and so forth.

CONDITION 3 Trials are mutually independent.

To satisfy this condition, we must assume that each of the ten tires will have no effect on the wear of any of the other nine tires. This could be accomplished by putting the ten tires on ten different trucks.

An important aspect of this problem that applies to most problems involving statistical analysis is the idea that basic assumptions may not be totally satisfied. Consequently, the results of the probability calculation are approximations. In most cases, an analyst must either choose a robust model that is relatively insensitive to slight departures from basic assumptions or be satisfied that the resulting calculation is close enough for decision-making purposes.

We can calculate the expected value and variance of a binomial random variable using the formulas for the expected value of a sum of n random variables and the variance of a sum of n mutually independent random variables. First note that the binomial distribution provides a count of the number of successes in n mutually independent Bernoulli trials. Second, recall that the Bernoulli distribution returns the value 1 for each success and the value 0 for each failure. Thus, the binomial random variable X is the sum of n independent Bernoulli random variables, and we have the relation

$$X = \sum_{i=1}^{n} Y_i$$

where Y_i denotes the ith Bernoulli random variable. Hence,

$$E(X) = E\left(\sum_{i=1}^{n} Y_i\right) = \sum_{i=1}^{n} E(Y_i) = \sum_{i=1}^{n} p = np$$

Similarly,

$$\text{Var}(X) = \text{Var}\left(\sum_{i=1}^{n} Y_i\right) = \sum_{i=1}^{n} \text{Var}(Y_i) = \sum_{i=1}^{n} (p - p^2) = np(1 - p)$$

Thus we have

$$E(X) = np \tag{4.7}$$

and

$$\text{Var}(X) = np(1 - p) \tag{4.8}$$

The geometric random variable is one of a number of elementary models used in the study of reliability. In such cases we would be interested in calculating the expected number of trials until component failure.

DEFINITION 4.3 The *geometric distribution* models a probability of mutually independent Bernoulli trials in which the probability of success is constant from trial to trial where the random variable of interest x is the number of Bernoulli trials until the first success (including the trial on which the first success occurs). Consequently, the geometric distribution has only one parameter, p, the probability of success on any given trial. The geometric distribution is given in functional form by

$$f(x|p) = p(1 - p)^{x-1} \qquad \text{for } x = 1, 2, 3, \ldots$$

We can justify this probability function by recognizing that the first success can occur on the xth trial only if all the previous $x - 1$ trials have resulted in failure.

The expected value and variance of the geometric random variable are given by Equations 4.9 and 4.10, respectively.

$$E(X) = \frac{1}{p} \tag{4.9}$$

$$\text{Var}(X) = \frac{1 - p}{p^2} \tag{4.10}$$

EXAMPLE 4.3 An electrical device will operate properly with probability 0.98 each time its switch is turned on.

(*a*) Calculate the expected number of times the device will be switched on before its first failure to operate properly.

(*b*) Calculate the probability that it will be switched on at least 100 times before it fails to operate.

SOLUTION

(*a*) This is a geometric random variable problem in which we are looking for the first failure to operate properly; therefore the parameter p must equal 0.02. Thus, the expected value is given by $1/p = 1/0.02 = 50$.

(*b*) What we want to compute is

$$\sum_{x=100}^{\infty} p(1-p)^{x-1} = 1 - \sum_{x=1}^{99} p(1-p)^{x-1}$$

In general, we have

$$\sum_{x=1}^{m} p(1-p)^{x-1} = 1 - (1-p)^m$$

so that the probability is given by

$$1 - [1 - (1-p)^{99}] = (1-p)^{99} = (0.98)^{99} = 0.1353$$

Although the presentation of the geometric random variable ends our introduction to some of the random variables that belong to the Bernoulli pyramid, you should recognize that the variables we have shown are only representative of a wider class of useful random variables that rely on the simple and basic assumption of Bernoulli trials. For instance, a next logical step in the development would be to investigate possible models for the number of Bernoulli trials until observing the rth success or until observing r successes *in a row*. The latter case could be used to describe random phenomena in which a sequence of at least r identical events must occur in a row before system failure.

The Bernoulli pyramid contains a special class of discrete random variables that are used extensively in business analysis problems. However, there are *many* more discrete random variables that are also very useful. We now present two of these for two different reasons. First, we will look at the hypergeometric random variable to illustrate the tricks that different sampling methods can play on us, and second, we will look at the Poisson random variable because it is so useful in the fields of quality control and simulation.

EXERCISES

4.2.1 Boogie North America does a quick check on the quality of all incoming shipments of stereo components by testing the top four components in each box. If Boogie North America is interested only in whether or not each component is defective, does this process constitute a binomial experiment (are the assumptions for use of the binomial random variable satisfied)?

4.2.2 Compute the following probabilities, which refer to tossing a coin for which the probability of a head on any one toss is 0.3.
(*a*) The probability of exactly two heads in three tosses
(*b*) The probability of at most two tails in four tosses
(*c*) The probability that the first head occurs before the fifth toss

4.2.3 Suppose that tossing a fair coin 10 times results in 10 heads. What is the probability that an eleventh toss will result in a head? Explain your answer.

4.3 THE HYPERGEOMETRIC DISTRIBUTION

When we use random variables for business decisions, we are interested in a certain population of values, whether these values be prices, quantities, or consumer statistics and opinions. In most cases, we determine which model is appropriate for describing how the values occur by the way in which we sample the population of values (or items). The hypergeometric distribution allows us to show how sampling methods can determine the proper model.

DEFINITION 4.4 The hypergeometric distribution models any experiment in which a random sample is taken without replacement from a finite population of size N where the random variable of interest counts the number of items in the sample that display a given characteristic. Thus, every item in the population either displays the characteristic of interest or does not. The hypergeometric distribution is given by

$$f(x|N, n, R) = \frac{\binom{R}{x}\binom{N-R}{n-x}}{\binom{N}{n}} \qquad \text{for } \max\{0, n - N + R\} \leqslant x \leqslant \min\{n, R\} \tag{4.11}$$

where R denotes the number of items in the population that display the characteristic of interest and we assume that $\max\{n, R\} \leqslant N$.

Notice that the upper limit on the domain of $f(x|N, n, R)$ is dependent on the relationship between n and R. Clearly, if $n \leqslant R$, the maximum number of type R items we could possibly choose in a sample of size n is given by n. Similarly, if $n \geqslant R$, the maximum number of type R items we could possibly choose in a sample of size n is given by R. The lower limit reflects the fact that if our sample size exceeds $N - R$, we must choose *at least* $n - (N - R)$ of the type R items.

EXAMPLE 4.4 A government agency decides to conduct a check on label specifications for a product being shipped by a company to local grocery stores. They choose five cans at random and without replacement from each crate. Suppose that in a particular crate there are 10 out of 30 cans whose contents do not meet their own label specifications, and let Y be the number of these in the sample.

(*a*) What type of random variable is Y?

(*b*) Find $\text{pr}(Y = 3)$.

(*c*) Find $\text{pr}(Y \geqslant 1)$.

SOLUTION

(*a*) Y is a hypergeometric random variable whose parameters are $N = 30, n = 5, R = 10$.

(*b*) From part (*a*), we have

$$\text{pr}(Y = 3) = f(3|N = 30, n = 5, R = 10) = \left\{\frac{\binom{10}{3}\binom{30-10}{5-3}}{\binom{30}{5}}\right\} = 0.160$$

(*c*)

$$\text{pr}(Y \geqslant 1) = 1 - \text{pr}(Y = 0) = 1 - \left\{\frac{\binom{10}{0}\binom{30-10}{5-0}}{\binom{30}{5}}\right\}$$

$$= 1 - \frac{\binom{20}{5}}{\binom{30}{5}} = 1 - 0.109 = 0.891$$

We can derive the hypergeometric probability function using the elementary combinatorial theorems in Chapter 2. For ease of explanation, let us assume that we want to control the quality of industrial components shipped in lots of size N. Since sampling is costly, we plan to minimize the number of defective components shipped per lot by choosing n components at random and without replacement. If the sample contains more than one defective component, the lot is rejected.

The hypergeometric random variable is the correct model for this situation. We need to determine the value of

$$f(x|N, n, R)$$

where R represents the actual number of defectives in the lot and x denotes the number of R defectives we choose in our sample size of n. If we have x defectives in our sample, we must also have $n - x$ nondefectives. We can choose the x defectives from the R defectives in the lot in

$$\binom{R}{x}$$

different ways and the $n - x$ nondefectives from the $N - R$ nondefectives in the lot in

$$\binom{N-R}{n-x}$$

different ways. This means that for every *one* of the $\binom{R}{x}$ ways of choosing x defectives, we have $\binom{N-R}{n-x}$ different ways to choose $n - x$ nondefectives to go with it, giving a total of

$$\binom{R}{x}\binom{N-R}{n-x}$$

different ways to choose x defective and $n - x$ nondefective components from a lot of size N. Dividing by the total number of ways to choose n components from N with no restrictions gives us the *probability* of choosing x defective components, or

$$f(x|N, n, R) = \frac{\binom{R}{x}\binom{N-R}{n-x}}{\binom{N}{n}}$$

The expected value and variance of a hypergeometric random variable are given by Equations 4.12 and 4.13, respectively.

$$E(X) = n\left(\frac{R}{N}\right) \tag{4.12}$$

$$\text{Var}(X) = \frac{nR(N - R)(N - n)}{N^2(N - 1)} \tag{4.13}$$

The hypergeometric random variable is the proper model for most survey experiments in which the response can be partitioned into one of two categories. This is because the same client is not allowed to be queried more than once. The next example illustrates its use.

EXAMPLE 4.5 The foreman for a machine shop wanted to choose the "best" two persons available to fill two positions in the shop. He narrowed the list to ten applicants, then chose two at random, since he was indifferent at that stage of the selection process. However, the ten applicants *did* include two who were better suited to the jobs than the eight others. What is the probability that the foreman chose the two best? What is the probability that he chose *at least* one of the two best?

SOLUTION For this example, $N = 10$, $n = 2$, and $R = 2$, and we need to calculate $f(2|N = 10, n = 2, R = 2)$. From Equation 4.11,

$$f(2|N = 10, n = 2, R = 2) = \frac{\binom{2}{2}\binom{8}{0}}{\binom{10}{2}} = \frac{1}{45} = 0.0222$$

The probability that he chooses at least one of the two best is given by

$$f(1|N = 10, n = 2, R = 2) + f(2|N = 10, n = 2, R = 2)$$

$$= \frac{\binom{2}{1}\binom{8}{1}}{\binom{10}{2}} + 0.0222 = 0.3777$$

We can now illustrate the role that sampling methods can play in determining the type of random variable that properly models an experiment. The next example shows this intimate connection.

EXAMPLE 4.6 Joe enjoys playing the carnival duck pond game. In this game an oval water trough contains 500 plastic ducks that pass before him, disappear momentarily, and reappear in an endless cycle.

Joe must pay \$1.00 to pick up four of the ducks, each of which has a number imprinted on the bottom side. If at least one of the ducks shows the number 10, he gets a big prize. However, Joe does not know that only 25 of the ducks have the number 10 on their bottom side. Since Joe is such an avid fan, the barker allows him to choose his four ducks in either of two different ways: with replacement or without replacement.

You are standing nearby and overhear the conversation. Without hesitation, you tell Joe which sampling method will maximize his chances for a big prize. What method did you recommend and why?

SOLUTION Assuming that the game characteristics and Joe interact in such a way that either method produces a random sample from the population of 500 ducks, we have with replacement

$$\begin{aligned}&\text{pr}\{\text{at least one number 10 duck in the four ducks chosen}\}\\&= 1 - \text{pr}\{\text{no number 10 ducks in the four ducks chosen}\}\\&= 1 - f\left(0 \mid n = 4, p = \frac{25}{500}\right) = 1 - \binom{4}{0}(0.05)^0(0.95)^4 = 0.1855\end{aligned}$$

and without replacement,

$$\begin{aligned}&\text{pr}\{\text{at least one number 10 duck in the four ducks chosen}\}\\&= 1 - \text{pr}\{\text{no number 10 ducks in the four chosen}\}\\&= 1 - f(0 \mid N = 500, n = 4, R = 25) = 1 - \left\{\frac{\binom{25}{0}\binom{475}{4}}{\binom{500}{4}}\right\} = 0.1860\end{aligned}$$

Thus, you tell Joe to choose sampling without replacement because in so doing he increases his probability of getting a big prize by 0.27%.

This solution uses both the binomial and the hypergeometric models, with the type of model dependent on the sampling method. Actually, there are other intimate connections between these two distributions, as is indicated by the closeness of the answers. To show the connection, let us first look at how $f(x \mid n, p)$ and $f(x \mid N, n, R)$ are related for the values $N = 1000$, $n = 4$, $R = 10$ as shown in Table 4.2.

From the numbers in Table 4.2 and Example 4.3, we could hypothesize that if the population size N is large, the binomial probability function is a good approximation to the hypergeometric probability function. This idea is supported by the fact that

Table 4.2 Comparison of the Binomial and Hypergeometric Probability Functions for a Particular Case with $N = 1000, n = 4, R = 10$

x	Binomial	Hypergeometric
0	0.9606	0.9605
1	0.0388	0.0389
2	0.0006	0.0005
3	0.0000	0.0000
4	0.0000	0.0000

$$\lim_{N \to \infty} f(x|N, n, R) = f(x|n, p) \tag{4.14}$$

whenever the ratio $R/N = p$ is held constant. That is, for a fixed value of R/N, the hypergeometric probability function converges to the binomial probability function as the population size becomes large.

EXERCISES

4.3.1 A box contains twelve screws, three of which are defective due to missing slots where the screwdriver is normally inserted. If a carpenter grabs four of the twelve at random for his pocket,
(*a*) What is the expected number of defective screws in his pocket?
(*b*) What is the probability that he has at least one defective screw in his pocket?

4.3.2 An accounting firm pools its audit process in an attempt to ensure that the audits are done in a purely professional manner. That is, given that they have twelve audits to execute, they place these audits in a pool and randomly assign them to their accountants. Suppose that Jim will execute three of the audits, that there are four automobile firms among the twelve audits, and that Jim hates to do an audit for an automobile firm. What is the probability that Jim will have to do at least one audit for an automobile firm?

4.3.3 A box contains three red balls and five black balls. John wants to calculate the probability of getting exactly one black ball among a random choice of three from the eight. To do this John uses the binomial distribution, but he samples without replacement. How far is his calculation from the true probability?

4.3.4 Calculate the variance for each of the random variables you used in Exercises 4.3.1 to 4.3.3.

4.4 MULTIVARIATE EXTENSIONS OF THE BINOMIAL AND HYPERGEOMETRIC DISTRIBUTIONS

We should mention two other important discrete distributions that are extensions of the binomial and hypergeometric distributions before we present the Poisson distribution. These are the multinomial and multivariate hypergeometric distributions, respectively, and they are fundamental models for certain useful statistical tests that we will present in Chapters 9 and 10.

DEFINITION 4.5 The multinomial distribution models any experiment that can be described as a sequence of n mutually independent trials in which exactly one of k outcomes, denoted by $E_1, E_2, \ldots, E_k$, occurs on each trial. We assume that $\text{pr}\{E_i\} = p_i$, $i = 1, 2, \ldots, k$, is constant from trial to trial. The multinomial random variables count the number of each type of outcome occurring during the n trials. Hence, the multinomial distribution has k parameters: The number of trials n and the probability of outcome E_i, with $i = 1, 2, \ldots, k-1$ (since $\sum_{i=1}^{k} p_i = 1$ implies that $p_k = 1 - \sum_{i=1}^{k-1} p_i$). The multinomial probability function is given by

$$f(\boldsymbol{x}|n, \boldsymbol{p}) = \frac{n!}{x_1!x_2!\ldots x_k!} p_1^{x_1} p_2^{x_2} \cdots p_k^{x_k} \tag{4.15}$$

where $\sum_{i=1}^{k} x_i = n$; $\sum_{i=1}^{k} p_i = 1$; $\boldsymbol{p}$ denotes the sequence $p_1, p_2, \ldots, p_{k-1}$; and $\boldsymbol{x}$ denotes the sequence $x_1, x_2, \ldots, x_{k-1}$.

The multinomial distribution gets its name from the fact that any given value for $f(\boldsymbol{x}|n, \boldsymbol{p})$ can be found to be one of the terms in the multinomial expansion of $(p_1 + p_2 + \cdots + p_k)^n$.

EXAMPLE 4.7 Paul the plumber knows that there are four categories into which problems with household drains can fall: there are the three most common problems and all others. If we call these T_1, T_2, T_3, and T_4, respectively, and assume that the probabilities for each type of problem on a particular job are 0.6, 0.2, 0.15, and 0.05, respectively, what is the probability that there will be five T_1 problems and three T_2 problems among the next twelve service calls he answers for household drain problems?

SOLUTION This is a multinomial random variable problem for which $n = 12$ (note that x_3 is T_3 and T_4 jobs combined) and

$$\begin{aligned} p_1 &= 0.6 \qquad \text{with } x_1 = 5 \\ p_2 &= 0.2 \qquad \text{with } x_2 = 3 \\ p_3 &= 0.2 \qquad \text{with } x_3 = 4 \end{aligned}$$

Thus, the probability is given by

$$\frac{12!}{5!3!4!}(0.6)^5(0.2)^3(0.2)^4 = 0.0276$$

The derivation of Formula 4.15 is a straightforward extension of the argument given for the binomial distribution. Thus, the probability that E_i occurs x_i times, $i = 1, 2, \ldots, n$, is given by

$$p_1^{x_1} p_2^{x_2} \cdots p_k^{x_k}$$

for any particular sequence of these events. To find the number of such sequences, first select x_1 positions in the sequence of n trial outcomes for the E_1's, then x_2 positions of the remaining $n - x_1$ trial outcomes for the E_2's, then x_3 positions out of $n - x_1 - x_2$ remaining for the E_3's. Thus, the number of sequences having exactly x_i occurrences of E_i, $i = 1, 2, \ldots, k$, is given by

$$\binom{n}{x_1}\binom{n - x_1}{x_2}\binom{n - x_1 - x_2}{x_2} \cdots \binom{n - x_1 - x_2 - \cdots - x_{k-2}}{x_{k-1}}\binom{x_k}{x_k}$$

$$= \frac{n!}{x_1!x_2! \cdots x_k!}$$

which is the number of mutually exclusive sequences of x_i outcomes of type E_i, $i = 1, 2, \ldots, k$, we can have in a sequence of n trials. Thus, the total probability over all such sequences is given by

$$\frac{n!}{x_1!x_2! \cdots x_k!} p_1^{x_1} p_2^{x_2} \cdots p_k^{x_k}$$

The following example illustrates use of the multinomial.

EXAMPLE 4.8 Suppose a manufacturing device contains three critical components that are electrically activated and that are arranged as follows:

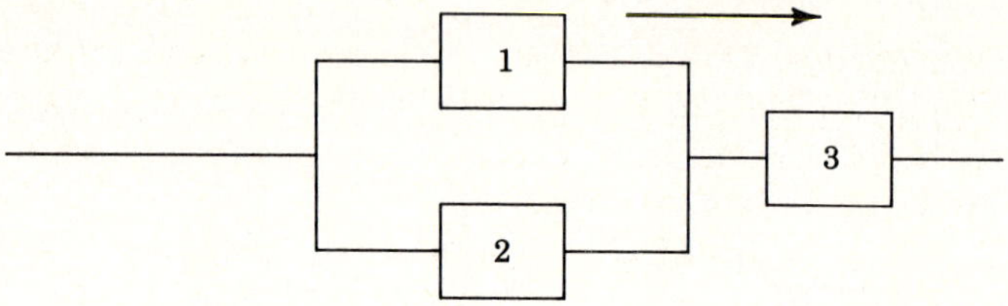

where the direction of the arrow indicates the component activation sequence. The device will fail to perform properly if components 1 *and* 2 fail or if component 3 fails. Each time the device fails, all components are replaced by new ones. If each component is assumed to work independently of the others with probability of failure 0.05, what is the probability that, among the last ten failures of the device, seven failures were due to component 3 failing, and one failure was due to all three components failing?

SOLUTION Failure of the device can occur in exactly one of the following three ways:

(i) Components 1 and 2 fail, but component 3 does not fail.
(ii) Component 3 fails, but component 1 or 2 or both 1 and 2 do not fail.
(iii) All three components fail.

Denoting these outcomes by E_1, E_2, and E_3, respectively, we have

$$\begin{aligned}
p_1 = \operatorname{pr}\{E_1 \mid \text{device fails}\} &= \frac{\operatorname{pr}(E_1)}{\operatorname{pr}(\text{device fails})} \\
&= \frac{(0.05)^2(0.95)}{\operatorname{pr}(E_1) + \operatorname{pr}(E_2) + \operatorname{pr}(E_3)} \\
&= \frac{0.002375}{0.052375} = 0.045346 \\
p_2 = \operatorname{pr}\{E_2 \mid \text{device fails}\} &= \frac{0.049875}{0.052375} = 0.952267 \\
p_3 = \operatorname{pr}\{E_3 \mid \text{device fails}\} &= \frac{0.000125}{0.052375} = 0.002387
\end{aligned}$$

From this information we can say that the desired probability is given by

$$f(X = 2, 7|n = 10, p = 0.045346, 0.952267)$$
$$= \frac{10!}{2!7!1!}(0.045346)^2 (0.952267)^7 (0.002387)^1 = 0.0013243$$

This problem reduces to a binomial problem if we ask for the probability that among the last ten failures of the device, at least eight were due to component 3 failing. In this case

$$p_1 = 0.952267 \qquad p_2 = 0.047733$$

and the desired probability is given by

$$\sum_{j=8}^{10} f(j|n = 10, p = 0.952267) = \sum_{j=8}^{10} \binom{10}{j} (0.952267)^j (0.047733)^{10-j}$$
$$= 0.986595$$

The hypergeometric distribution can be generalized in a similar manner to give a multivariate extension.

DEFINITION 4.6 The multivariate extension of the hypergeometric distribution is a model for any experiment in which a sample of size n is selected at random, without replacement, from a population of size N that can be thought of as the union of k mutually exclusive and collectively exhaustive subsets. The probability function maps the total number of outcomes corresponding to each of k subsets and is given by

$$f(\boldsymbol{x}|N, n, \boldsymbol{R}) = \left\{ \frac{\binom{R_1}{x_1}\binom{R_2}{x_2} \cdots \binom{R_k}{x_k}}{\binom{N}{n}} \right\} \tag{4.16}$$

for $\max\{0, n - (N - R_i)\} \leqslant x_i \leqslant \min\{n, R_i\} \qquad i = 1, 2, \ldots, k$

where $\boldsymbol{x}$ denotes the sequence $x_1, x_2, \ldots, x_{k-1}$ and $\boldsymbol{R}$ denotes the sequence $R_1, R_2, \ldots, R_{k-1}$.

$$\sum_{i=1}^{k} R_i = N$$

$$\sum_{i=1}^{k} r_i = n$$

$$R_i \leqslant N \qquad i = 1, 2, \ldots, k$$
$$n \leqslant N$$

The following example illustrates the use of $f(\boldsymbol{x}|N, n, \boldsymbol{R})$ in calculating probabilities of this type.

EXAMPLE 4.9 Jimmy the Creek (from an Indian tribe whose reservation is on a remote island in Quetico National Park) likes to play acey-deucy at the local dance hall on Saturday nights. In this game, you are dealt two cards face up and are to guess whether the next card dealt will fall between the first two (on an ordinal basis where deuces are lowest and aces are highest). What is the probability that Jimmy the Creek gets a maximum spread from the first two cards he is dealt if he receives his cards before anyone else?

SOLUTION This distribution models this problem because the 52 cards can be thought of as the union of the set of all aces ($R_1 = 4$), the set of all deuces ($R_2 = 4$), and everything else ($R_3 = 44$). Thus we have

$$f(X = 1, 1|N = 52, n = 2, \boldsymbol{R} = 4, 4) = \left\{ \frac{\binom{4}{1}\binom{4}{1}\binom{44}{0}}{\binom{52}{2}} \right\} = 0.0120663$$

The expected value and variance for the multivariate extensions are much the same as those for the hypergeometric and binomial distributions. For instance, the expected number of outcomes corresponding to subset i is given by

$$n\left(\frac{R_i}{N}\right) \tag{4.17}$$

for the multivariate hypergeometric distribution, whereas the expected number of type E_i outcomes is given by

$$np_i$$

where $p_i = \text{pr}\{E_i\}$, for the multinomial distribution.

EXERCISES

4.4.1 A company produces rubber bands in six different sizes, A, AA, B, BB, C and E, and markets them in boxes labeled "Assorted Rubber Bands." One box contains five, six, seven, eight, four, and six of the different sizes, respectively. Calculate the probability of

(*a*) Exactly three AA and two B sizes in a random sample of size seven from the box (assume sampling without replacement).

(*b*) One of each size in a random sample of size six from the box (assume sampling without replacement).
(*c*) One of each size in a random sample of size six from the box (assume sampling with replacement).
(*d*) The expected number of size E rubber bands in a random sample of size four from the box (assume sampling without replacement).

4.4.2 Company personnel can be described as follows:

	Age		
Sex	Less than 25 Years	At Least 25 but Less than 35 Years	At Least 35 Years
Male	4	6	15
Female	10	8	7

The union chooses a representative at random each year to sit on an employee relations board. Assume that the given numbers will apply for the next six years.
(*a*) Calculate the probability that exactly one male and one female under the age of 25 years are chosen for the board during the next four years.
(*b*) Write the expression for the probability of all six types of employees being chosen during the next six years.
(*c*) What is the expected number of females under the age of 35 years that will be chosen for the board during the next five years?

4.4.3 Refer to Exercise 4.4.2 and assume that we will choose five employees at random for a job satisfaction survey. If the sampling is without replacement, calculate
(*a*) The expected number of females under the age of 35 years that will be chosen for the survey.
(*b*) The probability of exactly one male under the age of 25 years and exactly one female at least 35 years of age being chosen for the survey.

4.4.4 Paul sells firewood for which the following composition probabilities apply for a given shipment:
- the probability of at least 50% ashwood is 0.1
- the probability of at least 50% oakwood is 0.2
- the probability of at least 50% softwood is 0.7

Calculate the probability that at least three of the next five shipments contain at least 50% softwood.

4.5 THE POISSON DISTRIBUTION

The Poisson distribution is one of the basic models used in process control and simulation studies. Typically we use the Poisson random variable whenever we need a model for the number of events occurring in a given region R. Typically, R represents a time interval of any length, an area of any size, or a volume of any size. If we were interested in modeling the number of telephone calls arriving at a switchboard per hour, then R would represent time in hours. If we were interested in modeling the number of bubble defects per square meter of sheet glass produced by a certain production line, then R would represent area in square meters. If we were to model the PCB level (to the nearest milligram) per liter of river water, then R would represent volume in liters.

Although the Poisson random variable is used with abandon in applied work, some important assumptions must be satisfied before it can be expected to be an adequate model for a given problem. These restrictions are outlined in the following definition.

DEFINITION 4.7 A Poisson distribution is a model for any experiment that possesses the following properties:

1. The number of events occurring in R_i is independent of the number of events occurring in R_j whenever R_i and R_j are disjoint regions.
2. The probability of exactly one event occurring in a *small* region is proportional to the size of the region, and the occurrence of the event is independent of the number of events occurring outside the region.
3. The probability of more than one event occurring in a *small* region is negligible.

The Poisson random variable counts the number of events occurring in a region. The Poisson probability function is given by

$$f(x \mid \mu R) = \frac{(\mu R)^x e^{-\mu R}}{x!} \qquad x = 0, 1, 2, \ldots \tag{4.18}$$

in which $\mu > 0$ represents the expected number of events occurring in a standard unit of region; R represents the number of standard units of region over which we want to calculate the probability of x events; and e represents Euler's constant, 2.71828 . . . .

The average number of events occurring per standard unit of region (as calculated from observation of a large number of such regions) is generally accepted as a good estimate for the parameter μ. The following examples will assume that such observation and subsequent estimation for μ have been conducted.

EXAMPLE 4.10 The average number of bubble defects per square meter of sheet glass produced by a production line is two. What is the probability that the next 50 cm by 80 cm sheet produced by the line contains fewer than two bubble defects?

SOLUTION The events of interest are bubble defects, the standard unit of region is 1 square meter, and $\mu = 2$. The region of interest represents 0.5×0.8 square meter, so that $R = 0.4$. Consequently,

$$\text{pr}\{\text{less than two bubble defects in 0.4 square meter}\}$$
$$= \sum_{x=0}^{1} \frac{[2(0.4)]^x e^{-2(0.4)}}{x!} = 0.8088$$

EXAMPLE 4.11 In the past, trucks have arrived at a loading dock at a rate of one every two hours. What is the probability that two trucks will arrive in the next hour?

SOLUTION The events of interest are truck arrivals at the loading dock, the standard unit of region is a time interval of two hours, and $\mu = 1$. The region, or time interval, of interest represents 0.5 standard unit of region, so that $R = 0.5$. Thus,

$$\text{pr}\{\text{two trucks arriving in one hour}\} = \frac{[1(0.5)]^2 e^{-1(0.5)}}{2!} = 0.0758$$

As you observed in these examples, calculating Poisson probabilities can be tedious. Tables have been constructed to help alleviate this problem. A representative set of values for the product μR and the corresponding distribution function values can be found in Tables A.3 and A.4. However, the calculation of Poisson probability function values, although tedious, is often more appealing than the calculation of binomial probability function values when the probability of success is small and the number of trials is large. The Poisson distribution is a good approximation to the binomial distribution under these circumstances and is often used because of ease of computation. Specifically, if $f(x|n, p)$ and $g(x|\mu R)$ represent the binomial and Poisson probability functions, respectively, then

$$f(x|n, p) \simeq g(x|\mu R) \tag{4.19}$$

when n is large, p is small, and we set $\mu R = np$.

A rule of thumb for the use of approximation 4.19 is to look for the value of n times p (or n times $1 - p$) to be less than 5; otherwise the normal distribution approximation is more suitable. Since the normal random variable will be discussed in the next chapter, we will delay our discussion of its use as an approximation to the binomial distribution.

We end this preliminary discussion of the Poisson by calculating its expected value and variance, but we will talk about it in later chapters. The expected value of a Poisson random variable X is computed as follows:

$$E(X) = \sum_{x=0}^{\infty} x \frac{(\mu R)^x e^{-(\mu R)}}{x!} = \sum_{x=1}^{\infty} \frac{(\mu R)^x e^{-(\mu R)}}{(x-1)!}$$

$$= (\mu R) \sum_{x=1}^{\infty} \frac{(\mu R)^{x-1} e^{-(\mu R)}}{(x-1)!}$$

If we let $y = x - 1$, this last expression becomes

$$(\mu R) \sum_{y=0}^{\infty} \frac{(\mu R)^y e^{-(\mu R)}}{y!}$$

But the series in this latter expression sums to 1 because it represents the probability of *at least* no events happening in the region R (this can be shown mathematically too). Thus

$$E(X) = \mu R \tag{4.20}$$

One can show in a similar fashion that

$$\text{Var}(X) = \mu R \tag{4.21}$$

Thus, the Poisson random variable has the rather unusual property that its expected value and variance are exactly the same.

EXERCISES

4.5.1 Customers arrive at a supermarket at a rate of three every five minutes. What is the expected number of customers during the next seven-minute period?

4.5.2 The secretaries at a bank like to go to Ted's Sandwich Shack for lunch. Service at Ted's proceeds at a rate of seven customers every five minutes. If four secretaries go to Ted's for lunch on a given day, what is the probability that it takes at most five minutes for them to be served?

4.5.3 Sheet metal is produced with one defect per square meter on the average. What is the probability that there will be at least two but not more than five defects in the next 3 square meters of sheet metal produced by the process?

4.5.4 The catalog for Cut-Rate Warehouse advertises merchandise at prices far below those offered by other retail outlets. However, the probability that the Warehouse has a catalog item in stock when a customer asks for it is 0.96, so that not all customers are satisfied with availability. What is the probability that at least 7 of the next 100 customers will not be satisfied with merchandise availability (that is, that the Warehouse will not have the item the customer wants in stock)?

4.6 SUPPLEMENTARY EXAMPLES

EXAMPLE 4.12 A large box contains both balsam and regular golf balls, with 60% being balsam. If five balls are chosen at random from the box,

(*a*) What is the probability that exactly one is balsam?
(*b*) What is the probability that at least three are regular?
(*c*) What is the probability that at most four are balsam?

SOLUTION Since the box is large and we have no estimate of the total number of golf balls it contains, the appropriate model is the binomial distribution with $p = 0.6$ (or $p = 0.4$).

(*a*)
$$\begin{aligned}\text{pr}\{\text{exactly one balsam}\} &= f(1|n = 5, p = 0.6)\\ &= \binom{5}{1}(0.6)^1(1 - 0.6)^4\\ &= 5(0.6)(0.0256) = 0.077\end{aligned}$$

(*b*)
$$\begin{aligned}\text{pr}\{\text{at least three regular}\} &= \text{pr}\{\text{at most two balsam}\}\\ &= F(2|n = 5, p = 0.6)\\ &= \sum_{x=0}^{2} f(x|n = 5, p = 0.6) = 0.317\end{aligned}$$

Alternatively,

$$\begin{aligned}\text{pr}\{\text{at least three regular}\} &= 1 - \text{pr}\{\text{at most two regular}\}\\ &= 1 - F(2|n = 5, p = 0.4)\\ &= 1 - 0.683 = 0.317\end{aligned}$$

(*c*)
$$\text{pr}\{\text{at most four balsam}\} = F(4|n = 5, p = 0.6) = 0.922$$

EXAMPLE 4.13 A production line produces transistor components. The process produces 3% defective when in control. Suppose that the process is in control and you inspect components at random from the line.

(*a*) What is the probability that the first defective component that appears is the tenth component inspected?

(*b*) How many components would you need to inspect (on the average) in order to encounter a defective component?

SOLUTION

(*a*) $f(10|p = 0.03) = (0.03)(1 - 0.03)^9 = 0.03(0.97)^9 = 0.0228$

Since each component represents a Bernoulli trial, it is reasonable to assume independence; the process being in control implies that p is constant at 0.03; and the variable of interest is the number of trials to the first defective.

(*b*) By the same reasoning as in (*a*), we have a geometric random variable with $p = 0.03$, and we want to determine its expected value. Thus

$$E(X) = \frac{1}{0.03} = 33.33 \text{ components}$$

EXAMPLE 4.14 Ball Bearings, Ltd., arrives at a manufacturing concern each month with a load of bearings offered at cut-rate prices. The catch is that the bearings are of five different sizes, and the manufacturing concern must live with what they get when the driver chooses at random from the cargo. Suppose that the bearing sizes are denoted by 51, 52, 53, 54, and 55, and assume that:

(i) The delivery truck carries

- 10 bearings of size 51
- 20 bearings of size 52
- 15 bearings of size 53
- 10 bearings of size 54
- 5 bearings of size 55

(ii) The manufacturing concern needs

- 2 bearings of size 51
- 4 bearings of size 52
- 3 bearings of size 53
- 1 bearing of size 54
- 0 bearings of size 55

and asks the driver to get ten bearings out of the truck. Calculate the probability that the company's needs will be met.

SOLUTION This problem is properly modeled using the extension to the hypergeometric random variable with

$$R_1 = 10 \quad R_2 = 20 \quad R_3 = 15 \quad R_4 = 10 \quad R_5 = 5$$
$$r_1 = 2 \quad r_2 = 4 \quad r_3 = 3 \quad r_4 = 1 \quad r_5 = 0$$

and $N = 60$ with $n = 10$. Thus, the desired probability is given by

$$\frac{\binom{10}{2}\binom{20}{4}\binom{15}{3}\binom{10}{1}\binom{5}{0}}{\binom{60}{10}} = 0.013$$

EXAMPLE 4.15 People pass through a library turnstile at a rate of 100 per hour during the evening hours of operation. Calculate:

(*a*) The probability that at least ten people pass through the turnstile during a ten-minute period during the evening hours of operation.

(*b*) The probability that at least two but not more than twelve people use the turnstile during a five-minute period during the evening hours of operation.

SOLUTION

(*a*) The standard unit of region is 60 minutes with $\mu = 100$. Thus, $R = 10/60$ and $\mu R = 16.7$, and we compute

$$\sum_{x=10}^{\infty} \left\{ \frac{e^{-16.7}(16.7)^x}{x!} \right\} = 1 - \sum_{x=0}^{9} \frac{e^{-16.7}(16.7)^x}{x!}$$
$$= 1 - 0.0318 = 0.9682$$

(*b*) Here $R = 5/60$ and $\mu R = 8.3$, and we compute

$$\sum_{x=0}^{12} \frac{e^{-8.3}(8.3)^x}{x!} - \sum_{x=0}^{1} \frac{e^{-8.3}(8.3)^x}{x!} = 0.9207 - 0.0023$$
$$= 0.9184$$

EXAMPLE 4.16 The promotions department in a local retail firm decides to interview 100 people at random in a major shopping center in the area to determine how many of the shoppers have heard the firm's radio advertisement. The radio station maintains that they reach 30% of the buying public with the type of advertisement schedule purchased by the company. Assuming that 300 of the 1000 people using the shopping center during the time when the interviews take place have heard the advertisement, discuss the problems in calculating probabilities like: The probability that at most 20 of the 100 in the sample have heard the retail company advertisement or the probability that at least 25 but not more than 40 of the 100 in the sample have heard the retail company advertisement.

SOLUTION We can use the hypergeometric random variable in this problem with $N = 1000$, $n = 100$, and $R = 300$. Thus, letting X denote the random variable, we get

$$\text{pr}\{X \leqslant 20\} = \sum_{x=0}^{20} \frac{\binom{300}{x}\binom{1000-300}{100-x}}{\binom{1000}{100}}$$

Using the binomial approximation, we get

$$\text{pr}\{X \leqslant 20\} \approx \text{pr}\{Y \leqslant 20\}$$

where Y is a binomial random variable with parameters $n = 100$ and $p = R/N = 0.3$. Thus,

$$\text{pr}\{X \leqslant 20\} \approx \sum_{x=0}^{20} \binom{100}{x} (0.3)^x (0.7)^{100-x}$$

The problem is still unwieldly, and we need another approximation in order to simplify the calculations. We will show how to do this in Chapter 5 when we introduce the normal approximation to the binomial (Section 5.4). The second problem is of the same type, namely

$$\begin{aligned}\text{pr}\{25 \leqslant X \leqslant 40\} &\approx \text{pr}\{25 \leqslant Y \leqslant 40\} \\ &= \sum_{x=0}^{40} \binom{100}{x} (0.3)^x (0.7)^{100-x} \\ &\quad - \sum_{x=0}^{24} \binom{100}{x} (0.3)^x (0.7)^{100-x}\end{aligned}$$

EXAMPLE 4.17 The output of a manufacturing process includes, on the average, four defective items per 10,000 produced. What is the probability that there will be at least two but not more than eight defective items in the next 10,000 items produced?

SOLUTION This is a binomial random variable problem with parameters $n = 10{,}000$ and $p = 4/10{,}000$, so that $np = 4$, and the Poisson approximation can be used to get

$$\sum_{x=2}^{8} \frac{e^{-4}(4)^x}{x!} = \sum_{x=0}^{8} \frac{e^{-4}(4)^x}{x!} - \sum_{x=0}^{1} \frac{e^{-4}(4)^x}{x!}$$

$$= 0.9786 - 0.0916$$

$$= 0.8870$$

EXAMPLE 4.18 Refer to Example 4.17. Suppose that a new safety device has been added to the production line and we are interested in studying its influence on quality (whether the average number of defective items per 10,000 has increased since the device was added to the line). If the next 10,000 produced include eight defective items, with all other production factors remaining the same, would you say that the average rate per 10,000 has increased?

SOLUTION We want to compute the probability

$$\binom{10{,}000}{8}\left(\frac{4}{10{,}000}\right)^8 \left[1 - \left(\frac{4}{10{,}000}\right)\right]^{10{,}000-8}$$

which is approximated by use of the Poisson distribution as

$$\frac{e^{-4}(4)^8}{8!} = 0.0298$$

The last figure is a measure of the likelihood of observing eight defective items given that the average number of defectives is still four per 10,000 items produced. The question is whether or not this qualifies as a rare event in the sense that the average rate has not really increased and we observed a high value of 8 only because of an unusual production run of 10,000 items. This question will be properly answered in Chapters 9 and 10, where concepts of statistical inference are introduced and applied to common problems in hypothesis testing.

Tabular Chapter Summary

Random Variable	Probability Function	Expected Value	Variance
Bernoulli	$f(x\mid p) = \begin{cases} p & \text{if } x = 1 \quad \text{(success)} \\ 1 - p & \text{if } x = 0 \quad \text{(failure)} \end{cases}$	p	$p - p^2$
Binomial	$f(x\mid n, p) = \binom{n}{x} p^x (1 - p)^{n-x} \quad x = 1, 2, 3, \ldots, n$	np	$np(1 - p)$
Geometric	$f(x\mid p) = p(1 - p)^{x-1} \quad x = 1, 2, 3, \ldots$	$\frac{1}{p}$	$\frac{1 - p}{p^2}$
Hypergeometric	$f(x\mid N, n, R) = \frac{\binom{R}{x}\binom{N-R}{n-x}}{\binom{N}{n}}$	$\frac{nR}{N}$	$\frac{nR(N - R)(N - n)}{N^2(N - 1)}$
	$\max\{0, n - N + R\} \leqslant x \leqslant \min\{n, R\}$		
Poisson	$f(x\mid \mu R) = \frac{(\mu R)^x e^{-\mu R}}{x!} \quad x = 0, 1, 2, \ldots$	μR	μR

4.7 SUMMARY

In Chapter 2, we discussed the basic concepts of probability and their use in determining the likelihood of occurrence for the outcome of a random experiment. Chapter 3 introduced the general idea of a random variable to help to describe the random experiments of Chapter 2. In doing so, it was necessary to associate probabilities with the possible values of the random variable by identifying its probability distribution. In this chapter we have acquainted you with some of the more common discrete probability distributions that can be used to describe the chance occurrence of many physical phenomena. Many statisticians prefer to label these distributions as natural distributions, since they can be used to describe events that occur in nature.

The simplest distribution (other than the degenerate case of a constant) is the Bernoulli. This can be used in describing experiments that must result in one of two possible outcomes. If we perform a series of independent identical Bernoulli trials and are interested in the total number of successes, the binomial distribution is appropriate. Switching the variable of interest from the number of successes to the number of trials needed to obtain a success gives rise to the geometric distribution. If the Bernoulli trials are dependent as a result of sampling without replacement from a finite categorical population, then the hypergeometric distribution becomes appropriate.

If our random variable is the type that counts the number of occurrences in a specified unit (time, area, distance), and if the assumptions of the Poisson distribution are satisfied, then it is the appropriate model to use. These assumptions are: (1) events occurring in nonoverlapping units are independent, (2) the expected number of occurrences is directly proportional to the size of the unit, and (3) the probability of more than one occurrence in a very small unit is zero for all practical purposes.

It is imperative that you become intimately familiar with the assumptions associated with each of these distributions. Doing so will enable you to recognize the situations in which each will apply in practice. If you use a distribution when one or more of its assumptions are violated, the resulting probability will be *at best* a reasonable approximation and in many situations may lead to disastrously wrong decisions.

For your convenience, we have summarized most of the distributions of this chapter and their pertinent attributes in a tabular format.

REVIEW EXERCISES

Level 1

4.1 In the manufacture of a certain article, it is known that one out of ten of the articles produced is defective. What is the probability that a random sample of four of the articles will contain
(*a*) No defectives?
(*b*) Exactly one defective?
(*c*) Exactly two defectives?
(*d*) Not more than two defectives?

4.2 A coin is tossed repeatedly. Find the probability that the fourth head appears on the eleventh toss (assume a fair coin).

4.3 A company buys a particular electronic component in lots of 10,000 for use in assembling one of its products. Their quality control inspector selects 100 items at random from each lot and rejects the lot if there are more than four defectives in the sample. Good lots contain 2% defectives.
(*a*) What is the exact probability distribution of the number of defectives in the sample if the lot is good?
(*b*) What is the probability that the inspector will reject a good lot?
(*c*) What is the probability that the inspector will accept a lot that has 5% defectives in it?

4.4 On a certain highway, chuckholes are known to occur randomly with an average of two per mile.
(*a*) What is the probability that you would encounter no more than five holes in a 3-mile stretch?
(*b*) What is the probability of driving at least 2 miles without encountering a chuckhole?

4.5 Ray's salvage has just purchased a truckload of damaged merchandise. Included in the purchase are ten color TVs, of which four are inoperable because of internal damage. Ray will not permit you to test a set before you purchase it. Hence, all sales are "as is" and are final. If you randomly select two sets,
(*a*) How many working sets should you expect to get?
(*b*) What is the standard deviation of the number of working sets for a random selection of two sets?

4.6 Suppose that Ray (from Problem 4.5) has been offered a boxcar load of transistor radios (the boxcar was involved in a derailment). There are 10,000 radios in the boxcar, and Ray feels that he can make a substantial profit if no more than 2000 have been damaged irreparably.

(*a*) If the quality satisfies his requirements for a profit, what is the probability that he buys them if he decides to buy when a sample of ten gives three or fewer defectives?

(*b*) If Ray selects 20 for testing, buying when six or fewer are defective, what is the chance that he will fail to buy the lot when it is to his advantage to buy?

4.7 In a machining process, parts are categorized as good (G), bad (B), and reworkable (R). In the past the process has yielded 80% G, 15% R, and 5% B. Find the probability that of the next ten parts:

(*a*) Seven will be good, two reworkable, and one bad.

(*b*) Eight or more will be good.

4.8 A professor prepares nine questions for an exam. Five of these questions will be chosen for the exam. A student knows the answers to five of the nine questions. If four correct answers are passing, what is the student's probability of passing? What is the average number of questions the student can be expected to answer?

4.9 A components manufacturer produces parts on a machine that yields a defective part with probability 0.01.

(*a*) If the manufacturer ships parts in boxes of ten, what is the probability that a box will contain fewer than two defectives?

(*b*) How many parts should be included to ensure, with probability of at least 0.98, that there are at least ten good parts in a box?

4.10 A truck dock has an arrival rate of 0.75 truck per hour. What is the probability that:

(*a*) Exactly six trucks arrive in an eight-hour shift?

(*b*) At least nine trucks arrive in an eight-hour shift?

4.11 Between 9:30 and 10:30 A.M., customers arrive at the Queen Mart on an average of five per hour. What is the probability that:

(*a*) Exactly six will arrive during that time tomorrow?

(*b*) Three or more will arrive between 9:30 and 10:00 tomorrow?

(*c*) No more than thirteen will arrive in total during that time over the next three days?

4.12 To determine which three soldiers to assign to an unpleasant detail, the sergeant puts the names of the eight candidates in a hat, stirs well, and draws out three. Of the eight candidates, two are women. Determine the probability that the detail will include exactly one woman.

4.13 Twenty persons were randomly selected in the student union and asked their political association and their feeling regarding the current administration's handling of the inflation problem. The results were as follows:

	Republican		Democrat		Independent	
	Male	Female	Male	Female	Male	Female
Favor	1	0	1	1	2	0
Not favor	1	2	3	2	3	4

If four persons were to be taken from this group to form a committee to gain more information on the matter, what is the probability of having:

(*a*) Two men and two women?
(*b*) Two men in favor?
(*c*) Four persons not in favor?
(*d*) Three Democrats and two independents?
(*e*) Two Republicans, one Democrat, and one independent?

4.14 For a binomial distribution with $n = 20, p = 0.3$, find

(*a*) $P(X = 7)$.
(*b*) $P(4 \leqslant X \leqslant 9)$.
(*c*) The expected value and variance of the number of successes.

4.15 A manufacturing representative has found that he averages three sales for each ten calls he makes. What is the probability that on his current business trip

(*a*) He will make one sale in his first three calls?
(*b*) He must make four calls before getting his first sale?
(*c*) He must make fifteen calls before he makes six sales?

Level 2

4.16 A part to a certain machine is known to break, on the average, once in four days. How many parts must be available so that there is *less than* one chance in 100 of having more breakages than parts available on any given day?

4.17 The Transit Taxi Company, operating a fleet of ten cars, serves three nearby small communities, Hudson, Stow, and Kent. The cars move freely among the cities depending on the requests for service. Assuming that the traffic pattern of the cars is random among the three communities, what is the probability that at the close of any given day there will be five cars in Hudson, two cars in Stow, and three cars in Kent?

4.18 The C-All Novelty Company buys Creepie Peepies from a Hong Kong supplier. Lots that have 10% defective Creepie Peepies are considered to be acceptable. C-All has just received a lot of 10,000 Creepie Peepies and is trying to decide whether they should keep it or return it to the supplier.

(*a*) If they select four at random, what is the probability that they will find exactly one defective if the lot is of acceptable quality? *Calculate* the answer directly.

(*b*) If they decide to select twenty at random and to reject the lot if more than four defectives are found, what is the probability that they will reject the lot by mistake?

(*c*) If the lot has 2500 defectives in it, what is the probability that they will accept it by mistake if they use the plan in (*b*)?

4.19 For the binomial distribution in Exercise 4.14, determine the expected value and variance of the proportion of successes.

4.20 The owner of Steve's No-Tell Motel is considering buying color television sets to rent to his overnight guests. He has ten rooms and believes that about half of his customers would be willing to rent sets. Finally he buys six sets. Assuming 100% occupancy at all times,

(*a*) What is the probability that there will be more requests than TVs?

(*b*) If his total cost per set per day is C, what rental must he charge in order to break even in the long run?

(*c*) What is the probability that a customer requesting a color television in Exercise 4.20 will receive one?

4.21 Two machines produce the same item. When in control, each produces defectives with a probability of 0.05. The items are packaged for shipment in boxes of 100 as they come from the machines, with each box having 50 items from each machine. On Monday, machine 1 went out of control and began producing defectives at a 20% rate. This was not detected until Wednesday. During the time that machine 1 was out of control, what are the expected value and standard deviation of the number of defectives per box?

Level 3

4.22 Derive the variance for a Poisson random variable.

4.23 Provide a mathematical argument supporting Equation 4.14.

4.24 Refer to the C-All Novelty Company in Exercise 4.18. They have just hired you as their quality control inspector for Creepie Peepies. They would like you to devise a sampling plan that will accept good lots at least 90% of the time, but accept lots having 25% defective no more than 10% of the time. Can you do so?

4.25 A particular sales manager feels that salespeople should sell a car to 20% of all potential customers. With this in mind, the manager hires a salesperson on a trial basis and observes the number of sales. If it takes the salesperson more than seven customers to make the first sale, the manager fires him or her.

(*a*) What is the probability that the manager retains a salesperson who is only half as good as the manager wants?

(*b*) What is the probability that the manager fires a good salesperson?

Chapter 5
Probability Distributions for Continuous Phenomena

KEY WORDS AND TERMS

Uniform Distribution
Exponential Distribution
Normal Distribution
Lognormal Distribution
Gamma Distribution
Beta Distribution
Wiebull Distribution

5.1 INTRODUCTION

Continuous random variables are used to model stochastic phenomena that reflect some type of measurement. The word *continuous* is appropriate, since for any two possible measurements, any value between the two could result from the next experiment or trial. The measurement could represent length, area, weight, time, or volume. Thus, for any experiment in which the range values represent measurements, a continuous random variable would be an appropriate model. In this chapter, we examine the most commonly used distributions for continuous phenomena.

5.2 THE UNIFORM DISTRIBUTION

Sometimes the simplest models turn out to be the most useful, and the uniform distribution is a case in point. It is a keystone in computer simulation studies, where it is used to generate random deviates (that is, range values) for more complicated or sophisticated distributions.

DEFINITION 5.1 The uniform distribution has a density function given by

$$f(x|a, b) = \begin{cases} \dfrac{1}{b - a} & a \leqslant x \leqslant b \\ 0 & \text{elsewhere} \end{cases} \tag{5.1}$$

where a and b are real numbers satisfying $a < b$. Thus, the uniform distribution can assume any one of the real numbers between its parameters a and b. Also, note that the uniform distribution has a constant value of $1/(b - a)$ for each point in this interval of real numbers. This property makes calculating probabilities for the random variable range values easy, as the following example illustrates.

EXAMPLE 5.1 In a simulation study of traffic patterns, a light is set to switch between red and green according to a uniform distribution whose parameters are 30 and 36 (seconds). Graph the density function, interpret the result, and calculate the probability that the next switch will take at least 34 seconds.

The probability density function is

$$f(x|a = 30, b = 36) = \begin{cases} \dfrac{1}{6} & 30 \leqslant x \leqslant 36 \\ 0 & \text{elsewhere} \end{cases}$$

This is illustrated by the graph in Figure 5.1. The interpretation of this graph is that all intervals of the same length lying entirely between 30 and 36 are equiprobable. That is, the probability is uniformly distributed between 30 and 36; hence the name uniform. The probability that the next switch takes at least 34 seconds is given by

$$\text{pr}(34 \leqslant X \leqslant 36) = \int_{34}^{36} \frac{1}{6}\, dx$$

This area can be calculated using either calculus or simple geometry. Using geometry and the formula for the area of a rectangle, we get

$$\frac{1}{6}(36 - 34) = \frac{1}{3}$$

Figure 5.1 Graph of a Uniform Density Function Whose Parameters are 30 and 36

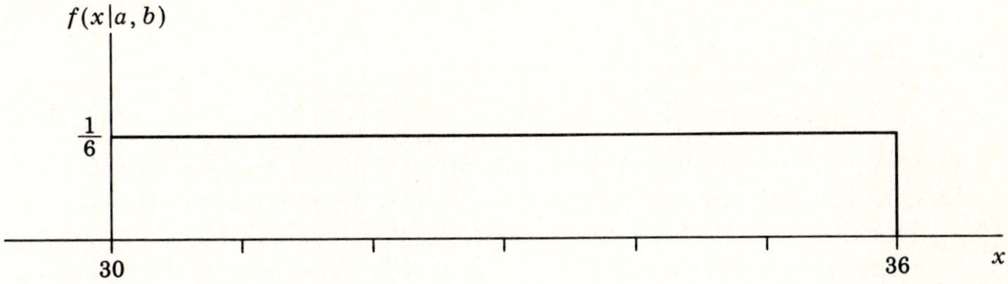

Using calculus, we get

$$\int_{34}^{36} \frac{1}{6}\,dx = \frac{1}{6}x\Big|_{34}^{36} = \frac{1}{6}(36) - \frac{1}{6}(34) = \frac{1}{3}$$

One frequently hears the phrase "random zero-one deviate" in the construction of computer simulation studies. This means that the computer is being asked to generate a range value from a uniform distribution whose parameters are zero and one (the standard uniform distribution).

The standard uniform distribution is also used to construct random sampling plans. Frequently the sampling plan is developed using a random number table. The table entries represent a collection of uniform zero-one deviates with the decimal point omitted.

EXERCISES

5.2.1 Suppose that X is a random variable uniformly distributed on the interval from 1 to 4. Calculate the probabilities that

(*a*) $\frac{1}{4} \leqslant X \leqslant \frac{3}{4}$

(*b*) $\frac{1}{4} \leqslant X \leqslant 5$

(*c*) $2 \leqslant X \leqslant 3$

(*d*) $2 \leqslant X \leqslant 5$

and calculate the expected value of X.

5.2.2 Suppose that X is a random variable having a uniform distribution defined over the interval from the real number a to the real number b. Calculate

(a) $\mathrm{pr}(a \leqslant X \leqslant b)$.

(b) $\mathrm{pr}\left(\frac{a+b}{2} \leqslant X \leqslant b\right)$.

(c) The expected value of X.

5.2.3 Pullin' Pete has found that the time it takes him to milk his cow is a random variable that appears to be uniformly distributed between 5 minutes and 12 minutes. Pete has a customer for fresh whole milk who stops by each morning promptly at 7:15. Having had a big night at the Hoedown last evening, Pete arose late this morning and didn't start pullin' until 7:05.

(a) What is the probability that the customer will not have to wait for Pete to finish pullin'?

(b) What is the average amount of time that Pete's pullin' each day? (Assume morning and evening milkings.)

5.2.4 Suppose that you want to use a random number table that was generated from a standard uniform distribution to select a five-digit number. What is the probability that your number will be between 15064 and 49352? (Random numbers generally are decimals with the decimal point omitted.)

5.3 THE NEGATIVE EXPONENTIAL DISTRIBUTION

The negative exponential distribution has many applications in the areas of queueing theory, waiting times, and reliability theory. It is connected mathematically with the Poisson distribution, discussed in the previous chapter, in the sense that if the Poisson assumptions are satisfied, then the time between events must be negative exponential in distribution.

DEFINITION 5.2 The *negative exponential distribution* has a density function given by

$$f(t|\beta) = \begin{cases} \beta e^{-\beta t} & t \geqslant 0 \\ 0 & \text{elsewhere} \end{cases} \tag{5.2}$$

where β is a positive number.

The parameter β is the reciprocal of the expected time until the event of interest occurs. The negative exponential distribution is usually used for a time or distance measurement, such as time or distance between events or time until failure. The connection between the Poisson distribution and the negative exponential distribution is that if X is Poisson with parameter μ, then the time between events T is exponential with parameter μ; that is, $\beta = \mu$. This idea is used in the following example.

EXAMPLE 5.2 Suppose that cars enter an intersection at a rate of five per fifteen seconds. What is the probability that the next time interval between cars entering the intersection is at least three seconds? [If T is negative exponential with parameter five, we need to calculate the probability that

$$\text{pr}(T' \geqslant 3 \text{ seconds})]$$

This problem has the same structure as the Poisson in that $\beta = \mu = 5$ and the standard time unit (STU) is 15 seconds. Therefore,

$$\text{pr}(T' \geqslant 3 \text{ seconds}) = \text{pr}\left(T \geqslant \frac{3}{15}\right)$$

$$= \int_{3/15}^{\infty} 5e^{-5t}\,dt$$

We could also have expressed β as the rate of occurrence per second, $\beta = 5/15$ per second. Then

$$\text{pr}(T' \geqslant 3) = \int_{3}^{\infty} \frac{1}{3}(e^{-t/3})\,dt = e^{-3/3} - 0 = e^{-1}$$

In either case, we have a problem, because there are no easy geometrical formulas to use in determining the area inside the curve. This is due to the fact that the function $\beta e^{-\beta t}$ is nonlinear, as shown in Figure 5.2 on page 110.

Fortunately, the integral of the negative exponential density function has what we call a *closed form*, meaning that it can be calculated mathematically using the theorems of calculus. In general,

$$\int_{a}^{b} \beta e^{-\beta t}\,dt = e^{-\beta a} - e^{-\beta b} \tag{5.3}$$

Figure 5.2 Graph of a Negative Exponential Distribution with Parameter β

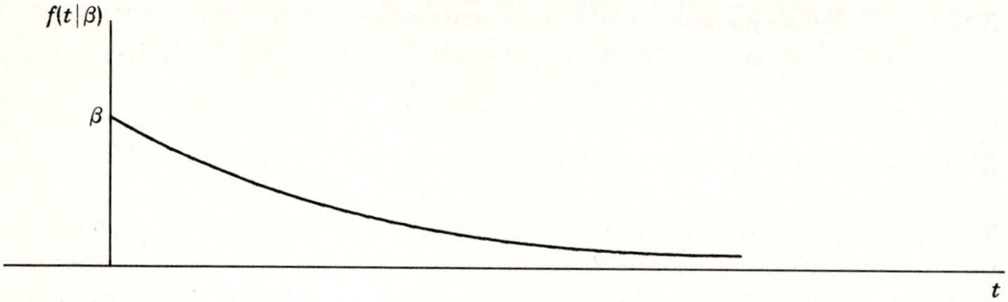

For the problem at hand, $\beta = 5$, $a = 3/15$, and $b = \infty$, giving $e^{-5(3/15)} - 0 = e^{-1}$. This still does not tell us much, since Euler's constant (e) is an irrational number. Table A.5 in the back of the book gives values of e^{-x} for selected values of x. Table A.5 gives $e^{-1} = 0.3679$, which is the probability that at least three seconds elapse between the next two cars entering the intersection.

The expected value and variance of an exponential random variable T whose density function has parameter β are given by

$$E(T) = \frac{1}{\beta} \tag{5.4}$$

$$\text{Var}(T) = \frac{1}{\beta^2} \tag{5.5}$$

EXERCISES

5.3.1 Customers arrive at a bank at a rate of three every five minutes. What is the probability that at least one minute elapses between the arrivals of the next two customers?

5.3.2 Vehicles pass through an intersection at a rate of five every two minutes. What is the expected time between the arrivals of the next two vehicles?

5.3.3 Repair points on a transmission line occur at a rate of one every 20 meters. What is the probability that there are at most 15 meters between repair points?

5.3.4 Calculate the standard deviation for each of the distributions you used in Exercises 5.3.1 to 5.3.3 and interpret your results.

5.4 THE NORMAL DISTRIBUTION

The normal distribution is very useful as a model for many business phenomena, such as processing times, assembly times, performance rates, income levels, consumption levels, and fill volumes. It is easier to use as a model (in a mathematical sense) than many other distributions, and its use is frequently justified as a good approximation for several other continuous distributions. In general, the distribution arises whenever the majority of a set of measurements cluster about a central point with the remainder of the measurements symmetrically distributed about the central point. The distribution is also known as the *Gaussian distribution.* The graph of its density function is shown in Figure 5.3. The normal curve in Figure 5.3 is the plot of the density function given in the following definition.

Figure 5.3 The Normal Curve

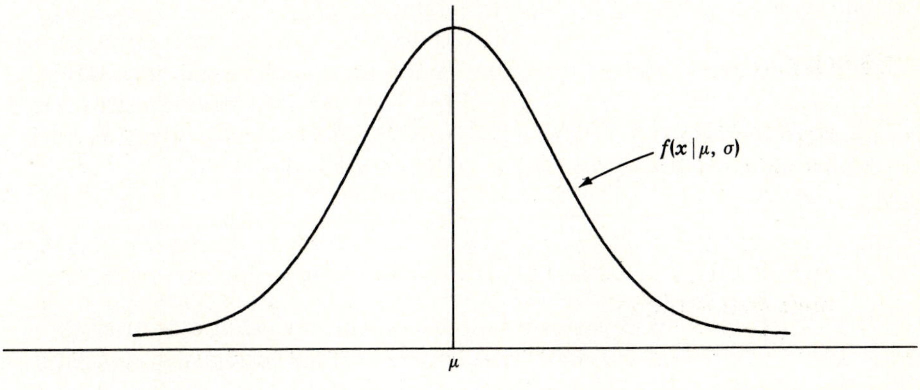

DEFINITION 5.3 The normal distribution has a density function given by

$$f(x|\mu, \sigma) = \frac{1}{\sigma\sqrt{2\pi}} e^{-(x-\mu)^2/2\sigma^2} \tag{5.6}$$

where $\pi = 3.14159\ldots$, μ is the expected value of X, and σ is the standard deviation of X.

Thus, like the uniform, the normal is a two-parameter distribution, and it has the property of being symmetric about a vertical line drawn through its expected value. Thus, if X is normal with expected value μ, then

$$\text{pr}\{\mu - \alpha \leqslant X \leqslant \mu\} = \text{pr}\{\mu \leqslant X \leqslant \mu + \alpha\}$$

for any positive real number α.

Now that you have the good news, the bad news is that the integral given by

$$\int_{x_1}^{x_2} \frac{1}{\sigma\sqrt{2\pi}} e^{-(x-\mu)^2/2\sigma^2}\, dx$$

does not have a closed form and thus must be numerically integrated using a computer. Now think of the fact that there are an infinite number of normal distributions (a different one each time the value of a parameter, μ or σ, is changed), each of which must be numerically integrated every time a new normal probability question arises. This is disconcerting when you consider the number of data sets in industry, nature, and scientific studies that can be adequately modeled using normal distributions. Fortunately, this difficulty is circumvented by the transformation given in the following statement:

If X is a normal random variable having mean μ and standard deviation σ, then $(X - \mu)/\sigma$ is a normal random variable having mean 0 and standard deviation 1; this is called the *standard normal* random variable. Traditionally, Z is used to denote this variable:

$$Z = \frac{X - \mu}{\sigma} \tag{5.7}$$

The significance of this statement lies in the fact that any normal probability statement can be transformed into an equivalent *standard*

Figure 5.4 The Correspondence Between Cumulative Probabilities and Deviates for the Standard Normal Distribution Table (see Appendix in the back of this book)

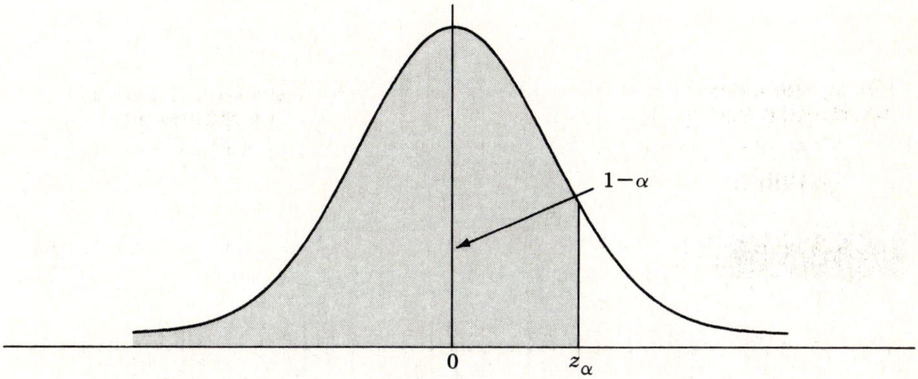

normal probability statement (a reflection of the fact that probabilities are invariant under scalar transformations). Thus, by constructing a table using numerical integration of the standard normal distribution, we have provided a means for numerically integrating any normal distribution. Such a table of values is called a *standard normal table*, and its use requires some practice. Table A.6 in the back of the book contains what we call *standard normal deviates* and their corresponding *cumulative probabilities.* The two numbers are related by the equation

$$\text{pr}(Z \leqslant z_\alpha) = 1 - \alpha \tag{5.8}$$

in which Z represents a standard normal random variable, z_α represents the deviate, and $1 - \alpha$ represents the cumulative probability. The correspondence is illustrated in Figure 5.4.

EXAMPLE 5.3 If Z is a standard normal random variable, determine the following:

(*a*) $\text{pr}(Z \leqslant 1.2)$
(*b*) $\text{pr}(Z > 1.4)$
(*c*) $\text{pr}(Z \leqslant -0.5)$
(*d*) $\text{pr}(Z > -0.85)$
(*e*) $\text{pr}(1.2 < Z \leqslant 1.4)$
(*f*) $\text{pr}(-0.85 < Z \leqslant -0.5)$
(*g*) $\text{pr}(-0.5 < Z \leqslant 1.4)$
(*h*) $\text{pr}(-1.96 < Z \leqslant 1.96)$

SOLUTION Parts (*a*) through (*h*) represent various circumstances that we may encounter when we use the standard normal table to evaluate a probability statement. Since all the statements involve the standard normal variable, we refer directly to Table A.6. Now let us proceed with the solution.

(*a*) The relationship

$$\text{pr}(Z \leqslant 1.2) = 1 - \alpha$$

indicates that $z_\alpha = 1.2$ for this problem, and we want to determine the value of $1 - \alpha$. The leftmost column of the table gives values of z_α out to one decimal place, and one must move along the row to the proper column to find the second decimal place. The body (as opposed to the margins) gives values for $1 - \alpha$. Thus, we locate the row headed by 1.2 and the column headed by 0.00 to obtain 0.8849 for the probability. Thus, $1 - \alpha = 0.8849$ and $\text{pr}(X \leqslant 1.2) = 0.8849$.

The next few parts are handled in much the same fashion.

(*b*) $\text{pr}(Z > 1.4) = 1 - \text{pr}(Z \leqslant 1.4) = 1 - 0.9192 = 0.0808$

(*c*) $\text{pr}(Z \leqslant -0.5) = 0.3085$

(*d*) $\text{pr}(Z > -0.85) = 1 - \text{pr}(Z \leqslant -0.85) = 1 - 0.1977 = 0.8023$

(*e*) $\text{pr}(1.2 < Z \leqslant 1.4) = \text{pr}(Z \leqslant 1.4) - \text{pr}(Z \leqslant 1.2)$
$= 0.9192 - 0.8849 = 0.0343$

We can justify this last result by referring to Figure 5.5, in which the horizontally lined area represents $\text{pr}(Z \leqslant 1.2)$ and the total lined area represents $\text{pr}(Z \leqslant 1.4)$. Note that $(Z \leqslant 1.2)$ is a subset of $(Z \leqslant 1.4)$, and since the desired area lies between 1.2 and 1.4, we exclude the area to the left of 1.2 by subtraction. Similar procedures are used to solve the remainder of the problem. Specifically,

(*f*) $\text{pr}(-0.85 < Z \leqslant -0.5) = \text{pr}(Z \leqslant -0.5) - \text{pr}(Z \leqslant -0.85)$
$= 0.3085 - 0.1977 = 0.1108$

(*g*) $\text{pr}(-0.5 < Z \leqslant 1.4) = \text{pr}(Z \leqslant 1.4) - \text{pr}(Z \leqslant -0.5)$
$= 0.9192 - 0.3085 = 0.6107$

and finally,

(*h*) $\text{pr}(-1.96 < Z \leqslant 1.96) = \text{pr}(Z \leqslant 1.96) - \text{pr}(Z \leqslant -1.96)$
$= 0.975 - 0.025 = 0.95$

Figure 5.5 Graphical Representation for Example 5.3(*e*)

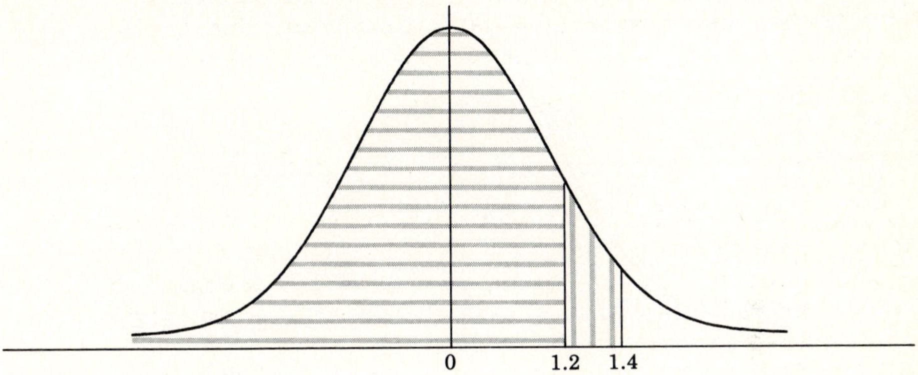

The standard normal table can be used in reverse, starting with a probability and then finding a z_α value in the margins. Remember, we are using the capital letter Z to represent a standard normal random variable and z_α to represent the specific value such that α is the area in the right-hand tail.

EXAMPLE 5.4 Find the value v for which

(*a*) $\text{pr}(-v \leqslant Z \leqslant +v) = 0.95$
(*b*) $\text{pr}(Z \leqslant v) = 0.995$
(*c*) $\text{pr}(Z > v) = 0.7995$

SOLUTION

(*a*) A good first step is to sketch the problem graphically. We do this in Figure 5.6*a* to illustrate the connection between the specified area (which is shaded) and the values we want to determine. Since the normal curve is symmetrical about zero, the area of 0.95 must be evenly divided, with 0.475 of the area over the interval $[-v, 0]$ and 0.475 of the area over the interval $[0, v]$. This means that 0.5 - 0.475 or 0.025 of the area must lie to the right of v, so that $v = z_{0.025} = 1.96$.

(*b*) We can refer to the definitional formula to find v in this example: If $\text{pr}(Z \leqslant v) = 0.995$, then $\text{pr}(Z > v) = 0.005$, so that (averaging)

$$v = z_{0.005} = 2.575$$

This is shown graphically in Figure 5.6*b*.

Figure 5.6 Graphical Representation for Example 5.5

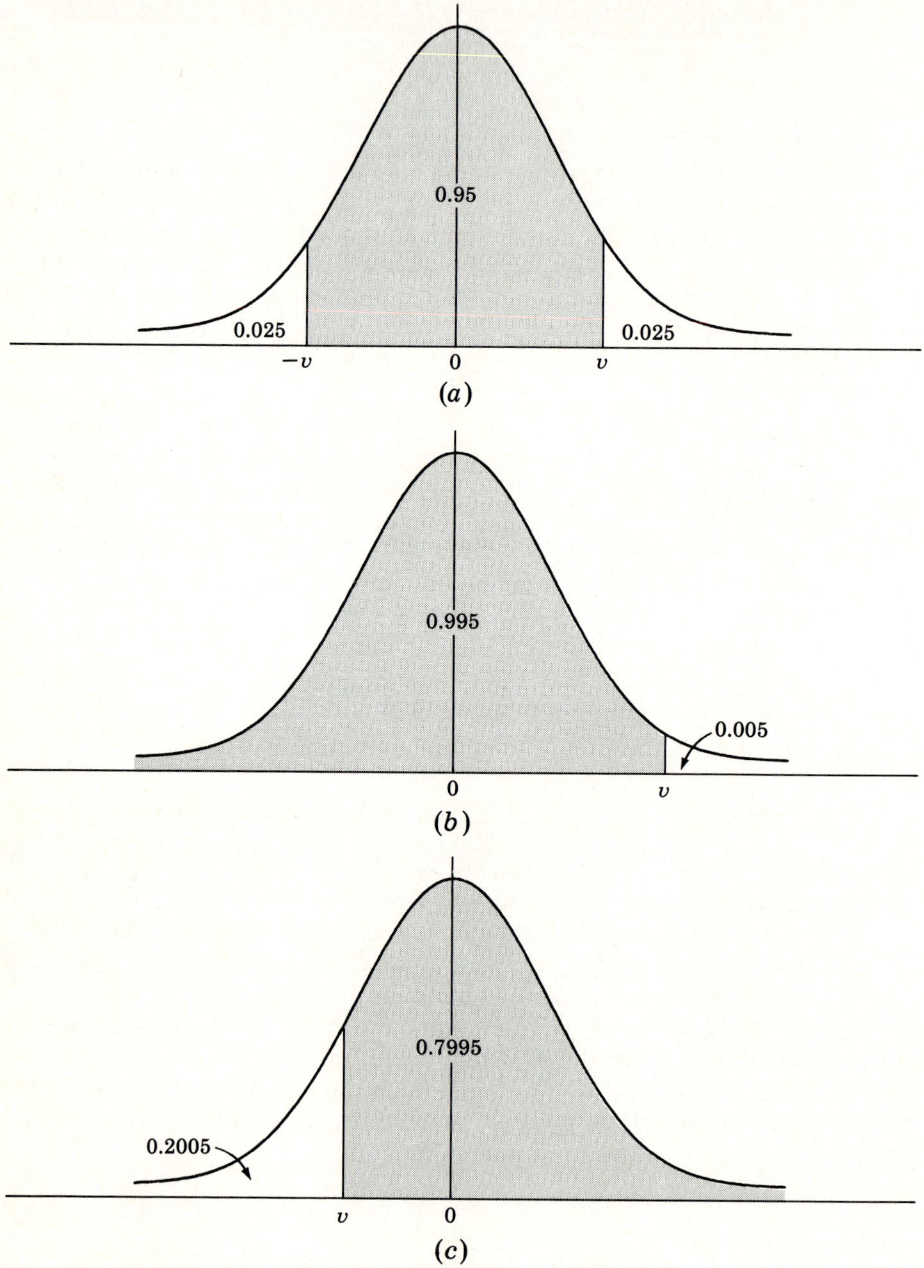

(c) In this case we note that since the area to the right of v is greater than 0.5, v must be to the left of zero. Graphically, this is shown in Figure 5.6c. Therefore, the area to the left of v is 0.2005 and

$$v = z_{0.7995} = -0.84$$

It should be emphasized that the values for Z represent the number of population standard deviations from the mean in a normal distribution. Thus, when we know a value of Z, we can find the area to the left of the value in Table A.6. This area represents the probability of being below the indicated number of standard deviations from the mean. The sign of Z indicates on which side of the mean we are. Correspondingly, if an area (probability) is specified, we can use the table to determine the distance from μ, in terms of the number of standard deviations, that will give rise to it.

The next step is to show how to use the transformation to calculate probabilities for "nonstandard" normal distributions. First, we will justify the method mathematically, then we will illustrate its use with an example.

Suppose that we have a normally distributed random variable X with mean μ and standard deviation σ, and we want to evaluate the probability statement

$$\text{pr}(x_1 < X \leqslant x_2)$$

where $x_1 < x_2$. Using the laws of algebra, we obtain

$$\begin{aligned}\text{pr}(x_1 < X \leqslant x_2) &= \text{pr}(x_1 - \mu < X - \mu \leqslant x_2 - \mu)\\ &= \text{pr}\left(\frac{x_1 - \mu}{\sigma} < \frac{X - \mu}{\sigma} \leqslant \frac{x_2 - \mu}{\sigma}\right)\\ &= \text{pr}\left(\frac{x_1 - \mu}{\sigma} < Z \leqslant \frac{x_2 - \mu}{\sigma}\right)\end{aligned}$$

where Z is the standard normal random variable in the previously mentioned transformation, Equation 5.7.

It is easy to show that the variable Z does have a mean of 0 and a variance of 1. These follow from the basic properties given in Chapter 3 for the expected value operator. We show the derivation for the variance here and leave the proof regarding the mean as an exercise. We have

$$\begin{aligned}\text{Var}(Z) &= \text{Var}\left(\frac{X - \mu}{\sigma}\right) = E\left[\left(\frac{X - \mu}{\sigma}\right)^2\right] = E\left[\frac{(X - \mu)^2}{\sigma^2}\right]\\ &= \frac{1}{\sigma^2}E[(X - \mu)^2] = \frac{1}{\sigma^2}(\sigma^2) = 1\end{aligned}$$

since σ^2 is a constant and $E[(X - \mu)^2]$ is defined to be the variance of X, or σ^2.

EXAMPLE 5.5 A certain type of drill bit lasts on the average 3.0 hours with a standard deviation of 0.5 hour. On the assumption that drill bit life is adequately described by a normal distribution, calculate the probability that a given bit will last over 2.3 hours but not more than 3.7 hours.

SOLUTION The problem is to calculate

$$\text{pr}(2.3 < X \leqslant 3.7)$$

in which X represents a normal random variable with mean 3 and standard deviation 0.5. Using the transformation of Equation 5.7 gives

$$\begin{aligned} \text{pr}(2.3 < X \leqslant 3.7) &= \text{pr}(2.3 - 3 < X - 3 \leqslant 3.7 - 3) \\ &= \text{pr}\left(\frac{2.3 - 3}{0.5} < \frac{X - 3}{0.5} \leqslant \frac{3.7 - 3}{0.5}\right) \\ &= \text{pr}(-1.4 < Z \leqslant 1.4) \\ &= 0.9192 - 0.0808 \\ &= 0.8384 \end{aligned}$$

in which Z is a standard normal random variable.

The first few times you use this procedure, you should include each step shown in order to convince yourself that the procedure is mathematically consistent (this will also reduce the possibility of error). In addition, you will find it helpful to sketch the curve and to shade in the area of interest.

Any variable you obtain using the standardization formula given in Equation 5.7 will have a mean of 0 and a variance of 1, but not all the resulting variables will have a normal distribution. The resulting variable has an exact normal distribution only if the original variable X is normally distributed. A powerful property of the normal family of distributions is that performing a scalar transformation on a normal random variable yields another normal random variable. In addition, if one constructs a linear combination of a set of normal random variables, the new variable also has a normal distribution. That is, if $X_1, X_2, \ldots, X_k$ are normal random variables and $a_1, a_2, \ldots, a_k$ are real numbers, then

$$L = \sum_{i=1}^{k} a_i X_i \tag{5.9}$$

will also have a normal distribution. The mean of L is

$$\mu_L = \sum_{i=1}^{k} a_i \mu_i \tag{5.10}$$

where $\mu_i = E(X_i)$, and if the X_is are *independent*, the variance of L is given by

$$\sigma_L^2 = \sum_{i=1}^{k} a_i^2 \sigma_i^2 \tag{5.11}$$

where $\sigma_i^2 = \text{Var}(X_i)$.

We now finish our discussion of the normal distribution by showing how it can be used as an approximation in evaluating binomial probability statements. Recall that the previous chapter presented a method of using the Poisson approximation to the binomial whenever $\min\{np, n(1-p)\} < 5$, where n (assumed to be large) is the number of trials and p is the probability of success for each of the n trials. The approximation method commonly used when this inequality is not satisfied employs the normal probability distribution. The first question to be answered is, If a normal distribution is appropriate for approximating the binomial probability (the value is not in the binomial table and it is computationally cumbersome), what normal distribution should be used in the approximation? The answer is easy to remember because one simply equates the expected values and standard deviations for the two distributions. Thus, if X is binomial with parameters n and p, then X has an expected value of np and a standard deviation of $[np(1-p)]^{1/2}$. This means that the normal distribution used in the approximation should be the one whose expected value is

$$\mu = np$$

and whose standard deviation is

$$\sigma = [np(1-p)]^{1/2}$$

(The fact that a normal distribution is appropriate is justified by the central limit theorem. This theorem will be discussed in detail in Chapter 8.)

The next step involves an adjustment called a correction for continuity, which acknowledges that we are approximating a discrete distribution (the binomial) by using a continuous distribution (the normal). Figure 5.7 illustrates why this is necessary. Notice that what we have

Figure 5.7 Graphical Representation of a Binomial Distribution with Parameters 10 and 1/2 and the Density Function Graph of its Normal Approximation

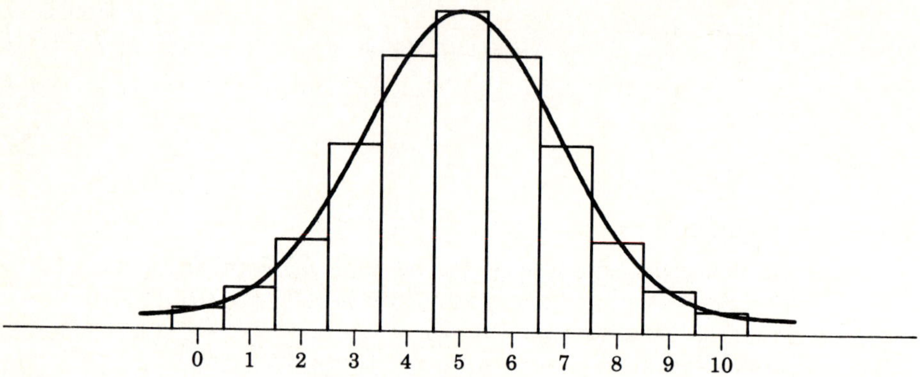

done is to construct rectangles above each number so that the area of each such rectangle equals the probability that the binomial random variable assumes that value. If we wanted to use the rectangles to compute the probability that $X = 7$, where X is binomial with parameters 10 and 1/2, we would simply check the height of the rectangle centered at the number 7. But if we use the normal approximation, asking the question

$$\text{pr}(X = 7) = ?$$

(where X is normal with mean 5 and variance 2.5) is nonsensical, since for continuous distributions the probability of any single point is zero! What we really want is an area under the normal curve that is approximately the same as that contained in the rectangle centered around 7. This approximation is given by

$$\text{pr}(6.5 < Y \leqslant 7.5)$$

since the base of the rectangle begins at 6.5 and ends at 7.5. This is illustrated in Example 5.6.

EXAMPLE 5.6 Suppose that X is a binomial random variable with parameters 25 and 0.3. Use the normal approximation to calculate

$$\text{pr}(5 \leqslant X \leqslant 8)$$

SOLUTION The proper normal random variable Y for the approximation has an expected value given by

$$np = 25 \times 0.3 = 7.5$$

and a standard deviation given by

$$\begin{aligned}[np(1-p)]^{1/2} &= (25 \times 0.3 \times 0.7)^{1/2} \\ &= 2.2913\end{aligned}$$

Thus,

$$\begin{aligned}\text{pr}(5 \leqslant X \leqslant 8) &\approx \text{pr}(4.5 < Y \leqslant 8.5) \\ &= \text{pr}\left(\frac{4.5 - 7.5}{2.2913} < Z \leqslant \frac{8.5 - 7.5}{2.2913}\right) \\ &= \text{pr}(Z \leqslant 0.44) - \text{pr}(Z \leqslant -1.31) \\ &= 0.6700 - 0.0951 \\ &= 0.5749\end{aligned}$$

EXERCISES

5.4.1 Given a normal distribution with $\mu = 36$ and $\sigma = 4$, find
(*a*) The area below 32.
(*b*) The area above 28.
(*c*) The area between 40 and 46.

5.4.2 Suppose that Y denotes a standard normal random variable. Find the deviate that has 40% of the area below it and the deviate that has 15% of the area above it.

5.4.3 Given a normal distribution with $\mu = 40$ and $\sigma = 6$, find
(*a*) The value with 40% of the area below it.
(*b*) The value with 15% of the area above it.

5.4.4 A soft drink machine is regulated so that it discharges an average of 0.213 liter with a standard deviation of 0.05 liter. If the cups used to dispense the drinks will hold 0.275 liter, then calculate (assuming a normal distribution for fill)
(*a*) The percentage of cups that will overflow.
(*b*) The cup size for which only 5% of the drinks will overflow.

5.4.5 We restate Example 4.16, which can be handled using the normal approximation to the binomial.

The promotions department in a local retail firm decides to interview 100 people at random in a major shopping center in the area to determine how many of the shoppers have heard the firm's radio advertisement. The radio station maintains that they reach 30% of the buying public with the type of advertisement schedule purchased by the company. If 300 of the 1000 people using the shopping center during the time when the interviewing takes place have heard the advertisement, calculate

(*a*) The probability that, at most, 20 of the 100 people in the sample have heard the advertisement.

(*b*) The probability that at least 25 but not more than 40 of the 100 in the sample have heard the advertisement.

5.5 SUPPLEMENTARY EXAMPLES

EXAMPLE 5.7 People are serviced at a bank teller's window at a rate of one every four minutes. What is the probability that the next person is served in less than three minutes?

SOLUTION Since the length of service is a measure of the time between services, we will use the negative exponential random variable as a model. In this case the standard time unit is four minutes and $\beta = 1$. Thus, letting T denote the random variable, we get

$$\text{pr}(0 \text{ minutes} \leqslant T \leqslant 3 \text{ minutes}) = \text{pr}\left(0 \leqslant T \leqslant \frac{3}{4}\right) = \int_0^{3/4} e^{-t}\, dt$$

$$= 1 - e^{-3/4} = 1 - 0.4724 = 0.5276$$

EXAMPLE 5.8 Refer to Example 5.7. What is the probability that a person is served in less than three minutes on at least four of the next six visits to the bank teller?

SOLUTION We use the binomial random variable with $n = 6$ and $p = 0.5276$ as previously calculated. Thus, the probability we want is given by

$$\sum_{x=4}^{6} \binom{6}{x}(0.5276)^x (0.4724)^{6-x} = 1 - \sum_{x=0}^{3} \binom{6}{x}(0.5276)^x (0.4724)^{6-x}$$

$$= 1 - 0.6032 = 0.3968$$

EXAMPLE 5.9 One of the production lines in a factory produces defective items at a rate of one every ten minutes. Calculate

(*a*) The probability that the time interval between the next two defective items is less than ten minutes.

(*b*) The probability that at least two defective items are produced in the next ten-minute time interval.

SOLUTION

(*a*) The standard time unit is ten minutes and $\beta = 1$. Thus, the probability we want is given by

$$\int_0^{10/10} e^{-t}\,dt = 1 - e^{-1} = 0.6321$$

(*b*) This is a Poisson random variable problem with $\mu = 1$ and $R = 1$. Thus, the probability we want is given by

$$\sum_{x=2}^{\infty} \frac{e^{-1}(1)^x}{x!} = 1 - \sum_{x=0}^{1} \frac{e^{-1}(1)^x}{x!} = 1 - (e^{-1} + e^{-1})$$

$$= 1 - \frac{2}{e} = 0.2642$$

EXAMPLE 5.10 Refer to Example 5.9 and calculate the probability that the next defective item is produced in less than ten minutes from now.

SOLUTION This and the previous exercise appear to be two different problems, but their solutions are exactly the same. This is because the negative exponential distribution is the *only* probability distribution that has the "memoryless" property given here.

If s and t are positive real numbers and T represents a negative exponential random variable whose parameter is β, then

$$\text{pr}(T \geqslant s + t \mid T \geqslant t) = \text{pr}(T \geqslant s)$$

Simple algebra and the axioms of Chapter 2 easily show how this peculiar property can be justified. Note that

$$\text{pr}(T \geqslant s + t \mid T \geqslant t) = \frac{\text{pr}(T \geqslant s + t \cap T \geqslant t)}{\text{pr}(T \geqslant t)} = \frac{\text{pr}(T \geqslant s + t)}{\text{pr}(T \geqslant t)}$$

$$= \frac{e^{-\beta(s+t)}}{e^{-\beta t}} = e^{-\beta s} = \text{pr}(T \geqslant s)$$

Thus, it makes no difference how long it has been since the last defective was produced; the probability that the next defective is produced in less than ten minutes will be unchanged no matter how long it has been since the last defective.

EXAMPLE 5.11 A manufacturer of picture tubes for black and white portable televisions finds that variable costs are \$25 per tube and that the tubes have an average failure rate of one every ten years. The manufacturer offers a warranty on these tubes of two years, during which time the tube will be replaced free of charge. The cost for replacing a tube is the cost of the tube plus \$15 for labor. If the manufacturer sells each tube for \$35, what is the expected profit per tube?

SOLUTION Since we are interested in time to occurrence in this situation, we will consider the negative exponential distribution. You should check the assumptions to see whether they seem reasonable. If they are, then we need to determine the probability that a tube will have to be replaced under warranty, that is, within two years after its sale. We have that $\beta = 1$ per ten years and

$$\begin{aligned}
\text{pr}(T < 2 \text{ years} \mid \beta = 1 \text{ per ten years}) &= \text{pr}\left(T < \frac{2}{10} \,\middle|\, \beta = 1\right) \\
&= \int_0^{0.2} e^{-t}\, dt = -e^{-t}\Big|_0^{0.2} \\
&= 1 - e^{-0.2} \\
&= 1 - 0.8187 = 0.1813
\end{aligned}$$

For each tube sold, the manufacturer realizes a \$10 profit if it outlasts the warranty and a \$30 loss if it doesn't, since

$$35 - 25 - 40 = -30$$

Therefore, the expected profit on a tube is

$$10(0.8187) - 30(0.1813) = \$2.748$$

EXAMPLE 5.12 A company is considering the purchase of a new machine that will reduce variable costs on a product by \$.05 per unit but increase fixed costs by \$200,000 a year. The current demand exceeds production. The production manager believes that the machine will average 6,000,000 units per year with a standard deviation of

1,100,000. She also states that the distribution of units produced is likely to be bell-shaped. Determine:

(*a*) The probability that the company would realize a loss from the purchase of the machine.

(*b*) The probability that the profit would be at least $190,000 per year.

SOLUTION

(*a*) First we need to determine the level of production that the machine must attain to break even. Dividing the increase in fixed costs by the decrease per unit in variable costs gives 4,000,000 units per year to break even and above that level to show a profit. To find the probability of a loss, we interpret bell-shaped to mean approximately normal. Thus

$$\begin{aligned}\text{pr(loss)} &= \text{pr}(X < 4{,}000{,}000)\\ &= \text{pr}\left(Z < \frac{4{,}000{,}000 - 6{,}000{,}000}{1{,}100{,}000}\right)\\ &= \text{pr}(Z < -1.82) = 0.0344\end{aligned}$$

(*b*) Realizing a profit of $190,000 per year implies that the machine must produce 190,000/0.05 units above its breakeven volume. Thus, it would have to make 3,800,000 more than the 4,000,000 required to break even. The probability of doing this is

$$\begin{aligned}\text{pr(profit} \geqslant \$190{,}000) &= \text{pr}(X \geqslant 7{,}800{,}000)\\ &= \text{pr}\left(Z \geqslant \frac{7{,}800{,}000 - 6{,}000{,}000}{1{,}100{,}000}\right)\\ &= \text{pr}(Z \geqslant 1.64) = 1 - 0.9495 = 0.0505\end{aligned}$$

EXAMPLE 5.13 Pokono Plastics has found that daily demand for a particular plasticizer is variable but averages 500 pounds with a standard deviation of 100 pounds. Lately they have been running short of this item; this has caused delays in production, and they feel that they must revise their ordering policy. It takes three days from the time an order is placed until they receive it, and the accounting department calculates that they should order a 30-day supply on each order. If they assume that daily usage is independent from one day to the next, how much should they order at this time to be 90% certain of not running out before the next order arrives? Assume that they will reorder in 30 days.

SOLUTION We cite a theorem, to be presented later, that states that the sums of independent, identically distributed random variables will have a distribution that is approximately normal. Thus, we need to find only the mean and the standard deviation for the total demand in question.

Since the next order will not be placed for 30 days and will not be received for 33 days (because of what is called the lead time), we must plan for a 33-day period. The demand over this time span is simply the sum of the daily demands and averages

$$E\left(\sum_{i=1}^{33} D_i\right) = \sum_{i=1}^{33} \mu_D = \sum_{i=1}^{33} 500 = 33(500) = 16{,}500 \text{ pounds}$$

Since day-to-day usage is assumed to be *independent*, we have

$$\mathrm{Var}\left(\sum_{i=1}^{33} D_i\right) = \sum_{i=1}^{33} \mathrm{Var}(D_i) = 33(100)^2 = 330{,}000$$

Thus, if we let L be the total demand, L will be approximately normally distributed with a mean of 16,500 and a standard deviation of $\sqrt{330{,}000}$, or 574.5 pounds.

Also, the company wishes to be 90% certain of not running out of this item before the next order arrives. This requires that they order more than the average demand over the time in question. In fact, they must order an amount such that

$$\mathrm{pr}(L \leqslant q) = 0.90$$

But since L has a normal distribution, we have

$$\mathrm{pr}(Z \leqslant z_{0.1}) = 0.90$$

This implies that q, when standardized, must correspond to $z_{0.1}$. Table A.6 gives $z_{0.1} = 1.28$ and, hence,

$$\frac{q - 16{,}500}{574.5} = 1.28$$

or

$$\begin{aligned} q &= 16{,}500 + 1.28(574.5) = 16{,}500 + 735.36 \\ &= 17{,}235.36 \text{ pounds} \end{aligned}$$

If orders must be placed in whole pounds, we would round upward to 17,236 pounds to ensure that the 90% service is satisfied.

5.6 SUMMARY

This chapter concludes our study of basic probability distributions. In Chapter 4, we studied some of the discrete distributions that could be used to model the random variations of natural phenomena. In this chapter we considered a few of the continuous distributions that may be used to describe certain phenomena that arise in nature.

The first distribution covered is also the simplest. The uniform or rectangular distribution is used extensively in the area of simulation. The next distribution we considered was the negative exponential. As with the uniform, it is easy to use. It is intrinsically related to the Poisson distribution of Chapter 4 and is basically used to model the time between occurrences.

The most commonly used distribution in statistical analysis is the normal. Its probability density function appears to be quite complicated, but it is one of the most versatile and easiest to use of all continuous distributions. It can be used to model a wide variety of phenomena where the outcomes vary symmetrically around a central point with rapidly decreasing probability as one moves farther away from the center in either direction. Its ease of use comes from the fact that all normal distributions have the same area within a specified number of standard deviations of the mean. Thus, only the standard normal table is required to evaluate probability statements for all normal distributions.

[We can extend our study of continuous distributions by presenting four that are somewhat more specialized in their areas of application. The lognormal distribution utilizes the standard normal table to evaluate probabilities and may be used to describe random variables that can assume only positive values. The gamma distribution is often used to model experiments that can assume only nonnegative values, of which the negative exponential is a special case. The beta distribution is extensively used in the study of order statistics (positions of observations in the array) and extreme values. The Weibull distribution is extremely useful in studying the reliability of systems and in warranty analysis.]

For ease in reference, the distributions of this chapter are summarized in the table on pages 128–129.

Tabular Chapter Summary

Random Variable	Density Function	Expected Value	Variance	Common Use(s)
Uniform	$f(x\|a, b) = \begin{cases} \frac{1}{b-a} & a \leqslant x \leqslant b \\ 0 & \text{elsewhere} \end{cases}$	$\frac{a+b}{2}$	$\frac{(b-a)^2}{12}$	To generate random deviates in simulation studies
Negative exponential	$f(t\|\beta) = \begin{cases} \beta e^{-\beta t} & t \geqslant 0 \\ 0 & \text{elsewhere} \end{cases}$	$\frac{1}{\beta}$	$\frac{1}{\beta^2}$	As a model for the time between events whose numbers per unit time are properly modeled using a Poisson random variable
Normal	$f(x\|\mu, \sigma) = \frac{1}{\sigma\sqrt{2\pi}} e^{-(x-\mu)^2/2\sigma^2}$ $-\infty < x < +\infty$	μ	σ^2	The foundation for most statistical methods encountered in business decision making
Lognormal	$f(x\|\mu, \sigma) = \begin{cases} \frac{1}{\sigma x\sqrt{2\pi}} e^{-(\ln x-\mu)^2/2\sigma^2} & x > 0 \\ 0 & \text{elsewhere} \end{cases}$	$e^{(1/2)\sigma^2+\mu}$	$(e^{\sigma^2+2\mu})(e^{\sigma^2}-1)$	A model for a process that can take on only positive values

Tabular Chapter Summary *(continued)*

Random Variable	Density Function	Expected Value	Variance	Common Use(s)
Gamma	$f(x\|\alpha, \beta) = \begin{cases} \dfrac{\beta^{\alpha} x^{\alpha-1} e^{-\beta x}}{\Gamma(\alpha)} & x \geqslant 0 \\ 0 & \text{elsewhere} \end{cases}$	$\dfrac{\alpha}{\beta}$	$\dfrac{\alpha}{\beta^2}$	Used in reliability to model time between failures and in statistical goodness-of-fit tests
Beta	$f(x\|\alpha, \beta) = \begin{cases} \dfrac{\Gamma(\alpha+\beta)}{\Gamma(\alpha)\Gamma(\beta)} [x^{\alpha-1}(1-x)^{\beta-1}] & 0 \leqslant x \leqslant 1 \\ 0 & \text{elsewhere} \end{cases}$	$\dfrac{\alpha}{\alpha+\beta}$	$\dfrac{\alpha\beta}{(\alpha+\beta)^2(\alpha+\beta+1)}$	A model for any process that returns values with finite upper and lower limits
Weibull	$f(x\|\alpha, \beta) = \begin{cases} \alpha\beta x^{\beta-1} e^{-\alpha x^{\beta}} & x > 0 \\ 0 & \text{elsewhere} \end{cases}$	$\dfrac{\Gamma(1+1/\beta)}{\alpha^{1/\beta}}$	$\dfrac{\Gamma(1+2/\beta) - \Gamma(1+1/\beta)^2}{\alpha^{2/\beta}}$	Models lengths of time to failure for many mechanical devices, biological plants, and animals

REVIEW EXERCISES

Level 1

5.1 For a standard normal distribution, determine:
(*a*) $\text{pr}(Z \geqslant 1.37)$
(*b*) $\text{pr}(Z \leqslant -0.45 \text{ or } Z \geqslant 1.26)$
(*c*) $\text{pr}(Z < 1.52)$
(*d*) z_0 such that $\text{pr}(Z \leqslant z_0) = 0.8438$
(*e*) z_0 such that $\text{pr}(Z \geqslant z_0) = 0.7357$
(*f*) z_0 such that $\text{pr}(-z_0 \leqslant Z \leqslant z_0) = 0.9544$

5.2 Most random number tables are constructed from the standard uniform distribution. If a number is selected at random from such a table, what is the probability that it will be between 0.2496 and 0.8721?

5.3 Customers arrive at a checkout counter at a rate of three every five minutes. What is the probability that
(*a*) No more than 12 minutes will elapse between customer arrivals?
(*b*) The time between the arrival of the third and fourth customers will be at least 6 minutes, but not more than 18 minutes?

5.4 A random variable has a negative exponential distribution with a mean of 5. Determine
(*a*) $\text{pr}(4 \leqslant T < 8)$
(*b*) $\text{pr}(10 \leqslant T \leqslant 16)$
If $V = T/2$, find
(*c*) $\text{pr}(2 \leqslant V < 4)$
(*d*) $\text{pr}(5 \leqslant V < 8)$
(*e*) Compare your answers for (*a*) and (*b*) with those for (*c*) and (*d*), and comment.

5.5 Pills Unlimited (PU) purchases 250-milliliter bottles from Purity Plastics, Limited. PU uses these bottles for packaging their basic headache remedy. The bottles arrive in lots of 100,000, and PU randomly samples 1000 bottles from each lot and inspects them for cracks, nicks, deformities, and neck size. Pills has found that, from a cost standpoint, lots with 5% or fewer defective are acceptable. After inspecting the sample bottles, PU accepts the lot if no more than 60 defective bottles have been found and rejects it otherwise. Determine:

(*a*) The probability that PU accepts a lot that has exactly 5% defective.
(*b*) The probability that PU accepts a lot that has only 4% defective.
(*c*) The probability that PU accepts a lot if it has 7% defective.
(*d*) The probability that PU rejects a lot with 8% defective.

5.6 The loan departments of a large system of commercial banks found that home loans issued by them over the past year were approximately normally distributed with a mean of $48,000 and a standard deviation of $8500. Assume that loan conditions at the bank remain essentially the same for next year.
(*a*) What proportion of new loans would you expect to be $39,000 or more?
(*b*) Ninety percent of next year's loans would be expected to be below what value?

5.7 Commuter trains arrive at a station once every 15 minutes on the average. What is the probability that you would have to wait between 12 and 21 minutes for a train?

5.8 The average number of hours per week that a student spends studying business statistics is 12 and the standard deviation is 3. If the distribution of study times is approximately normal,
(*a*) What is the probability that a randomly selected student studies this subject between 8 and 15 hours per week?
(*b*) What percentage of students study business statistics less than 6 hours per week?

5.9 Refer to Exercise 5.8 and assume that grades are to be assigned according to the amount of time spent studying the course. If 10% of all students are to receive A's and you want one, how many hours must you spend studying business statistics each week at the least? What must you assume to answer this question?

5.10 Structural defects in extruded plastic pipe are known to occur on the average of one every 50 feet. Determine the probability that:
(*a*) At least 75 feet will be extruded with no structural defects.
(*b*) Between 100 and 300 feet will be extruded between defects.
(*c*) Under what assumptions are your answers to (*a*) and (*b*) correct?
(*d*) Do you feel the assumptions in (*c*) are reasonable?

5.11 When a production process is in control, it produces defective items 3% of the time. When the process is out of control, this percentage increases to 10%. The quality control inspector selects 100 items at random and stops the process if the sample contains six or more defectives.

(*a*) What is the probability that the inspector fails to stop the process when it should be stopped?

(*b*) What is the probability that the inspector stops the process incorrectly?

5.12 Starting salaries for graduates with a bachelor's degree in business are known to be normally distributed with a mean of \$12,000 per year and a standard deviation of \$1380.

(*a*) What is the minimum salary that a graduate must receive to be in the top 20%?

(*b*) If there are 400 business majors in this year's graduating class, how many can be expected to earn between \$11,490 and \$12,600?

5.13 If it is known that the middle 90% of the values for a normal random variable are between 100 and 140, what are the mean and standard deviation?

5.14 Amalgamated Grain Producers owns wheat ranches in Oregon and North Dakota. The yields for the ranches are normally distributed, with the Oregon ranch having a mean of 500,000 bushels per year and a standard deviation of 80,000 bushels per year. The North Dakota ranch has an average yield of 300,000 bushels per year and a standard deviation of 60,000 bushels. If the Oregon yield for this year is anticipated to be 540,000 bushels, what must the North Dakota yield be if it is just as extreme?

Level 2

5.15 A manufacturer of TV picture tubes has found that a tube lasts an average of 2.5 years. If the manufacturer guarantees the tubes for one year, what percentage can the manufacturer expect to replace free of charge (using a negative exponential distribution)? If the tubes cost \$110 to produce and sell for \$140, what will be the expected net profit on 100 such tubes? Is the guarantee appropriate? Suggest three possible alternatives the manufacturer can take in order to break even (vary only one thing at a time).

5.16 Prove that if X is a random variable having a mean of μ and a standard deviation of σ, the random variable

$$Z = \frac{X - \mu}{\sigma}$$

has a mean of zero.

5.17 Suppose that the bottles used by Pills Unlimited have an average interior neck diameter of 1.85 cm and a standard deviation of 0.1 cm. The stoppers for these bottles are slightly tapered, with the bottom diameter having an average of 1.65 cm and a standard deviation of 0.08 cm. The top diameter averages 2.05 cm and has a standard deviation of 0.12 cm. If all diameters are normally distributed, when a stopper and bottle are randomly selected, what is the probability that
(*a*) The stopper will go completely inside the bottle?
(*b*) The stopper will not go into the bottle?

5.18 An elevator in a certain building has a listed capacity of 2000 pounds with a safety factor of 500 pounds. Assume that the weights of persons using this elevator are normally distributed with a mean of 165 and a standard deviation of 25 pounds. On a rainy Monday morning a number of people arrive late for work and crowd the elevator. If a total of 12 persons push their way into the elevator, what is the probability that the listed capacity will be exceeded? If two more people, making a total of 14, shove their way into the elevator, what is the probability of exceeding the actual capacity? Suppose the 14 people already aboard squeeze together to make room for you. Will you enter or not, and why?

5.19 Conglomerate Corporation has four major divisions. Sales for each division are normally distributed with means and standard deviations as listed in the table (in millions of dollars).

Division	A	B	C	D
Average yearly sales	95	150	300	120
Standard deviation of yearly sales	10.2	20.4	37.3	9.1

For any randomly selected year, what is the probability that Conglomerate's total sales
(*a*) Will exceed 700 million dollars?
(*b*) Will be less than 600 million dollars?
(*c*) Will be between 625 and 710 million dollars?

5.20 If a set of n independent random variables, each distributed normally with a mean of μ and a standard deviation of σ, are averaged, what will be the distribution of that average? That is, let X_i, $i = 1, 2, \ldots, n$, be independent random variables from a $N(\mu, \sigma^2)$ distribution and determine the distribution of

$$\overline{X} = \frac{1}{n} \sum_{i=1}^{n} X_i$$

5.21 Before commencing a new advertising campaign, a local electronics retailer found that weekly sales volume was approximately distributed as a normal distribution with a mean of \$25,000 and a standard deviation of \$6000. After the campaign had been in effect for eight weeks, the retailer checked the records for the last four weeks and found that sales had averaged \$30,562 over that period. Determine the probability of sales averaging this high or higher and indicate any assumptions you must make in doing so. On the basis of your probability, would you recommend that the retailer continue the new advertising campaign? Why or why not?

5.22 We have seen that a sum of independent random variables (observations) from the same normal distribution will be normally distributed with a mean equal to $n\mu$ and a standard deviation of $\sqrt{n}\,\sigma$. If σ is 25 and we wish the sample total to deviate from its expected value by no more than 100 with a probability of 0.90, how many observations should be taken?

Level 3

5.23 The Greystone Electric Company (GE) advertises that the time to failure of its 100-watt bulbs follows a negative exponential distribution with an average life of 750 hours. Consumer advocate Rolph Nater doubts the average life claim and decides to test some of GE's bulbs. To do so, he purchases a package of bulbs at each of 20 different stores scattered throughout the Washington area. Since a package contains two bulbs, he randomly selects one bulb from each package for testing. His results show that 14 of the bulbs burned out in the first 525 hours of life. Determine the probability of that many or more bulbs failing in the indicated time period if GE's claim is true. If you were Rolph, would you accuse GE of false advertising? Why or why not? What assumptions are necessary in addition to those of a negative exponential distribution?

5.24 Verify that the mean and variance for a negative exponential distribution are $1/\beta$ and $1/\beta^2$, respectively.

5.25 Refer to Exercise 5.22. If we want the sample average to deviate from the population mean μ by no more than 5, how many observations will be required for the 0.90 probability to be satisfied?

5.26 Refer to Exercise 5.25. Using the number of observations you indicated, we selected a sample and calculated its average to be 101.7. Determine the minimum and maximum values for an interval that would be expected to include the true value of μ with a probability of 0.90.

Chapter 6
Bayesian Decision Analysis

KEY WORDS AND TERMS

States of Nature
Payoff Table
Opportunity Loss Table
Maximin
Minimax
Maximum Expected Value
Minimum Expected Opportunity Loss
Expected Value of Perfect Information
Expected Value of Sample Information

6.1 INTRODUCTION

This chapter introduces traditional approaches to making decisions, using many of the basic tools and concepts presented in the first five chapters. Some of what is presented is traditionally known as *decision theory*, since much of what is done in the applications does not employ the type of inferential procedures used in the remainder of this text. Many of the more interesting problems in this area employ Bayes' rule in the solution procedure, which convinces some authors to classify the methods as Bayesian statistics.

Our sequence of presentation provides some basic building blocks that are necessary before one can understand the more useful and advanced techniques to be presented later. The first of these are called *payoff* and *opportunity loss tables*.

6.2 PAYOFF AND OPPORTUNITY LOSS TABLES

A basic premise for the construction of payoff and opportunity loss tables is that one can define a partition for the *states of nature* and the *states of action*. By this we mean that the decision will revolve around the existence of some type of stochastic phenomena that precipitate a condition or state of nature that is unknown to the decision maker at the time some decision or action must be taken. The tables require that all states of nature and all possible actions be considered.

EXAMPLE 6.1 Tea Ltd. of Halifax buys shipments of tea in sight unseen fashion. That is, some of the boxes in the shipment are of good quality and some are of bad quality. Define the states of nature and action set for the decision problem of whether to buy a particular box of tea.

SOLUTION The states of nature in this problem are simply that each box either is of good quality or is not (in which case it is of poor quality). The partition is given by

$$s_1 = \text{the box is of good quality}$$
$$s_2 = \text{the box is of poor quality}$$

in which we assume that all boxes can be classified as s_1 or s_2. The actions in this problem are to buy or not to buy, given by

$$a_1 = \text{buy the box}$$
$$a_2 = \text{don't buy the box}$$

which qualifies as a partition of the collection of possible actions.

The remainder of this chapter will employ the notation given here in that if there are n states of nature in the partition, they will be denoted by

$$s_1, s_2, \ldots, s_n$$

or, equivalently, by

$$s_j \qquad j = 1, 2, \ldots, n$$

If there are m possible actions, these will be denoted by

$$a_1, a_2, \ldots, a_m$$

or, equivalently, by

$$a_i \qquad i = 1, 2, \ldots, m$$

With this notation we can develop the payoff and opportunity loss tables for a decision problem.

DEFINITION 6.1 The payoff from pursuing action a_i when the state of nature is s_j is given by P_{ij}; it denotes the observed return, whether monetary or otherwise, experienced under the simultaneous occurrence of two conditions. That is, the payoff is conditioned by a_i and s_j.

Payoffs are often calculated with the aid of the formula

$$P_{ij} = r_{ij} - c_{ij}$$

in which r_{ij} is the revenue resulting from employing action a_i when the state of nature is s_j, and c_{ij} is the cost associated with employing action a_i when the state of nature is s_j. Payoff would mean the same thing as profit if c_{ij} included fixed costs. There are much more complicated formulas we could utilize if we wanted to include nonmonetary aspects of the problem as explicit variables in the model.

EXAMPLE 6.2 Suppose that Tea Ltd. of Halifax pays $1.20 per box of tea. If the box is good, they can sell it for $2.50; otherwise they can sell it for only $0.60. Construct a payoff table using this formula.

SOLUTION First note that:

- P_{11} implies that they purchase a good box of tea
- P_{12} implies that they purchase a poor box of tea
- P_{21} implies that they don't purchase a good box of tea
- P_{22} implies that they don't purchase a poor box of tea

Thus we have

i	j	r_{ij}	c_{ij}	$P_{ij} = r_{ij} - c_{ij}$
1	1	2.50	1.20	1.30
1	2	0.60	1.20	-0.60
2	1	0	0	0
2	2	0	0	0

and the payoff table looks like

	State of Nature	
P_{ij}	s_1 (good)	s_2 (poor)
a_1 (purchase)	1.30	-0.60
a_2 (don't purchase)	0	0

EXAMPLE 6.3 Jake Spud is the middleman for potato distribution in Aroostook County. He buys barrels of potatoes that are classified good, bad, or ugly, but pays for each shipment in bulk at an average rate of \$10 per barrel. If he can sell a good barrel for \$20, a bad barrel for \$12, and an ugly barrel for \$8, construct the payoff table for the buy–don't buy decisions.

SOLUTION Let a_1 indicate a buy action, a_2 indicate a don't buy action, s_1 indicate a good barrel, s_2 indicate a bad barrel, and s_3 indicate an ugly barrel of potatoes. Then we have

i	j	r_{ij}	c_{ij}	$P_{ij} = r_{ij} - c_{ij}$
1	1	20	10	10
1	2	12	10	2
1	3	8	10	-2
2	1	0	0	0
2	2	0	0	0
2	3	0	0	0

so that the payoff table is given by

	State of Nature		
P_{ij}	s_1 (good)	s_2 (bad)	s_3 (ugly)
a_1 (buy)	10	2	-2
a_2 (don't buy)	0	0	0

The next important concept to be introduced is what we call the *opportunity loss* L_{ij} associated with pursuing action a_i when the state of nature is s_j. In this case a_i may not be the optimum or maximum payoff for state s_j, and we will have lost the opportunity to attain a larger payoff.

DEFINITION 6.2 The opportunity loss is the difference given by

$$L_{ij} = \max \{P_{kj} \quad k = 1, 2, \ldots, m\} - P_{ij}$$

that is, the amount lost as a result of choosing action a_i rather than an action that maximizes the payoff for a fixed state of nature s_j.

Note that in Definition 6.2 we are maximizing over k for a fixed j. In simpler terms, this says to find the largest payoff in the column, then subtract every entry in that column from it. This is illustrated in Example 6.4.

EXAMPLE 6.4 Determine the opportunity loss tables for Examples 6.2 and 6.3.

SOLUTION In Example 6.2 we have

$$\begin{aligned} L_{11} &= \max \{P_{k1} \quad k = 1, 2\} - P_{11} \\ &= 1.30 - 1.30 = 0.00 \\ L_{21} &= \max \{P_{k1} \quad k = 1, 2\} - P_{21} \\ &= 1.30 - 0 = 1.30 \\ L_{12} &= \max \{P_{k2} \quad k = 1, 2\} - P_{12} \\ &= 0 - (-0.60) = 0.60 \\ L_{22} &= \max \{P_{k2} \quad k = 1, 2\} - P_{22} \\ &= 0 - 0 = 0.00 \end{aligned}$$

Thus, the opportunity loss table for Example 6.2 looks like

L_{ij}	s_1 (good)	s_2 (poor)
a_1 (purchase)	0.00	0.60
a_2 (don't purchase)	1.30	0.00

In Example 6.3 we have, for column 1,

$$\begin{aligned} L_{11} &= \max \{P_{k1} \quad k = 1, 2\} - P_{11} \\ &= 10 - 10 = 0 \\ L_{21} &= 10 - P_{21} \\ &= 10 - 0 = 10 \end{aligned}$$

For column 2,

$$\begin{aligned} L_{12} &= \max \{P_{k2} \quad k = 1, 2\} - P_{12} \\ &= 2 - 2 = 0 \\ L_{22} &= 2 - P_{22} \\ &= 2 - 0 = 2 \end{aligned}$$

Finally, for column 3,

$$\begin{aligned} L_{13} &= \max \{P_{k3} \quad k = 1, 2\} - P_{13} \\ &= 0 - (-2) = 2 \\ L_{23} &= 0 - P_{23} \\ &= 0 - 0 = 0 \end{aligned}$$

As a result, the opportunity loss table for Example 6.3 looks like

L_{ij}	s_1 (good)	s_2 (bad)	s_3 (ugly)
a_1 (buy)	0	0	2
a_2 (don't buy)	10	2	0

EXERCISES

6.2.1 Sungleam, Inc., has an opportunity to buy used mixing bowls in lots of size 25. The bowls are bought as a package deal, and Sungleam must live with what they get in terms of the appearance of the bowls. The bowls are purchased at a cut-rate price, but some are in horrible shape and will be next to impossible to sell unless their price is a very low \$0.50. The others will sell for \$2.50 each. Sungleam's sales records indicate that most shipments can be classified as poor, fair, or good according to the number of horrible bowls in the lot. A poor shipment has 20, a fair shipment 12, and a good shipment 4 horrible bowls, respectively. If Sungleam pays \$32.50 for each shipment of 25 bowls, determine:
(*a*) The actions and states of nature.
(*b*) The payoff table.
(*c*) The opportunity loss table.

6.2.2 A manufacturing process produces a special absorption agent in 1-inch squares, with each square monitored by a measuring device that calculates the agent's effectiveness as one of five values, 1.016, 1.523, 1.768, 1.922, and 2.107. Each production run will output the squares at a uniform agent effectiveness level independent of the number of squares produced, but the level varies from one run to the next. As a result of setup costs, market demands, and overhead variations, management is restricted to three volume levels for production runs: 10,000 squares, 20,000 squares, and 30,000 squares. The company costs are given in the following table.

	Agent Effectiveness				
Volume Level	1.016	1.523	1.768	1.922	2.107
10,000	0.68	0.71	0.74	0.77	0.80
20,000	0.65	0.70	0.75	0.80	0.85
30,000	0.60	0.66	0.72	0.76	0.84

If the selling price per square is

0.75 for a 1.016 agent effectiveness level
0.77 for a 1.523 agent effectiveness level
0.79 for a 1.768 agent effectiveness level
0.81 for a 1.922 agent effectiveness level
0.83 for a 2.107 agent effectiveness level

determine

(*a*) The actions and states of nature.
(*b*) The payoff table.
(*c*) The opportunity loss table.

6.3 COMMON DECISION CRITERIA

The previous section gave us some basic material we could use to explore some common criteria for choosing from among many actions. The most basic of these are the maximin and minimax procedures.

6.3.1 The Maximin Criterion

"The worst that can happen, will; therefore choose the action providing the best protection against what is inevitable." This statement is a proper reflection of the maximin criterion. The idea is to choose action a_i corresponding to P^*_{ij}, where

$$P^*_{ij} = \max \{ \min \{P_{ij}, \ j = 1, 2, \ldots, n\}, \ i = 1, 2, \ldots, m\}$$

(assuming that we have m actions and n states of nature).

This notation describes the rather simple procedure of finding the minimum payoff for each action, then selecting the action corresponding to the largest of these minimum payoffs.

EXAMPLE 6.5 Use the maximin criterion to choose an action for Examples 6.2 and 6.3.

SOLUTION The payoff table for Example 6.2 is

P_{ij}	s_1 (good)	s_2 (poor)
a_1 (purchase)	1.30	-0.60
a_2 (don't purchase)	0	0

Therefore,

$$\begin{aligned} &\max \{ \min \{P_{ij}, \ j = 1, 2\}, \ i = 1, 2\} \\ &= \max \{ \min \{1.30, -0.60\}, \min \{0, 0\}\} \\ &= \max \{-0.60, 0\} \\ &= 0 \end{aligned}$$

Since the value 0 occurs when the action is a_2, and a_2 is a don't buy decision, a manager taking the pessimistic viewpoint should get out of the tea business as it now stands, since the decision will always be "don't buy," and if you don't buy, you can't sell.

We will use a tabular procedure for Example 6.3 to show how this formula works.

P_{ij}	s_1 (good)	s_2 (bad)	s_3 (ugly)	$\min \{P_{ij} \quad j = 1, 2, 3\}$
a_1 (buy)	10	2	-2	-2
a_2 (don't buy)	0	0	0	0

$$0 = \max \{ \min \{P_{ij}, \quad j = 1, 2, 3\}, \quad i = 1, 2\}$$

So again, this philosophy says not to buy any potatoes, and Mr. Spud might as well go into the tea business.

6.3.2 The Minimax Criterion

The minimax criterion works with opportunity loss values seeking the minimum of the set of maximum loss values across all possible actions. Formally we choose that action a_i corresponding to L^*_{ij}, where, assuming m actions and n states of nature,

$$L^*_{ij} = \min \{ \max \{L_{ij}, \quad j = 1, 2, \ldots, n\}, \quad i = 1, 2, \ldots, m\}$$

That is, the action selected is the one that minimizes the maximum opportunity loss in the sense that we first calculate a maximum opportunity loss for each action, then select the action providing the smallest such maximum opportunity loss.

EXAMPLE 6.6 Use the minimax criterion to choose an action for Examples 6.2 and 6.3.

SOLUTION The opportunity loss table for Example 6.2 is

P_{ij}	s_1 (good)	s_2 (poor)
a_1 (purchase)	0.00	0.60
a_2 (don't purchase)	1.30	0.00

Therefore,

$$
\begin{aligned}
&\min \{ \max \{L_{ij},\ j = 1, 2\},\ \ i = 1, 2\} \\
&= \min \left\{ \max \{0.00, 0.60\}, \max \{1.30, 0.00\} \right\} \\
&= \min \{0.60, 1.30\} \\
&= 0.60
\end{aligned}
$$

Since the value of 0.60 occurs when the action is a_1, and a_1 is a buy decision, all boxes of tea should be purchased.

Again, we will use a tabular procedure for Example 6.3 to show how the formula works.

L_{ij}	s_1 (good)	s_2 (bad)	s_3 (ugly)	$\max \{L_{ij} \quad j = 1, 2, 3\}$
a_1 (buy)	0	0	2	2
a_2 (don't buy)	10	2	0	10

$$2 = \min \{ \max \{L_{ij} \quad j = 1, 2, 3\},\ \ i = 1, 2\}$$

So again the decision to buy gives the minimum of the maximum values.

One drawback of these two procedures is that they ignore the frequency of occurrence of each state; thus their use often makes little sense. A more reasonable method would employ a weighted average that acknowledges the relative frequency of each state of nature. This is exactly what the maximum expected value and minimum expected opportunity loss procedures do.

6.3.3 The Maximum Expected Value Procedure

The maximum expected value procedure calculates an expected payoff for each possible action, then selects the action that provides the maximum of the expected values. Formally, we choose an action a_i corresponding to

$$\max \{E(P|a_i) \qquad i = 1, 2, \ldots, m\}$$

where

$$E(P|a_i) = \sum_{j=1}^{n} P_{ij} \times \text{pr}(s_j)$$

and pr(s_j) is the probability of the occurrence of the jth state of nature.

EXAMPLE 6.7 Sellem has a franchise for Stellar Dorbo cookies and must order on a monthly basis in lots of size 25. That is, he must order by the box, and each box contains 25 packages of cookies. Each box costs him \$9.25 to order, and the sale of a box to a local merchant will return \$10.75. Mr. Sellem checks his sales volume records for the last 24 months and observes the following information:

Number of Boxes Sold	Number of Weeks of Sales Volume at This Level
3	5
4	6
5	9
6	4

How many boxes should he order if he uses the maximum expected value procedure to choose his action? (Note that unsold boxes cannot be returned.)

SOLUTION The states of nature are the set of demands for the product in numbers of boxes. Hence,

$$s_1 = 3$$
$$s_2 = 4$$
$$s_3 = 5$$
$$s_4 = 6$$

The set of possible actions is given by the number of boxes he is willing to order. From past data we can assume that he would be unwise to order less than three boxes (or suffer severe customer dissatisfaction) or more than six boxes. So

$$a_1 = 3$$
$$a_2 = 4$$
$$a_3 = 5$$
$$a_4 = 6$$

Next, we use the relative frequency definition of probability (Definition 2.5*b*) to get values for

$$\text{pr}(s_j) \qquad j = 1, 2, 3, 4$$

Doing this, we obtain

$$\text{pr}(s_1) = \frac{5}{24}$$

$$\text{pr}(s_2) = \frac{6}{24}$$

$$\text{pr}(s_3) = \frac{9}{24}$$

$$\text{pr}(s_4) = \frac{4}{24}$$

Remembering that action a_i means a supply of $i + 2$ boxes of cookies, and state of nature s_j means a demand for $j + 2$ boxes, we get payoff values given by

$$P_{ij} = \begin{cases} (10.75 - 9.25)(j + 2) - 9.25(i - j) & \text{if } i \geqslant j \\ (10.75 - 9.25)(i + 2) & \text{if } i < j \end{cases}$$

or equivalently,

$$P_{ij} = 1.5 \times (\min\{i, j\} + 2) - 9.25 \times \max\{0, i - j\}$$

Thus, we get the payoff table given by

P_{ij}	s_1 (demand of 3)	s_2 (demand of 4)	s_3 (demand of 5)	s_4 (demand of 6)
a_1 (order 3)	4.50	4.50	4.50	4.50
a_2 (order 4)	-4.75	6.00	6.00	6.00
a_3 (order 5)	-14.00	-3.25	7.50	7.50
a_4 (order 6)	-23.25	-12.50	-1.75	9.00

Let's check a value in the payoff table before we proceed. Suppose he takes action a_3 when the state of nature is s_2. This means he ordered $i + 2$ or five boxes when the demand is actually $j + 2$ or four boxes. Thus, he will sell four boxes and get stuck with one box. His return on the four boxes sold is

$$4 \times 10.75 = 43.00$$

but the cost of his order is

$$5 \times 9.25 = 46.25$$

giving a negative payoff of

$$43.00 - 46.25 = -3.25$$

The next step is to calculate the expected value associated with each action as given in the formula.

$$E(P|a_i) = \sum_{j=1}^{4} P_{ij} \times \text{pr}(s_j)$$

Doing this, we get

$$\begin{aligned}
E(P|a_1) &= 4.5\left(\frac{5}{24}\right) + 4.5\left(\frac{6}{24}\right) + 4.5\left(\frac{9}{24}\right) + 4.5\left(\frac{4}{24}\right) \\
&= 4.50 \\
E(P|a_2) &= -4.75\left(\frac{5}{24}\right) + 6\left(\frac{6}{24}\right) + 6\left(\frac{9}{24}\right) + 6\left(\frac{4}{24}\right) \\
&= 3.76 \\
E(P|a_3) &= -14\left(\frac{5}{24}\right) - 3.25\left(\frac{6}{24}\right) + 7.5\left(\frac{9}{24}\right) + 7.5\left(\frac{4}{24}\right) \\
&= 0.33 \\
E(P|a_4) &= -23.25\left(\frac{5}{24}\right) - 12.5\left(\frac{6}{24}\right) - 1.75\left(\frac{9}{24}\right) + 9\left(\frac{4}{24}\right) \\
&= -7.125
\end{aligned}$$

Thus

$$\begin{aligned}
\max\{E(P|a_i), i = 1, 2, 3, 4\} &= E(P|a_1) \\
&= 4.50
\end{aligned}$$

The proper action is to order three boxes (action a_1) per month, which will satisfy demand only 21% of the time (5/24) but will maximize the expected payoff in the long run.

The same decision can be obtained (often less laboriously) by using the opportunity loss table. In that case, we minimize the expected opportunity loss. The procedure is discussed in the next section.

6.3.4 The Minimum Expected Opportunity Loss Procedure

This procedure will always indicate the same action as the maximum expected value procedure of the last section. The difference is that the minimum expected opportunity loss method calculates expected values from an opportunity loss table and searches for a minimum value. Specifically, we choose a course of action a_i corresponding to

$$\min \{E(L|a_i) \quad i = 1, 2, \ldots, m\}$$

where

$$E(L|a_i) = \sum_{j=1}^{n} L_{ij} \operatorname{pr}(s_j)$$

and $\operatorname{pr}(s_j)$ is the probability of the occurrence of the jth state of nature.

EXAMPLE 6.8 Use the minimum expected opportunity loss procedure to determine an action for Example 6.7.

SOLUTION The loss table, given here, is determined by using the method of Section 6.2.

L_{ij}	s_1 (demand of 3)	s_2 (demand of 4)	s_3 (demand of 5)	s_4 (demand of 6)
a_1 (order 3)	0.00	1.50	3.00	4.50
a_2 (order 4)	9.25	0.00	1.50	3.00
a_3 (order 5)	18.50	9.25	0.00	1.50
a_4 (order 6)	27.75	18.50	9.25	0.00

Calculating the values for $E(L|a_i) \quad i = 1, 2, 3, 4,$ we get

$$E(L|a_1) = 0.00\left(\frac{5}{24}\right) + 1.50\left(\frac{6}{24}\right) + 3.00\left(\frac{9}{24}\right) + 4.50\left(\frac{4}{24}\right)$$
$$= 2.25$$

$$E(L|a_2) = 9.25\left(\frac{5}{24}\right) + 0.00\left(\frac{6}{24}\right) + 1.50\left(\frac{9}{24}\right) + 3.00\left(\frac{4}{24}\right)$$
$$= 2.99$$

$$E(L|a_3) = 18.50\left(\frac{5}{24}\right) + 9.25\left(\frac{6}{24}\right) + 0.00\left(\frac{9}{24}\right) + 1.50\left(\frac{4}{24}\right)$$
$$= 6.42$$

$$E(L|a_4) = 27.75\left(\frac{5}{24}\right) + 18.50\left(\frac{6}{24}\right) + 9.25\left(\frac{9}{24}\right) + 0.00\left(\frac{4}{24}\right)$$
$$= 13.88$$

Thus,

$$\min\ \{E(L|a_i),\ i = 1, 2, 3, 4\} = \min\ \{2.25, 2.99, 6.42, 13.88\}$$
$$= 2.25$$

indicating action a_1 (order three boxes of cookies), which is the same conclusion as before.

EXERCISES

6.3.1 Refer to Exercise 6.2.1. Determine the action to be taken using
(*a*) The maximin procedure.
(*b*) The minimax procedure.

6.3.2 Refer to Exercise 6.2.2. Determine the action to be taken using
(*a*) The maximin procedure.
(*b*) The minimax procedure.

6.3.3 Suppose that we have the following probabilities for the states of nature given in Exercise 6.2.1:

$$\text{pr}(s_1) = 0.36$$
$$\text{pr}(s_2) = 0.25$$
$$\text{pr}(s_3) = 0.39$$

Determine the action to be taken by
(*a*) The maximum expected value procedure.
(*b*) The minimum expected opportunity loss procedure.

6.3.4 Suppose that we have the following probabilities for the states of nature given in Exercise 6.2.2:

$$\text{pr}(s_1) = 0.03$$
$$\text{pr}(s_2) = 0.12$$
$$\text{pr}(s_3) = 0.16$$
$$\text{pr}(s_4) = 0.27$$
$$\text{pr}(s_5) = 0.42$$

Determine the action to be taken by
(*a*) The maximum expected value procedure.
(*b*) The minimum expected opportunity loss procedure.

6.4 THE EXPECTED VALUE OF PERFECT INFORMATION

We are now ready to expand on the last two sections with what we call an *expected value of perfect information*. By perfect information we mean an accurate prediction of the state of nature that will allow us to respond with the best action for that particular state. The first step in this process is to determine an expected payoff when we are provided with perfect information.

In the cookie problem of Example 6.7, we have

$$\text{pr}(s_1) = \frac{5}{24} \qquad \text{pr}(s_3) = \frac{9}{24}$$

$$\text{pr}(s_2) = \frac{6}{24} \qquad \text{pr}(s_4) = \frac{4}{24}$$

and a payoff table given by

P_{ij}	s_1 (demand of 3)	s_2 (demand of 4)	s_3 (demand of 5)	s_4 (demand of 6)
a_1 (order 3)	4.50	4.50	4.50	4.50
a_2 (order 4)	-4.75	6.00	6.00	6.00
a_3 (order 5)	-14.00	-3.25	7.50	7.50
a_4 (order 6)	-23.25	-12.50	-1.75	9.00

If we knew the state of nature was going to be s_1, we would choose a_1 as our action, to give us a payoff of 4.50. Likewise, we would choose a_2 if s_2 were the state of nature, a_3 for s_3, and a_4 for s_4. In each case, we would choose the action that would maximize the payoff for the given state of nature. But notice that even if we know exactly what the state of nature is going to be (perfect information), the state of nature will vary over time, with s_1 occurring 21% of the time, s_2 25% of the time, s_3 37% of the time, and s_4 17% of the time. This means that when we have perfect information on the state of nature, we can expect a payoff of

4.50 about 21% of the time
6.00 about 25% of the time
7.50 about 37% of the time
9.00 about 17% of the time

giving a long-run average payoff of

$$4.50(0.21) + 6.00(0.25) + 7.50(0.37) + 9.00(0.17) = 6.75$$

This example illustrates the fundamental aspects of calculating the expected payoff under perfect information. We can express the general solution mathematically as

$$\text{EPPI} = \sum_{j=1}^{n} \max \{P_{ij},\ i = 1, 2, \ldots, m\}\ \text{pr}(s_j)$$

where EPPI stands for the expected payoff under perfect information. For this problem, we have

$$\begin{aligned} \text{EPPI} &= \max \{4.50, -4.75, -14.00, -23.25\}\ \text{pr}(s_1) \\ &\quad + \max \{4.50, 6.00, -3.25, -12.50\}\ \text{pr}(s_2) \\ &\quad + \max \{4.50, 6.00, 7.50, -1.75\}\ \text{pr}(s_3) \\ &\quad + \max \{4.50, 6.00, 7.50, 9.00\}\ \text{pr}(s_4) \\ &= 4.50\left(\frac{5}{24}\right) + 6.00\left(\frac{6}{24}\right) + 7.50\left(\frac{9}{24}\right) + 9.00\left(\frac{4}{24}\right) = 6.75 \end{aligned}$$

The expected value of perfect information (EVPI) is given by

$$\text{EVPI} = \text{EPPI} - \text{MEV}$$

where MEV stands for the maximum expected value as given in Section 6.3.3. Therefore, the mathematical expression for EVPI must be given by

$$\text{EVPI} = \sum_{j=1}^{n} \max \{P_{ij},\ i = 1, 2, \ldots, m\}\ \text{pr}(s_j)$$

$$- \max \left\{ \left[\sum_{j=1}^{n} P_{ij}\ \text{pr}(s_j) \right],\ i = 1, 2, \ldots, m \right\}$$

In the previous problem, we have

$$\begin{aligned} \text{EVPI} &= 6.75 - 4.50 \\ &= 2.25 \end{aligned}$$

and the interpretation of this value is that it represents the absolute maximum to be paid for information on the states of nature. An equivalent expression given by using the opportunity loss table values is

$$\text{EVPI} = \min \{E(L|a_i) \quad i = 1, 2, \ldots, m\}$$

or the same value as the minimum expected opportunity loss as given in Section 6.3.4, since the EOL = 0 with perfect information.

Perfect information is impossible to obtain, but sample or imperfect information generally will be available in some form. Obviously, if one buys information, the upper bound on its value is the EVPI. A more realistic procedure is found in the next section, in which we discuss placing a value on sample information.

EXERCISES

6.4.1 Calculate the expected value of perfect information for Exercise 6.3.3 using:
 (*a*) The formula EVPI = EPPI − MEV.
 (*b*) The opportunity loss table.

6.4.2 Calculate the expected value of perfect information for Exercise 6.3.4, using the formula

$$\text{EVPI} = \sum_{j=1}^{n} \max \{P_{ij},\ i = 1, 2, \ldots, m\}\ \text{pr}(s_j)$$

$$- \max \left\{ \left[\sum_{j=1}^{n} P_{ij}\ \text{pr}(s_j) \right],\ i = 1, 2, \ldots, m \right\}$$

6.4.3 If the probabilities associated with the states of nature in Exercise 6.4.1 are changed to $\text{pr}(s_1) = 0.2$, $\text{pr}(s_2) = 0.6$, $\text{pr}(s_3) = 0.2$, what is the effect on the EVPI?

6.4.4 Change the pr(s_j) in Exercise 6.4.2 to pr(s_1) = 0.15, pr(s_2) = 0.10, pr(s_3) = 0.25, pr(s_4) = 0.35, and pr(s_5) = 0.15, and determine the new EVPI.

6.4.5 What do the previous exercises suggest to you about the relationship between the EVPI and the assignment of probabilities to the states of nature?

6.5 THE EXPECTED VALUE OF SAMPLE INFORMATION

The methods we have presented have built a framework from which we can present one of the more common methods used in decision analysis. The method uses Bayes' rule to help evaluate the long-term value of sample information through the calculation of an expected value. In the process, we will use the fundamental concepts of conditional and joint probabilities and will introduce two new terms that are typically used in Bayesian decision analysis.

One must first understand that we frequently do not have access to perfect information, but we usually *do* have an opportunity to sample either explicitly (test ten items from the shipment at random and see how many of the ten are defective) or intuitively (let an expert use his nose to judge the quality of a hold of carp). The sampling methods are part of the responsibility of quality control (or quality assurance) groups in many companies, as they want to control the quality of incoming raw materials or the quality of outgoing finished goods.

EXAMPLE 6.9 Jake sniffs out carp at Halifax Harbor in an attempt to control the quality of fish sold in Canadian fish shops. He scoops up a bucket full of carp at random from each hold, and if his nose tells him the fish is of poor quality, the hold is not purchased. Historically, one out of every eight holds is of poor quality and should not be purchased for human consumption. Furthermore, suppose that:

1. The government pays \$1200 per hold of carp
2. A hold of good-quality carp can be sold for \$2500
3. A poor-quality hold can be sold for \$600 (to a cat food company)
4. The probability that
   a. Jake rejects the hold given that the hold is good is 0.02
   b. Jake accepts the hold given that the hold is poor is 0.08

Calculate Jake's value to the government on a per hold basis (the maximum they should be willing to pay Jake, on a per hold basis, for his services).

SOLUTION As a first step, we construct the payoff table (which does not depend on Jake's nose).

	P_{ij}	State of Nature (Hold Quality)	
		s_1 (good)	s_2 (poor)
Action	a_1 (buy)	1300	-600
	a_2 (don't buy)	0	0

From this, we can calculate the expected value of perfect information:

$$\begin{aligned}
\text{EVPI} &= \text{EPPI} - \text{MEV} \\
&= [\max\{1300, 0\}\ \text{pr}(s_1) + \max\{-600, 0\}\ \text{pr}(s_2)] \\
&\quad - [\max\{1300\ \text{pr}(s_1) - 600\ \text{pr}(s_2), 0 \cdot \text{pr}(s_1) + 0 \cdot \text{pr}(s_2)\}] \\
&= 1300\left(\frac{7}{8}\right) + 0\left(\frac{1}{8}\right) - \max\left\{1300\left(\frac{7}{8}\right) - 600\left(\frac{1}{8}\right), 0\left(\frac{7}{8}\right) + 0\left(\frac{1}{8}\right)\right\} \\
&= 1137.50 - 1062.50 \\
&= 75.00
\end{aligned}$$

The probabilities for the states of nature are called *prior probabilities*, since they are calculated on the basis of historical record and do not depend on information due to sampling. That is, we can say, the probability that a hold at hand (no pun intended) is of good quality is 7/8 "prior" to any additional sample information due to Jake's nose.

Thus, the obvious question to answer next is, how do the probabilities of the states of nature change once we have Jake's decision? There are two cases: the event that Jake decides to accept the hold (denoted by A) and the event that Jake decides to reject the hold (denoted by R). First, suppose that Jake accepts the hold. Then we want to calculate the effect of this decision on the probability distribution associated with the states of nature,

$$\text{pr}(s_1 \mid A) \quad \text{and} \quad \text{pr}(s_2 \mid A)$$

We call these *posterior* probabilities, since they are the result of calculations *after* additional information has been considered. So, looking at a probability table, we have

	s_1 (good)	s_2 (poor)
A		x_2
R	x_1	
	7/8	1/8

and we could complete the table if we knew values for x_1 and x_2. But

$$x_1 = \text{pr}(s_1 \cap R) = \text{pr}(R|s_1)\,\text{pr}(s_1)$$
$$= (0.02)\left(\frac{7}{8}\right) = 0.0175$$

$$x_2 = \text{pr}(s_2 \cap A) = \text{pr}(A|s_2)\,\text{pr}(s_2)$$
$$= (0.08)\left(\frac{1}{8}\right) = 0.01$$

Thus, the joint probability table is given by

	s_1 (good)	s_2 (poor)	
A	0.8575	0.0100	0.8675
R	0.0175	0.1150	0.1325
	0.8750	0.1250	1.0000

From this, we obtain

$$\text{pr}(s_1|A) = \frac{\text{pr}(s_1 \cap A)}{\text{pr}(A)} = \frac{0.8575}{0.8675} = 0.9885$$

and

$$\text{pr}(s_2|A) = \frac{\text{pr}(s_2 \cap A)}{\text{pr}(A)} = \frac{0.0100}{0.8675} = 0.0115$$

Now, we can use the minimum expected opportunity loss procedure (or the maximum expected value procedure) with the new conditional probabilities. These are *posterior* probabilities, since they are the result of calculations after additional information has been considered.

The loss table is given by

L_{ij}	s_1 (good)	s_2 (poor)
a_1 (buy)	0	600
a_2 (don't buy)	1300	0

Using posterior probabilities for the states of nature, we get a posterior minimum expected opportunity loss, given by

$$\begin{aligned}&\min\{0\ \text{pr}(s_1\,|\,A)+600\ \text{pr}(s_2\,|\,A),\ 1300\ \text{pr}(s_1\,|\,A)+0\ \text{pr}(s_2\,|\,A)\}\\&=\min\{0(0.9885)+600(0.0115),\ 1300(0.9885)+0(0.0115)\}\\&=0(0.9885)+600(0.0115)\\&=6.90\end{aligned}$$

Since this value corresponds to a buy action, we conclude that if Jake's nose says to accept, we should buy the hold of carp.

Now, let's see what happens if Jake's nose says to reject the hold. In this case,

$$\text{pr}(s_1\,|\,R)=\frac{\text{pr}(s_1\cap R)}{\text{pr}(R)}=\frac{0.0175}{0.1325}=0.1321$$

and

$$\text{pr}(s_2\,|\,R)=1-\text{pr}(s_1\,|\,R)=1-0.1321=0.8679$$

so that

$$\begin{aligned}&\min\{0\ \text{pr}(s_1\,|\,R)+600\ \text{pr}(s_2\,|\,R),\ 1300\ \text{pr}(s_1\,|\,R)+0\ \text{pr}(s_2\,|\,R)\}\\&=\min\{0(0.1321)+600(0.8679),\ 1300(0.1321)+0(0.8679)\}\\&=\min\{520.75,\ 171.73\}=171.73\end{aligned}$$

Since 171.70 corresponds to a don't-buy action, we conclude that if Jake's nose says to reject, we should not buy the hold of carp.

At this point, we have reviewed the entire spectrum of what can happen in the sampling process and can properly evaluate its worth. Evidently, it must be worth something, since the action changes from buy to don't buy with a change in Jake's recommendation from accept to reject (the analysis can stop here if the action is the same no matter what Jake says). The final step is indicated by the following:

EVPI Given Jake Accepts	pr(Jake Accepts)
6.90	0.8675
EVPI Given Jake Rejects	pr(Jake Rejects)
171.73	0.1325

Thus, the expected value of perfect information *after sampling* is given by

$$6.90(0.8675) + 171.73(0.1325) = 28.75$$

Thus, the value of perfect information has dropped from 75.00 to 28.75 as a result of being allowed to employ Jake's nose. Since 75.00 is an upper bound on Jake's value, it follows that his services are worth no more than

$$75.00 - 28.75 = 46.25$$

That is, the government should pay Jake no more than $46.25 on a per hold basis for his services.

In general, the expected value of sample information, EVSI, is given by the equation

$$\text{EVSI} = \text{prior EVPI} - \text{posterior EVPI}$$

where prior EVPI is the expected value of perfect information prior to any sampling plan, using the prior probabilities, and posterior EVPI is the expected value of perfect information calculated on the premise that a given sampling plan is to be employed to assist in the choice of an action.

EXAMPLE 6.10 Let's take another look at Exercise 6.2.1, in which Sungleam, Inc., has an opportunity to buy used mixing bowls in lots of size 25. The bowls are bought as a package deal, and Sungleam must live with what they purchase in terms of the appearance of the bowls. The bowls are purchased at a cut-rate price of \$32.50 for 25, and some of them may be in horrible shape. These horrible bowls sell for \$0.50 each, and the others sell for \$2.50 each. Sungleam classifies each shipment as poor, fair, or good in quality, according to the number of horrible bowls in the 25. Poor shipments contain 20 horrible bowls, fair shipments contain 12 horrible bowls, and good shipments contain 4 horrible bowls, and all shipments can be placed in one of the three categories. Suppose the quality control plan is to choose three bowls at random from each box of 25 to help judge the overall quality of the 25. Then:

(*a*) Determine an action for each of the four possible sampling results (0, 1, 2, or 3 horrible bowls are found among the three chosen at random).

(*b*) If the action varies according to the sampling result, determine the maximum amount of money Sungleam should pay the quality control group for their inspection of the three bowls in each shipment.

SOLUTION You will get a good look at the use of Bayes' rule in this problem, compared with its use (shall we say, "incognito") in Example 6.9, where it was disguised.

(*a*) The first step in the analysis is to determine how the probabilities for the states of nature (the quality of the shipment) change once the sample information is presented. That is, we will now calculate the posterior probabilities given the prior probabilities for each possible sample outcome. Recall from Exercise 6.3.3 that for a given shipment,

$$\begin{aligned}
\text{pr(poor quality)} &= \text{pr}(s_1) = 0.36 \\
\text{pr(fair quality)} &= \text{pr}(s_2) = 0.25 \\
\text{pr(good quality)} &= \text{pr}(s_3) = 0.39
\end{aligned}$$

which, according to our agreement, are the prior probabilities. We now see how these change once we know the results of the inspection. Let X be a random variable that assumes the value k, $k = 0, 1, 2$, and 3, according to the number of horrible bowls in the sample of size three.

Case One $X = 0$ (no horrible bowls are found in the sample of size 3). What we want to know is $\text{pr}(s_i | X = 0)$, $i = 1, 2, 3$. Since the states of nature form a partition, this is a case for Bayes' rule, and we get

$$\text{pr}(s_i | X = 0) = \frac{\text{pr}(X = 0 | s_i)\,\text{pr}(s_i)}{\sum_{j=1}^{3} \text{pr}(X = 0 | s_j)\,\text{pr}(s_j)}$$

Thus, we have our solution for this case once we can find

$$\text{pr}(X = 0 | s_j) \qquad j = 1, 2, 3$$

In other words, we want to know probability that we observe no horrible bowls in a random sample of size three, given that there are 20, 12, or 4 horrible bowls in the 25. Assuming that the sampling is without replacement, this is a hypergeometric random variable problem.

1. The population size, N, is 25.
2. The size of the subset of interest (the number of horrible bowls in the 25), R, is either 20, 12, or 4.
3. The size of the random sample, n, is 3.

Using the hypergeometric probability function, we get

$$\text{pr}(X = 0 | s_1) = \frac{\binom{20}{0}\binom{25-20}{3-0}}{\binom{25}{3}} = \frac{10}{2300}$$

$$\text{pr}(X = 0 | s_2) = \frac{\binom{12}{0}\binom{25-12}{3-0}}{\binom{25}{3}} = \frac{286}{2300}$$

$$\text{pr}(X = 0 | s_3) = \frac{\binom{4}{0}\binom{25-4}{3-0}}{\binom{25}{3}} = \frac{1330}{2300}$$

Thus,

$$\begin{aligned}\text{pr}(X = 0) &= \sum_{j=1}^{3} \text{pr}(X = 0 | s_j)\,\text{pr}(s_j)\\ &= \frac{10}{2300} \times 0.36 + \frac{286}{2300} \times 0.25 + \frac{1330}{2300} \times 0.39\\ &= 0.25817\end{aligned}$$

Using the convenient tabular format, we obtain

State of Nature	Prior Probability $pr(s_j)$	Conditional Probability $pr(X = 0 \mid s_j)$	Posterior Probability $pr(s_j \mid X = 0) = \dfrac{pr(X = 0 \mid s_j)\,pr(s_j)}{0.25817}$
s_1 (poor)	0.36	$\frac{10}{2300}$	0.0061
s_2 (fair)	0.25	$\frac{286}{2300}$	0.1204
s_3 (good)	0.39	$\frac{1330}{2300}$	0.8735

Case Two $X = 1$ (one horrible bowl is found in the sample of size 3). In this case, we want to calculate $pr(s_j \mid X = 1)$, $i = 1, 2, 3$. Again, we use the hypergeometric probability function to get

$$pr(X = 1 \mid s_1) = \frac{\binom{20}{1}\binom{25-20}{3-1}}{\binom{25}{3}} = \frac{200}{2300}$$

$$pr(X = 1 \mid s_2) = \frac{\binom{12}{1}\binom{25-12}{3-1}}{\binom{25}{3}} = \frac{936}{2300}$$

$$pr(X = 1 \mid s_3) = \frac{\binom{4}{1}\binom{25-4}{3-1}}{\binom{25}{3}} = \frac{840}{2300}$$

Thus,

$$\begin{aligned} pr(X = 1) &= \sum_{j=1}^{3} pr(X = 1 \mid s_j)\,pr(s_j) \\ &= \frac{200}{2300} \times 0.36 + \frac{936}{2300} \times 0.25 + \frac{840}{2300} \times 0.39 \\ &= 0.27547 \end{aligned}$$

Again using the tabular procedure, we get:

State of Nature	Prior Probability	Conditional Probability $\text{pr}(X = 1 \mid s_j)$	Posterior Probability $\text{pr}(s_j \mid X = 1) = \dfrac{\text{pr}(X = 1 \mid s_j)\,\text{pr}(s_j)}{0.27547}$
s_1 (poor)	0.36	$\frac{200}{2300}$	0.1136
s_2 (fair)	0.25	$\frac{936}{2300}$	0.3693
s_3 (good)	0.39	$\frac{840}{2300}$	0.5171

Case Three $X = 2$ (two horrible bowls are found in the sample of size 3). In this case, we have, similarly,

$$\text{pr}(X = 2 \mid s_1) = \frac{950}{2300}$$
$$\text{pr}(X = 2 \mid s_2) = \frac{858}{2300}$$
$$\text{pr}(X = 2 \mid s_3) = \frac{126}{2300}$$

So,

$$\begin{aligned}\text{pr}(X = 2) &= \sum_{j=1}^{3} \text{pr}(X = 2 \mid s_j)\,\text{pr}(s_j) \\ &= \frac{950}{2300} \times 0.36 + \frac{858}{2300} \times 0.25 + \frac{126}{2300} \times 0.39 \\ &= 0.26333\end{aligned}$$

Thus, the table is:

State of Nature	Prior Probability	Conditional Probability	Posterior Probability
s_1 (poor)	0.36	$\frac{950}{2300}$	0.5647
s_2 (fair)	0.25	$\frac{858}{2300}$	0.3542
s_3 (good)	0.39	$\frac{126}{2300}$	0.0811

Case Four $X = 3$ (three horrible bowls are found in the sample of size 3). Again, using the hypergeometric, we obtain:

$$\text{pr}(X = 3|s_1) = \frac{1140}{2300}$$

$$\text{pr}(X = 3|s_2) = \frac{220}{2300}$$

$$\text{pr}(X = 3|s_3) = \frac{4}{2300}$$

So,

$$\begin{aligned}\text{pr}(X = 3) &= \sum_{j=1}^{3} \text{pr}(X = 3|s_j)\text{pr}(s_j)\\ &= \frac{1140}{2300} \times 0.36 + \frac{220}{2300} \times 0.25 + \frac{4}{2300} \times 0.39\\ &= 0.20303\end{aligned}$$

The resulting table is

State of Nature	Prior Probability	Conditional Probability	Posterior Probability
s_1 (poor)	0.36	$\frac{1140}{2300}$	0.8789
s_2 (fair)	0.25	$\frac{220}{2300}$	0.1178
s_3 (good)	0.39	$\frac{4}{2300}$	0.0033

We can now determine the course of action using the posterior probabilities for each case. We do this by using the minimum expected opportunity loss procedure with the posterior probabilities instead of the prior probabilities. Recall that the opportunity loss table is given by

L_{ij}	s_1 (poor)	s_2 (fair)	s_3 (good)
a_1 (buy)	10.00	0	0
a_2 (don't buy)	0	6.00	22.00

Thus, using posterior probabilities, we get the following.

Case One $(X = 0)$

$$E(L|a_1) = 10.00(0.0061) + 0(0.1204) + 0(0.8735) = 0.061$$

$$E(L|a_2) = 0(0.0061) + 6.00(0.1204) + 22.00(0.8735) = 19.939$$

Since $\min\{E(L|a_1), E(L|a_2)\} = 0.061$, the action is to buy the shipment when no horrible bowls are observed in the random sample of size three.

Case Two $(X = 1)$

$$E(L|a_1) = 10.00(0.1136) = 1.136$$

$$E(L|a_2) = 6.00(0.3693) + 22.00(0.5171) = 13.592$$

Since $\min\{E(L|a_1), E(L|a_2)\} = 1.136$, the action is to buy the shipment when one horrible bowl is observed in the random sample of size three.

Case Three $(X = 2)$

$$E(L|a_1) = 10.00(0.5647) = 5.647$$

$$E(L|a_2) = 6.00(0.3542) + 22.00(0.0811) = 3.909$$

and the action is to not buy the shipment when two horrible bowls are observed in the random sample of size three.

Case Four $(X = 3)$

$$E(L|a_1) = 10.00(0.8789) = 8.789$$

$$E(L|a_2) = 6.00(0.1178) + 22.00(0.0033) = 0.779$$

and the action is to not buy the shipment when three horrible bowls are observed in the random sample of size three.

Since the actions vary according to the sampling result, we know that the sampling process must have some value, and we move on to part (*b*).

(*b*) The expected value of sample information, EVSI, is given by

$$\text{Prior EVPI} - \text{posterior EVPI}$$

The prior EVPI was found in Exercises 6.3.3*b* and 6.4.1 to be \$3.60. To calculate the posterior EVPI, we need to calculate an EVPI using the posterior probabilities for each of the four cases, which we have already done, since we used the minimum expected opportunity loss procedure to determine the actions in part *a*. Thus, we can provide the following table for the determination of the posterior EVPI.

Case	k	$\text{EVPI}\|X = k$	$\text{pr}(X = k)$	$\text{pr}(X = k) \times (\text{EVPI}\|X = k)$
1	0	0.061	0.25817	0.016
2	1	1.136	0.27547	0.313
3	2	3.909	0.26333	1.029
4	3	0.779	0.20303	0.158
				$\Sigma = 1.516$

Thus, the posterior EVPI is \$1.52 rounded to the nearest cent. The value of perfect information has dropped from \$3.60 to \$1.52 as a result of the random sampling plan, and so the value of the plan must be

$$\$3.60 - \$1.52 = \$2.08$$

This means that the maximum amount of money Sungleam should pay the quality control group for their inspection of three bowls chosen at random from each shipment of size 25 is \$2.08.

EXERCISES

6.5.1 Use the probabilities calculated in Example 6.9 and the MEV approach to verify that the EVSI is $2.08.

6.5.2 Referring to Example 6.9, determine whether it would be worthwhile for Sungleam to randomly sample a single bowl from each lot.

6.5.3 What is the value to Sungleam of sampling two bowls at random from each lot?

6.5.4 Rework Example 6.8 using the MEV approach.

6.5.5 Morning Glory Winery ships table wine to most major cities in North America. They purchase their grapes in bulk volume equivalent to one barrel of wine. Each barrel equivalent is labeled either semisweet or dry by the vineyard producing it, indicating the type of wine the grapes will produce. However, the shipments are mislabeled too often, and Morning Glory has suffered losses in the past as a result of this problem. Fredrico the grape evaluator offers his services at $5.00 per barrel equivalent as a judge of the proper character of the resulting wine. If:

1. Fredrico accurately predicts the character of the wine with probability 0.95
2. Of the shipments, 38% are labeled semisweet and the other 62% are labeled dry.
3. The costs and returns are as follows:

	Label on Wine Shipment			
	Semisweet		Dry	
	Right Label	Wrong Label	Right Label	Wrong Label
Cost per barrel equivalent	300	300	400	400
Return per barrel equivalent	400	250	700	250

4. If semisweet grapes are labeled as dry 40% of the time, and dry grapes are labeled semisweet 40% of the time should Morning Glory employ Fredrico at his asking price per barrel equivalent?

6.6 SUPPLEMENTARY EXAMPLES

EXAMPLE 6.11 A grocer assigns the following probabilities to the demand for the *Daily Chronicle* as follows:

Demand in Number of Copies	Probability
$\leqslant 11$	0.00
12	0.18
13	0.32
14	0.22
15	0.28
$\geqslant 16$	0.00

If a copy sells for \$0.35 and costs \$0.25, determine:

(*a*) The number of copies the grocer should order if he *cannot* return unsold copies

(*b*) The number of copies the grocer should order if he *can* return unsold copies

SOLUTION

(*a*) We need only consider the demands of 12, 13, 14, and 15, since all other demands occur with zero probability. Let the jth state of nature represent a demand of $11 + j$ copies and the ith action represent the ordering of $11 + i$ copies. Since the grocer cannot return unsold copies, he must absorb the loss on the unsold copies, and so his cost for action i is $0.25(11 + i)$ independent of the state of nature. For state of nature j, the return is $0.35(11 + j)$. Hence, the payoff equation is given by

$$P_{ij} = \begin{cases} 0.35(11 + j) - 0.25(11 + i) & j \leqslant i \\ 0.10(11 + i) & j > i \end{cases}$$

and the payoff table is given by

P_{ij}	s_1 (demand is 12)	s_2 (demand is 13)	s_3 (demand is 14)	s_4 (demand is 15)
a_1 (order 12)	1.20	1.20	1.20	1.20
a_2 (order 13)	0.95	1.30	1.30	1.30
a_3 (order 14)	0.70	1.05	1.40	1.40
a_4 (order 15)	0.45	0.80	1.15	1.50

To determine the action, we use the maximum expected value procedure:

$$E(P|a_1) = 1.20(1) = 1.20$$

$$E(P|a_2) = 0.95(0.18) + 1.30(1 - 0.18) = 1.237$$

$$E(P|a_3) = 0.70(0.18) + 1.05(0.32) + 1.40(1 - 0.18 - 0.32)$$
$$= 1.162$$

$$E(P|a_4) = 0.45(0.18) + 0.80(0.32) + 1.15(0.22) + 1.50(0.28)$$
$$= 1.010$$

Since $\max\ \{E(P|a_i),\ \ i = 1, 2, 3, 4\} = E(P|a_2)$, action a_2 should be employed and the grocer should order 13 copies.

(*b*) If the grocer can return unsold copies, the payoff equation is given by

$$P_{ij} = \begin{cases} (0.10)(11 + j) & i \geqslant j \\ (0.10)(11 + i) & i < j \end{cases}$$

and the payoff table is given by

P_{ij}	s_1 (demand is 12)	s_2 (demand is 13)	s_3 (demand is 14)	s_4 (demand is 15)
a_1 (order 12)	1.20	1.20	1.20	1.20
a_2 (order 13)	1.20	1.30	1.30	1.30
a_3 (order 14)	1.20	1.30	1.40	1.40
a_4 (order 15)	1.20	1.30	1.40	1.50

Employing the maximum expected value procedure, we get

$$E(P|a_1) = 1.20(1) = 1.20$$

$$E(P|a_2) = 1.20(0.18) + 1.30(1 - 0.18) = 1.282$$

$$E(P|a_3) = 1.20(0.18) + 1.30(0.32) + 1.40(1 - 0.18 - 0.32)$$
$$= 1.332$$

$$E(P|a_4) = 1.20(0.18) + 1.30(0.32) + 1.40(0.22) + 1.50(0.28)$$
$$= 1.360$$

Since $\max \{E(P|a_i),\ \ i = 1, 2, 3, 4\} = E(P|a_4)$, action a_4 should be employed, and the grocer should order 15 copies of the *Daily Chronicle.*

EXAMPLE 6.12 Calculate the expected value of perfect information for Example 6.11*a* and *b*.

SOLUTION

(*a*) $\text{EVPI} = 1.20 \text{ pr}(s_1) + 1.30 \text{ pr}(s_2) + 1.40 \text{ pr}(s_3) + 1.50 \text{ pr}(s_4)$
$= 0.216 + 0.416 + 0.308 + 0.420$
$= 1.360$

From the example, $\text{MEV} = E(P|a_2) = 1.237$, and so the expected value of perfect information is given by

$$\text{EVPI} = \text{EPPI} - \text{MEV} = 1.360 - 1.237 = 0.123$$

(*b*) The EPPI is the same, but MEV $= E(P|a_4) = 1.360$, and so

$$\text{EVPI} = 1.360 - 1.360 = 0$$

which coincides with our intuition concerning the value of perfect information when there is no penalty for overestimating the demand.

6.7 SUMMARY

This chapter has shown how some of the basic probability axioms of Chapter 2, the probability distributions of Chapters 4 and 5, and the concept of expected value can be used to evaluate sample information when it is intended to aid the managerial decision-making process. The common criteria of maximin and minimax have limited value, since they do not consider the probabilities for the states of nature. The maximum expected value and minimum expected opportunity loss procedures relax this limitation by indicating an action after consideration of the probabilities. Last, the expected value of sample information is used to show how additional information on the states of nature can be evaluated.

Probably one of the most important assumptions in the latter part of the chapter is that the states of nature and their probabilities are defined and accurate. Frequently, historical records are used to get reliable estimates for these probabilities and reasonable choices for the states of nature.

REVIEW EXERCISES

Level 1

6.1 Guido Gambacci operates a bakery in Vancouver. Each day he must decide on the number of Italian rum cakes to make for the next day. A rum cake sells for \$18.00 when fresh but for only \$5.00 if it is a day old. Day-old cakes always sell. The cost of making an Italian rum cake is \$12.00. Guido has five regular customers for fresh Italian rum cake, but the probability that a particular customer will request one on any given day is only 0.5.

(*a*) Construct the payoff table for Guido.

(*b*) What will be Guido's decision if he uses the maximin criterion?

(*c*) What is his decision according to the minimax decision?

6.2 Refer to Exercise 6.1 and determine:
(*a*) The prior distribution for Guido's states of nature.
(*b*) His decision if he uses the MEV criterion.
(*c*) The value of perfect information for Guido (use the payoff table).
(*d*) If Guido could call each customer at a cost of 20 cents per call plus 15 cents for his time for each call, should he do so before making his decision?

6.3 Refer to Exercise 6.1 and
(*a*) Determine Guido's decision using the minimum EOL criterion.
(*b*) Find the EVPI using the opportunity loss table.

6.4 Suppose Guido (of Exercise 6.1) estimates that the loss associated with the inability to provide a fresh cake when it is requested (the ill-will cost of understocking) is $2.00. Construct his payoff table in this case and determine:
(*a*) The maximin decision.
(*b*) The MEV decision.
(*c*) The minimax decision.
(*d*) His new EVPI.
(*e*) Should he call the customers in this case?

6.5 Your instructor has two sections of this class. Your class is scheduled to take an exam today, and the other class is to take its exam over the same material tomorrow. The two exams have been prepared and are being given randomly to the two classes. One of the following is true: both exams are hard, only one is hard, or both are easy. Feeling somewhat unprepared, you ask the instructor if you may take the exam with tomorrow's class and he gives you the following options: You can take today's exam with no penalty, or you can take the exam with the other class and incur a penalty of half a grade point; for example, if you earn a B (3 points) on tomorrow's exam, your actual grade will become C+ or 2.5 points. Your payoff table is

	States		
Actions	Both Hard	One Hard	Both Easy
Take now	1.0	2.0	3.0
Take tomorrow	2.0	2.0	2.5

(*a*) What is your decision using minimax?
(*b*) What is your decision using maximin?

(*c*) From past experience you believe the probabilities for both hard, one hard, and both easy to be 0.5, 0.3, and 0.2, respectively. What decision will maximize your expected grade?

(*d*) What is it worth to you (in quality points) to know the exact number of difficult exams?

6.6 Refer to your dilemma in Exercise 6.5. Realizing that the exams have been assigned randomly to classes, what is it worth to you to find out the level of difficulty of tomorrow's exam?

6.7 A merchant has two units of an item with a unit cost of $300. She is trying to decide whether to give a regular money-back guarantee on the items or a double-your-money-back guarantee. If she gives the regular guarantee, the items will sell for $600 each, whereas with the double guarantee she can charge $750 each. The items are either both good, both defective, or one of each, and the quality cannot be determined by simple inspection.

(*a*) Construct the merchant's payoff table.

(*b*) Determine the maximin decision.

(*c*) Determine the minimax decision.

(*d*) What decision maximizes the expected payoff?

6.8 Refer to the merchant in Exercise 6.7.

(*a*) What is it worth to her to know the quality of both items?

(*b*) Suppose she can send one of the items out for testing at a cost of $75 before she decides. Is it worthwhile for her to do so?

(*c*) Should she consider sending both items out for testing?

6.9 Give two examples of situations in which the minimax criterion would be preferred over the MEV criterion.

6.10 A manufacturing firm is considering whether or not to expand plant capacity by purchasing one of two machines, one of which has twice the capacity of the other but costs over 50% more. The firm's existing capacity is not quite sufficient to meet the current demand for its product. Sales appear to have a high correlation with general economic conditions. If conditions improve, the high-capacity machine will be required; if they remain stable, the low-capacity machine will suffice; and if a recession is on the way, the current capacity will do. After a careful analysis of revenues and costs under the various combinations, the following payoff table was determined (in millions of dollars).

Buy	Economic Conditions		
	Improve	Same	Decline
High	2.5	1.5	-1.0
Low	1.6	2.0	0.0
None	1.2	1.2	0.8

(*a*) What are the maximin and minimax decisions?

(*b*) If the firm believes the economy will improve with a probability of 0.35, remain the same with a probability of 0.40, and decline with a probability of 0.25, what decision will maximize the expected value?

6.11 The President of the United States has an economic advisor whom some people consider to be infallible. If the advisor offers his services to the firm in Exercise 6.10, what is the maximum amount the firm should be willing to pay for the advisor's forecast of the economy?

6.12 Checking past forecasts of the economic advisor of Exercise 6.11, the firm finds the following frequencies: When the economy improved, he predicted that it would do so 85% of the time and predicted stable conditions and declines 10% and 5% of the time, respectively. When the economy remained stable, he had predicted improvement, stability, and decline with frequencies 15%, 60%, and 25%, respectively. In a declining economy, his predictions have been improvement, stability, and decline with probabilities 0.10, 0.20, and 0.70. With this additional information, how much will his forecast be worth to the firm?

6.13 A manufacturer of precision machine parts, the demand for which is highly seasonal, rents additional equipment during the busy season. Two machines are available for rent. One machine can be rented from Ace Rental at a cost of $2500 a month. This machine is guaranteed to operate perfectly in that Ace Rental will pay all costs of reworking any defective parts that the machine produces. A second machine can be rented from Machine Rental at a monthly cost of $1800. Although the lower rental price looks attractive, Machine Rental gives no guarantee concerning defective parts produced. The cost for reworking defective parts is $10 for each part. Based on past experience, the probability distribution for the percentage of defective parts has been

Percentage Defective	Probability
0.04	0.5
0.08	0.3
0.12	0.2

(*a*) Set up a payoff matrix (showing conditional *costs*). Clearly define the states of nature and alternative actions if a machine is needed for one peak month to produce 1000 parts.

(*b*) Which machine should the company rent based on expected costs?

(*c*) What is the expected value of perfect information?

6.14 You are contemplating introducing a new product and determining whether or not to invest a large amount of money in advertising. Your return depends on market conditions, which you believe will be poor with probability 0.5, good with probability 0.4, and excellent with probability 0.1. The following table shows your estimate of profits under each advertising strategy in $10,000.

	Market Conditions		
Actions	Poor	Good	Excellent
High-cost advertising	-100	80	400
Medium-cost advertising	-80	100	250
Low-cost advertising	-60	125	175
Forget product	0	0	0

(*a*) What is the maximin decision?

(*b*) What is the minimax decision?

(*c*) What is the decision using the EMV criterion?

6.15 (*a*) Refer to the above problem. If you could hire Dr. D'Mico, who claims to be infallible as a market researcher, what is the maximum you should consider paying him?

(*b*) Looking at past occasions when you have used Dr. D'Mico, you find that he is not infallible. Instead, the following patterns emerge: When the market has been poor, he has predicted poor 60% of the time, good 30% of the time, and excellent 10% of the time; whereas when it has been good, he has predicted each condition, respectively, as 20%, 50%, and

30%; and when it was excellent, the percentages were 10%, 20%, and 70%. Dr. D'Mico offers to give you his predictions this time for $100,000. Should you accept his offer?

Level 2

6.16 Prove that maximizing the expected value is equivalent to minimizing the expected opportunity loss.

6.17 Determine the optimum number of phone calls that Guido Gambacci, the baker in Exercise 6.1, should make each day before making his Italian rum cakes.

6.18 Suppose you are offered a ticket in a raffle that has six prizes: one $5000 prize, two $2000 prizes, and three $1000 prizes. The cost of the ticket is $1.00, and there will be exactly 20,000 tickets.
(*a*) Construct your payoff table.
(*b*) Since the raffle is being conducted by a charitable group, will you buy the ticket?
(*c*) Would you buy the ticket using maximin?
(*d*) Would you buy the ticket using MEV?

6.19 For many of you, your answers to parts (*b*) and (*d*) in Exercise 6.18 will appear contradictory. Explain this apparent contradiction.

6.20 An assumption of the MEV criterion of optimality is that the usefulness of the payoffs is linearly related to their values. Exercise 6.18 indicates that this may not be true. Construct another example in which your expected value would be positive, but you would refuse to play.

6.21 One way of handling the contradiction in Exercise 6.20 is to assign a *utility* value to each payoff, then construct a new table of utilities. One approach to doing so is to assign 0 to the least desirable payoff and 1 to the most desirable, with all other payoffs receiving utilities between 0 and 1 subject to the following restrictions (let u_{ij} be the utility of x_{ij}):

1. If $x_{ij} \leqslant x_{mn}$, then $u_{ij} \leqslant u_{mn}$ for all *ij* and *mn*.
2. $x_{ij} \leqslant x_{mn}$ and $x_{mn} \leqslant x_{kr}$ implies $u_{ij} \leqslant u_{kr}$.

(*a*) Use the concepts just presented to construct a utility table for your payoff table in Exercise 6.18.
(*b*) A utility table is analyzed in the same manner as a payoff table. Therefore, determine the decision that maximizes your expected utility.

6.22 Use your example from Exercise 6.20 and construct and analyze your utility table for it as in Exercise 6.21.

6.23 You are a degree candidate with a B average in this course. You are given a choice of whether or not to take the final exam. If you don't take the exam, your average will be reduced by 1 point. If you take the exam and it is hard, your grade will drop to a D; if it is average, your grade will remain at B; and if it is easy, you can raise your grade to an A.
(*a*) Using the four-point system (A = 4, B = 3, C = 2, D = 1, F = 0), set up the payoff table for this situation.
(*b*) What decision would you make if you are a pessimist?
(*c*) What decision would you make if you are an optimist?
(*d*) Determine the opportunity loss table.
(*e*) In checking fraternity files, you find that in the past your instructor has given hard exams 70% of the time, average exams 20% of the time, and easy exams 10% of the time. What action would you choose according to the expected value criterion?
(*f*) What is the point value of perfect information?
(*g*) The proportions of difficult questions on hard, average, and easy exams are, respectively, 0.6, 0.3, and 0.2. If you were able to see one question selected at random from the final to determine whether or not the question was difficult, how much would this privilege be worth to you in point value?

6.24 Refer to Exercise 6.23. Assume that you have been interviewing for jobs and have an offer from a company in which you are highly interested. The actual offer, though, is contingent on your grade in this course. If you get an A, the offer is for $30,000 a year in the Honolulu office; a B will get you $15,000 in the Minneapolis office; a C earns you $11,500 and a position as manager of the one-person office in Buzzard's Roost, Utah; and with any other grade, you can forget it. Construct the payoff table in light of this additional information and use it to answer the questions of Exercise 6.23.

6.25 Construct a utility table for Exercise 6.24 and find the decision that maximizes your expected utility.

6.26 A linear utility function gives the same decisions in terms of maximizing expected value as does the initial payoff. Utility functions that are linear indicate that the person is indifferent to risk. On the same graph, sketch utility curves for a person who is
(*a*) A pessimist (risk evader).
(*b*) An optimist (risk taker).

(c) Graph your utility curves for Exercises 6.21 and 6.25 and comment.

Level 3

6.27 A bank has repossessed two automobiles valued at \$2000 and \$3500, respectively. The bank intends to auction off both automobiles with the use of sealed bids. If you and one other person are the only people submitting bids and each of you has \$4000 available, what should each of you submit for bids if each uses the maximin criterion? (If you bid x on the \$2000 auto, then your bid on the \$3500 auto is not more than $4000 - x$.)

6.28 Retail outlets occasionally confront the Bayesian decision problem of a one-time order for a specialty item that has value for only one month or so during a particular season. Suppose that p_i denotes the relative frequency probability for a demand for i items as calculated from sales records for similar items sold by the outlet. Also, suppose that it is possible to order i items, where $0 < i \leqslant n$. Determine an optimal ordering policy based on the MEV criterion if W is the wholesale price per unit, R is the retail price per unit, and S is the salvage value per unit.

6.29 Fine Tuned, manager of the Magic Light bookshop, wants to place a one-time order for *Uranus Worshipers: Bandog Cult of the Occult World.* Ms. Tuned can order any number between 0 and 500 of the books and figures that x of them will be sold at the retail price of \$4.00, with probability $\text{pr}(X = x)$, in which X is binomial with parameters 500 and $p = 0.60$. She can buy each copy for \$2.00, and has arranged to sell any unsold copies to regional law-enforcement agencies for 75% of the retail price. How many books should she order? [*Hint*: See Exercise 6.28.]

Chapter 7
Sample Data and Statistics

KEY WORDS AND TERMS

Population
Random Sample
Statistic
Stratified Sampling
Cluster Sampling
Mean
Median
Range
Variance
Standard Deviation
Histogram

7.1 INTRODUCTION

A common problem in business decision making arises when we need information about a collection of objects but find that the cost of obtaining the information is prohibitive. For instance, suppose we need to know the average shelf life of current inventory. If the inventory is large, the cost of checking records for each item might be high enough to cancel the benefit of having the information. On the other hand, a hunch about the average shelf life might not be good enough for decision-making purposes. This means we must arrive at a compromise that involves selecting a small number of items and calculating an average shelf life as an estimate of the average shelf life of *all* items in inventory. This is a compromise, since the measurements for a sample from the inventory will produce only an estimate of the value we want, *but* at a substantial savings.

What we would like to know is how "good" the estimate is and how much more will it cost to make it "better." Information of this type is intimately related to how the test items are selected and how the measurements on the selected items are manipulated mathematically to estimate a value for the overall inventory. This chapter begins our study of methods for obtaining measurements so that we can arrive at valid conclusions. We also will be able to calculate the accuracy and confidence we can expect from the methods used to obtain the measurements. This chapter begins the investigation with common methods of measuring data location and dispersion.

7.2 STATISTICS AND POPULATIONS

Our presentation of sample statistics requires a knowledge of some definitions of various aspects of the environment in which measuring and calculation is done. The most fundamental of these is the population.

DEFINITION 7.1 The total set of objects for which information is desired and from which items are to be selected for measurement is called the *population.*

The word *objects* in this definition could mean physical items, such as boxes of cereal, nuts and bolts, people, or machines, or it could mean the set of possible measurements for physical items, such as the weights of all the cereal boxes filled by a certain production line or the weights of the people who step out of an elevator. In any case, we must be able to determine whether or not a selected item is from the population of interest. If we want to estimate the tons of steel in inventory produced over the last month, we need to know how to tell whether a quantity of steel in inventory was produced in the last month. If this identification is not always possible, reliable population measurements may be difficult to obtain.

DEFINITION 7.2 A *sample* is any subset of items from the population. If the population is finite, a sample of size n is called a *random sample* whenever all samples of size n are equiprobable. If the population is infinite, a sample is called a random sample if the chance variation of each observation is governed by the same probability distribution.

DEFINITION 7.3 Any measurement obtained from a set of items selected from the population of interest is called a *statistic.*

We use sampling to strike a balance between the reliability or accuracy of statistics gathered and the cost of obtaining that accuracy. For instance, the cost of the staff needed to gather data on the weight of cereal boxes may be prohibitive, and yet sampling the population of such weights introduces the question of how reliably the sample statistics represent the population.

Although there are many different ways of obtaining a sample from a population, all such methods work toward reducing the cost of data collection while providing reliable results. Some of these are discussed in the next section.

EXERCISES

7.2.1 A factory has three production lines, and management wants an estimate of the overall fraction of the items produced that are defective. For convenience the production of only one of the three lines is sampled in order to get the estimate. What is the population that is sampled? Explain.

7.2.2 A bridge commission wants to estimate the average weight of trucks traveling on a bridge that connects two major metropolitan areas. What criteria can you list that could be used to identify a member of the population of interest that should be weighed as part of the sample?

7.2.3 Many populations can be identified from the cards in an ordinary 52-card deck. List several ways in which the deck can be partitioned into subpopulations from which a sample of size five could be taken.

7.3 COMMON SAMPLING METHODS

There *are* some rather curious and innovative methods of sampling that do not lend themselves easily to analysis. Among these would be the pick and hope method, a classification that includes all intuitive sampling routines in which one selects according to an unstructured pattern and hopes that the results are representative of the population. However, we suggest choosing a variation of one of the following three alternatives, which represent the most popular methods.

7.3.1 Random Sampling

Random sampling is probably the most popular sampling method used in decision making today. Many decisions are made, for instance, by choosing a number out of a hat or a numbered bead from a barrel, and both of these methods are attempts to achieve a random choice from a set of items. But true random sampling is difficult to achieve without the aid of a computer or a random number table (whose values are usually generated by computer). The values in Table 7.1 are from a random number table, and the following example shows how to use it.

EXAMPLE 7.1 Suppose there are 5000 items in an inventory and we want to choose 50 of these at random. To do this we assign one of the numbers 1 through 5000 to each of the items in inventory. We then proceed through the random number table, *starting anywhere*, and systematically choose 50 *different* four-digit numbers. For simplicity, we will start with the upper left-hand value in Table 7.1

Table 7.1 Table of 200 Random Numbers

67323	57839	61114	62192	47547	58023	64630	34886	98777	75442
09255	13986	84834	20764	72206	89393	34548	93438	88730	61805
36304	74712	00374	10107	85061	69228	81969	92216	03568	39630
15884	67429	86612	47367	10242	44880	12060	44309	46620	55105
18745	32031	35303	08134	33925	03044	59929	95418	04917	57596
72934	40086	88292	65728	38300	42323	64068	98373	48971	09049
17626	02944	20910	57662	80181	38579	24580	90529	52303	50436
27117	61399	50967	41399	81636	16663	15634	79717	94696	59240
93995	18678	90012	63645	85701	85269	62263	68331	00389	72571
67392	89421	09623	80725	62620	84162	87368	29560	00519	84545
04910	12261	37566	80016	21245	69377	50420	85658	55263	68667
81453	20283	79929	59839	23875	13245	46808	74124	74703	35769
19480	75790	48539	23703	15537	48885	02861	86587	74539	65227
21456	13162	74608	81011	55512	07481	93551	72189	76261	91206
89406	20912	46189	76376	25538	87212	20748	12831	57166	35026
09866	07414	55977	16419	01101	69343	13305	94302	80703	57910
86541	24681	23421	13521	28000	94917	07423	57523	97234	63951
10414	96941	06205	72222	57167	83902	07460	69507	10600	08858
49942	06683	41479	58982	56288	42853	92196	20632	62045	78812
23995	68882	42291	23374	24299	27024	67460	94783	40937	16961

and move down each successive column, looking at the first four digits in each table entry. If the entry gives a number greater than 5000, subtract 5000 from it to get the number. If the four-digit number we obtain (either directly from the table or by using the subtraction procedure) is already on the sample list, move on to the next table entry. This system gives the following 50 numbers that indicate the items to be included in a random sample of size 50 from a population of 5000 items in inventory.

1732*	491	783	1226	1111*
925	3145*	1398	2028	3483*
3630	1948	2471*	2579*	37
1588	2145	1742*	1316	3661*
1874	3940*	3203	2091	3530
2293*	986	4008	741	3829
1762	3654*	294	2468	96*
2711	1041	1139*	4694*	4001*
4399*	4994	1867	668	962
1739*	2399	3942*	1888*	3756

*Number is Table 7.1 entry minus 5000.

It should be understood that such systematic procedures produce results that are consistent with the definition of a random sample given in Chapter 2. That is, our 50 values represent one of the $\binom{5000}{50}$ possible samples of size 50 from the inventory, each of which has a probability of $1/\binom{5000}{50}$ of being chosen.

7.3.2 Stratified and Cluster Sampling

One characteristic of random sampling methods that some people find disturbing is that these methods allow the so-called oddball sample to occur. This term refers to the possibility that the sample information may not be representative—for example, its average may be much smaller than that for the population—simply because, albeit by chance, the smallest values in the population were chosen as the sample. The advantage of random sampling is that we can easily calculate the probability of such an odd event. In fact, this type of calculation provides the foundation for much of what is done in elementary statistical decision analysis. Although corresponding calculations for stratified and cluster sampling are more difficult to obtain, these methods offer either some assurance that subpopulations will be represented in the sample (with stratified sampling) or substantial reductions in sampling costs (with cluster sampling).

Stratified sampling can be used whenever the population can be partitioned into smaller subpopulations, each of which is homogeneous according to the particular characteristic of interest. If there are k subpopulations and we let N_i denote the size of subpopulation i, N denote the overall population size, and n denote the sample size, then we select a stratified sample whenever we choose

$$n_i = n\left(\frac{N_i}{N}\right)$$

items at random from subpopulation i, $i = 1, 2, \ldots, k$.* The next example illustrates the use of stratified sampling.

EXAMPLE 7.2 The manager of Wine When Wanted, Inc., wants to know the average length of time between the production of his wine and its sale at retail outlets. When he looks at the records, he finds that very dry, dry, semisweet, sweet, and very sweet wines account for 12%, 32%, 42%, 12%, and 2% of sales, respectively. If he wants a sample of size 200 and decides on a stratified sample in which the

*There are many types of stratified sampling, and the reader is urged to explore standard sampling texts in this regard. The type we outline is a proportional allocation stratified sample.

strata are defined by the character of the wine, he should check the records on 24, 64, 84, 24, and 4 bottles, respectively, of each type of wine, chosen at random from each stratum.

Cluster sampling can be used whenever the population is homogeneous but can be partitioned. In many applications the partitioning is a result of physical distance. For instance, in the insurance industry, there are small "clusters" of employees in field offices scattered about the country. In such a case a random sampling of employee work habits might require travel to many of the "clusters" or field offices in order to get the data. Totally sampling each one of a small number of clusters chosen at random can eliminate much of the cost associated with the data requirements of management. The next example illustrates how this can be accomplished.

EXAMPLE 7.3 The management of an accounting firm wants to administer a standardized test to estimate the average performance level for accountants in its employ. The firm has 15 offices scattered throughout the country, and it is discovered that the recommended random sample size of 200 test results requires data from most of the 15 offices. Collecting these data would consume more time and money than management is willing to spend, and so they decide to use cluster sampling. Each office contains 40 accountants, and so five field offices are selected at random from the 15 and all 40 employees in each of the five field offices selected are tested in order to obtain the 200 test results.

The homogeneous assumption is an important one for cluster sampling. That is, the results may not be representative unless we can assume, referring to Example 7.3, that the performance level on the test would be about the same for most accountants in the firm's employ.

Our discussion of random, stratified, and cluster sampling methods does not do justice to the wide variety of methods available for general applications of statistical analysis. However, we anticipate that these methods will be suitable for most situations that require representative sampling from a population of interest.

Now that we have an indication of how to collect the data, we need to know how to manipulate it in order to obtain information on the original population. The most common measures of data population characteristics are given in the next two sections.

EXERCISES

7.3.1 Explain how you would use a random number table to select 12 cards at random from an ordinary deck of 52 cards.

7.3.2 Provide an example of a situation in which a cluster sample is more appropriate than a stratified sample.

7.3.3 The management of a chain of retail stores scattered throughout the country wants to estimate the average purchase amount per customer visit. Discuss how the sampling for this estimate might be done and the conditions under which it would be most appropriate.

7.3.4 Stratified sampling is to be used to provide a sample of size 50 under the following conditions: Population size of 10,000 and five strata of sizes 2053, 1925, 2085, 2018, and 1919. How many items should be sampled at random from each stratum?

7.4 MEASURES OF LOCATION

Without question, the most widely used measure of location is the arithmetic mean, or average value. Typically, the mean is denoted by placing a small horizontal bar over a suitable letter of the alphabet, as in $\overline{X}$, pronounced "X bar." For instance, the average for the set of data

$$\{7, 4, 3, 5, 6, 3, 7, 4\}$$

is determined by adding all values and dividing by the total number of observations. In general the formula is given by

$$\overline{X} = \left(\sum_{i=1}^{n} X_i \right) / n \tag{7.1}$$

which for these data gives (using $\overline{x}$ for a specific value)

$$\overline{x} = \frac{7+4+3+5+6+3+7+4}{8} = 4.875$$

At this point we should note the following definition, which distinguishes the arithmetic mean *of a sample* from the arithmetic mean *of the population.* Recall that the population mean is given by

$$\sum_{i=1}^{n} x_i \, \mathrm{pr}(X = x_i)$$

for a finite population of size N. The arithmetic mean of the population is calculated by letting $\mathrm{pr}(X = x_i) = 1/N$.

DEFINITION 7.4 The arithmetic mean calculated from a set of sample statistics is called the *sample average.*

Note that the mean always displays the following two characteristics:

1. It is unique (that is, there is one and only one mean for a given set of data points).
2. It may or may not be one of the original data points (for instance, the result 4.875 given in the previous example is not one of the values from which the mean was computed).

There are many situations in which sample statistics are provided in tabular form. For instance, suppose that data on the size of insurance claims in a given year for a particular insurance company are provided in the following tabular form:

Size of Claim			
Lower Limit	Class Mark	Upper Limit	Number of Claims
0	2,999.995	5,999.99	45
6,000	9,499.995	12,999.99	35
13,000	18,999.995	24,999.99	15
25,000	37,499.995	49,999.99	5

Thus, 45 out of the 100 claims were less than $6000, 35 were at least $6000 but less than $13,000, and so forth. In such a case, the data are said to be in *grouped* form, that is, they are categorized by numerical value. Here the data are separated into four *classes* with the *lower class limit* for each class given in the first column, the *class mark* for each class given in the second column, and the *upper class limit* for each class given in the third column. A class frequency is displayed in the Number of Claims column. Note that we do not know the individual amount of each claim. The sample average calculated for the grouped data is given by

$$\overline{x} = \frac{(2999.995 \times 45) + (9499.995 \times 35) + (18{,}999.995 \times 15) + (37{,}499.995 \times 5)}{100}$$

$$= 9399.99$$

in which the class mark for each class is multiplied by the frequency for that class. In general the formula for calculating the sample average for grouped data is given by

$$\overline{X} = \frac{\sum_{k=1}^{c} m_k f_k}{n} \tag{7.2}$$

where we assume that there are c classes with class marks $m_1, m_2, m_3, \ldots, m_c$ whose respective frequencies are $f_1, f_2, f_3, \ldots, f_c$ (so that $f_1 + f_2 + f_3 + \cdots + f_c = n$).

Another common measure of location is the *median*. To find the median for a set of observations, we first order the data points from smallest to largest, then find the middle value. For instance, if the values are

$$1, 3, 5, 1, 6$$

then reordering from smallest to largest gives

$$1, 1, 3, 5, 6$$

so that the median, or middle value, is 3. Technically, the median is any number greater than at most half of the values and less than at most half of the values. Thus, 2 could not be the median because 2 is less than three of the values (which is more than one-half of them). Similarly, 4 could not be the median because 4 is greater than three of the values (which is more than one-half of them). Thus, 3 is the only number that will satisfy our requirements.

This implies that the median value will be unique for any set of values consisting of an *odd* number of data points. On the other hand, notice that the median for the following set of ordered data

$$\{3, 4, 4, 5, 6, 7\}$$

is given by *any* number *between* 4 and 5. This will be the case whenever there are an *even* number of data points to start with and the two middle values are not the same. Although any number between 4 and 5 satisfies our requirements for the median value, we agree to compute the median value as one-half the magnitude of the two middle values. Thus for an even number of values, the median is the average of the two middle values (whether the two middle values are the same or not).

The median has two important characteristics that we should note:

1. The median may or may not be unique.
2. The median may or may not be one of the data values.

We also can calculate the median for grouped data. We first locate the class in which the median for the *ungrouped* data would fall. For our data on insurance claims, this would be the *second* class, since less than half the values are in the first class and more than half the values are in the first two classes. The class in which the ungrouped median falls is called the *median class*, mc. In general the formula for the grouped median m is given by

$$m = \text{mclb} + \text{wmc}\left[\frac{(n/2) - \sum_{i=1}^{\text{mc}-1} f_i}{f_{\text{mc}}}\right] \tag{7.3}$$

where mclb is the median class lower bound*, wmc is the width of the median class, n is the total number of data points, and f_{mc} is the frequency of the median class.

For the insurance claims information, we have

$$\text{wmc} = 12{,}999.99 - 5999.99 = 7000$$

$$\frac{n}{2} = 50$$

$$\sum_{i=1}^{mc-1} f_1 = \sum_{i=1}^{2-1} f_i = f_1 = 45$$

$$f_{mc} = f_2 = 35$$

Thus,

$$m = 5999.995 + 7000\left(\frac{50 - 45}{35}\right) = 6999.995$$

As a final calculation we will compare grouped and ungrouped means and medians for a set of data that represent the size in thousands of dollars of 35 randomly selected claims from the population of last year's claims for a particular insurance company.

These measurements are given in Table 7.2 in ungrouped form and in Table 7.3 in grouped form.

Table 7.2 Claims Amounts for Insurance Claims Sample ($1000) (Ungrouped Form)

1.7	2.2	3.7	5.2	8.2
1.3	6.5	5.1	1.2	0.3
3.5	1.2	0.8	6.8	6.9
3.4	5.5	3.3	4.1	2.8
8.2	3.9	7.8	1.1	2.7
6.5	4.4	4.5	2.4	1.2
2.9	1.1	1.9	6.1	1.3

*The class bound is halfway between the upper limit of one class and the lower limit of the next class. For our table these are: −.005, 5999.995, 12,999.995, and so forth.

Table 7.3 Claims Amounts for Insurance Claims Sample ($1000) (Grouped Form)

Class	Class Limits	Class Mark	Frequency (f_i)
1	0.1-1.59	0.845	9
2	1.6-3.09	2.345	7
3	3.1-4.59	3.845	8
4	4.6-6.09	5.345	3
5	6.1-7.59	6.845	5
6	7.6-9.09	8.345	3

Table 7.4 shows the mean and median for the grouped and ungrouped data, where the grouped mean is given by

$$\overline{X} = \frac{\sum_{i=1}^{c} m_i f_i}{n}$$

$$\overline{x} = \frac{0.845 \times 9 + 2.345 \times 7 + 3.845 \times 8 + 5.345 \times 3 + 6.845 \times 5 + 8.345 \times 3}{35}$$

$$= 3.716$$

and the grouped median is calculated by observing that

$$\text{mc} = 3$$

$$\text{wmc} = 4.59 - 3.09 = 1.5$$

$$\text{mclb} = 3.095$$

$$\frac{n}{2} = 17.5$$

$$\sum_{i=1}^{\text{mc}-1} f_i = \sum_{i=1}^{2} f_i = 9 + 7 = 16$$

$$f_{\text{mc}} = f_3 = 8$$

so that

$$m = 3.095 + 1.5\left(\frac{17.5 - 16}{8}\right) = 3.376$$

Procedures for calculating the grouped mean and grouped median as estimates for the ungrouped mean and median are important because statistics are often only available in grouped form. However, it is impor-

Table 7.4 Measures of Location for the Grouped and Ungrouped Data in Tables 7.2 and 7.3

	Grouped	Ungrouped
Mean	3.716	3.706
Median	3.376	3.350

tant to note that the accuracy of this estimation is highly dependent on the number of classes used and the way in which the classes are constructed. We will discuss this problem in detail in Section 7.6.

We have deliberately avoided discussing a third common measure of location, the *mode*, which, in short, is the most frequently occurring data value in the sample. As a meaningful measure of location its value is questionable because a set of data may have no mode, one mode, or more than one mode and because, in a sense, there is no generally accepted definition of a mode. Also, one can proceed with further analysis given the mean or median, whereas the mode usually is an end in itself. For instance, if we are given the sample average, we can proceed with an analysis that will allow us to estimate how closely it approximates the population mean. A comparable analysis with the mode is difficult to obtain.

Now some final comments concerning the mean and median. First, the mean is unique, whereas the median may not be unique. Second, the median is affected less by extreme values in the sample than is the mean. Third, the grouped median usually is affected less by choice of classes than is the grouped mean.

The next step in the analysis of data is an introduction to common measures of dispersion. Notice that up to this point we have restricted our comments to measures of central tendency (location) and have not discussed measures of variability. In terms of describing data, both are essential in making decisions.

EXERCISES

7.4.1 Find the arithmetic mean and median for the following sets of data:

(*a*) $\{2, 2, 2, 2, 2\}$
(*b*) $\{6, 12\}$
(*c*) $\{1, 2, 115\}$
(*d*) $\{6, 10, 10, 10\}$
(*e*) $\{5, 10, 3000\}$

7.4.2 Find the arithmetic mean and median for the following set of data:

7	2	7	2	2
3	5	1	2	3
5	2	8	8	9
4	5	3	1	8
2	9	8	1	7

7.4.3 The numbers in Exercise 7.4.2 are presented here in grouped form. Calculate the grouped mean and median.

Class	Class Limits	Class Mark	Frequency
1	0.5-1.99	1.245	3
2	2.0-3.49	2.745	9
3	3.5-4.99	4.245	1
4	5.0-6.49	5.745	3
5	6.5-7.99	7.245	3
6	8.0-9.49	8.745	6

7.4.4 The numbers in Exercise 7.4.2 are presented here in grouped form. Calculate the grouped mean and median.

Class	Class Limits	Class Mark	Frequency
1	0.6-1.59	1.095	3
2	1.6-2.59	2.095	6
3	2.6-9.59	6.095	16

7.5 MEASURES OF DISPERSION

An example is the best way to illustrate why we need measures of variability or dispersion in addition to measures of location in order to distinguish one population from another.

Suppose we have data on the hours worked in a given week for 30 hobbits* and want to compare these data with similar hours-worked

*Hobbits are *very* industrious little beings with furry feet.

data for 20 hobbits in the same factory 15 years ago. Tables 7.5 and 7.6 contain the past and present hours-worked data, respectively.

The medians for the two hobbit groups indicate that the hours worked per week were predominantly less than the population mean 15 years ago, whereas the current hours worked per week are predominantly greater than the population mean. Thus, although the hours averaged (per hobbit) has not changed significantly, the medians indicate that hobbit work patterns are different than they were in the past. We can expand on this claim by measuring the dispersion of each set of statistics about its respective sample average.

The sample averages and medians for each set of data can be seen in Table 7.7. Notice that the average values for the sets of data are about the same.

Table 7.5 Data on Hours Worked per Week for 20 Hobbits of 15 Years Ago

40.0	55.0	67.5	37.5	50.0
62.5	105.0	85.0	87.5	45.0
120.0	70.0	90.0	65.0	102.5
42.5	52.5	125.0	95.0	75.0

Table 7.6 Data on Hours Worked per Week for 30 Current Hobbits

55.0	86.0	65.0	50.0	60.0	90.0
88.0	80.0	85.0	60.0	62.5	50.0
92.0	62.5	90.0	75.0	62.5	90.0
80.0	60.0	97.5	67.5	90.0	80.0
70.0	77.5	80.0	62.5	80.0	60.0

Table 7.7 Sample Averages and Medians for Table 7.5 and Table 7.6 Data

	Table 7.5 Data	Table 7.6 Data
Sample Average	73.625	73.617
Median	68.750	76.250

There are three common measures of dispersion or variability: the *range*, the *variance*, and the *standard deviation*. The simplest of the three measures is the range R, which is the difference between the largest and smallest values in the sample. Technically, if the sample size is n, then

$$R = X_{(n)} - X_{(1)} \tag{7.4}$$

where $X_{(n)}$ represents the largest measurement in the sample and $X_{(1)}$ represents the smallest measurement in the sample.

The respective ranges for the Table 7.5 and Table 7.6 data are 87.5 and 47.5, which gives us the added information that the number of hours worked per week is less variable now than it was 15 years ago. However, the range is highly susceptible to extreme values, so this leaves us wondering whether the high variability 15 years ago was due to only one or two hobbits who worked an unusually small or large number of hours compared with the others in their group. In such a case, we could say that, even though the ranges were quite different, the variability for the two samples would be comparable. That is, the gain in computational ease that results when we use the range to measure dispersion is offset by its susceptibility to extreme values. Thus, we may encounter situations in which it would be desirable to have the following method of measuring the dispersion of a set of data values.

The most common method of measuring variability is the sample *variance* s^2, given by

$$s^2 = \frac{\sum_{i=1}^{n} (X_i - \overline{X})^2}{n - 1} \tag{7.5}$$

where n denotes the number of values in the sample, X_i, $i = 1, 2, 3, \ldots, n$ denotes the ith value, and $\overline{X}$ is the sample average.

The variance estimates the *population variance* σ^2 (pronounced "sigma squared"), which is given by (for discrete populations)

$$\sigma^2 = \sum_{i=1}^{m} (y_i - \mu)^2 \, \mathrm{pr}(y_i)$$

where m represents the number of distinct values in the population, y_i, $i = 1, 2, 3, \ldots, m$ denotes the set of m possible values in the population, $\mathrm{pr}(y_i)$, $i = 1, 2, 3, \ldots, m$ denotes the probability that a random choice from the m values results in the value y_i, and μ (mu) denotes the population mean, given by

$$\mu = \sum_{i=1}^{m} y_i \, \mathrm{pr}(y_i)$$

Here we are assuming that the population of possible values is finite; that is, that it can be listed in a counting fashion. Such a population is said to be *discrete*. In our example of hours worked per week, the population is discrete (assuming no fractional hours of work) and is given by

$$\{0, 1, 2, \ldots, 168\}$$

because a hobbit could choose to not work (giving a zero value) but could not work more than the total number of hours in a week, 168. Here

$$\begin{aligned} y_1 &= 0 \\ y_2 &= 1 \\ y_3 &= 2 \\ &\vdots \\ y_{169} &= 168 \end{aligned}$$

so that $m = 169$. Further, $\mathrm{pr}(y_{30})$ would denote the probability that a hobbit works 29 hours in any given week. Sometimes we need to estimate these probabilities from past data, sometimes there are mathematical procedures and counting methods that guide us, and at other times the probabilities are obvious. For instance, if company policy limits the number of hours per week per hobbit to 40, then

$$\mathrm{pr}(y_i) = 0 \qquad \text{for } i \geqslant 42$$

Clearly, the individual probabilities may be difficult to obtain, which is one of the reasons we need estimators for the population variance.

The sample variances for Tables 7.5 and 7.6 have been calculated and are, respectively, 709.52 and 190.18. Thus, we can see that the hours worked 15 years ago were highly variable about their average value, in comparison with the current situation. Perhaps the union has tried to put the hours worked per week per hobbit on a more equitable basis than in the past.

The *sample standard deviation* is our last measure of dispersion. It is simply the square root of the sample variance. The sample standard deviation s is used to avoid the interpretive difficulties of the sample variance, which is expressed in squared units.

We also can compute the range, variance, and standard deviation for grouped data as we did for the mean and median. Briefly, the calculation for the range is given by

$$R = \text{UL}_c - \text{LL}_1 \tag{7.6}$$

where c denotes the number of classes, UL_c is the upper limit of the cth class, LL_1 is the lower limit of the first class, and the classes are assumed to be labeled according to a smallest to largest class-mark sequence.

The calculation for the grouped variance is given by

$$s^2 = \frac{\sum_{i=1}^{c} (m_i - \overline{X})^2 f_i}{n - 1} \tag{7.7}$$

where c denotes the number of classes, $\overline{X}$ denotes the grouped mean, m_i is the mark of the ith class, $i = 1, 2, 3, \ldots, c$, f_i is the frequency of the ith class, $i = 1, 2, 3, \ldots, c$, and n is the total number of data points. The grouped standard deviation is the square root of the grouped variance.

The data from Table 7.6 are expressed in grouped form in Table 7.8 to illustrate the calculation of R, s^2, and s.

First, we calculate the grouped range:

$$R = 100.0 - 47.6 = 52.4$$

Second, we need to find the grouped mean and grouped variance. They are given by

$$\overline{x} = \frac{(52.8 \times 3) + (63.3 \times 10) + (73.8 \times 3) + (84.3 \times 8) + (94.8 \times 6)}{30}$$

$$= 75.2$$

Table 7.8 Data on Hours Worked per Week for 30 Current Hobbits (Grouped Form)

Class	Class Limits	Class Mark	Frequency (f_i)
1	47.6- 58.0	52.8	3
2	58.1- 68.5	63.3	10
3	68.6- 79.0	73.8	3
4	79.1- 89.5	84.3	8
5	89.6-100.0	94.8	6

and

$$s^2 = \frac{(52.8 - 75.2)^2 \times 3 + (63.3 - 75.2)^2 \times 10 + (73.8 - 75.2)^2 \times 3}{30 - 1}$$

$$+ \frac{(84.3 - 75.2)^2 \times 8 + (94.8 - 75.2)^2 \times 6}{30 - 1}$$

$$= \frac{1505.28 + 1416.10 + 5.88 + 662.48 + 2304.96}{29}$$

$$= 203.27$$

Consequently,

$$s = \sqrt{203.27} = 14.257$$

A complete contrast of the two data sets used in our comparison is given in Table 7.9.

From this information we observe that

1. The average number of hours worked per week per hobbit is about the same as it was 15 years ago.
2. More hobbits were working less than the average number of hours 15 years ago, whereas more hobbits are working more than the average number of hours currently.
3. There was about twice as much spread in the number of hours worked per week in the past compared with the current hours-worked patterns.
4. The number of hours worked per week per hobbit is less variable now than it was in the past.

In fact, using the empirical law of Section 3.6, we can estimate that approximately 68% of the current hobbits work 73.617 ± 13.791 hours

Table 7.9 Measures of Location and Dispersion for Tables 7.5 and 7.6

	Data for 20 Hobbits 15 Years Ago	Data for 30 Hobbits	Grouped Data for 30 Hobbits
Average	73.625	73.617	75.200
Median	68.750	76.250	75.550
Range	87.500	47.500	52.400
Variance	709.520	190.180	203.270
Standard deviation	26.637	13.791	14.257

Table 7.10 Comparison of Population Dispersion Estimates for Two Hobbit Groups

Approximate Percentage of Hobbits	Hours-Worked Interval 15 Years Ago	Current Hours-Worked Interval
68%	46.988 to 100.262	59.826 to 87.408
95%	20.351 to 126.899	46.035 to 101.199
99%	00.000* to 153.536	32.244 to 114.990

*The value of zero is used here because the negative lower limit given by the formula has no practical relevance.

per week, whereas the comparable interval 15 years ago was 73.625 ± 26.637. Similar comparisons can be made at the 95% and 99% levels as shown in Table 7.10.

This completes our introduction to common measures of location and dispersion. Although our discussion has been extensive, we have restricted the presentation to methods that describe the sample observations. These methods do not calculate how accurately the sample represents the characteristics of the population from which it was drawn. This *very* important concept will be explored in the next chapter, which deals with estimation methods.

EXERCISES

7.5.1 Find the range, variance, and standard deviation for the following sets of data:

(*a*) {2, 2, 2, 2, 2}
(*b*) {1, 2, 3, 4, 5, 100}
(*c*) {1, 100, 2, 88, 3, 75}

7.5.2 Find the range, variance, and standard deviation for the following set of data:

72	63	23	25	11
37	12	98	29	17
52	88	77	71	18
45	31	35	83	28
92	81	42	22	51

7.5.3 The numbers in Exercise 7.5.2 are presented here in grouped form. Calculate the grouped range, variance, and standard deviation.

Class	Class Limits	Class Mark	Frequency
1	11-25.9	18.45	7
2	26-39.9	32.95	5
3	40-54.9	47.45	4
4	55-68.9	61.95	1
5	69-83.9	76.45	5
6	84-98.9	91.95	3

7.6 HISTOGRAMS

Visual displays add a valuable dimension to the description of data sets and populations from which the data are produced. Histograms (a type of bar graph) are a popular method of producing visual displays that are heavily used in most management information systems.

It should be understood that the treatment of the data in the manner we are about to describe is independent of the random sample assumption. Histograms are simply a convenient method of reducing data to a form that gives us information about the underlying process that generated the data.

The construction of a histogram uses some of the concepts outlined in our discussions of grouped data. We can see how these concepts are used by providing a detailed analysis of some computer installation data representing the number of jobs processed over a 200-day period. The initial results of data collection are contained in Table 7.11 (see page 198).

The first step in the construction is to order the data beginning with the smallest and ending with the largest value. This is done in Table 7.12 (see page 199).

The second step is to choose class limits for the histogram. The choice of these limits is critical in that it determines the value of the resulting histogram. The point is well taken if we examine the information obtained in two extreme cases, one of which uses one class whose end points are 0.5 and 388.5 and the other of which uses 388 classes beginning at 0.5 and ending at 388.5. In between these two extremes there should be a more reasonable choice for the class limits. First note that you must make three decisions that together will determine the classes for your histogram. These are to

Table 7.11 Number of Jobs Processed by a Computer Installation on a Per Day Basis over a 200-Day Period

72	78	62	16	40	124	16	112	79	87
257	26	44	18	71	69	106	163	76	54
55	25	98	17	99	42	89	194	9	31
66	59	77	92	31	85	22	131	155	8
294	60	91	30	35	55	92	146	155	158
346	64	138	128	19	88	68	105	157	229
210	34	9	53	1	21	72	84	100	93
388	106	44	140	153	69	77	106	89	2
86	30	41	157	106	29	21	123	80	78
45	70	71	30	180	2	197	4	72	102
40	41	73	16	48	70	151	17	52	27
22	34	10	26	34	180	112	151	33	57
95	98	51	117	130	24	8	103	41	29
85	107	56	140	104	74	6	174	157	52
32	84	23	13	133	132	131	198	64	73
93	50	8	144	122	123	38	212	54	197
61	48	91	228	139	21	105	116	67	54
137	57	157	198	75	184	24	84	22	30
376	25	56	154	134	91	112	160	36	46
98	118	126	142	100	155	150	161	62	60

1. Decide on the number of intervals of equal size you should use
2. Determine a first class lower limit
3. Determine a last class upper limit

A poor choice of any one of the values can change the appearance of the histogram drastically from that which would have resulted had the values been chosen with more care. It is difficult to provide a standard for determining the number of classes that will apply to all situations, but the following suggestions may be of some value in this regard.

One guideline for choosing the number of classes of *equal* size c between the smallest and largest amount observed is to use the value suggested by Sturges,*

$$c = 1 + 1.433 \ln n \tag{7.8}$$

*H. A. Sturges, "The Choice of a Class Interval," *J. American Statist. Association*, 21 (1926), 65–66.

Table 7.12 Computer Data from Table 7.11 Arranged from Smallest to Largest Entry

1	21	31	48	62	77	92	112	139	161
2	21	32	50	64	77	93	112	140	163
2	22	33	51	64	78	93	112	140	174
4	22	34	52	66	78	95	116	142	180
6	22	34	52	67	79	98	117	144	180
8	23	34	53	68	80	98	118	146	184
8	24	35	54	69	84	98	122	150	194
8	24	36	54	69	84	99	123	151	197
9	25	38	54	70	84	100	123	151	197
9	25	40	55	70	85	100	124	153	198
10	26	40	55	71	85	102	126	154	198
13	26	41	56	71	86	103	128	155	210
16	27	41	56	72	87	104	130	155	212
16	29	41	57	72	88	105	131	155	228
16	29	42	57	72	89	105	131	157	229
17	30	44	59	73	89	106	132	157	257
17	30	44	60	73	91	106	133	157	294
18	30	45	60	74	91	106	134	157	346
19	30	46	61	75	91	106	137	158	376
21	31	48	62	76	92	107	138	160	388

in which ln denotes the natural logarithm. For our example c equals 8.592, so that nine classes can be used in the histogram. Table 7.13 gives the suggested number of classes for a small selection of sample sizes.

Table 7.13 Suggested Number of Histogram Classes for a Small Selection of Sample Sizes

Sample Size	Suggested Number of Classes
25	6
50	7
75	7
100	8
150	8
200	9
500	10
1000	11

Table 7.14 Class Limits, Frequencies, and Relative Frequencies for Table 7.11 Data

Class	Class Limits	Frequency	Relative Frequency
1	0.0- 43.2	55	0.275
2	43.3- 86.4	57	0.285
3	86.5-129.6	40	0.200
4	129.7-172.8	30	0.150
5	172.9-216.0	11	0.055
6	216.1-259.2	3	0.015
7	259.3-302.4	1	0.005
8	302.5-345.6	0	0.000
9	345.7-388.8	3	0.015

Note that Table 7.13 only provides guidelines and that we assume the resulting histogram will be used for visual interpretation. That is, from a *management information standpoint*, the histogram will tend to reflect the overall visual character of the population as well as can be expected, given the data of the sample. The suggested number of intervals can follow another suggestion to advantage when the objective is a *statistical* analysis of the type to be discussed in Chapter 10. At that point we will also provide a method for using *unequal* class widths so that the analysis can provide more reliability than that available using equal class widths.

The choice of where to begin the first class and where to end the last class relies on the use of common sense. In our example we can let the first-cell lower limit be 0 and the last-cell upper limit be 388.8 to get the values in Table 7.14.* Notice the relative frequency column that has been added to our usual procedure. This column contains the results of dividing the respective frequencies by 200, the total number of data values in the sample.

Plotting the relative frequency for each class as a bar graph gives the histogram shown in Figure 7.1.

The plot, unlike the measures of location and dispersion of the previous sections, gives the manager a visual display of the range of the

*The limits should be close to the largest and smallest values. The choices 0.0 and 388.8 give us a uniform class width of 43.2 (other choices give repeating decimals for the class width).

Figure 7.1 Histogram Based on Nine Intervals for the Data of Table 7.11

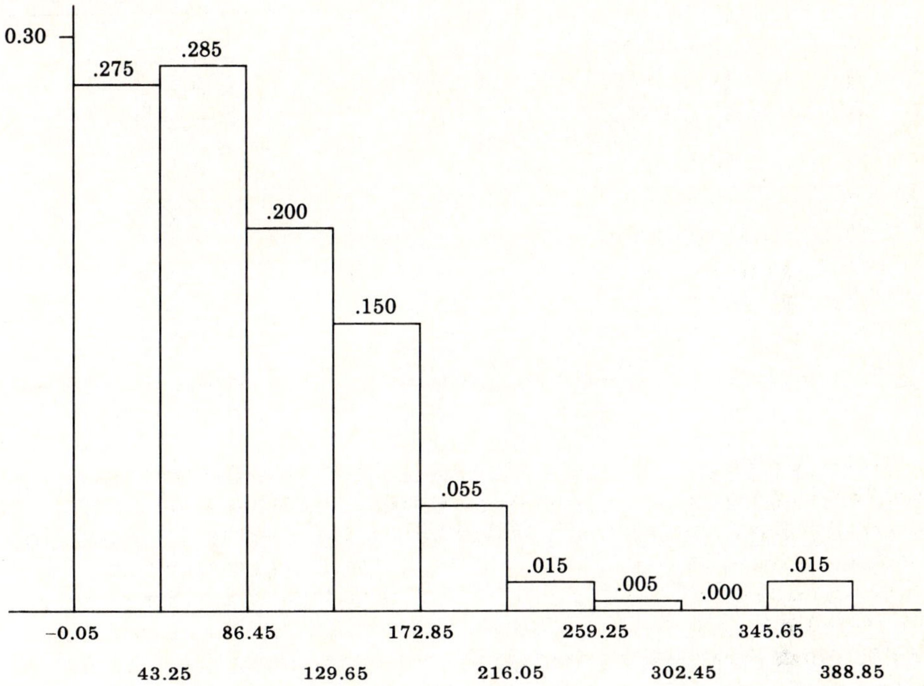

data, its central tendency, and the character of the dispersion throughout its range of values.

EXERCISES

7.6.1 Construct a histogram for the following data:

14.9	36.9	13.0	17.3	20.8
89.3	28.6	63.6	59.8	31.3
69.7	35.0	24.2	83.3	73.3
17.5	24.9	72.5	88.0	65.9
48.7	22.1	34.2	67.4	51.2

7.6.2 Construct a histogram for the following data:

5	2	4	9	5	2
4	5	6	5	1	4
2	7	1	1	7	6
7	7	4	7	3	6
4	7	7	3	9	7
9	6	9	9	8	7

7.6.3 Construct a histogram for the following data:

8.42	7.52	7.41	9.47	3.52
4.91	3.97	0.35	6.99	0.49
0.06	8.27	8.37	0.98	4.27
3.97	4.91	0.49	3.97	2.76
6.07	9.51	3.36	3.33	6.47
0.72	2.21	8.72	1.66	7.24

7.7 SUPPLEMENTARY EXAMPLES

The following 64 values are used in the supplementary examples:

98	47	92	96	36	10	7	178
37	85	30	37	74	47	43	221
76	24	68	72	11	84	80	260
115	62	44	110	49	21	117	6
269	39	6	148	86	58	154	49
231	77	82	185	23	96	191	35
192	15	20	223	61	33	227	78
153	54	58	261	73	70	264	20

EXAMPLE 7.4 Calculate the arithmetic mean, median, range, and variance for the given 64 values.

SOLUTION The arithmetic mean is given by

$$\bar{x} = \frac{5868}{64} = 91.6875$$

The median is halfway between the thirty-second and thirty-third values, which are 72 and 73, respectively. Thus the median is (72 + 73)/2 or 72.5.

The smallest value is 6 and the largest is 269, and so the range is 269 - 6 or 263.

The variance formula is

$$s^2 = \frac{\sum_{i=1}^{n} (X_i - \bar{X})^2}{n - 1}$$

We can improve this formula, from a computational standpoint, as we did in Chapter 3 for the calculation of the variance of a random variable. An equivalent expression is

$$s^2 = \frac{\sum_{i=1}^{n} X_i^2 - n(\bar{X})^2}{n - 1}$$

For the given information,

$$\sum_{i=1}^{n} X_i^2 = 885{,}034.00$$

and

$$n(\bar{X})^2 = 538{,}022.24$$

where $n = 64$, so that

$$s^2 = 5508.12$$

Thus, the sample variance is 5508.12 units squared, and the sample standard deviation is 74.22 units.

EXAMPLE 7.5 Construct a histogram for the given 64 values.

SOLUTION From Table 7.13 we see that seven classes are appropriate for 64 sample values. Let the first-class lower limit be 4.6 and the seventh-class upper limit be 270.5. In this case, the classes for the histogram, the frequency for each class, and the relative frequency for each class are as follows:

Class	Class Limits	Frequency	Relative Frequency
1	4.6- 42.5	18	0.2812
2	42.6- 80.5	20	0.3125
3	80.6-118.5	11	0.1719
4	118.6-156.5	3	0.0469
5	156.6-194.5	4	0.0625
6	194.6-232.5	4	0.0625
7	232.6-270.5	4	0.0625

The histogram is given by Figure 7.2.

EXAMPLE 7.6 Calculate the grouped median, the grouped mean, the grouped range, and the grouped variance for the information in Example 7.5.

SOLUTION We must determine the class marks for the classes of Example 7.5. These are as follows:

Figure 7.2 Histogram Based on the Data Given Above

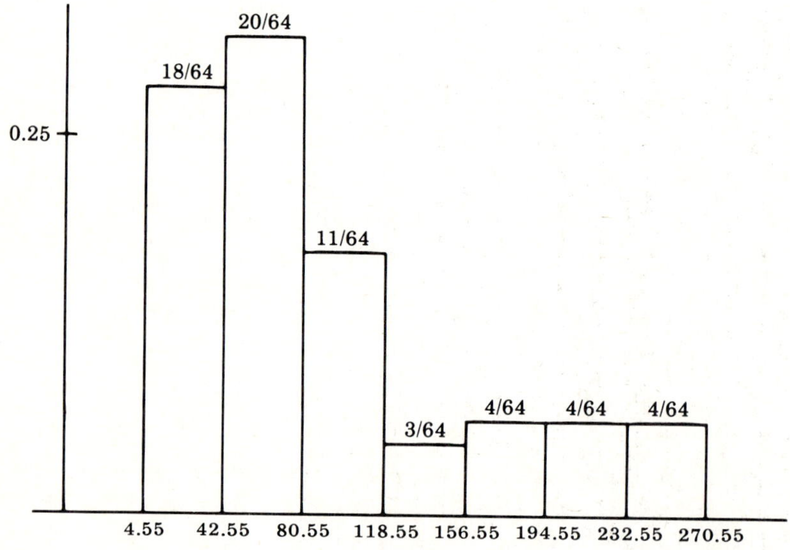

Class	Class Limits	Class Mark	Frequency
1	4.6- 42.5	23.55	18
2	42.6- 80.5	61.55	20
3	80.6-118.5	99.55	11
4	118.6-156.5	137.55	3
5	156.6-194.5	175.55	4
6	194.6-232.5	213.55	4
7	232.6-270.5	251.55	4

In this example,

$c = 7$

$m_7 = 251.55$ (mark of the seventh class)

$m_1 = 23.55$ (mark of the first class)

$\text{mc} = 2$ (median class)

$\text{mclb} = 42.55$ (median class lower limit)

$\text{wmc} = 38.0$ (width of the median class)

$n = 64$ (number of data values)

$f_{\text{mc}} = 20$ (frequency of the median class)

$$\sum_{i=1}^{\text{mc}-1} f_i = f_1 = 18$$

and so

$$m = 42.55 + 38.0\left[\frac{(64/2) - 18}{20}\right] = 69.15$$

$$\bar{x} = \frac{\sum_{i=1}^{7} m_i f_i}{64}$$

$$= \frac{23.55(18) + 61.55(20) + \cdots + 251.55(4)}{64}$$

$$= \frac{5725.20}{64}$$

$$= 89.46$$

$$R = 270.5 - 4.6$$

$$= 265.9$$

and finally

$$s^2 = \frac{\sum_{i=1}^{c} (m_i - \overline{X})^2 f_i}{n - 1} = \frac{\sum_{i=1}^{7} (m_i - 89.46)^2 f_i}{63}$$

$$= \frac{(23.55 - 89.46)^2 (18) + (61.55 - 89.46)^2 (20)}{63}$$

$$\frac{+ \cdots + (251.55 - 89.46)^2 (4)}{63}$$

$$= \frac{298{,}163.43}{63} = 4732.75$$

EXAMPLE 7.7 Suppose we have the following groups of values that correspond to items in a population we want to sample.

Group	Values
1	{90, 20, 14, 62, 82, 20, 96, 5, 76, 61}
2	{92, 26, 86, 85, 60, 3, 31, 3, 97, 13}
3	{26, 69, 97, 91, 83, 86, 44, 78, 92, 58}
4	{74, 82, 24, 58, 2, 99, 28, 61, 17, 15}
5	{30, 24, 2, 37, 33, 85, 88, 64, 26, 22}
6	{53, 43, 57, 46, 42, 53, 57, 51, 42, 46}
7	{75, 80, 96, 86, 72, 97, 71, 73, 82, 83}
8	{13, 11, 12, 4, 16, 1, 13, 3, 10, 4}
9	{27, 23, 25, 26, 25, 26, 26, 27, 28, 21}
10	{65, 65, 64, 64, 66, 62, 65, 65, 60, 62}

(*a*) If the groups correspond to strata, show how to collect a stratified sample of size 20.

(*b*) If the groups correspond to clusters show how to collect a cluster sample of size 20.

SOLUTION

(*a*) Since each group contains 10 values, a stratified sample of size 20 requires a random choice of two values from each group. Suppose we get 20 random numbers from 1 to 10 from a computer and they are, in the sequence of output, 8, 1, 4, 2, 9, 6, 6, 7, 6, 4, 2, 8, 7, 8, 2, 8, 6, 2, 9, and 1. If we use these numbers in pairs as they appear in the computer output to indicate the values to be chosen from each group, then

- The eighth and first values from group 1, or 5 and 90, are in the sample.
- The fourth and second values from group 2, or 85 and 26, are in the sample.

  .

  .

  .
- The ninth and first values from group 10, or 60 and 65, are in the sample.

In this case the stratified sample of size 20 is given by

5, 90, 85, 26, 92, 86, 99, 28, 85, 37,
43, 51, 71, 73, 11, 3, 26, 23, 60, 65

Note that we require two different random numbers from 1 to 9 in each pair, so that we have $\binom{10}{2}$ ways to choose two values at random from each stratum. Thus, there are $\binom{10}{2}^{10}$ or $(45)^{10}$ possible stratified samples of size 20.

(*b*) We need choose only two groups at random from the ten to get a cluster sample of size 20 from this population of 100 values. Thus, there are $\binom{10}{2}$ or 45 possible cluster samples of size 20. To get one of these, suppose the first two distinct random numbers from 1 to 10 output by the computer are 2 and 8. Then *all* the values in group 2 and *all* the values in group 8 would be in the cluster sample, or

92, 26, 86, 85, 60, 3, 31, 3, 97, 13,
13, 11, 12, 4, 16, 1, 13, 3, 10, 4

7.8 SUMMARY

We have reviewed some of the most common methods of collecting, describing, and presenting sample statistics. The statistics we examined were the mean and median (which are measures of central tendency, or location, of the data) and the variance, standard deviation, and range (which are measures of the spread, or dispersion, of the data about its mean). We then introduced histograms (a type of bar graph) as a method of visually characterizing the data. We should note that there are other valuable statistics and graphical methods that can be used to characterize data. For instance, there are statistics that measure the *skewness* of data (its tendency to "tail off" in one direction), and others that measure the *kurtosis* of data (its tendency to peak or lie flat). The advanced exercises in this chapter provide some of these other statistics that are useful when analyzing data. There are also other graphical means of data characterization. For instance, the frequency polygon (a type of line graph) is an alternative to the histogram as a management information tool. Finally, we should note that the median is a special case of a more general class of descriptions called *percentiles*. Formulas for calculating the percentiles for grouped and ungrouped data are given in the advanced exercises.

7.9 TABULAR CHAPTER SUMMARY

Table 7.15 contains the formulas for computing the measures of location and dispersion for ungrouped data that were discussed in the chapter. We assume n data values represented by $X_1, X_2, \ldots, X_n$ of which $X_{(i)}$ is the ith smallest (that is, X_1 is the smallest and $X_{(n)}$ is the largest).

Table 7.16 contains the formulas for computing the statistics presented in Table 7.15 for grouped data. We assume n data values and c cells with midpoints $m_1, m_2, \ldots, m_c$ and respective frequencies $f_1, f_2, \ldots, f_c$.

Table 7.15 Statistical Formulas for Ungrouped Data

Statistic	Notation	Formula
Mean	$\overline{X}$	$\dfrac{\sum_{i=1}^{n} X_i}{n}$
Median	m	$X_{((n+1)/2)}$ if n is odd
		$\frac{1}{2}[X_{(n/2)} + X_{((n/2)+1)}]$ if n is even
Range	R	$X_{(n)} - X_{(1)}$
Variance	s^2	$\dfrac{\sum_{i=1}^{n} (X_i - \overline{X})^2}{n-1}$
Standard deviation	s	$\left[\dfrac{\sum_{i=1}^{n} (X_i - \overline{X})^2}{n-1}\right]^{1/2}$

Table 7.16 Statistical Formulas for Grouped Data

Statistic	Notation	Formula
Mean	$\overline{X}$	$\dfrac{\sum_{i=1}^{c} m_i f_i}{n}$
Median	m	$\text{mclb} + \text{wmc}\,\dfrac{(n/2) - \sum_{i=1}^{\text{mc}-1} f_i}{f_{\text{mc}}}$
Range	R	$\text{UL}_c - \text{LL}_1$
Variance	s^2	$\dfrac{\sum_{i=1}^{c} (m_i - \overline{X})^2 f_i}{n - 1}$

REVIEW EXERCISES

Level 1

7.1 Give an example of a situation in which stratified sampling would be more appropriate than cluster or random sampling.

7.2 Calculate the mean and median for the following sets of data:
(*a*) {1, 2, 3}
(*b*) {−10, 2, +10}
(*c*) {1, 2, 1000}
(*d*) {7, 8, 6, 7}

7.3 Calculate the standard deviation for each of the sets of data in Exercise 7.2.

7.4 Construct a histogram for the following 64 values.

5.6	68.4	9.9	48.9	10.9	27.2	24.0	39.6
18.0	51.1	19.0	55.4	21.2	18.0	14.3	34.9
24.0	45.0	14.3	50.0	39.6	19.5	15.8	13.9
35.1	38.0	21.1	15.8	40.5	27.2	26.1	13.9
46.0	26.1	27.8	27.8	51.1	38.0	35.7	18.9
58.1	19.5	35.7	30.8	55.4	45.0	48.9	19.0
63.0	7.4	36.5	40.5	44.9	45.0	10.9	14.5
63.0	7.4	46.0	50.0	35.1	30.8	21.1	18.6

7.5 Calculate the mean and median for the 64 values in Exercise 7.4.

7.6 Calculate the range, variance, and standard deviation for the 64 values in Exercise 7.4.

7.7 Determine the class limits and class marks corresponding to the histogram of Exercise 7.4 and calculate the grouped median, grouped mean, and grouped variance on the basis of your results.

7.8 Construct a histogram for the following 90 values (note the negative signs).

-0.03	0.03	-0.02	0.02	0.04	0.00
0.01	0.10	0.07	-0.01	0.12	0.04
0.01	0.07	0.01	0.01	0.09	0.04
0.00	0.06	0.00	0.02	0.08	0.06
0.01	0.02	0.01	-0.01	0.06	0.08
0.23	0.05	0.05	0.02	0.05	0.01
0.06	-0.01	0.08	0.04	0.10	-0.07
0.05	0.07	-0.01	0.07	0.10	0.18
0.09	0.13	-0.01	0.11	0.02	0.04
0.02	0.04	0.00	0.10	0.05	0.01
0.06	0.09	-0.01	0.01	0.07	0.12
0.09	0.06	0.06	0.04	0.04	-0.25
0.02	0.02	0.03	-0.08	0.08	0.09
0.07	0.01	0.03	0.06	0.07	0.06
-0.01	0.01	0.03	0.03	0.06	-0.02

7.9 Calculate the mean, median, and variance for the 90 values in Exercise 7.8.

7.10 Determine the class limits and class marks corresponding to the histogram of Exercise 7.8 and calculate the grouped median, grouped mean, and grouped variance on the basis of your results.

7.11 Explain how you would conduct a stratified sample of size 50 for a population with strata of sizes 25, 75, 310, 40, and 500.

7.12 The following information represents typical gaps in traffic (measured in seconds) for metropolitan freeways. Construct a histogram for this information.

0.01	1.33	2.46	3.94	6.23	8.29	10.95	16.88
0.05	1.37	2.46	4.22	6.27	8.30	11.12	16.89
0.09	1.41	2.54	4.23	6.34	8.44	11.36	17.91
0.10	1.42	2.56	4.25	6.38	8.47	11.37	17.99
0.12	1.53	2.68	4.25	6.51	8.59	11.41	18.01
0.25	1.57	2.69	4.51	6.52	8.78	11.58	18.75
0.36	1.63	2.73	4.57	6.66	8.79	11.60	18.80
0.37	1.67	2.78	4.62	6.73	8.83	11.71	19.05
0.42	1.76	2.82	4.62	6.74	9.00	11.79	19.63
0.46	1.76	2.90	4.85	6.81	9.07	11.87	21.22
0.50	1.77	2.98	4.88	6.90	9.26	11.89	22.13
0.50	1.79	3.02	5.00	6.91	9.31	12.01	22.32
0.50	1.81	3.06	5.12	7.08	9.53	12.50	23.17
0.57	1.86	3.09	5.27	7.17	9.55	12.61	24.83
0.62	1.88	3.10	5.38	7.18	9.71	12.62	25.62
0.86	1.95	3.36	5.46	7.33	9.79	12.65	26.51
0.95	2.00	3.38	5.62	7.57	9.88	13.81	26.82
1.01	2.02	3.39	5.70	7.61	9.89	13.97	27.21
1.04	2.25	3.57	5.79	7.87	9.91	14.38	28.75
1.11	2.25	3.57	5.80	7.88	10.21	14.51	29.79
1.22	2.31	3.60	5.93	7.89	10.33	15.04	30.35
1.23	2.38	3.60	5.97	7.90	10.46	15.29	31.22
1.27	2.39	3.77	6.11	7.94	10.51	15.36	32.81
1.31	2.43	3.88	6.17	8.19	10.52	15.81	33.93
1.32	2.45	3.89	6.19	8.26	10.91	15.94	34.05

7.13 Calculate the mean and median for the values in Exercise 7.12.

7.14 Determine the class limits and class marks corresponding to the histogram of Exercise 7.12 and calculate the grouped median, grouped mean, and grouped variance on the basis of your results.

7.15 Find three different nonnegative integers that would represent a population with the property that the mean and variance are the same.

Level 2

7.16 Sometimes we would like to compare the dispersions of two different types of data. In such cases it is useful to use the dimensionless statistic called the *sample coefficient of variation* u, given by

$$u = \frac{s}{\overline{X}}$$

in which the sample standard deviation is computed by the formula

$$\left(\frac{1}{n}\right)\left[\sum_{i=1}^{n} (X_i - \overline{X})^2\right]^{1/2}$$

Compare the dispersions for the data sets in Exercises 7.4 and 7.8 using this method.

7.17 Calculate the sample coefficient of variation for the four data sets given in Exercise 7.2.

7.18 A measure of the degree of asymmetry about the sample average is provided by the *sample coefficient of skewness* v, given by

$$v = \frac{\left(\frac{1}{n}\right) \sum_{i=1}^{n} (X_i - \overline{X})^3}{s^3}$$

which assumes the formula for s given in Exercise 7.16. If the histogram is skewed to the right, then $v > 0$, and vice versa. In addition, $v = 0$ if the histogram is symmetric, but the condition $v = 0$ does not necessarily imply symmetry. Calculate v for the data in Exercise 7.4.

7.19 Refer to Exercise 7.18 and calculate the sample coefficient of skewness for the data in Exercise 7.8.

7.20 A measure of the peakedness of a histogram is provided by the *coefficient of kurtosis* w, given by

$$w = \frac{\left(\frac{1}{n}\right) \sum_{i=1}^{n} (X_i - \overline{X})^4}{s^4}$$

which assumes the formula for s given in Exercise 7.16. Samples from a normal population typically exhibit a value of w close to 3.0. Calculate w for the data in Exercise 7.8.

7.21 A procedure called *coding* can provide an easy way to compute measures of location and dispersion for grouped data *when the classes are of equal width*. Concisely, coding is the linear transformation of the class marks given by

$$y_i = \frac{m_i - a}{b}$$

in which y_i is the new class mark for class i, m_i is the original class mark for class i, a is one of the original class marks, and b is the uniform class width. Alternatively, we can make a more judicious choice for a and b as follows: For an even number of classes c, choose a as the boundary between class $(c/2)$ and class $(c/2) + 1$, and let b equal one-half the uniform class width. For an odd number of classes, let a equal the class midpoint for class $(c + 1)/2$ and b equal the uniform class width.

When coding is used, we can get the true grouped mean using the formula

$$\overline{X} = b\overline{Y} + a$$

where $\overline{Y}$ represents the grouped mean as calculated using the coded class marks. Use this formula to calculate the true grouped mean for the following grouped data.

Class Limits	Class Marks	Frequency
2.01- 3.50	2.755	6
3.51- 5.00	4.255	7
5.01- 6.50	5.755	5
6.51- 8.00	7.255	15
8.01- 9.50	8.755	10
9.51-11.00	10.255	2

7.22 Calculate, using coding, the grouped mean for the following grouped data.

Class Limits	Class Marks	Frequency
00.1-15.0	7.55	7
15.1-30.0	22.55	15
30.1-45.0	37.55	6
45.1-60.0	52.55	10
60.1-75.0	67.55	2

7.23 When coding is used, we can get the true grouped variance using the formula

$$s^2 = b^2 s_y^2$$

where s_y represents the grouped variance as calculated using the coded class marks. Use this formula to calculate the true grouped variance for the grouped data in Exercise 7.21.

7.24 Calculate, using coding, the true grouped variance for the grouped data in Exercise 7.22.

7.25 Calculate the grouped mean and grouped variance for the data in Exercises 7.21 and 7.22 *without using coding* and compare the results with your answers to Exercises 7.21 to 7.24.

Level 3

7.26 Let the number of values in a sample be denoted by n. The pth percentile for the sample, $0 \leqslant p \leqslant 100$, is any value y that satisfies both of the following two conditions:

1. The number of values in the sample less than y is not more than

$$n\left(\frac{p}{100}\right)$$

2. The number of values in the sample greater than y is not more than

$$n\left(1 - \frac{p}{100}\right)$$

Thus, the median is the 50th percentile. Find the 20th, 50th, and 90th percentiles for the following data.

0.33	0.49	0.88	0.92	0.92	0.95	1.06	1.07	1.09	1.15
1.20	1.24	1.30	1.30	1.36	1.50	1.79	1.88	2.01	2.19
2.42	2.98	3.33	3.58	3.71	3.76	3.92	3.98	4.27	4.33
4.40	4.46	4.64	4.94	4.98	5.12	5.16	5.50	5.66	6.18
6.54	6.56	6.81	6.98	7.86	7.88	8.15	8.23	8.29	8.59

7.27 For grouped data the pth percentile is given by

$$B_{i+1} + w_{i+1} \frac{n(p/100) - \sum_{j=0}^{i} f_j}{f_{i+1}}$$

where i is the greatest integer for which (defining $f_0 = 0$)

$$\frac{\sum_{j=0}^{i} f_j}{n} \leqslant \frac{p}{100}$$

B_{i+1} is the lower bound of the $(i+1)$st class, w_{i+1} is the width of the $(i+1)$st class, and f_j is the frequency of the jth class. Construct a histogram for the data in Exercise 7.26 and calculate the 20th, 50th, and 90th grouped percentiles on the basis of your results.

7.28 Develop a formula for calculating the sample coefficient of skewness (given in Exercise 7.18) for grouped data and apply it to the histogram of Exercise 7.8.

7.29 Prove that the coding formulas given in Exercises 7.21 and 7.23 are correct.

7.30 Let $y_i = (m_i - a)/b$ and find the relationship between the coefficient of skewness in Exercise 7.28 and its coded counterpart.

Chapter 8
Estimators and Sampling Distributions

KEY WORDS AND TERMS

Sampling Distribution
Unbiased Estimator
Efficient Estimator
Sufficient Estimator
Asymptotically Unbiased Estimator
Asymptotic Efficiency
Central Limit Theorem
Chi-Square Distribution
Student's t Distribution
Fisher's F Distribution

8.1 INTRODUCTION

The calculations for the grouped median, mean, and variance given in Chapter 7 are a proper introduction for the material in this chapter. This is because these calculations are actually estimators for the ungrouped median, mean, and variance. That is, if the data are provided in grouped form, we need to estimate (since we cannot directly calculate) the statistics for the data that were used to get the grouped information. We acknowledge this estimation by noting that if m_1 is the midpoint of the ith cell and n_i is the number of observations that fall in the ith cell, then the product

$$n_i \times m_i$$

is an estimate for the sum of the observations in the ith cell. This means that the series

$$s_1 = \sum_{i=1}^{c} n_i m_i$$

where c is the number of cells, is an estimate for the series

$$s_2 = \sum_{i=1}^{n} X_i$$

in which $x_1, x_2, \ldots, x_n$ represent the original data or observations. Thus, if s_1 is an estimate for s_2, then s_1/n is an estimate for s_2/n. We need only note that s_1/n is the grouped mean and s_2/n is the sample average to understand how the grouped mean is an estimate for $\bar{x}$.

There are many other ways in which we could estimate $\bar{x}$ from grouped data. The variety of choices brings us to the question of how to determine whether or not an estimation procedure is better than some other procedure. However, when we attempt to answer this question, we should understand that we are most often trying to estimate a *population* value rather than a sample value. Thus, we will assume that the raw, or ungrouped, data are at our disposal and that we are interested in manipulating the information to get an estimate of a population parameter of interest. Specifically, we want to examine the characteristics of typical estimation procedures wherein a sample statistic is used to estimate a population parameter. A case in point familiar to all of us is the use of the sample statistic $\bar{x}$ to estimate the normal distribution parameter μ, or the population mean.

Before proceeding, the reader should understand that although other population statistics may be of interest, the population parameter values are of *more* interest because they allow us to completely describe the population from which the sample was taken. In general, we assume that a certain probability distribution adequately describes the population under study. This distribution will always have parameters, or values that must necessarily be estimated before any interval probabilities of the type discussed in Chapter 3 can be calculated. These same parameters are then said to be the population parameters. Thus, it is of prime importance to study how we can provide a good guess for the parameters when the guess is based on our sample information. The material of this chapter provides the framework for such a study.

8.2 ESTIMATING PARAMETERS

Generally speaking, we can talk about two categories of estimation methods, *point* estimation and *interval* estimation. A point estimate is the result of using the sample information to get a single predictor (one real number) for the population parameter. The sample average and variance, $\overline{X}$ and s^2, respectively, are point estimates for the normal population parameters μ and σ^2. An interval estimate is the result of using the sample information to get an interval of real numbers that we think should contain the population parameter. Thus the interval $[\bar{x} - s, \bar{x} + s]$ is an interval estimate for the normal population parameter μ. All good interval estimates are probability statements that show how probable or likely it is that the interval will contain the population parameter of interest. Intuitively this should make sense because a sample statistic is a random variable. We can understand this latter point if we recognize that each time a random sample of size n is drawn from a population of values, there is no guarantee that a statistic calculated using the n values will be any given fixed real number (here we assume that the population size exceeds the sample size). Thus, if we know the distribution of the sample statistic, we can provide a probability statement in the form of an interval estimate. The distributions associated with sample statistics are referred to as *sampling distributions*. When we pass judgment on a particular estimator, we will be examining the features of its sampling distribution.

A perfect estimator is one whose sampling distribution has zero variance, that is, one whose sampling distribution is concentrated entirely at one point. Such an estimate would never be wrong and so would always equal the parameter it estimates. Although such perfect estimators are rare, they can occur if all items in the population produce the same sample value, that is, if the population distribution has no variance.

We need to provide some guidelines for an estimator, or to state the characteristics that are generally accepted as desirable. Perhaps the most fundamental of these desirable qualities is that of *unbiasedness*, which is defined as follows.

DEFINITION 8.1 An estimator whose expected value equals the parameter to be estimated is an *unbiased estimator*.

Remember that an expected value is an expression of the long-term average result. This means that whereas a perfect estimator gives the correct prediction every time, a nonperfect but unbiased estimator gives the correct prediction only on the average. Thus, an unbiased estimator gives predictions that are greater than or less than the parameter to be estimated, but the error in prediction will balance out as the number of estimations increases. The following example gives a familiar unbiased estimator.

EXAMPLE 8.1 The sample average $\overline{X}$ is an unbiased estimator for the population expected value μ. We can show this using the rules for expected value given in Chapter 3. The general form of the sample average is

$$\overline{X} = \frac{\sum_{i=1}^{n} X_i}{n}$$

in which X_i represents the ith observation from the population. The value of this ith observation will change from one sample of size n to another and is a random variable whose expected value is μ for any given value of i, $i = 1, 2, \ldots, n$. Thus,

$$E(\bar{X}) = E\left(\frac{\sum_{i=1}^{n} X_i}{n}\right) = \frac{1}{n}E\left(\sum_{i=1}^{n} X_i\right) = \frac{1}{n}\sum_{i=1}^{n} E(X_i)$$

$$= \frac{1}{n}\sum_{i=1}^{n} \mu = \frac{1}{n}(n\mu) = \mu$$

so that $\bar{X}$ is an unbiased estimator of μ.

Another desirable property relates to the variance of the estimator in comparison with the variance of other possible estimators. This second property is called *efficiency* and is defined as follows.

DEFINITION 8.2 An estimator is *efficient* if (a) it is unbiased and (b) its sampling distribution has the smallest variance among all unbiased estimators.

Note that in this definition we restrict ourselves to unbiased estimators in order to avoid the ambiguous situation of trying to balance bias against variance when comparing one estimator with another. This property is also known as *minimum variance* in many textbooks.

A third important property an estimator can have is called *sufficiency*.

DEFINITION 8.3 An estimator that uses all the information about the parameter contained in the sample is called a *sufficient estimator.*

Perhaps the most valuable guideline that would indicate sufficiency is the explicit use of all sample points in the calculation of the estimator. For instance, the sample average and standard deviation are sufficient estimators because they each have formulas that show the use of every sample point. Beyond this guideline the determination gets more involved. For instance.

$$\frac{X_{(n)} - X_{(1)}}{6}$$

(the sample range divided by 6, or for that matter by any constant) is a sufficient estimator for the population variance because it is impossible to determine the sample range without considering (utilizing) all the data values in the sample.

The value of sufficiency lies in the fact that

1. Sufficiency is a necessary condition for efficiency (if it is not sufficient, it can't be efficient).

2. It is difficult to test for efficiency in an estimator.

Regarding the second point, we recognize that frequently we have no idea of the minimum variance an unbiased estimator can have, but can provide only a lower bound.*

The previous properties are called *small-sample* properties of estimators. We now list what we call *large-sample* properties, which are what we look for in an estimator if it is either biased or not efficient. In order to present these properties in a concise and informative manner, we will use the following notation:

- θ (theta) will always represent the population parameter to be estimated.
- $\hat{\theta}$ will always represent an estimator for the population parameter θ.

The first large-sample property is given in Definition 8.4.

DEFINITION 8.4 $\hat{\theta}$ is an *asymptotically unbiased* estimator of θ if $\lim_{n\to\infty} E(\hat{\theta}) = \theta$.

Thus, an asymptotically unbiased estimator has an expected value that tends toward the parameter to be estimated as the sample size increases. Fundamental limit properties tell us that an unbiased estimator is also asymptotically unbiased, since $E(\hat{\theta}) = \theta$ independent of the value of n, which is the sample size.

In order to introduce the concept of asymptotic efficiency, we will use the definition for mean-squared error in conjunction with the definition for consistency.

DEFINITION 8.5 The expected value given by $E[(\hat{\theta} - \theta)^2]$ is called the *mean-squared error.*

Note that whereas the variance measures the dispersion of the distribution of $\hat{\theta}$ around its expected value, the mean-squared error measures the dispersion of $\hat{\theta}$ around the *true* value of the parameter.

DEFINITION 8.6 $\hat{\theta}$ is a *consistent* estimator of θ if $\lim_{n\to\infty} \text{pr}\{\theta - \epsilon \leqslant \hat{\theta} \leqslant \theta + \epsilon\} = 1$ for any $\epsilon > 0$.

Thus, consistency means that the sampling distribution tends to become concentrated at the true parameter as the sample size increases. This

*One such bound is called the Cramer-Rao lower bound. See C. R. Rao, *Linear Statistical Inference and Its Applications* (New York: Wiley, 1965), pp. 265ff.

means that a consistent estimator guarantees that an increase in sample size will result in better estimates because the bias and the mean-squared error both decrease. In fact, one of the reasons we have introduced the mean-squared error, MSE, is because it provides an easy way to determine consistency, since if

$$\lim_{n\to\infty} \text{MSE}(\hat{\theta}) = 0$$

then $\hat{\theta}$ is a consistent estimator of θ.

This brings us to the last property of estimators to be discussed in this chapter.

DEFINITION 8.7 $\hat{\theta}$ is an *asymptotically efficient* estimator of θ if

(*a*) $\lim_{n\to\infty} |E(\hat{\theta})| < \infty$

(*b*) $\lim_{n\to\infty} \text{Var}(\hat{\theta}) < \infty$

(*c*) $\hat{\theta}$ is a consistent estimator.

(*d*) No other consistent estimator of θ has a smaller *asymptotic variance.**

Although all this is a bit difficult to digest, it does illustrate the care required in developing good estimation procedures. Surely, any estimator that has most of these properties would be valuable. This is the case with a certain class of estimators called *maximum likelihood* estimators, which are

1. Asymptotically unbiased
2. Asymptotically efficient
3. If unbiased, then efficient

In addition, their sampling distributions tend toward the normal distribution as the sample size increases. The most familiar maximum likelihood estimator, MLE, is $\overline{X}$, which is an MLE of μ. Another is

$$\hat{\sigma}^2 = \frac{\sum_{i=1}^{n} (X_i - \overline{X})^2}{n}$$

which is an MLE of the variance of a population.

*The asymptotic variance calculation is $\left(\frac{1}{n}\right)\lim_{n\to\infty} \text{Var}(\sqrt{n}\,\hat{\theta})$.

The method of moments also provides a valuable procedure for estimating parameters when maximum likelihood estimators are difficult to determine.

EXERCISES

8.2.1 In the construction industry it is a common practice to estimate the variance of the time to project completion using the formula

$$\left[\frac{b - a}{6}\right]^2$$

in which a denotes the shortest time to completion experienced in the past and b denotes the largest time to completion experienced in the past. Discuss the good and bad aspects of using this type of estimator for the population variance.

8.2.2 The median is often used to estimate the population mean. Discuss the good and bad aspects of using this type of estimator for the population mean.

8.3 THE DISTRIBUTIONS OF SUMS OF RANDOM VARIABLES

We need to know much more than the expected value and variance of sums of random variables to efficiently produce bounds on estimation error. If possible, we would especially like to know the actual distribution of these sums. At first, it might seem strange that the distribution of sums of random variables should be so important in this area, and, indeed, in all of statistical inference. However, we need only consider the material presented in the last section to see how these sums enter the picture. Recall that efficiency is one of the qualities we look for in an estimator and that sufficiency is a necessary condition for efficiency. An estimator will be sufficient whenever all the sample values are required for its determination. Clearly, this last requirement is met whenever the estimation formula explicitly contains each sample value, and this is the case whenever the formula contains the sum of the sample values. Since each sample value is a random variable, their sum is a random variable. Look at some of the examples in the last two subsections and you can begin to understand how important it is to provide information on the distribution of sums of random variables.

8.3.1 The Central Limit Theorem

One of the most powerful tools available to us in this area is the information provided by the central limit theorem. Actually, this theorem provides us with an approximation (albeit a very good one) that relies on conceding a small number of assumptions in return for a wealth of information. We now provide a statement of the theorem followed by a less technical résumé of its content.

THE CENTRAL LIMIT THEOREM Let $X_1, X_2, \ldots, X_n$ denote a series of statistically independent random variables with means $\mu_1, \mu_2, \ldots, \mu_n$ and finite variances $\sigma_1^2, \sigma_2^2, \ldots, \sigma_n^2$. Then the distribution of the sum given by

$$\sum_{i=1}^{n} X_i$$

approaches a normal distribution independent of the distributions of the random variables that compose the sum.

This normal distribution has mean

$$\mu_n = \sum_{i=1}^{n} \mu_i \tag{8.1}$$

and variance

$$\sigma_n^2 = \sum_{i=1}^{n} \sigma_i^2 \tag{8.2}$$

Notice that the process of choosing n data values from the same population (which is the case in most practical applications) produces a series of n random variables. If we use random sampling, we satisfy the requirement of independent random variables *and* we get n random variables with the same expected value μ and variance σ^2. From Equations 8.1 and 8.2, the sum of the sample values has an expected value given by

$$\mu_n = n\mu \tag{8.3}$$

and a variance given by (assuming that the population is infinite)

$$\sigma_n^2 = n\sigma^2 \tag{8.4}$$

Thus, the assumption of random sampling from the same population in conjunction with the central limit theorem lets us say that the distribution of $\overline{X}$, the sample average, converges to a normal distribution as the sample size increases. The larger the sample size, the better the normal approximation. Equivalently, we could say that the distribution of (again assuming an infinite population)

$$\frac{(\overline{X} - \mu)\sqrt{n}}{\sigma} \tag{8.5}$$

approaches the standard normal as the sample size increases.

This last result comes from the fact that if

$$\sum_{i=1}^{n} X_i$$

is normal with expected value $n\mu$ and $n\sigma^2$, then

$$\frac{\sum_{i=1}^{n} X_i - n\mu}{\sqrt{n\sigma^2}}$$

must be normal with expected value 0 and variance 1 (see Section 5.4). Multiplying the numerator and denominator of this last fraction by $1/n$ gives the result shown in Equation 8.5. We can now hint at the power of the central limit theorem.

EXAMPLE 8.2 In a large city a candidate for mayor wants to try to judge his popularity by taking a preliminary sample poll of the voters. An agency does this for him and discovers that 320 of the 688 polled would vote for the candidate at the time of the poll. Discuss this situation from the viewpoint of the central limit theorem.

SOLUTION We assume random sampling with replacement from a Bernoulli population whose parameter is p, the proportion of voters in the city that would vote for the candidate at the time of the poll. Thus, we have the following:

1. A sequence of Bernoulli random variables, $X_1, X_2, \ldots, X_n$, each having the same parameter p
2. $E(X_i) = p$ and $\text{Var}(X_i) = p(1 - p)$ for $i = 1, 2, \ldots, n$
3. $E\left(\sum_{i=1}^{n} X_i\right) = np$ and $\text{Var}\left(\sum_{i=1}^{n} X_i\right) = np(1 - p)$

The central limit theorem, Equation 8.5, tells us that

$$\frac{(\overline{X} - \mu)\sqrt{n}}{\sigma}$$

or

$$\frac{(\overline{X} - p)\sqrt{n}}{\sqrt{p(1 - p)}}$$

is approximately standard normal in distribution. We can use this last result in conjunction with the material in Section 5.4 to tell us how close our estimate of 320/688 is to the actual proportion of voters in the city who would vote for the candidate at the time of the poll. This important aspect of estimators, called a *confidence interval*, is developed in the next chapter.

8.3.2 The Chi-Square Distribution

Before we present the chi-square distribution, we need to introduce the concept of *degrees of freedom*. Mathematically, the degrees of freedom for a sum of squares will be determined by the number of variables minus the number of independent linear restrictions placed on the variables. One linear restriction placed on the sample variance calculation

$$s^2 = \frac{\sum_{i=1}^{n} X_i^2 - \frac{(\Sigma X_i)^2}{n}}{n - 1}$$

is that imposed by $\overline{X}$. To see this, suppose we are ready to calculate ΣX_i^2 knowing that $\overline{X}$ is 50. If $n = 3$, then ΣX_i must be 150. Of the three variables X_1, X_2, X_3, any two may assume any value within their domain, but once they do so, the value for the third variable is dictated by the restriction $\Sigma X_i = 150$. For example, if we let $X_1 = 54$ and $X_2 = 44$, then X_3 must be 52 or else $\overline{X}$ could not be 50. In general s^2 has $n - 1$ degrees of freedom.

Confidence intervals for the variance of a random variable can be found once the sampling distribution of

$$s^2 = \frac{1}{n - 1} \sum_{i=1}^{n} (X_i - \overline{X})^2$$

can be determined. We will not adopt a general approach to the problem, however, but instead will focus on the useful special case in which we assume that the sample of n values is drawn from a normal population of values. Simple algebra allows us to develop the rationale as follows:

$$(n-1)s^2 = \sum_{i=1}^{n} (X_i - \overline{X})^2 = \sum_{i=1}^{n} (X_i - \mu + \mu - \overline{X})^2$$

$$= \sum_{i=1}^{n} [(X_i - \mu) - (\overline{X} - \mu)]^2$$

$$= \sum_{i=1}^{n} [(X_i - \mu)^2 - 2(X_i - \mu)(\overline{X} - \mu) + (\overline{X} - \mu)^2]$$

$$= \sum_{i=1}^{n} (X_i - \mu)^2 - 2(\overline{X} - \mu) \sum_{i=1}^{n} (X_i - \mu) + n(\overline{X} - \mu)^2$$

$$= \sum_{i=1}^{n} (X_i - \mu)^2 - 2(\overline{X} - \mu)n(\overline{X} - \mu) + n(\overline{X} - \mu)^2$$

$$= \sum_{i=1}^{n} (X_i - \mu)^2 - n(\overline{X} - \mu)^2$$

Thus,

$$\frac{(n-1)s^2}{\sigma^2} = \sum_{i=1}^{n} \left(\frac{X_i - \mu}{\sigma}\right)^2 - \left[\frac{(\overline{X} - \mu)\sqrt{n}}{\sigma}\right]^2$$

From Section 5.4 we know that each of the n terms in the sum is the square of a standard normal random variable, as is the right-hand term. One can show (using advanced mathematical techniques) that the sum of the squares of n independent standard normal random variables is a chi-square distribution with n degrees of freedom. The assumption of random sampling gives us the independence of the n standard normal distributions, so that the first term on the right follows a chi-square distribution with n degrees of freedom, written χ_n^2. In the same way

$$\left[\frac{(\overline{X} - \mu)\sqrt{n}}{\sigma}\right]^2$$

is chi-square with *one* degree of freedom. Thus, it is not surprising that since

$$\frac{(n-1)s^2}{\sigma^2} + \left[\frac{(\overline{X} - \mu)\sqrt{n}}{\sigma}\right]^2 = \sum_{i=1}^{n} \left(\frac{X_i - \mu}{\sigma}\right)^2$$

[that is, $(n-1)s^2/\sigma^2 + \chi_1^2 = \chi_n^2$], then

$$\frac{(n-1)s^2}{\sigma^2}$$

should be a chi-square random variable with $n - 1$ degrees of freedom. This is, in fact, the case, although what we have provided does not constitute a proof. (We would need to establish the independence of $(n-1)s^2/\sigma^2$ and $[(\overline{X} - \mu)\sqrt{n}/\sigma]^2$ and use the property that if $\chi_1^2 + \chi_2^2$ are independent chi-square random variables having r_1 and r_2 degrees of freedom, respectively, then $\chi_1^2 + \chi_2^2$ is a chi-square random variable with $r_1 + r_2$ degrees of freedom. A proof of this result follows directly from Cochran's theorem, which is stated in Chapter 12.)

The chi-square distribution does not have a closed form, and so tables for the calculation of areas under the curve have been included in Table A.7 in the back of the book. The notational form for the values in Table A.7 conforms to the following format:

$$\text{pr}\{\chi_{n-1}^2 \geqslant \chi_{n-1,\alpha}^2\} = \alpha$$

in which χ_{n-1}^2 denotes a chi-square random variable with $n - 1$ degrees of freedom. This means that $\chi_{n-1,\alpha}^2$ is a value that a random variable with a chi-square distribution with $n - 1$ degrees of freedom will exceed with probability α. Thus, α represents a right-hand tail area. The chi-square distribution is skewed right and nonnegative in domain, so that all possible values of χ_{n-1}^2 are nonnegative. Note that the chi-square table is set up a little differently than the standard normal table in that the distribution abscissa values are in the body of the table rather than along the perimeter. This facilitates an efficient presentation of the information, since it provides for the interaction of three rather than two variables—in a chi-square table we need to list areas, abscissas, *and* degrees of freedom. Areas are listed along the top, and degrees of freedom are listed down the leftmost column. The next two examples illustrate use of the table.

EXAMPLE 8.3 Use the χ^2 table to find

(*a*) $\chi_{15,0.9}^2$

(*b*) $\chi_{15,0.99}^2$

(*c*) $\chi_{8,0.05}^2$

SOLUTION

(*a*) $\chi_{15,0.90}^2 = 8.55$ (intersection of row 15 and the column headed by 0.90)

(*b*) $\chi_{15,0.99}^2 = 5.23$ (intersection of row 15 and the column headed by 0.99)

(c) $\chi^2_{8,\,0.05} = 15.51$ (intersection of row 8 and the column headed by 0.05)

EXAMPLE 8.4 Use Table A.7 to find values for y and w that satisfy the given probability statements.

(a) $\text{pr}\{\chi^2_6 \geqslant y\} = 0.95$

(b) $\text{pr}\{\chi^2_{20} < y\} = 0.01$

(c) $\text{pr}\{\chi^2_{18} \geqslant 7.01\} = w$

(d) $\text{pr}\{\chi^2_{10} < 2.56\} = w$

SOLUTION

(a) $y = \chi^2_{6,\,0.95} = 1.64$

(b) If $\text{pr}\{\chi^2_{20} < y\} = 0.01$, then $\text{pr}\{\chi^2_{20} \geqslant y\} = 0.99$, so $y = \chi^2_{20,\,0.99} = 8.26$

(c) Since $\text{pr}\{\chi^2_{18} \geqslant 7.01\} = w$, then $\chi^2_{18,\,w} = 7.01$, and we scan row 18 of the table until we find 7.01. The value is found in the second column, which is headed by 0.99. Thus, $w = 0.99$.

(d) If $\text{pr}\{\chi^2_{10} < 2.56\} = w$, then $\text{pr}\{\chi^2_{10} \geqslant 2.56\} = 1 - w$, so that $\chi^2_{10,\,1-w} = 2.56$. Scanning row 10 of the table, we find the value in the second column, which is headed by 0.99. Thus $1 - w = 0.99$ or $w = 0.01$.

The chi-square sampling distribution is developed here with our presentation of the central limit theorem in order to emphasize its importance as a tool in statistical analysis. Two more useful sampling distributions are provided in the next section.

EXERCISES

8.3.1 Find

(a) $\chi^2_{5,\,0.025}$

(b) $\chi^2_{9,\,0.99}$

(c) $\chi^2_{17,\,0.05}$

(d) $\chi^2_{17,\,0.95}$

8.3.2 Find values for y that satisfy the given probability statements.

(a) $\text{pr}\{\chi^2_{14} \geqslant y\} = 0.99$

(*b*) $\text{pr}\{\chi_9^2 \geqslant y\} = 0.01$

(*c*) $\text{pr}\{\chi_{25}^2 \geqslant y\} = 0.95$

8.3.3 Find values for y that satisfy the given probability statements.

(*a*) $\text{pr}\{\chi_9^2 < y\} = 0.01$

(*b*) $\text{pr}\{\chi_{25}^2 < y\} = 0.95$

(*c*) $\text{pr}\{\chi_{14}^2 < y\} = 0.99$

8.3.4 Find values for w that satisfy the given probability statements.

(*a*) $\text{pr}\{\chi_{17}^2 \geqslant 10.09\} = w$

(*b*) $\text{pr}\{\chi_{23}^2 \geqslant 35.17\} = w$

(*c*) $\text{pr}\{\chi_8^2 \geqslant 20.09\} = w$

8.3.5 Find values for w that satisfy the given probability statements.

(*a*) $\text{pr}\{\chi_{23}^2 < 10.20\} = w$

(*b*) $\text{pr}\{\chi_8^2 < 13.36\} = w$

(*c*) $\text{pr}\{\chi_{17}^2 < 33.41\} = w$

8.4 OTHER SAMPLING DISTRIBUTIONS

It is important to note that $\overline{X}$ and s^2 undergo a linear transformation from which we establish a sampling distribution. The general form of the transformation is given by

$$y' = ax + b$$

and Table 8.1 shows how the linear transformation applies to the two previous cases.

Table 8.1 Linear Transformations on Sample Statistics and the Corresponding Sampling Distributions

Statistic (x)	a	b	$ax + b$	Sampling Distribution
$\overline{X}$	$\frac{\sqrt{n}}{\sigma}$	$\frac{-\mu\sqrt{n}}{\sigma}$	$\frac{(\overline{X} - \mu)\sqrt{n}}{\sigma}$	Standard normal
s^2	$\frac{n-1}{\sigma^2}$	0	$\frac{(n-1)s^2}{\sigma^2}$	Chi-square with $n - 1$ degrees of freedom

The next sample distribution to be presented is a nonlinear combination of the sample average and the unbiased estimator of the population variance. This step actually simplifies the estimation of the population mean μ if the sampling is from a normal population. This sampling distribution is called the t distribution, and it is discussed in the following subsection.

8.4.1 Student's *t* Distribution

Observation of Table 8.1 reveals that each transformation employs the population parameter that is being estimated. This should be somewhat disconcerting, since it seems to indicate that we know the parameters we are trying to estimate! We can work around this problem in some special cases, but many situations require an alternative distribution and procedure. A distribution developed by the brewmaster W. S. Gossett gives us this type of flexibility *when we are interested in estimating a mean for a normal population.* Thus, we start with the assumption that we have a random sample from a normal population whose mean and variance are unknown. However, we look at the transformation

$$\frac{(\overline{X} - \mu)\sqrt{n}}{\sigma}$$

and see that it utilizes μ and σ, which we do not know, since they are population (rather than sample) values. The next chapter will illustrate that we do not need to know μ to develop the confidence interval; however, σ is required. When the sample size is large, the substitution of s for σ does not significantly alter the distribution from the standard normal (we will discuss why shortly). However, for small sample sizes the alteration cannot be ignored, and the true distribution of

$$t_{n-1} = \frac{(\overline{X} - \mu)\sqrt{n}}{s} \qquad (8.6)$$

(in which $s = [\Sigma(X_i - \overline{X})^2/(n-1)]^{1/2}$) must be utilized. The statistic given in expression 8.6 has a sampling distribution called Student's t distribution with $n - 1$ degrees of freedom; it is the ratio of two independent random variables, one of which has a standard normal distribution. The other is the square root of a chi-square random variable divided by its degrees of freedom. We will not establish the required independence but can show how expression 8.6 conforms to the t distribution format. Simply,

$$t_{n-1} = \frac{(\overline{X} - \mu)\sqrt{n}}{s} = \frac{(\overline{X} - \mu)\sqrt{n}/\sigma}{(s^2/\sigma^2)^{1/2}}$$

$$= \frac{(\overline{X} - \mu)\sqrt{n}/\sigma}{\{[(n-1)s^2/\sigma^2]/(n-1)\}^{1/2}}$$

If we are sampling from a normal population, the numerator in this last ratio has a standard normal distribution and $(n - 1)s^2/\sigma^2$ in the denominator has a chi-square distribution with $n - 1$ degrees of freedom.

The t distribution looks much like the standard normal distribution in that it is symmetrical and has an expected value of 0. However, its variance is always greater than 1. This variance is inversely related to the sample size and approaches 1 as the t distribution approaches the standard normal distribution (this explains the close agreement between the two for large sample sizes). The t distribution table is set up like the chi-square table, and the same abscissa notation is employed,

$$\text{pr}\{t_v \geqslant t_{v,a}\} = a$$

The following two examples illustrate the use of Table A.8, which is required because there is no closed form for the t distribution.

EXAMPLE 8.5 Use Table A.8 to find

(*a*) $t_{14,0.05}$

(*b*) $t_{14,0.95}$

(*c*) $t_{30,0.01}$

(*d*) $t_{20,0.90}$

SOLUTION

(*a*) $t_{14,0.05} = 1.761$ (intersection of row 14 and the column headed by 0.05)

(*b*) $t_{14,0.95} = -1.761$ (intersection of row 14 and the column headed by 0.05 and symmetry about 0)

(*c*) $t_{30,0.01} = 2.457$ (intersection of row 30 and the column headed by 0.01)

(*d*) $t_{20,0.90} = -1.325$ (intersection of row 20 and the column headed by 0.10 and symmetry about 0)

EXAMPLE 8.6 Use Table A.8 to find values for y and w that satisfy the given probability statements.

(*a*) $\text{pr}\{t_8 \geqslant y\} = 0.95$

(*b*) $\text{pr}\{t_{18} < y\} = 0.01$

(*c*) $\text{pr}\{t_{30} \leqslant 1.697\} = w$

(*d*) $\text{pr}\{t_{18} \geqslant -1.330\} = w$
(*e*) $\text{pr}\{t_{28} < 1.313\} = w$
(*f*) $\text{pr}\{t_{14} < -2.624\} = w$

SOLUTION

(*a*) $y = t_{8,\,0.05} = -1.860$

(*b*) If $\text{pr}\{t_{18} < y\} = 0.01$, then $\text{pr}\{t_{18} \geqslant y\} = 0.99$, and so $y = t_{18,\,0.99} = -2.552$.

(*c*) Since $\text{pr}\{t_{30} \geqslant 1.697\} = w$, then $t_{30,\,w} = 1.697$, and we scan row 30 of the table until we find 1.697. We find the value in the column headed by 0.05. Thus $w = 0.05$.

(*d*) Similarly we know that $t_{18,\,w} = -1.330$, and so we scan row 18 of the table and find the value 1.330 in the column headed by 0.10. Thus, using symmetry about 0, $w = 0.90$.

(*e*) If $\text{pr}\{t_{28} < 1.313\} = w$, then $\text{pr}\{t_{28} \geqslant 1.313\} = 1 - w$, and so $t_{28,\,1-w} = 1.313$. Using the same procedure as in parts (*c*) and (*d*), we get $1 - w = 0.10$ or $w = 0.90$.

(*f*) If $\text{pr}\{t_{14} < -2.624\} = w$, then $\text{pr}\{t_{14} \geqslant -2.624\} = 1 - w$, and so $t_{14,\,1-w} = -2.624$. Hence, $1 - w = 0.99$ or $w = 0.01$.

The final normal related distribution we will discuss is Fisher's F distribution, which is presented in the next subsection.

8.4.2 Fisher's *F* Distribution

If $\chi^2_{v_1}$ and $\chi^2_{v_2}$ are *independent* chi-square random variables with v_1 and v_2 degrees of freedom, respectively, then Fisher's F distribution is given by the ratio

$$F_{v_1,\,v_2} = \frac{\chi^2_{v_1}/v_1}{\chi^2_{v_2}/v_2}$$

The shape of the density function of the distribution is similar to that of the chi-square distribution, and, in fact,

$$\lim_{v_2 \to \infty} F_{v_1,\,v_2}$$

is a chi-square random variable. Fisher's F distribution is heavily used in analysis-of-variance procedures and in the fundamental comparison of two sample variances. This is because, according to our chi-square development, if we choose $\upsilon_1 + 1$ items at random from a normal population whose variance is σ_1^2 and get the unbiased variance estimator s_1^2, then $\upsilon_1 s_1^2/\sigma_1^2$ has a chi-square distribution with υ_1 degrees of freedom. Similarly, a random choice of $\upsilon_2 + 1$ items at random from a second normal population whose variance is σ_2^2, giving an unbiased variance estimator s_2^2, makes $\upsilon_2 s_2^2/\sigma_2^2$ conform to a chi-square distribution with υ_2 degrees of freedom. If these two random variables are independent, the ratio

$$\frac{(\upsilon_1 s_1^2/\sigma_1^2)/\upsilon_1}{(\upsilon_2 s_2^2/\sigma_2^2)/\upsilon_2}$$

that is

$$\frac{s_1^2/\sigma_1^2}{s_2^2/\sigma_2^2}$$

must have an $F_{\upsilon_1, \upsilon_2}$ distribution. Typically we would say that this ratio has "an F distribution with υ_1 and υ_2 degrees of freedom," meaning that υ_1 is the degrees of freedom for the chi-square random variable in the numerator and υ_2 is the degrees of freedom for the chi-square random variable in the denominator. Note that if we assume that $\sigma_1^2 = \sigma_2^2$, the latter ratio is simply s_1^2/s_2^2, a very convenient and heavily used ratio.

Before we show how to use the tables (yes, this distribution does not have a closed form either), we should note that a tabular procedure for this distribution must employ values for υ_1, υ_2, the right tail area, and the abscissa value. Because of this complication, we choose to reduce the number of tabulated right tail area values to the two values $\alpha = 0.05$ [given in Table A.9(a)] and $\alpha = 0.01$ [given in Table A.9(b)]. Actually, we have provided the equivalent of four tabulated right tail area values because of the fundamental relationship that if

$$\text{pr}\{F_{\upsilon_1, \upsilon_2} \geqslant f_{\upsilon_1, \upsilon_2, \alpha}\} = \alpha$$

then

$$f_{\upsilon_1, \upsilon_2, \alpha} = \frac{1}{f_{\upsilon_2, \upsilon_1, 1-\alpha}} \qquad (8.7)$$

This means that we also have $\alpha = 0.95$ and $\alpha = 0.99$ at our disposal. The next few examples will illustrate how to use Equation 8.7 and how to use Tables A.9(a) and A.9(b).

EXAMPLE 8.7 Find

(*a*) $f_{20,\,4,\,0.05}$

(*b*) $f_{4,\,20,\,0.05}$

(*c*) $f_{20,\,4,\,0.95}$

(*d*) $f_{14,\,20,\,0.99}$

SOLUTION

(*a*) $f_{20,\,4,\,0.05} = 5.80$ (Table A.9(*a*) at the intersection of the column headed by 20 and the row headed by 4)

(*b*) $f_{4,\,20,\,0.05} = 2.87$ (Table A.9(*b*) at the intersection of the column headed by 4 and the row headed by 20)

(*c*) $f_{20,\,4,\,0.95}$ cannot be found in the tables, but we can calculate it using Equation 8.7 as follows:

$$f_{20,\,4,\,0.95} = \frac{1}{f_{4,\,20,\,(1-0.95)}} = \frac{1}{f_{4,\,20,\,0.05}} = \frac{1}{2.87} = 0.35$$

(*d*) $f_{14,\,20,\,0.99}$ cannot be found in the tables, but we can calculate it using Equation 8.7 as follows:

$$f_{14,\,20,\,0.99} = \frac{1}{f_{20,\,14,\,(1-0.99)}} = \frac{1}{f_{20,\,14,\,0.01}}$$

$$= \frac{1}{3.51} \qquad \text{[use Table A.9}(b)\text{]}$$

$$= 0.28$$

EXAMPLE 8.8 Find values for y and w that satisfy the given probability statements.

(*a*) $\text{pr}\{F_{10,7} \geqslant y\} = 0.05$

(*b*) $\text{pr}\{F_{10,7} < y\} = 0.01$

(*c*) $\text{pr}\{F_{6,\,30} \geqslant 3.47\} = w$

(*d*) $\text{pr}\{F_{13,\,24} < 0.41322\} = w$

SOLUTION

(*a*) Since $\text{pr}\{F_{10,\,7} \geqslant y\} = 0.05$, then

$$y = f_{10,\,7,\,0.05} = 3.64$$

(*b*) If $\text{pr}\{F_{10,\,7} < y\} = 0.01$, then $\text{pr}\{F_{10,\,7} \geqslant y\} = 0.99$, so that

$$y = f_{10,\,7,\,0.99} = \frac{1}{f_{7,\,10,\,(1-0.99)}} = \frac{1}{f_{7,\,10,\,0.01}} = \frac{1}{5.20} = 0.19$$

(*c*) Since $\text{pr}\{F_{6,\,30} \geqslant 3.47\} = w$, then $f_{6,\,30,\,w} = 3.47$ so that $w = 0.05$ or $w = 0.01$. If $w = 0.01$, then $f_{6,\,30,\,w} = 3.47$. Hence $w = 0.01$.

(*d*) If $\text{pr}\{F_{13,\,24} < 0.41322\} = w$, then $\text{pr}\{F_{13,\,24} \geqslant 0.41322\} = 1 - w$, and so $f_{13,\,24,\,1-w} = 0.41322$. Since none of the values in Tables A.9(*a*) and (*b*) is less than 1.00, we know we must use Equation 8.7 to get

$$f_{24,\,13,\,w} = \frac{1}{f_{13,\,24,\,1-w}} = \frac{1}{0.41322} = 2.42$$

so that $w = 0.01$ or $w = 0.05$. If $w = 0.01$, then $f_{24,\,13,\,w} = 3.59$. If $w = 0.05$, then $f_{24,\,13,\,w} = 2.42$. Hence $w = 0.05$.

This concludes our presentation of the basic sampling distributions. They will be used throughout the next few chapters as we discuss the fundamental properties and implications of interval estimation and hypothesis testing.

EXERCISES

8.4.1 Find

(*a*) $t_{5,\,0.05}$

(*b*) $t_{7,\,0.95}$

(*c*) α such that $\text{pr}\{t_{17,\,1-\alpha} \leqslant t_{17} \leqslant t_{17,\,\alpha}\} = 0.95$ (*Hint*: draw a picture)

8.4.2 Find values for y that satisfy the given probability statements.

(*a*) $\text{pr}\{t_7 \geqslant y\} = 0.95$

(*b*) $\text{pr}\{t_{17} < y\} = 0.01$

8.4.3 Find values for w that satisfy the given probability statements.

(*a*) $\text{pr}\{t_{29} \geqslant 1.311\} = w$

(*b*) $\text{pr}\{t_{17} \geqslant -1.740\} = w$

(*c*) $\text{pr}\{t_{27} < 1.314\} = w$

(*d*) $\text{pr}\{t_{13} < -1.771\} = w$

8.4.4 Find

(*a*) $f_{15,\,3,\,0.05}$

(*b*) $f_{3,\,15,\,0.05}$

(*c*) $f_{9,\,24,\,0.95}$

(*d*) $f_{24,\,9,\,0.99}$

8.4.5 Find values for y and w that satisfy the given probability statements.
(*a*) $\text{pr}\{F_{15,6} \geqslant y\} = 0.05$
(*b*) $\text{pr}\{F_{6,15} < y\} = 0.05$
(*c*) $\text{pr}\{F_{5,29} \geqslant 3.73\} = w$
(*d*) $\text{pr}\{F_{19,30} < 0.3521\} = w$

8.5 SUPPLEMENTARY EXAMPLES

EXAMPLE 8.9 Show that the estimator

$$\hat{\mu} = \frac{\sum_{i=1}^{n} X_i}{n+1}$$

is not an unbiased estimator for the population mean μ.

SOLUTION We calculate the expected value of $\hat{\mu}$ as follows:

$$E(\hat{\mu}) = \frac{E\left(\sum_{i=1}^{n} X_i\right)}{n+1} = \frac{\sum_{i=1}^{n} E(X_i)}{n+1} = \frac{\sum_{i=1}^{n} \mu}{n+1} = \frac{n\mu}{n+1}$$

Since $E(\hat{\mu}) \neq \mu$, $\hat{\mu}$ is a biased estimator of μ.

EXAMPLE 8.10 Argue that the estimator $\tilde{\mu} = (X_1 + X_n)/2$ is not an efficient estimator for the population mean μ whenever $n \geqslant 2$.

SOLUTION Since X_1 is the first and X_n the last value in the unordered sample, the estimator $\tilde{\mu}$ uses only two values from the sample. This means that not all the data collected are utilized (either implicitly *or* explicitly) in determining the estimator. Therefore, it cannot be sufficient, which implies that it cannot be efficient.

EXAMPLE 8.11 The number of clients entering a bank during a certain hour of the day can be modeled using a Poisson distribution. Show how the central limit theorem would be applied to give us the distribution for the average number of clients entering the bank during this hour.

SOLUTION Let $X_1, X_2, \ldots, X_n$ represent the number of clients entering the bank during the hour of interest on n different days selected at random for observation. The expected value and variance of the population are both μ. The central limit theorem says that

$$\sum_{i=1}^{n} X_i$$

will be approximately normal in distribution with mean $n\mu$ and variance $n\sigma^2$, where μ is the population mean and σ^2 is the population variance. Thus, $\overline{X}$, the sample average, is approximately normal in distribution with expected value μ and variance σ^2/n. Since $\sigma^2 = \mu$ for a population that can be modeled using a Poisson random variable, $\overline{X}$ is approximately normal in distribution with expected value μ and variance μ/n.

EXAMPLE 8.12 Find values for y that satisfy the given probability statements.

(*a*) $\text{pr}\{F_{11,\,10} \geqslant y\} = 0.99$

(*b*) $\text{pr}\{t_{21} < y\} = 0.025$

(*c*) $\text{pr}\{\chi^2_{15} < y\} = 0.05$

SOLUTION

(*a*) Since $\text{pr}\{F_{11,\,10} \geqslant y\} = 0.99$ is not given in the tables, we must convert the problem using Equation 8.7 to obtain

$$y = f_{11,\,10,\,0.99} = \frac{1}{f_{10,\,11,\,(1-0.99)}} = \frac{1}{f_{10,\,11,\,0.01}} = \frac{1}{4.54} = 0.22$$

(*b*) If $\text{pr}\{t_{21} < y\} = 0.025$, then $\text{pr}\{t_{21} > -y\} = 0.025$ or

$$y = -t_{21,\,0.025} = -2.080$$

(*c*) If $\text{pr}\{\chi^2_{15} < y\} = 0.05$, then $\text{pr}\{\chi^2_{15} \geqslant y\} = 0.95$ or

$$y = \chi^2_{15,\,0.95} = 7.26$$

EXAMPLE 8.13 Find values for $\chi^2_{11,\,\alpha}$ and $\chi^2_{11,\,1-\alpha}$ such that

$$\text{pr}\{\chi^2_{11,\,1-\alpha} \leqslant \chi^2_{11} < \chi^2_{11,\,\alpha}\} = 0.95$$

SOLUTION Since $\text{pr}\{\chi^2_{11} < \chi^2_{11,\,1-\alpha}\} = \alpha$ and $\text{pr}\{\chi^2_{11} \geqslant \chi^2_{11,\,\alpha}\} = \alpha$, then

$$\text{pr}\{\chi^2_{11,\,1-\alpha} \leqslant \chi^2_{11} < \chi^2_{11,\,\alpha}\} = 1 - 2\alpha$$

or $1 - 2\alpha = 0.95$, so that $\alpha = 0.025$. Thus,

$$\chi^2_{11,\,\alpha} = \chi^2_{11,\,0.025} = 21.92$$

and

$$\chi^2_{11,\,1-\alpha} = \chi^2_{11,\,1-0.025} = \chi^2_{11,\,0.975} = 3.82$$

This means that we have shown that

$$\text{pr}\{3.82 \leqslant \chi^2_{11} < 21.92\} = 0.95$$

8.6 SUMMARY

This chapter presented the most fundamental sampling distributions used in business decision analysis today. It should be understood that we *must* use the sampling distribution principle for all generally accepted estimation procedures, since all such estimators are random variables whose values will change with each new set of sample data. That is, although each estimation procedure may give one unique statistic, this statistic represents a range value for a random variable whose distribution is called a sampling distribution. These sampling distributions are required whenever we need to produce an interval estimate rather than a point estimate. Independent of the type of estimation, we should always use "good" estimators. Small-sample properties for good estimators are unbiasedness and efficiency, whereas large-sample properties for good estimators are asymptotic unbiasedness, consistency, and asymptotic efficiency.

The central limit theorem provides an approximate sampling distribution for the sample average that is independent of the distribution that describes the population from which the sample is drawn. Therein lies its great versatility: it is an effective tool that provides bounds for a wide range of population parameters. The chi-square random variable, Student's t distribution, and Fisher's F distribution are used in the construction of three other useful sampling distributions. All three rely on the assumption of random sampling from a normal population. The next step is to introduce some of the basic principles of statistical inference. Much of what we will present next is dependent on the sampling distributions developed in this chapter.

REVIEW EXERCISES

Level 1

8.1 Find values for y that satisfy the given probability statements.
(*a*) $\text{pr}\{F_{2,\,14} \geqslant y\} = 0.05$
(*b*) $\text{pr}\{t_{13} \geqslant y\} = 0.001$
(*c*) $\text{pr}\{\chi^2_{11} < y\} = 0.025$

8.2 Find values for w that satisfy the given probability statements.
(a) $\text{pr}\{\chi^2_{19} < 6.84\} = w$
(b) $\text{pr}\{t_{13} < 1.771\} = w$
(c) $\text{pr}\{F_{12,\,14} \geqslant 3.80\} = w$

8.3 Illustrate that z_α and $t_{n,\alpha}$ are in close agreement (independent of the value of α) for large sample sizes.

8.4 The central limit theorem says that the sample average will be approximately normal in distribution *independent* of the underlying population from which the sample is drawn. We will now illustrate this point by using the uniform distribution as the underlying population. To do this we have selected 50 samples of size five from a uniform distribution defined on the interval from 0 to 10. Compute the sample average for each of these samples of size 5 to get the 50 sample means. Use the histogram procedure of Chapter 7 to construct a histogram for the 50 values you obtain (note the similarity of the resulting histogram to a normal distribution).

Sample Number	Sample Results	Sample Number	Sample Results
1	3.5 0.7 5.3 3.9 4.9	26	7.9 6.3 2.4 9.5 6.1
2	5.6 6.2 3.3 4.4 4.2	27	7.6 1.5 6.2 7.8 3.1
3	3.6 4.0 9.8 3.2 3.2	28	8.6 6.1 7.4 3.6 7.8
4	5.7 6.2 0.5 2.6 0.6	29	9.7 5.2 3.7 2.2 7.8
5	0.7 3.9 9.3 7.4 0.8	30	7.3 2.1 2.1 8.8 0.2
6	5.2 8.7 2.4 8.4 8.2	31	7.4 4.2 6.5 4.5 1.1
7	6.1 6.1 8.7 1.1 5.3	32	8.1 9.5 0.0 5.6 4.9
8	5.8 6.1 6.1 2.0 8.2	33	3.7 1.3 1.7 3.3 3.3
9	7.6 7.0 4.2 3.5 1.3	34	8.5 9.4 6.9 8.1 7.8
10	1.8 8.2 8.1 9.3 2.9	35	3.6 1.0 6.5 0.9 2.7
11	0.2 8.0 1.2 5.8 5.0	36	3.7 5.0 6.6 7.6 2.2
12	7.6 7.3 2.5 3.1 0.3	37	1.9 4.9 1.4 3.0 0.1
13	1.1 6.9 9.6 6.5 7.9	38	5.3 2.4 1.4 3.4 8.5
14	8.4 6.1 5.9 6.3 9.2	39	3.4 2.4 1.2 9.1 4.5
15	6.5 3.8 6.5 3.5 0.7	40	4.6 9.2 5.1 4.7 1.8
16	7.8 3.3 4.1 2.2 9.1	41	6.5 6.2 2.8 4.9 5.2
17	6.8 2.1 6.7 1.3 3.0	42	5.2 4.2 9.2 4.2 2.4
18	5.8 5.1 2.6 7.5 6.9	43	4.6 5.6 4.9 5.5 3.6
19	6.5 0.3 1.2 4.5 0.2	44	1.6 2.5 5.5 0.7 1.9
20	6.9 2.0 3.0 0.6 5.1	45	2.2 6.9 9.5 1.8 7.0
21	6.1 9.0 0.6 0.8 0.6	46	8.8 1.3 9.5 7.4 1.4
22	3.7 6.9 7.0 1.1 1.2	47	4.8 7.4 8.5 8.9 6.3
23	1.1 7.3 1.2 3.5 7.7	48	2.5 6.2 9.5 5.7 2.8
24	1.3 7.3 0.8 1.6 6.0	49	4.9 7.4 7.6 1.8 9.3
25	7.9 0.1 8.1 5.7 5.7	50	4.9 1.7 3.6 3.3 7.8

8.5 Refer to Exercise 8.4. What are the expected value and variance for the normal approximation (using the central limit theorem) of a sample average for a sample of size five from a uniform population with parameters 0 and 10?

8.6 The expected value and variance of a uniform distribution defined on the interval from a to b are given by

$$\frac{a+b}{2} \quad \text{and} \quad \frac{(b-a)^2}{12}$$

respectively. Suppose a random sample of size four is drawn from a uniform distribution with $a = 3$ and $b = 7$. What is the probability that the sample average is between 5 and 6?

8.7 Find values for w that satisfy the given probability statements.
(*a*) $\text{pr}\{2.18 \leqslant \chi_8^2 < 13.36\} = w$
(*b*) $\text{pr}\{1.319 \leqslant t_{23} < 2.500\} = w$
(*c*) $\text{pr}\{2.94 \leqslant F_{20,\,9} < 4.81\} = w$

8.8 Find values for y that satisfy the given probability statements.
(*a*) $\text{pr}\{F_{10,\,16} < y\} = 0.99$
(*b*) $\text{pr}\{F_{16,\,10} < y\} = 0.05$
(*c*) $\text{pr}\{t_{20} < y\} = 0.95$

8.9 Refer to Exercise 8.6. Suppose a random sample of size 16 is drawn from a uniform distribution with $a = 3$ and $b = 7$. What is the probability that the sample average is between 5 and 6? Compare your answer with that of Exercise 8.6 and explain why the two answers are different.

8.10 Find a value for w where $\text{pr}\{0.3509 \leqslant F_{10,\,15} < 3.8000\} = w$.

8.11 Refer to Exercise 8.6. Suppose a random sample of size *one* is drawn from a uniform distribution with $a = 3$ and $b = 7$. What is the probability that the sample average is between 5 and 6? Be careful about how you formulate your solution and compare your answer with those of Exercises 8.6 and 8.9.

8.12 The expected value and variance for the sample average are μ and σ^2/n when we assume that the sampling is at random from an infinite population or at random with replacement from a finite population. In either case we assume that μ and σ^2 denote the population mean and variance, respectively. When the sampling is without replacement from a finite population of size N, the expected value and variance for the sample average are

$$\mu \quad \text{and} \quad \frac{\sigma^2}{n}\left(\frac{N-n}{N-1}\right)$$

respectively. We can still use the central limit theorem to approximate the distribution of the sample average in the latter case. That is, we say that

$$\frac{\overline{X}-\mu}{\sqrt{(\sigma^2/n)[(N-n)/(n-1)]}} = \left[\frac{(\overline{X}-\mu)\sqrt{n}}{\sigma}\right]\left[\frac{(N-1)}{(N-n)}\right]^{1/2}$$

is approximately standard normal in distribution when the sampling is at random without replacement from a finite population of size N. Let $\mu = 50$, $\sigma = 1$, and $N = 5$. Use this information to approximate the probability that the sample average for a random sample of size three (without replacement) from this finite population will be greater than 50.

8.13 Refer to Exercise 8.12. Suppose current inventory contains 500 items whose average shelf life μ is 42 days with a standard deviation σ of 8.4 days. Suppose a random sample of size 49 is selected, without replacement, from this inventory. What is the probability that the sample average is between 40 and 50 days?

8.14 Refer to Exercise 8.13. What is the probability that the sample average is no more than 10 days away from the population mean of 42 days?

8.15 Refer to Exercise 8.12. Suppose current inventory contains 500 items whose average shelf life μ is 42 days with a standard deviation σ of 8.4 days. Suppose a random sample of size 16 is selected, without replacement, from this inventory. What is the probability that the sample average is between 40 and 50 days? Compare your answer to the one for Exercise 8.13 and to the one you get by using the central limit theorem *without* the correction for sampling without replacement from a finite population.

Level 2

8.16 Discuss how large a sample you would require before using z_α values in place of $t_{\upsilon,\alpha}$ values and explain why practitioners are inclined to make this substitution.

8.17 Find values for $f_{12,\,15,\,\alpha}$ and $f_{12,\,15,\,1-\alpha}$ such that

$$\text{pr}\{f_{12,\,15,\,1-\alpha} \leqslant F_{12,\,15} < f_{12,\,15,\,\alpha}\} = 0.98$$

8.18 Find two different ordered pairs (a, b) such that

$$\text{pr}\{a \leqslant t_{10} \leqslant b\} = 0.90$$

8.19 Find two different ordered pairs (a, b) such that

$$\operatorname{pr}\{a \leqslant \chi^2_{17} \leqslant b\} = 0.90$$

8.20 Show that $F_{1, v} = t_v^2$.

8.21 The time between arrivals at a loading dock can be modeled using an exponential distribution. Show how to apply the central limit theorem to give an approximate distribution for the average amount of time between arrivals at the loading dock.

8.22 The tourist center at the major port of entry for a popular summer vacation area reported an average of 1863 tourists per week entering the area during the season. The standard deviation of the number of tourists was reported to be 310. The minister of tourism then decided the area should offer a special vacation package to increase the number of tourists. In order to support his idea, he estimated that tourism at the center would increase by at least 20% (over the first month of next year's season), on the average, if voters would approve the special package he was proposing. What is the probability of averaging the estimated number of tourists over the stated time period given that the package makes no difference? (Assume that the standard deviation remains stable and that weekly tourism volume is normally distributed.)

8.23 Refer to Exercise 8.22. If the number of tourists *did* increase by 20% over the reported level, would you continue the special package proposed by the minister? (Support your answer with a statistical argument.)

Level 3

8.24 Prove Equation 8.7.

8.25 Illustrate that the minimum distance between a and b when we require

$$\operatorname{pr}\{a \leqslant t_v \leqslant b\} = 1 - 2\alpha$$

occurs when $-a = b = t_{v, \alpha}$. This will be the case for any symmetric distribution centered at zero.

Chapter 9
Introduction to Statistical Inference

KEY WORDS AND TERMS

Confidence Interval
Confidence Level $(1 - \alpha)$
Null Hypothesis (H_0)
Alternate Hypothesis (H_a)
Type I Error
Type II Error
Significance Level (α)
Operating Characteristic (β)
Power $(1 - \beta)$
Statistical Test

9.1 INTRODUCTION

Up to this point we have been investigating a sequence of tools that we will now use as a basis for making decisions in situations of uncertainty. Almost all decision-making processes have an element of uncertainty in them, some more than others. For example, suppose you have just been offered a job and must decide whether or not to accept it. You must consider whether it's really the line in which you'll be happy and whether or not the salary is right. Another question might be, "Will a better offer come next week or next month?" You must make your decision now and in most instances you'll never know whether your decision is the "right" one. In fact, either accepting *or* rejecting the offer could be a bad decision.

Another common situation in which we must make a decision is the purchase of household goods. For simplicity, let us consider the purchase of a refrigerator. There are various sources of information that can help us make such a decision. We might consult *Consumer Reports* for their recommendation of the "best" brand for our particular needs. (But even the best brand occasionally produces a "lime," something a bit greener than a lemon.) After selecting a brand and size, we can then embark on a comparison-shopping trip. After, say, fifteen stores, we decide, "That's it!" and buy the refrigerator at the store that offers the lowest price/best service warranty combination. But how do we know that the sixteenth store wouldn't have given us a better deal or that *Consumer Reports* didn't just happen to have obtained exceptionally good refrigerators for their test samples?

Of these two situations, the first decision had to be based on a very limited amount of information. In the second case, we were able to obtain a sample of prices, and we used information about quality and performance based on someone else's sample. Even if we checked every store in the immediate area, we still wouldn't know whether or not one of them was going to feature our refrigerator model in a sale next week. In both these situations, we gather all the information we can in an attempt to minimize the chance of making a bad decision.

The concept of *statistical inference* employs most of the characteristics just outlined. However, the major difference between the previous situations and the decision-making process known as statistical inference lies in (1) the type of information obtained, (2) the availability of the information, and (3) the use of the information in making a final decision. In this chapter we introduce the basic concepts of statistical inference in a simplified but general form.

9.2 INTERVAL ESTIMATES FOR PARAMETERS

One branch of the inferential process is concerned with estimating values for parameters. The main difference between the estimation approach and the hypothesis-testing approach is our willingness to assume an initial value. In hypothesis testing we will assume a value and use the sample to decide whether or not that value appears reasonable. In the estimation approach we calculate a point estimate from the sample and use it to obtain a set of values for the parameter that is reasonable with respect to the sample. Note that the end results are basically the same. In this book we present interval estimation first because the student generally finds it easier to follow and the transition to hypothesis testing is simple.

Using the criteria given for estimators in Chapter 8, we can select an estimator for a given parameter, take a random sample from the population, and evaluate the estimator to obtain a point estimate of the unknown parameter. We know that the probability of our estimate being equal to the true value is zero for most situations. Thus, it will be more useful if we can state how probable it is that an interval based on our estimate will contain the parameter to be estimated. That is, we can specify a range of possible values for the parameter and be $(1 - \alpha) \times 100\%$ confident that our range or interval will include the true value. Interval estimates are known as *confidence intervals* and are defined formally as follows:

DEFINITION 9.1 A *confidence interval* for a parameter θ is a set of values, determined by the sample, that is expected to contain the true value of θ with a probability of $1 - \alpha$.

DEFINITION 9.2 The *confidence level*, denoted by $1 - \alpha$, of an interval estimate is the probability that the interval will contain the unknown value.

The true meaning of Definition 9.2 is that in repeated sampling, if we calculate intervals of level $1 - \alpha$ from each sample, we can expect that $(1 - \alpha) \times 100\%$ of all the intervals will contain the true value. On the other hand, we expect also that $\alpha \times 100\%$ of the intervals will not contain the true value. That is, $100\alpha\%$ of the time we expect to obtain an oddball or nonrepresentative sample, with the result being an exceptionally poor estimate or guess. The same interpretation will be given to α in hypothesis testing, as well it should be, since it is the same α.

A confidence interval for any parameter is simply a probability statement about a sample value or variable. As such, its construction requires certain things. These requirements are listed here, and you should recognize that the last two are necessary for the evaluation of any probability statement.

Requirements for constructing a confidence interval for a parameter θ.

1. A statistic (estimator) that can be used to estimate θ or some function of θ
2. The type of distribution that governs the chance variation of the estimator in (1)
3. Values for all parameters other than θ for that family of distributions

9.2.1 Two-sided Interval Estimates

We wish to use requirements (1), (2), and (3) to determine values LL (lower limit) and UL (upper limit)—occasionally only one will be of interest—such that

$$\text{pr}\{[\text{LL}, \text{UL}] \text{ covers the true value of } \theta\} = 1 - \alpha \tag{9.1}$$

This statement generally is written in the form

$$\text{pr}\{\text{LL} \leqslant \theta \leqslant \text{UL}\} = 1 - \alpha \tag{9.2}$$

Equation 9.2 is the more common form for a confidence interval, but it is often misleading to the student, since it tends to imply that the variable is θ, whereas in fact the variable is the interval [LL, UL]. We will adhere to the less common but more precise form given in Equation 9.1. The following example will help to illustrate the concepts underlying the construction of an interval estimate.

EXAMPLE 9.1 The Chemtrol Company is considering a change in their process for making high-impact plastic. They hope the change will increase its mean compressive strength. Their chemists have assured them that the standard deviation of compressive strengths will be unchanged by the alteration in the process and have provided the quality control group with eight pieces of the new plastic for testing purposes. From their records the quality control group has determined that the standard deviation is 50 pounds per square inch (psi). Testing of the sample blocks gave an average compressive strength of 1032.4 psi. The company would like a range of values that can be expected to include the true mean compressive strength with a probability of 0.95.

SOLUTION The parameter to be estimated is the mean μ, and the first requirement is to obtain a point estimate of it. In Chapter 8, it was shown that the sample mean $\overline{X}$ is an unbiased and minimum-variance estimator of μ, that is, $E(\overline{X}) = \mu$, and its values are expected to be closer to the true mean than those of any other estimator. Thus, we have

$$\overline{X} = 1032.4 \text{ psi}$$

as our point estimate of μ.

The second requirement is to obtain a distribution for the values of $\overline{X}$. If the distribution of compressive strengths is normal, then $\overline{X}$ will have a normal distribution, since it is a linear combination of normal variables. If not, the central limit theorem tells us that the distribution of $\overline{X}$ will be approximately normal.

Third, the parameters for the normal distribution are $\mu_{\overline{X}}$ and $\sigma_{\overline{X}}$. We already know that $\mu_{\overline{X}} = E(\overline{X}) = \mu$, and in Chapter 8 we found that $\sigma_{\overline{X}} = \sigma/\sqrt{n}$ for an infinite population (compressive strengths would be continuous). Therefore, we know that $\overline{X}$ is at least approximately normal with mean μ and standard deviation $\sigma/\sqrt{n} = 50/\sqrt{8} = 17.678$ psi.

Checking in the standard normal table (Table A.6) we find that

$$\text{pr}[-1.96 \leqslant Z \leqslant +1.96] = 0.95$$

We know also that $(\overline{X} - \mu)/\sigma_{\overline{X}}$ is equivalent to a Z variable. Thus, by substitution, we have

$$\text{pr}\left[-1.96 \leqslant \frac{\overline{X} - \mu}{\sigma_{\overline{X}}} \leqslant 1.96\right] = 0.95$$

Since probabilities are invariant under transformations inside the brackets, we may solve the inequality for μ. That is,

$$-1.96 \leqslant \frac{\overline{X} - \mu}{\sigma_{\overline{X}}} \leqslant 1.96$$

implies

$$-1.96\sigma_{\overline{X}} \leqslant \overline{X} - \mu \leqslant 1.96\sigma_{\overline{X}}$$

and

$$\overline{X} - 1.96\sigma_{\overline{X}} \leqslant \mu \leqslant \overline{X} + 1.96\sigma_{\overline{X}}$$

which gives the more common form (Equation 9.2)

$$\text{pr}[\overline{X} - 1.96\sigma_{\overline{X}} \leqslant \mu \leqslant \overline{X} + 1.96\sigma_{\overline{X}}] = 0.95$$

Since the values for $\overline{X}$ and $\sigma_{\overline{X}}$ are known, the upper and lower limits are numbers, namely UL and LL, respectively. Thus, the statement may be rewritten as

$$\text{pr}\{[\overline{X} - 1.96\sigma_{\overline{X}}, \overline{X} + 1.96\sigma_{\overline{X}}] \text{ contains } \mu\} = 0.95$$

Substituting the numerical values gives

$$\text{LL} = 1032.4 - 1.96(17.678) = 997.751 \text{ psi}$$

and

$$\text{UL} = 1032.4 + 1.96(17.678) = 1067.049 \text{ psi}$$

Therefore, a 95% confidence interval for the mean compressive strength of plastic made by the new process is

$$[997.751, 1067.049] \text{ psi}$$

This is represented graphically in Figure 9.1.

Figure 9.1 Illustration of a 95% Confidence Interval for the Mean Compressive Strength of Plastic Made by the New Process

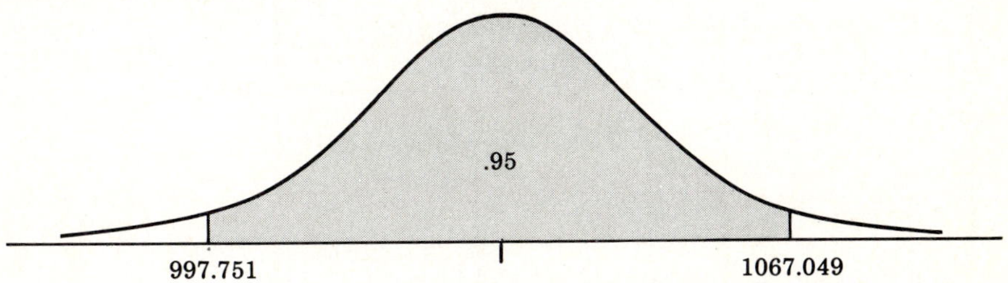

The construction of the interval in Example 9.1 can be explained intuitively as follows. We know that normally distributed random variables will yield values within 1.96 standard deviations of the mean 95% of the time. Thus, we can be 95% confident that our value for $\overline{X}$ will be within $\pm 1.96\sigma_{\overline{X}}$ or ± 34.649 psi of the true mean compressive strength. If our value happened to fall at the lower end of this range, μ would be 34.649 above, or $1032.4 + 34.649 = 1067.049$. On the other hand, if $\overline{X}$ is as extreme in the opposite direction, μ would be 34.649 psi below 1032.4, or 997.751 psi. If $\overline{X}$ is less than $1.96\sigma_{\overline{X}}$ away from μ, this interval will include the true value. You should also note that if $\overline{X}$ is more than 1.96 standard deviations away from μ, the interval will not contain the true value; there is 5% chance of this occurring.

For an arbitrary confidence level of $1 - \alpha$ we can derive an interval estimate for μ when σ is known. The procedure is identical to that used in Example 9.1, but we repeat it for the general case to reinforce the concepts. Considering our three requirements, we have for a random sample of size n:

1. $\overline{X}$ is our estimator for μ.
2. $\overline{X} \sim N(\mu, \sigma_{\overline{X}})$, or at least approximately so.
3. $\sigma_{\overline{X}} = \sigma/\sqrt{n}$ if the population is infinite.

   or

3′. $\sigma_{\overline{X}} = \dfrac{\sigma}{\sqrt{n}} \sqrt{\dfrac{(N-n)}{(N-1)}}$ if $N(<\infty)$ is the population size.

Therefore, it follows that

$$\frac{\overline{X} - \mu}{\sigma_{\overline{X}}} \sim N(0, 1)$$

and that

$$\text{pr}\left(-z_{\alpha/2} \leqslant \frac{\overline{X} - \mu}{\sigma_{\overline{X}}} \leqslant +z_{\alpha/2}\right) = 1 - \alpha$$

where $z_{\alpha/2}$ represents the value from Table A.6 that has $\alpha/2$ of the area to the right of it. That is, it is the z value such that $\text{pr}(Z > z_{\alpha/2}) = \alpha/2$. Multiplying all parts of the inequality by $\sigma_{\overline{X}}$ gives

$$\text{pr}(-z_{\alpha/2}\,\sigma_{\overline{X}} \leqslant \overline{X} - \mu \leqslant +z_{\alpha/2}\sigma_{\overline{X}}) = 1 - \alpha$$

Subtracting $\overline{X}$ from all terms leaves

$$\text{pr}(-\overline{X} - z_{\alpha/2}\,\sigma_{\overline{X}} \leqslant -\mu \leqslant -\overline{X} + z_{\alpha/2}\,\sigma_{\overline{X}}) = 1 - \alpha$$

Finally, upon multiplying by -1, which reverses the inequalities, we obtain

$$\text{pr}(\overline{X} + z_{\alpha/2}\sigma_{\overline{X}} \geqslant \mu \geqslant \overline{X} - z_{\alpha/2}\sigma_{\overline{X}}) = 1 - \alpha$$

or, in conventional form,

$$\text{pr}(\overline{X} - z_{\alpha/2}\sigma_{\overline{X}} \leqslant \mu \leqslant \overline{X} + z_{\alpha/2}\sigma_{\overline{X}}) = 1 - \alpha$$

Thus, the bounds for a two-sided $1 - \alpha$ confidence interval for μ when σ is known are given by

$$\text{LL} = \overline{X} - z_{\alpha/2}\sigma_{\overline{X}} \qquad \text{and} \qquad \text{UL} = \overline{X} + z_{\alpha/2}\sigma_{\overline{X}} \tag{9.3}$$

The form to be used for $\sigma_{\overline{X}}$ will be dictated by the population size in any particular situation.

EXAMPLE 9.2 Merle's Machine Shop purchases precision 5-inch bolts in boxes of 100 each. These bolts are supposed to average 5 inches in length, but the true mean may deviate from box to box; the standard deviation has been relatively stable at 0.02 inch. When each box arrives, Merle selects 10 bolts at random, measures each with a micrometer, and calculates the sample mean. For the most recent box, it was found that $\overline{X}$ equals 4.982 inches. What is a 98% confidence interval for the true average length of the bolts in this box?

SOLUTION Since we are estimating μ with σ known, the limits on the interval are given by Equation 9.3. In this case we have a finite population with $N = 100$ and $n = 10$. Thus

$$\sigma_{\overline{X}} = \frac{0.02}{\sqrt{10}}\sqrt{\frac{100 - 10}{100 - 1}} \qquad \text{or} \qquad \sigma_{\overline{X}} = 0.006$$

For a 98% confidence interval we have $\alpha = 1 - 0.98 = 0.02$ and $z_{\alpha/2} = z_{0.01} = 2.326$. Therefore, our limits are

$$\text{LL} = 4.982 - 2.326(0.006) = 4.968$$

and

$$\text{UL} = 4.982 + 2.326(0.006) = 4.996$$

Our confidence interval for the mean length of the bolts in this box is

$$\text{pr}\{[4.968, 4.996] \text{ contains } \mu\} = 0.98$$

Since this interval does not include the value 5.00, we would conclude that the bolts in this box are too short on the average.

The results obtained in Equation 9.3 are contingent on the fact that σ is a known value. If this is not the case, we use s as a point estimate of σ, in which case the statistic $(\overline{X} - \mu)/s_{\overline{X}}$ has a t distribution with $n - 1$ degrees of freedom. Thus, when we replace σ with s, we need to use Table A.8 in the back of the book rather than Table A.6.

As we mentioned earlier, it is also possible to construct one-sided interval estimates, confidence intervals with a finite bound on one side only. We illustrate this in the next section, but this material may be omitted without loss of continuity if so desired.

9.2.2 One-sided Interval Estimates

In many practical applications only a single bound will be of interest. For example, if we are trying to estimate the proportion of items that are defective, we would obtain an upper bound only, since the smaller the proportion the better. On the other hand, if we are trying to establish a guarantee for a product to help improve sales, the main concern is the probability of replacement while under warranty. This is directly related to the product's average life, and hence we might prefer to determine only a lower bound for μ in this case.

The procedure for determining a single bound follows the same pattern as the two-sided procedure. The difference lies in the placement of the area associated with α. In the two-sided case α was split equally between the two tails, whereas in this case we place the total α in the tail of interest. (Other divisions are possible if both bounds are of interest but one is considered more critical than the other; for example, if $\alpha = 0.05$, we could put 0.04 in one tail and 0.01 in the other.) For an upper bound only, we want to find an interval $(-\infty, \text{UL}]$ such that

$$\text{pr}\{(-\infty, \text{UL}] \text{ covers } \theta\} = 1 - \alpha$$

where θ represents the parameter of interest. If the specific parameter is the population mean μ, then we have that

1. $\overline{X}$ is our estimator for μ.
2. $\overline{X} \sim N(\mu, \sigma_{\overline{X}})$
3. $\sigma_{\overline{X}} = \sigma/\sqrt{n}$ if the population is infinite and the sample is random.

Thus, for an upper bound for μ when σ is known, we must determine UL such that

$$\text{pr}\{(-\infty, \text{UL}] \text{ covers } \mu\} = 1 - \alpha$$

Referring to Equation 9.2, we see that LL becomes $-\infty$ (or whatever physical bound is logical), and UL is altered by merely using z_α rather than $z_{\alpha/2}$. Thus, the upper limit UL is given by

$$\text{UL} = \overline{X} + z_\alpha \sigma_{\overline{X}} \tag{9.4}$$

Correspondingly, for a lower bound only, our probability statement of interest is

$$\text{pr}\{[\text{LL}, +\infty) \text{ covers } \mu\} = 1 - \alpha$$

with LL given by

$$\text{LL} = \overline{X} - z_\alpha \sigma_{\overline{X}} \tag{9.5}$$

As before, if σ is not known, we replace it with the sample standard deviation s and use the t table to find the limiting number of standard deviations. We illustrate this variation and a one-sided upper limit in the following example. *Note* that the use of t assumes normality.

EXAMPLE 9.3 In contract negotiations with the Merit Fabrication Company, the union is demanding a piecework clause in the contract. The company and the union have agreed to accept a clause stipulating the payment of a per piece bonus for each item produced above the average per shift. Before ratifying the contract, the union is demanding that a specific number be stated in the contract and agrees to allow the company to use the upper limit of a 60% confidence interval for μ.

From production records for the past year the company randomly selects 25 shifts and determines the output for each shift. The mean and standard deviation of the output for these 25 shifts were 128.65 and 15.36 pieces, respectively. Use these to determine the number of pieces that should be produced on a shift before the piecework bonus is paid.

SOLUTION Since we are estimating μ, our estimator is $\overline{X}$, and, by the central limit theorem, $\overline{X}$ is approximately normal in distribution. Thus, if σ were known, the statistic $(\overline{X} - \mu)/\sigma_{\overline{X}}$ would have a standard normal distribution. Since σ is not known, we use its estimate, s, instead. From Chapter 8 we know that this substitution changes the distribution and that the resulting statistic, $(\overline{X} - \mu)/s_{\overline{X}}$, will have a t distribution with $n - 1$ degrees of freedom where $s_{\overline{X}} = s/\sqrt{n}$.

Since we need only an upper bound for μ, we use Table A.8 and find that $t_{24,\,0.4}$ equals 0.256. Substituting this value for z_α and the sample results into Equation 9.4 gives

$$\text{UL} = 128.65 + 0.256(3.072) = 129.436 \text{ pieces}$$

Thus, Merit can be 60% confident that the interval $(-\infty, 129.436]$ contains the true mean output per shift, and the piecework bonus should be paid for all pieces in excess of this value. Note that the physical limit in this case is 0 and the actual interval should be (0, 129.436). This interval is illustrated in Figure 9.2.

In the next section we begin a discussion of the other major approach to statistical inference, hypothesis testing. The only change is that now we assume a value for the parameter and evaluate how extreme the sample results are under this assumption. As you will see, the concepts of interval estimation and hypothesis testing are intrinsically related. Actually, there is a one-to-one correspondence between confidence intervals and hypothesis tests.

EXERCISES

9.2.1 The Hedson Public School System has a complete audit of their records once every year with interim sample audits once every quarter (three months). In a sample audit, 36 accounts are randomly sampled and checked for errors in the account balance, with the amount of the error being recorded. A sample audit was completed recently and gave an average error of −$22.51. The last complete audit gave a standard deviation of error amounts equal to $15.92. Assume that the standard deviation is unchanged and determine a 90% confidence interval for the true average error at this time. The Hedson System has an average of 5000 open accounts at any particular time. Thus, you may ignore the finite population correction factor.

9.2.2 The Council for Consumer Education (CCE) is attempting to estimate the average fat content per pound of hamburger sold by a local supermarket chain. To do so, they purchased 1 pound of hamburger on each of ten randomly selected days. The hamburger was then fried and the amount of fat poured off and weighed. The results, in ounces of fat, are

$$5.6,\ 4.1,\ 4.7,\ 3.0,\ 3.2,\ 5.2,\ 4.9,\ 5.8,\ 3.8,\ 3.7$$

Figure 9.2 One-sided 60% Confidence Interval for the Mean Output per Shift in Example 9.3

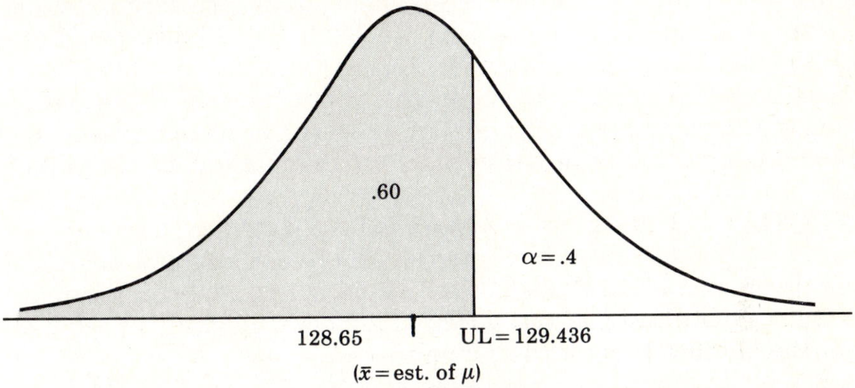

What is a 95% upper bound on the average fat content of a pound of hamburger sold by this supermarket?

9.2.3 Harkin's Department Stores is contemplating the acquisition of an independent store. Before making a firm offer, they would like to see an estimate of the store's average monthly sales volume along with 95% bounds on this average volume. To obtain their estimate, they selected 16 months at random from the store's records (it has been in operation for 25 years); the resulting mean and standard deviation were 34.2 and 3.8 (in thousands of dollars), respectively. Within what range might Harkin's expect the true average monthly sales volume to fall? What problems might Harkin's encounter with what they did?

9.2.4 A professional organization is considering changing its dues structure from a flat fee of $15 per member to 0.07% of the member's annual salary. Before doing so, they would like to obtain an estimate of the members' average salary. The organization currently has 2000 members, and the dues committee decides to poll 100 members, selected at random. The average salary for the 100 members who were polled is $22,580. National statistics indicate that the standard deviation of salaries for this profession is $3200. What are 92% bounds on the average salary of the organization's members?

9.3 THE STATISTICAL HYPOTHESIS

In Chapter 6 we encountered a situation involving several decision alternatives, with the resulting decision being determined by the state of nature that is in effect at the time. The statistician's task generally is to decide what the state of nature is. Ordinarily, a statistician's decision choices can be translated into statements about the probability distribution of the outcomes for some random process. These statements, in statistical jargon, are known as *hypotheses*, defined formally as follows.

DEFINITION 9.3 A *statistical hypothesis* is a statement about the nature of the distribution of interest. The hypothesis may specify the family or type of distribution that is believed to apply or, assuming a family of distributions, may specify individual members of the family by stipulating values for the parameters.

Normally, but not always, the statistician will dichotomize the decision set into two hypotheses, one called the *null hypothesis* and the other called the *alternative hypothesis.* The null hypothesis, denoted by H_0, is the one that is assumed to be true until there is sufficient evidence to the contrary. The alternative hypothesis, denoted H_a, is generally the one of interest and most often corresponds to the conclusion the decision maker would prefer to reach. However, before the decision maker can conclude H_a, the sample evidence must indicate it beyond a reasonable doubt. It should be noted that the choice of H_0 and H_a is not fixed, but is dependent on what is of interest to the individual or company and thus depends entirely upon the statistician's point of view. This perhaps alarming situation is actually quite logical, and to illustrate this point we consider the following situations.

EXAMPLE 9.4 I propose that you and I engage in a game of chance. One of us will toss a coin (mine); each time a tail occurs, I will pay you \$2, whereas if a head results, you pay me \$2. Being of a suspicious nature, you ask to examine the coin first and find that it is not of the two-headed variety. Being knowledgeable in statistics, you realize that simply having a tail does not an honest coin make, and so you request a few sample tosses, 20 to be exact, before deciding to play. Before allowing you your 20 tosses, I demand that you state your hypotheses in advance (standard statistical procedure, or SSP). Let us investigate what you might state.

(A) The *Gambler*: By gambler, we mean an individual who is willing to take a chance as long as he or she has an even shot at winning. Thus, if the coin is fair, the gambler is willing to play; he or she will refuse to play only if the evidence "strongly" indicates that the coin is biased against him or her. If you are a gambler, your hypotheses will take the form

$$H_0\colon\ p_H \leqslant 0.5 \qquad \text{vs.} \qquad H_a\colon\ p_H > 0.5$$

(B) The *Almost-Sure-Thing*: We now assume that you prefer to play only if it appears that the odds are in your favor. In this case, you will play only if there is "strong" evidence that the probability of a head is less than 0.5. Thus, your hypotheses would be

$$H_0\colon\ p_H \geqslant 0.5 \qquad \text{vs.} \qquad H_a\colon\ p_H < 0.5$$

(C) *Head or Tail*: This time we change the game slightly. Instead of stipulating your outcome, I allow you to choose the one you prefer (heads or tails). In this instance, you are interested in detecting whether the coin is biased in either direction. If it is not biased, you would be indifferent, but if it is biased, you prefer the more likely outcome. Hence, your hypotheses would be

$$H_0\colon\ p = 0.5 \qquad \text{vs.} \qquad H_a\colon\ p \neq 0.5$$

Example 9.4 illustrates how the hypotheses may change with one's viewpoint even though the situation remains basically the same. The following two examples indicate other forms that the hypotheses may take and relate more directly to the business environment.

EXAMPLE 9.5 Let us assume that we have been hired by our local branch bank to determine whether they should have more tellers available during peak hours. They have been receiving many complaints from their customers about excessive waiting times. Since waiting times, to some extent, are dictated by the number of customers present, we first need to determine the distribution of customer arrivals. Checking assumptions, we find that a Poisson distribution might be appropriate, and we decide to take a sample to test our belief. Our hypotheses would be of the form

$$H_0\colon\ X \sim P(\mu) \qquad H_a\colon\ X \nsim P(\mu)$$

where X denotes the number of arrivals. That is, we are testing to see whether the Poisson family appears to be appropriate.

EXAMPLE 9.6 A different situation occurs when we consider a manufacturing process producing items that are either defective or nondefective. In this case, we are interested in the overall quality of the output, which translates into the percent of items produced that are defective. Since the population is dichotomous and essentially infinite, with each item being independent of any other, the binomial distribution would be an appropriate model for the number of defectives occurring in a given random sample. Here the unknown variable is the quality of the items, or stated statistically, the value of the parameter p. If we assume that 3% defectives is acceptable, then we select a random sample to test the hypothesis set

$$H_0:\quad p \leqslant 0.03 \qquad H_a:\quad p > 0.03$$

where H_0 states that the quality is acceptable and H_a indicates that it is unacceptable.

We will consider hypotheses about general models or families of distributions in Chapter 10, but for the present we will confine ourselves to hypotheses of the *parametric* type as shown in Examples 9.4 and 9.6.

To stress the importance of the statement of H_0 and H_a, we use the following situation. Our legal process is based primarily on protection of the individual, and thus the accused is assumed innocent until "proven" guilty beyond a reasonable doubt. Suppose, then, that you are an accused person on trial for murder. As such, you want the jury's null and alternative hypotheses to be

$$H_0\text{: Innocent} \qquad \text{vs.} \qquad H_a\text{: Guilty}$$

and you would want them to accept H_0 if there is *any* reasonable doubt.

Now change your role to that of a member of the jury. In this capacity you are faced with a new set of values. You would prefer not to turn a guilty person back into society, but on the other hand you don't want to burden your conscience with the possibility of having convicted an innocent person. As a jury member, your selection of hypotheses is simplified, since society dictates that you choose the set corresponding to the accused's point of view.

Since we are all actors upon the stage of life, let us consider one last part in this simple drama. Assume that you are prescient and know that you are to be the next victim of the accused. What hypothesis set do you want the jury to use now? Obviously, your choice of hypotheses would be

$$H_0\text{: Guilty} \qquad \text{vs.} \qquad H_a\text{: Innocent}$$

because you would want the jury to accept H_0 if any doubt existed.

Thankfully, statisticians are not faced with such momentous decisions. They are more apt to have to determine whether or not a particular process is adhering to some quality standard. The choice of hypotheses and of what constitutes a reasonable doubt is determined by the economics of the problem. Usually, a statistician will determine what conclusion is more economically desirable and formulate the alternative hypothesis to coincide with it. Let us explain the reasons for letting the alternative correspond to the conclusion of interest.

Let us return to the jury trial and consider what is actually being said with each verdict. If the jury decides that the accused is innocent, what they really are saying is *not* that he is innocent, but that they haven't been given sufficient evidence to conclude otherwise. That is, the prosecutor did not prove guilt beyond a reasonable doubt. Hence, they are making a weak and somewhat wishy-washy decision. On the other hand, if they return a verdict of guilty, they are saying that a reasonable doubt no longer exists and that the evidence "proves" the guilt of the accused. Thus, a verdict of guilty is a much stronger decision than a verdict of innocent, which is actually no decision at all.

In statistics, we prefer to base our conclusion on the stronger decision whenever possible. Since rejecting the null hypothesis and concluding the alternative is the stronger decision, we will more often state H_a first. There will be occasions where this will not be true, but in those cases the hypotheses generally correspond to the first set in a series of decisions. There are other good reasons for selecting what we want to conclude to correspond to the alternative hypothesis. These reasons relate to the two types of decision errors we can make in any hypothesis-testing situation. These *very* important types of error are discussed in the next section.

EXERCISES

9.3.1 Reconsider Exercise 9.2.1 and determine an applicable hypothesis set that might be of interest for that situation. Explain why you stated the hypotheses as you did.

9.3.2 Consider Example 9.1 and indicate what hypotheses might be of interest to Chemtrol. Justify your selection of H_0 and H_a.

9.3.3 State an hypothesis set that would be appropriate for Exercise 9.2.2 and justify your selection.

9.3.4 What hypotheses would you state for the situation in Exercise 9.2.4 and why?

9.4 TYPES OF STATISTICAL ERROR

Statistical decisions are always based on a set of sample data, and hence never prove anything. There is always the possibility of obtaining an oddball, or nonrepresentative, sample that might cause us to make a wrong decision. Wrong decisions can occur in two ways: (1) we can reject the null hypothesis H_0 when it is true, or (2) we can accept it when it is false. These two types of errors are so important in statistics that they must be formally defined.

DEFINITION 9.4 A Type I error is said to occur when the null hypothesis is rejected due to chance alone (the sample was an extreme or oddball one).

DEFINITION 9.5 A Type II error is said to occur when the null hypothesis is accepted due to chance alone (the sample is now oddball with respect to H_a being true).

To illustrate and clarify the two types of errors, we consider the following example.

EXAMPLE 9.7 A manufacturer of a very expensive piece of equipment has two rush orders on his desk for this particular item. Each order calls for 10 pieces. Currently there are 20 units available, but the manufacturer does not know if all 20 are in perfect working order. He breaks the 20 into two lots of 10 each for shipment and, since a complete inspection of this item is both costly and time-consuming, decides to randomly inspect two units from each lot. Let us assume that if 10% or fewer of the units in a rush order are operating improperly, it is acceptable for shipment. Otherwise, the lot is considered to be of poor quality and is retained for 100% inspection. Further, suppose that one lot has exactly one defective in it and that the other has exactly four.

The manufacturer has decided that if any defectives occur in a sample of two pieces, the lot will be held back for total inspection. Now consider the first lot, which has only one defective (this is acceptable). The probability that the manufacturer will obtain the one defective in the sample is given by (hypergeometric distribution)

$$\text{pr}\{X = 1 \mid N = 10, R = 1, n = 2\} = \frac{\binom{1}{1}\binom{9}{1}}{\binom{10}{2}} = \frac{9}{45}$$

If this happens, the manufacturer will postpone shipment until the lot has been totally inspected and will have made a bad or incorrect decision. That is, he will have rejected the hypothesis H_0: $p \leqslant 0.1$ by chance alone, since the lot was actually acceptable.

The second lot, on the other hand, is bad and should be retained for further inspection. However, if the sample shows no defectives, the manufacturer will make the wrong decision and ship the lot. The chance of his doing this is

$$\text{pr}\{X = 0 \mid N = 10, R = 4, n = 2\} = \frac{\binom{4}{0}\binom{6}{2}}{\binom{10}{2}} = \frac{15}{45}$$

In this case, he would accept H_0 in error, since the alternative H_a: $p > 0.1$ is actually true (in fact, $p = 0.4$).

As you can see, an individual could make two "bad" decisions without being a poor decision maker, because luck dealt two somewhat unlikely samples. We summarize the possible decisions in Table 9.1.

Since our decisions are based on sample observations, each of the decision/state combinations has a probability associated with it for a given situation. The probability of committing a Type I error is denoted by α and is called the *significance level* of a test; that is,

$$\alpha = \text{pr}\{\text{reject } H_0 \mid H_0 \text{ true}\}$$

It follows that the probability of correctly accepting H_0 is $1 - \alpha$, that is

$$1 - \alpha = \text{pr}\{\text{accept } H_0 \mid H_0 \text{ true}\}$$

This is called the confidence level. Moving to the case in which H_a is true, we use β to denote the probability of a Type II error, sometimes calling it the *operating characteristic* of a test, that is,

$$\beta = \text{pr}\{\text{accept } H_0 \mid H_a \text{ true}\}$$

Table 9.1 Decision Situations for Hypothesis Testing

	True State	
Decision	H_0 true	H_a true
H_0	Correct Decision (No error)	Wrong Decision (Type II error)
H_a	Wrong Decision (Type I error)	Correct Decision (No error)

The *power* of a test is its ability to detect the alternative and is simply $1 - \beta$, or

$$\text{Power} = 1 - \beta = \text{pr}\{\text{reject } H_0 \mid H_a \text{ true}\}$$

To illustrate these probabilities, we revisit the manufacturer of Example 9.7.

EXAMPLE 9.8 The manufacturer's test consisted of selecting two items at random and rejecting H_0 if any defectives occurred. Using that fact, we found the probability of his rejecting the first lot incorrectly to be

$$\text{pr}\{X = 1 \mid N = 10, R = 1, n = 2\} = \frac{9}{45}$$

Thus, for this test, $\alpha = 9/45$, and correspondingly, $1 - \alpha = 36/45$. For the second lot, which was bad, the probability that he would mistakenly accept it was found to be

$$\beta = \text{pr}\{X = 0 \mid N = 10, R = 4, n = 2\} = \frac{15}{45}$$

and the power of his procedure to detect bad lots with four defectives in them is

$$\text{Power} = 1 - \beta = \text{pr}\{X \geqslant 1 \mid N = 10, R = 4, n = 2\} = \frac{30}{45}$$

Example 9.8 reinforces the fact that statistics do *not* prove things. They only tend to support or tend to contradict. The manufacturer, not knowing the actual quality, is always liable to make the incorrect decision because a statistic indicates the wrong action. Generally, a nominal value of α (and often β) will be specified before the data are actually collected. The value selected indicates the risk of rejecting the null hypothesis incorrectly that the decision maker is willing to run. In specifying β we must also state a particular alternative value for the parameter being tested. By doing so we are saying that we are willing to accept the null hypothesis incorrectly $\beta \times 100\%$ of the time when the stated alternative is actually true. After we have stated the hypotheses, and selected a level of α (and possibly β), the next step is to construct a test that will guarantee that α (and β) will not be exceeded. We will explore this procedure later in the chapter. Before ending this section, we revisit the coin-tossing proposition.

EXAMPLE 9.9 Let's return to the coin-tossing proposition of Example 9.4. Suppose that in the first two cases given you will refuse to participate if: (A) 13 or more heads occur; (B) 8 or more heads occur. For case C you decide to reject the honesty of the coin and choose the more frequently occurring side if either less than 8 heads or more than 12 occur.

(*a*) Determine α for each of the three cases.

(*b*) If the coin is actually biased so that $p_H = 0.3$, what will be the probability of accepting H_0 in each case?

SOLUTION

(*a*) Since coin tossing is the classical example for binomially distributed random variables, we easily find our probabilities in Table A.1 or A.2.

(A) The *Gambler*: The hypotheses are

$$H_0\colon\ p_H \leqslant 0.5 \qquad \text{vs.} \qquad H_a\colon\ p_H > 0.5$$

and you have decided to reject H_0 if $X \geqslant 13$. Alpha will be maximized when p_H is at the boundary of the set of null hypothesis values, that is, at $p_H = 0.5$. (You should think about that and verify it for yourself.) Thus, the binomial distribution of interest has $n = 20$, $p = 0.5$ and the desired probability is

$$\alpha = \text{pr}\{X \geqslant 13 \mid n = 20, p = 0.5\} = 0.1316$$

(B) The *Almost-Sure-Thing*: In this case your hypotheses are

$$H_0\colon\ p_H \geqslant 0.5 \qquad \text{vs.} \qquad H_a\colon\ p_H < 0.5$$

and you refuse to play if 8 or more heads occur. This translates to rejecting H_0 if 7 or fewer heads occur in your sample of 20 tosses. Referring to Table A.2 we find that

$$\alpha = \text{pr}\{X \leqslant 7 \mid n = 20, p = 0.5\} = 0.1316$$

This α is the same as the gambler's, but the situation is different. For the gambler the α of 0.1316 says that there is a 13.16% chance he or she won't play the game when the coin is actually honest, but for the almost-sure-thing 0.1316 indicates that there is only a 13.16% chance that he or she *will* play when the coin is honest.

(C) The *Choice*: This time your hypotheses are

$$H_0\colon\ p_H = 0.5 \qquad \text{vs.} \qquad H_a\colon\ p_H \neq 0.5$$

with H_0 being rejected if $X < 8$ or $X > 12$. This probability is

$$\begin{aligned}\alpha &= \text{pr}\{X < 8 \text{ or } X > 12 \mid n = 20, p = 0.5\} \\ &= \text{pr}\{X \leqslant 7 \mid n = 20, p = 0.5\} + \text{pr}\{X \geqslant 13 \mid n = 20, p = 0.5\} \\ &= 0.1316 + 0.1316 = 0.2632\end{aligned}$$

(*b*) The value used for p_H becomes 0.3, and we determine the probability of accepting H_0 rather than rejecting it.

(A) $\text{pr}\{\text{accept}\} = \text{pr}\{X < 13 \mid n = 20, p = 0.3\}$

$= \text{pr}\{X \leqslant 12 \mid n = 20, p = 0.3\} = 0.9987$

(B) Here accepting H_0 corresponds to making a Type II error and our $\text{pr}\{\text{accept } H_0\} = \beta$. We have

$$\beta_{0.3} = \text{pr}\{X \geqslant 8 \mid n = 20, p = 0.3\} = 0.2277$$

The notation $\beta_{0.3}$ represents the value of β when $p = 0.3$.

(C) $$\begin{aligned}\beta_{0.3} &= \text{pr}\{8 \leqslant X \leqslant 12 \mid n = 20, p = 0.3\} \\ &= \text{pr}\{X \leqslant 12 \mid n = 20, p = 0.3\} \\ &\quad - \text{pr}\{X \leqslant 7 \mid n = 20, p = 0.3\} \\ &= 0.9987 - 0.7723 = 0.2264\end{aligned}$$

EXERCISES

9.4.1 Refer to Example 9.9. The Gambler's hypotheses are H_0: $p_H \leqslant 0.5$ vs. H_a: $p_H > 0.5$, and we calculated α using $p_H = 0.5$. The values of 0.1, 0.2, 0.25, and 0.4 also satisfy H_0. Verify that if α is calculated using any of those values, the resulting probability is smaller than when we used $p_H = 0.5$.

9.4.2 Suppose that when the manufacturer of Example 9.7 divided the items into the two lots, three of the five defectives went into the first lot and two into the second lot. Determine α and the power for the test in each of these cases.

9.4.3 The Chemtrol Company of Example 9.1 knows that the plastic produced by their current process has an average compressive strength of 950 psi. The new process is supposed to increase this average. If Chemtrol has decided to convert to the new process if the mean of the eight test blocks exceeds 975, what is their chance of making a Type I error?

9.4.4 Refer to Exercise 9.4.3. If the new process actually increases the strength to 970 psi, what is the probability that Chemtrol will make the change? What is it if the psi increases to 980? to 990? to 1000?

9.5 THE STATISTICAL TEST

We have already presented the idea of a statistical test in our earlier examples. The concept itself is predicated on the idea that a random sample should be representative of the population itself, and thus that statistics calculated from the sample would be expected to be close in value to the corresponding population parameters. As we all know, the idea of *close* is quite relative. For example, if a moon shot is within one-half mile of its intended point of impact, it would probably be considered quite close, whereas if an airline pilot lands only within one-half mile of the runway, he'll probably never fly again. Therefore, our construction of a statistical test should utilize a measure of closeness that will be the same for all situations. At this point, we formally define a statistical test.

DEFINITION 9.6 A *statistical test* is a decision rule that indicates when to accept the null hypothesis, when to reject the null hypothesis, and, in some instances, when to reserve judgment.

As previously stated, we require a measure that will convey the same information about product quality as it will about the accuracy of next year's sales forecast. The measure that does this was first presented in Chapter 2 and forms the basis for all statistical inferences or decision making. Now you know why there is such an emphasis on probability in a statistics course and why those seemingly inane and useless problems with black balls and white balls are simplifications to help you understand this most important concept before using it in a decision-making situation.

In essence, there is only one statistical test or decision rule for any set of hypotheses. The manner in which it is stated may vary with one's assumptions or with the type of tables that are available, but it always can be reduced to the following statement:

> Reject H_0 if the *probability* of obtaining a sample as extreme as that observed (or one more extreme) is smaller than you are willing to attribute to chance alone.

If α has been stated in advance, it identifies the minimum probability with which one is willing to believe that the resulting outcome is due to chance alone. Consequently, this statement can be rewritten as:

Reject H_0, at significance level α, if pr(a sample as extreme or more so than that observed | H_0 true) $< \alpha$. (9.6)

The types of sample outcome that constitute extremity are designated by the alternative hypothesis. That is, the alternative hypothesis indicates the direction of extremity. As with confidence intervals and as indicated in previous examples, hypothesis tests can be either one-sided or two-sided. The statement of the alternative hypothesis dictates which it is to be. To illustrate this point as well as the basic hypothesis test, we return to our coin-tossing problem.

EXAMPLE 9.10 Refer to Exercise 9.9 and state the statistical test corresponding to each case if an α of 0.15 is to be used.

SOLUTION The Gambler's hypotheses are

$$H_0\colon\ p_{\mathrm{H}} \leqslant 0.5 \qquad \text{vs.} \qquad H_{\mathrm{a}}\colon\ p_{\mathrm{H}} > 0.5$$

Thus, if H_0 is true, we expect to observe 10 or fewer heads in the 20 tosses, whereas for H_{a} we expect more than 10. That is, H_{a} indicates that we should reject only for excessive numbers of heads. Hence, for an α of 0.15 our decision rule or test is

Reject H_0 (and refuse the game) if
$\mathrm{pr}\{X \geqslant \text{observed number of heads} \mid n = 20, p = 0.5\} < 0.15$

where X is a binomial random variable with the indicated parameter values. Note that this is identical to the decision rule given in Example 9.9, since, by checking in the cumulative binomial table, we find that

$$\mathrm{pr}\{X \geqslant 12 \mid n = 20, p = 0.5\} = 0.2517$$

and

$$\mathrm{pr}\{X \geqslant 13 \mid n = 20, p = 0.5\} = 0.1316$$

which implies that the only way our probability can be less than 0.15 is for the observed number of heads to be greater than or equal to 13.

(B) The risk-avoiding individual's hypotheses are

$$H_0\colon\ p_{\mathrm{H}} \geqslant 0.5 \qquad \text{vs.} \qquad H_{\mathrm{a}}\colon\ p_{\mathrm{H}} < 0.5$$

The alternative in this case indicates that the risk avoider should reject H_0 only for small numbers of heads. For an α of 0.15 the test is

Reject H_0 (and play the game) if
$\text{pr}\{X \leqslant \text{observed number of heads} \mid n = 20, p = 0.5\} < 0.15$

Again checking Table A.2 we find that this coincides with the decision rule, reject H_0 if $X \leqslant 7$, given in Example 9.9.

(C) For the chooser, the hypotheses are

$$H_0\colon\ p_{\text{H}} = 0.5 \qquad \text{vs.} \qquad H_a\colon\ p_{\text{H}} \neq 0.5$$

which implies rejection of H_0 if either too few or too many heads occur in the sample. For the stated α of 0.15, the test is

Reject H_0 if either
$\text{pr}\{X \leqslant \text{observed number of heads} \mid n = 20, p = 0.5\} < 0.075$

or

$\text{pr}\{X \geqslant \text{observed number of heads} \mid n = 20, p = 0.5\} < 0.075$

We have split the stated α equally between the two extremes since we are indifferent to which one occurs. Referring to Table A.2, we find that the corresponding test would be to reject H_0 if either

$$X \leqslant 6 \qquad \text{or} \qquad X \geqslant 14$$

Note what is actually being implied by the tests in Example 9.10 if H_0 is true:

(A) The Gambler feels that 12 heads in 20 tosses of an honest coin is not unreasonable, but that 13 is unreasonable—it is too far above the expected number 10.

(B) The risk avoider believes that 8 is not unreasonable but that 7 or fewer is.

(C) The chooser feels that any value between 7 and 13 inclusive is reasonable, whereas values outside that range are unreasonable.

Also, the stated α of 0.15 is only *nominal* (in name only), since for our discrete distribution there is no value for which the cumulative probability will be exactly 0.15. The exact αs for each of the cases are: (A) 0.1316; (B) 0.1316; (C) 0.1154.

At this point, we can observe that a statistical test is actually nothing more than a probability statement. As such, there are certain things that are required for its construction in any specific situation. These are:

1. A statistic that can be used to measure deviation from H_0 and that will have a known expected value whenever H_0 is true
2. The family or type of distribution that can be used to describe the chance variation of the statistic when H_0 is true
3. The specific member of the family in (2); that is, values for the parameters when H_0 is true

These are the same conditions that are required for constructing a confidence interval. The first condition is obvious, since all tests are based on comparing what is observed with what was expected to occur under H_0. Since the statistic is calculated from a random sample, it is a random variable and, as such, has a probability distribution associated with it. In order to be able to calculate probabilities for its values, we must know what type of distribution to use (condition 2) and which member of the family applies (condition 3), as is indicated in the following example.

EXAMPLE 9.11 Relating our requirements to the coin-tossing gamble, for any of the three situations, the number of heads occurring in the sample should be close to $np = 20(1/2)$ or 10 if H_0 is true. Deviations from 10 can occur in two ways: (*a*) by chance (or random variation) or (*b*) by the coin's not being fair, in which case the alternative, H_a, is true. Therefore, for the first condition, we considered X, the number of heads, for our statistic.

Since X is the result of performing a fixed number of Bernoulli trials, each of which is independent (the coin has no memory), with the probability of a head remaining constant (same coin each time), we recognized that X has a binomial distribution. This satisfies our second requirement.

The only thing remaining is to determine which member of the binomial family applies. The parameters of the binomial are n and p; for our example, n was specified to be 20 and H_0 stipulates that $p = 0.50$. Thus, we have a random variable with a completely specified distribution under H_0, and we were able to calculate the probability associated with any outcome.

The coin-tossing gamble is a simplified example, but it is an easy one to follow and the approach is quite logical as well as intuitively appealing. Before concluding this section, we consider another example that is more practical in nature and requires assumptions that are not as easily verified.

EXAMPLE 9.12 One of the products of a company is precision 5-centimeter (cm) bolts. The first step in the production process is to cut steel rods into 5-cm sections. The setting on the cutting machine is known to drift occasionally, with the result that pieces are either too long or too short. Those that are too long may be cut a second time, but the short ones are scrap, and it is important to detect shifts in this direction, particularly if the average length drops to 4.98 cm or less. It is known that the distribution of bolt lengths is approximately normal with a standard deviation of 0.04 cm independent of the average length.

State the hypotheses that would be of interest to the company and construct a test of the hypotheses for them if they wish to stop the process incorrectly no more than 10% of the time.

SOLUTION The company is interested primarily in detecting shifts to the left of 5 cm. Thus, their hypotheses are

$$H_0\colon\ \mu \geqslant 5.00 \qquad \text{vs.} \qquad H_a\colon\ \mu < 5.00$$

Since they do not want to stop the machine by mistake more than 10% of the time, the value for α is 0.10. To construct the test we note the following:

1. $\overline{X}$ is an unbiased estimator for μ.
2. $\overline{X} \sim N(\mu_{\overline{X}}, \sigma_{\overline{X}})$, since the bolt lengths are approximately normal initially.
3. $E[\overline{X} | H_0] = 5.00$ cm and

$$\sigma_{\overline{X}} = \frac{\sigma}{\sqrt{n}} = \frac{0.04}{\sqrt{n}}$$

Therefore, our three requirements are determined and the test can be stated as follows:

> Select a random sample of size n and reject H_0 if
> $\text{pr}\{\overline{X} \leqslant \text{observed value of the sample mean} \mid H_0\} < 0.10$.

The probability statement is evaluated using the normal distribution, and the condition stipulates that the value to be used for μ is 5.00 cm.

We can restate the actual test in three additional but equivalent forms. The first form is merely the standardization required to use Table A.6; the test becomes:

> Reject H_0 (and readjust the machine) if
> $$\text{pr}\left\{Z \leqslant \frac{\overline{x} - 5.00}{0.04/\sqrt{n}}\right\} < 0.10$$

But this probability will be less than 0.10 only if $(\bar{x} - 5.00)/(0.04/\sqrt{n}) < -z_{0.1}$, since $\mathrm{pr}\{Z \leqslant -z_{0.1}\} = 0.10$. Thus, the second way of stating the test is:

Reject H_0 if

$$\frac{\bar{x} - 5.00}{0.04/\sqrt{n}} < -z_{0.1} = -1.282$$

If we solve this last form for $\bar{x}$, we find that the statistic will be less than $-z_{0.1}$ only if the observed $\bar{x}$ is more than the indicated number of standard deviations below μ. That is, the test also may be stated as:

Reject H_0 if

$$\bar{x} < 5.00 - 1.282\left(\frac{0.04}{\sqrt{n}}\right)$$

We did not specify the sample size to be used in Example 9.12, since we will investigate the effect of varying the sample size in the next section. Also, as stated earlier, the statement of the alternative hypothesis *and* the selection of a value for α together determine the actual test to be used. The alternative hypothesis indicates what type of values for the statistic are to be considered extreme, and α indicates how extreme they must be in order to reject H_0. These concepts and their relationship to the power of the test are discussed in more detail in the next section.

EXERCISES

9.5.1 Refer to Exercise 9.2.2. Assuming that the government requires that hamburger have no more than 25% fat content by weight on the average, identify the appropriate hypotheses that a government inspector would use and construct a test procedure, indicating how you arrive at your test.

9.5.2 Assume that the standard deviation of the ounces of fat in a pound of hamburger is known to be 1.00 ounce for the supermarket chain in Exercise 9.2.2. Use your test from Exercise 9.2.2 to test your hypotheses. State both the decision and the conclusion you reach.

9.5.3 In the coin-tossing examples, assume that I permit you to toss the coin 50 times before making a decision. Construct the appropriate tests for the three cases under this condition when the nominal α is 0.15.

9.5.4 Stage Supplies sells feathers for use in variety shows. They package their feathers in 1-pound bags and sell them in lots of 25 bags each to stores specializing in theatrical supplies. Al's Theatrical Supply has just received a shipment of feathers from Stage Supplies and would like to check the weights. Feathers are quite bulky, and it is impossible to weigh the total lot at once. In fact, it is difficult to weigh even a single bag of them. Thus, because of the time and effort involved, Al must limit himself to weighing only five randomly selected bags. His results (in pounds) are 0.95, 1.02, 0.91, 0.94, 1.00. If he uses an α of 0.05, what conclusion should he reach about this shipment of feathers?

9.6 POWER AND THE STATED ALTERNATIVE

The power of a test was defined in Section 9.4 to be the ability of the test to detect (reject) a false null hypothesis. This ability is related directly to the value of the alternative used. Thus, if deviations from the null hypothesis in one direction are more critical than deviations in the other direction, the alternative should be stated with that in mind. The alternative hypothesis indicates the type of extreme values that tend to contradict H_0. If we let θ denote the parameter of interest and θ_0 its value as specified by H_0, we have the following possibilities for H_a:

$$H_a\colon\ \theta < \theta_0 \qquad H_a\colon\ \theta > \theta_0 \qquad H_a\colon\ \theta \neq \theta_0$$

The first alternative indicates that we are concerned only with deviations from H_0 on the low side, whereas the second implies that large values for θ are the ones we want to detect. The third possibility states that we will reject H_0 for large deviations in either direction. This was the rationale used in constructing the hypotheses for the three situations presented in the coin-tossing examples, for which θ corresponds to p_H.

One measure of the *goodness* of a test is its power, or the probability that it will lead one to reject H_0 when H_a is true. An ideal test would never cause one to make a wrong conclusion. It would lead to acceptance of H_0 when H_0 is true and to rejection of H_0 when H_a is true; that is, both α and β would equal zero. We illustrate this concept graphically in Figure 9.3 for each of the three alternatives. The graph of the probability of accepting H_0 for all possible values of the parameter θ is also known as the *operating characteristic* (OC) curve of the test. We can obtain the power curve by graphing $1 - \beta$ (1 − OC), and the ideal would be the reverse of the curve depicted in Figure 9.3 for each hypothesis set.

It should be apparent that the only way to obtain an ideal test is to sample the entire population. Doing so will ensure that the correct decision is made. As was indicated earlier, this is usually impossible, for any number of reasons. Therefore, we must settle for a random sample and trust that it is representative of the population. We can evaluate a test by constructing its OC curve or its power curve and comparing it with the ideal. The more nearly it approaches the ideal, the better the test. We will look at some possible tests and their corresponding OC curves in the following example.

EXAMPLE 9.13 Refer to Example 9.12, in which the company produces precision 5-cm bolts. The setting on the machine that cuts the steel rods into 5-cm sections is known to drift occasionally, with drifts to the low side being more critical. In Example 9.12 we formulated the hypotheses and constructed a general decision rule (test). The hypotheses are:

$$H_0\colon\ \mu \geqslant 5.00 \qquad \text{vs.} \qquad H_a\colon\ \mu < 5.00$$

with H_0 being rejected if

$$\text{pr}\{\overline{X} \leqslant \overline{x}_n \mid H_0\} < \alpha$$

where $\overline{x}_n$ is the mean resulting from a sample of n items. Construct OC curves for each of the following three specific (n, α) combinations.

(*a*) $n = 4, \alpha = 0.05$ (*b*) $n = 16, \alpha = 0.05$ (*c*) $n = 16, \alpha = 0.10$

Figure 9.3 Graph of $P_a = \text{pr}\{\text{Accept } H_0 | \theta\}$ for an Ideal Test for Each of Three Possible Alternatives

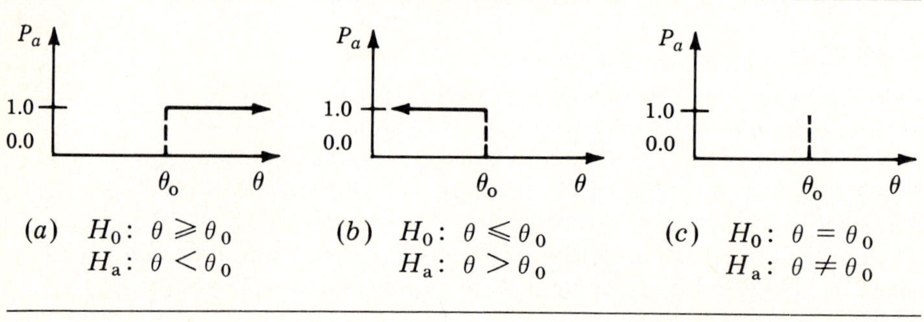

SOLUTION For all three tests the hypotheses are

$$H_0\colon\ \mu \geqslant 5.00 \qquad H_a\colon\ \mu < 5.00$$

Considering the requirements for constructing the tests, we determined that: (1) $\overline{X}_n$ is an unbiased estimator for μ; (2) the distribution of $\overline{X}_n$ is approximately normal; and (3) the standard deviation of $\overline{X}_n$ is $\sigma/\sqrt{n} = 0.04/\sqrt{n}$, and $E[\overline{X}|H_0] = 5.00$. With this information we can construct and evaluate the indicated tests.

(*a*) ($n = 4, \alpha = 0.05$). For this case, the basic decision rule is to

Reject H_0 if
$\text{pr}\{\overline{X} \leqslant \text{observed value} \mid \mu = 5.00\} < 0.05$

or, equivalently, if

$$\overline{X}_4 \leqslant \mu_0 - z_{0.05}\frac{\sigma}{\sqrt{n}} = 5.00 - 1.645\left(\frac{0.04}{2}\right) = 4.9671$$

The probability of committing a Type II error when $\mu = 4.98$ with this test is

$$\beta_{4.98} = \text{pr}\{\overline{X}_4 > 4.9671 | \mu = 4.98\} = \text{pr}\left\{Z > \frac{4.9671 - 4.98}{0.04/\sqrt{4}}\right\}$$
$$= \text{pr}\left\{Z > \frac{-0.0129}{0.02}\right\} = \text{pr}\{Z > -0.645\} = 0.7406$$

If $\mu = 4.96$, then

$$\beta_{4.96} = \text{pr}\{\overline{X}_4 > 4.9671 | \mu = 4.96\} = \text{pr}\{Z > 0.36\} = 0.3594$$

Similarly, for $\mu = 4.94$, $\beta_{4.94} = \text{pr}[Z > 1.36] = 0.0869$. Using these values, we can make a rough sketch of the OC curve for this test:

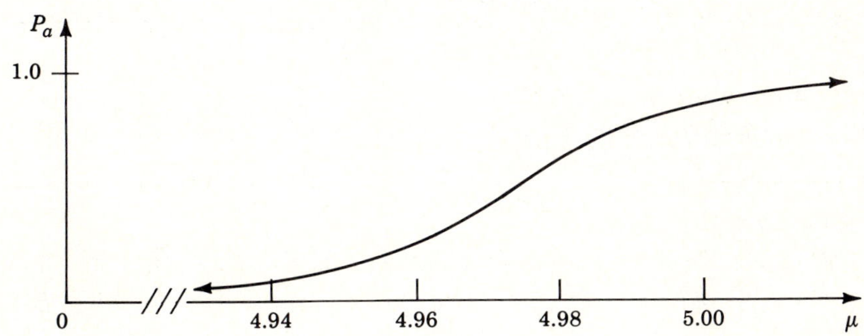

The power for each of the alternatives is $1 - \beta$, or 0.2594, 0.6406, and 0.9131, respectively.

(*b*) ($n = 16$, $\alpha = 0.05$). With the larger sample the decision rule becomes:

Reject H_0 if
$\text{pr}\{\overline{X}_{16} \leqslant \text{observed value} \mid \mu = 5.00\} < 0.05$

or if

$$\overline{X}_{16} \leqslant 5.00 - 1.645\left(\frac{0.04}{\sqrt{16}}\right) = 4.9836$$

With the additional information in the larger sample, the chance of a Type II error decreases accordingly. Using the same alternative values as in (*a*), we find:

$$\beta_{4.98} = \text{pr}\{\overline{X}_{16} > 4.9836 \mid \mu = 4.98\} = P\{Z > 0.36\} = 0.3594$$

$$\beta_{4.96} = \text{pr}\{\overline{X}_{16} > 4.9836 \mid \mu = 4.96\} = P\{Z > 2.36\} = 0.0091$$

$$\beta_{4.94} = \text{pr}\{\overline{X}_{16} > 4.9836 \mid \mu = 4.94\} = P\{Z > 4.36\} = 0.00$$

The power for each is 0.6406, 0.9909, and 1.00, respectively. The graph of the OC curve for this test is:

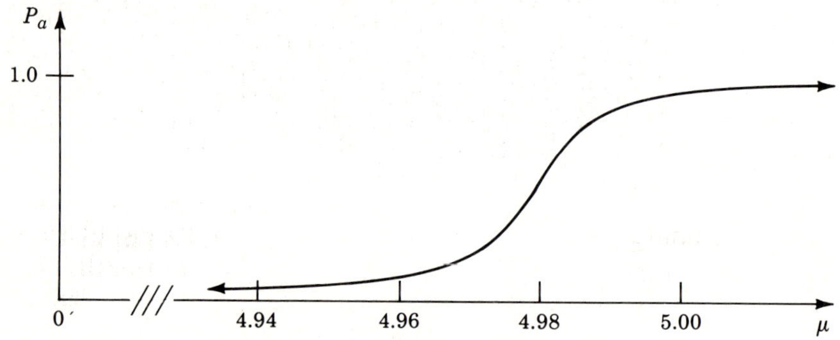

(*c*) ($n = 16$, $\alpha = 0.10$). The increase in α for this test indicates more willingness to reject H_0 by mistake or as a result of chance alone. The decision can be stated as:

Reject H_0 if
$\text{pr}\{\overline{X}_{16} \leqslant \text{observed value} \mid \mu = 5.00\} < 0.10$

or if

$$\overline{X}_{16} \leqslant 5 - 1.282\left(\frac{0.04}{\sqrt{16}}\right) = 4.9872$$

Calculating values of β as before gives

$$\beta_{4.98} = 0.2358 \qquad \beta_{4.96} = 0.0033 \qquad \beta_{4.94} = 0.00$$

with corresponding powers

$$\text{Power}(4.98) = 0.7642 \qquad \text{Power}(4.96) = 0.9967$$

$$\text{Power}(4.94) = 1.00$$

For case (*c*), the OC curve looks like

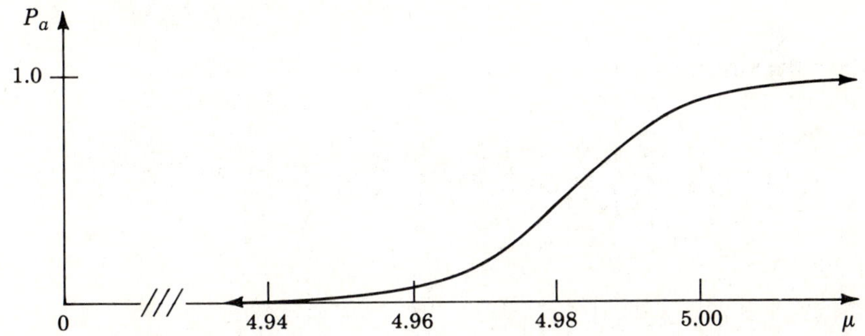

To determine which test to use, we compare the three OC curves with the ideal curve given in Figure 9.3*a*. The curves for the tests based on ($n = 16, \alpha = 0.05$) and ($n = 16, \alpha = 0.10$) are similar. The differences between the two occur primarily at $\mu = 5.00$ and $\mu = 4.98$, with $P_a(5.00)$ for $\alpha = 0.05$ being 0.95 and $P_a(4.98) = 0.3594$. For the test using $\alpha = 0.10$, these are, respectively, 0.10 and 0.2358. If the primary purpose is to detect shifts to 4.98, then we would prefer ($n = 16$, $\alpha = 0.10$), but if we are unwilling to (incorrectly) stop the machine for readjustment 10% of the time, then ($n = 16$, $\alpha = 0.05$) is to be preferred.

We could have used the power curves in Example 9.13 as easily as the OC curves by merely plotting $P_r = \text{pr}\{\text{reject } H_0 \mid \mu\}$, since $P_r = 1 - P_a$. Note that when α was increased with n remaining at 16, β decreased. Whenever n is held constant, an increase in α causes a decrease in β, whereas a decrease in α will cause β to increase. Thus, Example 9.13 illustrates the idea that it is possible to increase the power of a test by increasing the sample size with α fixed or by increasing α with the sample size held constant. Which way is preferred depends on the economics of the situation. If sampling is more costly than the time required for adjusting the machine, α may be changed. On the other hand, if sampling is inexpensive in comparison with down time, then we would select a larger sample. You also should note that if the two-sided alternative is used (that is, if detecting shifts in either direction is equally important), and both the sample size and α are held constant, the power for a specific value of the alternative will be lower. For example, if the alternative in Example 9.13 had been H_a: $\mu \neq 5.00$, the decision rule for ($n = 16$, $\alpha = 0.05$) would have been to reject H_0 at 0.05 if either

$$\text{pr}\{\overline{X}_{16} \leqslant \bar{x} \mid \mu = 5.00\} < 0.025 \quad \text{or} \quad \text{pr}\{\overline{X}_{16} \geqslant \bar{x} \mid \mu = 5.00\} < 0.025$$

Rewriting this gives the decision rule in terms of $\overline{X}$ values as

Reject H_0 at 0.05 if either

$$\overline{X}_{16} \leqslant 5.00 - 1.96\left(\frac{0.04}{\sqrt{16}}\right) = 4.9804$$

or

$$\overline{X}_{16} \geqslant 5.00 + 1.96\left(\frac{0.04}{\sqrt{16}}\right) = 5.0196$$

The power for $\mu = 4.98$ becomes

$$\begin{aligned} P_r &= 1 - \beta = 1 - P_a = 1 - \text{pr}\{4.9804 \leqslant \overline{X}_{16} \leqslant 5.0196 \mid \mu = 4.98\} \\ &= 1 - \text{pr}\{0.04 \leqslant Z \leqslant 3.96\} = 1 - 0.4840 = 0.516 \end{aligned}$$

as compared with 0.6406 for the alternative H_a: $\mu < 5.00$.

EXERCISES

9.6.1 Refer to Exercises 9.5.1 and 9.5.2 and construct the OC curve for your test.

9.6.2 If the sample size in Exercise 9.6.1 is increased to 25, with α remaining the same, construct the OC curve for the new test.

9.6.3 Sketch approximate power curves for your tests in Exercise 9.5.3.

9.6.4 Refer to Example 9.13 and construct power and OC curves when
(*a*) $n = 4$, $\alpha = 0.10$
(*b*) $n = 16$, $\alpha = 0.01$
(*c*) $n = 16$, $\alpha = 0.05$, and α is decreased to 0.03

9.7 RELATIONSHIP BETWEEN CONFIDENCE INTERVALS AND HYPOTHESIS TESTS

As we have indicated numerous times throughout this chapter, there is a one-to-one correspondence between hypothesis tests and confidence intervals. In both cases we are constructing intervals of acceptable or reasonable values. The difference is that in testing we assume a value for θ and construct a set of reasonable values for its estimator. The estimator approach pleads ignorance of the value of θ, obtains a value for the estimator, then calculates a set of values that corresponds to reasonable deviations of the estimator from θ. From both a practical and a theoretical point of view the procedures are identical. To illustrate this we consider the parameter μ and assume that σ is known, with n and α being fixed for both approaches.

If we are unwilling to specify a value for μ under H_0, then a $1 - \alpha$ confidence interval is given by Equation 9.3 as

$$\text{LL} = \overline{X} - z_{\alpha/2}\sigma_{\overline{X}} \qquad \text{and} \qquad \text{UL} = \overline{X} + z_{\alpha/2}\sigma_{\overline{X}}$$

What this says is that we expect the sample mean to be within $z_{\alpha/2}\sigma_{\overline{X}}$ units of the population mean with a probability of $1 - \alpha$. Correspondingly, if we specify a value for μ with H_0 and use a two-sided alternative, then a set of reasonable values for $\overline{X}$ when H_0 is true is given by the interval

$$\mu_0 - z_{\alpha/2}\sigma_{\overline{X}} \qquad \text{to} \qquad \mu_0 + z_{\alpha/2}\sigma_{\overline{X}}$$

In this case, we are $(1 - \alpha)100\%$ confident that $\overline{X}$ will be in the interval whenever H_0 is true.

To see this algebraically, note that for both intervals the distance from the center to either endpoint is the same, namely $z_{\alpha/2}\sigma_{\overline{X}}$. In both cases we start with the probability statement

$$\text{pr}\left\{-z_{\alpha/2} \leqslant \frac{\overline{X} - \mu}{\sigma/\sqrt{n}} \leqslant +z_{\alpha/2}\right\} = 1 - \alpha$$

and solve the inequality for whichever interval is desired. If μ_0 is specified, then

$$\begin{aligned} &\text{pr}\{\overline{X} \text{ is in } [\mu_0 - z_{\alpha/2}\sigma_{\overline{X}}, \mu_0 + z_{\alpha/2}\sigma_{\overline{X}}] \mid H_0\} \\ &= \text{pr}\{\mu_0 - z_{\alpha/2}\sigma_{\overline{X}} \leqslant \overline{X} \leqslant \mu_0 + z_{\alpha/2}\sigma_{\overline{X}} \mid \mu_0\} \\ &= \text{pr}\left\{-z_{\alpha/2} \leqslant \frac{\overline{X} - \mu_0}{\sigma_{\overline{X}}} \leqslant +z_{\alpha/2}\right\} = 1 - \alpha \end{aligned}$$

On the other hand, if $\overline{X}$ is known then

$$\begin{aligned}
&\text{pr}\{[\overline{X} - z_{\alpha/2}\sigma_{\overline{X}}, \overline{X} + z_{\alpha/2}\sigma_{\overline{X}}] \text{ contains } \mu_0 \mid H_0\} \\
&\quad = \text{pr}\{\overline{X} - z_{\alpha/2}\sigma_{\overline{X}} \leqslant \mu_0 \leqslant \overline{X} + z_{\alpha/2}\sigma_{\overline{X}} \mid H_0\} \\
&\quad = \text{pr}\{-z_{\alpha/2}\sigma_{\overline{X}} \leqslant -\overline{X} + \mu_0 \leqslant +z_{\alpha/2}\sigma_{\overline{X}} \mid H_0\} \\
&\quad = \text{pr}\{+z_{\alpha/2}\sigma_{\overline{X}} \geqslant \overline{X} - \mu_0 \geqslant -z_{\alpha/2}\sigma_{\overline{X}} \mid H_0\} \qquad \text{(multiplied by } -1) \\
&\quad = \text{pr}\left\{z_{\alpha/2} \geqslant \frac{\overline{X} - \mu_0}{\sigma_{\overline{X}}} \geqslant -z_{\alpha/2} \mid H_0\right\}
\end{aligned}$$

In conventional form this is

$$\text{pr}\left\{-z_{\alpha/2} \leqslant \frac{\overline{X} - \mu_0}{\sigma_{\overline{X}}} \leqslant z_{\alpha/2} \mid H_0\right\} = 1 - \alpha$$

since if H_0 is true, the variable in the middle is simply Z. Thus, both approaches lead to the same final result. These are illustrated graphically in Figure 9.4.

EXAMPLE 9.14 Refer to Example 9.1, in which the Chemtrol Company is considering a change in their process for making high-impact plastic. They were interested in determining whether the change would increase the mean compressive strength. The chemists have indicated that the standard deviation of compressive strengths will remain at 50 psi. Using the results of Example 9.1 and an α of 0.025, can it be concluded that the new process does the job?

SOLUTION The sample mean in Example 9.1 was 1032.4 psi, and the resulting 95% confidence interval for μ was [997.751, 1067.049]. The hypotheses of interest are (assuming that the current psi is 900)

$$H_0\colon\ \mu \leqslant 900 \qquad \text{vs.} \qquad H_a\colon\ \mu > 900$$

which indicates that in terms of a confidence interval for the true mean we want a lower bound only with an α of 0.025 or $1 - \alpha$ of 0.975. Since our two-sided 95% interval has 2.5% in each tail, we may simply ignore the upper bound to obtain a one-sided 97.5% interval for μ. Thus, our interval is $[997.751, \infty)$ and we are 97.5% confident that our value for $\bar{x}$ is no more than 34.649 psi above the true mean. This is shown graphically in Figure 9.5 (see page 278).

Note that the value of 900 for μ, specified by H_0, is far below the LL. Thus, the deviation of $\bar{x}$ from this value is too extreme to be reasonably attributed to chance, and we can reject H_0. Our conclusion is that the compressive strength appears to have been increased under the new process.

Figure 9.4 Representation of Confidence Interval and Hypothesis Test Results when (*a*) H_0 Is Accepted and (*b*) H_0 Is Rejected

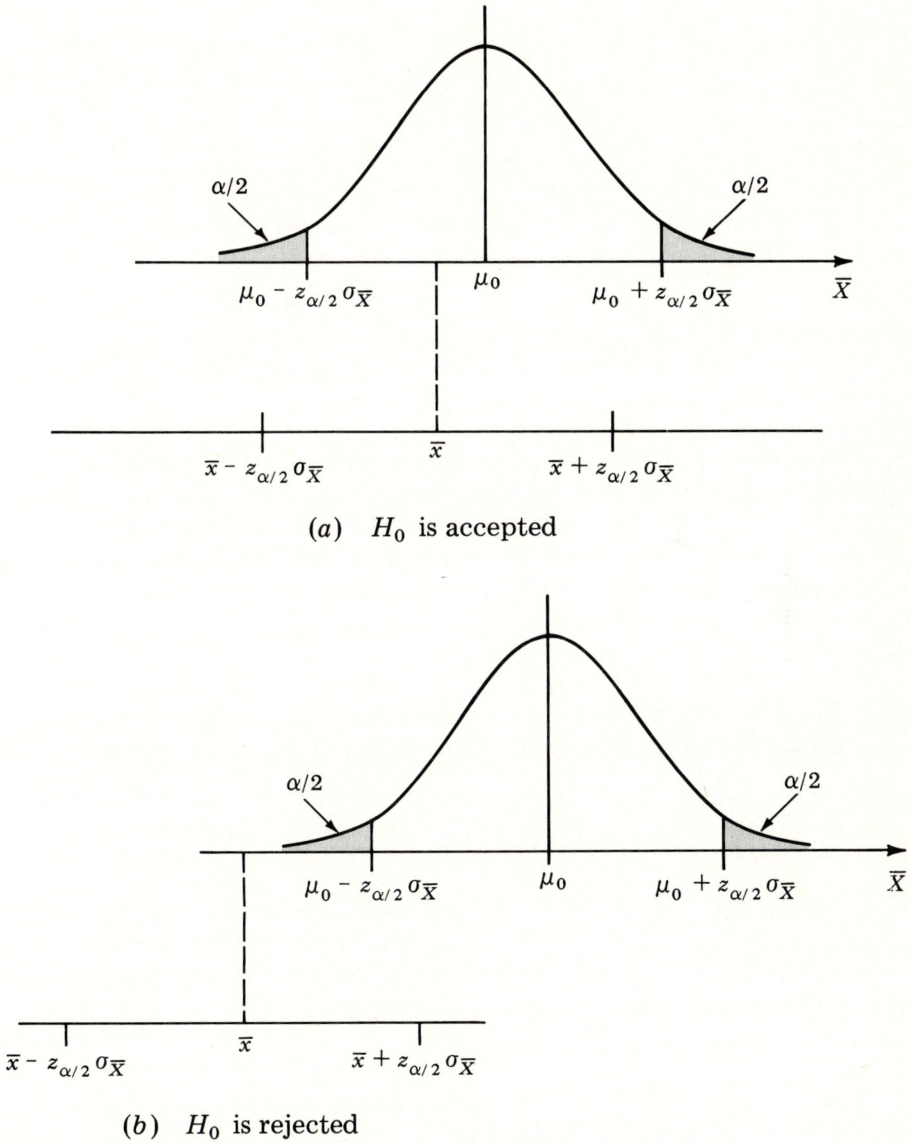

Figure 9.5 One-sided 97.5% Confidence Interval for Mean Compressive Strength of Plastic in Example 9.14

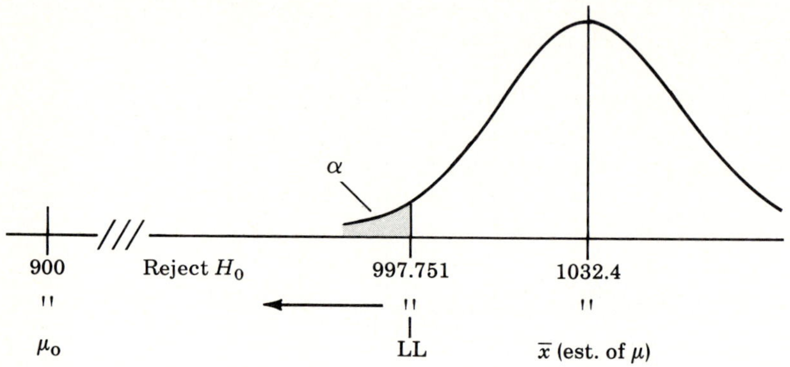

In general, when using a confidence interval in a hypothesis-testing situation, H_0 will be rejected if θ_0 is *not* in the interval and will be accepted if it is. In succeeding chapters, whenever we develop a test for a parameter, we will also present the corresponding procedure for constructing a confidence interval whenever it is appropriate.

EXERCISES

9.7.1 Refer to Example 9.3 and use the interval to test the company's claim that the average output per shift is 130 pieces. Use $\alpha = 0.4$.

9.7.2 Again refer to Example 9.3 and construct a 90% lower bound on the true average output per shift. How do you respond to the union's claim that the true average is 105 pieces per shift?

9.7.3 The owner of the store that Harkin's is considering acquiring in Exercise 9.2.3 claims that the store's sales have been averaging $40,000 over the past two years. On the basis of the results in Exercise 9.2.3, can you make any statement about this claim if α is 0.05? Why or why not?

9.8 SUPPLEMENTARY EXAMPLES

EXAMPLE 9.15 A 1975 survey of business graduates indicated that the average starting salary for a person holding a bachelor's degree and having no full-time work experience was $10,850 per year and the standard deviation was $1100. A random sample of one hundred

1979 graduates with the same qualifications had an average salary of \$15,725. Is this an indication that the average starting salary of business graduates with these qualifications increased between 1975 and 1979 for an α of 0.05? The 1979 sample standard deviation was quite similar to that of 1975.

SOLUTION This problem appears very straightforward at first, but there are some ramifications that we shall indicate later. From the information given, our hypotheses become

$$H_0\colon\ \mu \leqslant 10{,}850 \qquad \text{vs.} \qquad H_a\colon\ \mu > 10{,}850$$

Examining the three requirements for a test we find

1. Our estimator for μ is $\overline{X}$.
2. $\overline{X} \sim N(\mu, \sigma^2_{\overline{X}})$
3. $E[\overline{X}|H_0] = 10{,}850$, $\sigma_{\overline{X}} = 1100/\sqrt{100} = 110$

Thus, if H_0 is true, $\overline{X}$ is normally distributed with a mean of 10,850 and a standard deviation of 110. Using the basic test our decision rule is:

Reject H_0 if
$\text{pr}\{\overline{X} \geqslant 15{,}725|H_0\} < 0.05$

Calculating this probability, we find

$$\begin{aligned}\text{pr}\{\overline{X} \geqslant 15{,}725|H_0\} &= \text{pr}\left\{Z \geqslant \frac{15{,}725 - 10{,}850}{110}\right\}\\ &= \text{pr}\{Z \geqslant 44.318\} = 0\end{aligned}$$

Since $0 < 0.05$, we would reject H_0 and conclude that salaries of business graduates with these qualifications appear to have increased from 1975 to 1979.

At the beginning of the solution to Example 9.15 we indicated that there were some ramifications. These ramifications have to do with making direct comparisons over time. In this case, we should actually be comparing buying power of the two salaries—comparing "real" wages rather than actual wages. Comparing actual dollar values across time can be very misleading because of the consistent inflation in our economy. We will deal with this particular problem in more detail in Chapter 17. For the moment we simply reconsider Example 9.15 taking into account the increase in the cost of living.

EXAMPLE 9.16 Let us reconsider the results of Example 9.15. According to our calculations the 1979 sample mean is so far above \$10,850 that it is an impossible occurrence if H_0 is true. Checking statistics on inflation during the period 1975 to 1979, we find that the cost of living increased by approximately 10% per year. Use this additional information to reevaluate the previous conclusion.

SOLUTION If the cost of living increased by 10% per year for the four years from 1975 to 1979, then the cost of living in 1979 is $100\%(1.1)^4 = 146.41\%$ of what it was in 1975. That is, it takes \$1.46 in 1979 to buy the same goods that \$1.00 bought in 1975. Therefore, converting the 1979 average salary to its equivalent in 1975 dollars, we find that \$15,725 in 1979 dollars corresponds to $15{,}725/1.4641 = \$10{,}740.39$ in 1975 dollars. In terms of real earnings, we must then accept H_0, since 10,740.39 is not even on the correct side of μ_0 as indicated by H_a. That is,

$$\begin{aligned}\text{pr}\{\overline{X} \geqslant 10{,}740.39 \mid H_0\} &= \text{pr}\left\{Z \geqslant \frac{10{,}740.39 - 10{,}850}{110}\right\} \\ &= \text{pr}\{Z \geqslant -0.996\} \\ &= 0.8413 > 0.05\end{aligned}$$

Graphically, this is

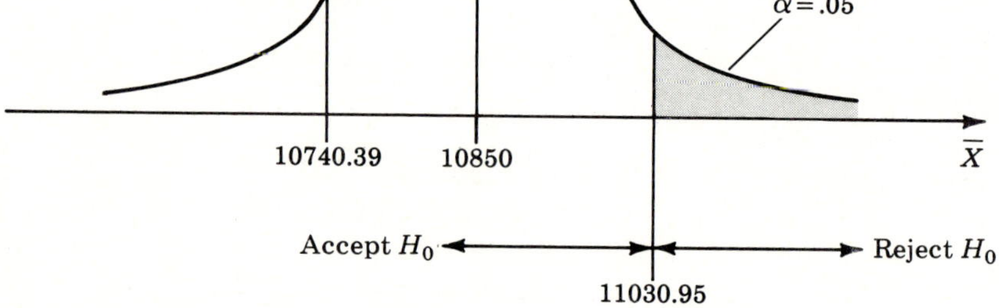

We also could have stated our test as

Reject H_0 if
$\overline{X} > \mu_0 + z_{0.05}\,\sigma_{\overline{X}} = 10{,}850 + 1.645(110) = 11{,}030.95$

where 11,030.95 is the maximum value one could expect for a sample mean from this population with 95% confidence. Any value greater than this implies rejection of H_0, whereas lower values lead to acceptance of H_0.

The main point to be made by Examples 9.15 and 9.16 is that you should not attempt to compare values that are not logically comparable. That is, don't compare oranges to kumquats.

EXAMPLE 9.17 Precision Bearings, Inc., recently completed repair work on a machine that produces Teflon bearings that are to have an average diameter of 2 inches. Precision is confident that the variance of the diameters, 0.000016 inch, is unchanged by the repairs, but they think that the average diameter may have changed. They have available the results of three random samples taken by different inspectors. These are:

Sample Number	Mean	n
1	2.001	20
2	1.998	10
3	2.004	15

Provide them with an estimate of the average diameter being produced and a corresponding 99% confidence interval.

SOLUTION To obtain a point estimate, we combine the three sample means by calculating their weighted average using the sample sizes as weights. We do this because the sample sizes indicate the amount of information contained in each. The resulting estimate is still unbiased, since

$$E\left[\frac{n_1\overline{X}_1 + n_2\overline{X}_2 + n_3\overline{X}_3}{n_1 + n_2 + n_3}\right]$$

$$= \frac{1}{n_1 + n_2 + n_3} E[n_1\overline{X}_1 + n_2\overline{X}_2 + n_3\overline{X}_3]$$

$$= \frac{1}{n_1 + n_2 + n_3} [n_1 E(\overline{X}_1) + n_2 E(\overline{X}_2) + n_3 E(\overline{X}_3)]$$

$$= \frac{1}{\Sigma n_i} [n_1 \mu + n_2 \mu + n_3 \mu]$$

$$= \frac{1}{\Sigma n_i} [\mu \Sigma n_i] = \mu$$

Thus, our estimate is $\frac{90.06}{45} = 2.0013$.

The fact that this is still a sample mean indicates that a normal distribution is applicable by the central limit theorem. Therefore, we have:

1. $\overline{X}_G = \Sigma\, n_i\overline{X}_i/\Sigma\, n_i$ is an unbiased estimator of μ
2. $\overline{X}_G \sim N\left(\mu, \sigma^2_{\overline{X}_G}\right)$
3. $\sigma^2_{\overline{X}_G} = \dfrac{\sigma^2}{\Sigma\, n_i}$

where $\overline{X}_G$ is the combined or grouped mean. You should show that (3) is true. Our 99% confidence interval is given by (Equation 9.3)

$$\text{LL} = \overline{X}_G - z_{\alpha/2}\sigma_{\overline{X}_G} = 2.0013 - 2.576(0.0006) = 1.9998$$

and

$$\text{UL} = \overline{X}_G + z_{\alpha/2}\sigma_{\overline{X}_G} = 2.0013 + 2.576(0.0006) = 2.0028$$

or

$$\text{pr}\{[1.9998, 2.0028] \text{ contains } \mu\} = 0.99$$

In Example 9.17 we assumed that the variance was unchanged. This assumption is unnecessary because we could have combined the sample variances by calculating their weighted average using the respective degrees of freedom as weights. This is known as *pooling the variances* and will be discussed in Chapter 11. If we had done so, we would have used the t_{42} distribution rather than $N(0, 1)$ for our table value.

9.9 SUMMARY

In this chapter we introduced the concepts of statistical inference—using sample data to attempt to infer something about the distribution of the population. The purpose was to acquaint the student with the fundamental reasoning that underlies all statistical procedures and to help tie together the material in the preceding chapters. We have stressed that any statistical test (or interval estimate) is simply a probability statement, and we will continue to stress this fact in succeeding chapters as we consider the standard test procedures.

The first step in any test is to identify what it is that you want to test, that is, to identify the hypotheses of interest. In stating the hypotheses, it must be realized that the stronger decision is to reject H_0; thus, in most cases the alternative hypothesis will correspond to the desired conclusion. Great care should be taken in selecting the hypothesis set, for occasionally a nonsensical conclusion can be reached if H_0 and H_a are reversed. There are three basic types of alternatives that may be stated, and all are strict inequalities. Which one is used depends on where you want the power of your test to be, that is, what you want to conclude.

For any test there are two types of error that can occur. The null hypothesis can be rejected due to chance alone (Type I error), or it can be accepted by chance when it should be rejected (Type II error). The value selected for α indicates the risk of committing a Type I error that you are willing to take. If β is specified in advance as well, this defines the risk of committing a Type II error, with both α and β being controlled by the sample size for a given set of hypotheses. The selection of values for α and β will (generally) be dictated by the economics of the problem. That is, they may be selected so as to balance the expected costs associated with each of the errors.

There is a one-to-one correspondence between statistical tests and confidence intervals, with both requiring the same components for their construction. These are:

1. An estimator of the parameter of interest
2. The family of distributions that govern the chance variation of the estimator (or of a function of the estimator)
3. Values for the parameters of the family that identify which member of the family is applicable

Once the hypotheses have been stated and these three items identified, the test consists of selecting a random sample, evaluating the estimator, calculating the probability of a value as extreme as that observed or more so when H_0 is true, and making a decision based on the magnitude of that probability. If you master this basic concept, you will find it easier to understand why future assumptions will be so important and why certain procedures follow the patterns they do. The concepts of this chapter are summarized in Table 9.2 on page 284.

REVIEW EXERCISES

Level 1

9.1 As an auditor, you discover that the cash "over and short" account for your client has consistently indicated that errors in billing exist. A sample of nine closing balances indicates that the average error is −\$12.00 and the standard deviation is \$3.30. Rather than check all the closing balances, you wish to provide the client with an estimate with 95% bounds for the true mean error. What do you tell the client?

Table 9.2 A Summary of Terms and Notation in Chapter 9

Term	Notation	Interpretation
Confidence interval	pr{[LL, UL] contains θ} $= 1 - \alpha$	The probability that an interval estimate will contain the true value of θ is $1 - \alpha$
Confidence level	$1 - \alpha$	$(1 - \alpha) \times 100\%$ of all such intervals will contain θ
Null hypothesis	H_0	The hypothesis assumed to be true initially
Significance level	$\alpha =$ pr{Type I error}	The probability of rejecting H_0 by chance alone
Operating characteristic	$\beta =$ pr{Type II error}	The probability of accepting H_0 when H_a is true
Power	$1 - \beta$	The probability of rejecting H_0 when H_a is true
Statistical test		The decision rule that identifies the sample outcomes for which H_0 is to be rejected

9.2 What is the effect on the length of the confidence interval if n is doubled while α is held constant?

9.3 A certain machine produces an average of 12.3 units per hour. An expensive new machine is being considered as a replacement. It is felt that the added expense is justified if the new machine can average more than 14 units per hour. If the manufacturer of the new machine will allow a 40-hour test period before you must make your final decision, what should your hypotheses be?

9.4 In a test-marketing experiment for a particular line of products, a company realized average weekly sales over a nine-week period of \$2100 in the store selected. If the company believes that standard deviation of weekly sales for this line in similar stores is \$350, determine a 90% confidence interval for the true average weekly sales for this line in this store. What assumptions must you make to justify your answer?

9.5 A bottling company purchases large quantities of 1-liter (L) bottles for use in their operation. These bottles are supposed to have an average capacity of 1.05 L and a standard deviation of 0.04 L. If the bottles are oversized, they will appear to be underfilled, and customer complaints result. If they are undersized, they overflow in the filling line, which creates clean-up problems and minor stoppages. What hypotheses should the packaging materials inspector use? Construct a test that will reject acceptable shipments of bottles only 10% of the time.

9.6 Refer to Exercise 9.5. A shipment of 25,000 bottles arrived this morning. The inspector selected a random sample of 50 from the shipment and found the average capacity to be 1.06 L. Should the inspector accept the shipment or return it?

9.7 Refer to Exercise 9.6 and identify other factors in addition to the test results that might influence the inspector's decision.

9.8 If a shipment of bottles in Exercise 9.5 has an actual mean capacity of 1.058 L, what is the probability that your test will detect the oversize? What is the probability of detecting a mean capacity of 1.044 L?

9.9 Construct the OC curve for your test in Exercise 9.5.

9.10 In an attempt to improve sales, a company embarked on a new advertising campaign 12 weeks ago. Prior to the new campaign sales had been averaging $10,000 per week and the standard deviation was $2000. Over the past 12 weeks sales have averaged $11,000. Using an α of 0.05, would you recommend a continuation of the campaign?

9.11 Refer to Exercise 9.10 and indicate what assumptions are necessary for your test to be valid.

9.12 The Harper Company has been receiving numerous complaints from their customers concerning the number of defective items in their purchases. Due to a backlog of orders Harper has been operating above peak capacity but allowing their facilities to produce defectives at a 10% rate. With the increase in the number of complaints, they have instructed the workers to keep closer surveillance on the equipment to try to lower the percentage of defectives. A recent sample of 25 items yielded one defective. Using $\alpha = 0.01$, would you conclude that their efforts have been successful?

9.13 Refer to Exercise 9.12 and assume that the sample size is 42 with two defectives. What is your conclusion at $\alpha = 0.05$?

9.14 If Harper decides to increase the sample size to 200, what is the maximum number of defectives they can observe in the sample and still conclude that the plan is working when $\alpha = 0.05$?

9.15 Refer to Exercise 9.3. If the standard deviation of the hourly output for the new machine is known to be 1.4 units, how large must the average hourly output be over the test period before the company will purchase the machine when $\alpha = 0.02$?

Level 2

9.16 Refer to Examples 9.7 and 9.8. If it costs the manufacturer $200 to inspect each piece of equipment before shipping, what are his expected sampling costs for the two lots with his current sampling plan?

9.17 Refer to the coin-tossing examples. Since you appear to be somewhat leery of my current proposal, here is an alternative. I have an urn that contains 20 balls; some are black and some are white, but the number of each is unknown to you. One of us will select a ball at random from this urn. If the ball is white, you win $2.00; if it is black, you pay me $2.00. Before you decide you may select four balls at random from the urn.
(*a*) State the hypotheses that you will test.
(*b*) Are you a gambler or a risk avoider?
(*c*) State the test that you will use.
(*d*) Determine α for your test if you sample without replacement.

9.18 If the urn in Exercise 9.17 actually has 12 black balls in it, what is β for your test with nonreplacement sampling? What is β if you sample with replacement?

9.19 Again refer to Exercise 9.17. If you sample with replacement, what is the statistical difference between this proposition and the coin-tossing one? What differences are there other than statistically?

9.20 Dandy Don, the Lipton Tea Taster, claims that he can tell, from one sip, which has been put into the cup first, the water or the tea. To test him, we prepared 15 cups of tea with the first ingredient determined by the toss of an honest coin: heads meant tea first, tails meant water first. Dandy identified 10 of the 15 cups correctly. Is this evidence that he has discriminating taste at $\alpha = 0.10$?

9.21 Refer to Exercise 9.2.4. The organization is considering the change in dues because the current $30,000 per year income is insufficient. They need at least an additional \$5000 per year from dues to avoid operating in the red. On the basis of the data in Exercise 9.2.4, would the change appear to solve their financial problems? Use $\alpha = 0.05$.

9.22 You should have concluded in Exercise 9.21 that the proposed change won't help the organization. Make a counterproposal on the basis of the sample information and provide 95% bounds on the new income from the dues of a single member.

9.23 The City Council is contemplating putting an increase in the city income tax on the ballot. Before doing so, they would like to have some idea as to the possibility of its successful passage. They call you in as a consultant to tell them how to get an estimate of the necessary information. What would you recommend to the Council?

9.24 The City Council in Exercise 9.23 followed your advice and polled 100 voters selected at random from the register at the Board of Elections. The results showed 46 people in favor of the tax increase.

(*a*) What are 95% limits on the true proportion who favor the increase?

(*b*) If the City Council indicates that they will put the issue on the ballot as long as there is a reasonable chance that at least 50% of the voters will vote for it, what do you recommend to them?

9.25 It has been shown that the occurrence of typographical errors can be reasonably described by a Poisson distribution. You currently need an accurate typist who will average fewer than two errors per page. Typist Tanya just walked into your office to apply for the job. You give her four pages of material to type, and she completes them in acceptable time with an average of 1.75 errors per page. If your nominal α is 0.10, will you show her the door?

Level 3

9.26 If the standard deviation σ is known for a particular process, determine a formula for calculating the sample size required to estimate the mean to within ϵ units of its true value with a probability of at least 0.95. Assume that the underlying distribution is normal.

9.27 What is the effect on the sample size in Exercise 9.26 if the probability level is changed from 0.95 to 0.90? From 0.95 to 0.99?

9.28 A local fast food restaurant claims that it has the fastest service in town; their average service rate is 2 customers per minute, and their service times have a negative exponential distribution. To test their claim, you make two random visits to the restaurant and find that the respective times for service are 5 minutes and 3.5 minutes. Assuming that the negative exponential distribution is correct, what do you say about their claim at $\alpha = 0.10$? Do you need to make any additional assumptions? If so, what?

9.29 Refer to Exercise 9.22 and construct a 95% confidence interval for the total revenue accruing each year from member dues if your proposal is used.

9.30 Tudor Arms, Inc., makes ammunition for shotguns. Their quality control department uses the following procedure in testing the shells to determine whether the ammunition is acceptable for shipping. They select a random sample of ten shells and test them, accepting the lot if no defectives are found, rejecting the lot if two or more are found, and taking another sample of ten if one is found. If a second sample is taken, they accept the lot if no defectives are found in that sample and reject it otherwise. If defective shells occur with a 0.05 probability when the process is working properly, determine:

(*a*) Their hypotheses and the α for their test procedure

(*b*) Their probability of a Type II error when the process is producing defectives at a 10% rate

(*c*) The expected number of shells they test each time when the process is operating properly

9.31 Suppose you wish to test the hypotheses

$$H_0\colon\ \mu \leqslant 100 \qquad \text{vs.} \qquad H_a\colon\ \mu > 100$$

for a normal distribution and desire that α be no more than 0.05 with β no larger than 0.1 when μ is actually 105. If $\sigma = 10$, how large a sample must you take to satisfy these requirements with your test? What is your test?

Chapter 10
One-Sample Inferences

KEY WORDS AND TERMS

Distribution-free (DF)
Nonparametric
Asymptotic Relative Efficiency (ARE)
Robust
Binomial Test
Z Test
t Test
χ^2 Test
Sign Test
Wilcoxon Signed Rank
Goodness of Fit
Kolmogorov-Smirnov (KS)
Runs Test

10.1 INTRODUCTION

In Chapter 9 we developed the basic concepts of statistical inference, or making a decision about the distribution of a population based on the information contained in a sample. Such decisions are based on the probability of occurrence of the sample, and they require the decision maker to be able to determine that probability. Thus, we must be able to define a statistic that measures what we want and for which we can determine the family of distributions that governs its chance variation. Once the family has been determined, we must identify which member of the family is applicable.

In this chapter we will develop certain inferential procedures that may be used in one-sample situations. This is a broad categorization that will be broken down into four basic subgroups within the chapter: (1) inferences about proportions; (2) inferences about parameters of normal populations; (3) procedures with less restrictive assumptions; and (4) inferences about the family or type of distribution. Whenever appropriate we will continue to utilize these basic subgroups in succeeding chapters.

Every statistical procedure has certain assumptions associated with its use. We will stress this fact throughout, and with each procedure we will identify whatever assumptions are necessary for its development and application. The assumptions will be necessary for determining what type of distribution to use in calculating the probability associated with a value of our statistic. For convenience we will construct our tests to correspond with the format of the table that will be used. This means that in most cases we will not actually calculate a probability to be compared with α. Instead we will determine, from a table, the maximum and/or minimum value (of the statistic) corresponding to α, then compare the statistic with that value. If the statistic is more extreme (farther out in the tail) than the tabulated value, this will imply that the probability associated with the sample is less than α and hence that H_0 should be rejected.

Our rationale for placing the more restrictive procedures, those based on the normal family, first is that they are the more commonly used. The distribution-free or nonparametric procedures are generally easier to use and to understand, but beginning with them might tend to place undue emphasis on them.

10.2 COMPARING PROCEDURES

As you will see, we often have more than one procedure for testing the same thing. Generally, it is the underlying assumptions that differentiate among the tests available for a given hypothesis set. If the assumptions for a procedure are violated, the main effect on the results is that any probabilities calculated under the assumptions are at best approximations. If the approximations are poor, the decisions and conclusions based on them may be the opposite of what they should be. If at any time you are uncertain about the validity of an assumption, drop back to a procedure that does not require that particular assumption. You will sacrifice some power (ability to reject H_0 when it is false) if the assumption is valid, but you will be more confident of your results if it is not.

All our procedures will assume that at least a certain level of measurement can be associated with the sample observations. The various levels were defined in Chapter 1, but since they are inherent to our developments in this and succeeding chapters, we will review them briefly.

1. Nominal measurement—this implies that we can categorize each sample observation as belonging to one of k mutually exclusive sets.
2. Ordinal measurement—this requires us to be able to take each observation and identify its relative position with respect to all other sample observations.
3. Interval measurement—this requires us to be able to associate a precise (within reason, with no gross rounding) numerical measurement with each observation. In such cases the numerical values will identify both positions and distances between observations.

Many texts tend to ignore levels of measurement in the assumptions for various procedures. We grant that they are less critical for large samples but stress that they may be a deciding factor for smaller samples.

To compare procedures, we must have a criterion for making such comparisons. The criterion that is cited most often is the concept of asymptotic relative efficiency (ARE). Before we discuss this concept further, it is necessary to define certain terms that will be used. Statistical procedures can be classified into two broad categories, distribution-free (also known as nonparametric) and parametric procedures.

DEFINITION 10.1 A statistical procedure is said to be *distribution-free* (DF) if the distribution of the sample statistic on which the procedure is based can be obtained without assuming a specific family of distributions for the population(s) in general.

DEFINITION 10.2 A statistical procedure is called *parametric* if it assumes a specific family of distributions for the population under consideration.

As the definitions indicate, the assumptions underlying the parametric procedures are more restrictive. Thus, one advantage of using distribution-free procedures is that they are more widely applicable. DF procedures also are easier to understand and simpler computationally (for smaller n) than their parametric counterparts. The major criticism of the DF procedures is that they are less powerful than parametric tests if the assumptions of the parametric test are satisfied, since the DF statistic may not utilize all the information available in the sample. Although this criticism is true, it is irrelevant to some extent, since if the parametric assumptions are satisfied, the parametric procedure should be used. The following definitions provide the standard basis for comparing test procedures.

DEFINITION 10.3 A statistical test is said to be *consistent* for a specified alternative H_a if the power approaches 1 as the sample size increases without bound, that is, if

$$\lim_{n \to \infty} \text{pr}\{\text{reject } H_0 \mid H_a \text{ true}\} = 1.0$$

In simplified form Definition 10.3 states that if a test is consistent, it is theoretically possible to select a large enough sample to ensure that the correct decision will be made. When comparing procedures we will require the property of consistency for all tests under consideration.

DEFINITION 10.4 Let D_1 and D_2 be consistent tests (decision rules) of H_0 versus H_a. Then D_1 is said to be more *efficient* than D_2 if the sample size n_1 required by D_1 to attain a specified power $1 - \beta$ is less than the sample size n_2 required by D_2. The limiting ratio n_1/n_2 of these sample sizes as n_1 approaches infinity and H_a approaches H_0 is called the *asymptotic relative efficiency* (ARE) of D_2 to D_1:

$$\text{ARE}(D_2 \text{ to } D_1) = \lim_{\substack{n_1 \to \infty \\ H_a \to H_0}} \frac{n_1}{n_2}$$

Violations of assumptions are most critical when we are using small samples, whereas ARE compares procedures strictly on a large-sample basis. Also, ARE generally is calculated under the assumption that the conditions of the more restrictive procedures are satisfied. The DF procedures are recommended whenever there is any doubt about the validity of the assumptions for the more restrictive procedure and when the sample sizes are small to moderate ($n \leqslant 30$). For samples in this size range the DF procedures are simpler, faster, and nearly as efficient as their parametric counterparts even if the parametric assumptions are justified. For $n > 30$, ranking procedures become quite cumbersome, and for very large n they become difficult even on a computer.

10.3 INFERENCES ABOUT PROPORTIONS

In Chapter 9 we briefly discussed inferences concerning the parameter p for a binomial distribution. In that particular instance we restricted ourselves to a specific example of an hypothesis test for which the binomial tables in the Appendix could be used. In this section we shall obtain general tests (decision rules) for both exact tests and large-sample approximations. In the large-sample case we will obtain interval estimates for p and develop a formula for determining the sample size required to estimate p with a specified maximum error for a confidence level of $1 - \alpha$.

10.3.1 Exact Tests

By *exact tests* we mean cases in which the binomial distribution with parameters n and p_0 is tabulated. Many texts would consider this as a small-sample situation, but what is relevant is not so much the sample size as the completeness of the available tables. The hypotheses of interest take the same basic forms here that they took in Chapter 9. In the general case these possible hypothesis sets, as dictated by the alternative of interest, are:

1. H_0: $p \leqslant p_0$ $\quad H_a$: $p > p_0$
2. H_0: $p \geqslant p_0$ $\quad H_a$: $p < p_0$
3. H_0: $p = p_0$ $\quad H_a$: $p \neq p_0$

The first set indicates that we are interested only in detecting shifts of p to the right (increases in the population proportion). Set 2 implies that we wish to detect decreases in p, and set 3 is applicable if we want to detect any change in the population parameter. Recall that the basic test for any hypothesis set is given by the decision rule (Equation 9.6), which says to

Reject H_0 at significance level α if
pr {a sample as extreme or more so than that observed | H_0 is true} $< \alpha$.

For a binomial distribution the random variable is the number of successes X. Thus, the sample will be only as extreme as X is. What constitutes possible extremity is indicated by H_a.

Considering the first hypothesis set, we see that if H_0 is true, we expect a value for X of np_0 or less, since, for a binomial random variable,

$$E[X] = np$$

If H_0 is true, then $p \leqslant p_0$, and correspondingly

$$E[X|H_0] \leqslant np_0$$

Since X is a random variable, what we observe should be in the neighborhood of np_0 or less. On the other hand, if H_a is true, then $E[X|H_a] > np_0$, and we should observe values for X in excess of np_0. The value for α simply indicates how far in the direction of H_a X must be before it is too extreme (unlikely) for H_0 to be believable. Summarizing, for the hypothesis set

$$H_0\colon\ p \leqslant p_0 \qquad H_a\colon\ p > p_0$$

the test is:

Reject H_0 if
pr $\{X \geqslant$ observed number of successes $| \, n, p_0\} < \alpha$

Following the same line of reasoning for

$$H_0\colon\ p \geqslant p_0 \qquad H_a\colon\ p < p_0$$

gives the decision rule:

Reject H_0 if
pr $\{X \leqslant$ observed number of successes $| \, n, p_0\} < \alpha$

For the two-sided case of set 3 we merely combine the previous two decision rules, replacing α with $\alpha/2$ in each. That is, when no preference for a particular tail is indicated, we split the total α in such a way that the two tails are nearly equiprobable under H_0. We say nearly since the distribution is discrete and will be symmetric only if $p_0 = 0.5$. With the two-sided alternative we expect a value of np_0 for our random variable and wish to reject H_0 for extreme deviations of X from this value in either direction. Hence our test is:

Reject H_0 either if

$$\text{pr}\,\{X \geqslant \text{the observed number of successes} \mid n, p_0\} < \frac{\alpha}{2}$$

or if

$$\text{pr}\,\{X \leqslant \text{the observed number of successes} \mid n, p_0\} < \frac{\alpha}{2}$$

Before proceeding, we summarize our results in Table 10.1 letting n_s be the observed number of successes in the sample.

In the Table we have labeled our stated α as only nominal, since the binomial distribution is discrete. The tests have exact significance levels that are $\leqslant \alpha$, with the exact α being the actual tabulated probability; that is, the exact α, say α_E, is (for set 1)

$$\alpha_E = \text{pr}\,\{X \geqslant n_s | n, p_0\} \tag{10.1}$$

EXAMPLE 10.1 The government has ordered Wetstone Rubber Products, Inc., to recall all the items produced over the last three years for one of their main product lines. The company would like to assess the effects of this recall on its total sales volume in the consumer market. The vice president for customer relations has assured

Table 10.1 Summary of Hypotheses and Corresponding Tests for the Binomial Parameter p (Exact Test)

Hypotheses		Test for Nominal α
1. $H_0: p \leqslant p_0$	$H_a: p > p_0$	Reject H_0 if $\text{pr}\,\{x \geqslant n_s \vert n, p_0\} < \alpha$
2. $H_0: p \geqslant p_0$	$H_a: p < p_0$	Reject H_0 if $\text{pr}\,\{x \leqslant n_s \vert n, p_0\} < \alpha$
3. $H_0: p = p_0$	$H_a: p \neq p_0$	Reject H_0 if either $\text{pr}\,\{x \geqslant n_s \vert n, p_0\} < \alpha/2$ or $\text{pr}\,\{x \leqslant n_s \vert n, p_0\} < \alpha/2$

the chairman of the board that Wetstone will experience no more than an additional 10% attrition of return customers. Being somewhat skeptical of this statement, the chairman asks Market Research to test the customer relations claims and report to him in the morning. Normal attrition of return customers runs at approximately 20%. Since time is limited, Market Research selects five company stores at random and for each store randomly selects, from customer records, the names of ten people who made purchases there in the last three years. These people then were called and, without knowing to whom they were talking, asked if they would continue to purchase Wetstone products. Of the 50 people sampled, 21 indicated that they would not buy another Wetstone product. What should Market Research report to the chairman in the morning if they use a nominal α of 0.05? What is the exact α of the test?

SOLUTION Since Customer Relations has stated that the attrition rate, normally 20%, will increase by no more than an additional 10%, the hypotheses of interest are

$$H_0\colon\ p \leqslant 0.30 \qquad H_a\colon\ p > 0.30$$

For $n = 50$ and $\alpha = 0.05$ the decision from Table 10.1 is:

Reject H_0 at a nominal α of 0.05 if
$\text{pr}\{X \geqslant n_s \mid 50, 0.3\} < 0.05$

Checking in Table A.2, we find that

$$\text{pr}\{X \geqslant 21 \mid 50, 0.3\} = 0.0478$$

Since this value is less than 0.05, it implies rejection of H_0. On the basis of this sample, Market Research should conclude that attrition has increased by more than an additional 10%. They actually have two options: (1) they can conclude that this is merely an extreme or unlikely event and let it go at that, or (2) they can reject H_0 on the basis that the sample is too extreme to have reasonably come from a population with $p = 0.3$. If they decide to reject H_0, the probability that they will have committed a Type I error is $\alpha_E = 0.0478$.

Whenever the sample size or p_0 is such that the exact probability cannot be found in the tables, we must resort to an approximation. The hypotheses and tests are unchanged when that occurs. The test merely becomes an approximate rather than an exact test, since the exact probability associated with the sample outcome is unobtainable. In the next subsection we look at two approximation methods given earlier in

Chapters 4 and 5. The only assumptions that are required for the exact tests given in Table 10.1 are:

1. The sample has been selected randomly.
2. The binomial distribution is the appropriate model for describing the chance variation of the number of successes.

If you have forgotten the characterization of a binomial random variable, we suggest you review Section 4.2.

10.3.2 Approximate Tests

To approximate the binomial probabilities required for the tests in Section 10.3.1, we have two options, and which one we select will be contingent on the degree of asymmetry in the original binomial distribution. That is, the deciding factors will be the sample size and how far away from 0.5 the hypothesized value p_0 is. The binomial distribution is symmetric only for $p = 0.5$. For large values of n (say $n > 50$), the approximations will be quite good. We may, by necessity, also be required to use the approximations when $n < 50$. If the stated conditions are satisfied, the approximations will still be reasonably good.

CASE I n *Large and* p$_0$ *Near the Extremes*

In this situation, p_0 near the extremes means that p_0 is either close to 0 or close to 1, and the applicable distribution to use for approximating binomial probabilities is the Poisson. The condition given in Chapter 4 for the Poisson approximation is either $np < 5$ or $n(1 - p) < 5$. For testing purposes this translates to $np_0 < 5$ or $n(1 - p_0) < 5$, and if H_0 is true the appropriate member of the Poisson family has $\mu = np_0$. We can assume without loss of generality that $p_0 < 1 - p_0$ at this point and in the future development. Thus, to evaluate the statement

$$\text{pr}\{X \geqslant n_s | H_0\}$$

we have

$$\text{pr}\{X \geqslant n_s | n, p_0\} = \text{pr}\{X \geqslant n_s | \mu = np_0\} \tag{10.2}$$

Applying this result to the decision rules of Table 10.1 and summarizing, we obtain Table 10.2.

EXAMPLE 10.2 In the past, Andrew Arthurson, Ltd., a public auditing firm, has found that they uncover major accounting errors in the books of 2% of their clients when they perform their annual audits.

Table 10.2 A Summary of Tests for p_0 When the Poisson Distribution Is the Appropriate Approximation

Hypotheses		Approximate Test for Nominal α
1. H_0: $p \leqslant p_0$	H_a: $p > p_0$	Reject H_0 if $\text{pr}\{X \geqslant n_s \mid \mu = np_0\} < \alpha$
2. H_0: $p \geqslant p_0$	H_a: $p < p_0$	Reject H_0 if $\text{pr}\{X \leqslant n_s \mid \mu = np_0\} < \alpha$
3. H_0: $p = p_0$	H_a: $p \neq p_0$	Reject H_0 if either $\text{pr}\{X \geqslant n_s \mid \mu = np_0\} < \alpha/2$ or $\text{pr}\{X \leqslant n_s \mid \mu = np_0\} < \alpha/2$

This year a random sample of 150 of their 8850 clients included six who had major accounting errors in their books. Is this an indication at $\alpha = 0.05$ that the percentage of their clients with major errors in the books is higher this year than previously?

SOLUTION The indicated hypotheses are

$$H_0\colon\ p \leqslant 0.02 \qquad H_a\colon\ p > 0.02$$

Since the population is finite in this case and sampling is assumed to be without replacement from a dichotomous population (either a client's records have a major error or they don't), the appropriate distribution would be the hypergeometric, with parameters $N = 8850$, $R = 177$ (if H_0 is true), and $n = 150$. Thus H_0 should be rejected if

$$\text{pr}\{X \geqslant 6 \mid N = 8850, R = 177, n = 150\} < 0.05$$

To evaluate this statement exactly is a nearly impossible task. We recall that the binomial distribution gives very close approximations to the hypergeometric whenever $n/N < 0.05$, and we find this ratio to be $150/8850 = 0.0169$ for this example. Thus we may approximate the given probability with

$$\text{pr}\{X \geqslant 6 \mid n = 150, p = 0.02\}$$

if H_0 is true.

Referring to Table A.2, we find that neither an n of 150 nor a p of 0.02 is listed. This forces us to approximate the approximation, and when we calculate np_0 we find that the Poisson distribution is appropriate. Thus, our decision rule finally becomes

Reject H_0 at a nominal α of 0.05 if
$\text{pr}\{X \geqslant 6 \mid \mu = 150(0.02)\} < 0.05$

Referring to Table A.4, we obtain

$$\text{pr}\{X \geqslant 6 \mid \mu = 3.0\} = 0.0839$$

Since $0.0839 \geqslant 0.05$, we conclude that on the basis of this sample there is not sufficient evidence to indicate rejection of H_0. That is, 6 or more out of 150 is not too unreasonable when only 2% of the accounts have major errors.

Before we go to our other approximation, we pause to consider the chance of making an error of the second kind. That is, we consider the calculation of β when a specific value in the alternative set is given. Letting the specific alternative be denoted by p_1, we have, in general, that

$$\beta_{p_1} = \text{pr}\{X \text{ is less extreme than } x_\alpha \mid p = p_1\} \tag{10.3}$$

where x_α is the least extreme value possible such that

$$\text{pr}\{X \text{ is as extreme or more so than } x_\alpha \mid p = p_0\} < \alpha$$

Thus, we need to use the alternative form of our test and use the tables to find x_α, where x_α is simply the least extreme value that can cause H_0 to be rejected. We apply this concept to the data of Example 10.2.

EXAMPLE 10.3 Determine x_α for Andrew Arthurson, Ltd., and use it to find both the probability of a Type II error and the power of the test if the true proportion of client accounts having major errors this year is 0.03.

SOLUTION We are still forced to use the Poisson approximation. On referring to Table A.4, we find (only large values of X are considered extreme)

$$\text{pr}\{X \geqslant 6 \mid \mu = 3.0\} = 0.0839$$

and

$$\text{pr}\{X \geqslant 7 \mid \mu = 3.0\} = 0.0335$$

Thus, for Arthurson, $x_\alpha = x_{0.05} = 7$, and an equivalent test is to reject H_0 if $n_s \geqslant 7$. The exact α for our test is $\alpha_E = 0.0335$.

The probability of committing a Type II error when $p_1 = 0.03$ is

$$\beta_{0.03} = \text{pr}\{X < 7 \mid \mu = 4.5\} = 0.8311$$

Correspondingly, the power of the test is

$$\text{Power} = 1 - \beta_{0.03} = 1 - 0.8311 = 0.1689$$

Note that the ability of the test in Examples 10.2 and 10.3 to differentiate between 0.02 and 0.03 is very small. The only ways to improve this ability are:

1. Let α increase, thus increasing the chance of a Type I error
2. Increase the sample size with α remaining fixed

Which we would use depends on the cost of committing a Type I error and the cost of increasing the sample size.

CASE II Both np_0 *and* $n(1 - p_0) \geqslant 5$

This condition is the one imposed in Chapter 4, when we used a member of the normal distribution to obtain an approximate value for a binomial probability. To evaluate the probability statement

$$\text{pr}\{X \leqslant k | n, p\}$$

we first adjust the statement to correct for continuity by rewriting it to correspond with the appropriate class boundaries. After this adjustment, we find the approximate probability using a normal distribution with $\mu = np$ and $\sigma = \sqrt{np(1 - p)}$. That is,

$$\text{pr}\{X \leqslant k | n, p\} = \text{pr}\{X \leqslant k + 0.5 | n, p\} = \text{pr}\left\{Z \leqslant \frac{(k + 0.5) - np}{\sqrt{np(1 - p)}}\right\}$$

Making this adjustment in our previous decision rules, we summarize the results in Table 10.3 (page 300). For illustration, let us reconsider Examples 10.2 and 10.3 with an increased sample size.

EXAMPLE 10.4 Suppose that Andrew Arthurson, Ltd., increased their sample size from 150 to 300 and observed 12 clients with major accounting errors. Determine a test at $\alpha = 0.05$ based on the number of major errors observed rather than the probability of occurrence specifically. Also determine the power of the test when $p_1 = 0.03$.

SOLUTION With the increase in n from 150 to 300, np_0 becomes 6 and $n(1 - p_0)$ is 294. Both of these are greater than 5, implying the normal approximation with (under H_0) $\mu = 300(0.02) = 6.0$ and $\sigma = \sqrt{(300)(0.02)(0.98)} = \sqrt{5.88} = 2.425$. The basic decision is:

Table 10.3 A Summary of Tests for p_0 When a Normal Distribution Is the Appropriate Approximation

Hypotheses	Approximate Test for Nominal α
1. $H_0: p \leqslant p_0$ $\quad H_a: p > p_0$	Reject H_0 if $\text{pr}\left\{Z \geqslant \frac{(n_s - 0.5) - np_0}{\sqrt{np_0(1-p_0)}}\right\} < \alpha$
2. $H_0: p \geqslant p_0$ $\quad H_a: p < p_0$	Reject H_0 if $\text{pr}\left\{Z \leqslant \frac{(n_s + 0.5) - np_0}{\sqrt{np_0(1-p_0)}}\right\} < \alpha$
3. $H_0: p = p_0$ $\quad H_a: p \neq p_0$	Reject H_0 if either $\text{pr}\left\{Z \geqslant \frac{(n_s - 0.5) - np_0}{\sqrt{np_0(1-p_0)}}\right\} < \frac{\alpha}{2}$ or $\text{pr}\left\{Z \leqslant \frac{(n_s + 0.5) - np_0}{\sqrt{np_0(1-p_0)}}\right\} < \frac{\alpha}{2}$

Reject H_0 if
$$\text{pr}\left\{Z \geqslant \frac{(n_s - 0.5) - np_0}{\sqrt{np_0(1-p_0)}}\right\} < \alpha$$

Since $\text{pr}\{Z \geqslant z_\alpha\} = \alpha$, we have

$$\frac{(n_s - 0.5) - np_0}{\sqrt{np_0(1-p_0)}} = z_\alpha$$

and solving for n_s gives

$$n_s = (np_0 + 0.5) + z_\alpha\sqrt{np_0(1-p_0)}$$

Substituting our specific values of $n = 300$, $p_0 = 0.02$, and $\alpha = 0.05$ gives $n_s = 10.489$, which implies that our decision rule becomes:

Reject H_0 at $\alpha = 0.05$ if
$n_s \geqslant 10.489$, or $n_s \geqslant 11$

since our variables are discrete. Thus, since $12 > 11$, H_0 would be rejected.

To determine the power we calculate

$$\begin{aligned} \text{pr}\{X \geqslant 11 | n = 300, p_1 = .03\} &= \text{pr}\{X \geqslant 11 - 0.5 | \mu = 9, \sigma = 2.955\} \\ &= \text{pr}\left\{Z \geqslant \frac{10.5 - 9}{2.955}\right\} \\ &= \text{pr}\{Z \geqslant 0.508\} \\ &= 0.3057 \end{aligned}$$

where we have interpolated between $z = 0.50$ and $z = 0.51$. Note that power has increased with the larger sample size but is still somewhat on the low side.

In the next subsection we apply the normal approximation to obtain approximate confidence intervals for a population proportion. An increase in sample size results in an increase of power, but the increase in Example 10.4 was not enough to obtain a reasonably powerful test. Therefore, we also will consider the determination of the sample size required to give a confidence interval of prespecified length and tests that have a prespecified power as well as α.

10.3.3 Interval Estimates and the Determination of *n*

In this section we restrict ourselves to consideration of cases in which n is large and the conditions for the normal approximation are satisfied empirically. By empirically we mean that the observed value indicates that np and $n(1 - p)$ are $\geqslant 5$. This translates to $X \geqslant 5$ and $n - X \geqslant 5$, where X is the observed number of successes. Referring to Section 9.2, we wish to determine an LL and a UL for p such that

$$\text{pr}\{[\text{LL, UL}] \text{ covers } p\} \geqslant 1 - \alpha$$

By the normal approximation we have

$$\frac{X - np}{\sqrt{np(1 - p)}} \sim N(0, 1) \qquad \text{or} \qquad \frac{X - np}{\sqrt{np(1 - p)}} \doteq Z$$

Thus, as before, we start with

$$\text{pr}\left\{-z_{\alpha/2} \leqslant \frac{X - np}{\sqrt{np(1 - p)}} \leqslant z_{\alpha/2}\right\} \doteq 1 - \alpha$$

With p as the unknown, we have a problem: the standard deviation of X is a function of p. To circumvent this problem, we merely replace p in our standard deviation with its estimate, that is, with x/n.

Making the indicated replacement and solving the equation for p yields the standard form

$$\text{pr}\left\{\frac{x}{n} - z_{\alpha/2}\sqrt{\frac{(x/n)[1-(x/n)]}{n}} \leqslant p \leqslant \frac{x}{n} + z_{\alpha/2}\sqrt{\frac{(x/n)[1-(x/n)]}{n}}\right\} = 1-\alpha \tag{10.4}$$

or

$$\text{LL} = \frac{x}{n} - z_{\alpha/2}\sqrt{\frac{(x/n)[1-(x/n)]}{n}}$$

$$\text{UL} = \frac{x}{n} + z_{\alpha/2}\sqrt{\frac{(x/n)[1-(x/n)]}{n}} \tag{10.5}$$

To obtain one-sided interval estimates, those with a single bound, we need only make the adjustments indicated in Chapter 9.

EXAMPLE 10.5 The Office for Consumer Safety (OCS) has received numerous complaints about another of Wetstone's (Example 10.1) product lines. In order to estimate the proportion of items in this line that have safety defects, OCS randomly selects 2500 items and tests them for safety defects. Of the 2500, eighty-two were found to be defective. What is a 99% upper bound on the percentage of items in the product line that have safety defects?

SOLUTION Our point estimate of p is $x/n = 82/2500 = 0.0328$. The resulting estimate for the standard deviation of the estimate is

$$\hat{\sigma}_{x/n} = \sqrt{\frac{(x/n)[1-(x/n)]}{n}} = \sqrt{\frac{0.0328(0.9672)}{2500}} = 0.0036$$

Since only an upper bound is required, our equation for UL becomes

$$\text{UL} = \frac{x}{n} + z_{\alpha}\sqrt{\frac{(x/n)[1-(x/n)]}{n}}$$

Substituting the numerical values, this bound is

$$\begin{aligned}\text{UL} &= 0.0328 + z_{0.01}(0.0036)\\ &= 0.0328 + 2.326(0.0036)\\ &= 0.0411\end{aligned}$$

Therefore, we have an interval for p given by

$$[0, 0.0411]$$

The lower bound for our interval is 0, since logically p cannot be less than 0.

In most of the examples so far we have selected a confidence level and sample size and pretty much let the width of the interval fall where it may. However, in many cases we prefer to plan or design our experiment in such a way that we can specify in advance that the resulting estimate will be no more than a given distance epsilon (ϵ) from the true value with a probability of $1 - \alpha$. The width of the interval is dictated by the confidence level and the standard error (deviation) of the estimator. That is, in the general two-sided case for p, the width w is given by UL − LL or

$$w = \text{UL} - \text{LL} = \frac{x}{n} + z_{\alpha/2}\,\sigma_{x/n} - \left(\frac{x}{n} - z_{\alpha/2}\,\sigma_{x/n}\right)$$
$$= 2z_{\alpha/2}\,\sigma_{x/n} = 2\epsilon$$

If we specify ϵ and α in advance, we may solve for the required sample size. Doing so gives

$$\epsilon = z_{\alpha/2}\,\sigma_{x/n} = z_{\alpha/2}\,\sigma_x / \sqrt{n}$$

since

$$\sigma_{x/n} = \sqrt{\frac{p(1-p)}{n}} = \frac{\sigma_x}{\sqrt{n}}$$

Upon solving this form we obtain

$$n = \left(\frac{z_{\alpha/2}\,\sigma_x}{\epsilon}\right)^2 = \left(\frac{z_{\alpha/2}\sqrt{p(1-p)}}{\epsilon}\right)^2 \tag{10.6}$$

There is still a problem, since σ_x is a function of p, which is the parameter we are trying to estimate. We have a way around this difficulty, though, for we can substitute the *maximum* value of σ_x to obtain the *largest sample size* that would be required to achieve the specified values. This maximum value occurs when $p = 0.5$. Using this in Equation 10.6 gives

$$n = \left(\frac{z_{\alpha/2}\sqrt{(1/2)(1/2)}}{\epsilon}\right)^2$$

or

$$n = \left(\frac{z_{\alpha/2}}{\epsilon}\right)^2\left(\frac{1}{4}\right) \tag{10.7}$$

We stress that the sample size yielded by Equation 10.7 is the largest that would possibly be required. If an alternative bound on p is available—for example, if the proportion of defectives cannot logically exceed 0.3—then that bound may be used to obtain a smaller sample size. If we denote this alternative bound on p by p_B, Equation 10.7 becomes

$$n = \left(\frac{z_{\alpha/2}}{\epsilon}\right)^2 p_B(1 - p_B) \qquad (10.8)$$

In general p_B is the closest value to 0.5 that is believed possible. When calculating a sample size we *always round upward* to the next integer value if rounding is required.

EXAMPLE 10.6 Referring to Example 10.5, assume that OCS wants to estimate the proportion of items having safety defects to within 0.01 of the true value. What size of sample would allow them to do so?

SOLUTION We make some minor alterations to our formulas, since they pertain to two-sided intervals and OCS desires only a single bound. We merely use z_α in the formulas rather than $z_{\alpha/2}$.

If OCS has absolutely no idea of the unknown proportion, they would use Equation 10.7 and obtain, for $\epsilon = 0.01$ and $1 - \alpha = 0.99$, a sample size of

$$n = \left(\frac{2.326}{0.01}\right)^2 \left(\frac{1}{4}\right) = 13{,}525.69 \text{ or } 13{,}526 \text{ items}$$

This is a very large sample and could be quite costly to obtain. Suppose OCS is confident that the true proportion is no larger than 0.10. Using this value for p_B in Equation 10.8 gives

$$n = \left(\frac{2.326}{0.01}\right)^2 (0.1)(0.9) = 4869.2484 \text{ or } 4870 \text{ items}$$

Just as we stipulated the maximum error ϵ for an estimate and calculated the sample size required to ensure that ϵ is not exceeded, we may specify a maximum value of β as well as α for a given alternative p_1. Doing so in the large n case gives two simultaneous equations in x_α and n. For the hypothesis set

$$H_0\colon p \leqslant p_0 \qquad H_a\colon p > p_0$$

these equations are

$$x_\alpha = (np_0 + 0.5) + z_\alpha \sqrt{np_0(1 - p_0)}$$

and

$$x_\alpha = (np_1 + 0.5) - z_\beta \sqrt{np_1(1 - p_1)}$$

The simultaneous solution of these equations is

$$n = \left(\frac{z_\alpha \sqrt{p_0(1 - p_0)} + z_\beta \sqrt{p_1(1 - p_1)}}{p_1 - p_0} \right)^2 \tag{10.9}$$

If the alternative is the other one-sided form, the only effect would be the use of $-z_\alpha$ and $-z_\beta$ in Equation 10.9. This would not change the value of n, since the term inside the parentheses is squared anyway. For a two-sided alternative we replace z_α by $z_{\alpha/2}$.

EXAMPLE 10.7 OCS of Examples 10.5 and 10.6 has decided that if the proportion defective in that product line is less than 4%, no action will be taken. But if they conclude that it is 4% or more, a total recall of this product line will be demanded. If OCS would like to fail to demand a recall test (to reject H_0) by chance with a probability of no more than 0.05 and to demand a recall by chance, when $p = 0.02$, with a probability of no more than 0.10, how large a random sample will they need to take?

SOLUTION The hypotheses of interest in this situation are

$$H_0\colon\ p \geqslant 0.04 \qquad H_a\colon\ p < 0.04$$

Identifying the elements for using Equation 10.9, we have $\alpha = 0.05$, $\beta = 0.10$, $p_0 = 0.04$, and $p_1 = 0.02$. Using Table A.6 gives $z_\alpha = 1.645$ and $z_\beta \doteq 1.282$. Therefore, the necessary sample size is

$$n = \left(\frac{-1.645\sqrt{0.04(0.96)} - 1.282\sqrt{0.02(0.98)}}{0.02 - 0.04} \right)^2$$

$$= \left(\frac{-0.3224 - 0.1795}{-0.02} \right)^2 = 629.76 \text{ or } 630 \text{ items}$$

The basic assumptions underlying all the procedures in this section are that:

1. A random sample is selected.
2. A binomial model is applicable.

In addition, to use the Poisson approximation, we assume that:

3. The binomial table is inadequate and either $np_0 < 5$ or $n(1 - p_0) < 5$.

If the normal approximation is to be used, the third assumption becomes

3′. The binomial table is inadequate and both $np_0 \geqslant 5$ and $n(1 - p_0) \geqslant 5$.

EXERCISES

10.3.1 Refer to Example 10.6 and assume that the Office for Consumer Safety feels that the sample size of 13,526 is too large but is unwilling to place a bound on p. Instead they decide to change ϵ from 0.01 to 0.02. What sample size will they need? Assume that α is maintained at 0.01.

10.3.2 What is the resulting change in sample size if OCS drops its required confidence level from 0.99 to 0.95?

10.3.3 Determine 99% upper bounds for OCS based on the two sample sizes calculated in Example 10.6. Assume that the resulting point estimate in both cases is 0.0328, and verify that ϵ is no larger than was specified.

10.3.4 Most manufacturing companies keep a close watch on product quality by periodically selecting random samples and checking for defectives. If too many defectives occur, they conclude that the process is out of control and stop it to search for a cause. The Pogo Company makes gidgets, an inexpensive but popular item, and feels that their quality is acceptable if the percentage of defectives does not go above 10%. Determine a sampling plan for Pogo, that is, (n, x_α), that has $n = 20$ and $\alpha = 0.15$.

10.3.5 Refer to Exercise 10.3.4 and determine the probability that your plan will fail to detect a shift in the quality level to 0.20; to 0.25; to 0.4.

10.3.6 Refer to Exercise 10.3.4. Pogo is unhappy with the power of your sampling plan. They have decided to change α to 0.10 from 0.15 and would like a new sampling plan that will detect shifts in the quality level to 0.20 with a probability of at least 0.80. Determine a new sampling plan for Pogo that will satisfy these requirements.

10.4 INFERENCES ABOUT NORMAL PARAMETERS

As with the binomial parameter p, we first encountered inferences about the location parameter for a normal distribution in Chapter 9. In this section we formalize those results for the two cases involving known and unknown σ's. In addition, we will look at inferences involving the variance of a normal distribution. All the developments of this section are contingent on the following assumptions being satisfied.

1. The observations constitute a random sample.
2. The underlying distribution is normal.
3. The numerical values associated with the observations represent an interval level of measurement.

Assumption 2 may be relaxed when the parameter of interest is μ, since the central limit theorem can be applied to obtain an approximately normal distribution for the estimator $\overline{X}$. The third assumption can be critical for small sample sizes but is less important as n increases. In general, procedures based on normal distributions are robust, which is defined as follows:

DEFINITION 10.5 An inferential procedure is said to be statistically *robust* if it is insensitive to slight violations of the assumptions on which it is based.

The procedures of this section become more robust as n gets larger.

10.4.1 Inferences for μ, Variance Known

The additional assumption implied by "variance known" is that the population variance σ^2 is a known value. In practice, this is unlikely to be the case, but we cover it here for its informative value. We wish to make inferences regarding the population mean and, recalling the requirements for doing so as given in Chapter 9, we have:

1. $\overline{X}$ is an efficient estimator of μ.
2. $\overline{X}$ has at least an approximately normal distribution.
3. $\mu_{\overline{X}} = \mu$, $\sigma_{\overline{X}} = \sigma/\sqrt{n}$ if the population is infinite.

For our development we will assume that the hypothesis set

$$H_0\colon\ \mu \leqslant \mu_0 \qquad H_a\colon\ \mu > \mu_0$$

is the one of interest. The same logic would hold for either of the other two sets. Applying the results of assumptions 1, 2, and 3 to the basic decision rule of Chapter 9 (Equation 9.6), our test for this hypothesis set is:

Reject H_0 at significance level α if
$\text{pr}\{\overline{X} \geqslant \bar{x} | H_0\} = \text{pr}\{\overline{X} \geqslant \bar{x} | \mu = \mu_0, \sigma_{\overline{X}} = \sigma/\sqrt{n}\} < \alpha$

As we saw in Chapter 9, this decision rule can be transformed into two alternative but equivalent forms. The first of these determines the most extreme value of $\overline{X}$ that is reasonable under H_0 for the specified α, whereas the second determines the maximum number of standard deviations that is reasonable for the same conditions.

Denoting the most extreme value of $\overline{X}$ as $\overline{x}_\alpha$, we need to determine the value of $\overline{x}_\alpha$ such that

$$\text{pr}\{\overline{X} \geqslant \overline{x}_\alpha | H_0\} = \alpha$$

Converting to the standardized form, this is

$$\text{pr}\left\{\frac{\overline{X} - \mu_0}{\sigma_{\overline{X}}} \geqslant \frac{\overline{x}_\alpha - \mu_0}{\sigma_{\overline{X}}}\right\} = \alpha$$

But $(\overline{X} - \mu_0)/\sigma_{\overline{X}} = Z$ and $\text{pr}\{Z \geqslant z_\alpha\} = \alpha$. Equating and solving for $\overline{x}_\alpha$ gives

$$\frac{\overline{x}_\alpha - \mu_0}{\sigma_{\overline{X}}} = z_\alpha$$

$$\overline{x}_\alpha = \mu_0 + z_\alpha \sigma_{\overline{X}}$$

Therefore, an alternative form of our basic decision rule is to

Reject H_0 if
$\overline{X} > \mu_0 + z_\alpha \sigma_{\overline{X}}$ (10.10)

For the second alternative we note that the only possible way the observed value for $\overline{X}$ can exceed $\overline{x}_\alpha$ is if the observed number of standard deviations from μ exceeds z_α. That is, rejecting H_0 if $\overline{X} > \overline{x}_\alpha$ is equivalent to rejecting H_0 if

$$\frac{\overline{X} - \mu_0}{\sigma_{\overline{X}}} > \frac{\overline{x}_\alpha - \mu_0}{\sigma_{\overline{X}}} = \frac{(\mu_0 + z_\alpha \sigma_{\overline{X}}) - \mu_0}{\sigma_{\overline{X}}}$$

$$= \frac{z_\alpha \sigma_{\overline{X}}}{\sigma_{\overline{X}}} = z_\alpha$$

That is, we

Reject H_0 if
$\dfrac{\overline{X} - \mu_0}{\sigma_{\overline{X}}} > z_\alpha$ (10.11)

These two decision rules are illustrated in Figure 10.1.

Figure 10.1 Graphs Illustrating Decision Rules 10.10 and 10.11

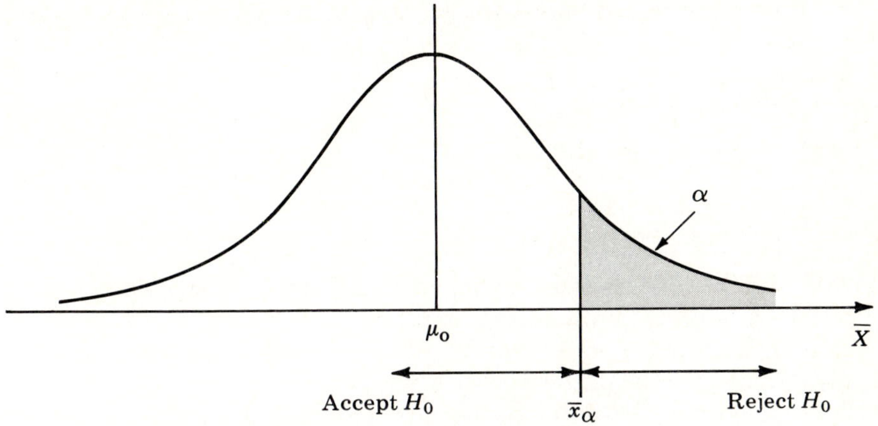

(a) Graph of Decision Rule Using x_α

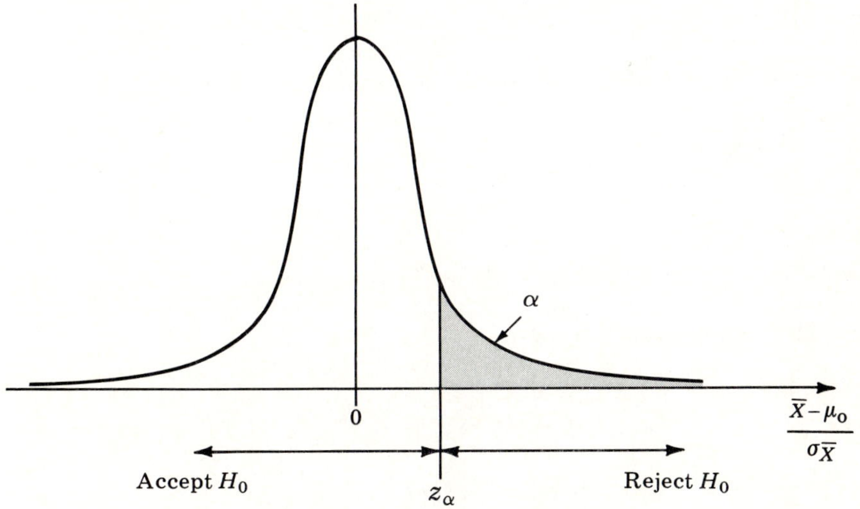

(b) Graph of Decision Rule Using z_α

We recommend that the student draw an appropriate sketch for the decision rule each time with the rejection region shaded. Doing so makes an incorrect decision less likely. We summarize the decision rules corresponding to the three basic hypothesis sets for μ in Table 10.4 for both $\bar{x}_\alpha$ and z_α.

When the hypotheses of interest correspond to the third set in Table 10.4*b*, we can combine the double decision rule into a single one by utilizing the fact that the standard normal distribution is symmetric about 0. The single decision rule becomes:

Reject H_0 if

$$\left| \frac{\bar{X} - \mu_0}{\sigma_{\bar{X}}} \right| > z_{\alpha/2} \qquad (10.12)$$

EXAMPLE 10.8 The marketing department of Hillebrand Enterprises provides the production department with a monthly sales forecast. Production has been complaining for years that Marketing is always overly optimistic with their forecasts. Marketing, on the other hand, says that they use statistical procedures in making their forecasts and that there is no bias whatsoever in them. They finally compromise

Table 10.4 Summaries of Tests for μ With a Known Variance When the Decision Rule Is Based on (*a*) $\bar{x}_\alpha$ and (*b*) z_α

Hypothesis Set		Decision Rule
(*a*) A Summary of Decision Rules Based on $\bar{x}_\alpha$ When σ is Known		
1. H_0: $\mu \leqslant \mu_0$	H_a: $\mu > \mu_0$	Reject H_0 if $\bar{X} > \mu_0 + z_\alpha \sigma_{\bar{X}}$
2. H_0: $\mu \geqslant \mu_0$	H_a: $\mu < \mu_0$	Reject H_0 if $\bar{X} < \mu_0 - z_\alpha \sigma_{\bar{X}}$
3. H_0: $\mu = \mu_0$	H_a: $\mu \neq \mu_0$	Reject H_0 if either $\bar{X} > \mu_0 + z_{\alpha/2} \sigma_{\bar{X}}$ or $\bar{X} < \mu_0 - z_{\alpha/2} \sigma_{\bar{X}}$
(*b*) A Summary of Decision Rules Based on z_α When σ is Known		
1. H_0: $\mu \leqslant \mu_0$	H_a: $\mu > \mu_0$	Reject H_0 if $\frac{\bar{X} - \mu_0}{\sigma_{\bar{X}}} > z_\alpha$
2. H_0: $\mu \geqslant \mu_0$	H_a: $\mu < \mu_0$	Reject H_0 if $\frac{\bar{X} - \mu_0}{\sigma_{\bar{X}}} < -z_\alpha$
3. H_0: $\mu = \mu_0$	H_a: $\mu \neq \mu_0$	Reject H_0 if either $\frac{\bar{X} - \mu_0}{\sigma_{\bar{X}}} > z_{\alpha/2}$ or $\frac{\bar{X} - \mu_0}{\sigma_{\bar{X}}} < -z_{\alpha/2}$

by agreeing to let Accounting take a random sample of 12 previous forecasts and run a statistical test on the errors at $\alpha = 0.05$. The sample results (in thousands of units) are (forecast error = actual sales - sales forecast):

1.4 -0.3 2.6 1.8 1.2 -1.1 1.7 3.1 0.9 2.1 1.5 -0.8

Assuming that the standard deviation of the forecast errors is 2.0 (in thousands), test Marketing's claim of no bias.

SOLUTION Assuming that Accounting is indifferent, they should construct their hypotheses as

$$H_0\colon \mu = 0 \qquad H_a\colon \mu \neq 0$$

That is, if there is no bias in the forecast errors, then the average error should be zero. Being uninfluenced by Production's claim, we prefer to test for any bias, not just a positive one; thus, we use the two-sided alternative. Using the decision rule indicated in Table 10.4*a*, Accounting should reject H_0 if either

$$\overline{X} > 0 + 1.96\left(\frac{2.0}{\sqrt{12}}\right) = 1.132$$

or

$$\overline{X} < 0 - 1.96\left(\frac{2.0}{\sqrt{12}}\right) = -1.132$$

The sample value for $\overline{X}$ is

$$\overline{x} = \frac{1}{12}\sum_{i=1}^{12} x_i = \frac{1}{12}(14.1) = 1.175$$

Since $1.175 > 1.132$, we reject H_0 and conclude that on the basis of this sample it appears that Marketing's forecasts have a definite bias.

Note that if, in Example 10.8, the alternative had been stated to reflect Production's opinion, we would still have rejected H_0. This should be obvious, since for the alternative $H_a\colon \mu > 0$ the critical $\overline{X}$ value, $\overline{x}_\alpha$, is 0.950, because the value for z_α is 1.645, not 1.96. Thus, the sample data actually support the contention of the production department.

If our main interest is in estimating the population mean and the variance happens to be known in advance, we solve the probability statement

$$\text{pr}\{-z_{\alpha/2} \leqslant \frac{\overline{X} - \mu}{\sigma_{\overline{X}}} \leqslant z_{\alpha/2}\} = 1 - \alpha$$

for μ. We did this in Chapter 9 and obtained limits of

$$\text{LL} = \overline{x} - z_{\alpha/2}\sigma_{\overline{X}}$$

and

$$\text{UL} = \overline{x} + z_{\alpha/2}\sigma_{\overline{X}}$$

If a single bound is desired, we simply adjust the limit of interest by using z_α rather than $z_{\alpha/2}$.

EXAMPLE 10.9 Having concluded that Marketing's forecasts are biased, Hillebrand Enterprises would like to obtain a 95% confidence interval for the true average bias in the forecasts. Use the data of Example 10.8 to provide them with one.

SOLUTION The data yield a sample mean of 1.175 and the population variance is indicated to be 4.0. Thus, a 95% interval estimate for μ is given by

$$\text{LL} = 1.175 - 1.96\left(\frac{2}{\sqrt{12}}\right) = 0.043$$

$$\text{UL} = 1.175 + 1.96\left(\frac{2}{\sqrt{12}}\right) = 2.307$$

and we have an interval of

$$[0.043, 2.307]$$

The interpretation of this interval is that there is a 95% chance that this interval contains the true error, where 0.043 represents 43 units and 2.307 implies 2307 units.

An appealing aspect of the unrealistic assumption of a known variance is that it facilitates power calculations and enables us to determine the sample sizes required for a given power and α quite easily. We consider sample size determination in Section 10.4.3 and simply review power in the current discussion. The power of a test is defined in Chapter 9 to be the probability that H_0 will be rejected if it is false, that is,

$$\text{Power} = \text{pr}\{\text{reject } H_0 \mid H_a \text{ true}\}$$

The calculation of a value for the power of a test requires that a specific alternative value be stated. If we let μ_a represent that value, then the power at μ_a becomes

$$P_r(\mu_a) = \text{pr}\{\text{reject } H_0 \mid \mu = \mu_a\}$$

where $P_r(\mu_a)$ is the notation used in Chapter 9 for the rejection probability at μ_a.

EXAMPLE 10.10 The advertising department for the Pogo Company contracted for a series of new television commercials for its gidgets in order to increase sales. The new commercials have increased advertising costs by \$3200 per week. Prior to the running of the new commercials, sales were averaging \$68,000 per week and had a standard deviation of \$14,200. Pogo has decided to evaluate the effect of the commercials once they feel that sales have stabilized by selecting a random sample of 10 weeks' sales. If Pogo uses $\alpha = 0.05$, what is the probability that the commercials will be continued if sales have increased to an average of \$75,000? Assume that the standard deviation of sales is unaffected by the new commercials.

SOLUTION Pogo should consider continuing the commercials only if the true average sales exceeds the original sales plus the cost of the commercials. Hence, the hypotheses are

$$H_0\colon\ \mu \leqslant 71{,}200 \qquad H_a\colon\ \mu > 71{,}200$$

Determining $\bar{x}_\alpha$ for the test, we have

$$\bar{x}_\alpha = 71{,}200 + 1.645\left(\frac{14{,}200}{\sqrt{10}}\right) = 78{,}586.76$$

with the resulting decision rule being to:

Reject H_0 if
$\bar{X} > \$78{,}586.76$

The probability of continuing the commercials is the probability that H_0 will be rejected. For $\mu_a = 75{,}000$ this is

$$\begin{aligned} P_r(75{,}000) &= \text{pr}\{\bar{X} > 78{,}586.76 \mid \mu = 75{,}000,\ \sigma_{\bar{X}} = 4490.43\} \\ &= \text{pr}\left\{Z > \frac{78{,}586.76 - 75{,}000}{4490.43}\right\} \\ &= \text{pr}\{Z > 0.80\} = 0.2119 \end{aligned}$$

These results are represented graphically in Figure 10.2.

10.4.2 Inferences for μ (σ Unknown)

This is a more realistic situation than the previous one, and to handle it we need make only a minor change in our previous results. The major difficulties that we will encounter arise in power calculations, where our table is not oriented to determining specific probabilities, and in the determination of sample sizes, for which a value is required for the standard deviation. For power calculations we will attempt only to estimate the value, and for calculating n we will use a prior estimate of σ.

Our assumptions remain as stated at the beginning of the section. Violations of the assumptions are slightly more critical when σ^2 is unknown, but procedures based on the t distribution are still robust. With σ^2 unknown we must estimate it with the sample variance, as given in Chapter 7. The formula, when using ungrouped data, is

$$s^2 = \frac{1}{n-1}\left\{\sum_{i=1}^{n} X_i^2 - \frac{(\Sigma X_i)^2}{n}\right\}$$

The substitution of s for σ in the standard formula

$$\frac{\overline{X} - \mu}{\sigma/\sqrt{n}}$$

Figure 10.2 Illustration of the Power of the Test in Example 10.10 when $\mu_a = 75{,}000$

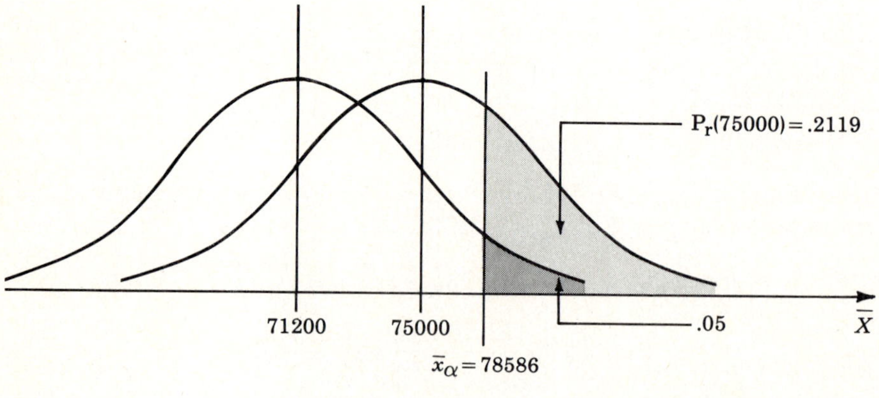

changes the distribution from an $N(0, 1)$ to a t_{n-1}; that is,

$$\frac{\overline{X} - \mu}{s/\sqrt{n}} \sim t_{n-1}$$

Once this substitution has been made and the t distribution identified, the logical development is identical to that of the previous section. We leave the algebra to the student and summarize the results in Table 10.5.

EXAMPLE 10.11 Refer to Example 10.8 and assume that the standard deviation of the forecast errors is unknown. Evaluate the sample results with this in mind.

SOLUTION The sample values are

1.4 −0.3 2.6 1.8 1.2 −1.1 1.7 3.1 0.9 2.1 1.5 −0.8

Table 10.5 A Summary of Tests for the Normal Mean When σ Is Unknown and the Test Is Based on (*a*) $\bar{x}_\alpha$ and (*b*) t_α

Hypothesis Set		Decision Rule for a Stated α
		(*a*) Decision Rules for Tests of μ Based on $\bar{x}_\alpha$
1. H_0: $\mu \leqslant \mu_0$	H_a: $\mu > \mu_0$	Reject H_0 if $\overline{X} > \mu_0 + t_{n-1,\,\alpha} s_{\overline{X}}$
2. H_0: $\mu \geqslant \mu_0$	H_a: $\mu < \mu_0$	Reject H_0 if $\overline{X} < \mu_0 - t_{n-1,\,\alpha} s_{\overline{X}}$
3. H_0: $\mu = \mu_0$	H_a: $\mu \neq \mu_0$	Reject H_0 if either $\overline{X} > \mu_0 + t_{n-1,\,\alpha/2} s_{\overline{X}}$ or $\overline{X} < \mu_0 - t_{n-1,\,\alpha/2} s_{\overline{X}}$
		(*b*) Decision Rules for Tests of μ Based on t_α
1. H_0: $\mu \leqslant \mu_0$	H_a: $\mu > \mu_0$	Reject H_0 if $\dfrac{\overline{X} - \mu_0}{s_{\overline{X}}} > t_{n-1,\,\alpha}$
2. H_0: $\mu \geqslant \mu_0$	H_a: $\mu < \mu_0$	Reject H_0 if $\dfrac{\overline{X} - \mu_0}{s_{\overline{X}}} < -t_{n-1,\,\alpha}$
3. H_0: $\mu = \mu_0$	H_a: $\mu \neq \mu_0$	Reject H_0 if $\left\lvert \dfrac{\overline{X} - \mu_0}{s_{\overline{X}}} \right\rvert > t_{n-1,\,\alpha/2}$

Previously, we calculated $\bar{x}$ to be 1.175. The sample variance is

$$s^2 = \frac{1}{12 - 1}\left\{35.31 - \frac{(14.1)^2}{12}\right\} = \frac{1}{11}\{35.31 - 16.57\} = 1.704$$

and

$$s = 1.305$$

The hypotheses of interest are

$$H_0\colon\ \mu = 0 \qquad H_a\colon\ \mu \neq 0$$

and for $\alpha = 0.05$, we reject H_0 if either

$$\bar{X} > 0 + 2.201\left(\frac{1.305}{\sqrt{12}}\right) = 0.829$$

or

$$\bar{X} < 0 - 2.201\left(\frac{1.305}{\sqrt{12}}\right) = -0.829$$

Since $\bar{x} = 1.175$, we still reject H_0 and conclude that we have observed either a very rare event or else there is a bias in the forecast errors.

If we want to obtain an estimate on the power of the test, we use Table A.8 to obtain an approximate probability for the statement (letting $\mu_a = 1.0$):

$$\begin{aligned} P_r(1.0) &\doteq 1 - \mathrm{pr}\{-0.829 \leqslant \bar{X} \leqslant +0.829 \mid \mu = 1.0\} \\ &\doteq \mathrm{pr}\{\bar{X} < -0.829 \mid \mu = 1.0\} + \mathrm{pr}\{\bar{X} > +0.829 \mid \mu = 1.0\} \\ &\doteq \mathrm{pr}\left\{t_{11} < \frac{-0.829 - 1.0}{0.377}\right\} + \mathrm{pr}\left\{t_{11} > \frac{0.829 - 1.0}{0.377}\right\} \\ &\doteq \mathrm{pr}\{t_{11} < -4.85\} + \mathrm{pr}\{t_{11} > -0.45\} \end{aligned}$$

The 0.377 in the above statement is our value for $s_{\bar{X}}$.

Referring to Table A.8, we find that

$$\begin{aligned} 0.0 < \mathrm{pr}\{t_{11} < -4.85\} = \mathrm{pr}\{t_{11} > 4.85\} &< \mathrm{pr}\{t_{11} > 4.437\} \\ &= 0.0005 \end{aligned}$$

Also, since

$$\mathrm{pr}\{t_{11} > -0.45\} = 1 - \mathrm{pr}\{t_{11} > +0.45\}$$

and

$$0.30 = \text{pr}\{t_{11} > 0.54\} < \text{pr}\{t_{11} > 0.45\} < \text{pr}\{t_{11} > 0.26\} = 0.40$$

we have that

$$0.60 \leqslant \text{pr}\{t_{11} > -0.45\} \leqslant 0.70$$

Therefore, the approximate power for our test when $\mu = 1.0$ is between 0.6 and 0.7.

The same substitution is required in calculating a confidence interval when the standard deviation is obtained from the sample data. The equation to be solved becomes

$$\text{pr}\{-t_{n-1,\,\alpha/2} \leqslant \frac{\overline{X} - \mu}{s_{\overline{X}}} \leqslant +t_{n-1,\,\alpha/2}\} = 1 - \alpha$$

Solving for μ inside the braces but omitting the algebra, we obtain

$$\begin{aligned} \text{LL} &= \bar{x} - t_{n-1,\,\alpha/2}s_{\overline{X}} \\ \text{UL} &= \bar{x} + t_{n-1,\,\alpha/2}s_{\overline{X}} \end{aligned} \tag{10.13}$$

where $s_{\overline{X}} = s/\sqrt{n}$ if the population is infinite and $s_{\overline{X}} = (s/\sqrt{n})\sqrt{(N-n)/N}$ if the finite population correction factor is used. One-sided intervals are handled in the usual manner.

EXAMPLE 10.12 A professor would like to estimate the average quality points earned per student for all students who have taken statistics from her to date. To do so she selects a random sample of 30 students from her grade records of the past ten years and calculates the quality points earned by each. The sample average and standard deviation are 2.23 and 0.58, respectively. Determine a 99% confidence interval for the average quality points earned by her students.

SOLUTION The population is finite, but over a ten-year span the professor probably has had between 1000 and 2000 students. Thus, $30/N < 0.05$ and we ignore the finite population corrector. The assumption of interval data does not seem unreasonable in this case. Though recent evidence suggests that grades are not normally distributed, the distribution would not be likely to deviate from normality enough to cause a problem when $n = 30$.

In light of this consideration, we use Equation 10.13 with $\bar{x} = 2.23$, $s_{\overline{X}} = 0.58/\sqrt{30} = 0.106$, and $t_{n-1,\,\alpha/2} = t_{29,\,0.005} = 2.756$. Substitution gives

$$LL = 2.23 - 2.756(0.106) = 1.938$$

and

$$UL = 2.23 + 2.756(0.106) = 2.522$$

Therefore, the professor can be 99% confident that the interval [1.938, 2.522] will contain the true average quality points earned by students in her course.

We now turn briefly to a consideration of the effect of the sample size on the results of this section. The derivation of the formulas will follow the same pattern as in Section 10.3.3.

10.4.3 Determination of Sample Size

As we saw in dealing with the population proportion, the width of a confidence interval and the power of a test for a specified α are influenced by the choice of sample size. The same relationship holds for inferences about means of normal distributions (in fact, for *all* statistical inferences). If the error bound ϵ is specified in estimating the population mean, then we have

$$UL - LL = (\bar{x} + z_{\alpha/2}\sigma_{\bar{X}}) - (\bar{x} - z_{\alpha/2}\sigma_{\bar{X}})$$

or

$$2\epsilon = 2z_{\alpha/2}\sigma_{\bar{X}} = 2z_{\alpha/2}\frac{\sigma}{\sqrt{n}}$$

Solving for n, we obtain

$$n = \left(\frac{z_{\alpha/2}\sigma}{\epsilon}\right)^2 \tag{10.14}$$

If the desired test is to have a specified power for a given alternative, it is a simple matter to solve for the sample size required. In that case we have, for the decision rule that implies rejection of H_0 whenever $\bar{X} > \bar{x}_\alpha$,

$$\text{pr}\{\bar{X} > \bar{x}_\alpha | H_0\} = \alpha$$

and, simultaneously,

$$\text{pr}\{\bar{X} \leqslant \bar{x}_\alpha | H_a\} = \beta$$

These probability statements imply that $\bar{x}_\alpha$ must be such that both

$$\bar{x}_\alpha = \mu_0 + z_\alpha \frac{\sigma}{\sqrt{n}}$$

and

$$\bar{x}_\alpha = \mu_a - z_\beta \frac{\sigma}{\sqrt{n}}$$

are true. Thus, we have two equations in two unknowns, and we solve them simultaneously to obtain

$$n = \left(\frac{(z_\alpha + z_\beta)\sigma}{\mu_a - \mu_0}\right)^2 \tag{10.15}$$

If our alternative is of the two-sided form, we replace z_α with $z_{\alpha/2}$. The formulas for sample size when the parameter of interest is μ are summarized in Table 10.6.

EXAMPLE 10.13

(*a*) Determine the sample size required to estimate a population mean within 3.5 units with a probability of 0.90. The standard deviation is not known for certain, but prior estimates indicate that it is unlikely to be larger than 28.

(*b*) Determine the sample size required to detect the alternative hypothesis, H_a: $\mu \neq \mu_0$, with a probability of 0.95 when $\mu_0 = 45$, $\mu_a = 40$, $\alpha = 0.02$, and $\sigma = 8$.

Table 10.6 Summary of Sample-Size Formulas When ϵ Is Specified or When μ_a and β Are Specified

Alternative or Estimate	Equation for n
Two-sided interval	$n = \left(\frac{z_{\alpha/2}\sigma}{\epsilon}\right)^2$
One-sided interval	$n = \left(\frac{z_\alpha\sigma}{\epsilon}\right)^2$
H_a: $\mu < \mu_0$ or H_a: $\mu > \mu_0$	$n = \left(\frac{(z_\alpha + z_\beta)\sigma}{\mu_a - \mu_0}\right)^2$
H_a: $\mu \neq \mu_0$	$n = \left(\frac{(z_{\alpha/2} + z_\beta)\sigma}{\mu_a - \mu_0}\right)^2$

SOLUTION

(*a*) We have that $\alpha = 0.10$, $\sigma \leqslant 28$, and $\epsilon = 3.5$. Therefore, the required sample size will be

$$n = \left(\frac{z_{\alpha/2}\sigma}{\epsilon}\right)^2 = \left(\frac{1.645(28)}{3.5}\right)^2$$
$$= (13.16)^2 = 173.18 \quad \text{or} \quad 174 \text{ items}$$

(*b*) The pertinent values are $1 - \beta = 0.95$, $\alpha = 0.02$, $\sigma = 8$, $\mu_a = 40$, and $\mu_0 = 45$, and the hypothesis is two-sided. From Table 10.6 our formula is

$$n = \left(\frac{(z_{\alpha/2} + z_\beta)\sigma}{\mu_a - \mu_0}\right)^2 = \left(\frac{(2.326 + 1.645)(8)}{40 - 45}\right)^2$$
$$= (6.3536)^2 = 40.37 \quad \text{or} \quad 41 \text{ observations}$$

All the equations in Table 10.6 require a value for σ. If the exact value is unknown, a prior estimate can be used if one is available. If none is available, the relationship between the range and σ for a normal distribution can be used. This relationship utilizes the fact that approximately 99% of the values should fall between $x_{(n)}$ and $x_{(1)}$ for large samples, and approximately 95% should fall in this range for smaller samples. (99% and 95% correspond to $\pm 3\sigma$ and $\pm 2\sigma$, respectively). That is, we can find a rough approximation for σ by using either

$$\sigma \doteq \frac{R}{6} \quad \text{or} \quad \sigma \doteq \frac{R}{4} \tag{10.16}$$

Which one you use will depend on the costs of sampling and how conservative you want to be, since $R/6$ will lead to smaller n's than will $R/4$.

10.4.4 Inferences for a Normal Variance

In this section our interest changes from the location of the distribution to its dispersion. The basic measure of dispersion is the variance, which also indicates the degree of uniformity of the population or process. That is, a smaller value for σ^2 indicates that the population values are more nearly uniform than a large value does.

The assumptions required for determining the distribution on which we base our inferential procedures are the same as those required for μ, namely:

1. A random sample from the population of interest.
2. Underlying normal distribution.

3. The numerical values associated with the sample data represent an interval level of measurement.

If these assumptions are satisfied, then we know from Chapter 8 that

$$s^2 = \frac{1}{n-1}\left\{\sum_{i=1}^{n} X_i^2 - \frac{(\Sigma X_i)^2}{n}\right\}$$

is an unbiased estimator of σ^2 and should be close to the true value. Second, from Chapter 8 we also know that

$$\frac{(n-1)s^2}{\sigma^2} \sim \chi^2_{n-1}$$

when our assumptions are correct. Thus, we have the three ingredients for constructing either interval estimates or test statistics: an estimator, its type of distribution, and the family member.

To construct an interval estimate, we proceed as before. Our initial probability statement is

$$\text{pr}\{\chi^2_{n-1,\ 1-\alpha/2} \leqslant \chi^2_{n-1} \leqslant \chi^2_{n-1,\ \alpha/2}\} = 1 - \alpha$$

Since $(n-1)s^2/\sigma^2$ is another form for the χ^2_{n-1} variable, we substitute this in the probability statement to obtain

$$\text{pr}\left\{\chi^2_{n-1,\ 1-\alpha/2} \leqslant \frac{(n-1)s^2}{\sigma^2} \leqslant \chi^2_{n-1,\ \alpha/2}\right\} = 1 - \alpha$$

Dividing within the braces by $(n-1)s^2$ gives

$$\text{pr}\left\{\frac{\chi^2_{n-1,\ 1-\alpha/2}}{(n-1)s^2} \leqslant \frac{1}{\sigma^2} \leqslant \frac{\chi^2_{n-1,\ \alpha/2}}{(n-1)s^2}\right\} = 1 - \alpha$$

Taking reciprocals and rewriting (reciprocals switch the directions of the inequalities), we have

$$\text{pr}\left\{\frac{(n-1)s^2}{\chi^2_{n-1,\ \alpha/2}} \leqslant \sigma^2 \leqslant \frac{(n-1)s^2}{\chi^2_{n-1,\ 1-\alpha/2}}\right\} = 1 - \alpha \qquad (10.17)$$

which is the conventional form for a $1 - \alpha$ confidence interval. Thus, our bounds are given by (using Table A.7 to obtain the χ^2 values)

$$\text{LL} = \frac{(n-1)s^2}{\chi^2_{n-1,\ \alpha/2}} \quad \text{and} \quad \text{UL} = \frac{(n-1)s^2}{\chi^2_{n-1,\ 1-\alpha/2}} \qquad (10.18)$$

and we are $(1 - \alpha) \times 100\%$ confident that the interval [LL, UL] contains σ^2. For one-sided intervals we calculate the bound of interest using α rather than $\alpha/2$.

For constructing the decision rule for a testing situation, our basic null hypothesis is

$$H_0\colon\ \sigma^2 = \sigma_0^2$$

As before, our test statistic is $(n - 1)s^2/\sigma^2$, and if H_0 is true, then

$$E\left[\frac{(n-1)s^2}{\sigma_0^2}\middle| H_0\right] = \frac{(n-1)}{\sigma_0^2}E[s^2 \mid H_0] = \frac{(n-1)}{\sigma_0^2}(\sigma_0^2) = n - 1$$

Large deviations from $n - 1$ in either direction indicate that our statistic may be more extreme than we are willing to attribute to chance alone. For our three basic hypothesis sets our decision rules are as follows:

$$H_0\colon\ \sigma^2 \leqslant \sigma_0^2 \qquad H_a\colon\ \sigma^2 > \sigma_0^2$$

If H_a is true, we expect our statistic to be significantly greater than $n - 1$. Thus, we

Reject H_0 if

$$\frac{(n-1)s^2}{\sigma_0^2} > \chi^2_{n-1,\ \alpha}$$

$$H_0\colon\ \sigma^2 \geqslant \sigma_0^2 \qquad H_a\colon\ \sigma^2 < \sigma_0^2$$

For this alternative it follows that we expect the statistic to be less than $n - 1$, and we

Reject H_0 if

$$\frac{(n-1)s^2}{\sigma_0^2} < \chi^2_{n-1,\ 1-\alpha}$$

$$H_0\colon\ \sigma^2 = \sigma_0^2 \qquad H_a\colon\ \sigma^2 \neq \sigma_0^2$$

For the two-sided alternative we wish to detect deviations in either direction. Therefore, we

Reject H_0 if either

$$\frac{(n-1)s^2}{\sigma_0^2} < \chi^2_{n-1,\ 1-\alpha/2} \qquad \text{or} \qquad \frac{(n-1)s^2}{\sigma_0^2} > \chi^2_{n-1,\ \alpha/2}$$

EXAMPLE 10.14 When in control, a beer bottling machine fills bottles with an average of 12 ounces. Periodically, random samples of ten bottles each are selected and the average fill and sample standard deviation are calculated. The current sample has $\bar{x} = 11.98$ ounces and $s = 0.16$ ounce.

(*a*) Construct a 90% confidence interval for the variance of the fill amounts.

(*b*) If the company desires the uniformity of the fill amounts, when the process is in control, to be such that the standard deviation is no more than 0.1, would you conclude that it is in control?

SOLUTION

(*a*) We have a random sample, obviously interval data, and it is quite reasonable to assume an underlying normal distribution for the fill amounts. Using Equation 10.18, we have

$$\text{LL} = \frac{(10 - 1)(0.16)^2}{16.92} = 0.0136$$

and

$$\text{UL} = \frac{(10 - 1)(0.16)^2}{3.33} = 0.0692$$

Hence, we are 90% confident that the interval, [0.0136, 0.0692] contains the true variance of the fill.

(*b*) If the process is in control, then $\sigma \leqslant 0.1$. In terms of σ^2, our hypotheses are

$$H_0\colon\ \sigma^2 \leqslant 0.01 \qquad H_a\colon\ \sigma^2 > 0.01$$

We reject H_0 if

$$\frac{(n - 1)s^2}{\sigma_0^2} > \chi^2_{n-1,\alpha} = \chi^2_{9,\,0.05} = 16.92$$

The value of the statistic is

$$\frac{(10 - 1)(0.16)^2}{(0.10)^2} = \frac{9(0.0256)}{0.01} = 23.04$$

Thus, we reject H_0 and conclude that this sample indicates the machine to be out of control. Note that all we really needed to do for our test was to check to see whether 0.01 was in the confidence interval of part (*a*). This is so because the lower bound corresponds to an $\alpha/2$ of 0.05, the same value as α for our test. Since 0.0136 represents the smallest reasonable value for σ^2 according to the sample, 0.01 is not reasonable, and H_0 is rejected. (See Figure 10.3.)

The assumption of normality is more critical in this development, since the calculation of sample variances requires squaring the observations. But the procedures are still robust, and slight deviations will not significantly affect the results. Power calculations for tests of variance are much more difficult than those for tests of means and are beyond the scope of this text. Rough approximations for the power against a specific alternative may be obtained in the same manner as was done for the t distribution. If we want a confidence interval for σ rather than for σ^2, we merely take square roots inside the probability statement. That is, a $(1 - \alpha) \times 100\%$ confidence interval for σ would be given by

$$\text{pr}\{[\text{LL}, \text{UL}] \text{ contains } \sigma\} = 1 - \alpha$$

where

$$\text{LL} = \sqrt{\frac{(n-1)s^2}{\chi^2_{n-1,\,\alpha/2}}} \qquad \text{UL} = \sqrt{\frac{(n-1)s^2}{\chi^2_{n-1,\,1-\alpha/2}}} \tag{10.19}$$

Figure 10.3 A Graphical Representation of the Test for Example 10.14(b)

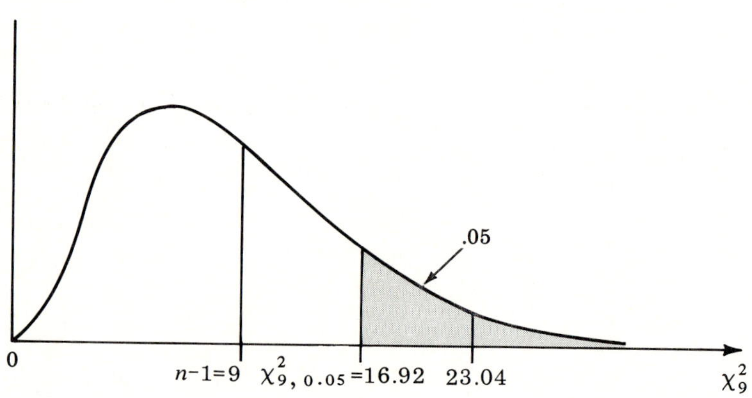

In the next section we relax the assumptions of normality and interval data to obtain more general tests for location. Doing so results in some loss of power for our procedures but makes the results more widely applicable.

EXERCISES

10.4.1 Refer to Example 10.10 and determine whether Pogo should continue using the commercials if a sample of six weeks gave average weekly sales of \$79,250 and an α of 0.10 is used.

10.4.2 Refer to Example 10.10 and determine the sample size required to detect a mean of \$77,500 with a detection probability of 0.90 and an α of 0.05.

10.4.3 Refer to Example 10.10, in which it was assumed that the standard deviation of sales was unchanged by the new commercials. If Pogo's sample in Exercise 10.4.1 had a standard deviation of \$22,456, would you conclude that this assumption was reasonable at $\alpha = 0.10$?

10.4.4 A school district, concerned about the publicized decline in the competence of high school graduates, has decided to test to see whether the decline is indicated in its system. Records of 15 years ago indicate that the average score on a standardized college entrance exam of high school graduates from the district was 570. A random sample of 24 of the district's 1394 seniors were given the same exam. The mean and variance of their test scores were 562 and 696, respectively. Would you conclude at $\alpha = 0.05$ that this indicates a decline in this district?

10.4.5 Refer to Exercise 10.4.4 and determine a 99% confidence interval for the standard deviation of scores for this year's students on that exam.

10.4.6 Refer to Exercise 10.4.4 and construct a 90% confidence interval for the average score of this year's students on that exam. Does your answer agree with the results of Exercise 10.4.4?

10.4.7 Comment on the reasonableness of the assumptions required for the procedure used in Example 10.10.

10.5 LESS RESTRICTIVE PROCEDURES

In many applications the assumptions underlying the procedures of Section 10.4 will be questionable. For large sample sizes (say $n > 30$), violations of the assumptions will be less critical. But for small sample sizes, which is where use of the distribution-free (DF) procedures is recommended, the assumptions should play the major role in determining which procedure to select. We restrict our consideration to tests for location, since there is no widely used DF test for a single variance.

10.5.1 The Sign Test

One of the simplest, safest, and most easily understood tests is the sign test. It is based on the binomial distribution, and as such it has readily available tables and is intuitively appealing. It is not the most powerful, but its efficiency is surprisingly high considering the simplistic nature of the information used. This information is merely the algebraic sign of the difference between two observations.

The hypothesis to be tested is that the median of the population is equal to some specified value, say M_0; the basic hypotheses are

$$H_0\colon\ M = M_0 \qquad H_a\colon\ M \neq M_0 \tag{10.20}$$

Another way of stating the null hypothesis is, since the median is the fiftieth percentile, that $\text{pr}\{X > M_0\} = \text{pr}\{X < M_0\} = 0.5$ or, by subtracting the median from each value, $\text{pr}\{X - M_0 > 0\} =$ $\text{pr}\{X - M_0 < 0\} = 0.5$. The latter statement leads directly to a statement based on the sign of the difference, pr{difference is positive} = pr{difference is negative} = 0.5. This will be the case if the null hypothesis is true and the following assumptions are satisfied.

Assumptions of the Sign Test

1. The observations are from an underlying continuous distribution, or $\text{pr}\{X = M\} = 0$. The basic assumption is still satisfied if the population is discrete but the median is not an observable value.
2. The observations constitute a random sample.
3. The observations are independent, which is satisfied by randomness if the population is infinite or by sampling with replacement if it is not.
4. At least nominal data are available.

Under these assumptions we may restate the hypotheses in Equation 10.20 as

$$H_0\colon\ p_+ = 0.5 \qquad H_a\colon\ p_+ \neq 0.5 \tag{10.21}$$

where $p_+ = \text{pr}\{X - M_0 > 0\}$. Thus, we intuitively base our test on the observed number of positive differences. If the null hypothesis is true, the number of +'s and the number of −'s should be the same, differing only because of random variation. The logical rejection region corresponds to situations in which there are either too many or too few +'s. If the true median is less than that specified in H_0, we should observe more −'s than +'s, and if it is greater than M_0, we should observe more +'s than −'s. See Figure 10.4. As usual, our measure of too many or too few is the probability of occurrence when H_0 is true.

When the observed value is too extreme, the probability of occurrence is less than α and H_0 is rejected. For the sign test we shall actually calculate the probability of occurrence and compare it with a stated level of significance, rather than determining x_α. The procedure will be as follows:

1. State the hypotheses and the nominal α to be used for testing.
2. Select a random sample from the population under consideration.
3. Subtract the hypothesized value, M_0, from every observation, recording only the sign of the difference.
4. Count the number of plus signs, say n_+, and the number of minus signs, say n_-.

Depending on whether the hypotheses are one-sided or two-sided, the decision rules or test criteria that arise are those given in Table 10.7 on page 328.

Note that each decision rule has been stated in terms of what H_a indicates should be smaller. To be able to calculate the indicated probabilities, we must know the distribution of the number of positive differences. Under our assumptions, we have a finite number of independent

Figure 10.4 Illustration of Possible Configurations of Data Points Observed When the (*a*) True Median Equals M_0, (*b*) True Median Is Less than M_0, (*c*) True Median Is Greater than M_0. Note that when M_0 is subtracted from each observation there will be (*a*) 5 +'s, 5 −'s, (*b*) 1 +, 9 −'s, (*c*) 8 +'s, 2 −'s.

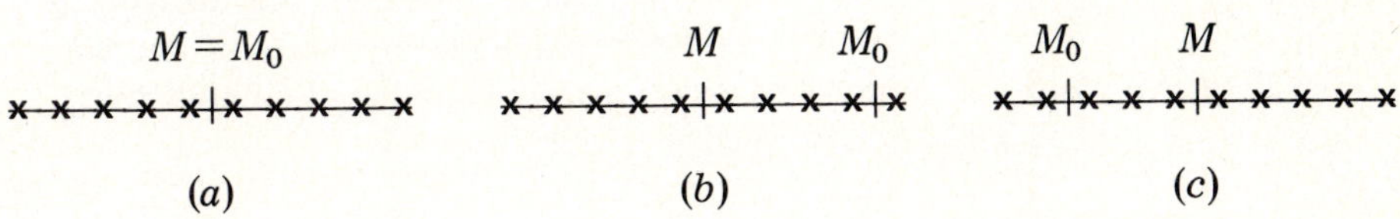

Table 10.7 A Summary of Hypothesis Sets and Decision Rules for Using the Sign Test for the Median

Hypothesis Set		Decision Rule for the Stated α
1. H_0: $M \leqslant M_0$	H_a: $M > M_0$	Reject H_0 if $\text{pr}\{X \leqslant n_- \mid n, 0.5\} < \alpha$
2. H_0: $M \geqslant M_0$	H_a: $M < M_0$	Reject H_0 if $\text{pr}\{X \leqslant n_+ \mid n, 0.5\} < \alpha$
3. H_0: $M = M_0$	H_a: $M \neq M_0$	Reject H_0 if $\text{pr}\{X \leqslant \min(n_+, n_-) \mid n, 0.5\} < \alpha/2$

trials from a very large dichotomous population with $p_+ = p_- = 1/2$. We recognize this as a binomial situation. Since any observation has a 50% chance of falling above the median, $p = 1/2$. Theoretically, no observations should equal M_0 and n should be the sample size. Practically, due to imprecise measurements, values will occasionally equal M_0, resulting in a difference of 0. When this occurs we will discard the 0 differences and reduce the sample size correspondingly to $n = n_+ + n_-$ (the number of nonzero differences). Thus, the probabilities indicated in Table 10.7 would be determined according to the binomial distribution as

$$\text{pr}\{X \leqslant n_+ \mid p = \frac{1}{2}, n = n_+ + n_-\} = \sum_{x=0}^{n_+} \binom{n}{x} \left(\frac{1}{2}\right)^x \left(\frac{1}{2}\right)^{n-x} \qquad (10.22)$$

You will recall that, for appropriate values of n, these probabilities can be found by using Table A.2 in the Appendix.

EXAMPLE 10.15 Consider a situation in which we want to compare family incomes in Fairfield with those reported by the Census Bureau. Realizing that the distribution of incomes tends to be highly skewed and that the mean is highly susceptible to extreme values, we decide to employ a distribution-free test of the median income, which was reported by the Census Bureau to be \$9500. For our test, we will use a nominal α of 0.10.

SOLUTION Knowing that people are highly sensitive about giving their exact incomes, we decide to simply ask them whether the family income is above or below \$9500, recording a + for an "above" answer and a - for a "below" answer. The obvious procedure is a sign test, and the hypotheses to be tested at $\alpha = 0.10$ are

$$H_0\colon M = 9500 \qquad H_a\colon M \neq 9500$$

The results of interviews with 20 randomly selected families were 16 above 9500 and 4 below. Thus, we have 16 +'s and 4 –'s, and H_0 will be rejected if Equation 10.22 is less than $\alpha/2 = 0.05$. We have for Equation 10.22

$$\begin{aligned} \text{pr}\{X \leqslant \min(4, 16) | p = 0.5, n = 20\} &= \text{pr}\{X \leqslant 4 | p = 0.5, n = 20\} \\ &= \sum_{x=0}^{4} \binom{20}{x} \left(\frac{1}{2}\right)^x \left(\frac{1}{2}\right)^{20-x} \\ &= 0.0059 \end{aligned}$$

Since the probability, 0.0059, is less than the $\alpha/2$ of 0.05, we reject H_0 and conclude that incomes in Fairfield appear to differ from those reported nationally.

In Example 10.15 we used a nominal α of 0.10. This is only nominal, since the distribution of +'s is discrete and the stated α will not generally correspond to one of the possible values that the variable can assume. The actual or exact α for Example 10.15 is 2(0.0059) = 0.0118, since the test was two-sided.

Note: If we can assume that the population with which we are concerned has a symmetric distribution, then the mean and median are identical. In that case the test of the hypothesis H_0: $M = M_0$ is equivalent to testing H_0: $\mu = \mu_0$, and the sign test provides an alternative to the one-sample t test discussed in Section 10.4.2 but with much less restrictive assumptions.

When the sample size increases to the extent that n cannot be found in Table A.2, the probability is calculated using the normal approximation to the binomial, as was presented in Section 10.3.2.

10.5.2 The Wilcoxon Signed-Rank Test

A test that is similar to the sign test but that has somewhat more restrictive assumptions is the Wilcoxon signed-rank test for population location. The test is based on ranking the differences of the observations minus the hypothesized median (or mean). It is surprisingly powerful, having an ARE of 0.955 relative to the one-sample t test even when the underlying population is normal and in most cases an ARE greater than 1.00 when compared with the t with nonnormal populations. In most situations in which the assumptions of the signed-rank test are satisfied, it is either the most powerful test or a close second. These assumptions are as follows.

Assumptions of Wilcoxon's Signed-Rank Test

1. The underlying distribution is continuous and symmetric. (Continuity may be relaxed if the population is either infinite or very large.)
2. The sample data are of at least an ordinal scale.
3. The observations constitute a random sample and are independent of one another.

Under these assumptions we can test hypotheses about the population median or mean, since the distribution is assumed to be symmetric. The basic hypotheses for the two-sided case are

$$H_0: M = M_0 \qquad H_a: M \neq M_0$$

As with the the sign test we state a recommended procedure to use. The steps to be followed are:

1. State the hypotheses and select the nominal α at which they are to be tested.
2. Select a random sample.
3. Subtract the hypothesized value M_0 from every observation, recording both the sign and the difference.
4. Rank the *absolute values* of the differences from low to high, assigning 1 to the lowest nonzero absolute difference, 2 to the next lowest, and so on up to n for the highest (discard zero differences).
5. Calculate T_+ and T_-, where

$$T_+ = \text{sum of the ranks of the positive differences}$$

and (10.23)

$$T_- = \text{sum of the ranks of the negative differences}$$

Our decision rules for the two-sided and various one-sided alternatives are based on the magnitudes of T_+ and T_-. If M_0 is the true median, the high and low ranks should be evenly divided among the positive and negative differences, differing only as a result of random variation, and T_+ and T_- should be approximately equal. If $M < M_0$, there should be more negative differences and correspondingly more of the high ranks should be assigned to them. Thus T_- should be greater than T_+, since their sum must equal $n(n + 1)/2$, which is the sum of the first n integers. In this case n is the number of nonzero differences. Conversely, by the same reasoning, if $M > M_0$, then T_+ should exceed T_-. Table A.10 in the Appendix gives critical values for the smaller sum corresponding to nominal α's of 0.05, 0.025, 0.01, and 0.005. The decision rule, then, is to reject H_0 if the sum that H_a indicates to be smaller is too small. We summarize the hypothesis sets and decision rules in Table 10.8.

Table A.10 is indexed by n, the total number of +'s and −'s, and by a nominal α. The value of n identifies the row and α identifies the column in which the critical value $T_{n,\alpha}$ will be found. The probability is only nominal, since T is a discrete random variable. The true α for a test that rejects H_0 usually will be somewhat less than the stated α.

EXAMPLE 10.16 Experience indicates that scores on a job aptitude test have been symmetrically distributed with a median of 52. A group of 12 applicants recently took the test with the following results:

61 48 44 57 76 30 65 59 22 72 88 39

Is there any evidence that these applicants have a superior aptitude for the job at $\alpha = 0.05$? (Because of the preponderance of extreme values, it is doubtful that the scores are normally distributed.)

SOLUTION Since the scores are symmetrically distributed, a test for the median is also a test for the mean. Since there is doubt about the normality of the distribution, the signed-rank test is preferable to a t test. The hypotheses indicated are

$$H_0\colon M \leqslant 52 \qquad H_a\colon M > 52$$

and H_0 will be rejected at $\alpha = 0.05$ if

$$T_- \leqslant T_{n,\,0.05}$$

where n will be the number of nonzero differences. Our procedure is to subtract 52 from each score, to rank the absolute differences, then

Table 10.8 A Summary of Hypothesis Sets and Decision Rules for Using the Wilcoxon Signed-Rank Test for Location

Hypothesis Set		Decision Rule for the Stated α
$H_0\colon M \leqslant M_0$	$H_a\colon M > M_0$	Reject H_0 if $T_- \leqslant T_{n,\alpha}$
$H_0\colon M \geqslant M_0$	$H_a\colon M < M_0$	Reject H_0 if $T_+ \leqslant T_{n,\alpha}$
$H_0\colon M = M_0$	$H_a\colon M \neq M_0$	Reject H_0 if $\min(T_+, T_-) \leqslant T_{n,\alpha/2}$

to add the ranks corresponding to the negative differences. These steps are summarized in the following table.

Score (X)	$X - 52$	$\lvert X - 52\rvert$	Rank of $\lvert X - 52\rvert$	Negative Ranks	Positive Ranks
61	+9	9	5		5
48	-4	4	1	1	
44	-8	8	4	4	
57	+5	5	2		2
76	+24	24	10		10
30	-22	22	9	9	
65	+13	13	6.5		6.5
59	+7	7	3		3
22	-30	30	11	11	
72	+20	20	8		8
88	+36	36	12		12
39	-13	13	6.5	6.5	
				$T_- = 31.5$	$T_+ = 46.5$

We have also calculated T_+ to show that $T_+ + T_- = n(n+1)/2$. Here $T_+ + T_- = 46.5 + 31.5 = 78 = 12(13)/2$. From Table A.10 we have, for $n = 12$ and $\alpha = 0.05$, that $T_{12,\,0.05} = 17$. Since $T_- = 31.5$ is not less than or equal to 17, we accept H_0 and conclude that there does not appear to be sufficient evidence that this group is superior.

In Example 10.16 there were two absolute differences with a value of 13. Theoretically, ties cannot occur, but in practice they do, either from sampling with replacement from a finite population or from using an imprecise measuring instrument on a continuous population. The general practice for handling ties in ranking procedures is to assign them the average of the ranks that would have been assigned to each of the tied values. Our two 13s should have received the ranks 6 and 7, but to avoid biasing the statistic with an arbitrary assignment they were each assigned a rank of 6.5. The next value, 20, then received the rank of 8, and so on.

Note: If there are any doubts about the symmetry of the population, the sign test is preferred over the signed-rank test. This is also the case if there are serious questions about the precision of the measurements.

For situations in which the sample size is too large for Table A.10, we may use the normal approximation to the distribution of T. It can be shown, using arithmetic series, that the mean and variance of T are

$$E(T) = \frac{n(n+1)}{4} \quad \text{and} \quad \sigma_T^2 = \frac{n(n+1)(2n+1)}{24}$$

Thus, for large samples, the null hypothesis is rejected if

$$z = \frac{T - [n(n+1)/4]}{\sqrt{n(n+1)(2n+1)/24}} \tag{10.24}$$

is less than the appropriate lower-tail z value from the standard normal table. Whether one uses T_+, T_-, or $\min(T_+, T_-)$ depends on the hypotheses being tested. For two-sided tests the critical z will be $-z_{\alpha/2}$, whereas for both one-sided tests it is $-z_\alpha$. The approximation *should not be used* if extremely small values of α are selected such as $\alpha <$ 0.005.

EXERCISES

10.5.1 In Example 10.15 it was reported that the median family income nationwide is \$9500. The town of Skunk Hollow has applied to HUD for federal funds, claiming that the town is economically depressed. Of a random sample of 16 families selected by HUD, 3 reported incomes above \$9500, 12 reported incomes below \$9500, and 1 indicated \$9500. Using $\alpha = 0.03$, what should HUD decide about the funding?

10.5.2 In 1977, a certain class of stocks on the NYSE was reported to have a median PE ratio of 12. A random sample of 15 of the stocks in this class recently gave the following PE ratios:

17 15 11 10 3 21 14 8 16 17 19 36 14 13 9

Would it appear that PE ratios for this class of stocks have increased since 1977? Use $\alpha = 0.10$ and evaluate the assumptions associated with your procedure.

10.5.3 The HiFi Corporation advertises that their electron tubes have a median operating life of at least 1000 hours. A random sample of 25 such tubes showed 15 failing before 1000 hours and 10 after 1000 hours of operation. Using $\alpha = 0.05$, can it be concluded that this sample contradicts their advertising claim?

10.5.4 The EPA has stipulated that the psi for clean air standards is to be no more than 100 on the average. A random sample of 20 days for the city of Smogville showed the following psi readings:

111 84 56 122 158 104 92 75 116 187

128 69 90 119 133 101 42 124 169 102

Is this evidence that the city's air is dirtier than the clean air standards recommend at $\alpha = 0.05$? (Assume the distribution of psi's is symmetric.)

10.6 TESTS FOR GOODNESS OF FIT AND RANDOMNESS

All the tests we have considered so far are concerned with values for the parameters of a distribution. In this section we turn our focus to testing to see whether a particular family of distributions adequately describes the sample data. In reality the tests of the preceding three sections are also tests for goodness of fit, but by convention they generally are not identified as such. To see this, all that is necessary is to realize that the hypothesis H_0: $\mu = \mu_0$ of Section 10.4.1 can be rewritten as

$$H_0:\ X \sim N(\mu_0, \sigma^2)$$

since we assume that the distribution is normal and that σ^2 is known. If H_0 is rejected, we are saying that the $N(\mu_0, \sigma^2)$ does not reasonably describe the sample data.

In this section we provide methods for testing the reasonableness of the normality assumption and for testing the appropriateness of other families as well. The final test of this section provides a "quick and dirty" (or rough) check on the randomness assumption.

10.6.1 The Chi-Square Test

The test presented in this section provides a method for testing whether the sample data agree with some specified distribution or family of distributions. The procedure compares observed frequencies with the frequencies that would be expected for the same categories if the null hypothesis is true. Basically, we extend the one-sample sign test to more than two categories and apply the central limit theorem. The null hypothesis of interest is that the specified distribution gives the probability of an observation's being in any particular category. In general, this would be of the form

$$H_0:\ F(x) = F_0(x) \qquad H_a:\ F(x) \neq F_0(x)$$

where $F(x) = \text{pr}\{X \leqslant x\}$ is the cumulative distribution function (cdf).

We can state a more workable form of the null hypothesis by transforming H_0 into an equivalent statement about a multinomial distribution with p_{oi}, the probability of the ith category, determined by F_0. For example, suppose we have hypothesized that the sample came from a normal distribution with a mean of 100 and a standard deviation of 20. If we defined categories by (1) $X < 80$, (2) $80 \leqslant X < 100$, (3) $100 \leqslant X < 120$, and (4) $X \geqslant 120$, then, using Table A.6 in the Appendix, we could rewrite H_0 in the equivalent form

$$H_0\colon\ p_1 = 0.1587,\ p_2 = 0.3413,\ p_3 = 0.3413,\ p_4 = 0.1587$$

and simply test it as a multinomial distribution. We obtain the p_i by calculating the probability of occurrence of the ith category assuming a normal distribution with $\mu = 100$ and $\sigma = 20$. For example,

$$\begin{aligned} p_1 &= \text{pr}\{X < 80 \mid \mu = 100, \sigma = 20\} \\ &= \text{pr}\left\{Z < \frac{80 - 100}{20}\right\} = \text{pr}\{Z < -1\} = 0.1587 \end{aligned}$$

and

$$\begin{aligned} p_2 &= \text{pr}\{80 \leqslant X < 100 \mid \mu = 100, \sigma = 20\} \\ &= \text{pr}\left\{\frac{80 - 100}{20} \leqslant Z < \frac{100 - 100}{20}\right\} = \text{pr}\{-1 \leqslant Z < 0\} = 0.3413 \end{aligned}$$

Thus we construct the test under the assumption that the null hypothesis is in the form

$$H_0\colon\ p_1 = p_{o1},\ p_2 = p_{o2}, \ldots, p_c = p_{oc} \tag{10.25}$$

where p_{oi} is the probability that an observation will belong to the ith class and c equals the number of classes.

Let O_i denote the observed frequency of the ith class—that is, the number of sample observations belonging to it—and let $E_i = np_{oi}$ (n is the sample size) be the expected frequency under H_0. Then we wish to compare O_i with E_i for all c categories simultaneously. Since

$$\sum_{i=1}^{c} (O_i - E_i) = \sum_{i=1}^{c} O_i - \sum_{i=1}^{c} E_i = n - n = 0$$

is always true, we cannot use the sum of the deviations as a measure of the total deviation of the sample from H_0. As we did in defining a measure of dispersion, we square the deviations before summing, to obtain

$$\sum_{i=1}^{c} (O_i - E_i)^2 \tag{10.26}$$

This statistic equals zero only when $O_i = E_i$ for all i, indicating perfect agreement with H_0, and increases only with increasing discrepancies between the sample distribution and that specified in the null hypothesis.

To construct the test we need a statistic with a known or tabulated distribution. The statistic of Equation 10.26 does not meet that condition, but it can be transformed or standardized intuitively as follows. (The formal proof for two categories is left to the student as an exercise.) If we consider only two categories, say the ith and all others grouped together, O_i becomes a binomial random variable with mean $E_i = np_i$ and standard deviation $\sqrt{np_i(1 - p_i)}$. Thus, for large n, the statistic $(O_i - E_i)/\sqrt{np_i(1 - p_i)}$ is approximately a standard normal variable. Also, since p_i varies between 0 and 1 exclusive $(0 < p_i < 1)$ and will be nearer 0 than 1, $1 - p_i$ can be roughly approximated by 1 and $(O_i - E_i)/\sqrt{E_i}$ is approximately an $N(0, 1)$ variable. Thus, upon squaring,

$$\frac{(O_i - E_i)^2}{E_i} = Z_i^2$$

will be approximately distributed as a chi-square variable with one degree of freedom. Summing over all categories gives a chi-square variable with $c - 1$ degrees of freedom:

$$\sum_{i=1}^{c} \frac{(O_i - E_i)^2}{E_i} \sim \chi^2_{c-1} \tag{10.27}$$

One degree of freedom has been lost as a result of the restriction placed on the O_i, namely

$$\sum_{i=1}^{c} O_i = n$$

From our previous considerations, small values for the left-hand side of Equation 10.27 would indicate agreement with H_0, whereas larger values tend to disagree.

Therefore we can construct our test procedure under the following assumptions.

Assumptions for χ^2 Goodness-of-Fit Test

1. A random sample is selected.
2. Either the population is infinite or the sampling is with replacement.
3. At least nominal (categorical) data are obtainable.

The general form of the hypotheses is

$$H_0\colon\ F(x) = F_0(x) \qquad H_a\colon\ F(x) \neq F_0(x)$$

which are rewritten as

$$H_0\colon\ p_1 = p_{o1},\ p_2 = p_{o2},\ \ldots,\ p_c = p_{oc}$$

$$H_a\colon\ \text{some } p_i \neq p_{oi},\ \text{where } \sum_{i=1}^{c} p_{oi} = 1$$

With these assumptions, the test procedure is outlined in the following steps.

1. State the hypotheses and specify α.
2. Define the categories to be used and select n such that $np_{oi} \geqslant 5$ for all i ($np_{oi} \geqslant 5$ is the condition for using the normal approximation to a binomial).
3. Select a random sample from the population being considered and count the number of observations O_i belonging to each category.
4. Calculate $\displaystyle\sum_{i=1}^{c} \frac{(O_i - E_i)^2}{E_i}$, where $E_i = np_{oi}$.
5. Reject H_0 if

$$\sum_{i=1}^{c} \frac{(O_i - E_i)^2}{E_i} > \chi^2_{c-1,\alpha} \tag{10.28}$$

where $\chi^2_{c-1,\alpha}$ is found in Table A.7 of the Appendix. The following example illustrates this procedure.

EXAMPLE 10.17 Equal Employment Opportunity (EEO) has reported that the work force in the United States is composed of 10% professional, 10% clerical, 30% skilled, 15% service, and 35% semiskilled/laborers. A random sample of 100 Akron residents indicated 15 professional, 15 clerical, 40 skilled, 10 service, and 20 semiskilled/laborers. At $\alpha = 0.10$, does the work force in Akron appear to be consistent with the EEO report for the nation?

SOLUTION The hypotheses indicated by the EEO report are

$$H_0\colon\ p_1 = 0.10,\ \ p_2 = 0.10,\ \ p_3 = 0.30,\ \ p_4 = 0.15,\ \ p_5 = 0.35$$

and

$$H_a\colon\ \text{some } p_i \text{ differ from } p_{oi}$$

The significance level is specified to be 0.10. Thus we will reject H_0 if

$$\sum_{i=1}^{5} \frac{(O_i - E_i)^2}{E_i} > \chi^2_{5-1,\,0.10} = 7.78$$

It will be computationally advantageous to use the following tabular format for calculating

$$\sum_{i=1}^{c} \frac{(O_i - E_i)^2}{E_i}$$

(1) Category	(2) O_i	(3) $E_i = np_{oi}$	(4) $O_i - E_i$	(5) $(O_i - E_i)^2$	(5)/(3) $\frac{(O_i - E_i)^2}{E_i}$	$\sum_{i=1}^{j} \frac{(O_i - E_i)^2}{E_i}$
1	15	10	5	25	2.5	2.5
2	15	10	5	25	2.5	5.0
3	40	30	10	100	3.33	8.33
4	10	15	-5	25	1.67	10.00
5	20	35	-15	225	6.43	16.43
	100	100	0		16.43	

Figure 10.5 A Graphical Representation of the Decision Rule for Example 10.17

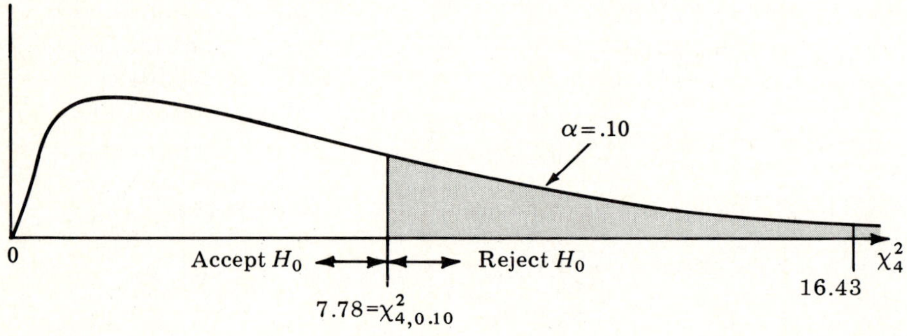

Since $16.43 > 7.78$, we reject H_0 and conclude that Akron's work force appears to deviate from that of the nation as reported by EEO. This is shown in Figure 10.5.

Note that an extra column has been added to the table presented in Example 10.17. The last column on the right represents the partial sums (running totals) of $\Sigma[(O_i - E_i)^2/E_i]$. As we can see, H_0 could have been rejected without calculating all the values, since the partial sum at category 3 exceeded 7.78. All further calculations simply increased a value that already implied rejection of H_0.

As indicated earlier, the χ^2 goodness-of-fit test can be used for any specific family of distributions. Some other examples are the normal, the Poisson, and the uniform families. When testing a distribution other than a multinomial, the p_{oi}'s are calculated using the specified distribution. The following example illustrates this requirement.

EXAMPLE 10.18 In many waiting line (queuing) studies, the service times are assumed to have a negative exponential distribution. In order to check this assumption for a local bank, a management consultant collected service times during the same period for a set of randomly selected days. The frequency table representing the sample of service times is given in Table 10.9. Test the fit of the negative exponential distribution for these data at $\alpha = 0.05$.

SOLUTION The hypotheses to be tested are:

H_0: T has a negative exponential distribution

H_a: T does not have a negative exponential distribution

Since no specific member of the family is indicated in H_0, we must estimate the parameter β from the sample data. The cost of doing so

Table 10.9 Grouped Data for Bank Service Times of Example 10.18

Class	Time (Minutes)	f_i	m_i
1	$0 \leq T < 0.5$	17	0.250
2	$0.5 \leq T < 1.25$	19	0.875
3	$1.25 \leq T < 2.25$	25	1.750
4	$2.25 \leq T < 4.00$	26	3.125
5	$4.00 \leq T < 12$	13	8.000

will be one degree of freedom. The maximum likelihood estimator (Chapter 8) for β is $\hat{\beta} = 1/\overline{X}$. Calculating $\overline{X}$ for the data in Table 10.9 gives

$$\bar{x} = \frac{1}{n} \sum_{i=1}^{c} m_i f_i = \frac{1}{100}(249.875) = 2.4988$$

Thus, our estimate of β is $1/2.4988 = 0.4002$.

Calculating p_{o1} and p_{o2}, the probability of occurrence of classes 1 and 2, for a negative exponential distribution with $\beta = 0.4$ gives

$$p_{o1} = \text{pr}\{0 \leqslant T < 0.5 | \beta = 0.4\} = \int_0^{0.5} 0.4e^{-0.4t}\, dt = 1 - e^{-0.4(0.5)}$$

$$= 1 - e^{-0.2} = 1 - 0.8187 = 0.1813$$

$$p_{o2} = \text{pr}\{0.5 \leqslant T < 1.25 | \beta = 0.4\} = \int_{0.5}^{1.25} 0.4e^{-0.4t}\, dt$$

$$= e^{-0.4(0.5)} - e^{-0.4(1.25)} = 0.8187 - 0.6065 = 0.2122$$

Similar computations yield $p_{o3} = 0.1999$, $p_{o4} = 0.2047$, and $p_{o5} = \text{pr}\{T > 4 | \beta = 0.4\} = 0.2019$. With these results we restate the hypotheses as

$$H_0\colon\ p_1 = 0.1813,\quad p_2 = 0.2122,\quad p_3 = 0.1999,$$
$$p_4 = 0.2047,\quad p_5 = 0.2019$$

$$H_a\colon\ \text{some } p_i \text{ are not as specified}$$

The test is:

Reject H_0 at 0.05 if

$$\sum_{i=1}^{c} \frac{(O_i - E_i)^2}{E_i} > \chi^2_{(c-1)-1,\,\alpha} = \chi^2_{5-2,\,0.05} = 7.81$$

Using our tabular format for the computations, we have

Class	$O_i = f_i$	$E_i = np_{oi}$	$\frac{(O_i - E_i)^2}{E_i}$	Partial Sum	$\chi^2_{3,0.05}$
1	17	18.13	0.07	0.07	7.81
2	19	21.22	0.23	0.30	7.81
3	25	19.99	1.26	1.56	7.81
4	26	20.47	1.49	3.05	7.81
5	13	20.19	2.56	5.61	< 7.81
	100	100.00			

Since $5.61 < 7.81$, we are able to conclude that on the basis of these data there is no reason to doubt that the service times during that period have a negative exponential distribution.

Note that in Example 10.18 we did not conclude that the negative exponential family is appropriate but that there is no reason to think it inappropriate. In some cases the values of the parameters may be specified by H_0. If they are, the degrees of freedom are the number of classes minus one, $\text{df} = c - 1$. If any parameters are estimated from the sample data, there is a resulting loss of precision in the test. This loss is taken into account by a reduction of one degree of freedom for each parameter estimated. That is, the degrees of freedom for such a test are given by

$$\text{df} = c - 1 - \text{number of estimated parameters} \tag{10.29}$$

In Example 10.18 we estimated β, and thus our degrees of freedom equalled $(c - 1) - 1 = (5 - 1) - 1 = 3$.

If the data have been collected and categorized before you receive them, you should use classes coinciding with the original categories so long as $np_{oi} \geqslant 5$. If any E_i (that is, np_{oi}) are less than 5, adjacent classes should be combined until the condition is satisfied. Note that this combining of classes will reduce c accordingly.

If you are designing the experiment, you can improve the power of the test slightly and simplify computations at the same time by defining the classes (before the data have been grouped) so that all classes are equiprobable under H_0. That is, define class i so that

$$\text{pr}\{\text{class } i | H_0\} = \frac{1}{c}$$

The next section presents an alternative to the χ^2 test, but it may be used only with ungrouped data and is not recommended unless all parameters are specified by H_0.

10.6.2 The Kolmogorov-Smirnov Test

The disadvantages of the χ^2 test for goodness of fit are that some information is lost in the grouping process, it discretizes all data even if the underlying distribution is continuous, and it provides only an approximate test. In most situations in which large sample sizes are available, the approximation will be accurate, but for small sample sizes the χ^2 test cannot even be used. An alternative to the χ^2 test is the Kolmogorov-Smirnov (KS) test for goodness of fit. The KS procedure is more restrictive in its assumptions than the χ^2, but it uses the data in an unaltered form and is an exact test. The assumptions underlying the derivation of the distribution for its test statistic are:

1. Random sample
2. An underlying continuous distribution
3. Distribution completely specified by H_0
4. At least interval data

The rationale for the KS procedure is that the sample cumulative distribution function $F_s(x)$ should be a good approximation of the true cumulative distribution function $F(x)$. In terms of the hypotheses, we have

$$H_0\colon\ F(x) = F_0(x) \qquad H_a\colon\ F(x) \neq F_0(x)$$

Thus, if H_0 is true, $F_s(x)$ should be close to $F_0(x)$ at all values for x. In particular, it should be close at each of the values occurring in the sample. If the deviation of $F_s(x)$ is excessive at any point, this implies that a lack of fit exists. The actual statistic, D, compares $F_s(x_i)$ with $F_0(x_i)$ and uses the largest absolute difference,

$$D = \max_x |F_0(x) - F_s(x)|$$

where the maximum is taken over all x. In simplified form

$$F_s(x_i) = \frac{i}{n} \tag{10.30}$$

From a graphical standpoint $F_0(x)$ is a monotonically increasing continuous function, whereas $F_s(x)$ is a step function. They are illustrated graphically in Figure 10.6.

To calculate D we use the formula

$$D = \max_i \left\{ \left| F_0(x_i) - \frac{i}{n} \right|, \left| F_0(x_i) - \frac{i-1}{n} \right| \right\} \tag{10.31}$$

Figure 10.6 A Graphical Representation of $F_0(x)$ and $F_s(x)$

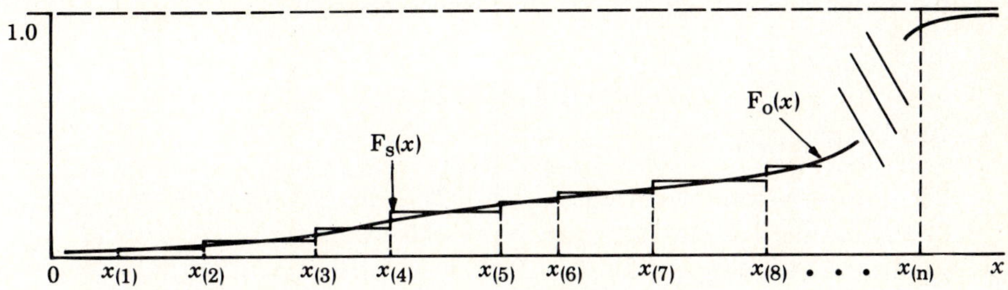

The formula in Equation 10.31 looks imposing, but it is really quite simple to use. The basic idea is that when the step occurs, we simply compare both the bottom and the top of the step with $F_0(x_i)$ and record the larger absolute difference. D is then the maximum of the larger absolute differences, that is, the largest of the larger. The decision rule for the hypotheses is to

$$\text{Reject } H_0 \text{ if } D > d_{n,\alpha} \tag{10.32}$$

where critical values, $d_{n,\alpha}$, are given in Table A.13. The distribution of the test statistic D is completely independent of the underlying distribution, assuming only continuity. Hence, the KS test is considered to be distribution-free.

EXAMPLE 10.19 Rather than the 100 observations that were used in Example 10.18, suppose we consider only the first 25 observations. Assume that these are:

1.14	2.68	0.36	0.69	0.80	1.92	4.37	3.86	0.55
0.62	2.13	0.25	0.43	1.38	0.96	2.20	4.28	5.76
0.80	2.72	3.06	4.15	0.32	1.88	0.40		

Test the assumption that these service times came from a negative exponential distribution with $\beta = 0.4$ using $\alpha = 0.05$.

SOLUTION The hypotheses to be tested are

$$H_0\colon F(t) = 1 - e^{-0.4t} \qquad H_a\colon F(t) \neq 1 - e^{-0.4t}$$

and we reject H_0 at $\alpha = 0.05$ if

$$D > d_{25,\,0.05} = 0.264$$

We have written $F_0(t)$ as $1 - e^{-0.4t}$ since for a negative exponential distribution

$$F(t) = \int_0^t \beta e^{-\beta x}\, dt = 1 - e^{-\beta t}$$

The computations required to calculate D are given in Table 10.10. The maximum of the larger absolute differences is at $i = 24$ and equals 0.1341. Since $0.1341 < 0.264$, we still conclude that there

Table 10.10 Computations for KS Test in Example 10.19

i (Rank)	$t_{(i)}$ = Ordered Values	$\frac{i}{n}$	$\frac{i-1}{n}$	$1 - e^{-0.4t_{(i)}}$	Larger|Diff|
1	0.25	0.04	0.00	0.0952	0.0952
2	0.32	0.08	0.04	0.1201	0.0801
3	0.36	0.12	0.08	0.1341	0.0541
4	0.40	0.16	0.12	0.1479	0.0279
5	0.43	0.20	0.16	0.1580	0.0420
6	0.55	0.24	0.20	0.1975	0.0425
7	0.62	0.28	0.24	0.2196	0.0604
8	0.69	0.32	0.28	0.2412	0.0788
9, 10	0.80	0.40	0.32	0.2738	0.1262
11	0.96	0.44	0.40	0.3189	0.1211
12	1.14	0.48	0.44	0.3662	0.1138
13	1.38	0.52	0.48	0.4242	0.0958
14	1.88	0.56	0.52	0.5286	0.0314
15	1.92	0.60	0.56	0.5361	0.0639
16	2.13	0.64	0.60	0.5734	0.0666
17	2.20	0.68	0.64	0.5852	0.0948
18	2.68	0.72	0.68	0.6577	0.0623
19	2.72	0.76	0.72	0.6631	0.0969
20	3.06	0.80	0.76	0.7059	0.0941
21	3.86	0.84	0.80	0.7865	0.0535
22	4.15	0.88	0.84	0.8099	0.0701
23	4.28	0.92	0.88	0.8195	0.1005
24	4.37	0.96	0.92	0.8259	0.1341
25	5.76	1.00	0.96	0.9001	0.0999

is no reason to doubt that a negative exponential distribution with $\beta = 0.4$ can be used to adequately describe the service times.

We emphasize that the KS test (as well as the χ^2) is not designed to differentiate between theoretical models. Instead it is merely a check to determine whether the proposed model adequately describes the sample data. It helps to answer the question, "Is the distribution that I would prefer to use reasonable for the population, or do the sample data exhibit such a drastic departure from this model that it must be rejected?" That is, as indicated in Chapter 9, the emphasis is on acceptance of H_0 because we wish to do further analysis under the proposed model.

10.6.3 The Runs Test

An assumption of all our test procedures has been that a random sample was selected. The runs test provides a rough check on the validity of this assumption. The test does not have high power, but it is extremely simple and easy to use even for large samples, and the efficiency may be quite adequate for most situations.

Assumptions of the Runs Test

1. A random sample has been selected.
2. The observations can be dichotomized.
3. At least nominal-level data are available.

The runs test specifically tests whether or not the order of occurrence of the observations follows a random pattern. For example, suppose an interviewer for a political preference poll was collecting sample data through "man on the street" interviews. The results of the interviews, as indicated by preference for either the Republican or the Democratic candidate, were, in order of occurrence, R R R R R R R R R R D D D D D D D D D D. If it were reported that of the twenty people interviewed, ten preferred the Democrat and ten preferred the Republican, the randomness of the sample might not be questioned. On the other hand, if the given sequence is reported, one becomes suspicious very quickly, since it seems quite unreasonable that the first ten should all prefer the Republican and the last ten should all prefer the Democrat. The runs test enables us to evaluate how unreasonable the sequence is. The procedure is quite simple:

1. State the hypotheses to be tested and the level of significance. Ordinarily the hypotheses will be H_0: random pattern of occurrence versus H_a: nonrandom pattern.

2. List the observations in order of occurrence.
3. Indicate to which of the two classes each observation belongs.
4. Count the number of runs R in the data (a *run* is defined to be an unbroken sequence of elements of the same type).

Nonrandomness in the data can be indicated in two ways. There can be too few runs; that is, elements of the same type tend to be grouped together, resulting in long runs, or too many runs, in which case an element of one type tends to be followed by an element of the opposite type, resulting in very short runs and cycling. Thus, our decision rule is given by

5. Reject H_0 at α if either

$$R \leqslant R_{n_1, n_2, \alpha/2} \qquad \text{or} \qquad R \geqslant R_{n_1, n_2, 1-\alpha/2} \tag{10.33}$$

Table A.12 in the Appendix gives upper and lower critical values for R at significance levels of 0.10, 0.05, and 0.025. The table is based on combinatorial formulas and thus is indexed by n_1, the number of elements of one type, and n_2, the number of elements of the other type, where $n_1 \leqslant n_2$. It is irrelevant which is identified as type 1 and which as type 2.

EXAMPLE 10.20 Test the randomness of occurrence of the political survey data given previously. Use $\alpha = 0.10$.

SOLUTION The observed sequence was

$$\underbrace{\text{R R R R R R R R R R}}_{1} \qquad \underbrace{\text{D D D D D D D D D D}}_{2}$$

with the indicated hypotheses being

$$H_0\text{: random pattern} \qquad H_a\text{: nonrandom pattern}$$

For the data, $n_1 = 10$ and $n_2 = 10$, and we reject H_0 at $\alpha = 0.10$ if

$$R \leqslant R_{10, 10, 0.05} \qquad \text{or} \qquad R \geqslant R_{10, 10, 0.95}$$

In Table A.12 we find $R_{10, 10, 0.05} = 6$ and $R_{10, 10, 0.95} = 16$. The observed value of R is 2 and the null hypothesis is rejected, since $2 < 6$. We conclude that the pattern of observation appears to be nonrandom.

The mean and variance of R are given by

$$\mu_R = \frac{2n_1 n_2}{n_1 + n_2} + 1 \quad \text{and} \quad \sigma_R^2 = \frac{2n_1 n_2 (2n_1 n_2 - n_1 - n_2)}{(n_1 + n_2)^2 (n_1 + n_2 - 1)} \tag{10.34}$$

As n_1 and n_2 approach infinity with their ratio remaining constant, the distribution of R approaches a normal distribution. Thus, for sample sizes too large for n_1 and n_2 to be included in Table A.12, we calculate

$$z = \frac{R - \mu_R}{\sigma_R} \tag{10.35}$$

and reject for the two-sided alternative if the absolute value exceeds $+z_{\alpha/2}$.

EXERCISES

10.6.1 There's a new game in town, and Marshall Matt Dillon doubts its honesty. The game is called craps and uses two six-sided dice. Mr. Dillon has temporarily shut down the game and confiscated the dice. While trying to decide whether or not to run the operator out of town on a rail, he has been absent-mindedly rolling one of the dice and jotting down the outcomes. He observed the following totals for the outcomes: one, 4; two, 7; three, 18; four, 19; five, 8; six, 4. Can you help him make his decision at $\alpha = 0.01$?

10.6.2 Pogo believes that a normal distribution having a mean of \$12,000 and a standard deviation of \$2400 can be used to model the weekly sales of one of the smaller product lines. They request your opinion of this belief and provide you with the following data for a random sample of weekly sales:

9926 15,234 12,316 10,722 6888 13,651 11,048

17,769 11,995 7838 12,009 14,272 14,111

11,502 9738 10,969 12,294 13,718 11,690 12,698

Using $\alpha = 0.05$, what do you tell them?

10.6.3 Refer to Exercise 10.6.2 and use the data given therein to test the randomness of the sample at 0.10. *Hint*: Determine the sample median and determine the number of runs above and below the median value.

10.7 SUPPLEMENTARY EXAMPLES

EXAMPLE 10.21 The production supervisor at the Pogo Company wants to estimate the percentage of downtime for a certain machine. To eliminate the cost of keeping constant records, he used a table of random numbers to select 400 times over a five-week period. Of the 400 times at which he checked the machine, it was down because of breakdowns, delays, and so on, a total of 36 times. What is his estimate of the percentage of downtime and what are 90% bounds on the estimate?

SOLUTION The machine either is down or it isn't, and since the check times were selected randomly and in advance, they should be independent with a constant probability of occurring. Thus, a binomial distribution with $n = 400$ and p unknown is appropriate, and the estimator for p is X/n. In this case we have

$$\frac{x}{n} = \frac{36}{400} = 0.09$$

or the machine is down 9% of the time.

For the confidence interval we note that $x > 5$ and use the normal approximation to the binomial. That is, using Equation 10.5, our bounds are given by

$$\text{LL} = 0.09 - 1.645\sqrt{\frac{0.09(0.91)}{400}} = 0.0665$$

$$\text{UL} = 0.09 + 1.645\sqrt{\frac{0.09(0.91)}{400}} = 0.1135$$

Thus, a 90% confidence interval for p is given by [0.0665, 0.1135].

The only assumption required in Example 10.21 that would appear to be questionable is that of a constant probability, since it might change if the time between two random checks is small. With the times randomly selected in advance any effect of this should be negligible.

EXAMPLE 10.22 The credit manager for Porsky's Department Store claims that the average balance on the statements mailed to credit customers is at least $69.00. To check this claim, an auditor selects a random sample of 28 statements from the credit files and finds that the average amount for the 28 is $64.32 and the standard deviation is $25.35. What would you say about the credit manager's claim using an α of 0.05?

SOLUTION We have a random sample and an interval level of measurement. The normality assumption is highly suspect because one would expect a distribution skewed to the right in this type of situation. However, in this instance the sample size of 28 is large enough that, with the central limit theorem, the violation of the normality assumption should have a very minimal effect. Therefore, since σ is not known, we will use the t distribution to test

$$H_0: \mu \geqslant 69.00 \qquad H_a: \mu < 69.00$$

That is, we assume that the credit manager is correct until the sample indicates otherwise. At α of 0.05 our test is:

Reject H_0 if

$$\bar{x} < \mu_0 - t_{n-1,\alpha}\left(\frac{s}{\sqrt{n}}\right) < 69 - 1.703\left(\frac{25.35}{\sqrt{28}}\right) < 69 - 8.16 = \$60.84$$

The value of $\bar{x}$ from the sample is \$64.32, which is not less than 60.84. Thus, we are able to conclude that there is no apparent reason to doubt the credit manager's claim on the basis of this sample.

If, in Example 10.22, it is felt that the distribution of statement amounts is so highly skewed that one might question the results of a t test, then we should use the sign test. Wilcoxon's test would not apply, since it assumes a symmetric population also. To use the sign test we would need the actual observations rather than the current summarized form.

EXAMPLE 10.23 The manager of Time Sharing, Incorporated, (TS) believes that interruptions to users of his computer facility are random and can be modeled by a Poisson distribution. To test this belief he has tabulated the number of service interruptions per week for the past 50 weeks. These are

Interruptions/Week	Number of Weeks
0	12
1	22
2	8
3	5
4	3

Is his belief supported at $\alpha = 0.05$?

SOLUTION The question here is one of goodness of fit, and since we are testing a discrete distribution, the χ^2 procedure is the obvious one to use. Thus, for the hypotheses

$$H_0\colon X \sim \text{Poisson} \qquad H_a\colon X \nsim \text{Poisson}$$

we reject H_0 if

$$\sum_{i=1}^{c} \frac{(O_i - E_i)^2}{E_i} > \chi^2_{c-2,\,\alpha}$$

We have $c - 2$ degrees of freedom, since μ is not specified and must be estimated from the sample data. The estimate for μ is $\overline{X}$, which, for these data, is 1.3.

Using Table A.3 we obtain a table of both observed and expected frequencies.

Interruptions	O_i	p_i	$E_i = np_i$	$\frac{(O_i - E_i)^2}{E_i}$	Partial Sum
0	12	0.273	13.65	0.20	0.20
1	22	0.354	17.70	1.04	1.24
2	8	0.230	11.50	1.07	2.31
3	5	0.100	5.00	0	2.31
4 or more	3	0.043	2.15	0.34	2.65

Since $2.65 < 7.81 = \chi^2_{3,\,0.05}$, we are able to conclude that a Poisson distribution appears to describe the distribution of interruptions adequately.

In Example 10.23 we could have combined the last two classes into one class of three or more because the expected number is less than 5. Since it was still in excess of 1, we did not do so. If we combine them, the resulting χ^2 value is 2.41, which is compared with a $\chi^2_{2,\,0.05}$ of 5.99.

EXAMPLE 10.24 A brewery indicates on its labels that each bottle of New Sloshingfroth contains 7% alcohol. A random sample of 18 bottles yielded approximate alcoholic contents of

6.1 6.5 7.2 6.9 7.0 7.1 6.8 6.6 6.7

7.4 6.4 7.2 7.3 6.3 7.5 7.0 6.0 6.2

If the distribution of alcoholic contents can be assumed only to be symmetric, can the brewery be accused of mislabeling its product? Use $\alpha = 0.10$.

SOLUTION The hypotheses of interest are

$$H_0\colon\ \mu = 7.0 \qquad H_a\colon\ \mu \neq 7.0$$

Since we can assume only symmetry, we eliminate the t test from consideration. Realizing that for a symmetric distribution the mean and median are identical, we consider the distribution-free alternatives. The indication of symmetry and data that are reasonably ordinal imply Wilcoxon's signed-rank test. Our test is to reject H_0 if

$$\min(T_+, T_-) \leqslant T_{n_+ + n_-,\, \alpha/2} = T_{16,\, 0.05} = 35$$

The differences, $x_i - 7.0$, the ranks of their absolute values, and the ranks corresponding to the positive differences are given in the following table.

$d_i = x_i - 7.0$	$R_{\lvert d_i \rvert}$	+'s	d_i	$R_{\lvert d_i \rvert}$	+'s	d_i	$R_{\lvert d_i \rvert}$	+'s
-0.9	15.0		-0.2	4.0		+0.3	6.5	6.5
-0.5	10.5		-0.4	8.5		-0.7	13.0	
+0.2	4.0	4.0	-0.3	6.5		+0.5	10.5	10.5
-0.1	1.5		+0.4	8.5	8.5	0	—	
0	—		-0.6	12.0		-1.0	16.0	
+0.1	1.5	1.5	+0.2	4.0	4.0	-0.8	14.0	

Therefore, $T_+ = 35 \qquad T_- = 101$ and $\min(T_+, T_-) = 35 \leqslant 35$

Hence H_0 is rejected, and we conclude that the labels appear to be incorrect.

In Example 10.24 we eliminated the t test because the problem indicated only symmetry, but with a sample of 18 from a symmetric population it is unlikely that lack of normality would be a problem. A more important consideration is probably the level of measurement. The alcoholic contents are given as approximate with no indication of how they were obtained. If they were measured precisely and rounded to the nearest tenth of a percent, the t test could probably be used. If the measuring instrument itself was crude, even ordinality might be questioned.

10.8 SUMMARY

In Chapter 9 we introduced the basic concepts of statistical inference, and in this chapter we have considered procedures that apply in specific situations involving a single sample. We considered inferences involving the parameter p, as well as inferences for the location and dispersion of the distribution. We concluded the chapter by considering two procedures that may be used to check whether a particular type of distribution seems reasonable and a test for randomness of the sample data.

For the binomial parameter we constructed tests based on exact binomial distributions and on the two approximations, Poisson and normal, when exact tables are not available. We also developed formulas for confidence intervals and sample size determination using the normal approximation.

In testing for location we have available tests that utilize the $N(0, 1)$ distribution (σ known) and the t_{n-1} distribution (σ unknown). Both these procedures assume normality and an interval level of measurement. If the normality assumption or the level of measurement is questionable, we have two alternatives available. The sign test requires only a nominal level of measurement and underlying continuity. Wilcoxon's signed-rank test requires symmetry and an ordinal level of measurement in addition to continuity. The choice of a procedure hinges on how valid the procedure's assumptions are for the situation at hand.

To test dispersion or variation, we looked only at the χ^2 test, which assumes normality and an interval level of measurement. Easy-to-use DF alternatives are not readily available in this instance. In checking the goodness of fit of a probability model, we looked at the χ^2 test and the Kolmogorov-Smirnov (KS) test. The χ^2 procedure groups the data into c mutually exclusive categories and compares observed frequencies with those expected under H_0. The KS test utilizes the actual observations and compares the observed cumulative probabilities with the theoretical values at each data point.

An assumption of all these procedures is that the sample is random. The runs test provides a "quick and dirty" check on this assumption. Though it is not a particularly powerful test, its ease and quickness of running make it appealing.

We summarize the procedures and their assumptions in Table 10.11, grouped by what the procedure tests. Remember that if any of the assumptions are violated, the probabilities associated with the inference are inexact and may be reasonable approximations at best.

Table 10.11 Summary of Tests and Associated Assumptions for Single Samples

Test For	Test	Assumptions
Goodness of fit	Kolmogorov-Smirnov (KS)	Random sample Underlying continuous distribution Interval data Completely specified distribution
	χ^2 test	Random sample Underlying continuous distribution or at least an infinite population At least nominal data
Randomness	Runs test	Underlying continuous distribution Data able to be dichotomized At least nominal level of measurement
Dispersion	χ^2 test	Random sample Underlying normal distribution At least interval level of measurement
Location	*t* test	Random sample Normal distribution (may be relaxed) At least interval level of measurement
	Z test	Random sample Normal distribution (may be relaxed) Population variance known At least interval level of measurement
	Wilcoxon signed-rank	Random sample Underlying continuous, symmetric distribution (continuity may be relaxed to infinite population) At least ordinal level of measurement
	Sign test	Random sample Continuous or at least infinite population At least nominal level of measurement
Proportions	Binomial	Random sample Independent trials Dichotomous outcomes per trial Constant probability of success from trial to trial

REVIEW EXERCISES

Level One

10.1 Refer to Example 10.21 and test the hypothesis that the percentage of time the machine is down is less than 10% at $\alpha = 0.10$.

10.2 Refer to Example 10.23 and test the hypothesis that the data came from a Poisson distribution with $\mu = 1.7$ using an α of 0.05.

10.3 Use the t test to evaluate the hypotheses of Example 10.24 with the data given there and an α of 0.10.

10.4 What are the advantages and disadvantages of using distribution-free procedures?

10.5 It has been reported that the distribution of grades in higher education today is 20% A's, 30% B's, 35% C's, 10% D's, and 5% F's or W's. In last term's statistics class of 250 students there were 30 A's, 55 B's, 105 C's, 35 D's, and 25 either failed or withdrew. Could the distribution of the grades for the statistics class be the same as that reported for higher education? Use $\alpha = 0.005$.

10.6 The Monroe Machine Shop uses a drill bit that has an average life of approximately 450 (in number of holes drilled). The manufacturer's rep for the Easy Tap Company has offered them a similar bit at the same price and claims that her bits will last longer on the average. She permits Monroe to test one dozen of the Easy Tap bits, with the following results in number of holes drilled:

462	396	501	504	427	462
438	493	481	441	472	513

Using an α of 0.05, what is your recommendation to Monroe?

10.7 Refer to Exercise 10.6 and evaluate the assumptions underlying the procedure you used.

10.8 What other tests could you have used in Exercise 10.6? Run those tests using the same α and explain any differences in your results.

10.9 The variance of the number of holes drilled for the bits that Monroe currently uses is 356. Are the lives of Easy Tap's bits as uniform as those of the current bits? Use an α of 0.10 and the data of Exercise 10.6.

10.10 When in control a production process produces items that are categorized as good, seconds, reworkable, and scrap in percentages of 80, 10, 7, and 3, respectively. A random sample of 200 items from the process had 140 good items, 25 seconds, 23 reworkable items, and 12 scrap items. At an α of 0.01, is this an indication that the process is out of control?

10.11 Explain the differences between the sign, Wilcoxon signed-rank, and t tests for location.

10.12 The results of the presidential elections, by party of the winning candidate, in order from 1900 through 1976 are: R R R D D R R R D D D D D R R D D R R D. Can we conclude at $\alpha = 0.10$ that the winning party occurs randomly?

10.13 In an advertising experiment customers were offered free samples of an item. The item was packaged in four different colors; everything else was the same, and each person could select the color of his or her choice. Of 160 people who took samples, 55 chose red, 42 chose blue, 34 chose yellow, and 29 chose brown. Does the color of the package appear to have any effect on choice? Use $\alpha = 0.05$.

10.14 In attempting to evaluate the risk associated with investing in Pitcairn Industries, an investor randomly selected 15 weeks and recorded the closing stock price for Pitcairn for the week. The closing prices were

15.5 13.25 16 10.5 18.75 16.25 11.5 10.75

14.00 14.25 16.5 9.75 12.25 11.25 8.50

Determine a 95% confidence interval for the variance of the weekly closing prices for Pitcairn stock.

10.15 It is desired to estimate the monthly food expenditures for a family of four to within \$15.00 of the true average with a probability of 0.95. If the standard deviation of expenditures is believed to be about \$35, how large a sample will be required?

Level Two

10.16 Show that

$$\operatorname{var}\left(\frac{X}{n}\right) = \frac{p(1-p)}{n}$$

10.17 Justify taking square roots of the endpoints in a confidence interval for σ^2 to obtain a confidence interval for σ.

10.18 Prove that Wilcoxon's signed-rank statistic has an expected value of $n(n + 1)/4$. *Hint*:

$$\sum_{k=1}^{n} k = \frac{n(n+1)}{2}$$

10.19 I just completed tossing an honest coin eight times and observed two runs. If I had no control over the outcomes of the tosses, how rare an occurrence is this?

10.20 Refer to Exercise 10.2 and Example 10.23. Explain why there is really no contradiction between the results of the two tests.

10.21 Approximate the power of the t test versus an alternative of $\mu_a = 465$ for the data in Exercise 10.6.

10.22 One way of testing for randomness of occurrence in a set of numerical observations is to find the median value for the data and then identify (by time of observation) each data point as being above or below the median. Do this for the data of Exercise 10.6 and count the number of runs above and below the median. Is there any evidence of nonrandomness in that sample at $\alpha = 0.10$?

10.23 It is desired to test the hypothesis set

$$H_0\colon\ \mu = 240 \qquad H_a\colon\ \mu < 240$$

with $\alpha = 0.05$ and a power of at least 0.9 for the alternative value of 220. How large a sample will be necessary if $\sigma = 40$ independent of the hypotheses?

10.24 A U.S. senator would like to obtain an estimate of constituents' opinions on a particular bill that will be coming to the floor of the Senate. How large a random sample will the senator's staff need to take to estimate the proportion in favor of the bill with an error of no more than 0.04 with a probability of 0.99?

10.25 When in control, a machining process is supposed to make steel cylinders with an inside diameter of 10 centimeters and a standard deviation of diameters of 0.002 centimeter. Changes in the diameter in either direction are critical. The quality control procedure is to randomly inspect nine cylinders periodically and calculate their average diameter. Within what limits must this average be for the inspector to conclude that the process is in control if $\alpha = 0.01$? Construct OC and power curves for the test.

Level Three

10.26 Prove that $\sigma_x = \sqrt{p(1-p)}$ is maximized at $p = 0.5$. (Requires simple derivatives.)

10.27 Prove that, for $c = 2$ in the χ^2 test for goodness of fit,

$$\sum_{i=1}^{2} \frac{(O_i - E_i)^2}{E_i} = \left(\frac{O_1 - np_1}{\sqrt{np_1(1-p_1)}} \right)^2$$

10.28 Prove that Wilcoxon's signed-rank statistic has a variance of $n(n+1)(2n+1)/24$. *Hint*:

$$\sum_{k=1}^{n} k^2 = \frac{n(n+1)(2n+1)}{6}$$

10.29 In a sequence of tosses of an honest coin resulting in n_1 heads and n_2 tails, what is the probability of observing exactly two runs if the sequence is actually random?

10.30 Test whether or not the following random sample could have come from a uniform distribution on the interval [3, 6], that is, $a = 3$, $b = 6$. Use $\alpha = 0.05$.

5.2	3.7	3.2	4.8	4.2	5.7	5.9	3.6
4.1	5.5	3.4	4.3	4.8	5.4	3.9	

References

1. Bradley, James V. *Distribution-Free Statistical Tests.* Englewood Cliffs, New Jersey: Prentice-Hall, Inc., 1968.
2. Gibbons, Jean Dickinson. *Nonparametric Statistical Inference.* New York: McGraw-Hill Co., 1971.
3. Lehmann, E. L. *Testing Statistical Hypotheses.* New York: John Wiley & Sons, Inc., 1959.
4. Lindgren, B. W. *Statistical Theory.* New York: The Macmillan Co., 1962.
5. Noether, Gottfried E. *Elements of Nonparametric Statistics.* New York: John Wiley & Sons, Inc., 1967.
6. Siegel, Sidney. *Nonparametric Statistics for the Behavioral Sciences.* New York: McGraw-Hill Co., 1956.

Chapter 11
Two-Sample Inferences

KEY WORDS AND TERMS

Paired Data (Randomized Block)
Paired *t* Test
Independent Samples *t* Test
F Test for Variances
Sign Test for Paired Data
Signed-Rank Test for Paired Data
Mann-Whitney *U* Test
Siegel-Tukey Test

11.1 INTRODUCTION

As director of purchasing for Amalgamated Motors Corporation (AMC), you must decide between two suppliers of a cylinder that is to have a diameter of 8 cm. To help you make the decision, each supplier has provided you with a sample of its cylinders. If you decide to use the samples, the immediate question to be resolved is, What are the primary attributes of interest? Suppose you have chosen to consider the mean diameter and the variation of the diameters between cylinders for each supplier. The next questions to arise are, What statistic do you use for each attribute? and Can you compare them between suppliers? If comparisons are possible, what assumptions are necessary for them to be valid, and what level of confidence can be associated with the results? The answers to these questions for specific situations are the objectives of this chapter.

In Chapter 10 we constructed inferential procedures that used the information contained in a single random sample. We now extend those procedures to situations in which two random samples are available. With one sample the emphasis is on trying to identify which member of a family appears to describe the chance variation in the population, or else whether or not a particular distribution seems reasonable. The two-sample case attempts to determine whether both samples could have come from the same population. If the characteristics of the samples differ by too much, we will decide that it is unreasonable to believe that they came from identical populations. The purpose of this chapter is to develop methods that will help us decide whether the samples differ by too much.

We vary the order slightly in this chapter by starting with inferences that assume normality. These are followed by tests for detecting differences between two binomial populations. As in Chapter 10, we conclude the chapter with the distribution-free procedures.

11.2 INFERENCES IN THE NORMAL FAMILY

This section is divided into three subsections. The first two are concerned with detecting differences in location for two normal distributions under two distinct methods of sampling. The first subsection considers the analysis of paired data, which are computationally easier to handle but more difficult to collect. The second subsection is applicable for detecting location differences using two independent random samples. The third subsection also uses two independent random samples, but it is concerned with comparing the variances.

11.2.1 Testing Location with Paired Data

The concept of *pairing* or *paired samples* relates to attempting to match the experimental subjects on some attribute before treatments are administered. The attribute on which the subjects are matched represents a random variable that is believed to affect the experimental results. The easiest way to understand why one would go to the trouble (and possibly expense) of pairing is to consider the following situation.

Suppose that I would like to evaluate two teaching methods. The two methods are (1) the current method and (2) an independent-study method. The same textbook is used for both methods and all students take a common exam at the end. To compare the methods I randomly select 15 students and assign them to the regular method, then randomly select 15 more and assign them to independent study. (Note that some control has occurred already, since both groups use the same text.) To compare the effects of the two methods, I calculate the average score for each group on the common exam.

Now let's assume that the regular class average was 76 and the independent-study average was 58, and that the difference between the two is highly significant. Can I conclude that the regular method is far superior to independent study for teaching statistics? Not necessarily, for suppose that by chance the students in the regular class had an average grade-point average (GPA) of 3.10, whereas the independent-study group had an average GPA of 2.20. (This would be rare if the samples were selected randomly, but it could happen.) To what would you attribute the difference in scores: teaching method or class differences? Most people would pick the class differences. Teaching method may well be the significant factor, but it would be impossible to convince anyone of that fact on the basis of these two samples.

From this example you can see that the object of pairing (or blocking, as it will be termed in Chapter 12) is to control the variability or effect of a secondary variable (or variables) that could significantly influence the responses. These secondary variables should average out through random sampling, but there is no guarantee that they will do so. To control this secondary variation as much as possible, we form pairs of subjects in which the two members of a pair are matched (as nearly as possible) with respect to the attributes of interest. For the previous example I could have selected a pair of students whose GPAs were both approximately 3.5, a pair whose GPAs were 2.4, and so forth. Then one member of each pair would be assigned randomly to one group and one to the other group. That is, I would ensure that I have one student whose GPA is 3.5, one whose GPA is 2.4, and so on, in each group. This guarantees that there will be no difference between the classes with respect to GPA. The treatments are then applied, and the resulting random sample is a set of pairs of measurements.

The data would appear as (x_{i1}, x_{i2}), where the first subscript, i, identifies the pair and the second identifies the treatment received. For analysis the data would be reorganized as in Table 11.1. Since the subjects were identical with respect to the controlled properties before experimentation, any differences within a pair can now be attributed to chance variation and factors (such as teaching methods) other than GPA.

Probably the best possible pairing is obtained in before/after-type experiments. In this case the subjects in both treatments are as nearly identical as one can expect. Psychologists often use identical twins in

Table 11.1 Sample Results for a Paired Experiment

Pair	Treatment 1	Treatment 2
1	x_{11}	x_{12}
2	x_{21}	x_{22}
.	.	.
.	.	.
.	.	.
n	x_{n1}	x_{n2}

their experiments in hopes of obtaining pairs with the same socioeconomic and genetic backgrounds. Marketing may attempt to evaluate two package designs by presenting both designs in identical displays and surroundings to an individual and having that individual choose between the two.

After the data have been obtained, the analysis is simple, since it follows the basic pattern of the one-sample t test of Chapter 10. The first step is to calculate the difference between the subjects within a pair for all pairs. For the hypothesis set,

$$H_0: \mu_1 = \mu_2 \qquad H_a: \mu_1 \neq \mu_2$$

H_0 indicates that the average response should be the same for both treatments. That is, the two measurements within a pair should be approximately the same, with any difference being a result of chance variation, if H_0 is true. Rewriting the hypotheses in the equivalent form,

$$H_0: \mu_1 - \mu_2 = 0 \qquad H_a: \mu_1 - \mu_2 \neq 0$$

indicates that the average difference should be near zero. Thus, for our first requirement in constructing a test, we have

$$\begin{aligned} E(X_{i1} - X_{i2}) &= E(X_{i1}) - E(X_{i2}) \\ &= \mu_1 - \mu_2 \end{aligned}$$

which equals 0 under H_0. If we let $D_i = X_{i1} - X_{i2}$ and consider

$$\overline{D} = \frac{1}{n} \sum_{i=1}^{n} D_i$$

then we obtain $E(\overline{D}|H_0) = 0$.

To find the distribution for the statistic $\overline{D}$, we must make the following assumptions:

1. A random sample of matched pairs of items with treatments assigned randomly within a pair
2. Underlying normal distributions
3. Interval level of measurement

With these assumptions, the average difference $\overline{D}$ will be normally distributed, since it represents a linear combination of normal variables. ($\overline{D}$ will be approximately normal according to the central limit theorem even if assumption 2 is violated.) An interval level of measurement is required, since we use actual differences, and grossly inaccurate measurements may give very misleading differences.

In summary, the distribution of $\bar{D}$ is given as

$$\bar{D} \sim N(\mu_1 - \mu_2, \sigma_{\bar{D}}^2)$$

or, if H_0 is assumed to be true, as

$$\bar{D} \sim N(0, \sigma_{\bar{D}}^2)$$

where $\sigma_{\bar{D}}^2 = \sigma_D^2/n$. The variance, σ_D^2, will be unknown, but we can estimate it by calculating the variance of the sample differences, that is, by

$$s_D^2 = \frac{1}{n-1}\left(\sum_{i=1}^{n} D_i^2 - \frac{(\Sigma D_i)^2}{n}\right) \tag{11.1}$$

where n represents the number of pairs in the sample. Substituting s_D for σ_D in the standardization formula gives

$$\frac{(\bar{D} - 0)}{s_D/\sqrt{n}} = \frac{\bar{D}\sqrt{n}}{s_D} \tag{11.2}$$

which has a t distribution with $n - 1$ degrees of freedom if H_0 is true. If H_0 is false, the formula still has a t distribution, but the distribution is shifted away from 0. Letting $\bar{d}$ represent the sample value, our decision rule is:

Figure 11.1 Graphical Interpretation of Paired t Test for H_0: $\mu_1 = \mu_2$ with Decision Rules Based on (a) $t_{n-1, \alpha/2}$ and (b) $\bar{d}_{\alpha/2}$

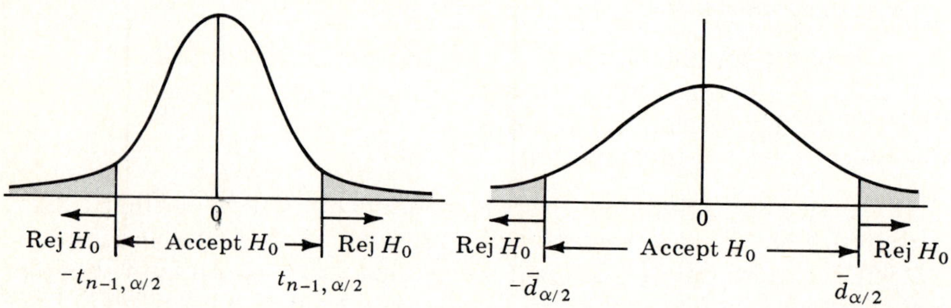

Reject H_0 if
$\text{pr}\{\bar{D} \text{ deviates from 0 by } \bar{d} \text{ or more} | H_0\} < \alpha$

That is, for our two-sided alternative,

Reject H_0 if

$$\text{pr}\{\bar{D} \geqslant |\bar{d}| \,\big|\, H_0\} = \text{pr}\left\{t_{n-1} \geqslant \left|\frac{\bar{d}\sqrt{n}}{s_D}\right|\right\} < \alpha$$

But this probability will be less than α only if

$$\left|\frac{\bar{d}\sqrt{n}}{s_D}\right| > t_{n-1,\ \alpha/2}$$

or, equivalently, if

$$|\bar{d}| > t_{n-1,\ \alpha/2}\frac{s_D}{\sqrt{n}} = \bar{d}_{\alpha/2}$$

This is illustrated in Figure 11.1.

Making the standard adjustment for the one-sided alternatives, we summarize the hypothesis sets and corresponding decision rules in Table 11.2.

Table 11.2 Summary of Hypothesis Sets and Decision Rules for the Paired t Test

Hypothesis Set		Decision Rule for the Stated α
$H_0\colon\ \mu_1 \leqslant \mu_2$	$H_a\colon\ \mu_1 > \mu_2$	Reject H_0 if $\bar{d} > t_{n-1,\alpha}\frac{s_D}{\sqrt{n}}$
$H_0\colon\ \mu_1 \geqslant \mu_2$	$H_a\colon\ \mu_1 < \mu_2$	Reject H_0 if $\bar{d} < -t_{n-1,\alpha}\frac{s_D}{\sqrt{n}}$
$H_0\colon\ \mu_1 = \mu_2$	$H_a\colon\ \mu_1 \neq \mu_2$	Reject H_0 if $|\bar{d}| > t_{n-1,\ \alpha/2}\frac{s_D}{\sqrt{n}}$

Before turning to an example, let us point out that the difference hypothesized under H_0 will not always be 0. The general form of the hypotheses is (for the two-sided alternative)

$$H_0: \mu_1 = \mu_2 + c_0 \qquad H_a: \mu_1 \neq \mu_2 + c_0 \tag{11.3}$$

where c_0 is a constant that will be zero in most cases. The change required in Equation 11.2 is the replacement of 0 by c_0. The decision rule for the hypotheses of Equation 11.3 is:

Reject H_0 if either

$$\bar{d} > c_0 + t_{n-1,\,\alpha/2}\frac{s_D}{\sqrt{n}} \qquad \text{or} \qquad \bar{d} < c_0 - t_{n-1,\,\alpha/2}\frac{s_D}{\sqrt{n}}$$

or, equivalently, in standardized form, if

$$\left|\frac{(\bar{d} - c_0)\sqrt{n}}{s_D}\right| > t_{n-1,\,\alpha/2}$$

Note that when we use the decision rule based on $\bar{d}$, we no longer combine the two statements by taking the absolute value. The reason is that the distribution of $\bar{D}$ as specified by H_0 is no longer symmetric about 0.

EXAMPLE 11.1 The marketing department of Pills Unlimited (PU) would like to advertise that a patient taking PU's biggest-selling drug will have an average reaction time more than 0.5 second faster than the average reaction time of a patient using their major competitor's drug. Before using the claim in its advertising, PU conducts the following experiment. Eleven people are randomly selected and one drug is administered. After the drug has taken effect, the subjects are given a timed reaction test. When enough time has passed for the effects of the first drug to wear off, the second drug is administered and the subjects' reaction times recorded while they are under its influence. To minimize carryover and time factors, the drug selected for the first dosage was random for each person. Each person then received the unused drug for the second dose. The reaction times, in seconds, for each subject with each drug are:

Subject:	1	2	3	4	5	6	7	8	9	10	11
Drug A:	6.0	3.8	5.1	2.9	3.6	5.1	7.3	2.7	4.5	4.6	5.8
PU's Drug:	4.1	3.7	5.2	2.3	2.8	4.5	6.7	1.9	3.4	4.6	6.0

Should the PU marketing department use the claim in its advertising? Let $\alpha = 0.05$.

SOLUTION The hypotheses of interest to PU are

$$H_0\colon\ \mu_{PU} \geqslant \mu_A - 0.5 \qquad H_a\colon\ \mu_{PU} < \mu_A - 0.5$$

Here $c_0 = 0.5$ and there are two reasons for selecting the stated alternative. First, the null hypothesis must stipulate a definite value in order to obtain a specific member of the normal family of distributions. Second, and more important, PU would like to minimize the chance of litigation that would result from false advertising. The decision rule for this hypothesis set is:

Reject H_0 if

$$\bar{d} < \bar{d}_\alpha = -c_0 - t_{n-1,\alpha}\frac{s_D}{\sqrt{n}}$$

Calculating the $d_i = X_{iP} - X_{iA}$, we get

d_i: -1.9 -0.1 $+0.1$ -0.6 -0.8 -0.6 -0.6 -0.8 -1.1 0 $+0.2$

with

$$\sum_{i=1}^{11} d_i = -6.2 \qquad \sum_{i=1}^{11} d_i^2 = 7.24$$

Thus, $s^2 = 0.3745$, $s_D/\sqrt{n} = 0.185$, and $\bar{d}_a = -0.5 - 1.812(0.185) = -0.835$. Our value for $\bar{d}$ is $\bar{d} = -6.2/11 = -0.564 \not< -0.835$. Therefore, we can't reject H_0 and must conclude that, on the basis of these data, we cannot state that the reaction time for PU's drug is more than 0.5 second faster than the time for drug A. That is, PU should not use the claim in its advertising.

The main point to remember in using the paired t test is that the sampling design dictates whether the observations are paired or not. In pairing, the pair identifications must be maintained to avoid trying to compare noncomparable responses. The purpose of pairing is to refine our estimate of the variance by removing from it variation between test items that is not really due to chance. This refinement costs us some degrees of freedom but hopefully results in a "cleaner" test of the effect of interest. Pairing is not always possible to achieve because of either the increased cost or the impossibility of obtaining matches on the attributes of interest. When this is the case, we must trust to randomization and select two independent random samples from the populations of interest.

Before we go to the analysis based on independent random samples, we indicate the procedure for obtaining an interval estimate of the difference between population means, that is, a confidence interval for $\mu_1 - \mu_2$. Our "best" estimator for $\mu_1 - \mu_2$ is $\overline{D} = \overline{X}_1 - \overline{X}_2$, which has a normal distribution (at least approximately) with mean $\mu_1 - \mu_2$ and variance σ_D^2/n. As with the test, σ_D^2 is estimated by s_D^2 and the standardized statistic, $(\overline{D} - \mu_D)\sqrt{n}/s_D$, has a t_{n-1} distribution. Thus, a $(1 - \alpha) \times 100\%$ confidence interval for $\mu_1 - \mu_2$ is given by

$$\begin{aligned} \text{LL} &= \bar{d} - t_{n-1,\,\alpha/2} \frac{s_D}{\sqrt{n}} \\ \text{UL} &= \bar{d} + t_{n-1,\,\alpha/2} \frac{s_D}{\sqrt{n}} \end{aligned} \tag{11.4}$$

One-sided bounds are obtained in the usual manner, and $s_{\bar{D}}^2 = (s_D^2/n)[(N - n)/N]$ if the population of differences is finite.

11.2.2 Testing Locations with Two Independent Samples

The hypotheses of interest in this case are identical to those of the previous section. The changes occur in the method of data collection and in the analysis procedure. The sampling is accomplished by selecting one random sample from the first population and a second random sample (independent of the initial one) from the other population. It also can be accomplished by selecting two independent random samples from the same population and applying one treatment to the subjects of the first sample and another to those of the second sample. Note that this was the initial proposal for evaluating the teaching methods.

Our objective will be to attempt to identify which of the situations depicted in Figure 11.2 is true.

Note that in Figure 11.2 we have shown the distributions to have identical variances. We did this because one of the assumptions required for constructing the distribution of our test statistic is that the populations have the same variance if the true values are unknown. If the population values are known (which is unrealistic), this assumption is unnecessary.

We will develop the test for the hypothesis set

$$H_0\colon \mu_1 \geqslant \mu_2 \qquad H_a\colon \mu_1 < \mu_2$$

and then summarize the tests for our three basic sets of hypotheses.

As we did for paired samples, we first rewrite the hypotheses in the form

$$H_0\colon \mu_1 - \mu_2 \geqslant 0 \qquad H_a\colon \mu_1 - \mu_2 < 0$$

Other equivalent forms of the hypotheses could be considered. For example,

$$H_0\colon \frac{\mu_1}{\mu_2} \geqslant 1 \qquad H_a\colon \frac{\mu_1}{\mu_2} < 1$$

Figure 11.2 Typical Independent Random Samples from Two Populations when (*a*) $\mu_1 = \mu_2$ and (*b*) $\mu_1 \neq \mu_2$ (in this case $\mu_1 < \mu_2$) (The numbers correspond to the samples.)

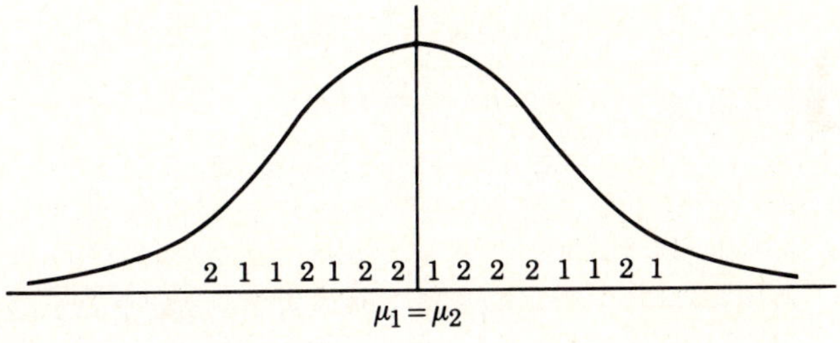

(*a*)

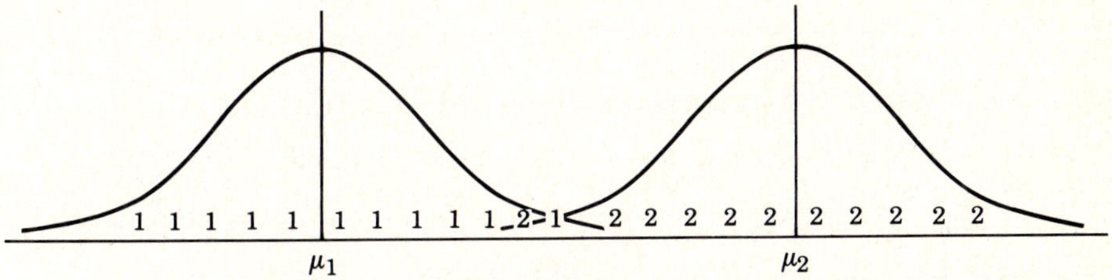

(*b*)

appears to satisfy our first requirement for developing a test, except that $E(\overline{X}_1/\overline{X}_2) \neq \mu_1/\mu_2$. Another problem with this statistic is the fact that its distribution is not readily available, since the ratio of two normal variables may not be normal. The statistic $\overline{X}_1 - \overline{X}_2$ satisfies both these requirements.

For the development of our test, the following assumptions are necessary:

1. Two independent random samples
2. Underlying normal distributions
3. $\sigma_1^2 = \sigma_2^2$ if they are unknown*
4. Interval level of measurement

If σ_1^2 and σ_2^2 are known, we can use the Z test, since we have a completely specified normal distribution. To see this, note that

$$\overline{X}_1 \sim N\left(\mu_1, \frac{\sigma_1^2}{n_1}\right) \quad \text{and} \quad \overline{X}_2 \sim N\left(\mu_2, \frac{\sigma_2^2}{n_2}\right)$$

Thus, regardless of whether or not H_0 is true, the difference of two normal variables is another normal variable, and we have

$$\overline{X}_1 - \overline{X}_2 \sim N\left(\mu_1 - \mu_2, \frac{\sigma_1^2}{n_1} + \frac{\sigma_2^2}{n_2}\right) \tag{11.5}$$

which, if H_0 is true, becomes

$$\overline{X}_1 - \overline{X}_2 \sim N\left(0, \frac{\sigma_1^2}{n_1} + \frac{\sigma_2^2}{n_2}\right)$$

If σ_1^2 and σ_2^2 are known, our decision rule (in standardized form) is to reject H_0 if

$$\frac{(\overline{X}_1 - \overline{X}_2) - 0}{\sqrt{\sigma_1^2/n_1 + \sigma_2^2/n_2}} < -z_\alpha$$

For the other possible hypotheses, the usual adjustments in the decision rules are made.

*If the population variances are known, then no problems arise and assumption 3 may be dropped. If they are unknown and unequal, we have what is known to statisticians as the Behrens-Fisher problem. A Russian mathematician, Y. V. Linnik, showed that an exact test for $H_0: \mu_1 = \mu_2$ exists in this case but did not develop a workable form for it. Approximate tests of H_0 are available when the variances are unknown and unequal. See, for example, Walpole and Myers.

The assumption that σ_1^2 and σ_2^2 are known is very unrealistic. When they are not known, we assume that they are equal, and the distribution of our statistic becomes

$$\overline{X}_1 - \overline{X}_2 \sim N\left(\mu_1 - \mu_2, \frac{\sigma^2}{n_1} + \frac{\sigma^2}{n_2}\right) \tag{11.6}$$

Rewriting $\sigma^2/n_1 + \sigma^2/n_2$ as $\sigma^2(1/n_1 + 1/n_2)$, we see that it is necessary to estimate only the single parameter σ^2. Since σ^2 is the same for both populations, we already have available two unbiased estimates of it: s_1^2 and s_2^2, the two sample variances. If we were forced to use only one, then we should use the sample variance based on the larger n, since it contains more information. Being economically minded statisticians, we realize that using only one would be wasteful, as this implies discarding the information contained in the other.

We avoid being wasteful by combining the two estimates of the same thing. Recalling the properties of estimators from Chapter 8, the new estimate should retain the property of unbiasedness and should have a tabulated distribution. In general, a simple average of s_1^2 and s_2^2 will not suffice, even though it is unbiased, because the sample sizes will be unequal. To obtain an estimate that gives equal weight to each observation, we "pool" the information contained in the two sample variances by taking a weighted average with the respective degrees of freedom as weights. Calling the pooled estimate (of σ^2) s_p^2, we have

$$s_p^2 = \frac{(n_1 - 1)s_1^2 + (n_2 - 1)s_2^2}{n_1 + n_2 - 2} \tag{11.7}$$

or equivalently,

$$s_p^2 = \frac{\sum_{j=1}^{2} \sum_{i=1}^{nj} (X_{ij} - \overline{X}_j)^2}{n_1 + n_2 - 2}$$

If the computational form for the numerator of a sample variance from Chapter 7 is used, the last form can be altered further to give

$$s_p^2 = \frac{\left(\sum_{i=1}^{n_1} X_{i1}^2 - (\Sigma X_{i1})^2/n_1\right) + \left(\sum_{i=1}^{n_2} X_{i2}^2 - (\Sigma X_{i2})^2/n_2\right)}{n_1 + n_2 - 2} \tag{11.8}$$

The form of s_p^2 to be used in a given situation will depend on the data at hand.

The estimate of σ is then $s_p = \sqrt{s_p^2}$. Substitution of this result for σ in the basic standardized statistic

$$\frac{(\overline{X}_1 - \overline{X}_2) - (\mu_1 - \mu_2)}{\sigma\sqrt{1/n_1 + 1/n_2}}$$

changes the distribution from a standard normal to a t with $n_1 + n_2 - 2$ degrees of freedom, that is

$$\frac{(\overline{X}_1 - \overline{X}_2) - (\mu_1 - \mu_2)}{s_p\sqrt{1/n_1 + 1/n_2}} \sim t_{n_1+n_2-2} \qquad (11.9)$$

Generalizing our basic hypothesis set to include nonzero differences under H_0, we have

$$H_0\colon\ \mu_1 - \mu_2 \geqslant c_0 \qquad H_a\colon\ \mu_1 - \mu_2 < c_0$$

Thus, we have a statistic with a known expected value and a completely specified distribution for its chance variation when H_0 is true. Following our previous pattern, the test is:

Reject H_0 at α if

$$\frac{(\overline{X}_1 - \overline{X}_2) - c_0}{s_p\sqrt{1/n_1 + 1/n_2}} < -t_{n_1+n_2-2,\,\alpha}$$

The equivalent test based on the observed difference between $\overline{X}_1$ and $\overline{X}_2$ is to reject H_0 if

$$(\overline{X}_1 - \overline{X}_2) < c_0 - t_{n_1+n_2-2,\,\alpha}\, s_p\sqrt{\frac{1}{n_1} + \frac{1}{n_2}}$$

The possible hypothesis sets and corresponding decision rules are given in Table 11.3 in standardized form. In Table 11.3,

$$s_{\overline{X}_1-\overline{X}_2} = s_p\sqrt{\frac{1}{n_1} + \frac{1}{n_2}} \qquad (11.10)$$

EXAMPLE 11.2 A company has decided to replace its old drill press with a new one and is considering two different makes. Since all costs of each press appear to be the same, the company would prefer simply to buy the faster one, if in fact there is a faster one. Before making the decision, each press is used on a trial basis, giving the following results.

	Machine 1	Machine 2
Number of hours:	10	8
Output/hour:	13.2	14.1
Standard Deviation:	1.4	2.3

Using $\alpha = 0.05$, does it matter which is purchased?

SOLUTION We wish to test

$$H_0\colon \mu_1 = \mu_2 \qquad H_a\colon \mu_1 \neq \mu_2$$

and we reject H_0 if

$$\left| \frac{(\overline{X}_1 - \overline{X}_2) - 0}{s_p \sqrt{1/n_1 + 1/n_2}} \right| > t_{n_1+n_2-2,\, \alpha/2} = t_{16,\, 0.025} = 2.120$$

We have

$$\left| \frac{(\overline{X}_1 - \overline{X}_2) - 0}{s_p \sqrt{1/n_1 + 1/n_2}} \right| = \left| \frac{(13.2 - 14.1) - 0}{1.848(0.474)} \right| = \left| \frac{-0.9}{0.876} \right| = 1.03$$

Table 11.3 Summary of Hypothesis Sets and Corresponding Standardized Decision Rules for the Two-Independent-Samples t Test

Hypothesis Set	Decision Rule for the Stated α
	Reject H_0 if:
1. $H_0\colon \mu_1 - \mu_2 \leqslant c_0 \quad H_a\colon \mu_1 - \mu_2 > c_0$	$\dfrac{(\overline{X}_1 - \overline{X}_2) - c_0}{s_{\overline{X}_1 - \overline{X}_2}} > t_{n_1+n_2-2,\, \alpha}$
2. $H_0\colon \mu_1 - \mu_2 \geqslant c_0 \quad H_a\colon \mu_1 - \mu_2 < c_0$	$\dfrac{(\overline{X}_1 - \overline{X}_2) - c_0}{s_{\overline{X}_1 - \overline{X}_2}} < -t_{n_1+n_2-2,\, \alpha}$
3. $H_0\colon \mu_1 - \mu_2 = c_0 \quad H_a\colon \mu_1 - \mu_2 \neq c_0$	$\left\lvert\dfrac{(\overline{X}_1 - \overline{X}_2) - c_0}{s_{\overline{X}_1 - \overline{X}_2}}\right\rvert > t_{n_1+n_2-2,\, \alpha/2}$

where

$$s_p^2 = \frac{(10-1)(1.4)^2 + (8-1)(2.3)^2}{10+8-2} = \frac{9(1.96)+7(5.29)}{16}$$

$$= \frac{54.67}{16} = 3.417$$

Since $1.03 < 2.120$, we cannot reject H_0. Hence, it cannot be concluded that one of the presses is faster, and the company needs some other criterion for making its decision.

If we are interested only in estimating the difference between the means of two normal distributions using independent random samples, our "best" point estimate is $\bar{x}_1 - \bar{x}_2$. As was shown earlier, if the assumptions for the test are satisfied, the distribution of

$$\frac{(\bar{X}_1 - \bar{X}_2) - (\mu_1 - \mu_2)}{s_{\bar{X}_1 - \bar{X}_2}}$$

is $t_{n_1+n_2-2}$. Therefore, a $(1-\alpha) \times 100\%$ two-sided confidence interval for $\mu_1 - \mu_2$ is given by

$$\begin{aligned} \text{LL} &= (\bar{x}_1 - \bar{x}_2) - t_{n_1+n_2-2,\,\alpha/2}\, s_p \sqrt{\frac{1}{n_1} + \frac{1}{n_2}} \\ \text{UL} &= (\bar{x}_1 - \bar{x}_2) + t_{n_1+n_2-2,\,\alpha/2}\, s_p \sqrt{\frac{1}{n_1} + \frac{1}{n_2}} \end{aligned} \tag{11.11}$$

We conclude this section with a test designed to detect differences in the scale parameters of two normal distributions, that is, a test for comparing the variances.

11.2.3 Comparing Variances for Two Normal Distributions

In this section we consider the problem of comparing the variances of two normal distributions. To this end we assume that independent random samples either are available or will be selected from the populations under consideration. Our basic hypothesis set for the following development is

$$H_0\colon\ \sigma_1^2 = \sigma_2^2 \qquad H_a\colon\ \sigma_1^2 \neq \sigma_2^2$$

To develop the distribution of the statistic to be used in making inferences, we need the following assumptions:

1. Independent random samples from each population
2. Underlying normal distributions
3. An interval level of measurement

From Chapter 8 we know that s_1^2 and s_2^2 are unbiased estimators for σ_1^2 and σ_2^2, respectively. In addition, we know that the variance of a random sample from a normal population is distributed as $\sigma^2 \chi^2_{n-1}/n - 1$. Thus, for our two estimators it follows that

$$s_1^2 \sim \frac{\sigma_1^2 \chi^2_{n_1-1}}{n_1 - 1} \qquad \text{and} \qquad s_2^2 \sim \frac{\sigma_2^2 \chi^2_{n_2-1}}{n_2 - 1}$$

It follows logically that the actual test statistic should be a function of s_1^2 and s_2^2. If we use the approach taken in comparing two means, it would seem intuitive that restating the hypotheses as

$$H_0\colon \sigma_1^2 - \sigma_2^2 = 0 \qquad H_a\colon \sigma_1^2 - \sigma_2^2 \neq 0$$

is the way to go. The obvious function for this restatement of H_0 is $s_1^2 - s_2^2$. This poses a problem distributionally, since, unlike the normal family, the difference of two χ^2 distributions does not have a distribution for which tables are readily available. An equivalent form of the hypotheses that quickly comes to mind is

$$H_0\colon \frac{\sigma_1^2}{\sigma_2^2} = 1 \qquad H_a\colon \frac{\sigma_1^2}{\sigma_2^2} \neq 1 \tag{11.12}$$

With the hypotheses in this form, we are in business, since the obvious function of s_1^2 and s_2^2 is s_1^2/s_2^2, and we know from Chapter 8 that the ratio of two independent χ^2 variables divided by their respective degrees of freedom has an F distribution. That is, the ratio

$$\frac{s_1^2/\sigma_1^2}{s_2^2/\sigma_2^2} = \frac{\chi^2_{n_1-1}/n_1 - 1}{\chi^2_{n_2-1}/n_2 - 1} \sim F_{n_1-1,\, n_2-1}$$

If H_0 is true, then $\sigma_1^2 = \sigma_2^2$, and we have

$$\frac{s_1^2}{s_2^2} \sim F_{n_1-1,\, n_2-1} \tag{11.13}$$

In constructing the decision rule, we note that if H_0 (as stated in Equation 11.12) is true, the ratio of the two sample variances should be close to 1. For the two-sided alternative, it follows that H_0 should be rejected if the ratio deviates from 1 by too much. By too much we mean, as always, if the probability of a deviation as large or larger than that observed is less than α. Thus, our decision rule for the alternative in Equation 11.12 is to:

Reject H_0 if either

$$\frac{s_1^2}{s_2^2} > F_{n_1-1,\, n_2-1,\, \alpha/2} \qquad \text{or} \qquad \frac{s_1^2}{s_2^2} < F_{n_1-1,\, n_2-1,\, 1-\alpha/2}$$

where $F_{n_1-1,\, n_2-1,\, 1-\alpha/2} = 1/F_{n_2-1,\, n_1-1,\, \alpha/2}$. Because F distributions have this reciprocal relationship, we can simplify this decision rule by simply using the larger of the two ratios, s_1^2/s_2^2 and s_2^2/s_1^2, for our test statistic. That is, we reject H_0 if either ratio gets too large or, equivalently, if

$$\frac{\max s^2}{\min s^2} > F_{n_{\text{num}}-1,\, n_{\text{den}}-1,\, \alpha/2} \tag{11.14}$$

In Equation 11.14 n_{num} is the sample size for the larger variance and n_{den} is the sample size for the smaller variance.

For one-sided alternatives we must use the variance indicated to be larger by H_a in the numerator.

A summary of the basic hypothesis sets and their decision rules is given in Table 11.4.

EXAMPLE 11.3 One of our assumptions in Section 11.2.2 was that the population variances were identical. We may use this test to check that assumption. We do so using an α of 0.02 and the data of Example 11.2.

SOLUTION In Example 11.2 we had $n_1 = 10$, $n_2 = 8$, $s_1 = 1.4$, and $s_2 = 2.3$. We wish to test the hypothesis set

$$H_0\colon\ \sigma_1^2 = \sigma_2^2 \qquad H_a\colon\ \sigma_1^2 \neq \sigma_2^2$$

Our decision rule is:

Reject H_0 if

$$\frac{\max s^2}{\min s^2} > F_{n_{\text{num}}-1,\, n_{\text{den}}-1,\, \alpha/2}$$

Table 11.4 Hypothesis Sets and Corresponding Decision Rules for Testing Variances for Two Normal Distributions

Hypothesis Set	Decision Rule for a Stated α
	Reject H_0 if
1. $H_0: \sigma_1^2 \leqslant \sigma_2^2 \quad H_a: \sigma_1^2 > \sigma_2^2$	$\dfrac{s_1^2}{s_2^2} > F_{n_1-1,\, n_2-1,\, \alpha}$
2. $H_0: \sigma_1^2 \geqslant \sigma_2^2 \quad H_a: \sigma_1^2 < \sigma_2^2$	$\dfrac{s_2^2}{s_1^2} > F_{n_2-1,\, n_1-1,\, \alpha}$
3. $H_0: \sigma_1^2 = \sigma_2^2 \quad H_a: \sigma_1^2 \neq \sigma_2^2$	$\dfrac{\max s^2}{\min s^2} > F_{n_{\text{num}}-1,\, n_{\text{den}}-1,\, \alpha/2}$

We have $\max s^2 = s_2^2 = (2.3)^2 = 5.29$, and hence

$$n_{\text{num}} - 1 = n_2 - 1 = 7$$

The smaller s^2 is s_1^2, which equals $(1.4)^2$, and

$$n_{\text{den}} - 1 = n_1 - 1 = 9$$

Therefore, the critical value, or maximum acceptable ratio under H_0, is $f_{7,\,9,\,0.01} = 5.61$.

The value of our statistic is $s_2^2/s_1^2 = 5.29/1.96 = 2.70$. Since $2.70 < 5.61$, we cannot reject H_0, and we conclude that we have no reason to doubt the assumption of equal variances for the populations of Example 11.2.

At times we may be interested in obtaining a confidence interval for the ratio of two variances; that is, we may want to construct a probability statement of the form

$$\text{pr}\left\{\text{LL} \leqslant \frac{\sigma_1^2}{\sigma_2^2} \leqslant \text{UL}\right\} = 1 - \alpha$$

Assuming normal distributions and two independent random samples, we know from our previous considerations that

$$\text{pr}\left\{f_{n_1-1,\, n_2-1,\, 1-\alpha/2} \leqslant \frac{s_1^2/\sigma_1^2}{s_2^2/\sigma_2^2} \leqslant f_{n_1-1,\, n_2-1,\, \alpha/2}\right\}$$

$$= \text{pr}\left\{f_{n_1-1,\, n_2-1,\, 1-\alpha/2} \leqslant \frac{\sigma_2^2 s_1^2}{\sigma_1^2 s_2^2} \leqslant f_{n_1-1,\, n_2-1,\, \alpha/2}\right\} = 1 - \alpha$$

From this we can obtain the required form, with

$$\text{LL} = \frac{s_1^2}{s_2^2}\left(\frac{1}{f_{n_1-1,\, n_2-1,\, \alpha/2}}\right) \quad \text{and} \quad \text{UL} = \frac{s_1^2}{s_2^2} f_{n_2-1,\, n_1-1,\, \alpha/2} \tag{11.15}$$

(The intervening algebra is left as an exercise.) For a one-sided interval we make the usual adjustments.

EXAMPLE 11.4 A company recently purchased a new machine and would like to obtain a 90% confidence interval for the ratio of the variance of an older machine to the variance of the new machine. Random samples of size 16 and size 11 were taken from the outputs of the new and old machines, respectively. The sample variances are $s_{\text{new}}^2 = 0.0016$ and $s_{\text{old}}^2 = 0.0080$. Construct the interval for the company.

SOLUTION We have $s_1^2 = s_{\text{old}}^2 = 0.0080$ and $s_2^2 = s_{\text{new}}^2 = 0.0016$, with $n_1 = 11$ and $n_2 = 16$. For $1 - \alpha = 0.90$, $\alpha/2 = 0.05$, which gives

$$f_{n_1-1,\, n_2-1,\, \alpha/2} = f_{10,\, 15,\, 0.05} = 2.54$$

and

$$f_{n_2-1,\, n_1-1,\, \alpha/2} = f_{15,\, 10,\, 0.05} = 2.85$$

Therefore, a 90% confidence interval for the ratio of the older machine's variance to that of the new machine is given by substituting these values into Equation 11.15. The results are

$$\text{LL} = \frac{0.0080}{0.0016}\left(\frac{1}{2.54}\right) = 1.97 \qquad \text{and} \qquad \text{UL} = \frac{0.0080}{0.0016}(2.85) = 14.25$$

These limits imply that

$$\text{pr}\left\{[1.97,\, 14.25] \text{ contains } \frac{\sigma_{\text{old}}^2}{\sigma_{\text{new}}^2}\right\} = 0.90$$

EXERCISES

11.2.1 It has been stated that accounting majors do better in statistics than other business students. To test this statement, eight accounting majors and nine other business students were randomly selected from last quarter's statistics class. Their grades on the common final exam are (total possible = 200):

Accounting:	152	139	176	182	121	86	137	149	
Other:	194	178	129	169	94	158	187	145	125

Is the statement supported at $\alpha = 0.10$?

11.2.2 In evaluating two types of rust inhibitors, 25 steel rods were selected. Half of each rod was coated with Type I rust inhibitor and the other half with Type II. The rods were then subjected to various soil, weather, and climactic conditions for one year. At that time the depth of corrosion on each half of each rod was measured. The results, coded by multiplying by 100, are given below.

Rod:	1	2	3	4	5	6	7	8	9
Type I Depth:	13.2	84	51	24	16.7	86	11.6	60	42
Type II Depth:	11.6	62	62	12	12.2	91	2.4	111	73
Rod:	10	11	12	13	14	15	16	17	18
Type I Depth:	54	76.1	57	18.7	71	25.4	14.7	39	33
Type II Depth:	50	30.3	22	26.9	60	17.6	1.1	22	37.5
Rod:	19	20	21	22	23	24	25		
Type I Depth:	54.1	67.2	46.3	19.8	36.5	27.4	54.9		
Type II Depth:	27.8	50.6	38.9	20.2	38.3	19.2	31.1		

Does there appear to be a difference in the effectiveness of the inhibitors at $\alpha = 0.01$?

11.2.3 Refer to Exercise 11.2.1 and construct a 90% confidence interval for the ratio of the variance in the scores of accounting majors to the variance in the scores of other business majors.

11.2.4 Use your confidence interval from Exercise 11.2.3 to test the assumption of equal variances required in Exercise 11.2.1. Explain the reason for your decision.

11.2.5 For Exercises 11.2.1 and 11.2.2, indicate what assumptions are required in each. Do you think that each of the assumptions is reasonable for the given situation? Why or why not?

11.3 TESTS OF PROPORTIONS

In this section we develop procedures for comparing two binomial distributions. Both procedures yield approximate tests, since they are based on the normal approximation to the binomial. If the only purpose of the test is to detect differences, then the two procedures are equivalent and you should use whichever one you find easier.

11.3.1 The Normal Approximation

The parameter of interest when we compare binomial distributions is the proportion of successes, p. We consider two dichotomous populations to be equivalent if the parameters p_1 and p_2 are equal. It makes no sense to test for or estimate the other parameter, which is the number of trials. Therefore, our basic hypothesis set is

$$H_0\colon\ p_1 = p_2 = p \qquad H_a\colon\ p_1 \neq p_2$$

The only assumptions necessary for developing the test are those of the binomial distribution (given in Chapter 4) and the restrictions required for the normal approximation.

The condition for approximating a binomial probability with a normal probability is that both np and $n(1 - p)$ be greater than or equal to 5. For the present situation, this condition translates to

$$n_1 p_1 \geqslant 5 \quad \text{and} \quad n_1(1 - p_1) \geqslant 5$$

and

$$n_2 p_2 \geqslant 5 \quad \text{and} \quad n_2(1 - p_2) \geqslant 5$$

With respect to the sample data, these become

$$x_1 \geqslant 5 \quad \text{and} \quad n_1 - x_1 \geqslant 5$$

and

$$x_2 \geqslant 5 \quad \text{and} \quad n_2 - x_2 \geqslant 5$$

Our best estimates of p_1 and p_2 are x_1/n_1 and x_2/n_2, respectively. Since both are unbiased estimators of their corresponding p values, they should be close to each other if H_0 is true. That is, under H_0

$$E\left[\frac{x_1}{n_1}\middle| H_0\right] = E\left[\frac{x_2}{n_2}\middle| H_0\right]$$

Using the basic form of a statistical test, we would reject H_0 at a significance level of α if

$$\text{pr}\left\{\text{a deviation as large or larger than } \left|\frac{x_1}{n_1} - \frac{x_2}{n_2}\right| \middle| H_0\right\} < \frac{\alpha}{2}$$

Under our conditions, x_1/n_1 and x_2/n_2 have approximately normal distributions with parameters $\mu_1 = p_1$, $\sigma_1^2 = p_1(1 - p_1)/n_1$ and $\mu_2 = p_2$, $\sigma_2^2 = p_2(1 - p_2)/n_2$, respectively. Thus, their difference will be approximately normal also. If H_0 is true, the parameters for the distribution are

$$\mu_{(x_1/n_1)-(x_2/n_2)} = 0$$

and

$$\sigma^2_{(x_1/n_1)-(x_2/n_2)} = \sigma^2_{x_1/n_1} + \sigma^2_{x_2/n_2} = \frac{p_1(1 - p_1)}{n_1} + \frac{p_2(1 - p_2)}{n_2}$$

$$= p(1 - p)\left(\frac{1}{n_1} + \frac{1}{n_2}\right)$$

At this point we have a problem because, even though $p_1 = p_2 = p$ if H_0 is true, p is still unknown. As we did in constructing a confidence interval for p, we use its estimate to obtain an estimate of the standard deviation. Since, under H_0, $p_1 = p_2 = p$, we can consider the two samples as one large sample from the common population, and our best estimate of p becomes $(x_1 + x_2)/(n_1 + n_2)$. The corresponding estimate of the standard deviation of the difference becomes

$$\hat{\sigma}_{(x_1/n_1)-(x_2/n_2)} = \sqrt{\left(\frac{x_1 + x_2}{n_1 + n_2}\right)\left(1 - \frac{x_1 + x_2}{n_1 + n_2}\right)\left(\frac{1}{n_1} + \frac{1}{n_2}\right)} \tag{11.16}$$

Letting $\hat{p} = (x_1 + x_2)/(n_1 + n_2)$, the statistic

$$\frac{(x_1/n_1) - (x_2/n_2)}{\sqrt{\hat{p}(1 - \hat{p})(1/n_1 + 1/n_2)}}$$

will have a distribution that is approximately $N(0, 1)$ if H_0 is true. Therefore, the basic decision rule becomes:

Reject H_0 if

$$\text{pr}\left\{Z > \left|\frac{(x_1/n_1 - x_2/n_2) - 0}{\sqrt{\hat{p}(1 - \hat{p})(1/n_1 + 1/n_2)}}\right|\right\} < \frac{\alpha}{2}$$

or, equivalently, if

$$\left|\frac{(x_1/n_1 - x_2/n_2) - 0}{\sqrt{\hat{p}(1 - \hat{p})(1/n_1 + 1/n_2)}}\right| > z_{\alpha/2}$$

For one-sided alternatives the usual adjustments are made.

The hypothesis sets and standardized decision rules are summarized in Table 11.5.

If you wish to test for a difference other than 0, that is, if H_0 is H_0: $p_1 = p_2 + c$ where c is a nonzero constant, then the zero in the decision rules of Table 11.5 is replaced with c. In addition, since the populations are no longer identical under H_0, the standard deviation is

Table 11.5 Hypothesis Sets and Decision Rules for Testing Two Binomial Distributions Using the Normal Approximation

Hypothesis Set	Decision Rule for a Stated α
	Reject H_0 at α if:
1. H_0: $p_1 \leqslant p_2$ H_a: $p_1 > p_2$	$\frac{(x_1/n_1 - x_2/n_2) - 0}{\sqrt{\hat{p}(1 - \hat{p})(1/n_1 + 1/n_2)}} > z_\alpha$
2. H_0: $p_1 \geqslant p_2$ H_a: $p_1 < p_2$	$\frac{(x_1/n_1 - x_2/n_2) - 0}{\sqrt{\hat{p}(1 - \hat{p})(1/n_1 + 1/n_2)}} < -z_\alpha$
3. H_0: $p_1 = p_2$ H_a: $p_1 \neq p_2$	$\left|\frac{(x_1/n_1 - x_2/n_2) - 0}{\sqrt{\hat{p}(1 - \hat{p})(1/n_1 + 1/n_2)}}\right| > z_{\alpha/2}$

altered by replacing p_1 with x_1/n_1 and p_2 with x_2/n_2 in the formula, to obtain

$$\hat{\sigma}_{(x_1/n_1)-(x_2/n_2)} = \sqrt{\frac{1}{n_1}\left(\frac{x_1}{n_1}\right)\left(1 - \frac{x_1}{n_1}\right) + \frac{1}{n_2}\left(\frac{x_2}{n_2}\right)\left(1 - \frac{x_2}{n_2}\right)}$$

EXAMPLE 11.5 Senator Jack S. Phogbound would like to know whether his popularity among his urban and rural constituents differs. He is trying to determine whether he should spend more time campaigning in one place than in another. As his campaign manager it is your responsibility to tell him where to go. To this end you have selected random samples of voters from the largest city in his state (Dogpatch) and from the rural area with the heaviest concentration of voters (Beaver Hollow). The voters in each sample were asked whether they preferred Jack S. or his opponent, Dick N. Tricky. Of the 200 Dogpatch voters polled, 112 preferred Jack S. and 88 preferred Dick N. In Beaver Hollow the 120 voters split 52 for Jack S. and 68 for Dick N. Using $\alpha = 0.10$, does it matter where Jack S. campaigns?

SOLUTION The hypothesis set of interest is the two-sided one in Table 11.5. That is, to detect a difference, the hypotheses are

$$H_0\colon\ p_1 = p_2 \qquad H_a\colon\ p_1 \neq p_2$$

Letting p_1 represent the proportion preferring the senator in Dogpatch and p_2 the corresponding proportion in Beaver Hollow, our decision is to:

Reject H_0 if

$$\left|\frac{(x_1/n_1 - x_2/n_2) - 0}{\sqrt{\hat{p}(1-\hat{p})(1/n_1 + 1/n_2)}}\right| > z_{\alpha/2}$$

From the sample data we have

$$\frac{x_1}{n_1} = \frac{112}{200} = 0.56 \qquad \text{and} \qquad \frac{x_2}{n_2} = \frac{52}{120} = 0.4333$$

The value for $\hat{p}$ is

$$\frac{x_1 + x_2}{n_1 + n_2} = \frac{112 + 52}{200 + 120} = \frac{164}{320} = 0.5125$$

Thus, the standard error (deviation) of the difference is

$$\sqrt{\hat{p}(1-\hat{p})\left(\frac{1}{n_1}+\frac{1}{n_2}\right)} = \sqrt{0.5125(0.4875)\left(\frac{1}{200}+\frac{1}{120}\right)} = 0.0576$$

and our statistic equals

$$\frac{(0.56 - 0.433) - 0}{0.0576} = 2.199$$

From the standard normal table, $z_{\alpha/2} = z_{0.05} = 1.645$. Hence, we reject H_0 and conclude that the senator's popularity among his urban and his rural constituents differs. You should tell him to do the majority of his campaigning in Beaver Hollow.

For one-sided alternatives and interval estimation the approach of this section is necessary. But if the alternative hypothesis is two-sided, we can use an equivalent procedure that is computationally easier. The procedure is based on the chi-square statistic and is presented in the next subsection.

11.3.2 The Chi-Square Test

The approach used in constructing the test for differences between two binomials in this section is similar to that taken in the goodness-of-fit test in Section 10.6.1. It also is a special case of the test of several binomials that will be presented in Chapter 13. The hypotheses to be tested are

$$H_0\colon\ p_1 = p_2 \qquad H_a\colon\ p_1 \neq p_2$$

If H_0 is true, our two random samples are actually the same as one combined sample from a single population. If this is the case, the observed numbers of successes are expected to differ from each other as a result of unequal sample sizes [since $E(X) = np$ for a binomial random variable] and of chance or sampling variation. To facilitate the analysis, the data are usually tabulated in a two-way table as shown in Table 11.6.

For H_0 true, our best estimate of the probability of a success p_s is given by

$$\hat{p}_s = \frac{x_1 + x_2}{n_1 + n_2} = \frac{n_s}{n}$$

Table 11.6 Organization of Sample Data for Testing H_0: $p_1 = p_2$ Using the χ^2 Test

	Population		
	I	II	Total
Number of successes	x_1	x_2	n_s
Number of failures	$n_1 - x_1$	$n_2 - x_2$	n_F
Sample size	n_1	n_2	$n_1 + n_2 = n$

Correspondingly, the estimate of the probability of a failure is

$$\hat{p}_F = \frac{n_1 + n_2 - (x_1 + x_2)}{n_1 + n_2} = \frac{n_F}{n}$$

Thus, the expected numbers of successes and failures from each sample are given by

$$E[S|\text{I}] = n_1\left(\frac{n_s}{n}\right) \qquad E[S|\text{II}] = n_2\left(\frac{n_s}{n}\right)$$

and

$$E[F|\text{I}] = n_1\left(\frac{n_F}{n}\right) \qquad E[F|\text{II}] = n_2\left(\frac{n_F}{n}\right)$$

Note in Table 11.6 that if one interior value is known, the remaining values are automatically determined by the marginal totals. That is, we have one degree of freedom.

The test statistic is identical to the one we used in Section 10.6.1 if we let $O_{s\text{I}} = x_1$, $O_{s\text{II}} = x_2$, $O_{F\text{I}} = n_1 - x_1$, and $O_{F\text{II}} = n_2 - x_2$. We merely compare what was observed with what is expected under H_0 for each cell in the table and calculate

$$\Sigma\Sigma \frac{(O - E)^2}{E}$$

By the same rationale used in Section 10.6.1 we find that the statistic is approximately distributed as a χ^2 distribution. As indicated, the distribution has one degree of freedom.

As before, rejection of H_0 is indicated only by large values of our statistic. Thus, the test is to:

Reject H_0 at α if

$$\Sigma\Sigma \frac{(O - E)^2}{E} > \chi^2_{1,\alpha} \tag{11.17}$$

EXAMPLE 11.6 Refer to Example 11.5 and use the χ^2 procedure to test the hypotheses of that problem at $\alpha = 0.10$.

SOLUTION The hypotheses are

$$H_0\colon\ p_1 = p_2 \qquad H_a\colon\ p_1 \neq p_2$$

and the value for $\chi^2_{1,\,0.10}$ from Table A.7 in the Appendix is 2.710. Note that, except for rounding, this value is identical to $(1.645)^2$, which is the square of $z_{\alpha/2}$ in Example 11.5. Using the layout of Table 11.6, we have

	Area		
Number For	Dogpatch	Beaver Hollow	Totals
Phogbound	112	52	164
Tricky	88	68	156
Totals	200	120	320

Our decision rule is to reject H_0 if the calculated value exceeds 2.710. We use the tabular format of Section 10.6.1 for our computations.

Cell	O	E	$O - E$	$\frac{(O-E)^2}{E}$	Partial Sum
PD	112	102.5	9.5	0.880	0.880
PB	52	61.5	-9.5	1.467	2.347
TD	88	97.5	-9.5	0.926	3.273
TB	68	58.5	9.5	1.543	4.816
	320	320.0	0.0		

Since 4.816 is greater than 2.710, we reject H_0 and conclude that Phogbound's popularity appears to differ by type of area.

To verify that the two procedures used in Examples 11.5 and 11.6 yielded identical results, we note that the table values for a χ^2 with one degree of freedom and a probability of α equal the squares of z values for a probability of $\alpha/2$. That is, $\chi^2_{1,\alpha} = z^2_{\alpha/2}$. The calculated value for z in Example 11.5 is 2.199, whereas the calculated χ^2 in Example 11.6 is 4.816. After rounding, $(2.199)^2 \doteq 4.816$.

EXERCISES

11.3.1 The Small Business Administration's (SBA) local office has received an application for a loan to open an exclusive clothing store in Rubbertown. A similar application by another individual was refused two years ago after a random sample of 300 residents living within a 5-mile radius of the proposed location yielded only 60 who said they would patronize such a store. Before ruling on this application, the SBA decided to select another random sample of residents to see whether the proportion has increased in the last two years. In the current sample of 400, 100 residents indicated that they would patronize the store. Can the SBA conclude, at $\alpha = 0.05$, that there has been an increase?

11.3.2 Refer to Exercise 11.3.1 and construct a 90% confidence interval for the true proportion of current residents who might patronize the store. Do the results here agree with your conclusion in Exercise 11.3.1?

11.3.3 In trying to evaluate which of two auditing methods detects a higher proportion of accounting errors, a CPA firm randomly selected two sets of ten accounts each. Each set was known to have 200 accounting errors in it. The two methods then were randomly assigned to the sets. The first method detected 150 of the 200 errors in its set, whereas the second method detected 162 errors. Is this evidence that the proportion of errors detected differs with the method used at $\alpha = 0.05$? Use the χ^2 procedure.

11.3.4 Refer to Exercise 11.3.3 and use the normal approximation to test the hypotheses at $\alpha = 0.05$. Do the results agree?

11.4 LESS RESTRICTIVE PROCEDURES

As in Chapter 10, we present alternatives to the procedures that assume normality. The distribution-free tests in the two-sample case are extensions of those presented in Chapter 10 for single samples. For situations involving paired data, we will extend the sign and signed-rank tests, whereas for detecting location differences with two independent samples we will develop a rank procedure known as the Mann-Whitney test. To test variances the Mann-Whitney test will be altered slightly.

11.4.1 Sign Test with Paired Data

The sign test of Section 10.5 can also be used with paired samples as a test for identical location parameters. As such, it provides a distribution-free alternative to the paired t test of this chapter. For this application the basic assumptions of the sign test are altered as follows:

1. The underlying populations are continuous.
2. The observations constitute a random sample of matched pairs.
3. The pairs are independent of each other.
4. At least nominal data are available; that is, within each pair it is possible to classify each member of the pair into one of two mutually exclusive categories.

Under these assumptions we are able to test for identical population medians (or means if we can assume symmetry). The procedure consists of subtracting the second observation in each pair from the first and recording the sign of the difference. The tests that follow will correspond to the differences being taken in this direction. If the direction is reversed, the decision rules corresponding to the one-sided alternatives must be altered to compensate for the reversal. The logical derivation then follows the line of reasoning presented in Section 10.5.

For hypotheses of the form (M denotes the population median)

$$H_0\colon\ M_1 = M_2 \qquad H_a\colon\ M_1 \neq M_2$$

we reject H_0 at the specified α if (n_+, n_- are the numbers of positive and negative differences, respectively)

$$\text{pr}\{X \leqslant \min(n_+, n_-) | n = n_+ + n_-, p = 0.5\} < \frac{\alpha}{2}$$

If the hypotheses are

$$H_0\colon M_1 \leqslant M_2 \qquad H_a\colon M_1 > M_2$$

we reject H_0 if

$$\text{pr}\{X \leqslant n_- | n = n_+ + n_-, p = 0.5\} < \alpha$$

The other possible hypotheses are

$$H_0\colon M_1 \geqslant M_2 \qquad H_a\colon M_1 < M_2$$

and H_0 is rejected if

$$\text{pr}\{X \leqslant n_+ | n = n_+ + n_-, p = 0.5\} < \alpha$$

The possible hypothesis sets and decision rules are summarized in Table 11.7, with the probabilities being calculated under a binomial distribution.

The changes required for the two-sample sign test are illustrated in the next example.

EXAMPLE 11.7 In an attempt to determine whether the color of the package design has an effect on sales, the advertising department of a particular company placed the product, packaged in two different colors, in side-by-side displays in various stores on 16 randomly selected days. Except for the color of the package the displays were identical. Each day the number of units with each package color sold was recorded, with the following results:

Table 11.7 Summary of Hypothesis Sets and Corresponding Decision Rules for the Two-Sample Sign Test

Hypothesis Set		Decision Rule for Stated α
		Reject H_0 if:
1. H_0: $M_1 \leqslant M_2$	H_a: $M_1 > M_2$	$\text{pr}\{X \leqslant n_- \| n, 0.5\} < \alpha$
2. H_0: $M_1 \geqslant M_2$	H_a: $M_1 < M_2$	$\text{pr}\{X \leqslant n_+ \| n, 0.5\} < \alpha$
3. H_0: $M_1 = M_2$	H_a: $M_1 \neq M_2$	$\text{pr}\{X \leqslant \min(n_+, n_-) \| n, 0.5\} < \frac{\alpha}{2}$

Day:	1	2	3	4	5	6	7	8	9	10	11	12	13	14	15	16
Color 1:	5	8	3	7	4	5	12	4	3	8	10	9	7	3	4	8
Color 2:	2	6	2	3	9	0	6	4	5	7	4	2	4	1	0	3

Can they conclude at $\alpha = 0.05$ that sales differ with the color of the package?

SOLUTION The hypotheses to be tested are

$$H_0\colon M_1 = M_2 \qquad H_a\colon M_1 \neq M_2$$

and since there is no indication of normal or even symmetric distributions, a sign test will be used. Calculating the differences corresponding to the first color minus the second, we observe 13 +'s, 2 −'s, and 1 zero. Discarding the zero and observing that $n_- < n_+$, our decision rule is to reject H_0 if

$$\text{pr}\{X \leqslant n_- | n = n_+ + n_-, p = 0.5\} < \frac{\alpha}{2}$$

Using Table A.2 we find

$$\text{pr}\{X \leqslant 2 | n = 15, p = 0.5\} = 0.0037$$

which is less than $\alpha/2$ of 0.025. Hence H_0 is rejected, and we conclude that sales appear to differ with package color.

Situations in which this procedure is applicable are identical to those for the paired t. The ideal experiment would be of the before/after type. Another way of satisfying the paired assumption is to match the experimental subjects on pertinent variables before treatment. The discussion and words of caution in Section 10.5 pertaining to the sign test are also applicable here.

11.4.2 The Wilcoxon Signed-Rank Test with Paired Data

If it is logical to add the assumptions of ordinal data and symmetric populations to those of the previous section, we can use the signed-rank test of Section 10.5 to test the hypothesis of identical locations (no treatment effect). It is assumed that the experimental design randomly assigns treatments within a pair. The ARE of the test relative to the paired-t test under the assumption of normality is 0.955. As the sample size decreases, the corresponding efficiency increases toward 1.00.

The procedure for the test is to calculate $X_{i1} - X_{i2}$ for each pair, where X_{i1} is the first observation in the ith pair and X_{i2} is the second, then to apply the steps indicated in Section 10.5. Once this is done, the following decision rules are applied to the appropriate hypotheses. For the two-sided alternative,

$$H_0\colon\ M_1 = M_2 \qquad H_a\colon\ M_1 \neq M_2$$

we reject H_0 at α if

$$T = \min(T_+, T_-) \leqslant T_{n,\alpha/2}$$

where n = number of nonzero differences. For hypotheses of the form

$$H_0\colon\ M_1 \leqslant M_2 \qquad H_a\colon\ M_1 > M_2$$

H_0 is rejected at α if

$$T_- \leqslant T_{n,\alpha}$$

The remaining possibility is

$$H_0\colon\ M_1 \geqslant M_2 \qquad H_a\colon\ M_1 < M_2$$

and we reject H_0 at α if

$$T_+ \leqslant T_{n,\alpha}$$

As with the one-sample signed-rank test, zero differences are discarded and tied differences are handled by the average rank procedure.

EXAMPLE 11.8 Let us reconsider the package design problem of Example 11.7. The data that were available were obviously of an ordinal scale, and it might not be unreasonable to assume that sales are symmetrically distributed. With these considerations we conduct the experiment on 13 more randomly selected days with the following results.

Day:	1	2	3	4	5	6	7	8	9	10	11	12	13
Color 1 Sales:	7	8	8	3	9	4	8	6	10	5	12	9	1
Color 2 Sales:	2	6	5	7	2	4	2	7	2	4	3	8	4

Can it be concluded that sales differ according to color using $\alpha = 0.10$?

SOLUTION The hypotheses of interest are

$$H_0\colon\ M_1 = M_2 \qquad H_a\colon\ M_1 \neq M_2$$

and α has been specified to be 0.10. Scanning the data, we see that there are 12 nonzero differences. Thus $n = 12$, and H_0 will be rejected if $T = \min(T_+, T_-) \leqslant T_{12,\,0.05} = 17$. Calculations for the differences and corresponding ranks are as follows:

Day:	1	2	3	4	5	6	7	8	9	10	11	12	13
d_i = Difference:	5	2	3	-4	7	0	6	-1	8	1	9	1	-3
$R_{\lvert d_i \rvert}$ = Rank:	8	4	5.5	7	10	-	9	2	11	2	12	2	5.5
Negative Ranks:				7				2					5.5

We find $T_- = 14.5$ and $T_+ = n(n + 1)/2 - T_- = 78 - 14.5 = 63.5$. Therefore $T = \min(T_+, T_-) = 14.5 < 17$, and H_0 is rejected. We conclude, at $\alpha = 0.10$, that sales appear to differ by color of package.

Note that in this example the zero difference was discarded. Also, the three 1s each received a rank of 2, the average of the first three ranks, and the two 3s received the rank of 5.5, the average of the fifth and sixth ranks.

Comments relating to the power and applicability of the procedure are identical to those in Section 10.5. The large-sample approximation of Section 10.5 can also be applied in this situation.

11.4.3 Mann-Whitney *U* Test for Identical Distributions

Sections 11.4.1 and 11.4.2 have provided us with procedures for handling paired data, where the subjects were matched on certain attributes before experimentation. In practice, matching is often impossible to accomplish, and we must resort to selecting two independent random samples. When this is true, the Mann-Whitney *U* test provides an alternative to the *t* test for two independent samples given in Section 11.2. The *U* test is a variation of a test known as the Wilcoxon rank-sum test, and the computational form of its statistic will use the rank sum. In reality, the procedure tests for any differences in the underlying distributions, but it is particularly sensitive to differences in location (in medians or means).

As with all statistical procedures, there are certain assumptions inherent in the U test. These are listed here.

1. Both samples have been randomly selected.
2. The populations are infinite or else sampling is with replacement. Thus, the samples are independent.
3. The populations are identical except for possible differences in location.
4. The data are of at least an ordinal scale.

In the second assumption, the condition that the population distributions be continuous is often imposed. The third assumption may be relaxed, but it is possible to reject H_0 with near certainty, even though it is true, for extreme violations of this condition when small samples are used. Most examples illustrating this last point are contrived and would not be expected to occur in practical situations.

Letting x_{i1}, $i = 1, 2, \ldots, n_1$, and x_{i2}, $i = 1, 2, \ldots, n_2$, denote the two sets of sample observations where $n_1 \leqslant n_2$, the test statistic U_1 is defined to be

$$U_1 = \sum_{i=1}^{n_2} u_{i1} \tag{11.18}$$

where u_{i1} equals the number of x_1's that are smaller than x_{i2}. Alternatively, we could use

$$U_2 = \sum_{i=1}^{n_1} u_{i2}$$

where u_{i2} is the number of x_2's that are smaller than x_{i1}. To facilitate the determination of the u_{i1}'s (or u_{i2}'s), we first rank the combined samples from low to high and identify the origin of each element in the array. Then, if H_0 is true and the assumptions are satisfied, all possible configurations of n_1 x_1's and n_2 x_2's are equally likely, with each having probability $1/C_{n_1}^{n_1+n_2}$. Under H_0 the X_1's and X_2's would be expected to be interspersed, and U_1 should be approximately equal to U_2, differing only because of unequal sample sizes and random variation. See Figure 11.3*a*. If the median of the X_1 distribution is less than that of the X_2 distribution, the X_1 values would be expected to be grouped at the low end of the array, resulting in a disproportionately large value for U_1. See Figure 11.3*b*. Conversely, if the X_2's are smaller than the

Figure 11.3 Possible Configurations of X_1's and X_2's Where Numbers Identify the Sample of Origin

(*a*) $M_1 = M_2$ and $U_1 = 23,\ U_2 = 26$	2 12 1 1 2 2 1 1 2 2 1 2 1 $M_1 = M_2$
(*b*) $M_1 < M_2$ and $U_1 = 55,\ U_2 = 1$	1 1 11 1 1 2 1 2 2 2 2 2 2 2 M_1 M_2
(*c*) $M_1 > M_2$ and $U_1 = 2,\ U_2 = 54$	2 2 22 2 2 1 1 2 1 1 1 1 1 1 M_2 M_1

X_1's, we expect U_1 to be disproportionately small. See Figure 11.3*c*. Note that $U_1 + U_2 = n_1 n_2$.

An alternative method for calculating U_1 and U_2 is to replace the observations with their ranks in the overall array and to calculate T_1 and T_2, where

$$T_1 = \text{sum of ranks associated with } X_1 \text{ values}$$

and

$$T_2 = \text{sum of ranks associated with } X_2 \text{ values}$$

Then

$$U_1 = n_1 n_2 + \frac{n_1(n_1 + 1)}{2} - T_1 \tag{11.19}$$

and

$$U_2 = n_1 n_2 + \frac{n_2(n_2 + 1)}{2} - T_2 \tag{11.20}$$

Summarizing, the overall procedure is as follows:

1. State the hypotheses to be tested and select an α.
2. Select random samples from each of the populations being compared.
3. Rank all the values as if they constituted one large sample of size $n_1 + n_2$.
4. Calculate U_1 and U_2.
5. (*a*) If the hypotheses in (1) are

$$H_0\colon M_1 = M_2 \qquad H_a\colon M_1 \neq M_2$$

we reject H_0 at the nominal α if $U = \min(U_1, U_2) \leqslant U_{n_1, n_2, \alpha/2}$.

(*b*) If the hypotheses are

$$H_0\colon\ M_1 \geqslant M_2 \qquad H_a\colon\ M_1 < M_2$$

we reject H_0 if $U_2 \leqslant U_{n_1, n_2, \alpha}$.

(*c*) If the hypotheses are

$$H_0\colon\ M_1 \leqslant M_2 \qquad H_a\colon\ M_1 > M_2$$

H_0 is rejected if $U_1 \leqslant U_{n_1, n_2, \alpha}$.

These hypothesis sets and their decision rules are summarized in Table 11.8.

Critical values for U are given in Table A.11 of the Appendix. The table is indexed by n_1, the smaller sample size; n_2, the larger sample size; and α, the nominal probability of being less than or equal to U. The exact probability will be less than or equal to that stated at the top of the column.

EXAMPLE 11.9 In order to determine whether a particular situation comedy show has an appeal factor that differs between men and women, seven men and eight women were randomly selected and asked to view the show in 15 separate rooms. At the conclusion each person was asked to rate the show on a scale of 0 to 100, with high scores indicating high appeal. The results of the ratings are as follows.

Women's Ratings:	24	48	15	26	37	82	33	41
Men's Ratings:	57	54	69	77	95	83	88	

Table 11.8 Summary of Hypotheses and Corresponding Tests for the Mann-Whitney U Test

Hypothesis Set		Decision Rule for Stated α
		Reject H_0 if:
1. $H_0\colon M_1 \leqslant M_2$	$H_a\colon M_1 > M_2$	$U_1 \leqslant U_{n_1, n_2, \alpha}$
2. $H_0\colon M_1 \geqslant M_2$	$H_a\colon M_1 < M_2$	$U_2 \leqslant U_{n_1, n_2, \alpha}$
3. $H_0\colon M_1 = M_2$	$H_a\colon M_1 \neq M_2$	$\text{Min}(U_1, U_2) \leqslant U_{n_1, n_2, \alpha/2}$

Can it be concluded that the show appeals differently to men and women at $\alpha = 0.10$?

SOLUTION The hypotheses indicated in the question are

$$H_0\colon M_M = M_W \qquad H_a\colon M_M \neq M_W$$

where M_M and M_W denote medians for men and women, respectively, and α is specified to be 0.10. The corresponding decision rule implies rejection of H_0 if

$$U = \min(U_1, U_2) \leqslant U_{7,8,0.05} = 13$$

Placing the data in an array gives the following.

Value:	15	24	26	33	37	41	48	54	57	69	77	82	83	88	95
Rank:	1	2	3	4	5	6	7	8	9	10	11	12	13	14	15
Sample:	W	W	W	W	W	W	W	M	M	M	M	W	M	M	M

Adding, we obtain

$$T_1 = \Sigma R_M = 80 \qquad \text{and} \qquad T_2 = \Sigma R_W = 40$$

Thus

$$U_1 = 7(8) + \frac{7(8)}{2} - 80 = 84 - 80 = 4$$

$$U_2 = 7(8) + \frac{8(9)}{2} - 40 = 92 - 40 = 52 = n_1 n_2 - U_1 = 56 - 4 = 52$$

and

$$U = \min(4, 52) = 4$$

Since $4 < 13 = U_{7,8,0.05}$, we reject H_0 and conclude that the show's appeal appears to differ according to the sex of the viewer. Note that it is assumed that interpersonal comparisons can be made in this instance.

If tied values occur, use the average rank procedure indicated previously. For sample sizes outside the range of Table A.11, use the Z statistic, given by $Z = (U - \mu_U)/\sigma_U$, with rejection of H_0 indicated for values of Z less than $-z_{\alpha/2}$ in the two-sided case and for values of z less than $-z_\alpha$ in both one-sided cases. The mean and the variance of U are given by

$$\mu_U = \frac{n_1 n_2}{2}$$

$$\sigma_U^2 = \frac{n_1 n_2 (n_1 + n_2 + 1)}{12}$$

The approximation will be good for both n_1 and n_2 greater than or equal to 25 and standard values of α. If n_1 is considerably less than n_2 and α is quite small, the approximation is a poor one.

The ARE of the U test relative to the comparable t test is 0.955 under normal conditions and has been shown to be never less than 0.864. For rectangular distributions the U test is just as powerful as the t test, giving an ARE of 1.00, and in many situations it has an ARE greater than 1.00, possibly approaching infinity.

Note: As mentioned at the beginning of this section, the U test is really a test of identical distributions. Thus, for it to be purely a test of location differences, the assumptions of equal variances and common shapes must be included with those stated earlier. If this is not the case (for example, if the two populations are skewed in opposite directions over the same range of possible values), the test may become less sensitive to differences in location.

11.4.4 The Siegel-Tukey Test

For the Mann-Whitney U test we assigned ranks to the data in the pooled sample corresponding to each observation's position in the array. Using a different ranking procedure, so long as it remains independent of whether the observation is an X_1 or an X_2, will not affect the distribution of the rank sums T_1 and T_2. Thus, the distribution of U_1 and U_2 is also independent of the method used to assign ranks. The Siegel-Tukey test uses this fact to assign the ranks in such a way that the U statistic becomes sensitive to differences in scale parameters, that is, in variances. As such, it provides an alternative to the F test for equal variances as presented in Section 11.2.3.

The assumptions of the test are similar to those of the Mann-Whitney U test. The only change in the assumptions is that the populations have distributions that may differ in variance but are otherwise identical. This assumption implies that the populations have the same median and mean. If this assumption is violated, the results of the test may be questionable. Thus in using this test rather than an F test, we substitute the assumption of identical locations for that of normal populations and interval data.

The hypotheses that we will be testing are the equality of variances and the one-sided alternatives. Except for the method of assigning ranks to the data, the procedure will correspond to that of Section 11.4.3. We will first place the data of the combined samples into an array. Next we rank the data from both ends toward the middle as follows:

1. Assign a rank of 1 to the smallest value and ranks of 2 and 3 to the largest and next largest values, respectively.
2. Assign ranks of 4 and 5 to the second and third smallest values, respectively, and ranks of 6 and 7 to the third and fourth largest values, respectively. Continue this pattern until all $n_1 + n_2$ ranks have been assigned.

After the ranks have been assigned, calculate T_1, T_2, U_1, and U_2 as before.

If the null hypothesis of equal variances is true, we would expect the two sets of observations to be interspersed throughout the array, yielding values for T_1 and T_2, and hence U_1 and U_2, of approximately the same magnitude. If $\sigma_1^2 < \sigma_2^2$, the X_1 values would be expected to be grouped in the middle of the array and to receive the high ranks. Thus, T_1 would be disproportionately large and U_1 correspondingly small. Conversely, for $\sigma_1^2 > \sigma_2^2$, the X_2 values are clustered in the middle, with T_2 becoming large and U_2 becoming small. These situations are illustrated in Figure 11.4.

As an example of the ranking procedure, we will assign ranks to the configurations in Figure 11.4. Ranking from both ends toward the middle, we obtain

$$1 \quad 4 \quad 5 \quad 8 \quad 9 \quad 12 \quad 13 \quad 14 \quad 11 \quad 10 \quad 7 \quad 6 \quad 3 \quad 2$$

for all three configurations. The corresponding values for T_1 and T_2 are: (*a*) $T_1 = 50$, $T_2 = 55$; (*b*) $T_1 = 63$, $T_2 = 42$; (*c*) $T_1 = 34$, $T_2 = 71$.

From the given considerations we construct the tests that are appropriate to the hypotheses of interest. If we are interested only in detecting a difference, our hypotheses become

$$H_0\colon\ \sigma_1^2 = \sigma_2^2 \qquad H_a\colon\ \sigma_1^2 \neq \sigma_2^2$$

and we reject H_0 at the stated α if

Figure 11.4 Possible Configurations Resulting When $n_1 = n_2 = 7$ and
(*a*) $\sigma_1^2 = \sigma_2^2$ (*b*) $\sigma_1^2 < \sigma_2^2$ (*c*) $\sigma_1^2 > \sigma_2^2$

1 2 2 1 1 2 1 2 2 1 2 1 1 2

(*a*)

2 2 1 2 1 1 1 2 1 2 1 1 2 2

(*b*)

1 1 1 2 2 1 2 2 2 2 1 2 1 1

(*c*)

$$U = \min(U_1, U_2) \leqslant U_{n_1, n_2, \alpha/2}$$

If the indicated hypotheses are

$$H_0\colon \sigma_1^2 \geqslant \sigma_2^2 \qquad H_a\colon \sigma_1^2 < \sigma_2^2$$

we reject H_0 if

$$U_1 \leqslant U_{n_1, n_2, \alpha}$$

For the remaining one-sided case

$$H_0\colon \sigma_1^2 \leqslant \sigma_2^2 \qquad H_a\colon \sigma_1^2 > \sigma_2^2$$

H_0 is rejected if

$$U_2 \leqslant U_{n_1, n_2, \alpha}$$

As usual, we summarize the hypotheses and decision rules in Table 11.9.

The following example illustrates the total procedure used in testing one of the hypotheses.

EXAMPLE 11.10 Your broker has advised you to consider buying stock in two unrelated companies. His report indicates that over the

Table 11.9 Summary of Hypotheses and Corresponding Decision Rules for the Siegel-Tukey Test

Hypothesis Set	Decision Rule for Stated α
	Reject H_0 if:
1. H_0: $\sigma_1^2 \leqslant \sigma_2^2$ H_a: $\sigma_1^2 > \sigma_2^2$	$U_2 \leqslant U_{n_1, n_2, \alpha}$
2. H_0: $\sigma_1^2 \geqslant \sigma_2^2$ H_a: $\sigma_1^2 < \sigma_2^2$	$U_1 \leqslant U_{n_1, n_2, \alpha}$
3. H_0: $\sigma_1^2 = \sigma_2^2$ H_a: $\sigma_1^2 \neq \sigma_2^2$	$\text{Min}(U_1, U_2) \leqslant U_{n_1, n_2, \alpha/2}$

past 20 years the average price per share has been the same for both companies. Stock A returns slightly higher dividends, and since you can only afford to invest in one, you would prefer stock A unless it is less stable than stock B. To facilitate making your decision, you randomly select nine weeks for A and eight weeks for B from the previous 20 years. The closing prices for A's nine weeks and B's eight weeks are given here.

A:	8	$11\frac{1}{2}$	$15\frac{1}{4}$	$10\frac{3}{8}$	$6\frac{3}{4}$	$9\frac{7}{8}$	10	$4\frac{1}{2}$	14
B:	12	$10\frac{1}{2}$	$8\frac{1}{4}$	$9\frac{1}{4}$	11	$8\frac{7}{8}$	$11\frac{1}{4}$	$10\frac{1}{4}$	

Using $\alpha = 0.05$, which stock should you purchase?

SOLUTION You are interested in the stability of the stocks, which can be measured by their respective variances; a small variance indicates a stable stock and a high variance indicates an unstable one. Since you prefer A unless it is more unstable than B, the hypotheses are

$$H_0: \sigma_A^2 \leqslant \sigma_B^2 \qquad H_a: \sigma_A^2 > \sigma_B^2$$

With $\alpha = 0.05$, $n_1 = 8$, and $n_2 = 9$, the decision rule implies rejection of H_0 if

$$U_B \leqslant U_{8, 9, 0.05}$$

In Table A.11 we find $U_{8, 9, 0.05} = 18$.

The computations required for calculating U_B are given in the following table:

Array:	$4\frac{1}{2}$	$6\frac{3}{4}$	8	$8\frac{1}{4}$	$8\frac{7}{8}$	$9\frac{1}{4}$	$9\frac{7}{8}$	10	$10\frac{1}{4}$
Rank:	1	4	5	8	9	12	13	16	17
R_b:				8	9	12			17

Array:	$10\frac{3}{8}$	$10\frac{1}{2}$	11	$11\frac{1}{4}$	$11\frac{1}{2}$	12	14	$15\frac{1}{4}$
Rank:	15	14	11	10	7	6	3	2
R_b:		14	11	10		6		

Thus $T_B = 87$ and

$$U_B = n_1 n_2 + \frac{n_1(n_1+1)}{2} - T_B = 8(9) + \frac{8(9)}{2} - 87 = 108 - 87 = 21$$

Since $21 > 18$, we can't reject H_0, and we conclude that on the basis of these samples it cannot be stated that $\sigma_A^2 > \sigma_B^2$ or that A is less stable. Thus, buy stock A.

For situations involving tied values we will use the average rank procedure. For sample sizes outside the range of Table A.11, use the normal approximation with $Z = (U - \mu_U)/\sigma_U$, where

$$\mu_U = \frac{n_1 n_2}{2} \quad \text{and} \quad \sigma_U^2 = \frac{n_1 n_2 (n_1 + n_2 + 1)}{12}$$

The decision rule corresponding to the two-sided alternative will be to reject H_0 if $Z < -z_{\alpha/2}$, and that corresponding to the one-sided alternatives will be to reject H_0 in both cases if $Z < -z_\alpha$.

EXERCISES

11.4.1 In a margarine-tasting experiment, 16 randomly selected persons were asked to taste each of two brands of margarine, Farkay and Crowned, and to state which was more nearly like butter. The brands were placed in unmarked dishes and were tasted by each individual in a random order. The 16 responses were as follows: F F C F F F C C Same F C F F C F F. Using $\alpha = 0.05$, does there appear to be any difference between the two brands in their similarity to butter? What type of experiment was this?

11.4.2 The marketing department of Consumer Products, Inc., provides the production department with quarterly sales forecasts in units at the beginning of each quarter. Production then uses these forecasts to set up the production schedule for the coming quarter. The marketing forecast and actual sales (in millions of units) for the last 12 quarters are given.

Quarter:	1	2	3	4	5	6	7	8	9	10	11	12
Forecast:	1.3	2.6	2.8	1.9	1.8	3.0	2.9	2.0	2.0	3.1	3.2	1.8
Actual:	1.1	2.0	2.5	2.0	1.3	2.1	2.5	2.3	1.3	2.1	2.3	2.6

Using $\alpha = 0.05$, could the production department accuse the marketing department of being overly optimistic? Do the assumptions of the signed-rank test seem reasonable in this case?

11.4.3 We wish to test for differences in the frequency of repair for two brands of copiers. Six copiers of brand *A* and six of brand *B* were selected at random, and the number of calls for repair service for each machine was observed. The data are as follows:

Brand *A*:	7	11	9	8	12	6
Brand *B*:	10	5	4	3	7	2

Test for a difference at $\alpha = 0.05$.

11.4.4 A quality control inspector has claimed that the night shift at Precision Die Products does not maintain as tight a control on its machine settings as the day shift. To test this claim, random samples of nine each were selected from the day shift and from the night shift, and the diameters of the pieces were measured. The results (in inches) are as follows:

Day:	4.990	5.020	4.995	4.998	5.001	5.015	4.996	4.991	5.003
Night:	4.891	4.898	5.021	5.018	5.019	4.992	5.024	4.895	5.021

Using $\alpha = 0.05$, does the inspector's claim appear to be supported?

11.4.5 Refer to Exercise 11.4.3 and test the assumption of equal variances at $\alpha = 0.05$.

11.4.6 Refer to Exercise 11.4.4 and test the assumption of identical location parameters.

11.4.7 Evaluate which of the procedures in Section 11.2 might be applicable for Exercises 11.4.1 through 11.4.6.

11.5 SUPPLEMENTARY EXAMPLES

EXAMPLE 11.11 In a survey of basic food prices in Cleveland, Ohio, and Portland, Oregon, a shopping list with fixed quantities of each of 15 common items, such as sugar, milk, and eggs, was made out. Workers were given copies of the list and sent to ten randomly selected stores in each city to purchase the items on the list. The average cost of the 15 items in the Cleveland stores was \$27.86, and the standard deviation was \$1.48. In Portland stores the cost of the items averaged \$26.42 with a standard deviation of \$1.16. Is it safe to conclude that food costs in Portland, Oregon, are less than those in Cleveland, Ohio, if $\alpha = 0.10$?

SOLUTION Since it is indicated that we would like to conclude that food costs in Portland are less than those in Cleveland, H_a is $\mu_p < \mu_c$. That is, our hypothesis set is

$$H_0\colon \mu_p \geqslant \mu_c \qquad H_a\colon \mu_p < \mu_c$$

The data obviously represent two independent random samples and are of an interval level of measurement. Thus, we have two possible procedures to consider for our test, the Mann-Whitney U test and the independent samples t test. Both assume equal variances, and so the only remaining criterion is the assumption about the underlying distributions. It is reasonable to think that the distributions of the costs do not deviate from normality by very much. Even though the sample sizes are on the small side, the central limit theorem should compensate for any lack of normality in the populations. With the added fact that the t test is robust, we opt for it.

Thus, our test is to reject H_0 if

$$\frac{(\bar{X}_p - \bar{X}_c)}{s_p\sqrt{1/n_p + 1/n_c}} < -t_{n_p+n_c-2,\,\alpha}$$

Checking in Table A.8 we find $-t_{18,\,0.10} = -1.33$. Doing the necessary arithmetic gives

$$s_p^2 = \frac{9(1.48)^2 + 9(1.16)^2}{18} = 1.768$$

and $s_p = 1.33$. The test statistic is equal to

$$\frac{26.42 - 27.86}{1.33\sqrt{1/10 + 1/10}} = -2.42$$

Since $-2.42 < -1.33$, we reject H_0 and conclude that food costs in Portland appear to be lower than those in Cleveland. The test is illustrated graphically in Figure 11.5.

Figure 11.5 Graphical Representation of the Test and Results for Example 11.11

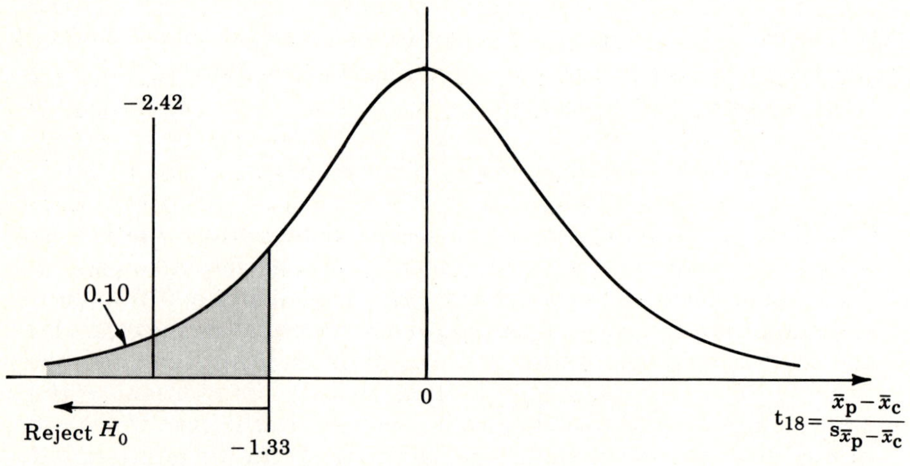

Note that if it had been deemed necessary to drop back to the Mann-Whitney procedure in Example 11.11, we would have had to retrieve the original observations. If they are unavailable, we have no choice but to use the t test.

In situations in which the choice of procedures has been narrowed to the sign test and the Wilcoxon signed-rank test, we must consider two things in our final decision. First and easiest to determine is the level of measurement that is represented by the numerical values associated with the observations. If the level is only nominal, the sign test is an automatic choice. If the level is at least ordinal, symmetry is the deciding factor. If you are uncertain about this property, you can get a "quick and dirty" check by determining the mean and median for each sample. If the mean and median are close in value for a given sample, it is probably not unreasonable to assume symmetry for that population. Recall that symmetry implies equality of the population mean and median.

EXAMPLE 11.12 The number of defective gidgets coming off each of two production lines was recorded by Pogo's quality control department on each of ten randomly selected days. These data are given in the following table. Assume that the daily output was the same for both machines, and test the contention that line II produces more defectives. Use $\alpha = 0.05$.

Day:	1	2	3	4	5	6	7	8	9	10
Line I:	172	165	191	184	174	142	180	169	161	200
Line II:	191	179	159	192	177	165	182	179	164	210

SOLUTION Since we want to compare the two populations with regard to location using paired samples, we have three candidates for the test procedure: the paired t, the signed-rank test, and the sign test. Since the sample size is small, our first consideration is normality. In this case the populations are obviously nonnormal, since the random variables are discrete, and we eliminate the t test.

Another obvious fact is the ordinality of the data as they are presented. Our last check is the symmetry assumption. Using the "quick and dirty" technique previously indicated, we have

$$\bar{x}_{\text{I}} = 173.8 \qquad \text{and} \qquad m_{\text{I}} = 173$$

while

$$\bar{x}_{\text{II}} = 179.8 \qquad \text{and} \qquad m_{\text{II}} = 179$$

Since the means and medians are quite close for both samples, we select the signed-rank test. Thus, our hypotheses are

$$H_0\colon\ \mu_{\text{I}} \geqslant \mu_{\text{II}} \qquad H_a\colon\ \mu_{\text{I}} < \mu_{\text{II}}$$

and we will reject H_0 if

$$T_+ \leqslant T_{n,\,\alpha} = T_{10,\,0.05} = 10$$

Calculating the differences and ranking their absolute values gives

d_i:	-19	-14	+32	-8	-3	-23	-2	-10	-3	-10
$R_{\lvert d_i \rvert}$:	8	7	10	4	2.5	9	1	5.5	2.5	5.5

Since $T_+ = 10 \leqslant 10$, we reject H_0 and conclude that on the basis of these data it appears that line II does produce more defectives than I.

In Example 11.12 the data could be considered to be interval, and even though the variable is obviously discrete and n is only 10, the central limit theorem should give a good approximation if symmetry is true. Thus, the t test should not be ruled out completely.

EXAMPLE 11.13 The variability in the ages of persons attending college during the 1950s and 1960s was relatively small. A study at that time indicated that the standard deviation was three years, based on a sample of 100 students. Currently, students are spreading their education over a longer period, and some educators have claimed that the variance of ages has more than doubled. To check this claim with respect to the students at a particular state university, 51 students were randomly selected. The standard deviation of their ages was 5.1 years. Does this sample support the claim at $\alpha = 0.05$?

SOLUTION The indicated hypotheses are

$$H_0\colon\ \sigma_1^2 \leqslant 2\sigma_2^2 \qquad H_a\colon\ \sigma_1^2 > 2\sigma_2^2$$

where σ_1^2 is the current variance. The only procedure that we have that appears to apply is the F test of two variances, and it must be altered for these hypotheses. To construct our test, we note the following:

$$\text{pr}\left\{\frac{s_1^2/\sigma_1^2}{s_2^2/\sigma_2^2} > f_{n_1-1,\,n_2-1,\,\alpha}\right\} = \alpha$$

But if H_0 is true, this is equivalent to

$$\text{pr}\left\{\frac{s_1^2/2\sigma_2^2}{s_2^2/\sigma_2^2} > f_{n_1-1,\,n_2-1,\,\alpha}\right\} = \alpha$$

or

$$\text{pr}\left\{\frac{s_1^2}{2s_2^2} > f_{n_1-1,\,n_2-1,\,\alpha}\right\} = \text{pr}\left\{\frac{s_1^2}{s_2^2} > 2f_{n_1-1,\,n_2-1,\,\alpha}\right\} = \alpha$$

Therefore, our decision rule is to:

Reject H_0 if

$$\text{pr}\left\{F_{n_1-1,\,n_2-1} > \text{observed}\left(\frac{s_1^2}{2s_2^2}\right)\right\} < \alpha$$

But this is equivalent to rejecting H_0 whenever

$$\frac{s_1^2}{s_2^2} > 2f_{n_1-1,\,n_2-1,\,\alpha}$$

For our samples $s_1^2 = (5.1)^2 = 26.01$, $s_2^2 = 9$, $n_1 = 51$, and $n_2 = 100$. The value from the F table is $f_{50,\,99,\,0.05} = 1.48$. Thus, we will reject H_0 if

$$\frac{s_1^2}{s_2^2} > 2(1.48) = 2.96$$

Our value is

$$\frac{26.01}{9} = 2.89 < 2.96$$

and H_0 cannot be rejected at $\alpha = 0.05$. Hence, we cannot conclude that the current variance of student ages is more than twice that in the 1950s. We might consider taking another sample, since the claim is supported at $\alpha = 0.10$.

The assumptions required in the preceding example if our procedure is to be valid are:

1. Interval data (this should pose no problems)
2. Random samples
3. Underlying normal distributions

The assumption of normality is probably not valid, since experience indicates that the distribution of ages of college students tends to be highly skewed to the right. That is, the majority of students are in the 18 to 25 age range, but there is a large number of students scattered through the upper ranges. The distribution of the ages is likely to resemble the graph in Figure 11.6.

11.6 SUMMARY

In this chapter we have extended the one-sample tests of Chapter 10 to situations in which we want to compare two distributions with respect to their location and scale parameters. For normal populations we have available two tests for comparing locations. The difference between the two relates to the sampling design. The paired t test requires that experimental subjects be matched on relevant attributes before treatments are applied; that is, a before/after type of experiment. The independent-samples t test requires that independent random samples be taken from each population. Since the variances are usually unknown, the assumption of equal variances is also required.

Figure 11.6 Possible Distribution of Ages of College Students

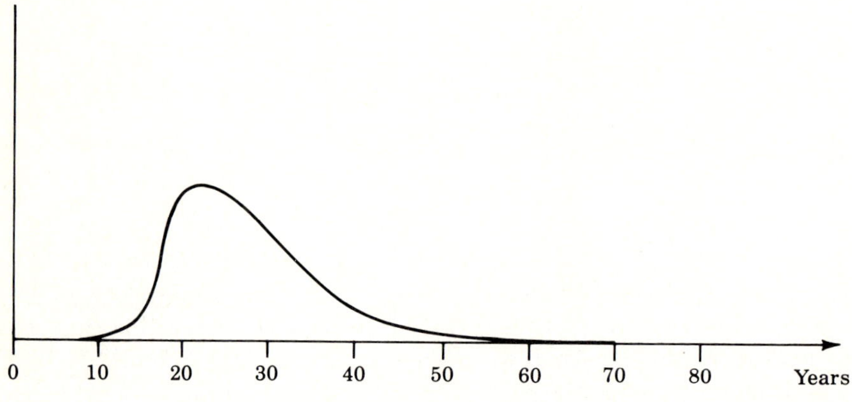

To compare binomial distributions when large samples are available, we extended the normal approximation of Chapters 5 and 10. For the two-sided alternative an equivalent and easier procedure is based on the χ^2 goodness-of-fit approach in Chapter 10. If a confidence interval for the difference in proportions is desired or the alternative is one-sided, the normal approximation should be used.

In Section 11.4 we presented the distribution-free alternatives to the normal-based tests for location and scale differences. For paired data the sign test and the signed-rank test are both alternatives to the paired *t* that assume neither interval data nor normality. The difference between the two is that the signed-rank test requires ordinal data and symmetric populations, whereas the sign test needs only nominal data and doesn't require symmetry. The Mann-Whitney *U* test is an alternative to the independent-samples *t* test. The difference in assumptions for the two is only the requirement of ordinal data and the dropping of the normality assumption for the *U* test. In comparing scale parameters (variances), we can choose between the *F* test and the Siegel-Tukey modification of the *U* test. The *F* test requires normal distributions and interval measurement, whereas the Siegel-Tukey uses ordinal data and requires identical locations.

If we believe that the assumptions for the normal-based tests are violated only slightly, they may still be used, since they are robust. For gross violations of assumptions, use of these tests would make our probabilities only approximations, and probably poor ones. In general, when in doubt, use a procedure with less restrictive assumptions. It is normally considered better to use a less powerful procedure that actually applies than a more powerful one that may not apply. The procedures of this chapter are summarized in Table 11.10 which appears on pages 408–409.

REVIEW EXERCISES

Level One

11.1 Explain the differences between the sign, signed-rank, and *t* tests for paired data.

11.2 Ten randomly chosen people were given a mathematics quiz to do in a specified amount of time. Some time later the same ten people were given a similar test but were allowed to take as much time as they wanted. Their scores are given here. Is this evidence that a time limit is a significant factor in a person's score on an exam? Use $\alpha = 0.05$.

Table 11.10 Summary of Tests and Associated Assumptions for Two-Sample Situations

Two-Sample Tests for Differences in	Test	Underlying Assumptions
Proportions with Independent Samples	Chi-square	Alternative is H_a: $p_1 \neq p_2$ Random sampling Underlying binomial distributions At least nominal data
	Normal approximation (Z)	Random sampling Underlying binomial distribution for the random variable of interest At least nominal data
Scale (Variances) with Independent Samples	Siegel-Tukey	Random samples Underlying continuous or at least infinite populations Common location parameters At least ordinal data
	F test	Random samples Underlying normal distributions At least interval data

Location Independent Samples	Mann-Whitney	Random samples Continuous (or infinite) distributions Common variances ($\sigma_1^2 = \sigma_2^2$) At least ordinal data
	t test	Random samples Underlying normal distributions Common variances ($\sigma_1^2 = \sigma_2^2$) At least interval data
Location Paired Data	Sign	Random sample of matched pairs Underlying continuous (or infinite) distributions At least nominal data
	Signed rank	Random sample of matched pairs Underlying continuous distributions Symmetric distributions At least ordinal data
	t test	Random sample of matched pairs Underlying normal distributions At least interval data

Person:	1	2	3	4	5	6	7	8	9	10
Timed Quiz:	50	83	75	65	90	98	73	80	88	69
Untimed Quiz:	65	80	85	60	95	95	80	95	97	69

11.3 Refer to Exercise 11.2 and identify the assumptions required for your test. Do you feel they are valid? What other tests might you consider?

11.4 In a consumer protection survey to determine whether two food chains in a certain area sold hamburger of the same quality, 1 pound was bought at each chain on each of ten randomly selected days. Each pound was then fried, and the amount of fat was poured off and weighed. The results, in ounces of fat, are as follows:

Store 1:	5.6	3.2	4.7	6.1	4.3	5.2	4.9	5.8	4.1	3.9
Store 2:	2.8	4.1	3.0	3.6	3.3	3.8	4.5	3.4	3.7	2.5

Can we conclude at $\alpha = 0.10$ that the fat content of the hamburger differs by store?

11.5 Refer to Exercise 11.4. The distribution of fat contents is likely to be skewed left because upper limits are imposed by government regulations. What effect would this have on your results? The test you used in Exercise 11.4 assumed that the variances were equal. Test this assumption at $\alpha = 0.10$, keeping in mind the additional information.

11.6 A tire manufacturer wishes to determine whether a new additive in the curing process will provide longer tire life (measured in terms of mileage). Ten types of tires are being considered. Two tires of each type were made, one with the additive and one without it. The tires were then tested under standard conditions, with the following results (rounded to the nearest 5000 miles):

Tire Type:	1	2	3	4	5	6	7	8	9	10
No Additive:	2.5	2.5	3.5	3.5	3.0	2.0	3.0	3.0	3.0	4.5
Additive:	3.5	2.0	4.0	3.5	4.5	2.5	3.5	2.5	3.5	4.0

(*a*) What test will you use to see whether there is an increase in tire life? Justify your choice.
(*b*) Run the test you selected at $\alpha = 0.05$.
(*c*) Run all other tests that might apply in this situation and explain any differences in your results.

11.7 In an attempt to determine whether there are differences between the grade point averages of males and females entering the College of Business, seven entering freshman males were selected and matched on age, socioeconomic background, high school background, and college entrance exam scores with seven entering freshman females. All the students were then advised to take the same courses through their first two years. Their GPAs upon entering the College of Business are given here. Use them to test for a difference in GPAs between males and females at $\alpha = 0.01$.

Pair:	1	2	3	4	5	6	7
Females' GPA:	2.75	3.17	2.36	3.53	3.14	2.27	2.69
Males' GPA:	2.25	2.87	2.56	3.23	3.04	2.07	2.44

11.8 Redo Exercise 11.7 using another test procedure. Which of the two tests you have used do you think is to be preferred? Why?

11.9 In testing for differences in grades between day and evening classes, 16 evening students and 21 day students were randomly selected and given identical exams. The evening group had an average of 63.4 and a variance of 256.8, whereas the day students had an average of 67.8 and a variance of 103.4. The t test for equality of means was run and indicated no difference. Using $\alpha = 0.10$, is there any reason to doubt the results of the t test?

11.10 Refer to Exercise 11.9 and construct a 90% confidence interval for the ratio of the evening class's variance to the variance of the day class.

11.11 A store manager wanted to evaluate whether or not a customer's tendency to read the unit pricing information available in the grocery store differs according to the customer's economic level. To do so the manager randomly selected four days over a one-month period and asked each customer on those days to fill out a questionnaire. The questionnaire asked whether the person reads the unit pricing information and whether family income was above or below \$16,000 annually. Names were not requested, and customers were asked not to fill out the questionnaire if they had done so previously. Of the 920 customers who filled out the questionnaires, 300 indicated incomes over \$16,000. Of the three hundred, 192 indicated that they read the unit pricing. For those with incomes less than \$16,000, the number who read the unit prices was 358. Do the proportions differ by income level at $\alpha = 0.05$?

11.12 Refer to Exercise 11.11. The store manager believed also that the proportions differ by sex of the customer. The third item on the questionnaire asked the sex of the customer. Of the 188 men in the sample, 92 indicated that they read the unit pricing. Construct a 95% confidence interval for the difference between the proportion of women and of men who read the unit pricing (for $p_w - p_m$).

11.13 In an effort to improve safety in its factory, a firm was experimenting with a safety training program for workers. A group of seven workers was selected at random and put through a safety training course. A second group of seven workers was also selected at random to be the control or comparison group and did not attend the training course. Two months after the training program a supervisor who did not know which employees had attended the course was asked to rank all 14 employees in terms of their safety behavior. A rank of 1 indicates the poorest safety behavior, and a rank of 14 indicates the best safety behavior. The resulting ranks for the employees were as follows:

6 2 7* 12* 9 14* 4* 3 1 5 8* 11 13* 10*

where the asterisks indicate those who were trained. At the 0.05 level of significance, can you conclude that the training was beneficial in improving safety?

11.14 The sales of new homes are intrinsically tied to interest rates. High interest rates mean that mortgage funds are in short supply and, when available, are very costly to the borrower. A real estate developer is interested in building a number of speculative homes in one of two communities. Knowing that interest rates may vary from region to region, he randomly selected seven financial institutions from each community and recorded their interest rates for a conventional \$35,000, 30-year first mortgage. These rates are:

Community A:	14.6	14.9	14.65	14.7	14.5	14.75	14.95
Community B:	14.8	15.0	14.6	14.9	15.1	15.2	15.0

The builder would prefer to build in Community B unless it would appear to be more difficult to sell there. What should she decide using an α of 0.05?

11.15 Justify your selection of a test procedure in Exercise 11.14.

11.16 A researcher found that people in low-income neighborhoods have less confidence in consumer protection legislation than people in high-income neighborhoods. The results (after correcting for location differences) in terms of responses to a particular question on a 10-point continuous scale are:

Low-Income:	5.0	6.0	5.5	4.0	4.5	5.0	
High-Income:	4.0	3.5	2.0	6.5	5.5	5.0	7.0

The researcher had randomly selected six people from the low-income area and seven from the high-income area and asked each to rate his or her feelings about the question. He then ran a Mann-Whitney U test and found that the medians appeared to differ. Test the equality of variances assumption of the U test at $\alpha = 0.10$, and indicate whether or not you agree with what the researcher did.

11.17 In an attempt to determine whether the proportions of people with allergies differ between Arizona and Ohio, 100 people were selected at random in each state. The people then were asked whether they had an allergy of any kind. In the samples 32 from Arizona and 18 from Ohio indicated allergies. Do the proportions appear to differ at $\alpha = 0.05$?

11.18 If you rejected H_0 in Exercise 11.17, could you conclude that the Arizona climate causes more allergic reactions to develop than the Ohio climate? Why or why not?

11.19 A company wishes to evaluate whether or not there is a significant difference in the average number of minutes of operating time on one gallon of fuel for two different types of engines. A random sample of engines is selected from each type and tests are performed, with the results shown. Is there a significant difference at the 0.05 level?

$$\text{Type } A\text{:}\quad \overline{X}_A = 22,\ s_A = 4,\ n_A = 3$$

$$\text{Type } B\text{:}\quad \overline{X}_B = 26,\ s_B = 2,\ n_B = 6$$

11.20 In a life-testing experiment on two brands of light bulbs, random samples of six bulbs each were selected for each brand and the times to failure were recorded. Test for a difference in median life between the two brands. (It is believed that time to failure is not normally distributed, but that the distributions should be otherwise similar.) Let $\alpha = 0.05$.

	Lives in Hours					
Brand A:	982	1024	956	1102	993	1000
Brand B:	951	894	1005	982	922	949

11.21 Two judges were asked to rate the full-bodied properties of 11 wines on a 1 to 10 scale. Use the results given here to test to see whether the judges differ in their tastes at $\alpha = 0.05$.

Wine:	1	2	3	4	5	6	7	8	9	10	11
Judge 1:	6	5	9	7	4	6	8	2	3	7	1
Judge 2:	4	3	6	8	2	4	9	1	1	5	1

11.22 In an attempt to determine whether there is a difference in the weathering properties of oil-base and latex exterior paints, the French Girl Paint Co. randomly selected 16 houses in various sections of the United States and painted one-half of one side of each house with exterior latex paint and the other half with exterior oil-base paint. After two years the company's chief chemist inspected each house and rated each half on a 1 to 10 scale (with a higher score indicating better weathering properties). The chemist did not know in advance which paint was used on which half. The ratings are as follows:

House:	1	2	3	4	5	6	7	8	9	10	11	12	13	14	15	16
Oil-Base:	8	5	7	4	3	5	8	9	7	5	6	6	8	4	6	3
Latex:	6	4	5	4	5	2	7	6	4	8	7	5	5	7	4	5

What conclusion should the French Girl Co. reach on the basis of these data if they use an α of 0.05 for their test?

11.23 Two different lathes are used to produce 2-inch cylinders for use in air compressors. It is suspected that the variance in machine A is greater than that in B. Random samples of five cylinders are taken from the output of each machine. Is the suspicion true? ($\alpha = 0.05$)

Machine A:	1.97	2.05	2.08	2.04	2.04
Machine B:	2.02	1.99	2.02	1.99	2.03

11.24 In evaluating the coloring ability of two types of red pigment used in plastic, ten different batches of plastic were split, and half of each batch was milled with each of the two pigments. The millman was then asked to rate each piece of plastic on a 0 to 100 scale as to color intensity. These ratings are as follows:

Batch:	1	2	3	4	5	6	7	8	9	10
Pigment 1:	86	92	54	63	39	77	43	94	88	69
Pigment 2:	54	76	47	75	22	69	51	72	81	54

Does there appear to be any difference in the intensity due to the pigments ($\alpha = 0.10$)?

11.25 For each of Exercises 11.19 through 11.24, indicate why you chose the procedure you did.

Level Two

11.26 Determine the decision rules for the alternatives H_a: $\mu_1 < \mu_2$ and H_a: $\mu_1 \neq \mu_2$ when the distributions are normal with known variances.

11.27 Prove that s_p^2 is an unbiased estimate of σ^2 if the population variances are equal.

11.28 A company has decided to purchase a new machine for their packaging department and is considering two different makes, A and B. The accounting department has made a cost study of the two machines and has determined that since A will be somewhat more expensive to operate, it must be able to produce a proportion of defectives more than 0.03 less than B's proportion to be feasible. If the company were able to obtain random samples of 400 pieces from the output of each machine, determine an expression for the difference in sample proportions that will make the probability that the company will purchase A by mistake no higher than 0.05.

11.29 The company in Exercise 11.28 was able to obtain their samples. Machine A's output had 15 defectives and B's had 32 defectives. Which machine should be purchased?

11.30 To estimate the standard deviation of the lengths of the items made by a particular machine, the quality control department selected five random samples at different times. The results of those samples are as follows:

Sample:	1	2	3	4	5
Sample mean:	7.02	7.03	6.99	7.04	6.95
s_i:	0.012	0.011	0.013	0.009	0.016
n_i:	11	16	21	9	6

Determine the best estimate of the standard deviation for the machine.

11.31 Use the data in Exercise 11.30 to estimate the average length of the items produced by this machine.

11.32 Try to develop a procedure that will compare two population medians by using the χ^2 test for two binomial distributions.

11.33 What will need to be assumed to use the test you developed in Exercise 11.32?

11.34 Determine a 99% confidence interval for the problem in Exercise 11.12.

11.35 Approximate the power of your test, based on the paired t test, against the specific alternative that a female's average GPA is 0.3 unit higher than a male's based on the data of Exercise 11.7.

Level Three

11.36 Construct a procedure for testing for identical populations using a runs test with two independent random samples such that the procedure will be sensitive to differences in location.

11.37 What assumptions will be necessary to use your procedure in Exercise 11.36?

11.38 Use the procedure you constructed in Exercise 11.36 to test the data of Exercise 11.4 at $\alpha = 0.10$.

A confidence interval can be constructed for the difference between two population medians based on the U test. The procedure consists of computing all the $n_1 n_2$ differences, $x_{i1} - x_{k2}$, $i = 1, 2, \ldots, n_1$, $k = 1, 2, \ldots, n_2$, and ranking them. An interval estimate with a confidence level of at least $1 - \alpha$ is then given by the $(U_{n_1, n_2, \alpha/2} + 1)$st values from either end of the array of differences.

11.39 Refer to Exercise 11.20 and determine an interval estimate for the difference in the median lives between light bulbs from the two brands. Take the differences as Brand 1 − Brand 2 and use a confidence level $\geqslant 0.90$.

11.40 Develop a procedure for finding an interval estimate of the difference in population medians based on the Wilcoxon signed-rank test.

References

1. Bradley, James V. *Distribution-Free Statistical Tests.* Englewood Cliffs, New Jersey: Prentice-Hall, Inc., 1968.
2. Gibbons, Jean Dickinson. *Nonparametric Statistical Inference.* New York: McGraw-Hill Co., 1971.
3. Lehmann, E. L. *Testing Statistical Hypotheses.* New York: John Wiley & Sons, Inc., 1959.
4. Lindgren, B. W. *Statistical Theory*. New York: The Macmillan Co., 1962.
5. Noether, Gottfried E. *Elements of Nonparametric Statistics.* New York: John Wiley & Sons, Inc., 1967.
6. Siegel, Sidney. *Nonparametric Statistics for the Behavioral Sciences.* New York: McGraw-Hill Co., 1956.

Chapter 12
c-Sample Inferences in the Normal Family (ANOVA)

KEY WORDS AND TERMS

Analysis of Variance (ANOVA)
Linear Model
Fixed Effects
Random Effects
Factor Levels
Crossed Design
Nested Design
Cochran's Theorem
Sum of Squares (SS)
Mean Square (MS)
Expected Mean Square
Replication
Randomized Block

12.1 INTRODUCTION

In Chapter 11, we considered methods that you could use to help you make a decision between two suppliers of a particular part. But what approach do you take if there are more than two possible suppliers, say five? Suppose you are in charge of advertising for a new product line and are trying to evaluate various packaging designs. The designs involve four different color schemes and five different package styles. How would you design a market survey to help you make the decision, and how would you analyze the data afterward? This chapter and Chapter 13 address questions of this type.

In the previous two chapters we considered situations in which we were interested in making a decision about either a single population or the equivalence of two populations. We now generalize those procedures so that we can handle an arbitrary but finite number of populations. The analyses for multiple samples have been separated into two chapters because of the amount of material involved. In this chapter we consider only populations whose distributions are members of the normal family. Chapter 13 develops alternatives to these procedures when the assumptions of normality and interval data are questionable.

In both this chapter and Chapter 13, the emphasis is on detecting whether there are differences, primarily with respect to location, among a set of members of the same family of distributions. For example, a typical decision-making situation might be one in which a company is considering the purchase of a large piece of equipment. It has received bids from a number of suppliers and through cost considerations has narrowed the choice to four products whose machines appear to be comparable. At this point the company's primary criterion is the average production rate. They prefer to buy the fastest machine if, indeed, one is faster than the others. If the production rates appear to be comparable, the company will use secondary criteria in making its decision. To use statistical analysis as an aid in making the decision would require sample data from each machine. These data could be obtained by using each machine on a trial basis, if the suppliers are agreeable, or from other companies that are using the machines currently.

Once the data are available, they could be analyzed using the procedures of Chapter 11 by running a series of tests comparing two machines at a time. This is known as *pairwise comparisons*. The drawbacks of making pairwise comparisons are: (1) a large number of tests may be required—in general, $\binom{n}{2}$ of them, or in the company's case six of them; (2) the more tests one runs, the higher the probability of rejecting at least once as a result of chance alone. The second disadvantage is more important, for if a stated α is used for all comparisons, the overall significance level, say α', is

$$\alpha' = \text{pr}\{\text{rejecting at least once} \mid H_0 \text{ true}\} \doteq \text{pr}\left\{X \geqslant 1 \,\middle|\, \binom{n}{2}, \alpha\right\}$$

That is, α' is approximately a binomial probability. It is approximate, since some tests may not be independent of all others, but it would be exact if all were independent. Thus, the overall α for the battery of tests would be approximately

$$\alpha' \doteq 1 - (1 - \alpha)^{\binom{n}{2}} \tag{12.1}$$

Note that this probability (and hence the chance of making a type I error) increases dramatically with n. For example, in the company's case, if an α of 0.05 is used for each comparison, the chance that the company will make a type I error overall is approximately

$$1 - (1 - 0.05)^{\binom{4}{2}} = 1 - (0.95)^6 = 1 - 0.7351 = 0.2649$$

or about 26%.

To avoid increasing the overall significance level, we prefer to test all populations simultaneously, making specific comparisons only after we have determined that some differences appear to exist. That is, initially we wish to test the hypothesis set

$$H_0\colon \mu_1 = \mu_2 = \cdots = \mu_c \qquad H_a\colon \text{some } \mu_j \text{ differ}$$

We will consider specific comparisons if H_0 is rejected. Tests that maintain the overall α at no more than the stated level are grouped under the general term of *analysis-of-variance* (ANOVA) procedures. From the name it would appear that we are trying to compare the variances of the populations. Actually, the ANOVA procedures test for equality of means, but they do so by calculating different estimates of the error variance and then comparing them.

12.2 GENERAL CONCEPTS

Before continuing, we need to explain some of the basic concepts and terminology that are germane to the analysis of variance. Among these are the idea of a linear model, the type of effects being considered, and the concept of an experimental design.

12.2.1 The Basic Model

The term *model* in ANOVA refers to how an observation arises— that is, what effects contribute to the particular response and how those effects go together. It was unnecessary to state the model explicitly when only one or two populations were of interest, but doing so simplifies the statement of the hypotheses and makes the overall approach easier to understand in the multisample case. In all analyses involving ANOVA we assume a linear model. That is, we assume that the response is a linear combination of the effects that contribute to it. The purpose of testing will be to determine which effects actually appear to contribute to the response.

For a one-factor model, such as the purchase of the equipment in our earlier example, the general representation is

$$X_{ij} = \mu_j + \epsilon_{ij}$$

where $i = 1, 2, \ldots, n_j$ and $j = 1, 2, \ldots, c$, X_{ij} is an observed response from population j, μ_j is the mean of the jth population, and ϵ_{ij} is a random error term. For our example, X_{ij} represents the ith observation from machine j, μ_j is the true average production rate for a machine from company j, and ϵ_{ij} represents the random variation in the actual production. We assumed the same model without stating it formally in the two-sample tests of location for the normal family. In that case c was equal to 2. In the one-sample test we were simply trying to decide whether or not X_i could be represented by the model $\mu_0 + \epsilon_i$.

As we proceed in the chapter and introduce more effects as possible contributors to our response, this model will be more convenient if we restate it in an alternative form. The preferred statement of the model is

$$X_{ij} = \mu + \alpha_j + \epsilon_{ij} \tag{12.2}$$

where $i = 1, 2, \ldots, n_j$ and $j = 1, 2, \ldots, c$. In Equation 12.2, μ represents the average of all the population means, α_j is the difference between μ and the mean of the jth population, and ϵ_{ij} is the random error term. With the model of Equation 12.2, our hypotheses take the equivalent form

$$H_0\colon\ \alpha_1 = \alpha_2 = \cdots = \alpha_c = 0 \qquad H_a\colon\ \text{some } \alpha_j \neq 0 \tag{12.3}$$

If the null hypothesis is true, the model of Equation 12.2 can be reduced to $X_{ij} = \mu + \epsilon_{ij}$. Thus, another way of stating our hypotheses is

$$H_0\colon\ X_{ij} = \mu + \epsilon_{ij} \qquad H_a\colon\ X_{ij} = \mu + \alpha_j + \epsilon_{ij}$$

Throughout this discussion we have been using the term *effect*. By this we mean the general variable of interest. This variable may be either quantitative or qualitative in nature. For our example, the effect of interest is the variable representing suppliers. In a marketing framework the effect of interest might be advertising, which could be quantitative if the variable is advertising expenditure or qualitative if the variable is advertising media. Terms that are used interchangeably with *effect* are *factor* and *treatment*. We explore the various types of effects in the next section.

Note: We assume linear models primarily because we will be assuming normal distributions. A unique property of the normal family is the fact that linear independence implies statistical independence and vice versa. We will require statistical independence in the construction of our test procedures.

12.2.2 Types of Effects

To discuss the two types of effects that are of interest to us, we need to identify the term *levels*. By the levels of an effect or factor we mean all possible values or categories that the effect (variable) can assume. In the evaluation of which machine to purchase, the supplier is the effect or factor and the levels would be the four suppliers in the final set under consideration. Since these four are the only suppliers of interest at this point, the number of levels is fixed at four, and we have no intention of trying to infer something about all possible suppliers. This is what is known as a *fixed effect*.

DEFINITION 12.1 An effect or factor is called a fixed effect if the levels being evaluated are the only ones of interest—if the population about which we wish to make a decision comprises only the levels used in the experiment. If all factors or effects in the experiment are fixed effects, the model is a fixed-effects model.

In many situations there will be far too many levels of interest for a given effect to include them all in the experiment. Thus, because of either economics or sheer physical limitation, we will be forced to use only a sample of the possible levels in the actual experiment. For example, if the marketing department wanted to evaluate the effect of advertising expenditures on an industrial product, there are any number of levels of expenditure that could be used. It would be prohibitive from a cost standpoint to try to include all of them in the actual experiment. Instead, levels might be defined in $10,000 increments and stratified, with one level being randomly chosen from each stratum. In this case advertising expenditure would be considered a *random effect*.

DEFINITION 12.2 A factor is considered to be a random effect if the levels included in the experiment are only a sample (preferably random) of the population of all levels of interest. If all the factors in the experiment are of the random-effects type, the model is a random-effects model.

Not all factors in a given model will necessarily be of the same type. If a model contains factors of both types, it is said to be a *mixed* model.

The test statistics in the chapter will pertain primarily to fixed-effects models. The formulas for calculating sums of squares and mean squares in the numerator are identical for both fixed-effects and random-effects models. But the mean square used in the denominator of the test statistic may change for a random-effects model. The terms *sum of squares* and *mean square* will be defined later. We have introduced both models so that you will be aware that the final step in the analysis (the tests) must be altered for a random-effects model.

12.2.3 Experimental Designs

The term *experimental design* relates to exactly how the data were collected. You first encountered different types of experimental designs in Chapter 11. Specifically, in testing two population means, we considered the situation in which two independent random samples were collected. This is considered a completely randomized experimental design. We also had the option of selecting a random sample of pairs of identical items, then randomly allocating treatments to the items within a pair. The paired design is a special case of a more general design known as a randomized block. Randomized block designs will be considered in more detail in Section 12.6.

In multifactor experiments there are two basic types of designs, the factorial (or crossed) design and the nested design. In a crossed design all levels of a given factor occur in combination with all levels of any other factor. This is illustrated in Figure 12.1 for a two-factor experimental design with three replications.

A *replication* is a repetition of the experiment. In Figure 12.1, the experiment is to be repeated three times. That is, there are three replications or three observations for each treatment or factor combination.

A factor B is said to be nested within another, A, if the levels of B occurring in combination with one level of A differ from the levels of B occurring in combination with any other level of A. Whereas the test statistic may vary depending on the fixed-effects or random-effects model, the calculation of the sums of squares changes if a factor is nested within another. The concept of nesting is illustrated in Figure 12.2.

In Figure 12.2 there appear to be six levels of B, but for analysis purposes there are really only two, since there are only two levels of B in combination with any level of A. Thus, in comparing for a B effect, we may compare only the levels of B within a given level of A. For example, B_1 may be compared only with B_2, B_3 with B_4, and B_5 with B_6. B_3 is not comparable with B_1, since the subjects received different levels of treatment A and differences between B_1 and B_3 cannot be separated from the differences between A_1 and A_2. That is, the B effect is confounded within the A effect. This is not the case in a

Figure 12.1 A Two-Factor Crossed Design With Three Replications

		Factor A		
		Level A_1	Level A_2	Level A_3
Factor B	Level B_1	× × ×	× × ×	× × ×
	Level B_2	× × ×	× × ×	× × ×

crossed design, since if all levels of B are in combination with all levels of A, the A effect averages out for each level of B.

Not all designs are completely crossed or completely nested. In many practical situations a design will have both crossed factors and nested factors at the same time. We will restrict our discussion to crossed designs for now and briefly touch on the computational changes necessary for a nested design in the chapter appendix.

EXERCISES

12.2.1 Indicate whether you would consider the following effects to be fixed or random and explain why:
(*a*) The evaluation of three proposed package designs
(*b*) The effect of various temperature levels on the yield of a chemical experiment
(*c*) The effect of different statistics textbooks on class performance
(*d*) The effect of machine operators on the output for a given company

12.2.2 Identify two possible experiments that you might construct in which the effects would be fixed.

12.2.3 Identify two experiments that you could design in which the effects would be random.

12.2.4 Design an experiment for a situation familiar to you in which all factors are crossed.

12.2.5 Identify an experiment that would be of interest to you in which at least one factor is nested.

Figure 12.2 A Two-Factor Design With B Nested Within A and Four Replications

Level A_1		Level A_2		Level A_3	
Level B_1	Level B_2	Level B_3	Level B_4	Level B_5	Level B_6
x	x	x	x	x	x
x	x	x	x	x	x
x	x	x	x	x	x
x	x	x	x	x	x

12.3 COCHRAN'S THEOREM

In this section we state without proof a theorem that is probably second only to the central limit theorem with respect to its importance in statistical analysis. The theorem itself is the basis and justification for the construction of most of our test procedures, not only in this chapter but also in Chapters 14 and 15. We state the theorem here in a somewhat simplified form.

COCHRAN'S THEOREM Let Q be a sum of squared random variables such that Q has a χ^2 distribution with ν degrees of freedom. Furthermore, assume that Q can be expressed as the sum of a set of Q_i, $i = 1, 2, \ldots, m$, where each Q_i represents a smaller sum of squares. Then the Q_i's will be independent, with each Q_i distributed as a χ^2 distribution with ν_i degrees of freedom, *if and only if* $\sum_{i=1}^{m} \nu_i = \nu$, that is, if and only if the degrees of freedom for the Q_i total the degrees of freedom for Q. The symbol ν is the Greek letter nu.

As stated earlier, one of the main uses of the χ^2 distribution is to determine the distribution of the variance of a random sample taken from a normal population. We will illustrate the use of Cochran's theorem by finding the distribution of s^2. Let X_i, $i = 1, 2, \ldots, n$, be a random sample from a $N(\mu, \sigma^2)$ distribution, and let

$$Q = \sum_{i=1}^{n} \left(\frac{X_i - \mu}{\sigma}\right)^2 = \frac{1}{\sigma^2} \sum_{i=1}^{n} (X_i - \mu)^2$$

Then, $(X_i - \mu)/\sigma$ is a $N(0, 1)$ random variable and, by the definition in Chapter 8, Q has a χ^2 distribution with n degrees of freedom. Also,

$$\begin{aligned}
\frac{1}{\sigma^2} \sum_{i=1}^{n} (X_i - \mu)^2 &= \frac{1}{\sigma^2} \sum_{i=1}^{n} (X_i - \overline{X} + \overline{X} - \mu)^2 \\
&= \frac{1}{\sigma^2} \sum_{i=1}^{n} (X_i - \overline{X})^2 + \frac{1}{\sigma^2} \sum_{i=1}^{n} (\overline{X} - \mu)^2 \\
&= \frac{1}{\sigma^2} \sum_{i=1}^{n} (X_i - \overline{X})^2 + \frac{n(\overline{X} - \mu)^2}{\sigma^2}
\end{aligned}$$

since it is easily shown that $\sum_{i=1}^{n} (X_i - \overline{X})(\overline{X} - \mu) = 0$. The first summation on the right-hand side is simply the numerator for s^2 and has $n - 1$

degrees of freedom. In the second term $\overline{X}$ is the only random variable and has no restrictions on it; hence, it has 1 degree of freedom. [This can also be seen from the fact that $\sqrt{n}\,(\overline{X} - \mu)/\sigma$ is a $N(0, 1)$ variable, and hence $n(\overline{X} - \mu)^2/\sigma^2$ is a χ^2 variable with 1 degree of freedom.] Thus, we have shown that the degrees of freedom on the right side sum to the degrees of freedom for Q,

$$n = (n - 1) + 1$$

By Cochran's theorem it follows that

$$\frac{1}{\sigma^2} \sum_{i=1}^{n} (X_i - \overline{X})^2 \sim \chi^2_{n-1}$$

Dividing both sides of the equation by $n - 1$ gives

$$\frac{1}{\sigma^2}\left(\frac{\sum_{i=1}^{n} (X_i - \overline{X})^2}{n - 1}\right) \sim \frac{\chi^2_{n-1}}{n - 1}$$

The term in parentheses is s^2 by definition, and so we have

$$\frac{s^2}{\sigma^2} \sim \frac{\chi^2_{n-1}}{n - 1}$$

or

$$\frac{(n - 1)s^2}{\sigma^2} \sim \chi^2_{n-1}$$

The separation of $\Sigma(X - \mu)^2$ into the other two sums of squares is called *partitioning.* This is an important aspect of analysis of variance. We will be partitioning an overall or total sum of squares into smaller sums of squares. Each of the smaller sums will reflect a different method of estimating the chance or error variance if our null hypotheses are true.

12.4 ANOVA FOR A ONE-FACTOR DESIGN

In this section we extend the two independent samples t test to situations involving more than two populations. Our model, in the form of Equation 12.2, is

$$X_{ij} = \mu + \alpha_j + \epsilon_{ij} \qquad i = 1, \ldots, n_j,\ j = 1, \ldots, c$$

and we wish to test the hypotheses

$$H_0: \alpha_j = 0 \text{ for all } j \qquad H_a: \text{ some } \alpha_j \neq 0$$

The sample data will be in the form shown in Table 12.1.

In order to construct our test, it is necessary to make the following assumptions:

1. Independent random samples from each population.
2. Each population has a normal distribution.
3. The population variances are identical.
4. Interval level of measurement.

The middle two assumptions actually relate to the random variable ϵ, which is our random-error term. They can be combined into the single assumption

$$\epsilon_{ij} \sim N(0, \sigma^2)$$

Assumption 3 is known generally as *homogeneity of variance.*

In addition to these assumptions, we will be assuming a fixed-effects model. This implies that the $\sum_j \alpha_j = 0$. In a one-factor design the test is identical for both the random- and fixed-effects models. The difference lies in the interpretation of the results and, theoretically, in the statement of the hypotheses.

To illustrate the basic idea underlying ANOVA, let us consider the following situation. Suppose that five samples have been selected and that all have the same sample size n. Assume further that all the information from the samples has been lost except for the five means and n.

Table 12.1 Typical Organization of Data for a One-factor Design

Sample or Treatment (A)				
A_1	A_2	A_3	. . .	A_c
x_{11}	x_{12}	x_{13}		x_{1c}
x_{21}	x_{22}	x_{23}		x_{2c}
⋮	⋮	⋮		⋮
$x_{n_1 1}$	$x_{n_2 2}$	$x_{n_3 3}$		$x_{n_c c}$

If H_0 is true and we want to estimate the population variance, how can we do it?

From Chapter 8, we know that $\sigma^2_{\overline{X}} = \sigma^2/n$. Thus, if we can estimate $\sigma^2_{\overline{X}}$ by itself, all we need to do to estimate σ^2 is to multiply the estimate of $\sigma^2_{\overline{X}}$ by n. Since we have available a random sample of 5 $\overline{X}$'s, we need only calculate their sample variance, $s^2_{\overline{X}}$, directly. That is, we calculate

$$s^2_{\overline{X}} = \frac{\sum_{1}^{5} (\overline{X}_{.j} - \overline{X}_{..})^2}{5 - 1}$$

where $\overline{X}_{..}$ represents the overall or grand mean, that is, the average of the five sample means, and $\overline{X}_{.j}$ is the mean for the jth sample. The dot merely indicates that we have averaged over the missing subscript(s). Therefore, if H_0 is true, an estimate of σ^2 is $ns^2_{\overline{X}}$. If H_0 is not true, $ns^2_{\overline{X}}$ includes an estimate of σ^2 but also contains a term representing the fact that the population means differ.

The numerator, $\Sigma\, (\overline{X}_{.j} - \overline{X}_{..})^2$, is called a *sum of squares* (SS), and the estimate, $s^2_{\overline{X}}$, is called a *mean square* (MS). The expected value of an s^2 is called its *expected mean square* and denoted EMS. The EMS identifies all elements that are being estimated by the particular MS. The EMS includes the items that dictate the test statistic. For our $ns^2_{\overline{X}}$, if we do not assume that H_0 is true, then its EMS is, for a fixed-effects model,

$$E\left[ns^2_{\overline{X}}\right] = \sigma^2 + \frac{n\, \Sigma\, \alpha_j^2}{c - 1}$$

To see this, we need to use only the properties of the expected value operator and simple substitution. We have

$$E\left[ns^2_{\overline{X}}\right] = E\left[\frac{\sum_j \sum_i (\overline{X}_{.j} - \overline{X}_{..})^2}{c - 1}\right] = E\left[\frac{\sum_j \sum_i [(\mu + \alpha_j + \overline{\epsilon}_{.j}) - (\mu + \overline{\epsilon}_{..})]^2}{c - 1}\right]$$

since

$$\overline{X}_{.j} = \frac{1}{n} \sum_i (X_{ij}) = \frac{1}{n} \sum_i (\mu + \alpha_j + \epsilon_{ij}) = \frac{1}{n}(n\mu + n\alpha_j + \sum_i \epsilon_{ij})$$

$$= \mu + \alpha_j + \frac{\sum_i \epsilon_{ij}}{n} = \mu + \alpha_j + \overline{\epsilon}_{.j}$$

The $\bar{\epsilon}_{.j}$ represents the average error in sample j. Similarly,

$$\bar{X}_{..} = \frac{1}{nc} \sum_j \sum_i (X_{ij}) = \frac{1}{nc} \sum_j \sum_i (\mu + \alpha_j + \epsilon_{ij})$$

$$= \frac{1}{nc}(nc\mu + n \sum_j \alpha_j + \sum_j \sum_i \epsilon_{ij}) = \mu + \frac{\sum_j \sum_i \epsilon_{ij}}{nc} = \mu + \bar{\epsilon}_{..}$$

since, for a fixed model, $\sum_j \alpha_j = 0$. Here $\bar{\epsilon}_{..}$ represents the average error in the total sample. Thus,

$$E\left[ns_{\bar{X}}^2\right] = E\left[\frac{\sum_j \sum_i (\alpha_j + \bar{\epsilon}_{.j} - \bar{\epsilon}_{..})^2}{c-1}\right] = E\left[\frac{\sum_j \sum_i \alpha_j^2 + \sum_j \sum_i (\bar{\epsilon}_{.j} - \bar{\epsilon}_{..})^2}{c-1}\right]$$

$$= E\left[\frac{n \sum_j \alpha_j^2 + n \sum_j (\bar{\epsilon}_{.j} - \bar{\epsilon}_{..})^2}{c-1}\right]$$

But the α_j's are constants, and $E\left[ns_{\bar{X}}^2\right]$ reduces to

$$E\left[ns_{\bar{X}}^2\right] = \frac{n \sum_j \alpha_j^2}{c-1} + nE\left[\frac{\sum_j (\bar{\epsilon}_{.j} - \bar{\epsilon}_{..})^2}{c-1}\right]$$

$$= \frac{n \sum_j \alpha_j^2}{c-1} + n\left(\frac{\sigma^2}{n}\right) = \sigma^2 + \frac{n \sum_j \alpha_j^2}{c-1}$$

since $\sum_j (\bar{\epsilon}_{.j} - \bar{\epsilon}_{..})^2/(c-1)$ is simply the sample variance of $\bar{\epsilon}_{.j}$, the average error in sample j, and σ^2 is our common error variance.

The $\sum \alpha_j^2$ will be zero if and only if H_0 is true. Therefore, if H_0 is true, we have that

$$E\left[ns_{\bar{X}}^2 \middle| H_0\right] = \sigma^2$$

In this development we have assumed that $n_j = n$ for all j, that is, we have assumed equal sample sizes. The general form, which allows for unequal sample sizes, is for the fixed-effects model

$$E[\text{MSA}] = \sigma^2 + \frac{\sum_j n_j \alpha_j^2}{c-1} \tag{12.4}$$

In Equation 12.4 the notation MSA indicates that the variance has been calculated by comparing the sample means of factor A with the overall mean.

If we can obtain an estimate of σ^2 that is independent of H_0, then for a test statistic we need to compare the MSA with that estimate. Suppose that our lost data have been recovered. Within a given sample the only reason that X_{ij} differs from $\overline{X}_{.j}$, if our model is correct, is because of the error variance. Therefore, each of the five sample variances provides an independent estimate of σ^2. Using the pooling concept of Section 11.2.2, we can combine the five estimates by taking their weighted average, with the weights being their respective degrees of freedom. Calling the numerator the sum of squares due to error (SSE), we have, allowing the n_j to differ,

$$\text{SSE} = \sum_j \sum_i (X_{ij} - \overline{X}_{.j})^2 \tag{12.5}$$

The degrees of freedom associated with the SSE is simply the sum of the individual degrees of freedom or the total sample size minus 1 for each different $\overline{X}_{.j}$. That is,

$$\text{df}_{\text{error}} = \sum_j n_j - c \tag{12.6}$$

Our estimate of σ^2 is given by MSE, where

$$\text{MSE} = \frac{\text{SSE}}{\sum_j n_j - c} \tag{12.7}$$

with

$$E(\text{MSE}) = \sigma^2 \tag{12.8}$$

What we have just done is partition the total sum of squares into two parts. The total sum of squares (SSTO) is the numerator for the sample variance calculated as if we had one large sample rather than c individual samples and is given by

$$\text{SSTO} = \sum_j \sum_i (X_{ij} - \overline{X}_{..})^2 \tag{12.9}$$

SSTO has $\Sigma\, n_j - 1$ degrees of freedom, since there are $\sum_j n_j$ variables and one restriction is imposed by $\overline{X}_{..}$. In order to determine distributions for our various statistics, we consider the formal partition of SSTO and apply Cochran's theorem. The partition is formed as follows and is illustrated in Figure 12.3.

$$\sum_j \sum_i (X_{ij} - \overline{X}_{..})^2 = \sum_j \sum_i [(X_{ij} - \overline{X}_{.j}) + (\overline{X}_{.j} - \overline{X}_{..})]^2$$

or

$$\begin{matrix} \sum_j \sum_i (X_{ij} - \overline{X}_{..})^2 & = & \sum_j \sum_i (X_{ij} - \overline{X}_{.j})^2 & + & \sum_j \sum_i (\overline{X}_{.j} - \overline{X}_{..})^2 \\ \text{SSTO} & = & \text{SSE} & + & \text{SSA} \end{matrix}$$

Figure 12.3 Graphical Representation of the Partitioning of SSTO

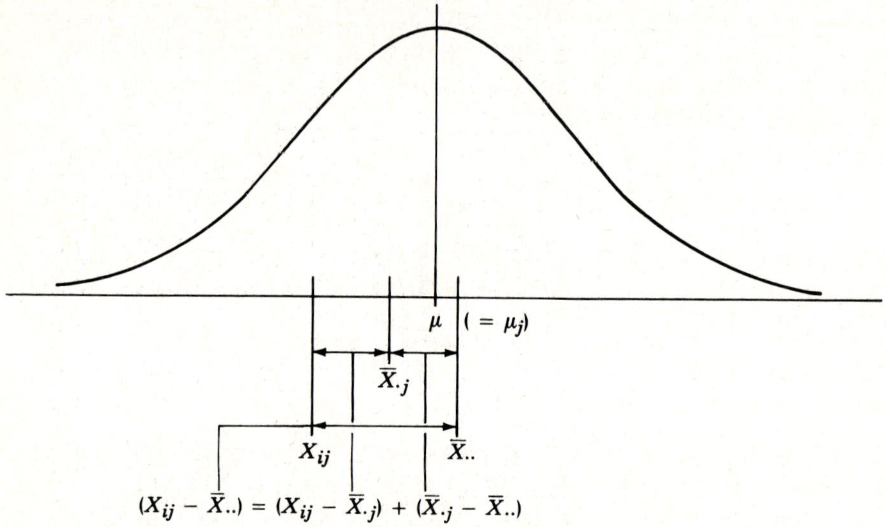

(*a*) Illustration of Partitioning of SSTO when H_0 is True

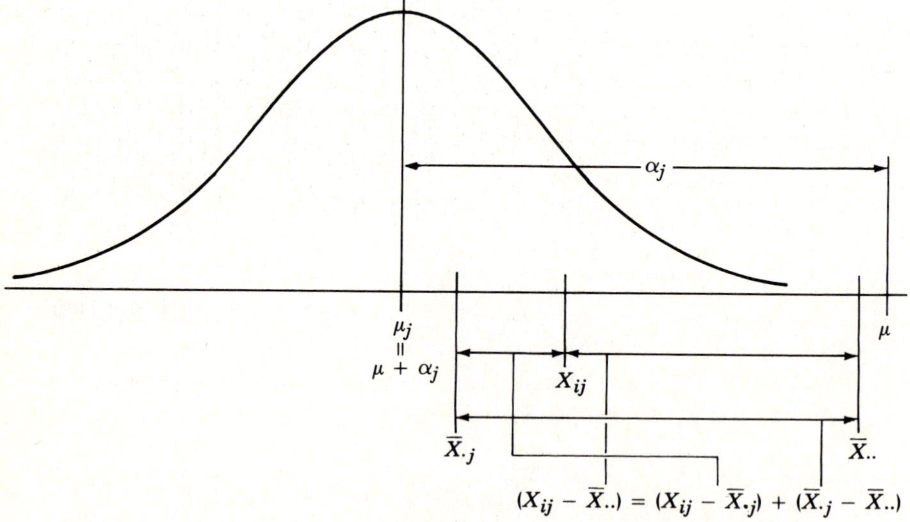

(*b*) Illustration of Partitioning of SSTO when H_0 is not True

If we can show that SSTO has a $\sigma^2\chi^2$ distribution and that the degrees of freedom on the right total the degrees of freedom on the left, then, by Cochran's theorem, we know that SSE and SSA will be independent and have $\sigma^2\chi^2$ distributions also. If H_0 is true, SSTO is merely the numerator for the variance of a random sample from a normal distribution. Thus, from Section 12.3, it follows that, if H_0 is true

$$\text{SSTO} \sim \sigma^2\chi^2_{\Sigma n_j - 1}$$

Since we have already shown that $\text{df}_E = \Sigma n_j - c$ and $\text{df}_A = c - 1$, and these total $\sum_j n_j - 1$, we know that

$$\text{SSE} \sim \sigma^2\chi^2_{\Sigma n_j - c} \quad \text{and} \quad \text{SSA} \sim \sigma^2\chi^2_{c-1}$$

If H_0 is true, Equations 12.4 and 12.8 are equivalent, that is, MSE and MSA are both estimates of σ^2. Therefore, under H_0, their ratio should be near 1:

$$E\left[\frac{\text{MSA}}{\text{MSE}}\bigg| H_0\right] = 1$$

If H_0 is not true, then the second term in Equation 12.8, $\Sigma n_j\alpha_j^2/(c-1)$ can only be positive, and our ratio is expected to exceed 1. Thus, we obtain the decision rule for the hypothesis set,

$$H_0\text{: } \alpha_j = 0 \text{ for all } j \qquad H_a\text{: some } \alpha_j \neq 0$$

as:

Reject H_0 at level α if
$$\text{pr}\left\{\frac{\text{MSA}}{\text{MSE}} \geqslant \text{observed } \frac{\text{MSA}}{\text{MSE}}\bigg| H_0\right\} < \alpha$$

But $\text{MSA} \sim \sigma^2\chi^2_{c-1}/(c-1)$ and $\text{MSE} \sim (\sigma^2\chi^2_{\Sigma n_j - c})/(\Sigma n_j - c)$ and are independent. Therefore, under H_0,

$$\frac{\text{MSA}}{\text{MSE}} \sim F_{c-1,\, \Sigma n_j - c}$$

and our decision rule becomes:

Reject H_0 at α if
$$\frac{\text{MSA}}{\text{MSE}} > f_{c-1,\, \Sigma n_j - c,\, \alpha} \qquad (12.10)$$

The formulas given for SSTO, SSE, and SSA are the definitional forms. Although they are very helpful in seeing exactly what is being compared, they are computationally laborious and tend to compound rounding errors. As with the variance for a single sample, there are equivalent forms available that simplify computations and require less rounding. They are given in Table 12.2. The CF term represents the correction factor for the overall mean, $\overline{X}_{..}$. If you are ever in doubt about the divisor in any of the terms, remember that you always divide by the number of items added before squaring.

It is general practice in statistics to summarize these results in what is called an analysis-of-variance (ANOVA) table. The ANOVA table for a one-factor experiment with a fixed-effects model, showing all items that are to be included, is given in Table 12.3.

Most introductory texts omit the E(MS) column, but it is probably the most important information in the table. It is included simply because it identifies how the F ratio is to be calculated. In the fixed-effects models all ratios use the MSE in the denominator, but that is not always true for random- or mixed-effects models. Most authors concern themselves with only the fixed-effects model; they do the student a disservice by not at least discussing the other models and the changes that are required for them.

Table 12.2 Computational Forms for the Sums of Squares in a One-factor Design

Sum of Squares		Computational Form
		$\text{CF} = \dfrac{\left(\sum_j \sum_i X_{ij}\right)^2}{\sum_j n_j}$
SSTO	=	$\sum_j \sum_i X_{ij}^2 - \text{CF}$
SSA	=	$\sum_j \dfrac{\left(\sum_i X_{ij}\right)^2}{n_j} - \text{CF}$
SSE	=	$\sum_j \sum_i X_{ij}^2 - \sum_j \dfrac{\left(\sum_i X_{ij}\right)^2}{n_j}$

Table 12.3 An ANOVA Table for a One-factor Fixed-effects Model

Source of Variation	Sum of Squares (SS)	Degrees of Freedom (df)	Mean Square (MS)	E(MS)	Calculated F (F Calc)
A	SSA	$c-1$	MSA	$\sigma^2 + \dfrac{\sum_j n_j\alpha_j^2}{c-1}$	$\dfrac{\text{MSA}}{\text{MSE}}$
Error	SSE	$\sum_j n_j - c$	MSE	σ^2	
Total	SSTO	$\sum_j n_j - 1$			

EXAMPLE 12.1 Three different methods are available for the assembly of a mechanical nutcracker. Motion and time study engineers for the company, Sweet Meats, Ltd., run an experiment to determine whether a difference exists in the average time required to complete the assembly using each method. If the data (in minutes) are as given, does it matter which method Sweet Meats chooses for its assembly line? Use $\alpha = 0.05$.

The experiment was conducted by randomly selecting 14 persons from the assembly department and randomly assigning methods to persons:

Method M_1	M_2	M_3
5.8	3.8	6.8
4.7	2.6	5.9
3.8	4.5	7.2
4.9	5.1	8.1
5.8		5.6

SOLUTION We have a completely randomized one-factor design with a fixed-effects model, since the three methods are the only ones of interest. Checking the assumptions for ANOVA, we find no reason to think that any are violated to an extent that would greatly affect our results. Sweet Meats' hypotheses are (for the model $X_{ij} = \mu + \alpha_j + \epsilon_{ij}$)

$$H_0\colon\ \alpha_j = 0\ \ j = 1, 2, 3 \qquad H_a\colon\ \text{some } \alpha_j \neq 0$$

and the decision rule is:

Reject H_0 if

$$\frac{\text{MSA}}{\text{MSE}} > f_{c-1,\, \Sigma n_j - c,\, \alpha} = f_{2,\, 11,\, 0.05} = 3.98$$

We will use the computational forms of Table 12.2. The necessary sums are:

$$\sum_j \sum_i X_{ij} = 74.6$$

$$\sum_j \sum_i X_{ij}{}^2 = 5.8^2 + 4.7^2 + \cdots + 8.1^2 + 5.6^2 = 425.14$$

$$\sum_i X_{i1} = 25.0 \qquad \sum_i X_{i2} = 16.0 \qquad \sum_i X_{i3} = 33.6$$

$$\text{CF} = \frac{(74.6)^2}{14} = 397.51$$

The resulting SS's are

$$\text{SSTO} = \sum_j \sum_i X_{ij}^2 - \text{CF} = 425.14 - 397.51 = 27.63$$

$$\text{SSA} = \sum_j \frac{\left(\sum_i X_{ij}\right)^2}{n_j} - \text{CF} = \left[\frac{(25)^2}{5} + \frac{(16)^2}{4} + \frac{(33.6)^2}{5}\right] - 397.51$$

$$= 414.79 - 397.51 = 17.28$$

$$\text{SSE} = \sum_j \sum_i X_{ij}^2 - \sum_j \frac{\left(\sum_i X_{ij}\right)^2}{n_j} = 425.14 - 414.79 = 10.35$$

Using the ANOVA format, we summarize our calculations in Table 12.4.

Table 12.4 ANOVA Table for Example 12.1

Source	SS	df	MS	E(MS)	F Calc
Methods	17.28	2	8.64	$\sigma^2 + \frac{\sum n_j \alpha_j^2}{2}$	9.19
Error	10.35	11	0.94	σ^2	
Total	27.63	13			

Since the calculated F of 9.19 exceeds 3.98, the table value, we reject H_0 and conclude that the methods appear to differ in their average assembly times.

Note that if there had been many different methods of assembly in Example 12.1, the model would have been a random-effects model. Since our design has only one factor (assembly methods), the analysis would not have changed, but the interpretation would be slightly different. In the random-effects model the levels of the factor are assumed to be selected at random. Thus, the α_j in the model are essentially random observations from the population of all α_j's. In this case the hypotheses are often written as

$$H_0\colon\ \sigma_A^2 = 0 \qquad H_a\colon\ \sigma_A^2 \neq 0$$

If we reject H_0, the conclusion then is that the population of methods contributes significantly to the total variance in X. Note, though, that the only way this can happen is for some of the α_j's themselves to be nonzero. The only change in the ANOVA table occurs in the $E(\text{MS})$ column. For a random-effects, one-factor model,

$$E(\text{MSA}) = \sigma^2 + n\sigma_A^2$$

$E(\text{MSE})$ is still equal to σ^2.

As we will see later, things become much more complicated as we go to higher-order models. By higher-order we mean models with more factors in them.

EXERCISES

12.4.1 Perform an analysis of variance on the following data at $\alpha = 0.05$.

A_1	A_2
15	24
18	23
12	26
14	18
11	19
	16

12.4.2 Analyze the data in Exercise 12.4.1 using a two independent samples t test at $\alpha = 0.05$ and compare your answers. Do the comparisons indicate anything to you?

12.4.3 Five students were randomly selected from the accounting majors at each of three universities: Akron, Kent State, and Cleveland State. The students were given a CPA exam and obtained the following scores (one student from C.S.U. became ill during the exam and did not complete it).

Akron	K.S.U.	C.S.U.
82	71	92
73	64	57
41	53	84
88	79	67
66	33	

Test for differences among the three schools using $\alpha = 0.05$.

12.4.4 Consider the given data on costs of operation in cents per mile for three different makes of trucks. Using a significance level of 0.05, test the hypothesis that mean costs of the three makes are equal.

Truck Type		
A	*B*	*C*
4	6	6
6	7	8
2	5	9
	6	7
	6	

12.4.5 Indicate which model, random effects or fixed effects, you feel is appropriate in each of Exercises 12.4.3 and 12.4.4, and why.

12.5 TWO-FACTOR DESIGNS WITH REPLICATION

Two-way ANOVA with replication is used when one is interested in testing two qualitative or quantitative variables simultaneously, both are variables of primary interest, and there may be an interaction effect present. The term *interaction* refers to a situation in which a particular level of one of the variables gives responses that could be considered out of character or pattern when it occurs in combination with a particular level of the other variable.

For example, suppose that one wished to determine whether there was a difference in students' scores due to professors and also due to textbooks. To determine this, one might have each of three professors teach three classes using a different one of three texts in each, then having all students take a common exam. An interaction between books and professors would be present if the students having professor 3 and book 2, for example, scored high on the exam, whereas those using book 2 with professors 1 and 2 scored about the same but significantly lower than those of professor 3, and the scores of all those using the other books were about the same regardless of professor.

Another way of looking at interaction is to recall that our model is linear and to consider what that implies. If there are individual effects due to professors or individual effects due to books, but no interaction between the two, we would expect that a graph of the mean responses for each professor versus the texts would show parallel relationships. Correspondingly, if the mean responses for each text are graphed versus the professors, we would expect parallel relationships in the absence of interaction. On the other hand, if an interaction is present, the relationships between mean responses within one factor and the levels of the other factor will no longer be parallel. That is, the graph will exhibit a lack of parallelism. Figure 12.4 illustrates various possibilities of these considerations. (Note that if no effects are present, all means should lie on one horizontal line.)

In this situation we will be working with the completely randomized type of design, with each treatment combination administered to n randomly selected subjects. Thus each response X_{ijk} can then be considered as a sum of the following components: overall mean, treatment 2 (or column) effect, treatment 1 (or row) effect, interaction effect, and random error.

That is,

$$X_{ijk} = \mu + \alpha_j + \beta_i + (\alpha\beta)_{ij} + \epsilon_{ijk}$$

where $i = 1, \ldots, r$ is the number of levels of treatment 1; $j = 1, \ldots, c$, the number of levels of treatment 2; and $k = 1, \ldots, n$, the number of observations per combination or cell.

Figure 12.4 Graphs of Expected Mean Responses for Various Combinations of Effects

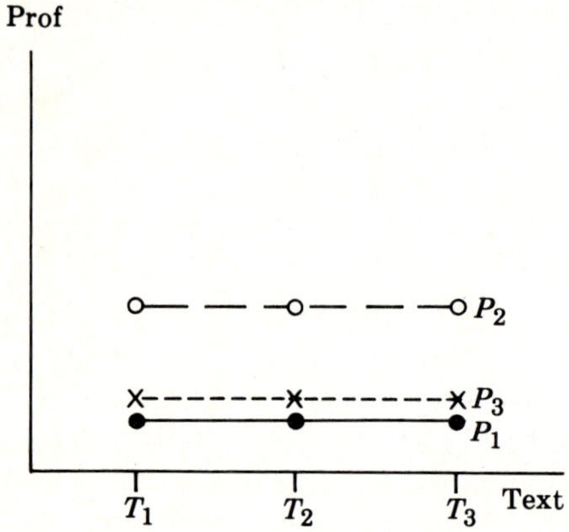

(*a*) An effect due to professor but not text or interaction

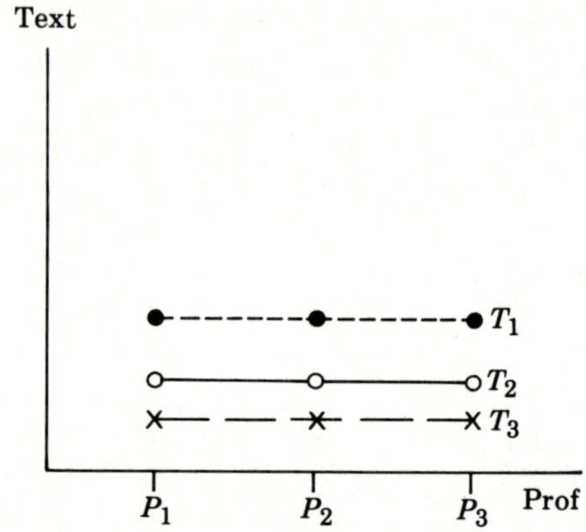

(*b*) An effect due to text but not professor or interaction

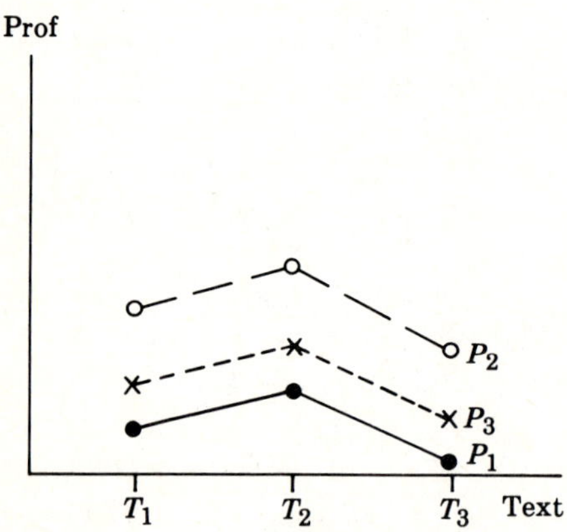

(*c*) Effects due to both text and professor but no interaction

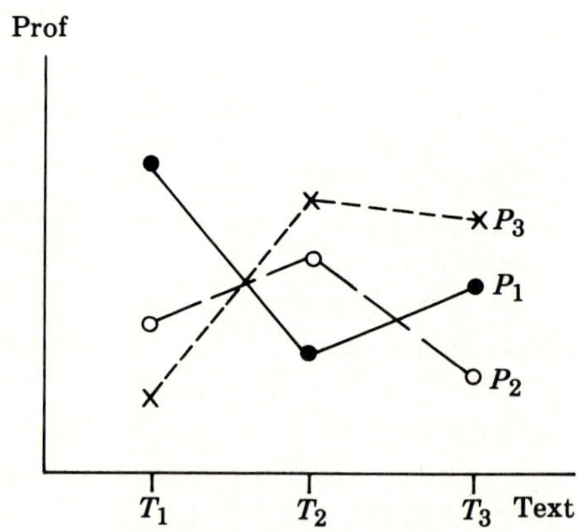

(*d*) Obvious effect due to interaction between professor and text

Suppose that we have three levels of treatment 2, two levels of treatment 1, and four observations per cell. Our design would then appear as follows:

		Treatment 2					
		Level 1		Level 2		Level 3	
Treatment 1	Level 1	x_{111}	x_{113}	x_{121}	x_{123}	x_{131}	x_{133}
		x_{112}	x_{114}	x_{122}	x_{124}	x_{132}	x_{134}
	Level 2	x_{211}	x_{213}	x_{221}	x_{223}	x_{231}	x_{233}
		x_{212}	x_{214}	x_{222}	x_{224}	x_{232}	x_{234}

The first subscript indicates the level of treatment 1 that was used, the second subscript indicates the level of treatment 2 that was used, and the third subscript identifies the observation in the cell (or the replicate).

The object, then, is to test to see whether there are any significant differences between the levels of treatment 1, the levels of treatment 2, and the various cell combinations after correcting for each treatment effect. To do this, we use the following breakup of the total sum of squares.

$$\begin{aligned} \text{SSTO} &= \text{SS}_{tr2} + \text{SS}_{tr1} + \text{SS}_{int} + \text{SS}_{error} \\ &= \text{SSA} + \text{SSB} + \text{SSAB} + \text{SSE} \end{aligned} \tag{12.11}$$

or

$$\begin{array}{ccccc} \Sigma\Sigma\Sigma(X_{ijk} - \bar{X}_{...})^2 & = & \Sigma\Sigma\Sigma(\bar{X}_{.j.} - \bar{X}_{...})^2 & + & \Sigma\Sigma\Sigma(\bar{X}_{i..} - \bar{X}_{...})^2 \\ \text{df} = rcn - 1 & = & c - 1 & + & r - 1 \end{array}$$

$$\begin{array}{cccc} + & \Sigma\Sigma\Sigma(\bar{X}_{ij.} - \bar{X}_{.j.} - \bar{X}_{i..} + \bar{X}_{...})^2 & + & \Sigma\Sigma\Sigma(X_{ijk} - \bar{X}_{ij.})^2 \\ + & (r-1)(c-1) & + & rc(n-1) \end{array}$$

Thus far we have assumed that the effects are additive. In addition, we need to assume that the subjects, before treatments, are selected from a normal population with variance equal to σ^2 and mean μ. Furthermore, we assume that the variances are all the same and independent of the treatments. That is, our assumptions are:

1. $X_{ijk} = \mu + \alpha_j + \beta_i + \alpha\beta_{ij} + \epsilon_{ijk}$

2. $\epsilon_{ijk} \sim N(0, \sigma^2)$ for all i, j, k

or

2′. $X_{ijk} \sim N(\mu, \sigma^2)$ before treatments

The procedure then consists of comparing the various mean squares with the error mean square. If the corresponding ratio is too large, this implies that the treatment means that gave rise to the numerator are further apart than one would expect due to chance variation. The results are summarized in an ANOVA table as follows:

Table 12.5 ANOVA Table for a Two-factor Design

Source	SS	df	MS	*F* Calc
Columns	SSA	$c - 1$	$\frac{\text{SSA}}{c-1}$	$\frac{\text{MSA}}{\text{MSE}}$
Rows	SSB	$r - 1$	$\frac{\text{SSB}}{r-1}$	$\frac{\text{MSB}}{\text{MSE}}$
Interaction	SSAB	$(c - 1)(r - 1)$	$\frac{\text{SSAB}}{(c-1)(r-1)}$	$\frac{\text{MSAB}}{\text{MSE}}$
Error	SSE	$rc(n - 1)$	$\frac{\text{SSE}}{rc(n-1)}$	
Total	SSTO	$rcn - 1$		

The computational forms for the various SS are given by:

$$\text{Correction factor} = \frac{(\Sigma\,\Sigma\,\Sigma\,X_{ijk})^2}{rcn} = \text{CF}$$

$$\text{SSA} = \sum_j \frac{\left(\sum_i \sum_k X_{ijk}\right)^2}{rn} - \text{CF}$$

$$\text{SSB} = \sum_i \frac{\left(\sum_j \sum_k X_{ijk}\right)^2}{cn} - \text{CF}$$

$$\text{SSTO} = \sum_i \sum_j \sum_k X_{ijk}^2 - \text{CF}$$

$$\text{SSE} = \sum_i \sum_j \sum_k X_{ijk}^2 - \sum_i \sum_j \frac{\left(\sum_k X_{ijk}\right)^2}{n}$$

$$\text{SSAB} = \text{SSTO} - \text{SSA} - \text{SSB} - \text{SSE}$$

Thus, to test for the various effects we have:

1. Interaction effects:

$$H_0\colon\ (\alpha\beta)_{ij} = 0 \text{ for all } i, j \qquad H_a\colon\ (\alpha\beta)_{ij} \neq 0 \text{ for some } i, j$$

   Reject H_0 at α if MSAB/MSE $> f_{(r-1)(c-1),\ rc(n-1),\ \alpha}$ and conclude that a significant interaction exists.

2. Row effects:

$$H_0\colon\ \beta_i = 0 \text{ for all } i \qquad H_a\colon\ \beta_i \neq 0 \text{ for some } i$$

   Reject H_0 at α if MSB/MSE $> f_{r-1,\ rc(n-1),\ \alpha}$ and conclude that a significant β effect exists.

3. Column effects:

$$H_0\colon\ \alpha_j = 0 \text{ for all } j \qquad H_a\colon\ \alpha_j \neq 0 \text{ for some } j$$

   Reject H_0 at α if MSA/MSE $> f_{c-1,\ rc(n-1),\ \alpha}$ and conclude that a significant α effect exists.

EXAMPLE 12.2 We want to determine whether there is any effect on the life of television tubes due to the size of the set, the manufacturer of the tube, or any specific combinations of the two. To determine this, two television sets of different size and three tube manufacturers were randomly selected. Then two tubes from each supplier were tested in each type of set, with the following results for tube lives in months.

Size of Set	Manufacturer A	B	C	$\sum_j \sum_k X_{ijk}$
1	22 16 (38)	8 13 (21)	9 10 (19)	78
2	19 23 (42)	16 15 (31)	16 13 (29)	102
$\sum_i \sum_k X_{ijk}$	80	52	48	$180 = \sum_i \sum_j \sum_k X_{ijk}$

Calculating the required sums yields

$$\sum_i \sum_j \sum_k X_{ijk}^2 = 2950$$

$$\sum_j \left(\sum_i \sum_k X_{ijk} \right)^2 = 80^2 + 52^2 + 48^2 = 11{,}408$$

$$\sum_i \left(\sum_j \sum_k X_{ijk} \right)^2 = 78^2 + 102^2 = 16{,}488$$

$$\left(\sum_i \sum_j \sum_k X_{ijk} \right)^2 = 180^2 = 32{,}400$$

$$\sum_i \sum_j \left(\sum_k X_{ijk} \right)^2 = 38^2 + 21^2 + 19^2 + 42^2 + 31^2 + 29^2 = 5812$$

Then, substituting into the formulas gives

$$\text{CF} = \frac{32{,}400}{12} = 2700 \qquad \text{where } r = 2,\ c = 3,\ n = 2$$

$$\text{SSA} = \frac{11{,}408}{4} - 2700 = 2852 - 2700 = 152$$

$$\text{SSB} = \frac{16{,}488}{6} - 2700 = 2748 - 2700 = 48$$

$$\text{SST} = 2950 - 2700 = 250$$

$$\text{SSE} = 2950 - \frac{5812}{2} = 2950 - 2906 = 44$$

$$\text{SSAB} = 250 - 152 - 48 - 44 = 250 - 244 = 6$$

Hence the ANOVA table is

Source	SS	df	MS	F Calc	F Table at 0.05
Columns (A)	152	2	76	10.37	5.14
Rows (B)	48	1	48	6.55	5.99
Interaction (AB)	6	2	3	0.41	5.14
Error	44	6	7.33		
Total	250	11			

Thus the tests are:

1. For interaction:

$$H_0\colon\ (\alpha\beta)_{ij} = 0 \text{ for all } i, j \qquad H_a\colon\ (\alpha\beta)_{ij} \neq 0 \text{ for some } i, j$$

   We cannot reject H_0, since $0.41 < 5.41$. We conclude that there appears to be no interaction.

2. Rows:

$$H_0\colon\ \beta_i = 0 \text{ for all } i \qquad H_a\colon\ \beta_i \neq 0 \text{ for some } i$$

   Reject H_0, since $6.55 > 5.99$, and conclude that there appear to be differences in the mean lives due to the size of the set.

3. Columns:

$$H_0\colon\ \alpha_j = 0 \text{ for all } j \qquad H_a\colon\ \alpha_j \neq 0 \text{ for some } j$$

   Reject H_0, since $10.37 > 5.14$, and conclude that tube life appears to differ between manufacturers.

In summary, both the size of the set and the tube manufacturer appear to have an effect on the expected life of the tube, but particular combinations of the two are of little or no consequence.

The justification for our construction of the tests in this section again comes from Cochran's theorem. If all null hypotheses are true, then instead of rc samples of size n, we have one large sample of size rcn from a normal population. Hence,

$$\frac{1}{\sigma^2} \sum_i \sum_j \sum_k (X_{ijk} - \overline{X}_{...})^2 \sim \chi^2_{rcn-1}$$

Since the degrees of freedom for the terms on the right-hand side of our partition add to $rcn - 1$, we have that each has a χ^2 distribution with the indicated degrees of freedom and is independent of all others. Therefore, if an individual H_0 is true, the ratio of its mean square to the mean square error will have the appropriate F distribution. As usual, H_0 is rejected if this ratio is too large.

The column for the E(MS)s was omitted from the ANOVA table in Example 12.2. The reason for this omission is that we constructed our test statistics as if we were using a fixed-effects model. Since both makes and suppliers were randomly selected, it is actually a random-effects model. In the one-factor experiment this didn't matter, but in the two-factor it does. To illustrate the difference, the $E(\mathrm{MS})$ for both the fixed-effects and random-effects models for a two-factor experiment are presented in Table 12.6.

As we can see from Table 12.6, we were lucky in Example 12.2: we accepted H_0 in the test for interaction and concluded that no interaction effects were present. Thus, we conclude that $\sigma_{AB}^2 = 0$.* In the fixed-effects model it does not matter whether we accept H_0 for the interaction test or not, since only σ^2 and the effect in question are present in the remaining $E(\mathrm{MS})$s. In the random-effects model, if the AB term is significant, the A and B factors are both tested against the interaction (AB) mean square. In *all* ANOVA tests you should begin testing with the highest-order interaction (the one next to the error term in the table) and work your way up the table, checking the $E(\mathrm{MS})$s for each effect each time.

If a model has both fixed and random effects in it for a factorial design, the fixed factor will have $n\sigma_{AB}^2$ as part of its $E(\mathrm{MS})$. The random effect will contain only error and its own effect in the $E(\mathrm{MS})$. If the population of levels for a factor is finite, then whenever its sample size occurs in the $E(\mathrm{MS})$, it is replaced by the finite population corrector; for example, for A, c is replaced by $1 - c/C$, where C is the number of levels in the population.

EXERCISES

12.5.1 An analyst for an automobile manufacturer wishes to test two antipollution devices and two brands of gasoline. For each combination, three randomly selected pollution readings were taken and the following data resulted. Use a significance level

*In analyzing the random-effects model, three approaches are used in practice whenever the interaction term is concluded to be insignificant. The purely theoretical approach guards against the possibility of making a Type II error when constructing the remaining test statistics. With respect to Table 12.6, it would do so by comparing MSA and MSB to MSAB even though it appeared to be insignificant. The other extreme attempts to build up the degrees of freedom for the denominator by comparing MSA and MSB to the MSE whenever MSAB is insignificant.

The second approach is theoretically incorrect and can create problems in interpreting the results. The third approach represents a compromise between the other two. If we conclude that AB is insignificant, this approach retests AB using a larger value for α (usually two to four times larger). If the retest fails to reject H_0, the procedure then either uses MSE for testing the main effects, or else pools SSAB with SSE and uses $(\mathrm{SSAB} + \mathrm{SSE})/(\mathrm{df}_{\mathrm{AB}} + \mathrm{df}_{\mathrm{E}})$ for testing the main effect. Only the first approach is theoretically correct.

Table 12.6 Expected Mean Squares for a Two-factor Design with Either Both Effects Fixed or Both Effects Random

Source	E(MS) (Fixed Effects)	E(MS) (Random Effects)
A	$\sigma^2 + \dfrac{rn \sum \alpha_j^2}{c-1}$	$\sigma^2 + n\sigma_{AB}^2 + rn\sigma_A^2$
B	$\sigma^2 + \dfrac{cn \sum \beta_i^2}{r-1}$	$\sigma^2 + n\sigma_{AB}^2 + cn\sigma_B^2$
AB	$\sigma^2 + n \dfrac{\sum \sum (\alpha\beta)_{ij}^2}{(r-1)(c-1)}$	$\sigma^2 + n\sigma_{AB}^2$
Error	σ^2	σ^2

of 0.01 and make a complete analysis of variance. For your information, marginal totals are given. Also, $\sum \sum \sum X_{ijk}^2 = 250$. (State the model and the type of effects present.)

	Device 1	Device 2	Totals
Gasoline 1	1	5	
	1	6	
	1	4	18
Gasoline 2	5	8	
	2	4	
	5	6	30
Totals	15	33	48

12.5.2 Suppose that both the antipollution devices and the gasolines used in the experiment were randomly selected. Does this change your conclusions in any way?

12.5.3 A firm has two different sources of raw material and two machines of different makes to produce a new product. The firm obtained three observations for each variable combination, with the following results. Use a significance level of 0.05 and analyze completely. Also, state the model and identify the type of effects present.

	Raw Materials A	B
Machine x	7	7
	6	5
	8	9
Machine y	4	3
	7	3
	4	3

12.5.4 A certain part produced by a company requires extensive machining. The company has three machines for doing this and six operators who can run the machines. In an attempt to determine whether there are differences in the average time required per part due to machines, operators, or particular combinations of the two, the company had each operator finish nine parts, three on each machine, and recorded the time for each part. The results (in minutes) are summarized here. State the model and type of effects and analyze at $\alpha = 0.05$.

$$\sum_i \sum_j \sum_k X_{ijk} = 3240$$

$$\sum_i \sum_j \sum_k X_{ijk}^2 = 210{,}620 \quad i = 1, 2, 3 \quad j = 1, 2, \ldots, 6 \quad k = 1, 2, 3$$

$$\sum_i \Big(\sum_j \sum_k X_{ijk}\Big)^2 = 3{,}524{,}400 \qquad \sum_j \Big(\sum_i \sum_k X_{ijk}\Big)^2 = 1{,}846{,}800$$

$$\sum_i \sum_j \Big(\sum_k X_{ijk}\Big)^2 = 622{,}290$$

Table 12.7 An Illustration of a Possible Randomized Block Design

		Treatments (Factor A)			
Blocks (Factor B)	B_1	T_4	T_1	$\cdots$	T_7
	B_2	T_8	T_c	$\cdots$	T_2
	$\vdots$	$\vdots$	$\vdots$	$\vdots$	$\vdots$
	B_r	T_{c-1}	T_6	$\cdots$	T_3

12.5.5 Refer to Exercises 12.5.1, 12.5.3, and 12.5.4. Evaluate each with respect to the ANOVA assumptions.

12.6 THE RANDOMIZED BLOCK DESIGN

Just as we extended the independent samples t test in the one-factor ANOVA, we can extend the paired t test. The extension of the paired t is called a *randomized block design*. The sampling concept is quite simple, and the analysis is simply that of a two-factor ANOVA *without* replication.

The idea of blocking, as in the paired t, is to control the variation of some secondary variable and hence to remove its effect from the error SS before testing. That is, the reason for blocking is to refine the SSE or to make it as nearly pure as a measure of error as possible. The concept is easily understood but often difficult to attain.

The sampling procedure is as follows: Select a random sample of blocks of items. Within each block the items or subjects are identical with respect to some attribute, the attribute being one of the levels of the blocking variable. Each block contains the same number of subjects as there are levels of the treatment. Then randomly assign treatment levels to the subjects within each block. Table 12.7 shows an example of a typical randomized block design. Note that within a given block, treatments occur in a random order.

For analysis purposes the data are rearranged into the standard two-factor table. That is, all observations corresponding to the same treatment are grouped together in a column (or row). This has been done in Table 12.8 for the design in Table 12.7.

Table 12.8 Rearranged Data for a Randomized Block Design

		Treatment (Factor A)				
		T_1	T_2	·	· ·	T_c
	B_1	X_{11}	X_{12}	·	· ·	X_{1c}
	B_2	X_{21}	X_{22}	·	· ·	X_{2c}
Blocks	·	·	·	·	· ·	·
(Factor B)	·	·	·	·	· ·	·
	·	·	·	·	· ·	·
	B_r	X_{r1}	X_{r2}	·	· ·	X_{rc}

As stated earlier, the analysis for a randomized block design is the same as that for a two-factor design with $n = 1$. The model is altered slightly to become

$$X_{ij} = \mu + \alpha_j + \beta_i + \epsilon_{ij} \tag{12.12}$$

Note that the interaction term is missing. This is because the randomized block analysis requires the additional assumption that the interaction term is 0, that $(\alpha\beta)_{ij} = 0$ for all i and j. The reason for this is that, without replication, it is impossible to measure the interaction and error terms separately, since there are no degrees of freedom available for error. All other assumptions are the same as for a two-way analysis. Summarizing, the assumptions for constructing the test in a randomized block design are

1. A random sample of blocks of items, with treatments being assigned randomly within a block
2. $\epsilon_{ij} \sim N(0, \sigma^2)$ for all i, j
3. Interval level of measurement
4. $\alpha\beta_{ij} = 0$ for all i, j

Since $n = 1$ in this case, the computational forms are simplified to give

$$\text{SSTO} = \sum_j \sum_i X_{ij}^2 - \text{CF} \qquad \text{CF} = \frac{(\Sigma\, \Sigma\, X_{ij})^2}{rc}$$

$$\text{SSTR} = \text{SSA} = \sum_j \frac{\left(\sum_i X_{ij}\right)^2}{r} - \text{CF}$$

$$\text{SSBL} = \text{SSB} = \sum_i \frac{\left(\sum_j X_{ij}\right)^2}{c} - \text{CF}$$

$$\text{SSE} = \text{SSAB} = \text{SSTO} - \text{SSA} - \text{SSB}$$

Our discussion has assumed that treatments correspond to columns and blocks to rows. This is not necessary. In identifying which is which for data that have already been collected, the variable of primary interest is the treatment and the secondary (or controlled variable) is the blocking one.

EXAMPLE 12.3 In evaluating the lives of four different brands of drill bits, a company felt that drill speed would have an effect on the life of a bit. They therefore selected five bits of each brand and randomly allocated a different speed to each bit in a specific brand. The lives in hours are as follows, with the data reorganized for computational ease.

Brand	Drill Speed				
	1500	2000	2500	3000	3500
A	4.0	3.0	2.0	1.0	1.0
B	3.0	3.5	2.5	1.2	0.5
C	3.5	2.5	3.0	1.8	0.2
D	4.5	3.0	2.5	1.5	0.3

(*a*) Construct the analysis-of-variance table and test to see whether brands differ at 0.05.
(*b*) Does it appear that blocking was necessary at $\alpha = 0.05$?

SOLUTION The model of interest is (note that blocks are columns)

$$X_{ij} = \mu + \alpha_j + \beta_i + \epsilon_{ij}$$

and we wish to test (*a*) H_0: $\beta_i = 0$ for all i; H_a: $\beta_i \neq 0$ for some i, and (*b*) H_0: $\alpha_j = 0$ for all j; H_a: $\alpha_j \neq 0$ for some j. The necessary computations follow. For blocks (speeds) we have the column totals of

$$\Sigma X_{i1} = 15 \quad \Sigma X_{i2} = 12 \quad \Sigma X_{i3} = 10 \quad \Sigma X_{i4} = 5.5 \quad \Sigma X_{i5} = 2.0$$

For treatments (rows), the totals are

$$\Sigma X_{1j} = 11 \quad \Sigma X_{2j} = 10.7 \quad \Sigma X_{3j} = 11 \quad \Sigma X_{4j} = 11.8$$

In addition, $\sum_j \sum_i X_{ij} = 44.5$ and $\sum_j \sum_i X_{ij}^2 = 128.81$. Thus our sums of squares become:

$$\text{CF} = \frac{(44.5)^2}{20} = 99.01$$

$$\text{SSTO} = 128.81 - \frac{(44.5)^2}{20} = 29.80$$

$$\text{SSA} = \frac{1}{4}(15^2 + \cdots + 2^2) - 99.01 = 26.80$$

$$\text{SSB} = \frac{1}{5}(11^2 + \cdots + 11.8^2) - 99.01 = 0.14$$

$$\text{SSAB} = 29.80 - 26.80 - 0.14 = 2.86$$

The ANOVA table (assuming that both factors are fixed effects) is

Source	SS	df	MS	E(MS)	F Calc
Speeds	26.80	4	6.70	$\sigma^2 + \frac{4\,\Sigma\,\alpha_j^2}{4}$	27.92
Brands	0.14	3	0.05	$\sigma^2 + \frac{5\,\Sigma\,\beta_i^2}{3}$	0.21
Error (AB)	2.86	12	0.24	σ^2	
Total	29.80	19			

(a) For testing H_0: $\beta_i = 0$ for all i; H_a: $\beta_i \neq 0$ for some i we reject H_0 if

$$\frac{\text{MSB}}{\text{MSE}} > f_{3,\ 12,\ 0.05} = 3.49$$

Since $0.21 < 3.49$, we cannot reject H_0. We conclude that, on the basis of these data, there appears to be no difference in the bit lives according to brand.

(b) For testing whether blocking was necessary, we note that blocking is necessary if evidence indicates that drill speed affects life. Thus, to test

$$H_0\colon\ \alpha_j = 0, \text{ all } j \qquad H_a\colon\ \alpha_j \neq 0, \text{ some } j$$

we reject H_0 if

$$\frac{\text{MSA}}{\text{MSE}} > f_{4,\ 12,\ 0.05} = 3.26$$

Since $27.92 > 3.26$, we reject H_0 and conclude that according to this test blocking appears to be necessary on drill speed.

In Example 12.3 we assumed that drill speed is a fixed effect. This is unrealistic, and we leave it to the student to correct the analysis to account for drill speed being a random effect. The degrees of freedom for a randomized block are given by the same formulas as in a two-factor model. The difference is that the interaction is used for the error term, since the degrees of freedom for true error would be

$$rc(n - 1) = rc(1 - 1) = 0$$

That is, we use for degrees of freedom error the term

$$\mathrm{df}_{\mathrm{error}} = \mathrm{df}_{AB} = (r - 1)(c - 1) \tag{12.13}$$

EXERCISES

12.6.1 Reevaluate the analysis in Example 12.3 assuming that speed is a random effect.

12.6.2 In gasoline efficiency tests, three different brands of gas were tested in three types of cars. The number of miles a car traveled on one gallon of gas was determined. Which brand of gas gives the best mileage? ($\alpha = 0.01$) State the model and type of effects that you assume.

Brand	Car *A*	Car *B*	Car *C*
Gas *A*	21	15	9
Gas *B*	21	19	20
Gas *C*	24	20	31

12.6.3 A manufacturer has three machines, A, B, C, that are used in a production process. Three operators are also qualified to run the machines. Believing that differences in output due to operators may exist, each machine is run by each operator, with the assignments being made randomly. The coded measurements of each operator/machine combination are:

	Machine		
Operator	*A*	*B*	*C*
1	5	6	11
2	1	8	9
3	9	4	10

Assuming normality, do there appear to be differences in machines at $\alpha = 0.05$?

12.6.4 Refer to Exercise 12.6.3 and test for operator differences at $\alpha = 0.05$.

12.6.5 Refer to Exercise 12.6.2 and test for differences in gasolines at $\alpha = 0.05$.

12.7 SUPPLEMENTARY EXAMPLES WITH COMPUTER OUTPUT

As you can see, the computations get rapidly out of hand as the size of the design or the number of replicates (observations per cell) increases. In such cases it is desirable to have the computations done by computer. There are a number of standard programs available in most university computer facilities. Among these are the BMD package (developed by the UCLA Health Sciences Computing Facility), the SPSS package (the Statistical Package for the Social Sciences, developed by the University of Chicago) and SAS (Statistical Analysis System, developed by North Carolina State University). In addition to these fairly universal programs, most universities have "in-house" programs that are also available for student use. All these programs are easy to use, but access procedures may vary from one system to another. You should check with your instructor to find what is available on your campus. The computer printouts presented here are from an in-house program using interactive APL at the University of Akron.

Before we look at the computer outputs, we will do two examples to illustrate a point that we mentioned earlier. Throughout our discussions we have stressed the fact that the MSE is an estimate of the error variance, σ^2, provided that the model is correct. The next two examples show what can happen if the model has been misstated.

EXAMPLE 12.4* In an attempt to analyze whether display width and/or shelf height has any effect on sales, a store randomly allocated a particular width (regular or wide) and a particular shelf to each of two weeks over a 12-week period. The store then recorded the number of cases of a certain item sold in each of the 12 weeks. The results are presented here. The store's assistant manager, who took a statistics course that covered only one-factor designs, recommended analyzing the results as two one-factor models. You know better, but you humor him by doing a one-way ANOVA on the width factor at $\alpha = 0.05$. What do you conclude?

*This example is an adaptation in part from J. Neter and W. Wasserman, "Sect. 17.4," *Applied Linear Statistical Models* (Homewood, Ill.: Richard D. Irwin Co., 1974).

Shelf	Display Width: Regular		Display Width: Wide		Row Totals
Bottom	47 43	(90)	46 40	(86)	176
Middle	62 68	(130)	67 71	(138)	268
Top	41 39	(80)	42 46	(88)	168
Column Totals	300		312		612

SOLUTION The model the assistant manager specified is a fixed-effects model (only two widths are of interest) with $X_{ij} = \mu + \alpha_j + \epsilon_{ij}$. The hypotheses to be tested are

$$H_0\colon\ \alpha_1 = \alpha_2 = 0 \qquad H_a\colon\ \text{both } \alpha_j \neq 0$$

with H_0 being rejected if

$$\frac{\text{MSA}}{\text{MSE}} > f_{1,\,10,\,0.05} = 4.96$$

The table gives both the row and the cell totals, in parentheses, as well as the column totals. We need only the column and grand totals for our one-way analysis. Using the computational forms for the SS, we have

$$\text{CF} = \frac{(612)^2}{12} = 31{,}212 \qquad \sum_j \sum_i X_{ij}^2 = 32{,}854$$

with

$$\text{SSTO} = 32{,}854 - 31{,}212 = 1642$$

$$\text{SSA} = \frac{(300)^2}{6} + \frac{(312)^2}{6} - 31{,}212 = 31{,}224 - 31{,}212 = 12$$

$$\text{SSE} = 32{,}854 - 31{,}224 = 1630$$

Thus, the ANOVA table is as follows:

Source	SS	df	MS	E(MS)	F Calc
Width (A)	12	1	12	$\sigma^2 + 6 \sum \alpha_j^2$	0.0736
Error	1630	10	163	σ^2 *	
Total	1642	11			

*The $E(\text{MSE}) = \sigma^2$ only if the model is correct.

Since $0.0736 < 4.96$, we cannot reject H_0 and must conclude that width appears to have no effect on sales.

In this case we know in advance that an incorrect analysis was performed. In practice you won't know that. An indication that the wrong model (in this case a one-factor rather than a two-factor) was used is the value of the calculated F. If H_0 is true, it should be close to 1. Rejection of H_0 is indicated only if the calculated F is significantly greater than 1. What do we conclude if it is unreasonably small, as the value in Example 12.4 is? We could attribute such a low value to chance variation (call it a miracle) and forget it. It may be simply chance, but before we conclude that, we should suspect that something may be wrong. We should double check all calculations, data collection and recording procedures, information relating to the original design that may have been omitted in the reporting of the data, and any other possibilities for mistakes. That is, try to find a logical reason for the value, and if all else fails, then call it a miracle. In our case we know the reason: The one-factor model is incorrect. While we have been assuming that the MSE is estimating σ^2, it is actually estimating (for Example 12.4) the expected value, given by

$$E(\text{MSE}) = \sigma^2 + 0.2 \sum_j \sum_i (\alpha\beta)_{ij}^2 + 0.4 \sum_i \beta_i^2$$

We leave this for you to verify as a Level Two exercise.

EXAMPLE 12.5 Analyze the data of Example 12.4 as it should be done using $\alpha = 0.05$ for all tests.

SOLUTION We still have a fixed-effects model, but for two factors it becomes

$$X_{ijk} = \mu + \alpha_j + \beta_i + \alpha\beta_{ij} + \epsilon_{ijk}$$

We wish to test the three sets of hypotheses:

(*a*) H_0: $\alpha\beta_{ij} = 0$ all i, j $\quad$ H_a: some $\alpha\beta_{ij} \neq 0$
(*b*) H_0: $\beta_i = 0$ all i $\quad$ H_a: some $\beta_i \neq 0$
(*c*) H_0: $\alpha_j = 0$ all j $\quad$ H_a: some $\alpha_j \neq 0$

The corresponding decision rules are:

(*a*) Reject H_0 if $\dfrac{\text{MSAB}}{\text{MSE}} > f_{2,\,6,\,0.05} = 5.14.$

(*b*) Reject H_0 if $\dfrac{\text{MSB}}{\text{MSE}} > f_{2,\,6,\,0.05} = 5.14.$

(*c*) Reject H_0 if $\dfrac{\text{MSA}}{\text{MSE}} > f_{1,\,6,\,0.05} = 5.99.$

Our SSTO and SSA will be unchanged from the one-factor model, but we need to calculate SSA, SSAB, and the new SSE. The row and cell totals are given in the table in Example 12.4, and the CF was calculated to be 31,212. Thus,

$$\text{SSA} = \frac{(176)^2}{4} + \frac{(268)^2}{4} + \frac{(168)^2}{4} - 31{,}212 = 1544$$

$$\text{SSAB} = \sum_j \sum_i \frac{\left(\sum_k X_{ijk}\right)^2}{n} - \sum_j \frac{\left(\sum_i \sum_k X_{ijk}\right)^2}{rn} - \sum_i \frac{\left(\sum_j \sum_k X_{ijk}\right)^2}{cn} + \text{CF}$$

$$= \left[\frac{(90)^2}{2} + \cdots + \frac{(88)^2}{2}\right] - 31{,}224 - 32{,}756 + 31{,}212 = 24$$

$$\text{SSE} = \sum_i \sum_j \sum_k X_{ijk}^2 - \sum_j \sum_i \frac{\left(\sum_k X_{ijk}\right)^2}{n} = 32{,}854 - 32{,}792 = 62$$

The ANOVA table for the two-factor analysis is

Source	SS	df	MS	E(MS)	F Calc
Width	12	1	12	$\sigma^2 + 6 \sum_j \alpha_j^2$	1.16
Height	1544	2	772	$\sigma^2 + 2 \sum_i \beta_i^2$	74.71
Interaction	24	2	12	$\sigma^2 + \sum_i \sum_j (\alpha\beta)_{ij}^2$	1.16
Error	62	6	10.33	σ^2	
Total	1642	11			

Therefore, for our tests we:

(*a*) Accept H_0, since $1.16 < 5.14$, and conclude that there is no apparent interaction effect.

(*b*) Reject H_0, since $74.71 > 5.14$, and conclude that shelf height appears to affect sales.

(*c*) Accept H_0, since $1.16 < 5.99$, and conclude that display width doesn't seem to matter.

In Example 12.5 we still accept the null hypotheses corresponding to display width, but this time the F value is at least reasonable; there are no obvious discrepancies. Also note that SSA and SSTO are the same for both models. If we did a one-factor analysis for shelf height, SSB also would have the same value. This is to be expected, since in either case we calculate SSA and SSB by comparing the corresponding column or row means. Computer printouts showing the ANOVA tables for Examples 12.4 and 12.5 are given in Tables 12.9 and 12.10, respectively.

In the printouts in Tables 12.9 and 12.10, the ANOVA table includes the subtotal corresponding to SSA + SSB + SSAB. This value is calculated by comparing the cell means; it corresponds to the SSA for a one-factor design with rc levels—that is, each cell represents a level.

Table 12.9 Computer Printout for the Analysis of Example 12.4

TABLE I

SOURCE	SUM SQ	D OF F	MEAN SQ	F RATIO
MEANS	12.0000	1	12.0000	.0736
WITHIN	1630.0000	10	163.0000	
TOTAL	1642.0000	11		

Table 12.10 Computer Printout for the Analysis of Example 12.5

TABLE II

SOURCE	SUM SQ	D OF F	MEAN SQ	F RATIO
ROWS	1544.0000	2	772.0000	74.7097
COLS	12.0000	1	12.0000	1.1613
INTER	24.0000	2	12.0000	1.1613
SUBTOT	1580.0000	5		
ERROR	62.0000	6	10.3333	
TOTAL	1642.0000	11		

EXAMPLE 12.6 A series of experiments was conducted to determine whether either the temperature of the bath, the time in the bath, or particular combinations of the two have any effect on the yield of a polymerization process. Five experiments were run at each temperature-time combination, with the following yields in grams:

Temperature	20 minutes	60 minutes
100°C	9,500 10,650 9,700 9,950 10,100	11,500 11,650 11,250 11,250 11,900
120°C	11,300 11,750 11,600 11,650 11,700	10,900 11,500 11,850 11,700 11,650

Analyze by ANOVA using $\alpha = 0.01$ for all tests.

SOLUTION We assume that the two temperatures and two times are the only ones of interest; therefore, we have a fixed-effects model. Thus,

$$X_{ijk} = \mu + \alpha_j + \beta_i + \alpha\beta_{ij} + \epsilon_{ijk}$$

The hypotheses to be tested are:

(*a*) H_0: $\alpha\beta_{ij} = 0$, all i, j $\quad H_a$: some $\alpha\beta_{ij} \neq 0$
(*b*) H_0: $\beta_i = 0$, all i $\quad H_a$: some $\beta_i \neq 0$
(*c*) H_0: $\alpha_j = 0$, all j $\quad H_a$: some $\alpha_j \neq 0$

Our corresponding decision rules are:

(*a*) Reject H_0 if $\dfrac{\text{MSAB}}{\text{MSE}} > f_{1,\,16,\,0.01} = 8.53.$

(*b*) Reject H_0 if $\dfrac{\text{MSB}}{\text{MSE}} > f_{1,\,16,\,0.01} = 8.53.$

(*c*) Reject H_0 if $\dfrac{\text{MSA}}{\text{MSE}} > f_{1,\,16,\,0.01} = 8.53.$

The computations were done by computer; the printout is given in Table 12.11. This program assumes a fixed-effects model, so that if some effects are random the F ratios on the printout should be ignored. From Table 12.11, we have the following decisions.

(*a*) For particular temperature/time combinations,

$$H_0:\ \alpha\beta_{ij} = 0, \text{ all } i, j \qquad H_a:\ \text{some } \alpha\beta_{ij} \neq 0$$

we reject H_0, since $29.66 > 8.53$. That is, the temperature and time factors appear to interact to give atypical yields for particular combinations of the two.

(*b*) For the temperature effect alone,

$$H_0:\ \beta_i = 0, \text{ all } i \qquad H_a:\ \text{some } \beta_i \neq 0$$

we reject H_0, since 30.4 exceeds 8.53. Thus, the yield appears to be affected by varying the temperature alone.

(*c*) For the time factor by itself,

$$H_0:\ \alpha_j = 0, \text{ all } j \qquad H_a:\ \text{some } \alpha_j \neq 0$$

we reject H_0, since $24.06 > 8.53$. We conclude that changing only the bath time affects the yield also.

Table 12.11 Printout for ANOVA for the Data of Example 12.6

TABLE II

SOURCE	SUM SQ	D OF F	MEAN SQ	F RATIO
ROWS	3321125.0000	1	3321125.0000	30.3993
COLS	2628125.0000	1	2628125.0000	24.0561
INTER	3240125.0000	1	3240125.0000	29.6579
SUBTOT	9189375.0000	3		
ERROR	1748000.0000	16	109250.0000	
TOTAL	10937375.0000	19		

Overall, it would appear that both the time in the bath and the temperature of the bath individually affect the yield, but appearances can be deceiving. One thing that is not deceiving is the fact that there appears to be a significant interaction between the two. Scanning the data, it would appear that for a 60-minute bath, changing the temperature from 100 to 120°C has no effect, whereas at 120°C, changing the bath time from 20 to 60 minutes has no effect. We make these comments to indicate that it may be difficult to explain significant main effects when a significant interaction is present. The resolution of this apparent contradiction is left to you in Exercise 12.22.

12.8 SUMMARY

In this chapter we have considered extensions of the analyses of Section 11.2. In addition, we looked at some of the things that we must consider in analyzing multifactor experiments. Among these are the concepts of fixed effects and random effects. A factor is a fixed effect if the only levels of interest are those used in the experiment. A factor is a random effect if the levels used in the experiment represent a sample of all possible levels. The main difference between the two is in the calculation of the F ratios. If all factors are fixed effects, then all other MSs are compared with the MSE. If some factors are random effects, then you must check the E(MS)s to determine how the F ratios should be calculated.

From the design standpoint we analyzed only crossed, or factorial, designs (the randomized block has both factors crossed). There are other possible designs, such as the nested design, which is discussed in the chapter appendix. The randomized block design is an attempt to refine the error variance by randomly selecting blocks of subjects (homogeneous within blocks) and randomly assigning treatment levels to subjects within each block. In trying to reduce the estimate of the error variance by blocking, we sacrifice degrees of freedom. We also could block on more than one variable at once. A one-factor experiment with blocking on two variables is called a Latin square, and one that blocks on three variables is a Graeco-Latin square. For treatments of Latin and Graeco-Latin squares, see Guenther (1) or Neter and Wasserman (2).

In terms of analysis, the one-factor analysis is an extension of the t test for two independent random samples. The randomized block analysis is an extension of the paired t test, but you should note that it cannot be used for a before/after type of experiment, since treatments cannot be randomly assigned within blocks.

For two-factor models with replication, the sums of squares for the main effects (A and B) compare the corresponding column and row means. The SSE is calculated by comparing items within each cell and then pooling the within-cell comparisons. The SSAB compares cell means after they have been corrected for any row and column differ-

Table 12.12 Computational Formulas for One- and Two-Factor Crossed Designs

One-Factor Model		
Source	Computational Form for SS	df
A	$\sum_j \frac{\left(\sum_i X_{ij}\right)^2}{n_j} - \text{CF}$	$c - 1$
Error	$\sum_j \sum_i X_{ij}^2 - \sum_j \frac{\left(\sum_i X_{ij}\right)^2}{n_j}$	$\sum_j n_j - c$
Total	$\sum_j \sum_i X_{ij}^2 - \text{CF}$	$\sum_j n_j - 1$
	$\left[\text{CF} = \frac{\left(\sum_j \sum_i X_{ij}\right)^2}{\sum n_j}\right]$	

Two-Factor Model		
Source	Formula for SS	df
A	$\frac{1}{rn} \sum_j \left(\sum_i \sum_k X_{ijk}\right)^2 - \text{CF}$	$c - 1$
B	$\frac{1}{cn} \sum_i \left(\sum_j \sum_k X_{ijk}\right)^2 - \text{CF}$	$r - 1$
AB	SSTO − SSA − SSB − SSE	$(r - 1)(c - 1)$
Error	$\sum_j \sum_i \sum_k X_{ijk}^2 - \frac{1}{n} \sum_j \sum_i \left(\sum_k X_{ijk}\right)^2$	$rc(n - 1)$
Total	$\sum_j \sum_i \sum_k X_{ijk}^2 - \text{CF}$	$rcn - 1$
	$\left[\text{CF} = \frac{1}{rcn}\left(\sum_i \sum_j \sum_k X_{ijk}\right)^2\right]$	

ences. In all cases SSTO is calculated by comparing the individual observations with the overall mean.

All our procedures assumed the following:

1. Random samples in some form
2. Observations in each cell represent a sample from a normally distributed population
3. An additive model
4. $\epsilon_{ij} \sim N(0, \sigma^2)$ for all i, j (k for two factors)

In addition, the randomized block design assumes

5. No interaction; that is, $\alpha\beta_{ij} = 0$ for all i, j, and a fixed effect, say A, implies $\sum_j \alpha_j = 0$.

Tables 12.12 and 12.13 summarize the formulas and appropriate test statistics. Rejection of H_0 is implied in all cases only if the statistic is too large. For a one-factor model the test statistic is MSA/MSE for both fixed effects and random effects.

REVIEW EXERCISES

Level One

12.1 Differentiate between fixed and random effects.

12.2 How did Cochran's theorem come into play in this chapter?

Table 12.13 Test Statistics for a Two-Factor Model

	Test Statistic for All Effects	
Factor	Fixed	Random
A	$\frac{\text{MSA}}{\text{MSE}}$	Varies
B	$\frac{\text{MSB}}{\text{MSE}}$	Varies
AB	$\frac{\text{MSAB}}{\text{MSE}}$	$\frac{\text{MSAB}}{\text{MSE}}$

12.3 A food company that wants to put a fruit salad on the market experiments to determine what levels they should use for the quantities of apples and bananas in the basic mix. Two levels are tested for each and replicated twice. What do you recommend? ($\alpha = 0.05$)*

		Bananas	
		B_1	B_2
Apples	A_1	21 23	19 21
	A_2	18 16	12 10

12.4 Evaluate the assumptions associated with your analysis in Exercise 12.3.

12.5 Photographic technicians run an experiment involving both time and temperature in developing a negative. For each experimental picture of the same object the gamma density is then measured. Do time and temperature interact ($\alpha = 0.05$)? Assume fixed effects.*

		Time	
		8 seconds	10 seconds
Temperature	60°	2.7 2.9	4.3 4.7
	70°	3.0 3.4	5.6 5.4

12.6 Evaluate your assumptions in Exercise 12.5.

12.7 Do you think a fixed-effects analysis is appropriate for Exercise 12.5? Redo your analysis assuming random effects and explain any differences.

*Our special thanks to K. Dunning for allowing us to use these problems.

12.8 You, as a market researcher, have undertaken to determine families' expected vacation expenditures. You have trained 10 interviewers, one under 25 and one over 25 from each of five different colleges, and, as a pilot study, had each one interview four families of similar size and income level. Based on the interviews with the 40 families, the interviewers presented the following classifications of family vacation expenditures.

	College 1	College 2	College 3	College 4	College 5
$\leqslant 25$	261	263	312	301	305
	239	258	308	274	333
	286	272	329	316	318
	255	248	289	290	342
> 25	289	265	307	317	315
	253	261	293	303	291
	252	282	298	307	276
	268	263	282	321	296

Using ANOVA, does there appear to be any interviewer bias due to college, age, or particular combinations of the two? The use of a computer is recommended for this exercise. Assume fixed effects.

12.9 In an attempt to determine whether there are differences in companies' R&D expenditures due to company size, type of industry, or both, six industries were selected (steel, oil, rubber, toy, aircraft, and auto) and three sales categories were defined ($1 billion or more, $500 million to $1 billion, and under $500 million). Then, three companies were randomly selected from each industry-sales category and their R&D expenditures as a percent of gross sales were recorded. The percentages were then transformed so as to satisfy the ANOVA assumptions. The summarized results are as follows:

$$\sum_i \sum_j \sum_k X_{ijk} = 1432 \qquad \sum_i \sum_j \sum_k X_{ijk}^2 = 68{,}965$$

$$\sum_j \Big(\sum_i \sum_k X_{ijk}\Big)^2 = 404{,}682.2 \qquad \sum_i \Big(\sum_j \sum_k X_{ijk}\Big)^2 = 721{,}672.767$$

$$\sum_i \sum_j \Big(\sum_k X_{ijk}\Big)^2 = 166{,}058.52 \qquad r = n = 3 \qquad c = 6$$

Analyze the data by ANOVA and indicate which differences exist, if any.

12.10 A school district concerned about equality of educational opportunity in its high schools compares standard college entrance test scores for students in each of its three schools. The students in each school were randomly selected.

School 1	School 2	School 3
510	600	650
480	450	420
550	540	580
500		

Using $\alpha = 0.05$, should they remain concerned?

12.11 A blood bank is considering the purchase of an additional autoanalyzer (which does blood types and several other tests from a blood sample). There are three models on the market that might be satisfactory. The analyst wishes to test the speed of each model in order to make a comparison based on this factor. The following data were collected from 12 blood samples allocated randomly to the three machines. Which autoanalyzer should the analyst recommend based on these random samples of the times required to process blood samples ($\alpha = 0.05$)? Assume that the times are normally distributed.*

#1	#2	#3
8	9	12
5	6	8
3	4	6
6	7	10

12.12 In an experiment evaluating four procedures for testing tensile strength, five materials were split into four equal segments each. One piece of each type, chosen at random, was tested by each procedure. Use the following results to test for differences

*This problem is an adaptation of an exercise in A. Hughes and D. Grawoig, *Statistics: A Foundation for Analysis* (Reading, MA: Addison-Wesley Pub. Co., 1971), p. 305.

between the procedures at $\alpha = 0.05$. The distributions are known to be normal.

Material	Procedure 1	2	3	4
A	10.1	11.4	9.8	12.2
B	14.3	13.7	11.4	15.6
C	9.2	10.3	9.3	10.9
D	20.4	17.6	15.4	20.1
E	6.7	7.0	6.9	7.2

12.13 An experiment is conducted on the melting point of hydroquinone. Two operators and two types of thermometers are used, with three observations taken for each operator-thermometer combination. The results are as follows. Test for differences using $\alpha = 0.05$.

	Thermometer 1	Thermometer 2
Operator 1	2 2 2	6 7 5
Operator 2	6 6 3	5 7 9

12.14 In an attempt to evaluate the protein content of three types of soya beans, three adjacent plots of land were selected in each of three sections of the country: the Midwest, Northwest, and South. In each section, each type of soya bean was randomly assigned to a plot and planted. The same type of fertilizer was used in all cases. The following data correspond to pounds of protein yielded per acre after coding.

	Type I	Type II	Type III
Midwest	4	-1	3
Northwest	6	0	1
South	3	2	2

(*a*) Identify the type of experiment.
(*b*) Present an analysis of variance and test the indicated hypothesis at $\alpha = 0.05$.
(*c*) What assumption had to be made for the test in part (*b*)?

12.15 In an advertising experiment, a researcher wanted to test product recognition time for four different television ads for a new product. Sixteen people were randomly selected, and four were then randomly assigned to view each ad. Three people failed to recognize the product and were dropped from the experiment. The results in seconds are given here.

Ad 1	Ad 2	Ad 3	Ad 4
4	7	10	11
3	5	8	12
5	8	9	10
	9		

(*a*) Present an analysis of variance table.
(*b*) Using $\alpha = 0.01$, can one conclude that the ads have different recognition times?

12.16 Complete the following analysis-of-variance table for a two-way design that had replication with fixed effects.

Source	SS	df	MS	*F* Calc
A		2		8.50
B		3		
AB	1200			
Error			50	
Total	5000	47		

12.17 Three types of calculators, A, B, and C, were checked to determine the number of seconds required to perform a certain computation. The results are given here. Using a significance level of 0.05, test the hypothesis of equality of mean times for the calculators.*

*Our special thanks to K. Dunning for allowing us to use this problem.

A	B	C
1	5	11
4	4	7
1	8	9
	9	9
	4	

12.18 Life tests of a product were run for four different manufacturers. Since the life of the product may be affected by the type of usage, three items of each brand were selected and randomly allocated to each of three different usages. The results, in coded units, are as follows:

	Manufacturer			
Usage	A	B	C	D
1	14	2	1	8
2	10	5	6	10
3	15	8	5	12

Perform the appropriate analysis and test for differences among the manufacturers at $\alpha = 0.05$.

12.19 For which of the preceding randomized block designs does it appear that blocking was necessary? Use $\alpha = 0.05$ and explain why or why not in each case.

Level Two

12.20 Prove that $\Sigma (X_i - \overline{X})(\overline{X} - \mu) = 0$.

12.21 Refer to Example 12.4 and show that the MSE used in it has an expected value of $\sigma^2 + 0.2 \sum_j \sum_i (\alpha\beta_{ij})^2 + 0.4 \sum_i \beta_i^2$.

12.22 Refer to Example 12.6 and analyze it as a random-effects model. Explain any differences in your conclusions.

12.23 Determine the expected mean squares for a two-factor model with replication if A is fixed and B is random.

12.24 Determine the expected mean squares for a two-factor replicated design with A random and B fixed.

12.25 Refer to Exercise 12.8 and analyze the results if all effects are assumed to be random. Which analysis do you think is the appropriate one?

12.26 Refer to Exercise 12.10 and determine a 95% confidence interval for the difference in mean test scores betweens schools 1 and 2.

12.27 For a random-effects model, it is possible to estimate the variance contribution of the factor means when H_0 is rejected by looking at the E(MS)s and then doing the necessary arithmetic. Derive the formula for doing so in a one-factor design with $n_j = n$ for all j.

12.28 Assume a random effects model in Exercise 12.17, and use your formula from Exercise 12.27 to estimate the variance contribution to computation time due to type of calculator.

12.29 What is the expected mean square for B in a random-effects model if the levels of A used in the experiment represent a random sample of five out of a population of 20?

Level Three

12.30 Prove that for a one-factor model, $E(\text{MSE}) = \sigma^2$.

12.31 Prove that $E\left[\sum_j \sum_i \alpha_j(\bar{\epsilon}_{\cdot j} - \bar{\epsilon}_{\cdot\cdot})\right] = 0$.

12.32 Refer to Exercise 12.21. Determine the expected value of the MSE in a general situation in which a one-factor model was used by mistake when a two-factor was applicable. (Assume fixed effects.)

12.33 The procedure for estimating the variance contribution of a factor in a one-way random-effects design can be extended to higher-order designs. Assume that the effects in Exercise 12.9 are random, and estimate the variance contribution of any that are significant.

12.34 Given the following set of summarized data:

Sample from Population	$\overline{x}_i$	s_i^2	n_i
1	32.4	10.8	15
2	24.7	8.6	22
3	33.6	9.2	13
4	32.9	11.7	27
5	25.2	10.1	31

Perform an analysis of variance to test for differences in the population means at $\alpha = 0.05$.

APPENDIX

12.9 OTHER DESIGNS

12.9.1 Higher-Order Designs

In this section we merely mention extending the analysis of the two-factor crossed design to handle a crossed design involving three factors. The same concepts carry through to even higher-order designs, but the arithmetic becomes increasingly more cumbersome. For a three-factor crossed design the model is given by

$$X_{ijkm} = \mu + \alpha_j + \beta_i + \alpha\beta_{ij} + \gamma_k + \alpha\gamma_{jk} + \beta\gamma_{ik} + \alpha\beta\gamma_{ijk} + \epsilon_{ijkm} \tag{12.14}$$

where $j = 1, \ldots, a$; $i = 1, \ldots, b$; $k = 1, \ldots, c$; $m = 1, \ldots, n$.

For testing purposes we will have seven sets of hypotheses, one for each subscripted effect other than the error term. The approach for determining the distributions of our test statistics follows the same pattern used in all the previous models. That is, SSTO is partitioned and Cochran's theorem applied.

The assumptions required for determining the distributions of the SSs for this model are:

1. Independent random samples of size n from each of the abc populations, that is, a random sample within each cell
2. ϵ_{ijkm} independently distributed as $N(0, \sigma^2)$
3. An interval level of measurement throughout

In addition, for the fixed-effects model it can be easily shown that all effects sum to zero over any of the subscripts i, j, or k.

Table 12.14 ANOVA Table for a Three-Factor Crossed Design with All Effects Fixed

Source	df	SS	MS	E(MS)	F Calc
A	SSA	$a - 1$	MSA	$\sigma^2 + bcn \frac{\Sigma\, \alpha_j^2}{a - 1}$	$\frac{\text{MSA}}{\text{MSE}}$
B	SSB	$b - 1$	MSB	$\sigma^2 + acn \frac{\Sigma\, \beta_i^2}{b - 1}$	$\frac{\text{MSB}}{\text{MSE}}$
AB	SSAB	$(a - 1)(b - 1)$	MSAB	$\sigma^2 + cn \frac{\Sigma\,\Sigma\, (\alpha\beta_{ij})^2}{(a - 1)(b - 1)}$	$\frac{\text{MSAB}}{\text{MSE}}$
C	SSC	$c - 1$	MSC	$\sigma^2 + abn \frac{\Sigma\, \gamma_k^2}{c - 1}$	$\frac{\text{MSC}}{\text{MSE}}$
AC	SSAC	$(a - 1)(c - 1)$	MSAC	$\sigma^2 + bn \frac{\Sigma\,\Sigma\, (\alpha\gamma_{jk})^2}{(a - 1)(c - 1)}$	$\frac{\text{MSAC}}{\text{MSE}}$
BC	SSBC	$(b - 1)(c - 1)$	MSBC	$\sigma^2 + an \frac{\Sigma\,\Sigma\, (\beta\gamma_{ik})^2}{(b - 1)(c - 1)}$	$\frac{\text{MSBC}}{\text{MSE}}$
ABC	SSABC	$(a - 1)(b - 1)(c - 1)$	MSABC	$\sigma^2 + n \frac{\Sigma\,\Sigma\,\Sigma\, (\alpha\beta\gamma_{ijk})^2}{(a - 1)(b - 1)(c - 1)}$	$\frac{\text{MSABC}}{\text{MSE}}$
Error	SSE	$abc(n - 1)$	MSE	σ^2	
Total	SSTO	$abcn - 1$			

Partitioning the total sum of squares, we obtain

$$\text{SSTO} = \text{SSA} + \text{SSB} + \text{SSAB} + \text{SSC} + \text{SSAC} + \text{SSBC} + \text{SSABC} + \text{SSE}$$

We assume that a computer would be used for the computations in a three-factor design. If not, then see Guenther (2), Chapter 5, for the appropriate computational forms.

Table 12.14 gives the ANOVA table for a three-factor crossed design with all fixed effects. For each of the seven tests the decision rule is to reject H_0 if the appropriate F_{calc} exceeds $f_{v_1, v_2, \alpha}$, where v_1 is the numerator degrees of freedom and v_2 is the degrees of freedom for error, given by $v_2 = abc(n-1)$.

For a random-effects model with a three-factor crossed design, all the E(MS)s change, and so, except for MSABC/MSE, all F_{calc}s may need to be altered. The expected mean squares for the random-effects model are given in Table 12.15. For numerical examples of three-factor models, we recommend Chapter 20 of Neter and Wasserman (3).

A problem can arise in constructing the test statistics for the main effects in the random model. Note that for factor A in Table 12.15, a

Table 12.15 Expected Mean Squares for a Three-Factor Crossed Design with All Effects Random

Factor	E(MS)
A	$\sigma^2 + n\sigma^2_{\text{ABC}} + nc\sigma^2_{\text{AB}} + nb\sigma^2_{\text{AC}} + nbc\sigma^2_{\text{A}}$
B	$\sigma^2 + n\sigma^2_{\text{ABC}} + nc\sigma^2_{\text{AB}} + na\sigma^2_{\text{BC}} + nac\sigma^2_{\text{B}}$
AB	$\sigma^2 + n\sigma^2_{\text{ABC}} + nc\sigma^2_{\text{AB}}$
C	$\sigma^2 + n\sigma^2_{\text{ABC}} + nb\sigma^2_{\text{AC}} + na\sigma^2_{\text{BC}} + nab\sigma^2_{\text{C}}$
AC	$\sigma^2 + n\sigma^2_{\text{ABC}} + nb\sigma^2_{\text{AC}}$
BC	$\sigma^2 + n\sigma^2_{\text{ABC}} + na\sigma^2_{\text{BC}}$
ABC	$\sigma^2 + n\sigma^2_{\text{ABC}}$
Error	σ^2

suitable test statistic does not exist if all interactions are significant. When this happens, A cannot be compared directly with either AB or AC (E[MSAB] does not contain σ^2_{AC}, and E[MSAC] does not contain σ^2_{AB}). One approach to resolving this is to form a linear combination of MSs for the denominator such that the E(MS) will contain all the terms in E(MSA) except $nbc\sigma^2_A$. In this case, the linear combination given by MSAC + MSAB − MSABC satisfies the criterion. The statistic, MSA/(MSAC + MSAB − MSABC) is called a *pseudo* or *quasi F* ratio and will have an approximate F distribution if H_0 is true. The degrees of freedom for the denominator are given by

$$df_{den} = \frac{(\text{MSAC} + \text{MSAB} - \text{MSABC})^2}{\dfrac{(\text{MSAC})^2}{(a-1)(c-1)} + \dfrac{(\text{MSAB})^2}{(a-1)(b-1)} + \dfrac{(\text{MSABC})^2}{(a-1)(b-1)(c-1)}}$$

For a general discussion of quasi-F ratios, see Section 20.5 in Neter and Wasserman (3).

12.9.2 Nested Designs

In this section we cover a two-factor design with the B-factor effects nested within the A effects. Our purpose is to acquaint the student with the computational changes necessary in nested designs. Our basic model will be

$$X_{ijk} = \mu + \alpha_j + \beta_i(\alpha_j) + \epsilon_{ijk} \qquad (12.15)$$

The notation $\beta_i(\alpha_j)$ means that the B effects are nested within the A effects and cannot be compared across levels of A. Note that this concept is exactly what we have with the error term. That is, the error effects in any model are nested within the highest-order interaction and we cannot compare observations between cells. The computations for determining SSB(A) follow the same idea as those for determining SSE [note this is actually SSE(AB)]. The idea is to calculate sums of squares for B within each level of A and then to pool across the A levels. That is,

$$\text{SSB(A)} = \sum_j \sum_i \sum_k (\overline{X}_{ij.} - \overline{X}_{.j.})^2$$

or in the computational form

$$\text{SSB(A)} = \frac{1}{n} \sum_j \sum_i \left(\sum_k X_{ijk}\right)^2 - \frac{1}{rn} \sum_j \left(\sum_i \sum_k X_{ijk}\right)^2 \qquad (12.16)$$

Table 12.16 Expected Mean Squares for the Two-Factor Model with *B* Nested in *A*

Source	E(MS) for A, B Fixed	E(MS) for A, B Random
A	$\sigma^2 + \frac{nr}{c-1} \sum_j \alpha_j^2$	$\sigma^2 + n\sigma^2_{B(A)} + nr\sigma^2_A$
B(A)	$\sigma^2 + \frac{n}{r-1} \sum_j \sum_i [\beta_i(\alpha_j)]^2$	$\sigma^2 + n\sigma^2_{B(A)}$
Error	σ^2	σ^2

The SSTO, SSA, and SSE are calculated exactly as before.

The table of expected mean squares for this model for both fixed effects and random effects is shown in Table 12.16. The degrees of freedom for a nested model are given by $df_A = c - 1$; $df_{B(A)} = c(r - 1)$, $df_E = rc(n - 1)$, $df_{TO} = rcn - 1$, where, in this case, r = number of levels of B within a level of A.

EXAMPLE 12.7 To evaluate methods for training new salespeople, a sales manager randomly selected six salespeople and three methods, then randomly assigned two salespeople to each method. At the end of the training period the salespeople were randomly assigned to territories. After three months the manager calculated average monthly sales records for each salesperson. Use these data, which are given here, to test for differences in both training methods and salesperson/territory combinations at $\alpha = 0.05$.

Method A_1		Method A_2		Method A_3	
Salesperson		Salesperson		Salesperson	
1	2	3	4	5	6
14	17	18	17	26	18
18	20	25	23	30	20
12	16	20	25	19	30

SOLUTION The salesperson/territory combinations are obviously nested within training methods, and we assume that both are random effects. The model is

$$X_{ijk} = \mu + \alpha_j + \beta_i(\alpha_j) + \epsilon_{ijk}$$

where $j = 1, 2, 3; i = 1, 2; k = 1, 2, 3$. Our hypotheses are, for the random-effects model,

1. $H_0\colon \sigma^2_{B(A)} = 0 \qquad H_a\colon \sigma^2_{B(A)} \neq 0$
2. $H_0\colon \sigma^2_A = 0 \qquad H_a\colon \sigma^2_A \neq 0$

For (1) we will reject H_0 if MSB(A)/MSE $> f_{3,\,12,\,0.05} = 3.49$. Since this is a random-effects model, the test for (2) is contingent upon the results of the first test. If H_0 is accepted in (1), then we may test versus the MSE; otherwise we must use MSB(A).

The necessary calculations are

Salespeople:	1	2	3	4	5	6
Totals:	44	53	63	65	75	68
Method totals:	97		128		143	
Overall total:	368					

$$\Sigma\,\Sigma\,\Sigma\, X^2_{ijk} = 7962$$

Then,

$$\text{SSTO} = 7962 - \frac{(368)^2}{18} = 438.44$$

$$\text{SSA} = \frac{1}{6}(97^2 + 128^2 + 143^2) - \frac{(368)^2}{18} = 183.44$$

$$\text{SSB(A)} = \frac{1}{3}(44^2 + 53^2 + \cdots + 68^2) - \frac{1}{6}(97^2 + 128^2 + 143^2)$$

$$= 7729.33 - 7707.00 = 22.33$$

$$\text{SSE} = 7962 - 7729.33 = 232.67$$

Thus, the ANOVA table is

Source	SS	df	MS	E(MS)
A	183.44	2	91.72	$\sigma^2 + 3\sigma^2_{B(A)} + 6\sigma^2_A$
B(A)	22.33	3	7.44	$\sigma^2 + 3\sigma^2_{B(A)}$
Error	232.67	12	19.39	σ^2
Total	438.44	17		

For our tests we first test hypothesis set (1) and cannot reject H_0, since $7.44/19.39 = 0.384 < 3.49$. There appears to be no significant contribution to the variance from salesperson/territory differences. Therefore, to test methods we will use the MSE. Our test for (2) is to reject H_0 if MSA/MSE $> f_{2,\,12,\,0.05} = 3.89$. Since $91.72/19.39 = 4.73 > 3.89$, we reject H_0 and conclude that there is a significant contribution to the variance due to the different methods; that is, the methods appear to differ. In this instance, we used the MSE in the denominator to build up our degrees of freedom, since MSB(A)/MSE was quite small and could not have been rejected at any reasonable value for α. Had we used MSA/MSB(A) the ratio would have been 12.33, which exceeds $f_{2,\,3,\,0.05} = 9.55$, and so we reach the same conclusion.

12.10 COMPARING TREATMENT MEANS AFTER H_0 IS REJECTED

In the fixed-effects model only, if we have rejected H_0 for a particular factor, we may be interested in trying to determine exactly which means are different and which appear to be similar. One method that we could use to do so would be to run independent-samples t tests on all possible pairs of the factor means. This approach suffers from the problem discussed in the introduction, namely, that the probability of rejecting for at least one pair increases dramatically with the number of comparisons. If you decide to do so anyway, the test statistic for comparing two column means μ_i and μ_j would be (MSE = error mean square)

$$\frac{\overline{X}_i - \overline{X}_j}{\sqrt{\text{MSE}(2/rn)}}$$

for a two-way design. Confidence intervals for the difference can be obtained from this statistic in the usual manner.

Two approaches for comparing means after H_0 has been rejected that keep the overall α less than or equal to the stated α are the Tukey method and Duncan's multiple-range procedure. These are presented briefly in the following sections. For both methods, the sample sizes must be equal for all means being compared.

12.10.1 The Duncan Multiple-Range Test

The rejection of the null hypothesis for a main effect in analysis of variance does not permit us to make any inferences about the individual means. We could, of course, perform t tests for the difference of means for all the pairs available, providing alpha can be adjusted. An alternative to such a procedure is to use the Duncan multiple-range test. This test is primarily concerned with the actual differences between various mean pairs and how each difference compares with a term known as the *least significant range* (LSR).

To use the Duncan multiple-range test, we must obtain the standard error of the mean, $s_{\overline{X}}$. For our situation $s_{\overline{X}}$ is given by

$$s_{\overline{X}} = \sqrt{\frac{\text{MSE}}{n_{\overline{X}}}} \tag{12.17}$$

where $n_{\overline{X}}$ is the number of observations on which each $\overline{X}$ is based. The MSE is, of course, the mean square error taken directly from the ANOVA table. The $n_{\overline{X}}$ is the common sample size for each of our samples. If we have unequal sample sizes, we would need to use an alternative procedure, such as the Scheffé multiple contrast test [see Guenther (2)].

The LSR is determined as

$$\text{LSR} = r s_{\overline{X}} \tag{12.18}$$

The r in Equation 12.18 is found in Table A.16, Standardized Ranges for Duncan's test. The proper values are obtained from the table based on the level of α, n_2, and the number of means inclusive between the ordered pairs for which we seek a difference. For example, the number of ordered pairs for $\overline{X}_v$ and $\overline{X}_w$, where $v > w$, would be equal to $(v - w) + 1$. The value for n_2 is equal to the degrees of freedom for error in the analysis-of-variance table.

$$n_2 = \text{df}_\text{E} \tag{12.19}$$

In using the Duncan multiple-range test, we are actually performing a type of t test because the values obtained from Table A.16 are functions of t values. Example 12.8 illustrates the procedure.

EXAMPLE 12.8 Students from four high schools were given a mathematical proficiency test to see whether the students' mathematical abilities were the same. Six students were randomly selected from each school and their scores on a 40-point scale were recorded. The ANOVA for equality of means rejected the null hypothesis at the 0.01 level of significance. The mean square error was 7.3085 and the sample means were $\bar{x}_{\cdot \text{east}} = 21.67$, $\bar{x}_{\cdot \text{west}} = 30.67$, $\bar{x}_{\cdot \text{north}} = 32.0$, and $\bar{x}_{\cdot \text{south}} = 22.5$. What can we conclude at the 0.01 level of significance?

SOLUTION Initially we will order the means according to their magnitude. Even though the identity of each mean is lost, it is necessary to return to this information for our conclusion. The ordered means are

$$\bar{x}_{(1)} = 21.67 \qquad \bar{x}_{(2)} = 22.5 \qquad \bar{x}_{(3)} = 30.67 \qquad \bar{x}_{(4)} = 32.0$$

and the standard error is

$$s_{\bar{X}} = \sqrt{\frac{7.3085}{6}} = \sqrt{1.218} = 1.104$$

and

$$n_2 = (6 - 1)4 = 20$$

From Table A.16 the ranges are

Number of Means Between	4	3	2
r	4.33	4.22	4.02

Thus, the LSR values are

	4	3	2
LSR	4.780	4.659	4.438

and evaluating the differences, we have

4 $\quad \bar{x}_{(4)} - \bar{x}_{(1)} = 32.0 - 21.67 = 10.33 > 4.780^*$

3 $\quad \bar{x}_{(4)} - \bar{x}_{(2)} = 32.0 - 22.5 = 9.5 > 4.659^*$

$\quad \bar{x}_{(3)} - \bar{x}_{(1)} = 30.67 - 21.67 = 9.0 > 4.659^*$

2 $\quad \bar{x}_{(4)} - \bar{x}_{(3)} = 32.0 - 30.67 = 1.33 < 4.438$

$\quad \bar{x}_{(3)} - \bar{x}_{(2)} = 30.67 - 22.5 = 8.17 > 4.438^*$

$\quad \bar{x}_{(2)} - \bar{x}_{(1)} = 22.5 - 21.67 = 0.83 < 4.438$

The significant values are denoted by an asterisk. This test would indicate that a difference exists between pairs 4 - 1, 4 - 2, 3 - 1, and 3 - 2, but not between 4 - 3 and 2 - 1. Interpreting these results with respect to our high schools, we could say that West and North were comparable schools with respect to student performance on the mathematical proficiency test and that they both were better than East and South, which were comparable to each other.

The Duncan multiple-range test enables us to further evaluate our data when the test of means is rejected. Its use, however, is restricted to samples of equal size and to pairwise comparisons. An alternative and more general approach is the Tukey method.

12.10.2 The Tukey Method of Multiple Comparisons

Tukey's method is more general than Duncan's because it can be used to compare more than two means at one time. Because of this, it is often preferred over Duncan's procedure. Tukey's method, similar to Duncan's, is based on the studentized range statistic, where, if Y_i ($i = 1, \ldots, r$) are independent and $N(\mu, \sigma^2)$ and s^2 is an unbiased estimate of σ^2, then

$$q_{r, n_2} = \frac{\max Y - \min Y}{s}$$

In our case r is the number of means to be compared and n_2 is the degrees of freedom error. Percentage points for q are given in Table A.17.

The procedure, as with Duncan's test, is to calculate all possible absolute differences between the sample means in question and then to compare each difference to

$$q_{r, n_2, \alpha} \sqrt{\frac{\text{MSE}}{n_{\overline{X}}}} \tag{12.20}$$

where $n_{\overline{X}}$ is the same as for Duncan's procedure.

EXAMPLE 12.9 We use Tukey's procedure on the data of Example 12.8 with $\alpha = 0.01$.

SOLUTION For Example 12.8 the ranked means are 21.67, 22.50, 30.67, and 32.00, which correspond to $\bar{x}_E$, $\bar{x}_S$, $\bar{x}_W$, and $\bar{x}_N$, respectively. In addition, MSE = 7.3085, $n_2 = 20$, $r = 4$, and $n_{\overline{X}} = 6$. Thus we need to use Table A.17a to obtain $q_{4,\,20,\,0.01}$. Doing so, we find that

$$q_{4,\,20,\,0.01} = 5.02$$

and our rejection value for all differences between means is given by

$$q_{4,\,20,\,0.01} \sqrt{\frac{\text{MSE}}{n_{\overline{X}}}} = 5.02 \sqrt{\frac{7.3085}{6}} = 5.54$$

Calculating all possible differences between means, we have

$$
\begin{aligned}
32.00 - 21.67 &= 10.33 > 5.54 \Rightarrow \text{Reject } H_0 \\
32.00 - 22.50 &= 9.50 > 5.54 \Rightarrow \text{Reject } H_0 \\
32.00 - 30.67 &= 1.33 < 5.54 \Rightarrow \text{Accept } H_0 \\
30.67 - 21.67 &= 9.00 > 5.54 \Rightarrow \text{Reject } H_0 \\
30.67 - 22.50 &= 8.17 > 5.54 \Rightarrow \text{Reject } H_0 \\
22.50 - 21.67 &= 0.83 < 5.54 \Rightarrow \text{Accept } H_0
\end{aligned}
$$

Our conclusions coincide with those of Example 12.8. That is, West and North appear comparable and have significantly higher scores than East and South, which are also comparable to each other.

Both Duncan's and Tukey's procedures can be used with higher-order designs. Tukey's procedure requires the rejection of H_0 before it is used and, while Duncan's does not, we do not recommend using Duncan's otherwise, since contradictory results can occur. In addition, we recommend their use only on means for main effects, since cell means may differ due to both main effects and various interaction effects. To compare interaction effects the cell means should be corrected for main effects first.

EXERCISES

12.10.1 For the following set of data, state the model and test all effects for significance at $\alpha = 0.05$. Assume a random-effects model.

B_1			B_2			B_3		
A_1	A_2	A_3	A_4	A_5	A_6	A_7	A_8	A_9
15	20	19	20	26	31	14	10	15
12	19	24	27	28	22	10	13	17
18	17	21	21	24	27	11	8	12
11	22	18	23	32	29	9	6	14

12.10.2 Assuming a fixed-effects model, analyze the following data at $\alpha = 0.01$ and state the model.

	A_1		A_2	
	C_1	C_2	C_3	C_4
B_1	8 5	7 10	14 22	11 9
	C_5	C_6	C_7	C_8
B_2	10 13	19 24	15 16	27 20

12.10.3 Refer to Exercise 12.10.2 and analyze differences between the means for factors A and B, where appropriate, at $\alpha = 0.05$ using Duncan's procedure.

12.10.4 Refer to Exercise 12.10.3 and use Tukey's procedure.

12.10.5 Refer to Review Exercise 12.8 and use Tukey's procedure to evaluate factor means for the main effects at $\alpha = 0.05$ where applicable.

12.10.6 Refer to Review Exercise 12.12 and evaluate the means for both materials and procedures using Duncan's test at $\alpha = 0.05$.

References

1. Davies, Owen L. *Design and Analysis of Industrial Experiments.* 2d. ed. London: Oliver and Boyd, 1963.
2. Guenther, William C. *Analysis of Variance.* Englewood Cliffs, New Jersey: Prentice-Hall, Inc., 1964.
3. Neter, John, and William Wasserman. *Applied Linear Statistical Models.* Homewood, Ill.: Richard D. Irwin, Inc., 1974.
4. Scheffé, Henry. *The Analysis of Variance.* New York: John Wiley & Sons, Inc., 1959.
5. Winer, B. J. *Statistical Principles in Experimental Design*. New York: McGraw-Hill Co., 1962.

Chapter 13
c-Sample Inference with Less Restrictive Assumptions

KEY WORDS AND TERMS

Test of Several Binomials
Test of Several Multinomials
Contingency Table
Kruskal-Wallis Test
Friedman's Test

13.1 INTRODUCTION

In Chapter 12 we developed tests for comparing multiple samples simultaneously under the restrictions that all samples came from normally distributed populations and that the numerical values associated with the outcomes were of an interval level of measurement. The procedures of Chapter 12 basically were extensions of the two-sample tests for normal distributions of Chapter 11.

In this chapter we develop procedures that are alternatives to the analysis-of-variance F tests for one-way and randomized-block designs. If we believe that the populations of interest are decidedly nonnormal or if the level of measurement is questionable, these procedures should be used. These tests are extensions of the distribution-free procedures presented in Chapter 11.

With a one-factor experimental design and data of an ordinal level, we use the Kruskal-Wallis procedure, which is an extension of the Mann-Whitney U test. For randomized-block designs with at least ordinal data within blocks, Friedman's test is the appropriate one. The Friedman procedure reduces to the sign test of Chapter 11 if only two treatments are being compared.

Distribution-free procedures that can be applied to higher-order experimental designs have been developed. In many cases they are cumbersome to use and beyond the scope of this text. If the need for them arises, see Bradley (1) or Gibbons (2).

13.2 TESTS FOR PROPORTIONS

We have divided this section into two parts, the first part concerned exclusively with comparing binomial distributions, and the second dealing with the general case of comparing several multinomials. The reason for the division is that most students find the binomial case much easier to understand. Once you have mastered the binomial case, the extension to multinomials will generally cause little, if any, difficulty.

13.2.1 Test of Several Binomials

In this section we present an analog of the ANOVA F test for a randomized design when the populations being considered are dichotomous. The assumptions of the procedure are similar to those of Section 11.3; they are restated below for ease in reference.

1. Independent random samples are selected from each of c populations.
2. The populations are very large or sampling is with replacement.
3. The level of measurement is at least nominal so that the data may be dichotomized.

We define the following notation. Let $n_{.j}$ = size of the sample from the jth population, $n_{i.}$ = total number of items of type i in the combined sample,

$$n = \sum_{j=1}^{c} n_{.j} = \sum_{i=1}^{2} n_{i.} = \text{total number of observations}$$

and O_{ij} = number of observations in the jth sample belonging to category i, $i = 1, 2$. Then, under the given assumptions, each sample of size $n_{.j}$ constitutes a random sample from a binomial distribution. Each observation from the jth population has a probability p_{ij} of belonging to the first category and a probability $1 - p_{ij}$ of belonging to the second category. To test whether the c populations are identical, the hypotheses of interest are

$$H_0\colon\ p_{11} = p_{12} = \cdots = p_{1c} \qquad H_a\colon \text{ some } p_{ij} \text{ differ} \tag{13.1}$$

The rationale for constructing the test is identical to that presented for testing two binomials. Thus, the statistic

$$\sum_{i=1}^{2} \sum_{j=1}^{c} \frac{(O_{ij} - E_{ij})^2}{E_{ij}}$$

has an approximate χ^2 distribution with $c - 1$ degrees of freedom. In this situation we determine the E_{ij} by first estimating the common probability of category i, assuming that H_0 is true, then multiplying by the number of observations from population j, since the expected value of a binomial variable is np. This is summarized in the equation

$$E_{ij} = n_{.j}\left(\frac{n_{i.}}{n}\right) = \frac{n_{.j}n_{i.}}{n} \tag{13.2}$$

We reason that if the populations are identical, the best estimate of p_i is given by the number of times i occurs divided by the total number of trials, as was indicated in Chapter 10. The corresponding test procedure for the hypotheses of Equation 13.1 follows.

1. Select random samples from each of the c populations being considered and count the number of observations of each type in each sample.

For convenience, organize the data in a tabular format with populations as columns and categories as rows, that is, in a $2 \times c$ table.

2. Calculate the E_{ij}'s to check that all are greater than or equal to 5. If any E_{ij}'s are less than 5, larger samples should be selected.
3. Calculate $\sum_{i=1}^{2} \sum_{j=1}^{c} \frac{(O_{ij} - E_{ij})^2}{E_{ij}}$
4. Reject H_0 at α if the probability associated with the value in step 3 is too small, that is, if

$$\sum_{i=1}^{2} \sum_{j=1}^{c} \frac{(O_{ij} - E_{ij})^2}{E_{ij}} > \chi^2_{c-1,\,\alpha} \tag{13.3}$$

EXAMPLE 13.1 In an effort to determine whether a difference exists in the percentages of defective items produced by three shifts, the Nite Lite Company randomly selected, at various times, 120 items during the first shift, 150 during the second shift, and 130 during the third shift. All items were inspected, and the number of defectives was recorded by shift with the following results: 8 on the first, 24 on the second, and 18 on the third. Can it be concluded at $\alpha = 0.05$ that the percentages differ?

SOLUTION The null hypothesis is that the percentage of defectives is the same for all shifts:

$$H_0\colon\ p_{D1} = p_{D2} = p_{D3} = p_D \qquad H_a\colon\ \text{some } p_{Dj} \neq p_D$$

Table 13.1 Breakdown of Defectives by Shift for the Nite Lite Company

	Shift 1	Shift 2	Shift 3	$n_{i.}$
Number Defective	8	24	18	50
Number Good	112	126	112	350
$n_{.j}$	120	150	130	$400 = n$

The decision rule for our test is to

Reject H_0 at $\alpha = 0.05$ if

$$\sum_{i=1}^{2} \sum_{j=1}^{3} \frac{(O_{ij} - E_{ij})^2}{E_{ij}} > \chi^2_{3-1,\,0.05} = 5.99$$

For clarity, in Table 13.1 the data are reorganized into a 2 X 3 table showing the number of good items, the number of defective items, the shift during which they were produced, and the totals. Using ij to denote the cell or category in Table 13.1, our calculations are as follows [for example, $E_{11} = 120(50/400) = 15$ and $E_{22} = 150(350/400) = 131.25$]:

Table 13.2 Computations for Example 13.1

ij	O_{ij}	E_{ij}	$O_{ij} - E_{ij}$	$\frac{(O_{ij} - E_{ij})^2}{E_{ij}}$	Partial Sum
11	8	15.00	-7.00	3.27	3.27
12	24	18.75	5.25	1.47	4.74
13	18	16.25	1.75	0.19	4.93
21	112	105.00	7.00	0.47	5.40
22	126	131.25	-5.25	0.21	5.61
23	112	113.75	-1.75	0.03	$5.64 < 5.99$
	400	400.00	0.00		

We have 2 degrees of freedom, since if numbers are specified for any two cells in Table 13.2, the remaining values are automatically determined by the marginal totals.

For this problem,

$$\sum_i \sum_j \left[\frac{(O_{ij} - E_{ij})^2}{E_{ij}}\right] = 5.64$$

which is less than 5.99. Therefore, we accept H_0 and conclude that the differences in the percentage defective between shifts appear to be due to chance variation.

As you can see, the extension from comparing two binomial distributions to comparing several was simple and direct. In the next subsection we extend from comparing binomials to comparing multinomials. This is also a direct and straightforward generalization.

13.2.2 Test for Several Multinomials (Contingency Tables)

It is logical to extend the analysis of the previous section to situations in which each item from a given population can assume any one of r categories ($r \geqslant 2$). We then wish to test to see whether the probability of an observation's belonging to any particular category is the same for all populations. That is, we wish to test the hypothesis that c multinomial populations, each having r categories, are identical. In parametric form, the hypotheses of interest are

$$H_0: \begin{cases} p_{11} = p_{12} = \cdots = p_{1c} = p_1 \\ p_{21} = p_{22} = \cdots = p_{2c} = p_2 \\ \quad \vdots \qquad\qquad\qquad \vdots \quad\ \ \vdots \\ p_{r1} = p_{r2} = \cdots = p_{rc} = p_r \end{cases} \qquad H_a: \text{at least two equations do not hold} \tag{13.4}$$

It will be simpler to state these hypotheses in terms of association between categories and populations or, equivalently, between row and column classifications. The restatement is

$$H_0: \text{no association between rows and columns} \qquad H_a: \text{some association} \tag{13.5}$$

We can see that the statements in Equations 13.4 and 13.5 are equivalent. If the populations are identical, then differences between cells in any one row would be due only to random sampling and differences in the sample sizes. That is, knowing from which population an item comes would not alter the probability of its belonging to any one category, or, symbolically,

$$\text{pr}\{X \text{ belongs to category } i | \text{population } j\} = \text{pr}\{X \text{ belongs to category } i\}$$

which is the definition of independence. Figure 13.1 gives a tabular layout showing the organization of the data. This representation is also called a *contingency table*, since independence implies that the likelihood of a category is not contingent on the population. Note the similarity to the randomized-block design of the two-way analysis of variance presented in Chapter 12. This test is a distribution-free alternative to two-way ANOVA when the data are nominal.

The test of the hypotheses in Equations 13.4 and 13.5 is based on the χ^2 distribution and follows the same intuitive derivation as the previous χ^2 tests. The underlying assumptions of the procedure follow.

1. Independent random samples are selected from each of the c populations; or

1′. One large sample of size n is selected, and the items are classified by population and category.

2. The populations are very large, or else sampling is with replacement.
3. The data are at least nominal and can be classified into r mutually exclusive categories.

Figure 13.1 Representation of Data for a Contingency Table With O_{ij} Denoting the Observed Frequency for the ijth Cell

	Population					
	1	2	3	. . .	c	Row Total
	O_{11}	O_{12}	O_{13}	. . .	O_{1c}	$n_{1.}$
	O_{21}	O_{22}	O_{23}	. . .	O_{2c}	$n_{2.}$
	O_{31}	O_{32}	O_{33}	. . .	O_{3c}	$n_{3.}$
Category	.	.	.		.	.
	.	.	.		.	.
	.	.	.		.	.
	O_{r1}	O_{r2}	O_{r3}	. . .	O_{rc}	$n_{r.}$
	$n_{.1}$	$n_{.2}$	$n_{.3}$	. . .	$n_{.c}$	$n = \sum_{i=1}^{r} n_{i.} = \sum_{j=1}^{c} n_{.j}$

The difference between the assumptions in (1) and (1′) is the distinction that is generally made between the test of several multinomials (Equation 13.4) and the test for association in a contingency table (Equation 13.5). The distinction is a moot one, and the procedure of analysis is the same in either case.

Then, if these assumptions are satisfied, the statistic

$$\sum_{i=1}^{r} \sum_{j=1}^{c} \frac{(O_{ij} - E_{ij})^2}{E_{ij}}$$

will be approximately distributed as a χ^2 variable with $(r-1)(c-1)$ degrees of freedom. The degrees of freedom, $(r-1)(c-1)$, is the number of cells to which frequencies may arbitrarily be assigned before all remaining cells become automatically determined by the marginal totals. Under the null hypothesis the best estimate of p_i is $n_{i.}/n$; that is, the number of items belonging to the ith category divided by the total number of observations. Thus, the expected frequency for the ijth cell is given by

$$E_{ij} = n_{.j}\hat{p}_i = n_{.j}\left(\frac{n_{i.}}{n}\right) \tag{13.6}$$

Following our previous pattern, the test procedure for the hypotheses of either Equation 13.4 or Equation 13.5 is given here.

1. State the hypotheses of interest (in either form) and select a value for α.
2. Select random samples from each of the c populations being compared and determine O_{ij}, the observed frequency in cell ij.

2′. Or select one large random sample from the combined population and determine O_{ij}.

3. Calculate the E_{ij}'s to check that all are greater than or equal to 5. If any E_{ij} are less than 5, either select a larger sample from the appropriate population(s) or combine adjacent categories until the condition is satisfied.
4. Calculate $\sum_{i=1}^{r} \sum_{j=1}^{c} \frac{(O_{ij} - E_{ij})^2}{E_{ij}}$
5. Reject H_0 at α if

$$\sum_{i=1}^{r} \sum_{j=1}^{c} \frac{(O_{ij} - E_{ij})^2}{E_{ij}} > \chi^2_{(r-1)(c-1),\,\alpha} \tag{13.7}$$

The analysis of an $r \times c$ table is illustrated in the following example.

EXAMPLE 13.2 In a civil rights compliance review of three large companies in the same industry and city, the review team randomly selected 100 cards from the personnel files of each company. They then classified each card as that of a white male, a black male, or a female. The resulting data are summarized in Table 13.3. Do the companies appear to differ in the composition of their staffs? Use $\alpha = 0.025$.

SOLUTION Using the form indicated in Equation 13.5, the hypotheses are

$$H_0\text{: composition of employees is independent of company}$$

and

$$H_a\text{: the composition is not independent of the company}$$

with α specified to be 0.025. Since none of the observed values are less than 5 and the $n_{.j}$'s are all equal, none of the expected values will be less than 5. (Note that if any O_{ij} had been < 5, we would have calculated the E_{ij} at this point.) Thus, our decision rule is to reject H_0 at $\alpha = 0.025$ if

$$\sum_{i=1}^{3}\sum_{j=1}^{3}\frac{(O_{ij}-E_{ij})^2}{E_{ij}} > \chi^2_{(3-1)(3-1),\,0.025} = 11.14$$

The necessary calculations are given in Table 13.4 on page 492. Since 11.51 already exceeds 11.14, we can stop, and H_0 is rejected at $\alpha = 0.025$. We conclude that the percentages of employees in each category appear to differ by company.

In all situations involving the chi-square family of distributions and nominal data, we stipulated that all expected frequencies were to be

Table 13.3 Classification of Employees for Example 13.2

	Company 1	Company 2	Company 3	Total
Number of females	16 (17)	8 (17)	27 (17)	51
Number of black males	36 (31)	33 (31)	24 (31)	93
Number of white males	48 (52)	59 (52)	49 (52)	156
Total	100	100	100	300

Table 13.4 Computations for Example 13.2

ij	O_{ij}	E_{ij}	$O_{ij} - E_{ij}$	$\frac{(O_{ij} - E_{ij})^2}{E_{ij}}$	Partial Sum
11	16	17	-1	0.06	0.06
12	8	17	-9	4.76	4.82
13	27	17	10	5.88	10.70
21	36	31	5	0.81	11.51 > 11.14
22	33	31	2		
23	24	31	-7		
31	48	52	-4		
32	59	52	7		
33	49	52	-3		
	300	300	0		

greater than or equal to 5. This is particularly critical for the test of Section 13.2.1. In the present section, this condition is occasionally relaxed to the extent that some of the E_{ij}'s may be less than 5 ("some" is often interpreted to mean less than 10%), but absolutely none may be less than 1. The approach of combining adjacent categories is illustrated in Example 13.3.

EXAMPLE 13.3 In an attempt to determine whether the choice of academic programs has changed over the years, the Registrar's office at Acorn U randomly selected 100 freshman registrations for each of

Table 13.5 Classification of Students by Major and Year of Enrollment

Area of Major	Year 1968	Year 1973	Year 1978	Totals
Business	17	24	34	75
Education	26	20	14	60
Arts	32	26	20	78
Science & Engineering	23	26	26	75
Nursing	2	4	6	12
Totals	100	100	100	300

the years in question and recorded the registrant's intended major. These data are given in Table 13.5. Is there any indication of a relationship between year of registration and the area chosen, using $\alpha = 0.10$?

SOLUTION The hypotheses in question are

$$H_0\text{: no relationship} \qquad H_a\text{: some relationship}$$

and α is specified at 0.10. Checking the E_{ij}'s first, we find that $E_{5j} = 4$ for all three years. This is not a critically low value, but we will regroup anyway for illustrative purposes. Nursing is probably most closely related to the science and engineering category. Thus we will combine those two categories to obtain the revised table, Table 13.6.

Our decision rule will be to reject H_0 if $\chi^2_{\text{calc}} > \chi^2_{6,\,0.10} = 10.645$. Since the sample sizes are the same for all years, each E_{ij} will be given by $n_{i.}/3$. Our calculations are given in Table 13.7 on page 494. Since $10.825 > 10.645$, we stop at this point and reject H_0. We conclude that the choice of a major area appears to be related to the year of registration.

In concluding this section, we point out that whenever possible you should attempt to identify the members of the populations of interest in advance. By doing so you can design the sampling procedure so that you select the same number of observations from each population. There are two advantages to this. First, all calculations are simplified because, since the $n_{.j}$ will all be equal, E_{ij} will be given by $n_{i.}/c$ for all j.

Table 13.6 Revised Classification of Students from Table 13.5

	Year			
Area of Major	1968	1973	1978	Totals
Business	17	24	34	75
Education	26	20	14	60
Arts	32	26	20	78
Science, Engineering, & Nursing	25	30	32	87
Totals	100	100	100	300

Table 13.7 Computations for Example 13.3

Cell	O_{ij}	E_{ij}	$O_{ij} - E_{ij}$	$\frac{(O_{ij} - E_{ij})^2}{E_{ij}}$	Partial Sum
1, 1	17	25	-8	2.560	2.560
1, 2	24	25	-1	0.040	2.600
1, 3	34	25	9	3.240	5.840
2, 1	26	20	6	1.800	7.640
2, 2	20	20	0	0.000	7.640
2, 3	14	20	-6	1.800	9.440
3, 1	32	26	6	1.385	10.825 > 10.645

Second, studies have indicated that the χ^2 test is more powerful when the $n_{.j}$ are equal. If it is impossible to design the sampling procedure to obtain equal $n_{.j}$, then the general formula for the E_{ij} must be used.

EXERCISES

13.2.1 In a comparison of the proportion of defectives produced by four different machines, the following data were obtained: Machine A had 25 defectives in 200 items, machine B had 10 in 120 items, C had 6 in 140 items, and D had 9 defectives in 160 items. Does the quality of the items produced appear to be the same for all machines? Let $\alpha = 0.05$.

13.2.2 A market research department wanted to test whether or not the type of advertising display appeals differently to people in different socioeconomic areas. For their experiment they selected one store in each of five different socioeconomic areas in a particular city. Then they placed the same product on two distinctly different displays situated next to each other. To minimize any effect due to location, they alternated the positions of the displays each day. After six days they tabulated the number of units sold from each display for each store. These data are presented here:

	Store				
	1	2	3	4	5
Display 1	152	112	56	37	63
Display 2	94	108	79	45	137

Do the displays appear to appeal differently to different people? Let $\alpha = 0.01$.

13.2.3 In an attempt to analyze whether or not the percentage return on stocks is related to type of industry, an investor randomly selected 50 stocks each from the utilities, rubber and chemical products, aerospace, steel, automotive, and fuel industries. He then classified the stocks from each industry by the percentage return on investment. This breakdown is shown in the following table.

	Industry					
Percent ROI	Utilities	Rubber & Chemicals	Aerospace	Steel	Auto	Fuel
0 to 4	4	26	20	15	18	10
4 to 8	29	19	28	27	19	24
8 or higher	17	5	2	8	13	16

Does it appear that a relationship exists at $\alpha = 0.01$?

13.2.4 The Cleveland Transit System (CTS) wishes to analyze customer attitudes toward its service. To do so, they randomly select people from various neighborhoods and ask questions about the service in that area. One of the questions asked the individual to indicate whether he or she believed the buses were consistently on time, 1 to 10 minutes late, or more than 10 minutes late. CTS then categorized the responses by neighborhoods to obtain the following table. Each table entry indicates the number of people belonging to that cell.

	East Side	West Side	South Side	Downtown
On time	12	18	22	25
Lateness: 1–10 min	28	36	26	14
Lateness: > 10 min	15	5	9	6

If CTS uses an α of 0.05, will they conclude that a customer's belief about how well buses maintain their schedules is contingent on the area of residence?

13.3 KRUSKAL-WALLIS TEST

In Chapter 11 we discussed the Mann-Whitney U test for unequal location parameters using two independent random samples. The Kruskal-Wallis test is a direct extension of the U test to situations involving c independent random samples. Thus, it provides a distribution-free alternative to the F test in the one-way analysis of variance presented in Chapter 12. Like the U test, the Kruskal-Wallis procedure is a test of identical distributions and is particularly sensitive to location differences. For $c = 2$ the test is identical to the rank-sum form of the U test.

Except for the obvious extension to cover c populations, the assumptions of the procedure are identical to those listed in Section 11.4.3. For ease of reference they are restated here with the indicated changes.

Assumptions of the Kruskal-Wallis Test

1. Independent random samples of size n_j are selected from each of the c populations of interest (or are selected before application of the c treatments).
2. The populations are infinite or sampling is with replacement.
3. The underlying distributions are identical with respect to shape and scale parameters.
4. No ties occur between observations in different samples.
5. The level of measurement is at least ordinal.

Under these assumptions, the hypotheses to be tested are

$$H_0: M_1 = M_2 = \cdots = M_c \qquad H_a: \text{some } M_j \text{ differ} \tag{13.8}$$

where M_j denotes the median for the jth population. If H_0 is true, we can determine the distribution of the Kruskal-Wallis statistic, H, by considering all possible permutations of the ranks of the r observations in the combined sample. To calculate H, we first rank all observations from 1 to n by increasing order of magnitude and calculate T_j, the sum of the ranks corresponding to the observations in the jth sample, for all $j = 1, 2, \ldots, c$. Then H is determined by the equation

$$H = \frac{12}{n(n+1)} \sum_{j=1}^{c} n_j \left(\overline{R}_j - \frac{n+1}{2}\right)^2 \tag{13.9}$$

where n_j is the number of observations in the jth sample, n is the total number of observations and equals $\sum_{j=1}^{c} n_j$, and $\overline{R}_j = T_j/n_j$.

The quantity $(n + 1)/2$ in the squared term of Equation 13.9 is the average of the ranks for the combined sample. If H_0 is true, each $\overline{R}_j$ has an expected value equal to $(n + 1)/2$. Thus, since H will equal zero only when all $\overline{R}_j = (n + 1)/2$, a value of zero for H indicates perfect agreement between the null hypothesis and the sample data. Conversely, large values for H will occur only when some of the $\overline{R}_j$'s differ significantly from $(n + 1)/2$. This happens if some samples receive a disproportionate share of either the high ranks or the low ranks. The occurrence of this would indicate that the observations corresponding to the aforesaid samples are grouped at the extremes of the array, which implies rejection of the null hypothesis of equal medians. In summary, our rejection region will correspond only to large values of the statistic H.

Equation 13.9 can be shown to be algebraically equivalent to the following:

$$H = -3(n+1) + \frac{12}{n(n+1)} \sum_{j=1}^{c} \frac{T_j^2}{n_j} \tag{13.10}$$

This is known as the computational form for H and involves fewer arithmetic operations.

Tables for the exact distribution of the H statistic are limited and exist only for very small sample sizes. When $c = 3$ and all $n_j \leqslant 5$, critical values corresponding to an α of 0.1, 0.05, or 0.01 can be found in *Owen's Handbook of Statistical Tables* (1962). In other situations, or if Owen's tables are not available, it has been shown that the distribution of H can be approximated with a χ^2 distribution with $c - 1$ degrees of freedom. As the sample sizes increase, the approximation improves and becomes exact in the limiting case.

As usual, we summarize this procedure and the test in the following steps.

1. State the hypotheses to be tested and select α.
2. Select random samples of size n_j from each of the c populations under consideration, or select a random sample of size n from a single population and randomly assign n_j items to receive the jth treatment, $j = 1, 2, \ldots, c$.
3. Construct an array of the combined set of observations and replace each observation with its corresponding position in the array.
4. Calculate H using Equation 13.10.
5. For hypotheses of the form

$$H_0\colon M_1 = M_2 = \cdots = M_c \qquad H_a\colon \text{some } M_j \text{ differ}$$

reject H_0 at α if

$$H > \chi^2_{c-1,\,\alpha} \tag{13.11}$$

where $\chi^2_{c-1,\,\alpha}$ is found in Table A.7 of the Appendix.

We illustrate this procedure with an example.

EXAMPLE 13.4 In attempting to determine whether academic salaries differ by geographical area, six associate professors were randomly selected from each of the following areas of the United States: Northeast, Southeast, Northwest, Southwest. Each professor was then asked his or her salary for the academic year; the results are

Table 13.8 Salaries of Associate Professors by Geographical Region of the United States in 1977

Northeast	Southeast	Northwest	Southwest
23.3	22.2	19.1	21.7
22.6	20.3	20.2	20.9
25.1	21.9	21.5	22.4
24.5	19.5	20.7	20.8
24.0	20.1		23.0
	21.4		

presented in Table 13.8 in thousands of dollars. Some professors refused to disclose their salaries, and the corresponding sample sizes were reduced accordingly. Note that refusals may affect the results and that conclusions reached after reducing the sample sizes may be questionable. We proceed with that in mind. Using $\alpha = 0.05$, can we conclude that salaries of associate professors differ by geographical area?

SOLUTION The hypotheses indicated by the question are

$$H_0\colon M_{NE} = M_{SE} = M_{NW} = M_{SW} \qquad H_a\colon \text{some differ}$$

with α specified at 0.05. Thus, we will reject H_0 if

$$H > \chi^2_{4-1,\,0.05} = \chi^2_{3,\,0.05} = 7.81$$

Placing the data into an array, we have, with areas identified,

19.1	19.5	20.1	20.2	20.3	20.7	20.8	20.9	21.4	21.5
NW	SE	SE	NW	SE	NW	SW	SW	SE	NW
21.7	21.9	22.2	22.4	22.6	23.0	23.3	24.0	24.5	25.1
SW	SE	SE	SW	NE	SW	NE	NE	NE	NE

Replacing each observation in the original table with its position in the array, we obtain the table of ranks and the corresponding rank sum for each sample.

	Northeast	Southeast	Northwest	Southwest
	17	13	1	11
	15	5	4	8
	20	12	10	14
	19	2	6	7
	18	3		16
		9		
T_j:	89	44	21	56

Squaring the rank sums gives

$$T_{\text{NE}}^2 = 7921 \qquad T_{\text{SE}}^2 = 1936 \qquad T_{\text{NW}}^2 = 441 \qquad T_{\text{SW}}^2 = 3136$$

with $n = \Sigma\, n_j = 5 + 6 + 4 + 5 = 20$. Then, using Equation 13.10,

$$\begin{aligned} H &= -3(n+1) + \frac{12}{n(n+1)} \sum_{j=1}^{c} \frac{T_j^2}{n_j} \\ &= -3(21) + \frac{12}{20(21)}\left(\frac{7921}{5} + \frac{1936}{6} + \frac{441}{4} + \frac{3136}{5}\right) \\ &= -63 + \frac{12}{420}(1584.20 + 322.67 + 110.25 + 627.2) \\ &= -63 + \frac{12}{420}(2644.32) = -63 + 75.552 = 12.552 \end{aligned}$$

Since $12.552 > 7.81$, we reject H_0 and conclude that salaries of associate professors appear to differ by geographical area, realizing that the refusals may have affected our results.

Our assumptions were such that, theoretically, ties should not occur. In practice, they do. If there are ties in the data, we will use the average rank procedure of Chapter 11. Ties within the same sample will not affect the value of H or its distribution, but those between samples will reduce the value of H, altering its distribution, and make the test less sensitive to actual population differences. If there are many between-sample ties, it is recommended that the user dichotomize the combined sample in terms of each observation's being above or below the overall median and construct a $2 \times c$ contingency table. If the sample sizes are large enough so that all $E_{ij} \geqslant 5$, the hypotheses may be tested using the procedure of Section 13.2.1.

The AREs of the Kruskal-Wallis test relative to the F test for a one-way ANOVA are identical to those of the Mann-Whitney U test with respect to the t test for two independent samples. In specific situations the ARE can be as high as infinity, but it is never lower than 0.864. If the populations are normally distributed, the ARE is 0.955, and if the distributions are uniform it is 1.000. If only two treatments are of interest, use the Mann-Whitney U test, since exact tables are available for U.

The Kruskal-Wallis test can be adjusted to test H_0 versus one-sided alternatives and has been generalized to handle factorial designs that test both main effects and interactions. For a thorough treatment of the appropriate changes in the procedure, see Bradley (1).

EXERCISES

13.3.1 In evaluating salaries for a particular job, the hourly rates for that job in cities of varying size were determined. The table shows the hourly rate for each city in the random sample as classified by the size of the city (1977 data).

Size of City (in 1000 persons)				
0–5	5–10	10–50	50–250	Over 250
2.89	2.94	3.47	3.88	3.72
3.22	3.32	3.26	3.56	4.10
2.96	3.17	3.39	3.65	3.98
3.15	2.92	3.18	3.70	3.84
3.06	3.12	3.44	3.52	4.08

Can we conclude that the size of the city has an effect on hourly rates for this job? Use $\alpha = 0.05$.

13.3.2 Acorn U gives entering MBA candidates a multiple-choice exam on introductory statistics in order to decide whether or not a student should be exempted from the first course in quantitative analysis. A random sample of 20 students with the following undergraduate majors showed the following scores on the exam.

Business	Science and Engineering	Education	Liberal Arts
72	69	62	29
48	92	31	40
64	77	47	59
83	86	26	36
51	58	50	42

Can it be concluded that the student's preparation in statistics differs by the area of his or her undergraduate major? Use $\alpha = 0.05$.

13.3.3 Discuss the assumptions of the ANOVA F test with respect to Exercise 13.3.1.

13.3.4 Would you consider using an F test for Exercise 13.3.2? Why or why not?

13.4 FRIEDMAN'S TEST

In using the sign test for paired data (Chapter 11), we calculated the within-pair differences and counted the resulting number of positive differences. Equivalently, we could have ranked the data within each pair, assigning a 1 to the smaller value and a 2 to the larger value, and used as a statistic the squared difference between the rank sums of the two treatments. Friedman's test uses a direct extension of the latter procedure to two or more treatments. It is equivalent to the sign test when only two treatments are being considered.

The procedure for Friedman's test consists of ranking the observations within each row (matching set) and adding the ranks associated with each treatment. Then the test statistic, denoted by S, is calculated by comparing the rank sums for each treatment with the average rank sum. The statistic is

$$S = \sum_{j=1}^{c} \left[T_j - \frac{n(c+1)}{2}\right]^2 \tag{13.12}$$

and will have a randomization distribution; that is, one that is determined by considering the results associated with all possible permutations of the c ranks within each matched set, provided the following assumptions are satisfied.

Assumptions for Friedman's Test

1. There are available n sets of c objects each, with the elements in each set being similar before treatments.
2. The assignment of treatments to objects within a set is random, or, if the same unit receives all treatments, there is no carryover effect. The latter part of the assumption can be approximately satisfied by administering the treatments in a random order.
3. There are no ties within sets (blocks) after treatments.
4. The level of measurement within each set (block) is at least ordinal.

If the null hypothesis of no treatment effects, H_0: $M_1 = M_2 = \cdots = M_c$, is true, then the ranks within each row should occur randomly.

Hence, the treatment rank sums T_j would be random also and should equal the average rank sum. The average is given by $n(c + 1)/2$, since the sum of the ranks, 1 to c, within each row is $c(c + 1)/2$, the number of rows equals n, and the total, $nc(c + 1)/2$, is divided by the number of treatments, c. If $S = 0$, T_j equals $n(c + 1)/2$ for all j and there is perfect agreement with the null hypothesis. Conversely, if, for example, one treatment yields appreciably higher responses, say treatment 1, then it will receive all the high ranks within each row. The high ranks will be c's, and the corresponding sum would equal nc. The resultant value for S would be larger than should be expected if H_0 is true. Thus, the null hypothesis is rejected for large values of S.

Equation 13.12 is algebraically equivalent to

$$S = \sum_{j=1}^{c} T_j^2 - \frac{c}{4}[n(c + 1)]^2 \qquad (13.13)$$

where c is the number of treatments, n is the number of matched sets, and $T_j, j = 1, 2, \ldots, c$, is the sum of the ranks assigned to subjects receiving treatment j. Equation 13.13 is computationally easier than Equation 13.12 and is recommended.

We summarize the overall procedure for performing Friedman's test in the following outline.

1. State the hypotheses of interest and select α.
2. Select a sample of n sets with c matched elements in each.
3. Randomly assign a treatment to each element in each set and record the results, taking care to maintain the matches.
4. Within each set, rank the results in ascending order, replacing each observation with its rank, and calculate S using Equation 13.13.
5. For the hypotheses

   $$H_0\colon M_1 = M_2 = \cdots = M_c \qquad H_a\colon \text{some } M_j \text{ differ}$$

   reject H_0 if

   $$S \geqslant S_{c,n,\alpha} \qquad (13.14)$$

   where values of $S_{c,n,\alpha}$ are given in Table A.14 of the Appendix for selected values of c and n at nominal α's of 0.10, 0.05, 0.01, and 0.005.

EXAMPLE 13.5 In evaluating four methods for training dogs, seven different breeds were selected. Within each breed litters consisting of four males, whose parentage was known to be common, were chosen, with each male in the litter being assigned to a different training

Table 13.9 Average Ratings of Dogs by Training Method and Breed

Breed	Method 1	Method 2	Method 3	Method 4
1	1.33	4.00	2.67	3.00
2	2.33	4.00	2.00	1.67
3	1.67	3.33	2.33	2.67
4	2.67	3.67	1.33	2.33
5	3.00	4.00	2.00	1.00
6	1.00	3.67	2.00	3.33
7	2.00	3.33	1.00	3.67

method. All dogs of the same breed worked with the same trainer. At the conclusion of the training period three independent judges were asked to rate the response of the dogs in each breed to a complicated series of commands. The three ratings for each dog were then averaged, and the results are recorded in Table 13.9. Using $\alpha = 0.05$, can it be concluded that the training methods differ?

SOLUTION The hypotheses to be tested at $\alpha = 0.05$ are

$$H_0\colon M_1 = M_2 = M_3 = M_4 \qquad H_a\colon \text{some } M_j \text{ differ}$$

The corresponding decision rule implies rejection of H_0 if $S \geqslant S_{c,n,\alpha} = S_{4,7,0.05} = 91$. To calculate the value of S for the data, we rank the observations within breeds and replace each observed value with its corresponding rank. The ranks under each method are then added to obtain the T_j's. This is accomplished in the following table.

	Breed	Method 1	Method 2	Method 3	Method 4
	1	1	4	2	3
	2	3	4	2	1
	3	1	4	2	3
	4	3	4	1	2
	5	3	4	2	1
	6	1	4	2	3
	7	2	3	1	4
T_j:		14	27	12	17

Substituting into Equation 13.13, we obtain

$$S = \sum_{j=1}^{c} T_j^2 - \frac{c}{4}[n(c+1)]^2$$

$$= (14^2 + 27^2 + 12^2 + 17^2) - \frac{4}{4}[7(5)]^2$$

$$= (196 + 729 + 144 + 289) - (35)^2 = 1358 - 1225 = 133$$

Since $133 > 91$, we reject H_0 and conclude that the methods appear to differ in effectiveness.

For values of c and n outside the range of Table A.14, the statistic $12S/nc(c+1)$ has an approximate chi-square distribution with $c - 1$ degrees of freedom. In such cases an approximate test can be obtained by rejecting the null hypothesis if

$$\frac{12S}{nc(c+1)} > \chi^2_{c-1,\alpha} \tag{13.15}$$

Friedman's test provides an alternative to the F test used in the randomized-block design of Chapter 12. When compared with the F test, with the condition of normality satisfied, its ARE varies depending on the number of treatments being compared. For $c = 2$ the ARE is 0.637, the same as the sign test, and it increases with the number of treatments to a maximum value of 0.955. If the underlying distributions are nonnormal, the actual ARE may be either above 0.955 or below 0.637, or between the two.

EXERCISES

13.4.1 In evaluating three additives to speed up the curing process of rubber, the research lab of Rubber Products selected a sample of crude rubber from each of seven different suppliers. Each sample was split into three equal parts, which were then randomly assigned to the three additives and milled under identical conditions. The curing times in minutes are as follows:

Supplier:	1	2	3	4	5	6	7
Additive 1:	24	22	27	31	23	30	28
Additive 2:	16	18	15	22	26	20	21
Additive 3:	32	28	30	26	29	25	30

Do the median curing times for the additives differ at $\alpha = 0.05$?

13.4.2 Refer to Exercise 13.3.2 and assume that the scores also can be identified with respect to the student's undergraduate average. We present those scores with this additional information:

	Area			
Average	Business	Science & Engineering	Education	Liberal Arts
< 3.00	72	69	62	29
3.00 - 3.25	51	58	50	42
3.25 - 3.50	48	92	31	40
3.50 - 3.75	64	77	47	59
$\geqslant 3.75$	83	86	26	36

Does it appear that statistics preparation differs by area of major at $\alpha = 0.05$ when this design is used?

13.4.3 What assumptions would have to be violated in Exercise 13.4.1 to eliminate the use of the F test? Do you feel that they are in this case?

13.5 SUPPLEMENTARY EXAMPLES

EXAMPLE 13.6 The Pogo Company is considering offering its employees a stock option plan and is interested in determining whether receptiveness to the plan is independent of the employee's job classification. To determine so, the Personnel Department randomly selected 200 employees from the personnel files and sent each a questionnaire detailing the stock option and the employee's attitude toward it in terms of yes, no, or undecided. The results of the survey

Table 13.10 Summary of Responses to Stock Option Plan for the Pogo Company

	Response		
Employee Category	Yes	No	Undecided
Factory	31	54	25
Maintenance	13	8	7
Clerical	20	12	8
Management	15	3	4

are given in Table 13.10. Is receptiveness to the plan independent of the employee's job category at $\alpha = 0.10$?

SOLUTION The hypotheses to be tested are

$$H_0\text{: Response is independent of category}$$

$$H_a\text{: Response is dependent on category}$$

Checking the expected values for cells (4, 2) and (4, 3), we find $E_{42} = 8.47$ and $E_{43} = 4.84$. E_{43} is less than 5 but only slightly so. Thus, we will not regroup and our decision rule is

Reject H_0 at $\alpha = 0.10$ if

$$\sum_i \sum_j \frac{(O_{ij} - E_{ij})^2}{E_{ij}} > \chi^2_{(4-1)(3-1),\, 0.10} = 10.64$$

Using the recommended tabular format, the computations are given in Table 13.11 on page 508.

Since the calculated value of 14.492 already exceeds the tabular value of 10.64, we stop at this point and reject H_0. Thus, on the basis of this sample it appears that an employee's attitude toward the stock option plan is related to job category.

In Example 13.6 we stated the hypotheses as H_0: no relationship versus H_a: some relationship but we could have stated them in parametric form. In the alternative form they become

Table 13.11 Computations for Example 13.6

i, j	O_{ij}	E_{ij}	$(O_{ij} - E_{ij})^2$	$\frac{(O_{ij} - E_{ij})^2}{E_{ij}}$	Partial Sum
1, 1	31	43.45	155.00	3.567	3.567
1, 2	54	42.35	135.72	3.205	6.772
1, 3	25	24.20	0.64	0.026	6.798
2, 1	13	11.06	3.76	0.340	7.138
2, 2	8	10.78	7.73	0.717	7.855
2, 3	7	6.16	0.71	0.115	7.970
3, 1	20	15.80	17.64	1.116	9.086
3, 2	12	15.40	11.56	0.751	9.837
3, 3	8	8.80	0.64	0.073	9.910
4, 1	15	8.69	39.82	4.582	14.492 > 10.645

$$H_0: p_{FY} = p_{MY} = p_{CY} = p_{MgtY} \quad \text{and} \quad p_{FN} = p_{MN} = p_{CN} = p_{MgtN}$$

and

$$H_a: \text{at least one equation is incorrect}$$

We used only two equations in H_0 because if they are true, the third one follows from them, since $p_U = 1 - (p_Y + p_N)$ for all categories. The χ^2 test is the obvious procedure for two reasons. First, the data are strictly nominal, since each observation merely represents yes, no, or undecided. Second, the hypotheses do not correspond to those of either the Kruskal-Wallis or the Friedman test.

EXAMPLE 13.7 A company is trying to evaluate four proposed locations for building warehouses. They have funds for building two warehouses and would like to know whether there are actually any differences among the four locations. In the evaluation the company had the president and each of the five vice presidents rate each location on a 1 to 20 scale with respect to market accessibility, transportation availability, legal restrictions, and tax structures. Their ratings are given in Table 13.12. Use $\alpha = 0.05$ to test for differences in the locations. The locations were presented to each person in a random order.

Table 13.12 Ratings of Possible Warehouse Locations for Example 13.7

	Rater					
Location	Pres.	VP1	VP2	VP3	VP4	VP5
1	13	6	10	12	11	5
2	9	12	7	8	13	9
3	8	5	14	10	7	14
4	5	16	8	9	16	7

SOLUTION Since each location was rated in a random order by each executive, we have a randomized-block design. The ratings themselves connote only relative position in the rater's opinion and convey nothing about actual differences between locations. Thus the data are at best ordinal and Friedman's test is indicated.

The hypotheses are

$$H_0: M_1 = M_2 = M_3 = M_4 \qquad H_a: \text{ some } M_i \text{ differ}$$

and our decision rule is

Reject H_0 if

$$\sum_{i=1}^{4} T_i^2 - \frac{4}{4}[6(4+1)]^2 \geqslant S_{4,\,6,\,0.05} = 76$$

Replacing the ratings in Table 13.12 by their ranks within each block, we obtain

Location	Pres.	VP1	VP2	VP3	VP4	VP5	T_i
1	4	2	3	4	2	1	16
2	3	3	1	1	3	3	14
3	2	1	4	3	1	4	15
4	1	4	2	2	4	2	15

Substituting into the computational form for S, we have

$$(256 + 196 + 225 + 225) - \frac{4}{4}[6(5)]^2 = 2$$

Since $2 < 76$, we cannot reject H_0, and we conclude that on the basis of these data there appear to be no significant differences among locations. That is, it appears to be irrelevant where the warehouses are built if the stated criteria are the only ones of interest.

In Example 13.7 the calculated value for S was 2, which is definitely less than the critical value of 76. In fact, the value of 2 is almost too small to be believable. Two is a very unlikely value in this situation, and while it definitely does not indicate rejection of H_0, it should at least lead you to double-check the data for errors. It also could indicate that the judges are weighting the indicated criteria very differently.

EXAMPLE 13.8 In an attempt to evaluate whether differences exist between types of stocks with respect to their annual yield (dividend as a percentage of price), a portfolio manager for the First Provincial Bank of Halifax randomly selected five companies from each of four types of industry. The four types were transportation, aerospace, utilities, and chemical. The annual yield for each of the companies is given in Table 13.13. Do there appear to be any differences in the annual yields by type of industry at $\alpha = 0.05$?

SOLUTION We have at least ordinal data and are interested in testing for location differences. Thus, the Kruskal-Wallis test is indicated for the hypotheses

$$H_0\colon M_1 = M_2 = M_3 = M_4 \qquad H_a\colon \text{some } M_j \text{ differ}$$

Table 13.13 Annual Yields for Each of Five Companies in Each of the Indicated Industry Types

Transportation	Aerospace	Utilities	Chemical
8.5	2.9	10.3	2.5
2.4	3.2	8.4	4.3
3.1	4.1	9.7	3.7
6.0	5.0	6.8	1.5
5.2	1.8	7.2	5.2

Our decision rule is:

Reject H_0 if
$H > \chi^2_{c-1,\alpha} = \chi^2_{3,\,0.05} = 7.81$

Replacing the yields in Table 13.13 by their ranks within the overall sample gives the following table.

	Transportation	Aerospace	Utilities	Chemical
	18	5	20	4
	3	7	17	10
	6	9	19	8
	14	11	15	1
	12.5	2	16	12.5
T_j:	53.5	34	87	35.5

Using the computational form for H,

$$H = -3(n+1) + \frac{12}{n(n+1)} \sum_{j=1}^{c} \frac{T_j^2}{n_j}$$

we have $n = 20$, $c = 4$, $n_j = 5$ for $j = 1, 2, 3, 4$, and

$$\sum_{j=1}^{c} T_j^2 = (2862.25 + 1156 + 7569 + 1260.25) = 12{,}847.50$$

Substituting these values gives

$$H = -3(21) + \frac{12}{20(21)}\left(\frac{12{,}847.5}{5}\right) = 10.41$$

Since $10.41 > 7.81$, we reject H_0 and conclude that annual yields (dividends as a percentage of price) differ by type of company.

In Example 13.8 we can actually consider the data to be of an interval level of measurement, and we should consider the ANOVA F test. We did not use the F test here because it assumes both normality and homogeneity of variance. Generally data that are percentages violate the normality assumption.

Table 13.14 A Summary of Tests for Chapter 13 and Their Classical (Parametric) Counterparts

	Several Multinomials (Association)	Location	
		Independent Samples	Matched Sets (Blocks)
	c samples (or 1 sample)	c samples	c treatments n blocks
Test	χ^2 test $df = (r - 1)(c - 1)$	Kruskal-Wallis (statistic is H)	Friedman's (statistic is S)
Assumptions	Nominal data Infinite populations rc mutually exclusive categories Random sample(s)	Random samples Equal variances and shapes Infinite populations Ordinal measurement	Randomized-block design Infinite populations Ordinal measurement within blocks
Classical Counterpart	F test for randomized blocks	F test for one-way ANOVA	F test for randomized blocks

A situation in which both normality and homogeneity of variance are violated is one in which the data represent binomial proportions. In this case, a square root transformation on the data stabilizes the variances and gives approximate normality.

13.6 SUMMARY

In this chapter we presented a further class of tests that are identified as distribution-free. All fall into the basic category of *c*-sample procedures. The χ^2 test of several binomials is a special case of the test of several multinomials, and both may be classified as tests of association between types of categories. The hypotheses are more easily stated and understood in the independence form than in the parametric formulation. In both cases the χ^2 test compares observed frequencies with those expected when H_0 is true. It does this by calculating the squared differences between observed and expected frequencies, dividing each by the expected value, and accumulating over all categories. That is, the χ^2 tests are based on the statistic

$$\sum_i \sum_j \frac{(O_{ij} - E_{ij})^2}{E_{ij}}$$

where O_{ij} is the observed frequency of cell *ij* and E_{ij} is the expected frequency of cell *ij* under H_0. The null hypothesis is rejected only if the statistic is too large when compared with the appropriate table value.

Both the Kruskal-Wallis test and Friedman's test are extensions of two-sample procedures for testing location differences. The Kruskal-Wallis test is an extension of the Mann-Whitney *U* test and has the same basic assumptions associated with it. Friedman's test relates to randomized-block designs and is an extension of the two-sample sign test for identical locations. The primary differences between the two are in the levels of measurement and the sampling designs. Kruskal-Wallis requires ordinal measurement throughout and independent random samples, whereas Friedman's requires ordinality only within blocks and utilizes subjects that have been matched before treatment.

The procedures of this chapter, their associated assumptions, and their classical counterparts are presented in tabular form in Table 13.14.

REVIEW EXERCISES

Level One

13.1 After a course is over, students often say, "If I'd had Professor So-and-So, I'd have gotten a better grade," implying that some instructors are easier than others. In an attempt to evaluate whether the proportion of high grades in statistics at Norka U differs by instructor, a researcher collected the following data:

	Instructor			
	Ms. Laffin	Mr. Barnem	Mr. Stunning	Mr. Cheek
No. of grades > C	15	36	20	10
No. of grades ⩽ C	30	44	32	20

Is the students' implication supported at $\alpha = 0.10$?

13.2 What are the differences between the Kruskal-Wallis test and Friedman's test?

13.3 In an attempt to evaluate self-teaching texts in statistics, eight groups of four students each were selected. The students in each group were matched in terms of their grades in an introductory mathematics course which was a prerequisite. Each student in a group was randomly assigned one of the four texts under consideration. At the end of a specified time period, all students were given a common exam. Group identification, text, and exam scores are as follows.

	Group							
	1	2	3	4	5	6	7	8
Text A	64	91	84	56	60	52	91	75
Text B	78	82	68	46	70	74	95	77
Text C	76	85	71	67	75	64	80	82
Text D	87	79	75	65	78	59	86	69

Do there appear to be any differences in the grades due to texts, using $\alpha = 0.10$?

13.4 Before the publicity about the disadvantages of aerosol products, the median weekly sales in number of units of the four types of deodorant produced by the Daisy-Fresh Co. were about the same. Recently the Marketing Department randomly selected 24 weeks from the past three years and then randomly selected 6 weeks from the 24 for each product and recorded its weekly sales in 10,000 units. Can we conclude that the median weekly sales are still equal at $\alpha = 0.01$?

Type of Deodorant			
Roll-On	Stick	Aerosol	Powder
3.8	3.1	2.0	2.9
3.2	2.7	1.5	1.9
2.5	4.2	2.1	3.3
2.8	3.0	0.8	2.2
3.6	2.3	1.3	4.0
2.4	3.4	0.4	3.5

13.5 If you wish to test for association in a 4 by 7 contingency table, how many degrees of freedom will the test statistic have?

13.6 What are the differences between the χ^2 test of several multinomials and the Kruskal-Wallis test?

13.7 In evaluating drill bits from various manufacturers, the Monroe Machine Shop obtained one-half dozen bits from each of six manufacturers. The life of each bit in number of holes drilled before wear-out was then recorded. A total of five of the bits were broken by accident, resulting in a reduction of the sample sizes for some manufacturers. The data collected are presented in tabular format on the following page.

Manufacturer					
A	B	C	D	E	F
462	498	532	526	540	499
396	486	504	481	508	471
427	510	512	493	490	519
501	429	461	515	513	507
454	473	485		482	500
	469	479			

Are the median lives the same for all manufacturers ($\alpha = 0.025$)?

13.8 In a test involving evaluation of four laundry detergents, five sheets were soiled with five different types of dirt (grease, regular dirt, chocolate, etc.). The sheets were then torn into four equal sections, and each section was washed in a different detergent. A homemaker was then asked to rank the four sections, which had been identified by numbers, as to cleanness, with a 1 corresponding to least clean and 4 to most nearly clean. Using these rankings, given below, are there any differences in the detergents' ability to clean at $\alpha = 0.10$?

	Dirt				
	1	2	3	4	5
Detergent 1	1	2	3	1	1
Detergent 2	4	3	4	4	3
Detergent 3	3	1	2	2	2
Detergent 4	2	4	1	3	4

13.9 In analyzing absenteeism, Rubber Products, Inc., checked their personnel records for the past year and constructed the following table. The table categorizes all employees by type of job and number of days absent during the past year.

Job Classification	X = Days Absent $X \leqslant 3$	$4 \leqslant X \leqslant 7$	$X > 7$
Management	80	17	3
Secretarial	50	30	25
Craftsman	150	65	35
Service	40	50	30
Laborer	130	138	57

Using $\alpha = 0.01$, can it be concluded that number of days absent is related to job classification?

13.10 The data in Exercise 13.9 also were broken down according to sex and number of days absent. The data showing the number of persons in each of these categories are as follows:

Sex	X = Days Absent $X \leqslant 3$	$4 \leqslant X \leqslant 7$	$X \geqslant 7$
Female	92	56	32
Male	397	243	80

Can we conclude that absenteeism is independent of sex at $\alpha = 0.025$?

13.11 Refer to Exercise 13.4. Suppose that rather than sampling a total of 24 weeks, Daisy Fresh's Marketing Department had sampled only six weeks and recorded the sales of each product in each week. Let the first row correspond to week 1, the second row to week 2, etc. Analyze the data for this design at $\alpha = 0.01$.

13.12 Of the two designs in Exercises 13.4 and 13.11, which do you think is better for this type of experiment and why?

13.13 To compare the effect of five different advertising designs on product recognition time, 24 people were selected at random. The participants were assigned randomly to view the different designs, with 4 viewing the first, 6 the second, 5 the third, 4 the fourth, and 5 the last. The experiment was terminated after 10 seconds if the product was not recognized in that time. The results, with the recognition time in seconds, were as shown in the following table (a 10 indicates nonrecognition).

Layout				
1	2	3	4	5
0.8	0.7	1.2	1.0	0.6
1.6	1.8	1.4	1.1	1.5
1.6	1.2	0.9	0.9	10.0
0.7	0.9	1.2	1.7	0.7
	1.6	1.3		1.4
	1.9			

Test for differences in median recognition time at $\alpha = 0.05$.

13.14 Refer to Exercise 13.13 and perform an analysis of variance using an F test at $\alpha = 0.05$.

13.15 Comment on the results of Exercises 13.13 and 13.14. Which procedure would you recommend and why?

Level Two

13.16 In situations in which the one-way ANOVA F test can be used, is the Kruskal-Wallis procedure also applicable? Is the converse true?

13.17 What are the differences between the χ^2 test for a contingency table and the ANOVA F test for a randomized-block design?

13.18 What advantage does the ANOVA procedure for a randomized-block design have over Friedman's test?

13.19 Refer to Example 13.7 and test to see whether vice presidents two through five appear to be using the same criteria weightings at $\alpha = 0.10$.

13.20 Refer to Example 13.8 and test for differences in the yields of transportation, aerospace, and chemical stocks at $\alpha = 0.05$.

Directed alternatives may be used with the χ^2 test. To do so, the direction must be stated by H_a in advance. The test procedure then has two stages. The first stage is to check the data to see whether the direction of H_a is satisfied. If H_0 is true, the probability of the correct direction is 0.5. If the direction is not as indicated by H_a, then H_0 is accepted. If it is as indicated, we proceed to the second stage and actually do the χ^2 test, but we use an α that is twice the stated significance level. We double α because we actually have run a series of two tests on the data. The approximate total α for the test then is

$$\alpha' = \text{pr}\{\text{correct direction}\}\ \text{pr}\{\text{reject } H_0 | \text{correct direction}\} \doteq \frac{1}{2}(2\alpha) = \alpha$$

13.21 Sociologists have indicated that there is an inverse relationship between the number of children an individual plans to have and the number of siblings that individual has. That is, the more siblings a person has, the fewer children he or she plans to have. Use the data given in the following table, which represent the responses of 242 randomly selected young adults, to test this hypothesis at $\alpha = 0.05$.

Anticipated Number of Offspring	Number of Siblings			
	0	1	2	3 or more
0	7	8	13	20
1	7	11	16	18
2	20	17	10	7
3	18	15	10	5
4 or more	15	14	7	4

13.22 Refer to Exercise 13.10 and test the hypothesis that females are absent more often than males at $\alpha = 0.05$.

13.23 Another test for equality of medians in the multisample situation is the median test. For this procedure we find the median for the combined sample and then identify each observation with respect to its position above or below the combined median. Then we construct a $2 \times c$ contingency table with classes identifying the sample of origin and its relative position to the median. The test is to reject H_0 at α if

$$\sum_{i=1}^{2} \sum_{j=1}^{c} \frac{(O_{ij} - E_{ij})^2}{E_{ij}} > \chi^2_{c-1,\,\alpha}$$

(*a*) State the hypotheses of interest for the median test.
(*b*) Identify what O_{ij} represents.
(*c*) Give an expression for determining E_{ij} in general.

13.24 What assumptions do you think are necessary for the median test of Exercise 13.23?

13.25 Dimwiddy's Department Stores is interested in evaluating three training methods for its salespersons. A group of 30 salespersons is randomly assigned, with 10 people given each training method. The results, in terms of number of sales during a specified period, are given here. Use the data to test for differences in training methods at $\alpha = 0.10$ with the median test of Exercise 13.23.

Method 1		Method 2		Method 3	
64	49	83	62	54	60
46	72	70	103	31	44
87	37	49	94	75	38
34	48	77	69	62	69
55	56	68	73	47	55

Level Three

13.26 Prove that the definitional and computational forms for the Kruskal-Wallis H statistic are equivalent.

13.27 In evaluating four suppliers of a particular component, a buyer randomly selected four boxes from a shipment by each supplier and determined the proportion of defectives in each box. These proportions are given in the following table.

Supplier			
1	2	3	4
0.07	0.04	0.09	0.03
0.03	0.02	0.04	0.05
0.06	0.05	0.08	0.11
0.08	0.06	0.10	0.07

Do the proportions appear to differ at $\alpha = 0.10$?

13.28 Prove that the definitional and computational forms for Friedman's S are equivalent.

13.29 Use the square root transformation on the data of Exercise 13.27 and analyze them using the Kruskal-Wallis procedure with $\alpha = 0.05$.

13.30 Analyze the transformed data of Exercise 13.29 using an F test with $\alpha = 0.05$.

References

1. Bradley, James V. *Distribution-Free Statistical Tests.* Englewood Cliffs, New Jersey: Prentice-Hall, Inc., 1968.
2. Gibbons, Jean Dickinson. *Nonparametric Statistical Inference.* New York: McGraw-Hill Co., 1971.
3. Lehmann, E. L. *Testing Statistical Hypotheses.* New York: John Wiley & Sons, Inc., 1959.
4. Lindgren, B. W. *Statistical Theory.* New York: The Macmillan Co., 1962.
5. Noether, Gottfried E. *Elements of Nonparametric Statistics.* New York: John Wiley & Sons, Inc., 1967.
6. Siegel, Sidney. *Nonparametric Statistics for the Behavioral Sciences.* New York: McGraw-Hill Co., 1956.

Chapter 14
Improving the Ability to Estimate—The Simple Case

KEY WORDS AND TERMS

Regression
Dependent Variable (Y)
Independent Variable (X)
Scatter Diagram
Least-Squares Estimation (LSE)
Residual ($Y_i - \hat{Y}_i$)
Pure Error
Lack of Fit
Coefficient of Determination (R^2)
Pearson's Product Moment Correlation (r_{XY})
Spearman's Rank Correlation (r_s)
Residual Plot

14.1 INTRODUCTION

EXAMPLE 14.1 Suppose you are faced with the following simplified situation. As a product sales manager you have been asked by production for an estimate of next month's sales, that is, a sales forecast. Replying that you will call back in an hour with the forecast, you call your secretary and ask him to bring in the sales data for the past six months. Ten minutes later he hands you the following information:

Month:	April	May	June	July	Aug.	Sept.
Unit Sales (1000s):	22	34	28	25	30	36

(*a*) If these are the only data available, what do you tell production?
(*b*) In a few minutes your secretary returns and hands you the advertising expenditures for this product for the same six months. These are (in \$1000s):

April	May	June	July	Aug.	Sept.
4	7	6	5	6	8

If October's advertising is budgeted at $7000, will this information change your forecast?

SOLUTION

(*a*) If the unit sales are the only data available and we assume that this is not a seasonal product, your best estimate for the sales of any randomly selected month is the mean or average monthly sales. Therefore, you assume that these data are representative and that October's sales represent a random observation, and you tell production to expect sales of approximately

$$\bar{x} = \left(\frac{1}{n}\right) \Sigma x_i = \left(\frac{1}{6}\right)(175) = 29.167 \text{ thousand units}$$

(*b*) Scanning the additional information, it appears that there is a relationship between advertising and sales. In fact, as advertising increases, so do sales, apparently. Since your hour is almost up, you call production and simply tell them to expect sales of approximately 34,000 units. That is, you use May's sales as your October forecast, since advertising is the same and you don't have time to do a more sophisticated analysis.

The situation in Example 14.1 is very simplified, but it illustrates the fact that information contained in a second variable may be useful for predicting values of a primary variable. That is, it may enable us to explain some of the total variation of the primary (response) variable and hence replace the simple mean with an estimator that is a function (linear for the moment) of the secondary variable.

In this chapter we continue our study of linear models from Chapter 12, but our emphasis is shifted. In Chapter 12 we wanted to determine whether some response variable is affected by (linearly related to) changing values for some other variable(s). That is, in ANOVA, the primary focus is on testing to see whether an effect exists. Regression still tests to see whether a relationship exists, but does so with the intent of estimating an equation for the relationship. The object of trying to identify an equation for the relationship is to use the information provided by the secondary variable(s) to obtain more precise estimates or predictions for the response variable.

Standard terminology in regression is to call the response variable the *dependent variable* and the secondary variables the *independent* (or *predictor*) *variables.* These terms are misnomers to some extent and usually are misleading to the beginning student. The word *dependent* in everyday usage connotes a cause-and-effect relation. Regression (and correlation) say nothing about cause and effect. They simply indicate the existence or nonexistence of a relationship. The relationship may be of a cause-and-effect nature, but determining this is outside the role of statistics. A classic example is the fact that a high correlation (indicating a definite relationship) was found to exist between the number of storks in a certain locale and the number of babies born in the same area. Few people today would conclude that the storks caused an increase in the birth rate. The term *independent variables* is even more misleading, since in most cases they are not statistically independent either of each other or of the response variable. The term arises from the mathematical concept of linear independence. The predictor variables must be linearly independent of each other in order to determine an equation for the relationship.

The regression approach requires the predictor variables to be quantitative rather than qualitative, but qualitative variables can be used as predictors if dummy (indicator or zero-one) variables are included in the model. For ANOVA, qualitative variables can be used with no difficulty.

Opportunities to use regression arise in all areas of business. In marketing regression can be used to estimate the relationship between the amount of money spent for advertising and the company's sales volume. An accountant may use regression to estimate the relationship between sales and total cost to obtain a value for the company's fixed costs. A personnel manager may evaluate testing procedures by regressing productivity on examination scores. A financial analyst might use regression to relate a company's growth rate to any number of variables for corporate planning purposes. The production manager might use a regressive procedure to try to determine the best size for a work crew to get the most productivity per worker-hour. These are only a few of the many possible business applications. There are also numerous possibilities in government, education, medicine, and any other area you care to cite.

14.2 THE SIMPLE LINEAR REGRESSION MODEL

The easiest method to use in terms of both understanding and computational ease is the simple linear model. This model assumes that a straight-line relationship exists between the two variables of interest. We restrict our discussion in this chapter to this type of model.

Our basic model is given by

$$Y_x = \beta_0 + \beta_1 x + \epsilon \tag{14.1}$$

where Y_x is the response variable, x is a nonrandom variable, ϵ is a random error term with a mean of zero, and β_0 and β_1 are parameters.

By *response variable* we mean the variable for which we wish to predict values. The term *nonrandom* means that x, the predictor variable, is such that we will know its value in advance. As in ANOVA, ϵ is a random variable representing error or chance fluctuations. β_0 and β_1 are parameters of the model, that is, numerical constants. If we take the expected values of both sides of Equation 14.1, the model can be restated as

$$E[Y_x] = E[\beta_0 + \beta_1 x + \epsilon]$$

or

$$\mu_{Y|x} = \beta_0 + \beta_1 x \tag{14.2}$$

The notation $Y|x$ indicates that $\mu_{Y|x}$ is the mean of the population of Y values corresponding to a given value of x. Thus, the line given in Equations 14.1 and 14.2 is actually a line connecting the means of a set of populations. Note that if β_1 is zero, this line becomes parallel to the x axis. We will explore these last two points later.

Before we attempt to estimate β_0 and β_1, let us consider some preliminary analyses that we should make. We are assuming a linear relationship between x and Y, but in many practical situations this will be an invalid assumption. Therefore, before we go to the trouble of fitting a straight line to the observed data, we should check to see whether or not a straight line makes any sense. The easiest way to check this is to construct a *scatter diagram*. That is, simply plot all pairs of (x, y) values on standard graph paper. This is a very rough check, but it should indicate if absolutely no relationship exists or if the relationship is decidedly nonlinear. Figure 14.1 on page 526 gives four examples of scatter diagrams.

The scatter diagram in Figure 14.1*a* appears to be a totally random scattering of points. That is, there appears to be no relation of any kind between x and Y; thus, fitting a straight line to the data would probably be a waste of time. In Figure 14.1*b* the points appear to be scattered randomly, but around a line with a positive slope. In Figure 14*c*, the points appear to be scattered in a definite pattern, but the pattern is certainly not linear in nature. Again a straight line, at least over the full range of data, would probably be a waste of time. The pattern in Figure 14.1*d* also appears to be nonlinear, but fitting a straight line still could be very useful. Therefore, if we obtain patterns

Figure 14.1 Some Examples of Possible Scatter Diagrams

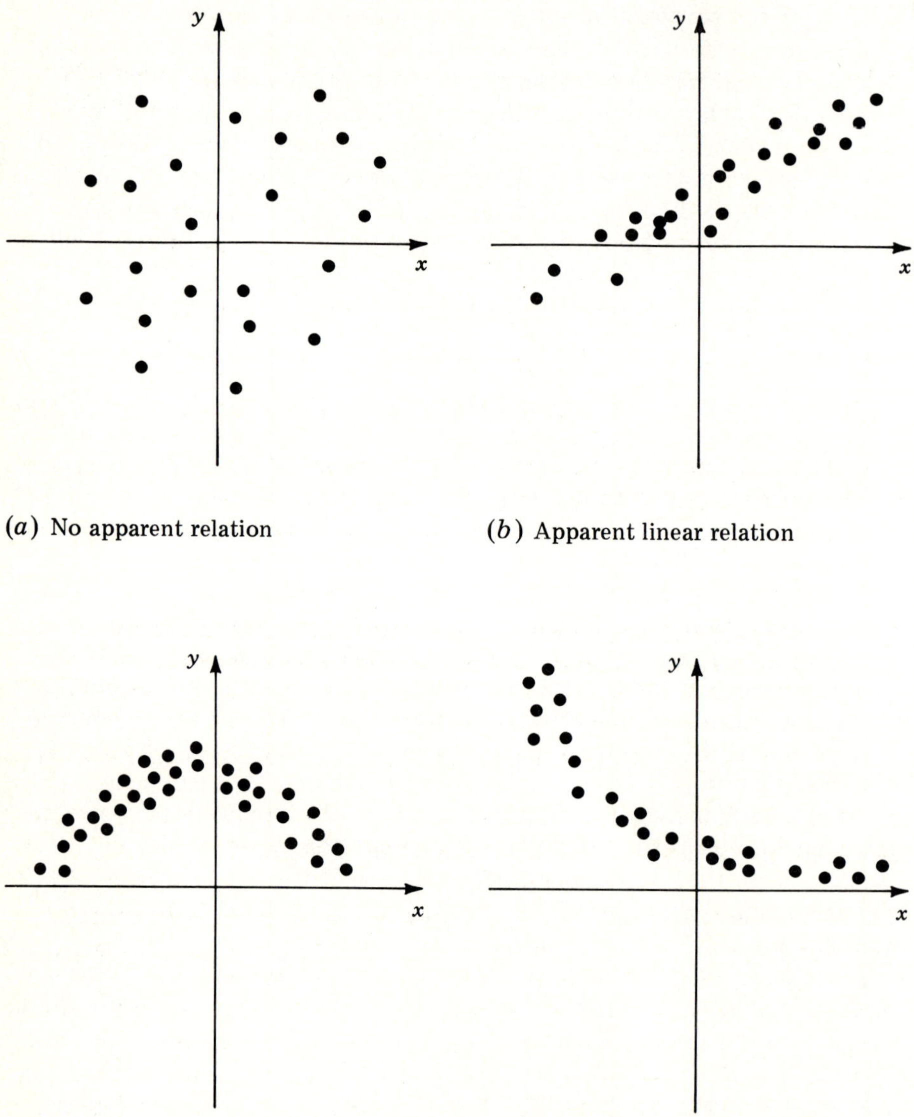

(*a*) No apparent relation

(*b*) Apparent linear relation

(*c*) Apparent nonlinear relation

(*d*) Apparent nonlinear relation

similar to those in Figure 14.1*b* and *d*, we need a method for determining estimates of β_0 and β_1. This is the topic of the next section.

EXERCISES

14.2.1 Construct scatter diagrams for the following sets of data and indicate the ones for which a straight-line relationship might be of some use.

(*a*) (x, y)	(*b*) (x, y)	(*c*) (x, y)	(*d*) (x, y)
-3, 5	0, 2	0, -4	5, 1
2, 11	0, 1	5, 0	4, 7
4, 6	4, 14	3, -4	0, 13
-1, 10	2, 7	0, -5	1, 14
5, 12	3, 12	-4, -2	2, 14
-3, 1	5, 16	-5, -1	3, 11
2, 0	3, 11	3, -3	5, 2
4, 9	4, 12	5, -1	4, 6
5, 3	2, 9	-5, 0	0, 15
-1, -1		1, -4	6, 2
		-1, -5	6, 0

14.2.2 Collect data on two variables for three different situations and plot your pairs of data points. Identify the ones for which you might use a straight line to describe the relationship. Possible sources of data are *Barron's*, the *Wall Street Journal*, your local newspaper, or any other publication that reports business data.

14.3 LEAST-SQUARES ESTIMATION OF THE MODEL

When the scatter diagram indicates that a straight line is not unreasonable, the next step is to determine what line to use. If β_0 and β_1 were known, we should have no problem, since obviously that would be the best line available and there would be no reason to take a sample. In practice, β_0 and β_1 will not be known, and we must be content to estimate them.

The simplest way is to fit a line to the points on the scatter diagram visually. That is, use a straightedge to draw a line that appears to give the "best" visual fit to the points as a whole. Then we can read an estimate of β_0 directly off the graph and find a corresponding estimate of β_1 by taking any two points on the line, say (x_1, y_1) and (x_2, y_2), $x_1 < x_2$, and calculating $(y_2 - y_1)/(x_2 - x_1)$, that is, the *rise* divided by the *run*. The problem with the visual approach is that for the same scatter diagram, ten different people will obtain ten different lines. In fact, if the same individual is given the same scatter diagram ten times, he or she is likely to come up with ten different lines. Thus, as you can see, we need some criterion for determining what constitutes a good fit.

Obviously, since the line is supposed to fit the data as a whole, we would like the total distance of all the points from the line to be as small as possible. We might express this as:

Make

$$\Sigma\ (Y_i - \hat{Y}_i)$$

as near to zero as possible, where $\hat{Y}_i$ is the point on the line corresponding to Y_i. The problem with this criterion is that more than one line will yield a value of exactly zero for this sum—for example, let $\hat{Y}_i = \overline{Y}$ for all i. (See Figure 14.2.) The next expression we might consider is

$$\Sigma\,|Y_i - \hat{Y}_i|$$

Figure 14.2 One Possible Line for which $\sum_{i=1}^{n} (y_i - \hat{y}) = 0$

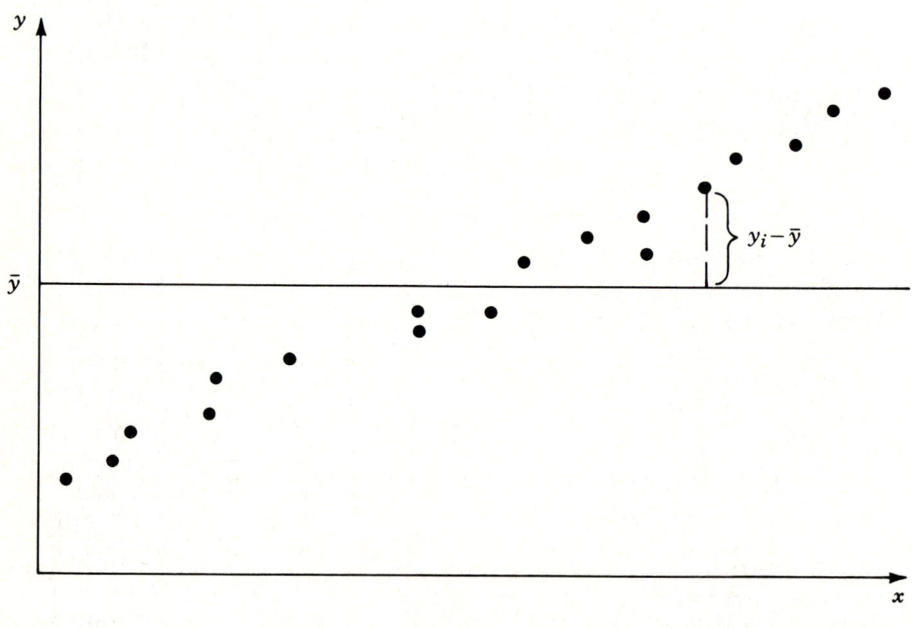

This is a legitimate measure of the fit, since, when minimized, it will yield a unique line and will equal zero only if all the observed values for Y are on the line. We discard this measure for two reasons: First, it is often difficult to work with absolute values, and second, their distributions are often almost impossible to determine. Finally, we come to the approach used in determining a measure of dispersion, namely, looking at the sum of the squared deviations.

The estimation procedure that minimizes the sum of the squared deviations of the observed values from the predicted values is called *L*east *S*quares *E*stimation (LSE). LSE is quite general and can be used for any model. If $\hat{Y}$ is the estimator of Y, then LSE says that we want to determine $\hat{Y}$ such that

$$\Sigma\,(Y - \hat{Y})^2$$

is a minimum when the sum is over all observed values of Y. Like the sum of the absolute differences, this will equal zero only if all the observed values are on the line. It increases as the actual deviations increase, and it yields a unique estimator when it is minimized. Since x is a nonrandom variable, its value will be known. Thus, to determine the estimator $\hat{Y}$, we need to determine only estimates for β_0 and β_1. That is, the LSE for $\hat{Y}$ coincides with the estimates for β_0 and β_1 that minimize

$$Q(b_0, b_1) = \sum_{i=1}^{n} (Y_i - b_0 - b_1 x_i)^2 \qquad (14.3)$$

The easiest way to determine the estimates of β_0 and β_1 that minimize $Q(b_0, b_1)$ is to take the partial derivatives with respect to each, equate them to zero, and solve them simultaneously. For those students who have a background in elementary calculus, we recommend they attempt it on their own. For the present we merely state the resulting equations. These are called the *normal equations* and are

$$\begin{aligned} nb_0 + b_1\,\Sigma\,x_i &= \Sigma\,Y_i \\ b_0\,\Sigma\,x_i + b_1\,\Sigma\,x_i^2 &= \Sigma\,x_i Y_i \end{aligned} \qquad (14.4)$$

In Equations 14.4 we have replaced β_0 and β_1 with b_0 and b_1, respectively, because b_0 and b_1 represent estimates of the corresponding parameters. That is, they are determined by the sample data. Solving the normal equations for b_0 and b_1 yields the formulas for our estimates. These are

$$b_1 = \frac{\sum_{i=1}^{n} x_i Y_i - (\Sigma x_i)(\Sigma Y_i)/n}{\sum_{i=1}^{n} x_i^2 - (\Sigma x_i)^2/n} \tag{14.5}$$

and

$$b_0 = \overline{Y} - b_1 \overline{x} \tag{14.6}$$

There are other alternative forms available for b_0 and b_1 (see the review exercises). We prefer the forms in Equations 14.5 and 14.6 because they involve less rounding and because the numerator and denominator for b_1 are values that will be required later. The numerator and denominator of b_1 are equivalent to

$$\Sigma (x_i - \overline{x})(Y_i - \overline{Y}) \quad \text{and} \quad \Sigma (x_i - \overline{x})^2$$

respectively. The denominator will be used often in our inferential procedures.

Once b_0 and b_1 have been calculated, the estimator for $\mu_{Y|x}$ (or predictor for Y_x) is given by

$$\hat{Y}_x = b_0 + b_1 x \tag{14.7}$$

The error term, ϵ, is no longer stated explicitly since its variability has been incorporated in the estimates b_0 and b_1. That is, b_0 and b_1 are random variables. To estimate $\mu_{Y|x}$ (or predict Y_x) we simply substitute the specific value of x into Equation 14.7.

EXAMPLE 14.2 Refer to Example 14.1 and determine the least-squares equation for predicting sales from advertising expenditures.

SOLUTION To facilitate calculations we reorganize the data into a tabular format.

Advertising (\$1000)	Sales (1000 units)		
x	y	xy	x^2
4	22	88	16
7	34	238	49
6	28	168	36
5	25	125	25
6	30	180	36
8	36	288	64
36	175	1087	226

Substituting these values into Equations 14.5 and 14.6 gives

$$b_1 = \frac{1087 - (36)(175)/6}{226 - (36)^2/6} = \frac{37}{10} = 3.7$$

and

$$b_0 = \frac{175}{6} - 3.7\left(\frac{36}{6}\right) = 29.167 - 22.2 = 6.967$$

Thus, our prediction equation is

$$\hat{Y}_x = 6.967 + 3.7x$$

where $\hat{Y}_x$ is sales in 1000s of units and x is advertising expenditure in \$1000. Using the equation, you determine the sales forecast by substituting the advertising expenditure for October, expressed in thousands of dollars, into the equation. That is, for a \$7000 expenditure, $x = 7$ and

$$\hat{Y}_7 = 6.967 + 3.7(7) = 32.867$$

or a sales forecast of 32,867 units.

Note that in Example 14.2 we made a point of identifying the units for x and Y. Without this identification a regression equation is useless, since another user would not know what value to use for x and would not be able to interpret the answer.

The values for b_0 and b_1 represent the intercept and slope of the line, respectively. On the practical side, b_0 represents the expected value for Y when $x = 0$ and b_1 is the change in Y for a one-unit change in x. For our example $b_0 = 6.967$, which implies that we expect the average sales with no advertising to be 6967 units. The problem with this interpretation of b_0 is that $x = 0$ is outside the range of our original x values. The value of 3.7 for b_1 implies that for each \$1000 spent on advertising, sales can be expected to increase by 3700 units on the average. Again this interpretation holds theoretically only within the range of the original x's. The reason for this restriction is that we have no information concerning the relationship between x and Y outside of that range. The relationship may change to nonlinear outside the range of experimentation, or, even though it still may be linear, the slope may change. For example, in Example 14.2, no matter how much is spent on advertising, we can't sell more than we can produce. On the other hand, there may be a saturation point for advertising, in which case any expenditures above the saturation point will have no effect on sales (they might possibly even be detrimental). In many applications in business and related areas it will be necessary to use the equation to extrapolate outside of the experimental range. Always use caution when doing so and remember that your equation may be inapplicable when used to extrapolate. See Figure 14.3.

So far we have assumed that a linear relationship is appropriate and have constructed our estimate of the model on this basis. Our purpose in doing so is to try to improve our accuracy in predicting Y by removing variation from the total sum of squares that can be considered as nonrandom. You should be aware that the line can always be calculated whether it is applicable or not. Thus, we need to be able to answer the following two questions: first, is the linear model applicable (or appropriate), and second, is the line useful (is it an improvement over the estimator $\overline{Y}$). In the next section we construct tests to help us answer these questions.

Figure 14.3 Illustration of a Possible Difficulty in Trying to Extrapolate with a Linear Model

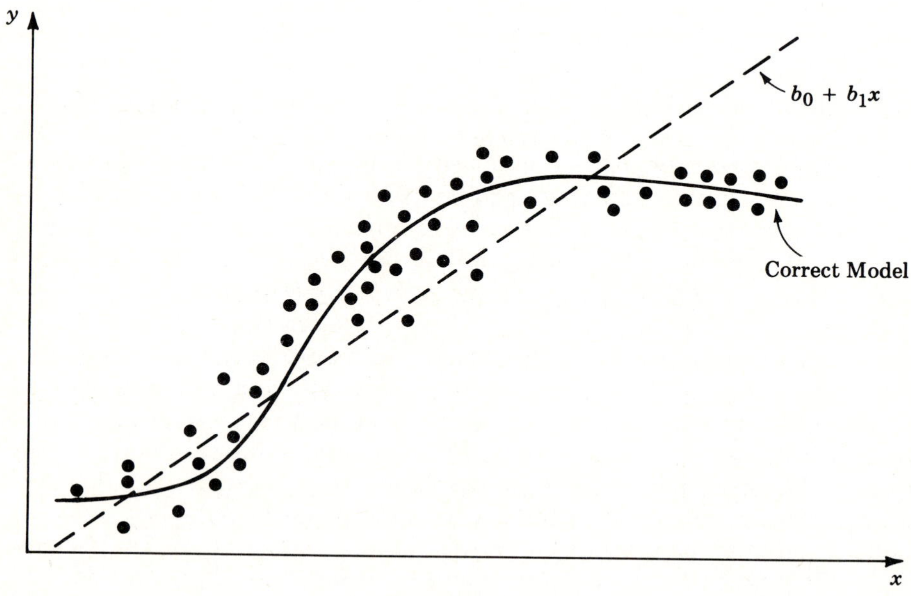

EXERCISES

14.3.1 Refer to Exercise 14.2.1 and visually fit a straight line to each of the sets of data given therein.

14.3.2 Determine the equations for the lines you constructed in Exercise 14.3.1.

14.3.3 Use the least-squares procedure to estimate a linear model for each of the sets of data in Exercise 14.2.1.

14.3.4 Use LSE to estimate a linear model for each of the sets of data you collected in Exercise 14.2.2.

14.3.5 Many instructors believe that there is a relationship between a student's score on the first exam and on future exams in the same course. The following data are from my student records. Use them to estimate a linear model for predicting a student's score on the second exam once the score on the first exam is known.

	Student									
	1	2	3	4	5	6	7	8	9	10
Exam 1 Score	72	90	68	75	72	65	84	90	75	84
Exam 2 Score	80	88	76	76	77	69	88	93	79	84

14.3.6 Refer to Exercise 14.3.5 and interpret your values for b_0 and b_1 in the context of the problem. Identify any restrictions that should be placed on your interpretations and on the equation in general.

14.4 INFERENCES FOR A SIMPLE LINEAR MODEL

As we indicated in the previous section, it is always possible to construct a straight line for the data. A straight line may be inappropriate, of no use, or both. In this section we present ways of testing the appropriateness of the model and whether or not it is useful. In addition, we derive procedures for constructing interval estimates for $\mu_{Y|x}$ and prediction intervals for Y_x. Since both tests and confidence intervals are probability statements, additional assumptions are required.

So far we have assumed only that x is a nonrandom variable and that ϵ is a random variable with an expected value of zero. For the procedures of this section to be exact, that is, for α to be as specified, we must be more specific about the distribution of ϵ. In particular, we must make the following assumptions to use $\hat{Y}$ to make statistical inferences.

1. $\epsilon \sim N(0, \sigma^2)$ for all values of x.
2. ϵ is independent of x.
3. ϵ_i and ϵ_j are independent for $i \neq j$.
4. The data are of an interval level of measurement.

The term *uncorrelated* is often substituted for *independence* in the assumptions, but the terms are equivalent when normality is assumed. A unique property of normal distributions is that the term uncorrelated

Figure 14.4 Graphical Representation of Regression Assumptions of Normality and Constant Variance

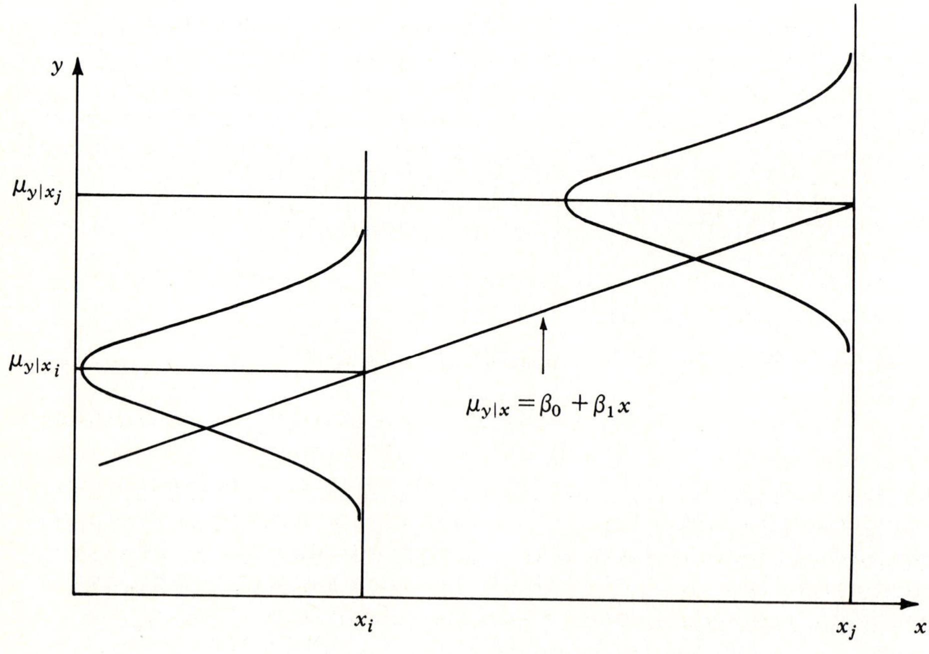

implies independence. Since the linear model expresses Y as a linear function of ϵ, we can restate assumption (1) as

1′. $Y_x \sim N(\mu_{Y|x}, \sigma^2)$

For a graphical illustration of assumption (1′), see Figure 14.4.

14.4.1 Is the Model Appropriate?

The first inference we consider is a test for the appropriateness of the model. That is, we would like to test the hypothesis set

$$H_0\colon \mu_{Y|x} = \beta_0 + \beta_1 x \qquad H_a\colon \mu_{Y|x} \neq \beta_0 + \beta_1 x$$

In words, H_0 specifies that the means of the Y populations are linearly related to x, whereas the alternative states that they are not. H_a does not state that there is no relationship; it states that if there is one, it is not linear. Note that this is a situation similar to the goodness-of-fit tests of Chapter 10 in that we prefer to accept the null hypothesis. Rejection of H_0 implies that we should try to formulate another model, and that can be a difficult thing to do.

To construct the test, we use the same approach that we took in ANOVA (the similarity will become obvious). In order to construct the test, we must be able to estimate the error variance independently of H_0. To do so requires replication for some of the x's; that is, we must have more than one observation on Y for some of the x's. Replication means the same thing here as it did in Chapter 12. We use a double subscript to indicate that there is more than one observation on Y for some x's. The subscript i will identify the particular observation for a specific x_j, and j will identify the x value. Thus, y_{ij} is the ith observation for Y corresponding to x_j.

We partition the total sum of squares for Y as follows, using the double subscript:

$$\sum_j \sum_i (Y_{ij} - \overline{Y})^2 = \sum_j \sum_i (Y_{ij} - \hat{Y}_j + \hat{Y}_j - \overline{Y})^2$$

$$\sum_j \sum_i (Y_{ij} - \overline{Y})^2 = \sum_j \sum_i (Y_{ij} - \hat{Y}_j)^2 + \sum_j \sum_i (\hat{Y}_j - \overline{Y})^2 \qquad (14.8)$$

since the cross-product term can be shown to be zero. If the model is correct—that is, if H_0 is true—the first term on the right arises only from error variation and hence can be used to measure the error variance. The second term arises from differences between the hypothesized model and the model estimated by $\overline{Y}$ and may be due either to chance or to the fact that β_1 is nonzero. These differences are illustrated in Figure 14.5 on page 536.

Figure 14.5 Graphical Representation of Components of $(y_i - \bar{y})$

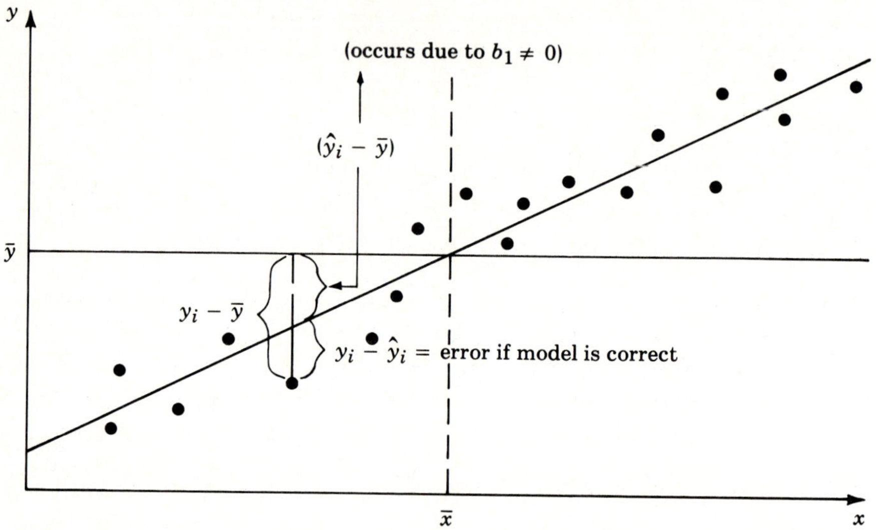

Neither of the terms on the right in Equation 14.8 is independent of the hypothesized model. Therefore, we need additional partitioning to obtain an estimate of σ^2 that is independent of the model. The fact that there are replications at some of the x_j's enables us to split the first term on the right in Equation 14.8. That is,

$$\sum_j \sum_i (Y_{ij} - \hat{Y}_j)^2 = \sum_j \sum_i (Y_{ij} - \overline{Y}_j)^2 + \sum_j \sum_i (\overline{Y}_j - \hat{Y}_j)^2 \qquad (14.9)$$

Note that the first term on the right in Equation 14.9 is precisely the SSE for a one-factor ANOVA from Chapter 12. Here we will call it the sum of squares for *pure error*, or SSPE. We use the term *pure error* because Y_{ij} differs from $\overline{Y}_j$ only due to chance variation regardless of the regression model, for the basic reason that the differences are calculated within a fixed value for x and then pooled. If the model is correct, the second term on the right in Equation 14.9 is also a measure of chance variation, since both $\overline{Y}_j$ and $\hat{Y}_j$ are estimating the same thing, namely $\mu_{Y|x}$. If the model is incorrect, $\overline{Y}_j$ still estimates $\mu_{Y|x}$, but $\hat{Y}_j$ is estimating something else. We have illustrated these concepts in Figure 14.6. As the graph indicates, $\Sigma\Sigma(\overline{Y}_j - \hat{Y}_j)^2$ measures the "lack of fit" between the observed sample means $\overline{Y}_j$ and the fitted model $\hat{Y}_j$.

Summarizing our results to this point, we have

$$\sum_j \sum_i (Y_{ij} - \overline{Y})^2 = \underbrace{\sum_j \sum_i (Y_{ij} - \overline{Y}_j)^2 + \sum_j \sum_i (\overline{Y}_j - \hat{Y}_j)^2}_{\sum_j \sum_i (Y_{ij} - \hat{Y}_j)^2} + \sum_j \sum_i (\hat{Y}_j - \overline{Y})^2$$

or

$$\text{SSTO} = \underbrace{\text{SSPE} + \text{SSLOF}}_{\text{SSE}} + \text{SSREG}$$

To construct the test, we need the distributions of the SSPE and SSLOF (sum of squares for *lack of fit*). If the linear model is correct, that is, if H_0: $\mu_{Y|x} = \beta_0 + \beta_1 x$ is true, then SSE is simply the numerator for the pooled variance of samples from normal distributions. As such, it will be distributed as $\sigma^2 \chi^2$ with $n - 2$ degrees of freedom.

Figure 14.6 Graphical Representation of Residual Components when the True Model is Nonlinear

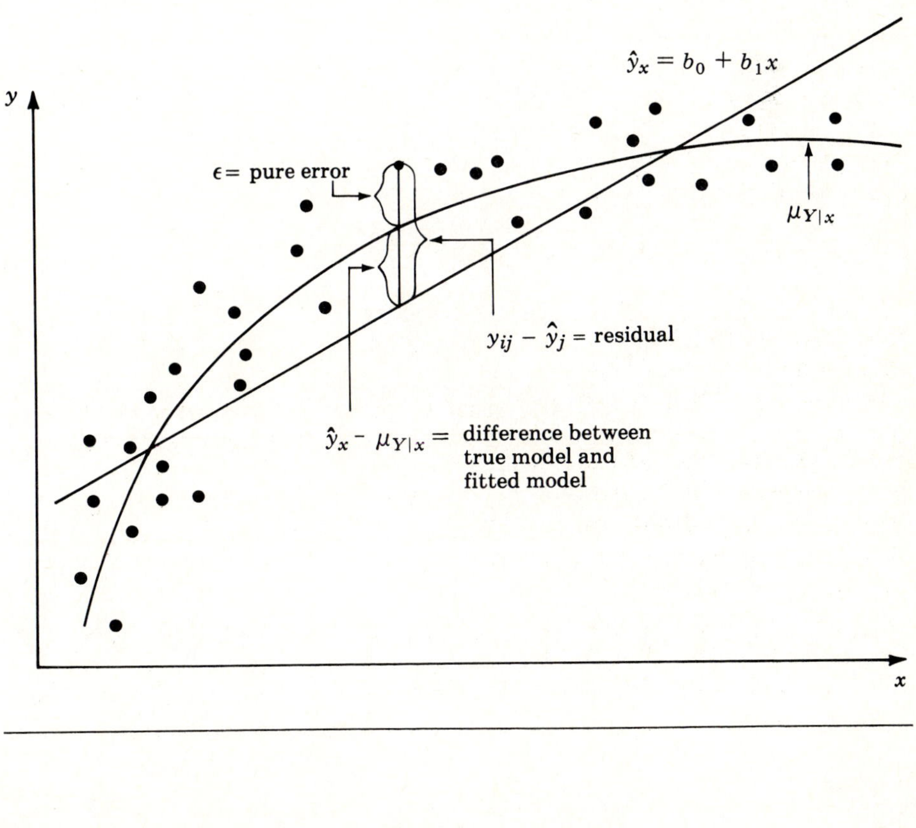

There are $n - 2$ degrees of freedom because we lose one for each of b_0 and b_1. Referring to Equation 14.9, we see that SSPE has $\sum_j n_j = n$ variables with one degree of freedom lost for each $\overline{Y}_j$, or for each different value of x; or SSPE has $n - c$ degrees of freedom, where c is the number of different x values. For SSLOF we have c variables with one degree of freedom lost for each of b_0 and b_1. That is, SSLOF has $c - 2$ degrees of freedom. Therefore,

$$\left(\frac{1}{\sigma^2}\right)\text{SSE} = \left(\frac{1}{\sigma^2}\right)\text{SSPE} + \left(\frac{1}{\sigma^2}\right)\text{SSLOF}$$

Since $(1/\sigma^2)\text{SSE} \sim \chi^2_{n-2}$ and $n - 2 = (n - c) + (c - 2)$, Cochran's theorem implies that SSPE and SSLOF are independent with each having a $\sigma^2\chi^2$ distribution. Converting each to a mean square by dividing by its respective degrees of freedom, we have that MSPE and MSLOF are estimates of σ^2 when H_0 is true. If H_0 is false, MSPE still estimates σ^2, but MSLOF estimates σ^2 plus a positive term representing the lack of fit of the model.

With these results, the decision rule for testing

$$H_0\colon\ \mu_{Y|x} = \beta_0 + \beta_1 x \qquad H_a\colon\ \mu_{Y|x} \neq \beta_0 + \beta_1 x$$

is to reject H_0 at α if

$$\frac{\text{MSLOF}}{\text{MSPE}} > f_{c-2,\, n-c,\, \alpha} \tag{14.10}$$

Table 14.1 Data for Example 14.3 (Sales in 1000s of Units and Advertising in \$1000)

Advertising	Sales	Advertising	Sales	Advertising	Sales
3	14	7	30	4	22
2	13	9	39	7	34
8	28	6	27	6	28
4	20	4	23	5	25
5	24	6	31	6	30
5	26	7	33	8	36

since, under H_0, MSLOF/MSPE has an $F_{c-2,\, n-c}$ distribution. As usual, critical values are found in Table A.9 of the Appendix. We reject H_0 only for large values of MSLOF/MSPE, since its expected value is 1 under H_0 and greater than 1 under H_a.

EXAMPLE 14.3 Having given production your sales forecast for next month, you decide that you really should be a bit more organized the next time. Since you have almost an hour until lunch, you call Winston, your secretary, and ask him to bring in the advertising expenditures and corresponding sales for the past 18 months. He brings you the data in Table 14.1. Use these data to test the appropriateness of the simple linear model. Use $\alpha = 0.05$.

SOLUTION The first step is to calculate the estimated line for these data. Calculating the required sums, we obtain (noting that there are only eight different values for x):

$$\sum_{j=1}^{8} n_j x_j = 102 \qquad \sum_{j=1}^{8} \sum_{i=1}^{n_j} y_{ij} = 483 \qquad \sum_{j=1}^{8} n_j x_j^2 = 636$$

$$\sum_{j=1}^{8} \sum_{i=1}^{n_j} y_{ij}^2 = 13{,}775 \qquad c = 8 \qquad n = \Sigma n_j = 18$$

$$\sum_{j=1}^{8} \sum_{i=1}^{n_j} x_j y_{ij} = 2941$$

Thus,

$$b_1 = \frac{2941 - (102)(483)/18}{636 - (102)^2/18} = \frac{2941 - 2737}{636 - 578} = 3.517$$

$$b_0 = \frac{483}{18} - 3.517\left(\frac{102}{18}\right) = 26.833 - 19.930 = 6.903$$

Therefore, the estimate of the linear model is

$$\hat{Y}_x = 6.903 + 3.517x$$

where Y is sales in thousands of units and x is advertising expenditure in \$1000s.

To test the aptness of the linear model, the hypotheses are

$$H_0\colon \mu_{Y|x} = \beta_0 + \beta_1 x \qquad H_a\colon \mu_{Y|x} \neq \beta_0 + \beta_1 x$$

The null hypothesis (and hence the model) will be rejected if MSLOF/MSPE $> f_{c-2,\, n-c,\, 0.05} = f_{6,\, 10,\, 0.05} = 3.22$. To facilitate the computations, we reorganize the data as in a one-factor design.

	Advertising Level (x_j)							
	2	3	4	5	6	7	8	9
	13	14	20	24	27	30	28	39
			23	26	31	33	36	
			22	25	28	34		
					30			
$\Sigma\, y_{ij}$:	13	14	65	75	116	97	64	39

Then,

$$\text{SSPE} = \sum_j \sum_i (y_{ij} - \bar{y}_j)^2 = \sum_j \sum_i y_{ij}^2 - \frac{\sum_j \left(\sum_i y_{ij}\right)^2}{n_j}$$
$$= 13{,}775 - 13{,}717.66 = 57.334$$

To calculate SSLOF, we find the $\hat{Y}_{x_j}$ by substituting the x_j's into our equation. Doing so gives

$$\hat{Y}_2 = 6.903 + 3.517(2) = 13.937$$
$$\hat{Y}_3 = 6.903 + 3.517(3) = 17.454$$
$$\hat{Y}_4 = 20.971 \qquad \hat{Y}_5 = 24.488 \qquad \hat{Y}_6 = 28.005$$
$$\hat{Y}_7 = 31.522 \qquad \hat{Y}_8 = 35.039 \qquad \hat{Y}_9 = 38.556$$

The corresponding $\bar{y}_j$'s are: $\bar{y}_2 = 13$, $\bar{y}_3 = 14$, $\bar{y}_4 = 21.667$, $\bar{y}_5 = 25$, $\bar{y}_6 = 29$, $\bar{y}_7 = 32.333$, $\bar{y}_8 = 32$, and $\bar{y}_9 = 39$, where the subscript refers to the value of x. Thus,

$$\text{SSLOF} = \sum_j \sum_i (\bar{Y}_j - \hat{Y}_j)^2 = \sum_j n_j(\bar{Y}_j - \hat{Y}_j)^2$$
$$= (13 - 13.937)^2 + (14 - 17.454)^2 + 3(21.667 - 20.971)^2$$
$$+ \cdots + 2(32 - 35.039)^2 + (39 - 38.556)^2 = 39.649$$

The mean squares are MSPE $= 57.334/10 = 5.7334$ and MSLOF $= 39.649/6 = 6.608$.

Therefore, the value of the test statistic is MSLOF/MSPE = 6.608/5.7334 = 1.153 < 3.22. Hence, you accept H_0 and conclude that there is no reason to doubt the appropriateness of the linear model.

The conclusion in Example 14.3 is that the relationship between sales and advertising appears to be linear. We stress that this result refers to the relationship within the advertising range of \$2000 to \$9000. It does not say anything about the relationship when the advertising expenditure is either below \$2000 or above \$9000. If H_0 had been rejected in Example 14.3, this would imply that the model should include some nonlinear terms in x, possibly x^2 or x^3. The new model could include the linear term also.

The test that we just constructed indicates whether or not the linear model appears to be appropriate but does not say anything about its usefulness. Since we can always construct a line, we need to be able to determine if it is useful. That is, we need to determine whether it is a better predictor than $\overline{Y}$ alone. This is the topic of the next section.

14.4.2 Is the Model Useful?

The question, "Is the model useful?" really asks whether it explains a significant amount of the variation between the y values. If the model does explain a significant portion of the total variation in Y, the sum of squares attributed to the error (chance) term is reduced accordingly. This results in a smaller value for our estimate of σ^2 and implies that using $\hat{Y}$ rather than $\overline{Y}$ will give us better precision in our predictions. To help illustrate this idea, we enlarge a section of Figure 14.5 as Figure 14.7 on page 542.

Recalling our initial partition of SSTO in Equation 14.8, we have (ignoring the subscript j and letting i go from 1 to n)

$$\sum_i (Y_i - \overline{Y})^2 = \sum_i (Y_i - \hat{Y}_i)^2 + \sum_i (\hat{Y}_i - \overline{Y})^2$$

We wish to test the hypotheses

$$H_0\text{: model is not useful} \qquad H_a\text{: model is useful}$$

From our previous discussion, these can be restated as

$$H_0\text{: } \mu_{Y|x} = \mu \text{ for all } x \qquad H_a\text{: } \mu_{Y|x} \neq \mu \text{ for all } x$$

Figure 14.7 Graphical Representation of the Components of a Total Deviation when the True Model is Linear

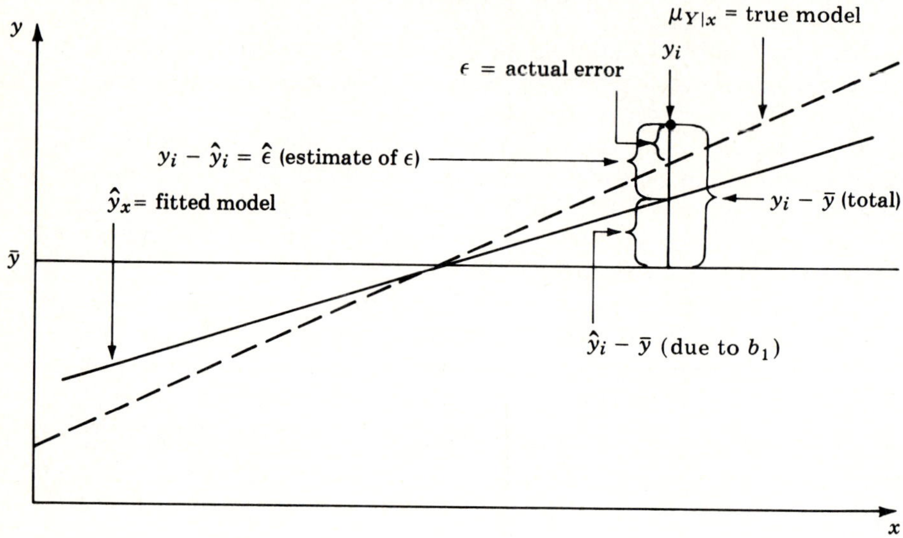

If H_0 is true, our best estimate for $\mu_{Y|x}$ is simply $\overline{Y}$, since β_1 would be zero and $\beta_0 = \mu$. But if H_0 is not true, then $\hat{Y}_x$ should be near $\mu_{Y|x}$ and differs from $\overline{Y}$ because $b_1 \neq 0$. Now, b_1 is a random variable, and so it may differ from zero by chance or it may differ from zero because the parameter that it estimates, β_1, is nonzero. Thus, our hypotheses may be rewritten as

$$H_0\colon\ \beta_1 = 0 \qquad H_a\colon\ \beta_1 \neq 0 \tag{14.11}$$

Under our assumptions, if H_0 is true, then $\Sigma\,(Y_i - \overline{Y})^2$ is the numerator of the variance for a random sample from a normal distribution. As such its distribution will be $\sigma^2\chi^2_{n-1}$. As we saw previously, this can be partitioned into

$$\begin{array}{ccccc} \Sigma\,(Y - \overline{Y})^2 & = & \Sigma\,(Y - \hat{Y})^2 & + & \Sigma\,(\hat{Y} - \overline{Y})^2 \\ \text{SSTO} & = & \text{SSE} & + & \text{SS}(b_1) \end{array}$$

SSE has $n - 2$ degrees of freedom, since there are n Y values and one degree of freedom is lost for each of b_0 and b_1. SS(b_1), on the other hand, has two random variables, b_1 and b_0, with a single restriction imposed by $\overline{Y}$; thus its degrees of freedom are 2 - 1 or 1. Since $(n-2)+1=n-1$, we can apply Cochran's theorem, finding that

$$\text{SSE} \sim \sigma^2 \chi^2_{n-2} \qquad \text{and} \qquad \text{SS}(b_1) \sim \sigma^2 \chi^2_1$$

provided that H_0 is true. As in Chapter 12, we need to determine the expected mean squares to see how the test should be constructed. These are:

$$E[\text{MSE}] = \sigma^2 \tag{14.12}$$

and

$$E[\text{MS}(b_1)] = \sigma^2 + \beta_1^2 \sum (x_i - \overline{x})^2 \tag{14.13}$$

Equation 14.12 holds regardless of H_0 if we have the correct model. If H_0 is true, Equation 14.13 reduces to σ^2 also, but if H_0 is false, $E[\text{MS}(b_1)] > \sigma^2$.

Since both MSE and $\text{MS}(b_1)$ are estimating the same thing under H_0, the ratio, $\text{MS}(b_1)/\text{MSE}$, should be close to 1 if H_0 is true. If H_a is true, this ratio should be greater than 1. Thus the obvious test is to reject H_0 if the ratio gets too large; that is, to reject H_0 if

$$\frac{\text{MS}(b_1)}{\text{MSE}} > f_{1,n-2,\alpha} \tag{14.14}$$

The F distribution is appropriate, since for H_0 true the statistic is the ratio of two independent χ^2 variables divided by their respective degrees of freedom.

There are also computational forms that we can use to determine the sums of squares in this case. These are presented in Table 14.2.

Table 14.2 Definitional and Computational Formulas for Calculating Sums of Squares for a Simple Linear Regression

Sum of Squares	Definitional Form	Computational Form
Regression[$\text{SS}(b_1)$]	$\sum (\hat{y}_i - \overline{y})^2$	$b_1 \left[\sum x_i y_i - \dfrac{(\sum x_i)(\sum y_i)}{n}\right]$
Error (SSE)	$\sum (y_i - \hat{y}_i)^2$	$\sum y_i^2 - b_0 \sum y_i - b_1 \sum x_i y_i$
Total (SSTO)	$\sum (y_i - \overline{y})^2$	$\sum y_i^2 - \dfrac{(\sum y_i)^2}{n}$

Table 14.3 The ANOVA Table for a Simple Linear Regression with Replication at Some x Values [$W_j = \eta_j - E(Y_j|H_0)$, where η_j is the expected value of Y at x_j for the correct model]

Source	SS	df	MS	E(MS)	F Ratio
Regression (b_1)	SS(b_1)	1	MS(b_1)	$\sigma^2 + \beta_1^2 \Sigma (x - \bar{x})^2$	$\frac{\text{MS}(b_1)}{\text{MSE}}$
Error	SSE	$n - 2$	MSE		
Lack of Fit	SSLOF	$c - 2$	MSLOF	$\sigma^2 + \frac{1}{c-2} \sum_j n_j W_j^2$	$\frac{\text{MSLOF}}{\text{MSPE}}$
Pure Error	SSPE	$n - c$	MSPE	σ^2	
Total	SSTO	$n - 1$			

As with the ANOVA procedures of Chapter 12, the results are usually summarized in an ANOVA table. Table 14.3 shows the general form. We have included, as part of the table, the terms that are relevant to testing the model for lack of fit. Remember that without replication at some of the x's these cannot be calculated.

EXAMPLE 14.4 Refer to Example 14.3, in which we concluded that the simple linear model seemed appropriate for the relationship between sales and advertising expenditures. You would now like to know whether your equation is useful; that is, whether it is better than $\overline{Y}$ as an estimator, at $\alpha = 0.05$.

SOLUTION In Example 14.3 we determined that the equation is $\hat{Y}_x = 6.903 + 3.517x$ with x in \$1000 and Y in 1000s of units. In addition, $\Sigma\, y = 483$, $\Sigma\, x = 102$, $\Sigma\, y^2 = 13{,}775$, $\Sigma\, x^2 = 636$, $\Sigma\, xy = 2941$, and $n = 18$.

The hypotheses to be tested are:

$$H_0\colon\ \beta_1 = 0 \qquad H_a\colon\ \beta_1 \neq 0$$

and H_0 will be rejected at $\alpha = 0.05$ if

$$\frac{\text{MS}(b_1)}{\text{MSE}} > f_{1,\, n-2,\, 0.05} = f_{1,\, 16,\, 0.05} = 4.49$$

Calculating the required sums of squares, we have (using the computational formulas)

$$\text{SSTO} = \Sigma\, y^2 - \frac{(\Sigma\, y)^2}{n} = 13{,}775 - \frac{(483)^2}{18} = 814.5$$

$$\begin{aligned}\text{SSE} &= \Sigma\, y^2 - b_0\, \Sigma\, y - b_1\, \Sigma\, xy \\ &= 13{,}775 - 6.903(483) - 3.517(2941) = 97.354\end{aligned}$$

$$\begin{aligned}\text{SS}(b_1) &= b_1\left[\Sigma\, xy - \frac{(\Sigma\, x)(\Sigma\, y)}{n}\right] = 3.517\left[2941 - \frac{(102)(483)}{18}\right] \\ &= 717.468\end{aligned}$$

The sum of SSE and $\text{SS}(b_1)$ differs slightly from SSTO as a result of rounding error. If we do not round to three places, $\text{SS}(b_1) = 717.52$ and $\text{SSE} = 96.98$. Rounding will normally be necessary when the values are calculated by hand. Usually the calculations are done on a computer so that rounding is minimal and will not be a problem in this situation. We use the unrounded values to construct the ANOVA table in Table 14.4 with the results of Example 14.3 included.

Since $118.4 > 4.49$, we reject H_0 and conclude that the equation is useful. In other words, the variation in advertising expenditures explains a significant amount of the variation in sales.

To relate the last two tests, we make the following comments (Test I refers to the test for aptness of the model, and Test II refers to the test for usefulness).

1. If both tests accept H_0, then the simple linear model is appropriate but it is no better than the estimator $\overline{Y}$. Therefore, you should attempt to find a new predictor and start over.
2. If Test I accepts H_0 and Test II rejects H_0 then the simple linear model is correct and is an improvement over $\overline{Y}$.
3. If Test I rejects H_0 and Test II accepts H_0, then the model is incorrect and is not useful. You might consider a higher-order model in x, since Test I indicates a nonlinear relationship. (The linear term in x should be retained, at least initially.)
4. If both tests reject H_0, then the model is incorrect but it is still useful within the given range of x values. You should try a higher-order model in x including the linear term, such as $Y_x = \beta_0 + \beta_1 x + \beta_2 x^2 + \epsilon$.

In our discussion we have continually referred to the amount of variation in Y that is explained by x. Another way of measuring the usefulness of the model is to use this concept.

Table 14.4 Analysis-of-Variance Table of Examples 14.4 and 14.3

Source	SS	df	MS	F Calc
Regression (b_1)	717.52	1	717.52	118.40
Error	96.98	16	6.06	
LOF	39.65	6	6.61	1.15
PE	57.33	10	5.73	
Total	814.50	17		

DEFINITION 14.1 The coefficient of determination for a regression model, denoted by R^2, is defined to be the ratio of the regression sum of squares to the total sum of squares,

$$R^2 = \frac{\text{SSReg}}{\text{SSTO}}$$

For the simple linear model, SSReg = SS(b_1). Thus R^2 (100%) indicates the percent of explained variation due to x for a simple linear model. The percent unexplained is simply $(1 - R^2)(100\%)$ or (SSE/SSTO)(100%); this represents the variation due to chance and other possible predictors that are not in the model. For Example 14.4, the percentage of variation explained by the linear model is

$$R(100\%) = \left[\frac{\text{SS}(b_1)}{\text{SSTO}}\right](100\%) = \left(\frac{717.52}{814.50}\right)(100\%) = 88.1\%$$

That is, 88.1% of the variation in sales is explained by the advertising expenditures.

In developing our test for usefulness, we considered only the alternative H_a: $\beta_1 \neq 0$ and used the F test. Since the numerator has only one degree of freedom, we also can use a t test. If the alternative is in any form other than $\beta_1 \neq 0$ or if we want a confidence interval for β_1, then the t distribution must be used. b_1 will have a normal distribution, since

$$b_1 = \frac{\Sigma (x_i - \bar{x})(Y_i - \bar{Y})}{\Sigma (x_i - \bar{x})^2} = \frac{\Sigma (x_i - \bar{x})Y_i}{\Sigma (x_i - \bar{x})^2} = \Sigma c_i Y_i$$

is a linear combination of normal random variables. Also, if the model is correct,

$$E(b_1) = \beta_1$$

and

$$\text{Var}(b_1) = \frac{\sigma^2}{\Sigma (x_i - \bar{x})^2} \tag{14.15}$$

where σ^2 is estimated by the MSE, which we denote by

$$s^2_{Y|x} = \text{MSE} \tag{14.16}$$

Therefore the statistic

$$\frac{b_1 - \beta_1}{s_{b_1}}$$

has a t distribution with $n - 2$ degrees of freedom if the model is correct. The estimated standard deviation for b_1 is given by

$$s_{b_1} = \sqrt{\frac{\text{MSE}}{\Sigma\,(x - \bar{x})^2}} = \frac{s_{Y|x}}{\sqrt{\Sigma\,(x - \bar{x})^2}} \qquad (14.17)$$

The value for $\Sigma\,(x - \bar{x})^2$ will have already been calculated. It is the denominator of b_1 in Equation 14.5. The test for a given hypothesis set follows the same pattern as our previous t tests. The possible hypothesis sets and their corresponding t tests are given in Table 14.5.

In Table 14.5, if $\beta_{01} = 0$, then the calculated value for t will be $\sqrt{F_{\text{calc}}}$, with the associated sign being the same as the sign of b_1.

To obtain a confidence interval, we solve the appropriate decision rule in Table 14.5 for β_1. For a two-sided interval the limits are

$$\text{LL} = b_1 - t_{n-2,\,\alpha/2}\, s_{b_1} \qquad \text{and} \qquad \text{UL} = b_1 + t_{n-2,\,\alpha/2}\, s_{b_1}$$

One-sided intervals are obtained in the usual manner.

EXAMPLE 14.5 It is logical to assume that any relationship between advertising and sales would be a positive one—otherwise, a lot of money is being spent foolishly on advertising. Use the results of

Table 14.5 A Summary of Possible Hypotheses and the Corresponding t Tests

Hypothesis Set		Decision Rule for a Stated α
H_0: $\beta_1 = \beta_{01}$	H_a: $\beta_1 \neq \beta_{01}$	Reject H_0 if $\left\lvert\frac{(b_1 - \beta_{01})}{s_{b_1}}\right\rvert > t_{n-2,\,\alpha/2}$
H_0: $\beta_1 \leqslant \beta_{01}$	H_a: $\beta_1 > \beta_{01}$	Reject H_0 if $\frac{(b_1 - \beta_{01})}{s_{b_1}} > t_{n-2,\,\alpha}$
H_0: $\beta_1 \geqslant \beta_{01}$	H_a: $\beta_1 < \beta_{01}$	Reject H_0 if $\frac{(b_1 - \beta_{01})}{s_{b_1}} < -t_{n-2,\,\alpha}$

Example 14.4 to test for a positive relationship between advertising expenditures and sales, and construct the appropriate one-sided confidence interval for β_1. Use $\alpha = 0.05$.

SOLUTION From Example 14.4, we have that $n = 18$, $b_1 = 3.517$, $\text{MSE} = s^2_{Y|x} = 6.06$, and $\Sigma (x - \bar{x})^2 = 58$ (from Example 14.3). Thus,

$$s_{b_1} = \sqrt{\frac{6.06}{58}} = 0.323$$

The hypotheses for testing for a positive relationship are

$$H_0\colon \beta_1 \leqslant 0 \qquad H_a\colon \beta_1 > 0$$

and H_0 is to be rejected at 0.05 if

$$\frac{(b_1 - 0)}{s_{b_1}} > t_{n-2,\,\alpha} = t_{16,\,0.05} = 1.746$$

We have that $(b_1 - 0)/s_{b_1} = 3.517/0.323 = 10.889 > 1.746$. Therefore, we reject H_0 and conclude that sales appear to increase with advertising. Note that, except for rounding, we get the same value for t from

$$+\sqrt{F_{\text{calc}}} = +\sqrt{118.40} = 10.881$$

For the one-sided confidence interval for β_1, we want a lower bound only. That is, our interval is given by $[\text{LL}, \infty)$ where

$$\text{LL} = b_1 - t_{n-2,\,\alpha} s_{b_1} = 3.517 - 1.746(0.323) = 2.953$$

Thus, we are 95% confident that the interval $[2.953, \infty)$ includes the true slope, β_1.

14.4.3 Using the Equation as an Estimator

In the previous two sections we determined methods for testing if the model is appropriate and whether or not it is useful. We now turn to its actual use once it is considered to be useful. We will use the equation in two basic ways. One way is to estimate and place bounds on the mean of a population of Y-values for a given x. The second way is to use the equation to forecast an actual value of Y at a given x. We use the term forecast in the second case because we are trying to predict the value of a random variable and this entails much more chance for error.

We consider first the problem of estimating and determining a confidence interval for $\mu_{Y|x}$, the mean of Y_x. As we saw earlier $\hat{Y}_x$ is an unbiased estimator for $\mu_{Y|x}$ if our model is correct; for example

$$E[\hat{Y}_x] = E[b_0 + b_1 x] = \beta_0 + \beta_1 x = \mu_{Y|x}$$

Letting x_p be the value of x that is to be used for predicting $\mu_{Y|x}$ we need only substitute x_p into our equation to obtain the point estimate. To obtain a confidence interval requires the distribution of $\hat{Y}_{x_p}$. Since

$$\hat{Y}_{x_p} = b_0 + b_1 x_p = (\overline{Y} - b_1 \bar{x}) + b_1 x_p = \overline{Y} + b_1 (x_p - \bar{x})$$

is a linear combination of normal variables and unbiased, it follows that $\hat{Y}_{x_p}$ is normally distributed with mean $\mu_{Y|x_p}$. For the variance we have ($\overline{Y}$ and b_1 are independent)

$$\begin{aligned} \mathrm{Var}(\hat{Y}_{x_p}) &= \mathrm{Var}[\overline{Y} + b_1 (x_p - \bar{x})] = \mathrm{Var}(\overline{Y}) + (x_p - \bar{x})^2 \,\mathrm{Var}(b_1) \\ &= \frac{\sigma^2}{n} + \frac{\sigma^2 (x_p - \bar{x})^2}{\Sigma (x - \bar{x})^2} \end{aligned} \tag{14.18}$$

Replacing σ^2 with its estimate, MSE $= s^2_{Y|x}$, implies that

$$\frac{\hat{Y}_{x_p} - \mu_{Y|x_p}}{s_{\hat{Y}}}$$

has a t distribution with $n - 2$ degrees of freedom. Setting up the probability statement and solving for $\mu_{Y|x_p}$ in the usual manner gives $(1 - \alpha)(100\%)$ bounds of

$$\mathrm{LL} = \hat{Y}_{x_p} - t_{n-2,\,\alpha/2} \sqrt{\mathrm{MSE}\left[\frac{1}{n} + \frac{(x_p - \bar{x})^2}{\Sigma (x - \bar{x})^2}\right]} \tag{14.19a}$$

and

$$\mathrm{UL} = \hat{Y}_{x_p} + t_{n-2,\,\alpha/2} \sqrt{\mathrm{MSE}\left[\frac{1}{n} + \frac{(x_p - \bar{x})^2}{\Sigma (x - \bar{x})^2}\right]} \tag{14.19b}$$

EXAMPLE 14.6 Construct a 95% confidence interval for the average monthly sales for all months with a \$7000 advertising expenditure. (Refer to Examples 14.3 and 14.4.)

SOLUTION Our equation is

$$\hat{Y}_x = 6.903 + 3.517x$$

and for \$7000 in advertising $x_p = 7.0$. Thus, the estimate for $\mu_{Y|7}$ is

$$\hat{Y}_7 = 6.903 + 3.517(7) = 31.522 \text{ (thousand units)}$$

For the confidence interval,

$$\sqrt{\text{MSE}}\sqrt{\frac{1}{n} + \frac{(x_p - \bar{x})^2}{\Sigma\,(x - \bar{x})^2}} = \sqrt{6.06}\sqrt{\frac{1}{18} + \frac{(7 - 5.667)^2}{58}}$$

$$= (2.462)\sqrt{0.0556 + 0.0306}$$

$$= 2.462(0.2936)$$

Therefore, 95% bounds on $\mu_{Y|7}$ are

$$\text{LL} = 31.522 - (2.12)(0.7228) = 29.990 \quad \text{(1000 units)}$$

and

$$\text{UL} = 31.522 + 2.12(0.7228) = 33.054 \quad \text{(1000 units)}$$

That is, we are 95% confident that the interval [29,990 to 33,054] will contain the true average sales when advertising is set at \$7000 per month.

In many situations we will be interested in using the relationship between x and Y to predict or forecast a single response at $x = x_p$, not to estimate the average response. That is, we will be interested in what is going to occur the next time, not what is going to occur on the average. In trying to predict Y_{x_p} we are attempting to forecast the value for a random variable rather than a constant. Thus, in determining the standard deviation for placing prediction limits on Y_{x_p} we must include the variability of Y_{x_p} also.

To obtain a point estimate for Y at $x = x_p$, we note that the "best" estimate of a variable's value is its expected value, that is, its mean. In this situation $E[Y_{x_p}] = \mu_{Y|x_p}$ is unknown and must itself be estimated. From our previous considerations we know that the "best" estimate of $\mu_{Y|x_p}$ is $\hat{Y}_{x_p}$ if the model is correct. Therefore, our point estimate or prediction for Y_{x_p} is $\hat{Y}_{x_p}$, the same estimator that we used for $\mu_{Y|x_p}$.

To determine a prediction interval for Y_{x_p}, we must have the distribution of $\hat{Y}_{x_p}$. We already know that, under our assumptions,

$$\hat{Y}_{x_p} \sim N\left[\mu_{Y|x_p}, \sigma^2\left(\frac{1}{n} + \frac{(x_p - \bar{x})^2}{\Sigma (x - \bar{x})^2}\right)\right]$$

Since Y_{x_p} is an observation from a normal distribution, it is also a normal random variable and is distributed as $N(\mu_{Y|x_p}, \sigma^2)$. Since we are interested in how far off our prediction might be, we consider the prediction error given by $Y_{x_p} - \hat{Y}_{x_p}$. As the difference of two normal random variables, this is also normally distributed and has a mean of zero. Since Y_{x_p} is a new random observation, it is independent of $\hat{Y}_{x_p}$, and hence the variance of the prediction error is given by

$$\sigma^2_{Y_{x_p} - \hat{Y}_{x_p}} = \sigma^2_{Y_{x_p}} + \sigma^2_{\hat{Y}_{x_p}} = \sigma^2 + \sigma^2\left[\frac{1}{n} + \frac{(x_p - \bar{x})^2}{\Sigma (x - \bar{x})^2}\right]$$

or

$$\sigma^2_{Y_x - \hat{Y}_x} = \sigma^2\left[1 + \frac{1}{n} + \frac{(x_p - \bar{x})^2}{\Sigma (x - \bar{x})^2}\right] \tag{14.20}$$

Replacing σ with its estimate $s_{Y|x}$ in the standardization formula gives a t distribution with $n - 2$ degrees of freedom. That is,

$$\frac{Y_{x_p} - \hat{Y}_{x_p}}{s_{Y_{x_p} - \hat{Y}_{x_p}}} \sim t_{n-2}$$

Therefore a $(1 - \alpha) \times 100\%$ two-sided prediction interval for Y_{x_p} is given by

$$\begin{aligned} \text{LL} &= \hat{Y}_{x_p} - t_{n-2,\,\alpha/2}\sqrt{\text{MSE}}\sqrt{1 + \frac{1}{n} + \frac{(x_p - \bar{x})^2}{\Sigma (x - \bar{x})^2}} \\ \text{UL} &= \hat{Y}_{x_p} + t_{n-2,\,\alpha/2}\sqrt{\text{MSE}}\sqrt{1 + \frac{1}{n} + \frac{(x_p - \bar{x})^2}{\Sigma (x - \bar{x})^2}} \end{aligned} \tag{14.21}$$

To find one-sided prediction limits, we make the standard adjustments.

Before turning to an example, let us point out that the width of the interval is influenced by the distance of x_p from $\bar{x}$, the mean of the original set of x values. As x_p moves farther away from $\bar{x}$, the intervals given by Equations 14.19 and 14.21 become wider. The intervals are narrowest at $x_p = \bar{x}$, since the last term under the radical is zero at that point and increases as the difference between x_p and $\bar{x}$ gets larger. Figure 14.8 gives graphical illustrations of both intervals. Note that the interval for $\mu_{Y|x_p}$ (solid curve) is always contained in the interval for Y_{x_p} (dashed curve).

EXAMPLE 14.7 Production actually asked you for a forecast of next month's sales, not the average sales for all months in which the advertising expenditure was \$7000. Your forecast is unchanged, but you would like to place a 95% lower bound on your sales forecast. Using the information in Example 14.6, do so.

SOLUTION Referring to Example 14.6, we see that the only differences between that lower bound and the desired one is a one under the radical and a change in the t value. The forecast is 31.522

Figure 14.8 Graphical Representations of Confidence Intervals (CI) for $\mu_{Y|xp}$ and Y_{xp}

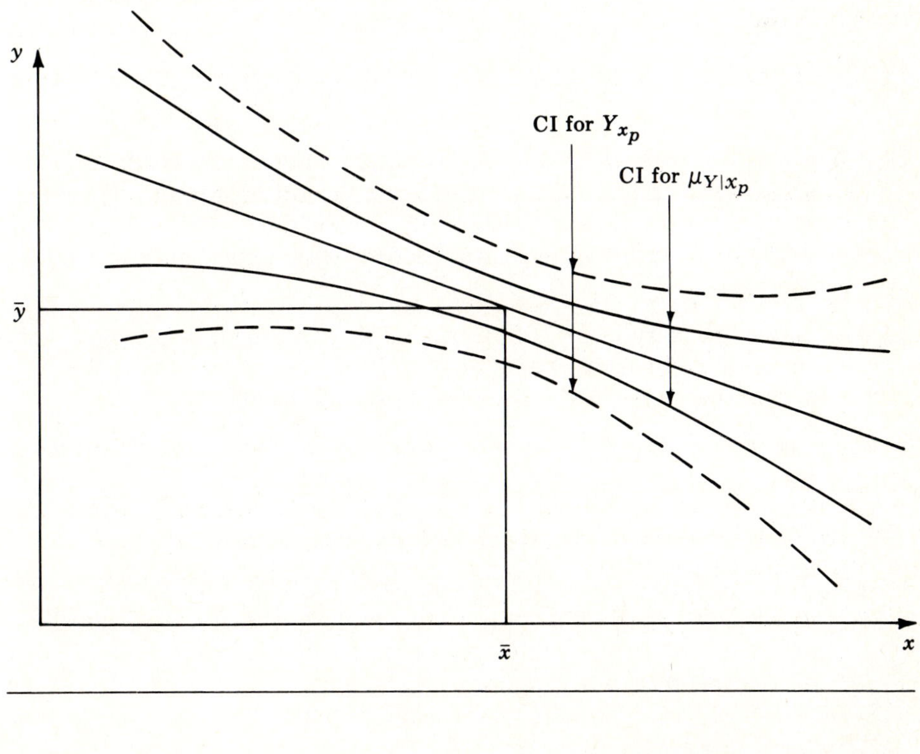

thousand units, $t_{16,\,0.05} = 1.746$, MSE = 6.06, $\Sigma\,(x - \bar{x})^2 = 58$, and $(x_p - \bar{x})^2 = (7 - 5.667)^2 = 1.777$. Thus, a 95% lower bound for the actual sales next month is given by

$$\hat{Y}_{x_p} - t_{n-2,\,\alpha}\sqrt{\text{MSE}}\sqrt{1 + \frac{1}{n} + \frac{(x_p - \bar{x})^2}{\Sigma\,(x - \bar{x})^2}}$$

$$= 31.522 - 1.746\sqrt{6.06}\sqrt{1 + \frac{1}{18} + \frac{1.777}{58}}$$

$$= 31.522 - 1.746(2.462)(1.0422) = 27.042 \text{ thousand units}$$

Our interpretation is that we are 95% confident that next month's sales will be at least 27,042 units.

EXERCISES

14.4.1 Refer to Exercise 14.2.1*a* and test to see if the simple linear model seems appropriate at $\alpha = .05$.

14.4.2 Refer to Exercise 14.2.1*b* and test to see if the simple linear model appears to be appropriate at $\alpha = .05$.

14.4.3 Refer to Exercise 14.4.1 and test the linear model for usefulness at $\alpha = .05$.

14.4.4 Refer to Exercise 14.4.2 and test the linear model for usefulness at $\alpha = .05$.

14.4.5 Refer to Exercise 14.3.5 and perform a regression analysis showing tests for both appropriateness and usefulness of the simple linear model. Use $\alpha = .05$ for both tests and indicate the percent of variation in exam 2 scores which can be explained by the exam 1 scores.

14.4.6 Refer to Exercise 14.2.2. Choose one of your data sets and test the model for usefulness at $\alpha = .05$. If possible also test the appropriateness of the simple linear model.

14.4.7 Refer to Exercise 14.4.5 and construct 95% confidence intervals for β_1, the average score on exam 2 of all students who might score an 80 on exam 1 and for your score on exam 2 if you got an 80 on exam 1. Interpret each of the intervals.

14.4.8 Refer to Exercise 14.4.6 and construct a 99% confidence interval for β_1. Pick a value for x_p and construct a 99% confidence interval for whichever you think is appropriate, $\mu_{Y|x_p}$ or Y_{x_p}. Interpret your intervals.

14.5 MEASURING THE STRENGTH OF THE RELATIONSHIP

Often in the results of studies in business and economics (and in all other fields, for that matter), you will see reported a statistic called the *correlation coefficient.* Correlation coefficients are dimensionless measures of the strength of the linear association between two variables. They are frequently reported in conjunction with regression analyses, but they are virtually useless in that instance, since they contribute no additional information to the problem. From a practical viewpoint, they could be useful in the planning of a regression analysis when a small pilot sample is selected to determine which variables to include in the full experiment.

There are two formulas for calculating the linear correlation between two variables that seem to occur more frequently than others. One is a parametric form requiring normal distributions for both X and Y if hypothesis tests are to be run. It is known as the Pearson product moment correlation. The other is a distribution-free alternative based on ranking procedures known as Spearman's coefficient. Pearson's is the more commonly used, so we turn our attention to it first.

14.5.1 Pearson's Correlation Coefficient

In studying the linear relationship between two normally distributed random variables, Pearson defined a measure, independent of the respective scales of measurement, of the strength of the linear relationship. For a bivariate population the theoretical value is called the *population correlation coefficient.* It is denoted by ρ_{XY} (pronounced rho) and is defined as

$$\rho_{XY} = \frac{\text{covariance }(X, Y)}{\sigma_X \sigma_Y}$$

As usual, in practice population values are unknown, and so they must be estimated from sample data. For a random sample of pairs of (X, Y) values, Pearson's sample correlation coefficient is denoted by r_{XY} and given by (computationally)

$$r_{XY} = \frac{\Sigma XY - (\Sigma X)(\Sigma Y)/n}{\sqrt{[\Sigma X^2 - (\Sigma X)^2/n]\,[\Sigma Y^2 - (\Sigma Y)^2/n]}} \tag{14.22}$$

If a simple linear regression analysis has been performed, then we can calculate r_{XY} by merely taking the square root of the coefficient of determination and associating the sign of b_1. That is,

$$r_{XY} = (\text{sign of } b_1)\sqrt{R^2} = (\text{sign of } b_1)\sqrt{\frac{\text{SS}(b_1)}{\text{SSTO}}} \tag{14.23}$$

If we note that $r_{XY} = b_1 s_x/s_y$, we can easily see that r_{XY} is measuring the same thing as b_1 but without having units associated with it. It is essentially measuring, in standardized form, how close the observed points (x_i, y_i) are to the line given by $\hat{Y}_x$. If no linear relationship exists, then the line is given by $\overline{Y}$ and r_{XY} should be near 0. If there is a strong positive linear relation, r_{XY} should be near +1, and if there is a strong negative linear relation, then r_{XY} can be expected to be close to −1.

We cite, for your information, some basic properties of r_{XY}.

1. $-1 \leqslant r_{XY} \leqslant +1$.
2. $r_{XY} = -1$ implies a line sloping downward to the right with all observations lying directly on the line (that is, a perfect negative linear relation between the observed x's and y's).
3. $r_{XY} = +1$ implies a perfect positive linear relation in the sample data.
4. r_{XY} near 0 implies little or no linear relation (but there may be a nonlinear relation between the two).
5. If X, Y are independent, $\rho_{XY} = 0$.
6. If X, Y are normal random variables, then $\rho_{XY} = 0$ implies that X and Y are independent.

To test whether r_{XY} is significantly different from zero, we first note that if b_1 is significantly different from zero, so is r_{XY}. That is, the following hypothesis sets are equivalent:

$$H_0\colon\ \beta_1 = 0 \qquad H_a\colon\ \beta_1 \neq 0$$

and

$$H_0\colon\ \rho = 0 \qquad H_a\colon\ \rho \neq 0$$

If only the value of r_{XY} and the corresponding sample size are known, then H_0 is rejected if

$$\left|\frac{r\sqrt{n-2}}{\sqrt{1-r^2}}\right| > t_{n-2,\,\alpha/2} \tag{14.24}$$

The statistic in Equation 14.24 is equivalent to the t statistic for b_1 in Table 14.5 if $\beta_{01} = 0$. Proof of the equivalence is left to the student as a Level 3 exercise. For one-sided alternatives, the decision rule becomes:

1. If H_a: $\rho > 0$, then reject H_0 if $r_{XY}\sqrt{n-2}/\sqrt{1-r_{XY}^2} > t_{n-2,\alpha}$.

2. If H_a: $\rho < 0$, then reject H_0 if $r_{XY}\sqrt{n-2}/\sqrt{1-r_{XY}^2} < -t_{n-2,\alpha}$.

EXAMPLE 14.8 Refer to Example 14.2. Calculate the correlation coefficient for those data and test for a positive linear relationship between sales and advertising at $\alpha = 0.01$.

SOLUTION In Example 14.2 we have already calculated that $\Sigma\, x = 36$, $\Sigma\, y = 175$, $\Sigma\, xy = 1087$, and $\Sigma\, x^2 = 226$. To determine r_{XY}, we also need $\Sigma\, y^2$. This is 5245. Thus the linear correlation between X and Y is

$$r_{XY} = \frac{1087 - (36)(175)/6}{\sqrt{[226 - (36)^2/6][5245 - (175)^2/6]}}$$

$$= \frac{37}{\sqrt{10(140.83)}} = \frac{37}{37.53} = 0.986$$

The hypothesis set to be tested is

$$H_0\colon\ \rho \leqslant 0 \qquad H_a\colon\ \rho > 0$$

with H_0 being rejected if

$$\frac{r_{XY}\sqrt{n-2}}{\sqrt{1-r_{XY}^2}} > t_{n-2,\alpha} = t_{4,0.01} = 3.747$$

The value for our statistic is

$$\frac{0.986\sqrt{6-2}}{\sqrt{1-(0.986)^2}} = \frac{0.986\sqrt{4}}{\sqrt{0.0278}} = 11.827$$

Since $11.827 > 3.747$, we reject H_0 and conclude that sales appear to increase linearly with advertising expenditures.

The correlation coefficient can be misleading occasionally. In many disciplines a correlation of 0.5 is considered to be quite good, but in reality the independent variable (X) is explaining only 25% of the total variation. For an r_{XY} of 0.3, which may be significantly different from zero, only 9% of the variation is being explained. Thus the question you must ask yourself is, "Even though X may be theoretically useful, is it useful from a practical viewpoint?"

14.5.2 Spearman's Correlation Coefficient

Spearman believed that the assumptions required by Pearson's product moment correlation—a bivariate normal distribution and interval measurement for both X and Y—are often unreasonable. Therefore, he derived an alternative procedure, with much less restrictive assumptions, by replacing the actual values in Pearson's formula, Equation 14.22, with their corresponding ranks within each individual set of values. The distribution of the correlation coefficient, denoted r_s, that results from these substitutions can be derived if the following conditions are satisfied.

Assumptions for Testing Spearman's Coefficient

1. A random sample of n units is selected.
2. A measurement on each variable, X, Y, is obtained for each unit in the sample.
3. The underlying distribution of each variable is continuous or at least infinite (to eliminate ties).
4. The data are at least ordinal.

The ARE of the test based upon r_s has a value of 0.912 when the assumption of normality is satisfied and a value of 1.000 if the distribution is uniform. For other distributions the ARE will vary.

The procedure for calculating r_s is summarized in the following steps. We assume a random sample of n pairs of observations has already been obtained.

1. Rank the n observations corresponding to X among themselves, replacing each value with its rank, R_{X_i}.
2. Rank the n Y_i's among themselves, replacing each with its rank, R_{Y_i}.

Then r_s is calculated using the equation,

$$r_s = \frac{\Sigma R_X R_Y - (\Sigma R_X)(\Sigma R_Y)/n}{\sqrt{[\Sigma R_X^2 - (\Sigma R_X)^2/n][\Sigma R_Y^2 - (\Sigma R_Y)^2/n]}}$$

which can be simplified to the form

$$r_s = \frac{\Sigma R_X R_Y - n(n+1)^2/4}{\sqrt{[\Sigma R_X^2 - n(n+1)^2/4][\Sigma R_Y^2 - n(n+1)^2/4]}} \tag{14.25}$$

If there are no ties in either the X's or the Y's, the denominator will reduce to

$$\Sigma R_X^2 - \frac{n(n+1)^2}{4} = \frac{n(n^2-1)}{12}$$

since ΣR_X^2 and ΣR_Y^2 will both equal $n(n+1)(2n+1)/6$, the sum of the squares of the first n integers.

There is an alternative method for calculating r_s that simplifies the arithmetic considerably. If there are no ties in the data, the resulting value will be identical with that obtained using Equation 14.25. When the number of ties is small and they are handled by the average rank procedure, the two methods yield slightly different values. If there is an excessive number of ties, Equation 14.25 should be used. The alternative procedure is to calculate the differences between corresponding X and Y ranks, $d_i = R_{X_i} - R_{Y_i}$, and then to use

$$r_s = 1 - \frac{6 \Sigma d_i^2}{n(n^2-1)} \tag{14.26}$$

Spearman's rank correlation coefficient has a maximum value of 1, which occurs when there is a perfect positive relationship. This can be seen from Equation 14.26, since, in such a situation, $R_{X_i} = R_{Y_i}$ and $d_i = 0$ for all i. Conversely, if there is a perfect negative relation, the smaller X values will correspond to the larger Y values, so that $R_{X_i} = n - R_{Y_i} + 1$. That is, if $R_{X_i} = i$, then $R_{Y_i} = n - i + 1$ and $d_i = 2i - (n+1)$, in which case it can be shown that $\Sigma d_i^2 = n(n^2-1)/3$ with a resultant r_s of -1. If there is no relation between X and Y, there should be no pattern between their respective ranks and we can expect the sum of the d_i^2 to be halfway between the extremes, that is, to equal $n(n^2-1)/6$, giving an r_s of zero. Thus, r_s is also restricted to the interval from -1 to $+1$ inclusive, and we can expect it to be close to the true value of ρ.

Utilizing these considerations, we can test for a relationship between X and Y. To do so, we:

1. State the hypotheses of interest and select α.
2. Select a random sample of n pairs of observations on X and Y.
3. Separately rank the X_i's and Y_i's, assigning a rank of 1 to the smallest X and also to the smallest Y, etc.
4. Calculate r_s using Equation 14.26 (14.25 if there is an excessive number of ties).
5. The decision rules for the appropriate hypotheses are given in Table 14.6 on page 560.

Table 14.6 Hypotheses and Corresponding Decision Rules Based on Spearman's Rank Correlation Coefficient

Hypothesis Set		Decision Rule for a Nominal
1. H_0: $\rho \leqslant 0$	H_a: $\rho > 0$	Reject H_0 if $r_s > r_{n,\alpha}$.
2. H_0: $\rho \geqslant 0$	H_a: $\rho < 0$	Reject H_0 if $r_s < -r_{n,\alpha}$.
3. H_0: $\rho = 0$	H_a: $\rho \neq 0$	Reject H_0 if $\lvert r_s \rvert > r_{n,\alpha/2}$.

Critical values for r_s are given in Table A.15 of the Appendix. The table is indexed by n, the number of pairs of observations, and α, the nominal level of significance. The number $r_{n,\alpha}$ in the body of the table is the largest value possible such that $\text{pr}(r \geqslant r_{n,\alpha}) \leqslant \alpha$ if no relationship exists. Since the distribution of r_s is symmetric about zero, we obtain lower-tail critical values by merely multiplying the corresponding upper-tail value by a negative one (−1).

The following example illustrates the computations required to determine the rank correlation coefficient and the test procedure. Both Equations 14.25 and 14.26 are used for comparative purposes.

EXAMPLE 14.9 It has long been contended that students who study more do better on statistics exams. To test this contention, 12 students were randomly selected and asked the number of hours per week spent studying statistics and what grade had been attained on the final exam. The results are as follows:

Student:	1	2	3	4	5	6	7	8	9	10	11	12
Hours/ week:	6.5	3.0	4.0	5.5	2.5	3.5	5.0	2.0	1.5	6.0	4.5	8.0
Exam grade:	93	62	72	86	50	76	83	48	40	95	79	88

(*a*) Calculate r_s using Equation 14.25.
(*b*) Calculate r_s using Equation 14.26.
(*c*) Test the contention at $\alpha = 0.05$.

SOLUTION The preliminary computations required for both (*a*) and (*b*) are given in the following table. The hours per week have been rearranged in ascending order, but the pairings are maintained.

(X) Hours/ Week	(Y) Exam Grade	R_X	R_Y	For Equation 14.25 R_X^2	R_Y^2	$R_X R_Y$	For Equation 14.26 $d_i = R_X - R_Y$	d_i^2
1.5	40	1	1	1	1	1	0	0
2.0	48	2	2	4	4	4	0	0
2.5	50	3	3	9	9	9	0	0
3.0	62	4	4	16	16	16	0	0
3.5	76	5	6	25	36	30	−1	1
4.0	72	6	5	36	25	30	+1	1
4.5	79	7	7	49	49	49	0	0
5.0	83	8	8	64	64	64	0	0
5.5	86	9	9	81	81	81	0	0
6.0	95	10	12	100	144	120	−2	4
6.5	93	11	11	121	121	121	0	0
8.0	88	12	10	144	100	120	+2	4
		78	78	650	650	645		10

(*a*) Using Equation 14.25, we have

$$r_s = \frac{\Sigma R_X R_Y - n(n+1)^2/4}{\sqrt{[\Sigma R_X^2 - n(n+1)^2/4][\Sigma R_Y^2 - n(n+1)^2/4]}}$$

$$= \frac{645 - 12(13)^2/4}{\sqrt{[650 - 12(13)^2/4][650 - 12(13)^2/4]}}$$

$$= \frac{645 - 507}{\sqrt{(650 - 507)(650 - 507)}} = \frac{138}{143} = 0.965$$

(*b*) Substituting from the last column on the right into Equation 14.26 gives

$$r_s = 1 - \frac{6(\Sigma d_i^2)}{n(n^2 - 1)} = 1 - \frac{6(10)}{12(143)} = 1 - \frac{5}{143} = 1 - 0.035 = 0.965$$

Since the two methods yield identical values, it is unnecessary to use both equations to calculate r_s. Use Equation 14.25 only if there are numerous ties in either variable.

(c) Since the contention is that more study time corresponds to higher grades, the hypotheses are

$$H_0: \rho \leq 0 \qquad H_a: \rho > 0$$

and H_0 will be rejected at $\alpha = 0.05$ if

$$r_s \geq r_{n,\alpha} = r_{12,\,0.05} = 0.497$$

Therefore we reject H_0, since $0.965 > 0.497$, and conclude that there appears to be a positive correlation between exam grades and study time.

For values of $n > 30$ the t distribution may be used to obtain an approximate test. In such situations the statistic $r_s\sqrt{n-2}/\sqrt{1-r_s^2}$ will be distributed as a t with $n - 2$ degrees of freedom. The decision rules correspond to those based on Pearson's r.

EXERCISES

14.5.1 Refer to Exercise 14.2.1*a*. Calculate both Pearson's and Spearman's linear correlation coefficients and compare the results.

14.5.2 Calculate Pearson's and Spearman's linear correlation coefficients for the data in Exercises 14.2.1*b* and 14.2.1*d*.

14.5.3 Test the correlations in Exercises 14.5.1 for significance at $\alpha = 0.05$.

14.5.4 Test the correlations of Exercises 14.5.2 for significance at $\alpha = 0.10$. Interpret the results of your tests.

14.5.5 Refer to Exercise 14.3.5 and calculate Spearman's correlation for the two exam scores. Test the belief that students who do well on the first exam also do well on the second exam at $\alpha = 0.05$.

14.5.6 Ten students were interviewed by each of two recruiters for similar positions. After each interview the recruiters rated the desirability of hiring each student on a 1 to 25 scale. The students and the corresponding ratings by each recruiter are given here.

Student:	1	2	3	4	5	6	7	8	9	10
Rating by First Recruiter:	21	14	23	17	9	7	19	25	13	12
Rating by Second Recruiter:	18	15	19	22	7	10	21	24	17	8

Do the recruiters appear to be using similar criteria in their judgments at $\alpha = 0.05$? Justify your choice of correlations.

14.6 SUPPLEMENTARY EXAMPLES

EXAMPLE 14.10 The Coast-to-Coast Trucking Company is trying to estimate its fixed costs per million ton-miles. CC has hired you as an accounting consultant and has provided the following data covering the last 14 months of operation.

Month:	1	2	3	4	5	6	7
Total cost ($1000):	280	220	340	375	356	430	380
Output (million miles):	2	1.5	4.0	4.1	3.6	5.8	4.2

Month:	8	9	10	11	12	13	14
Total cost ($1000):	310	435	330	360	440	420	415
Output (million miles):	2.9	7.1	3.2	3.7	6.9	5.6	4.8

Using simple linear regression, estimate both their fixed costs and their variable costs per million ton-miles. Test your model for usefulness at $\alpha = 0.05$.

SOLUTION The data were input into a timesharing computer terminal. The resulting printout is given in Table 14.7. From the printout we see that $b_0 = 208.901$ and $b_1 = 36.471$, with the model estimated by

$$\hat{Y} = 208.901 + 36.471x$$

for x in million ton-miles and $\hat{Y}$ in \$1000s.

In the equation, b_1 represents the change in total costs of a 1 million ton-mile change in output. That is, it represents a variable cost per million ton-miles of 36.471(\$1000) or \$36,471. As the Y intercept b_0 is the estimate of total cost when the output (x) is zero. Thus an estimate of the fixed costs is given by b_0 (\$1000) or \$208,901.

To test the model for usefulness, our hypotheses are

$$H_0\colon\ \beta_1 = 0 \qquad H_a\colon\ \beta_1 \neq 0$$

The null hypothesis will be rejected at $\alpha = 0.05$ if MSREG/MSE > $f_{1, 12, 0.05} = 4.75$. Referring to Table 14.7, we see that MSREG/MSE = 97.915 > 4.75, and so we reject H_0, concluding that output in million ton-miles appears to be a useful predictor of total cost.

We cannot formally test the linear model of Example 14.10 for appropriateness, since there are no replications. However, we can roughly check both the assumptions underlying our test and the linearity (appropriateness) of the model by a graphical procedure known as a *residual plot.* That is, we plot the residuals (differences between Y_i and $\hat{Y}_i$) by themselves to check normality and versus the corresponding x values to check for linearity, homogeneity of variance, independence of ϵ and x, and arithmetic errors. Figure 14.9 on page 566 shows representative plots for normality. (Note that if n is sufficiently large, the assumption $\epsilon \sim N(0, \sigma^2)$ can be tested using the chi-square test of Chapter 10.) Plots of the residuals versus x are given in Figure 14.10 on page 567.

If the normality assumption is satisfied, we expect the residuals to be dispersed symmetrically about zero, with the greatest concentration being at zero (see Figure 14.9*a*). If the plot is skewed in either direction (as in Figure 14.9*b*) or if the residuals are not concentrated at zero, the assumption of normality may not be justified.

If the model is correct and all assumptions are satisfied, we expect the residuals to be dispersed randomly about zero, with the maximum distance from zero being fairly uniform for all values of x (see Figure 14.10*a*). Figure 14.10*b* appears to exhibit a random scatter about zero, but the deviations appear to be increasing as x increases, implying that

Table 14.7 Computer Printout for Example 14.10

```
PLEASE ENTER THE MATRIX NAME.
EX149

OK, LET'S GO.
   1 □ 2 0 280
   2 □ 1.5 220
   3 □ 4.0 340
   4 □ 4.1 375
   5 □ 3.6 356
   6 □ 5.8 430
   7 □ 4.2 380
   8 □ 2.9 310
   9 □ 7.1 435
  10 □ 3.2 330
  11 □ 3.7 360
  12 □ 6.9 440
  13 □ 5.6 420
  14 □ 4.8 415
  15 □ END

YOUR DATA MATRIX EX149 HAS 2 COLUMNS.
WHICH OF THEM DO YOU WANT TO BE YOUR DEPENDENT VARIABLE
(OR PREDICTED VARIABLE)?  TYPE ONE NUMBER.
□:
        2
DO YOU WANT ALL THE REST TO BE USED TO PREDICT
THAT ONE?
Y
OK.  HERE GOES.

DEPENDENT VARIABLE:           Y;COST IN $1000
INDEPENDENT VARIABLE(S):      X;MILLION TON-MILES

MULTIPLE REGRESSION
-------------------

REGRESSION EQUATION: Y  DEPENDENT
                          'B'          STD. ERR        BETA
                        COEFFS          OF 'B'        COEFFS      F-VALUES
X                  36.47121789       3.68575510       .94383      97.91463
CONSTANT          208.90068982

STATISTICS FOR THE ENTIRE EQUATION

ANALYSIS OF VARIANCE
  SOURCE              SS          DF         MS                F          PROB.
REGRESSION      47930.99557        1    47930.99557      97.91463     .00000
ERROR            5874.21872       12      489.51823
TOTAL           53805.21429       13

      MULTIPLE R      MULT. R-SQR          DETER XTRSX
        .94383           .89082            1.0000000000
```

Figure 14.9 Examples of Residual Plots for Checking the Normality Assumption ($\hat{\epsilon}_i = y_i - \hat{y}_i$)

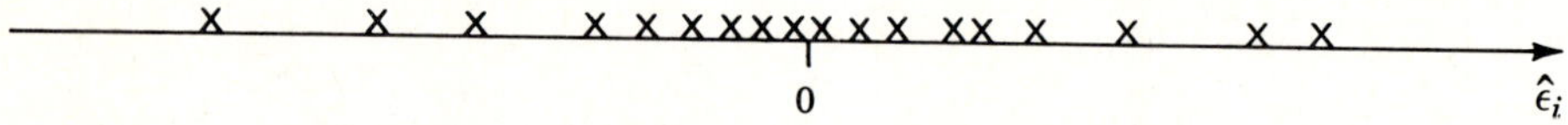

(*a*) No Obvious Deviation from Normality

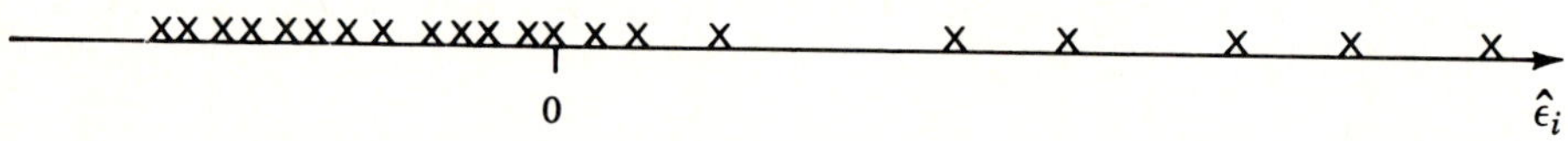

(*b*) Possible Deviation from Normality due to Right Skewness

the variance is not constant. In Figure 14.10*c* we have a constant variance, but the pattern does not appear to be random, which implies that higher-order terms in x should be considered; for example, try $Y = \beta_0 + \beta_1 x + \beta_2 x^2 + \epsilon$. The plot in Figure 14.10*d* indicates that a linear term in x is still present, and that is possible only if a computational error has been made.

EXAMPLE 14.11 The residuals for Example 14.10 are given in Table 14.8. Using them, construct residual plots to check the assumptions for Example 14.10.

SOLUTION The plots of the residuals (*a*) by themselves and (*b*) versus x are given in Figure 14.11 on page 568. Since the sample size is small, it is difficult to determine whether there are any discrepancies. There may be a slight deviation from normality according to Figure 14.11*a*, but the robustness of the F test should take care of it. Figure 14.11*b* indicates that there might be a nonlinear effect present, but we cannot say so for certain. We could try a quadratic term (x^2) just to see its effect.

Residual plots provide only a rough check of the assumptions. For us to be able to make any definitive conclusions using them, there usually must be a gross violation of the assumption in question, but gross violations are the critical ones. In general you should use residual plots to check for indications of possible problems in the analysis.

Figure 14.10 Examples of Plots of Residuals Versus x

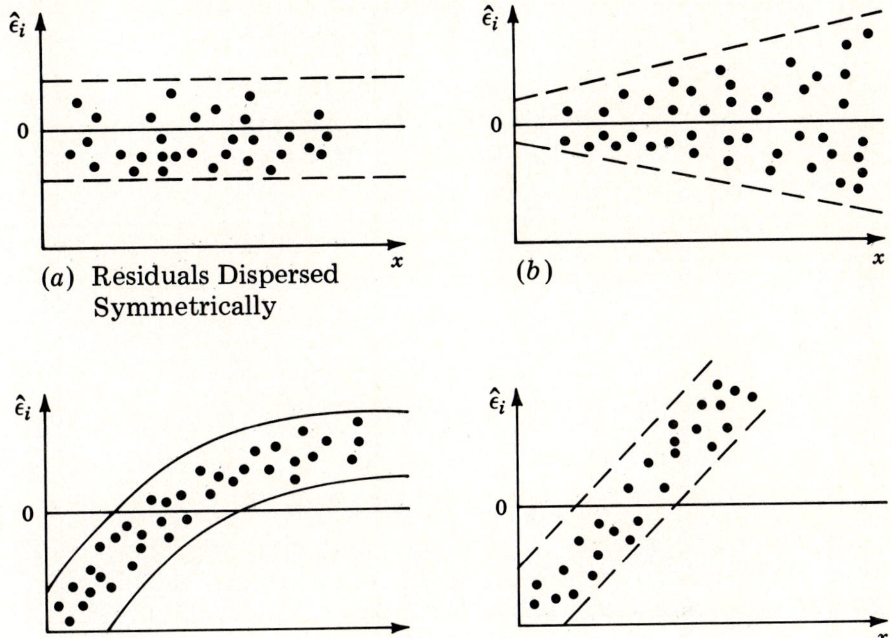

(a) Residuals Dispersed Symmetrically

(b)

(c)

(d)

Table 14.8 Table of Residuals (Y_i - Y_i) for Examples 14.10 and 14.11

TABLE OF RESIDUALS

ACTUAL-Y	PREDICT-Y	RESIDUALS	NORM-RES	CASE
280.00000	281.84313	¯1.84313	¯.08330	1
220.00000	263.60752	¯43.60752	¯1.97096	2
340.00000	354.78556	¯14.78556	¯.66827	3
375.00000	358.43268	16.56732	.74880	4
356.00000	340.19707	15.80293	.71425	5
430.00000	420.43375	9.56625	.43237	6
380.00000	362.07980	17.92020	.80995	7
310.00000	314.66722	¯4.66722	¯.21095	8
435.00000	467.84634	¯32.84634	¯1.48458	9
330.00000	325.60859	4.39141	.19848	10
360.00000	343.84420	16.15580	.73020	11
440.00000	460.55209	¯20.55209	¯.92891	12
420.00000	413.13951	6.86049	.31008	13
415.00000	383.96254	31.03746	1.40282	14

Figure 14.11 Residual Plots for Example 14.11

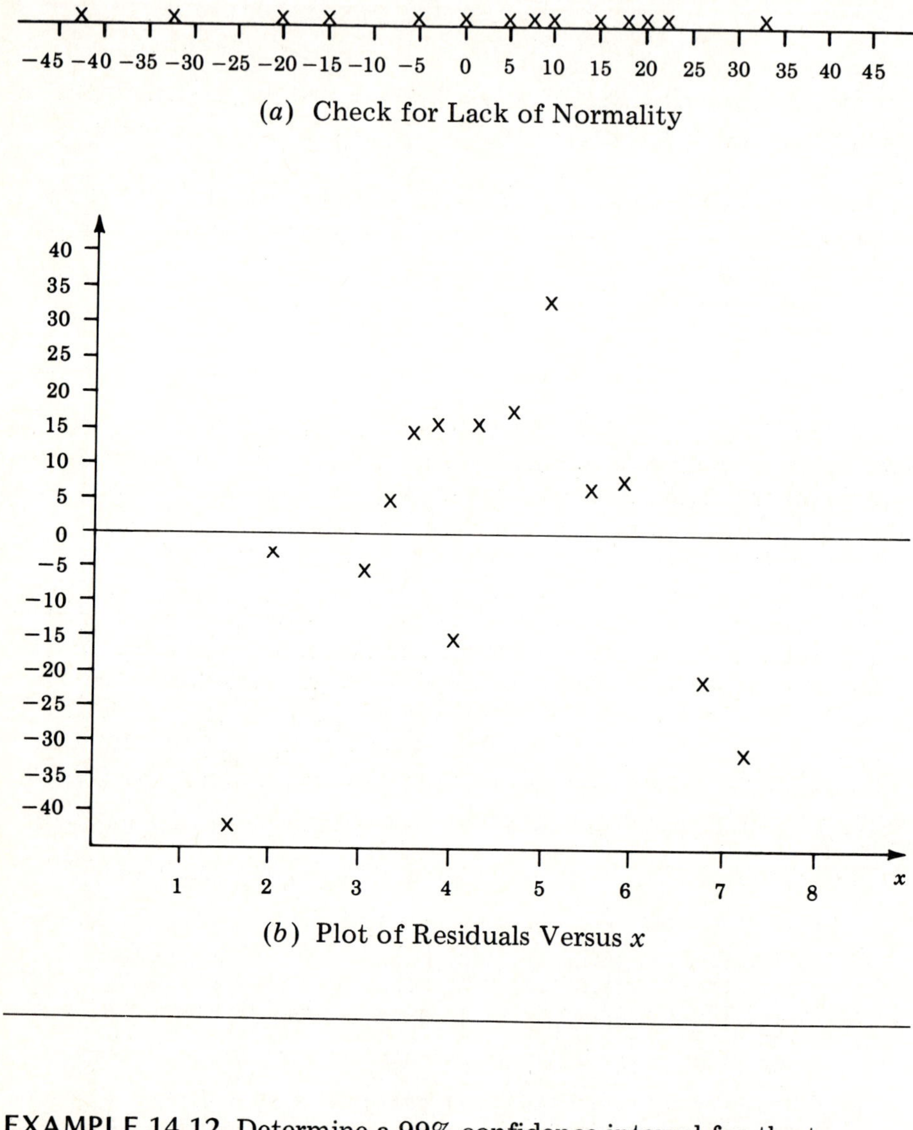

(*a*) Check for Lack of Normality

(*b*) Plot of Residuals Versus x

EXAMPLE 14.12 Determine a 99% confidence interval for the true variable cost per million ton-miles for Coast-to-Coast Trucking of Example 14.10.

SOLUTION Since b_1 is our estimate of the variable cost and also of β_1, we need to determine a 99% confidence interval for β_1. From the printout in Table 14.7 we obtain $b_1 = 36.471$ and $s_{b_1} = 3.686$. In Table A.8 of the Appendix we find $t_{12,\,0.005} = 3.055$. Thus, a 99% confidence interval for the variable cost (in \$1000s) per million ton-miles is

$$36.471 \pm 3.055(3.686) = [25.210, 47.732]$$

Thus Coast-to-Coast can be 99% sure that the interval [\$25,210, \$47,732] contains their true variable cost per million ton-miles.

Note that in Example 14.12 we could have calculated s_{b_1} directly if it were necessary. To do so requires calculating $\Sigma\,(x - \bar{x})^2$, which is 36.034. Thus s_{b_1} is

$$s_{b_1} = \frac{s_{Y|x}}{\sqrt{\Sigma\,(x - \bar{x})^2}} = \frac{22.125}{\sqrt{36.034}} = 3.686$$

EXAMPLE 14.13 The Police Department of Akersville is trying to estimate the required stopping distance for a standard passenger car under normal driving conditions. They selected six speeds and measured the distance required to come to a full stop from the point at which the brakes were applied. Since they felt that the relationship between speed and stopping distance might not be linear, they ran three tests at each speed. The data and the results of the computer run are given in Tables 14.9 (page 570) and 14.10 (page 571). A one-way ANOVA was performed also to obtain the SSPE. The results are given in Table 14.10*b*. Using the results in these tables, perform a complete analysis showing tests for linearity of the model and usefulness of the model and present residual plots. Use $\alpha = 0.01$ for testing linearity and 0.05 for testing usefulness.

SOLUTION From Table 14.10*a*, the estimate of the linear model is

$$\hat{Y}_x = -185.238 + 9.357x$$

with x = speed in mph and Y_x the stopping distance in feet. To test the linearity hypotheses of

$$H_0\colon\ \mu_{Y|x} = \beta_0 + \beta_1 x \qquad H_a\colon\ \mu_{Y|x} \neq \beta_0 + \beta_1 x$$

we reject H_0 if MSLOF/MSPE $> f_{4,\,12,\,0.01} = 5.41$. SSPE and MSPE are given in Table 14.10*b* as 1150 and 95.833, respectively. SSLOF is obtained by subtracting SSPE from SSE (in Table 14.10*a*) and is

$$\text{SSLOF} = \text{SSE} - \text{SSPE} = 28{,}142.857 - 1150.000 = 26{,}992.857$$

Table 14.9 Computer Input for Example 14.13

```
PLEASE ENTER THE MATRIX NAME.
EX1412

OK, LET'S GO.
    1 ⎕ 20   45
    2 ⎕ 20   55
    3 ⎕ 20   50
    4 ⎕ 30   95
    5 ⎕ 30   90
    6 ⎕ 30   85
    7 ⎕ 40  150
    8 ⎕ 40  140
    9 ⎕ 40  160
   10 ⎕ 50  230
   11 ⎕ 50  250
   12 ⎕ 50  240
   13 ⎕ 60  360
   14 ⎕ 60  375
   15 ⎕ 60  345
   16 ⎕ 70  515
   17 ⎕ 70  535
   18 ⎕ 70  525
   19 ⎕ END

YOUR DATA MATRIX EX1412 HAS 2 COLUMNS.
WHICH OF THEM DO YOU WANT TO BE YOUR DEPENDENT VARIABLE
(OR PREDICTED VARIABLE)?   TYPE ONE NUMBER.
⎕:
      2
DO YOU WANT ALL THE REST TO BE USED TO PREDICT
THAT ONE?
Y
OK.   HERE GOES.
```

Thus

$$\text{MSLOF} = \frac{26{,}992.857}{4} = 6748.214$$

and

$$\frac{\text{MSLOF}}{\text{MSPE}} = \frac{6748.214}{95.833} = 70.416$$

Since $70.416 > 5.41$, we reject H_0 and conclude that the relationship appears to be nonlinear.

Even if the relation is nonlinear, it may be useful over the range of x values used in the experiment. To test the usefulness of the linear prediction equation, we have

$$H_0\colon\ \beta_1 = 0 \qquad H_a\colon\ \beta_1 \neq 0$$

Table 14.10 Printout for Example 14.13 Showing (*a*) Regression Analysis and Residuals and (*b*) One-Factor ANOVA Table

---

```
DEPENDENT VARIABLE:         Y;STOPPING DISTANCE IN FEET
INDEPENDENT VARIABLE(S):    X;MILES PER HOUR

MULTIPLE REGRESSION
-------------------
```

REGRESSION EQUATION: Y DEPENDENT

	'B' COEFFS	STD. ERR OF 'B'	BETA COEFFS	F-VALUES
X	9.35714286	.57882123	.97073	261.33503
CONSTANT	¯185.23809524			

STATISTICS FOR THE ENTIRE EQUATION

ANALYSIS OF VARIANCE

SOURCE	SS	DF	MS	F	PROB.
REGRESSION	459669.64286	1	459669.64286	261.33503	.00000
ERROR	28142.85714	16	1758.92857		
TOTAL	487812.50000	17			

MULTIPLE R	MULT. R-SQR	DETER XTRSX
.97073	.94231	1.0000000000

TABLE OF RESIDUALS

ACTUAL-Y	PREDICT-Y	RESIDUALS	NORM-RES	CASE
45.00000	1.90476	43.09524	1.02756	1
55.00000	1.90476	53.09524	1.26599	2
50.00000	1.90476	48.09524	1.14677	3
95.00000	95.47619	¯.47619	¯.01135	4
90.00000	95.47619	¯5.47619	¯.13057	5
85.00000	95.47619	¯10.47619	¯.24979	6
150.00000	189.04762	¯39.04762	¯.93104	7
140.00000	189.04762	¯49.04762	¯1.16948	8
160.00000	189.04762	¯29.04762	¯.69261	9
230.00000	282.61905	¯52.61905	¯1.25464	10
250.00000	282.61905	¯32.61905	¯.77776	11
240.00000	282.61905	¯42.61905	¯1.01620	12
360.00000	376.19048	¯16.19048	¯.38604	13
375.00000	376.19048	¯1.19048	¯.02839	14
345.00000	376.19048	¯31.19048	¯.74370	15
515.00000	469.76190	45.23810	1.07865	16
535.00000	469.76190	65.23810	1.55553	17
525.00000	469.76190	55.23810	1.31709	18

(*a*) Regression Printout

TABLE I

SOURCE	SUM SQ	D OF F	MEAN SQ	F RATIO
MEANS	486662.5000	5	97332.5000	1015.6435
WITHIN	1150.0000	12	95.8333	
TOTAL	487812.5000	17		

(*b*) One-Factor Analysis of Variance

---

with rejection occurring if MSREG/MSE $> f_{1,\,16,\,0.05} = 4.49$. This ratio (from Table 14.10*a*) is 261.335, which exceeds 4.49. Thus H_0 is rejected; the linear term appears to be useful and should be retained in any higher-order models.

We already have rejected the linearity hypothesis, but we construct the residual plots anyway for comparison. They are presented in Figure 14.12. The graph in Figure 14.12*a* indicates a nearly bimodal and hence nonnormal distribution. This is probably the result of having thc wrong model. Figure 14.12*b* indicates that higher-order terms in x definitely should be considered but gives no evidence of a lack of homogeneity in the variances.

14.7 SUMMARY

In this chapter we have considered using the information contained in one variable to try to minimize errors associated with predicting values for another variable. Traditionally the predictor, x, is called the independent variable and the response variable, Y, is called the dependent variable. For the simple case presented here, it is assumed that the two variables are related in a linear fashion according to the model

$$Y_x = \beta_0 + \beta_1 x + \epsilon$$

where β_0 and β_1 are parameters, x is a nonrandom variable, and ϵ is a random variable with a mean of zero. The model may be rewritten in terms of a relationship between the means of the Y populations and x as

$$\mu_{Y|x} = \beta_0 + \beta_1 x$$

That is, the model actually specifies that the Y means lie on a straight line with an intercept β_0 and a slope β_1. The slope β_1 represents the change in the mean value of Y for a one-unit change in x.

β_0 and β_1 are never known in practice and must be estimated from sample data. Since the only way we can measure the accuracy of our estimates is to compare observed values with predicted values, we determine estimates of β_0 and β_1, denoted b_0 and b_1, so that

$$\Sigma\,(Y_i - \hat{Y}_i)^2$$

is minimized. This ensures that the sum of squared deviations of the actual values from the predicted values is the smallest possible. Thus, the name *L*east *S*quares *E*stimation (LSE) is associated with the procedure.

Figure 14.12 Residual Plots for Example 14.13

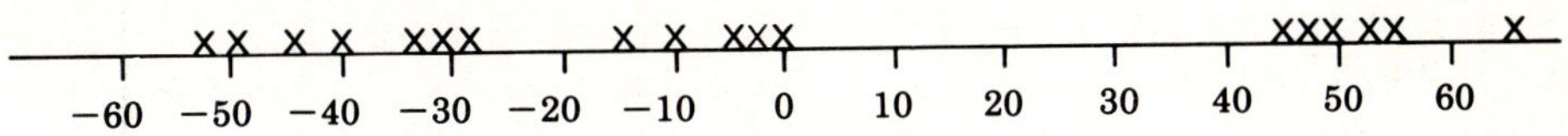

(a) Check for Normality

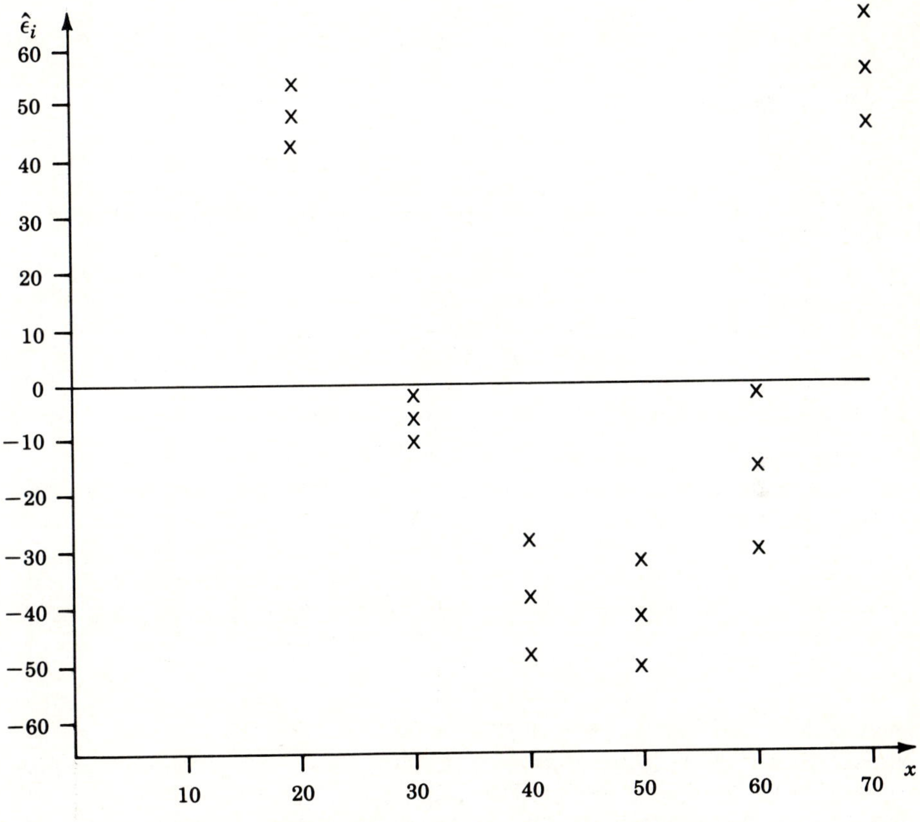

(b) Check for Homogeneity and Linearity

In the real world few relationships are truly linear, and so it is advisable to graph the (x, y) pairs on a scatter diagram first to see whether a linear model seems reasonable. If it is indicated, then LSE is used to obtain values for b_0 and b_1. Since the scatter diagram provides only a visual check for linearity, it is recommended that, if possible, the experiment be designed in such a way that the linearity assumption can be tested formally. To do so requires replication at some of the x values. In order to be able to test hypotheses about the model or to make probability statements about predictions, we assume that the random variable ϵ

1. Has a normal distribution
2. Is independent of the value of x
3. Has a constant variance over all values of x

(If these assumptions are violated, alternative procedures are available, but these are beyond the scope of this text. The interested reader should see Draper and Smith [1] and Neter and Wasserman [3].)

With these assumptions we partition the total sum of squares for Y and are able to apply Cochran's theorem to obtain their distributions. The partition is

$$\text{SSTO} = \text{SSE} + \text{SSREG}$$

and if there is replication, SSE is partitioned further to

$$\text{SSE} = \text{SSPE} + \text{SSLOF}$$

To test the linearity hypothesis, $H_0\colon \mu_{Y|x} = \beta_0 + \beta_1 x$, we reject H_0 if MSLOF/MSPE exceeds the appropriate F value. Rejection of the linearity hypothesis implies that the model is inappropriate and should be expanded to include higher-order terms in x. Note that an incorrect model still may be useful.

The model is useful if the error associated with predictions using it is smaller than the error associated with predictions based only on $\overline{Y}$. The error will be significantly smaller only if the SSE is significantly smaller. But the SSE becomes significantly smaller only if $\beta_1 \neq 0$. Thus the hypotheses for testing the model's usefulness are

$$H_0\colon \beta_1 = 0 \qquad H_a\colon \beta_1 \neq 0$$

We reject H_0 if MSREG/MSE exceeds the appropriate F value. This indicates that the model is a significant improvement over Y for predicting Y_x; that is, the model is useful. For all other hypotheses and confidence intervals concerning β_1, the t statistic

$$\frac{b_1 - \beta_1}{s_{b_1}}$$

is used for constructing the test. The degrees of freedom for the t distribution are $n - 2$.

Once we have determined that the model is useful, we need only substitute a value for the predictor, say x_p, into the equation for $\hat{Y}_x$ to obtain an estimate for $\mu_{Y|x_p}$ or a prediction for Y_{x_p}. A confidence interval for $\mu_{Y|x_p}$ or prediction limits for Y_{x_p} can be constructed using the t distribution. The two intervals differ because Y_{x_p} is a random variable, whereas $\mu_{Y|x_p}$ is a constant. Because of this difference, the variability of Y_{x_p} is included in the construction of the limits, and this results in a wider interval.

In addition to the value of b_1, we can measure the strength of the linear relationship between two variables by calculating a correlation coefficient. Pearson's product moment correlation uses the actual data and assumes that X and Y have a bivariate normal distribution for testing purposes. The usefulness of Pearson's correlation coefficient is questionable except for pilot experiments, since $r_{XY} = b_1(s_X/s_Y)$, and this contains no additional information about the relationship. Spearman's rank correlation coefficient, r_s, measures the same thing as Pearson's r_{XY} but uses the ranks of the X values (ranked among themselves) and Y values rather than the actual observations. Thus, Spearman's r_s has the advantages of requiring only ordinal data and of being distribution-free. Both correlation coefficients can assume values only between -1 and $+1$, with values near either extreme indicating a strong linear relation, and values near 0 indicating little or no linear relationship.

The computational formulas and test procedures are summarized in Tables 14.11 and 14.12 on pages 576–577.

Table 14.11 Summary of Formulas Used in Chapter 14

Parameter	Estimate	Computational Formula
β_1	b_1	$\dfrac{\Sigma\, x_i Y_i - (\Sigma\, x_i)(\Sigma\, Y_i)/n}{\Sigma\, x_i^2 - (\Sigma\, x_i)^2/n}$
β_0	b_0	$\overline{Y} - b_1 \overline{x}$
$\mu_{Y\|x}$	$\hat{Y}_x$	$b_0 + b_1 x$
$\sigma^2_{Y\|x}$	$s^2_{Y\|x}$	MSE
ρ	r_{XY}	$\dfrac{\Sigma\, X_i Y_i - (\Sigma\, X_i)(\Sigma\, Y_i)/n}{\sqrt{[\Sigma\, X_i^2 - (\Sigma\, X_i)^2/n]\,[\Sigma\, Y_i^2 - (\Sigma\, Y_i)^2/n]}}$
ρ	r_s	$1 - 6\, \Sigma\, d_i^2/n(n^2 - 1)$

Sum of Squares	Definitional Form	Computational Form	df
Total	$\Sigma\,(Y_i - \overline{Y})^2$	$\Sigma\, Y_i^2 - (\Sigma\, Y_i)^2/n$	$n - 1$
Regression	$\Sigma\,(\hat{Y}_i - \overline{Y})^2$	$b_1 \left[\Sigma\, x_i Y_i - \dfrac{(\Sigma\, x_i)(\Sigma\, Y_i)}{n}\right]$	1
Error	$\Sigma\,(Y_i - \hat{Y}_i)^2$	$\Sigma\, Y_i^2 - b_0\, \Sigma\, Y_i - b_1\, \Sigma\, x_i Y_i$	$n - 2$
Lack of fit	$\Sigma\,\Sigma\,(\overline{Y}_j - \hat{Y}_j)^2$	SSE − SSPE	$c - 2$
Pure error	$\Sigma\,\Sigma\,(Y_{ij} - \overline{Y}_j)^2$	$\Sigma\,\Sigma\, Y_{ij}^2 - \dfrac{\Sigma\left(\sum_i Y_{ij}\right)^2}{n_j}$	$n - c$

n is the total number of observations, c is the number of different x values, and n_j is the number of Y values corresponding to x_j.

Table 14.12 Summary of Inferential Procedures for Chapter 14

Hypotheses	Test
$H_0: \mu_{Y\|x} = \beta_0 + \beta_1 x$ $H_a: \mu_{Y\|x} \neq \beta_0 + \beta_1 x$	Reject H_0 if $\dfrac{\text{MSLOF}}{\text{MSPE}} > f_{c-2,\, n-c,\, \alpha}$
$H_0: \beta_1 = 0 \quad H_a: \beta_1 \neq 0$	Reject H_0 if $\dfrac{\text{MSREG}}{\text{MSE}} > f_{1,\, n-2,\, \alpha}$ or if $\left\|\dfrac{b_1}{s_{b_1}}\right\| > t_{n-2,\, \alpha/2}$ where $s_{b_1}^2 = \dfrac{s_{Y\|x}^2}{\Sigma (x - \bar{x})^2}$

For all other hypotheses about β_1, use the statistic $(b_1 - \beta_{01})/s_{b_1}$ with the t_{n-2} distribution, making the standard adjustments in the test.

Hypotheses	Test
$H_0: \rho = 0 \quad H_a: \rho \neq 0$	Reject H_0 if $\left\|\dfrac{r_{XY}\sqrt{n-2}}{\sqrt{1 - r_{XY}^2}}\right\| > t_{n-2,\, \alpha/2}$ (standard adjustment for a one-sided H_a) or Reject H_0 if $\|r_s\| > r_{n,\, \alpha/2}$

Confidence Intervals for	Formula
β_1	$b_1 \pm t_{n-2,\, \alpha/2} s_{b_1}$
$\mu_{Y\|x_p}$	$\hat{Y}_{x_p} \pm t_{n-2,\, \alpha/2} \sqrt{(\text{MSE})\left[\dfrac{1}{n} + \dfrac{(x_p - \bar{x})^2}{\Sigma (x - \bar{x})^2}\right]}$
Y_{x_p}	$\hat{Y}_{x_p} \pm t_{n-2,\, \alpha/2} \sqrt{\text{MSE}\left[1 + \dfrac{1}{n} + \dfrac{(x_p - \bar{x})^2}{\Sigma (x - \bar{x})^2}\right]}$

REVIEW EXERCISES

Level One

14.1 The General Stone Tire & Rubber Company wishes to predict annual sales of radial tires as a function of national automobile registrations. Data available for the past 10 years are as follows (x in millions, Y in thousands):

$$\Sigma x = 600 \qquad \Sigma Y = 800 \qquad \Sigma xY = 56{,}000$$
$$\Sigma x^2 = 46{,}000 \qquad \Sigma Y^2 = 74{,}000 \qquad n = 10$$

(*a*) Determine the appropriate regression equation.
(*b*) Determine the analysis-of-variance table and test whether there is a significant linear relationship between x and Y. Use a significance level of 0.05.

14.2 The following data represent the density (coded) of cigarette filters and the amount of tar and nicotine passing through in milligrams.

	Cigarette									
	1	2	3	4	5	6	7	8	9	10
Filter density:	1.5	2.0	1.6	1.4	2.1	2.2	1.9	2.6	1.8	1.7
Amount of residue:	20	30	25	35	15	11	18	08	22	17

Calculate a measure of the strength of the relationship and test to see whether filter density is useful as an indicator of the relative harmlessness of the cigarette (use $\alpha = 0.05$).

14.3 Refer to Exercise 14.2 and plot the data. Then
(*a*) Determine the prediction equation.
(*b*) Check the regression assumptions using residual plots.
(*c*) Estimate the average amount of residue ingested from smoking one Terrytown cigarette and place 90% bounds on your estimate. Terrytown cigarettes have a coded filter density of 1.75.

14.4 A weather observer has collected data on wind speed, X, and pollution count, Y, for a random sample of six days. Some resulting computations are the following (assume a normal distribution):

$$\Sigma X_i = 72 \qquad \Sigma X_i^2 = 976 \qquad \Sigma X_i Y_i = 3816$$

$$\Sigma Y_i = 360 \qquad \Sigma Y_i^2 = 24{,}400$$

What is the strength of the linear relationship between the two variables?

14.5 The following is a plot of the residuals from a regression equation against the independent variable X. What assumptions appear to be violated (if any)?

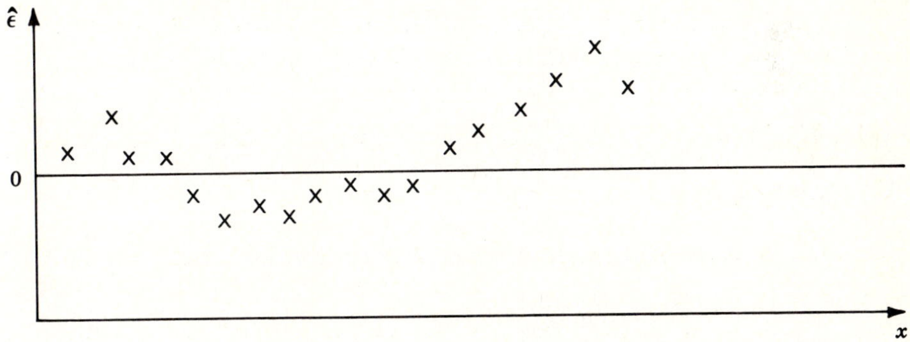

14.6 Two groups of consumers, each in a different marketing area, were asked to rank seven brands of coffee in order of preference. The results are given here. At the 0.05 level, is there a relationship between the rankings of consumers in the two areas?

	Ranking	
Brand	Group 1	Group 2
A	2	1
B	1	3
C	3	2
D	4	4
E	5	5
F	6	7
G	7	6

14.7 Refer to Exercise 14.1. If national automobile registrations is a useful predictor for radial tire sales, determine upper and lower bounds on radial tire sales for next year, when automobile registrations are expected to rise to 67 million ($\alpha = 0.10$).

14.8 Believing that there is a negative linear relationship between interest rates and the number of housing starts, a researcher in the building trades selected seven years at random and recorded the following results.

Year	Interest Rate (%)	Housing Starts (Millions)
1	8.0	50
2	8.5	42
3	7.2	71
4	6.3	86
5	5.4	80
6	7.5	78
7	4.8	91

(*a*) Assume a bivariate normal distribution and test this belief at $\alpha = 0.05$.
(*b*) What percentage of the fluctuation in housing starts may be explained by changing interest rates?
(*c*) Determine the least-squares linear regression equation, if your answer to (*a*) indicates that it is appropriate to do so, and obtain a 95% confidence interval for the average number of housing starts in years that have an average interest rate of 10.0%.
(*d*) The interest rate on home loans is expected to be 9.75% next year. Obtain 95% limits for the number of housing starts next year.

14.9 The following summarized data have been collected for age of trucks (x) and miles per gallon of gasoline (Y).

$$n = 15 \quad \Sigma (x_i - \bar{x})^2 = 84 \qquad \Sigma x_i = 45 \qquad \Sigma x_i^2 = 219$$
$$\Sigma Y_i = 60 \qquad \Sigma Y_i^2 = 313 \quad \Sigma x_i Y_i = 138$$

(*a*) Calculate the simple linear regression equation for predicting miles per gallon as a function of age.
(*b*) Construct the analysis-of-variance table and test the regression for significance at the 0.01 level.
(*c*) Determine a 99% confidence interval for the mileage of a four-year-old truck.
(*d*) Comment on your interval in (*c*).

14.10 It has often been stated that the stock market follows women's hemlines: as hemlines go up, so does the market. To determine whether there is any relationship between the two, 18 years were randomly selected and hemline lengths and the Dow Jones averages for the years determined. A correlation coefficient of 0.6 was calculated. Is the statement supported at $\alpha = 0.10$?

14.11 An industrial engineer collected data to determine whether there is a relationship between the speed at which a lathe is run and the variation in the runout of the wheel being machined by this lathe. The equation the engineer obtained is $\hat{Y} = 1521 - 42x$. Determine the strength of this relationship. Test the relationship for significance ($\alpha = 0.05$).

ANOVA TABLE

	SS	df	MS
Regression	15	1	15
Error	75	25	3
Total	90	26	

14.12 Show that Equation 14.5 can be rewritten as either

$$\frac{\Sigma\, x_i Y_i - n\bar{x}\bar{Y}}{\Sigma\, x_i^2 - n\bar{x}^2} \quad \text{or} \quad \frac{n\,\Sigma\, x_i Y_i - (\Sigma\, x_i)(\Sigma\, Y_i)}{n\,\Sigma\, x_i^2 - (\Sigma\, x_i)^2}$$

14.13 An investigator claimed that leadership ability is related to the amount of a certain hormone present in the blood. Six individuals were selected at random and ranked on the characteristic "leadership ability." The hormone content for each individual was determined from blood samples and recorded in milligrams per hundred milliliters of blood. Do these data support the contention at the 0.05 level?

Individual	Leadership Ability Rank	Hormone Content
A	1	131
B	4	174
C	6	186
D	5	200
E	3	189
F	2	156

14.14 For each of Exercises 14.2, 14.6, and 14.13, justify your choice of statistic for measuring the strength of the linear relation.

14.15 A marketing research experiment was conducted to study the relationship between the length of time a buyer needed to reach a decision and the number of alternative package designs of a product presented. Brand names were eliminated from the packages to reduce the effects of brand preferences. The buyers made their selections using the manufacturers' product descriptions on the packages as the only buying guide. The length of time necessary to reach a decision is recorded for 15 participants in the study.

Length of decision time (seconds)	5	8	8	7	9	7	9	8	9	10	10	11	10	12	9
Number of alternatives	2	2	2	2	2	3	3	3	3	3	4	4	4	4	4

(*a*) Is a prediction line appropriate for these data?
(*b*) Is the line a useful predictor? (Use $\alpha = 0.05$.)
(*c*) Find a 95% confidence interval for the average decision time when four alternatives are available.
(*d*) A new customer just walked into the display setup with two alternatives. What are 95% limits on the decision time?
(*e*) What percentage of the variability in decision time can be attributed to changing the number of alternatives?
(*f*) On the basis of these calculations, can we conclude that increasing the number of alternatives causes the decision time to increase?

14.16 It has been stated that long-run temperature forecasts are better indicators of fuel usage than are direct forecasts of heating oil sales. Before doing an extensive analysis of this relationship for the purpose of determining inventory levels, a regional distributor of heating oils has recorded the monthly sales volume and the average daily temperatures for nine randomly selected months. Using these data, the distributor calculated the correlation coefficient to be -0.9627.
(*a*) On the basis of this result, would you advise the distributor to do a more extensive analysis to obtain a linear prediction equation? (Use $\alpha = 0.01$.)
(*b*) What percentage of the variation in fuel usage appears to be explained by the temperature fluctuations?

14.17 The Novelty Company makes Dickey Birds. This is a product requiring a high degree of manual dexterity from the workers on the assembly line. In an effort to hire people who can do the job more quickly, and hence increase productivity, the company wants to give each job applicant a manual dexterity test to predict the average number of Dickey Birds that person can handle per hour. Eight workers were selected from the assembly line and given the manual dexterity test. Recorded below is the time (in minutes) each took to complete the test and the average number of Dickey Birds that worker has been handling per hour.

Worker:	1	2	3	4	5	6	7	8
Test time:	1	3	5	4	7	6	8	10
Number of Dickey Birds:	50	50	10	40	20	25	15	10

(*a*) Calculate the least-squares linear prediction equation.
(*b*) Test your equation for usefulness at $\alpha = 0.05$.
(*c*) Calculate a 90% confidence interval for the average number of Dickey Birds per hour that an applicant who just completed the test in five minutes can produce.

14.18 Use residual plots to check the regression assumptions of Exercise 14.17.

14.19 Occult Power, Ltd., believes that there is a linear relationship between the number of residential customers in a given area and the daily usage of electricity in megawatts (MW). To estimate the line, OP has collected the following data.

Customers (100,000s):	7	10	5	4	15	20	18	12
Average MW (1000s):	4	10	4	3	18	22	21	18

(*a*) Using LSE, estimate the parameters of the model and interpret your estimates.
(*b*) Is your equation a better (at $\alpha = 0.01$) predictor than the estimator $\overline{Y}$?
(*c*) What percentage of the variation in usage is explained by the number of residential customers?

14.20 Perform an analysis of the residuals for Exercise 14.19.

14.21 A neophyte statistician collected the following data and estimated the simple linear relation between x and Y to be $\hat{Y}_x = 125.58 - 11.88x$ with SSE = 524.66.

X:	2	2	2	7	7	5	5	10	10	10
Y:	94	98	96	49	53	74	76	2	0	1

Since you are no longer a neophyte, do you think that this was appropriate? Justify your answer using $\alpha = 0.01$.

14.22 Zoologic Thermometers, Inc., sells Amazon Chirp Bugs, found only in the upper Amazon River region of Brazil. In their advertising the company has stated that temperature can be determined by measuring the number of seconds between chirps of the bug. The FTC is now asking for proof. An experiment was performed by varying the temperature and measuring the time between chirps. The data obtained are:

Temperature:	15	20	25	30	35	43	46	51	59	62
Time:	21	23	28	33	33	36	34	39	43	41

Temperature:	66	68	71	74	75	76	77	79	81	84
Time:	34	44	40	33	47	42	37	30	25	47

(*a*) Calculate the straight-line equation.
(*b*) Test the equation for significance using $\alpha = 0.01$.
(*c*) Plot the residuals. Which assumptions do not appear to be satisfied?
(*d*) Calculate the coefficient of determination and the coefficient of correlation.

14.23 Refer to Exercise 14.4 and determine a 95% confidence interval for the average pollution count for all days with an average wind speed of 15 mph.

14.24 Tomorrow's weather forecast calls for an average wind velocity of 4 mph. Use the data of Exercise 14.4 to place 95% prediction limits on the pollution count for tomorrow.

14.25 Refer to Exercise 14.10. What percent of the variation in the DJI appears to be due to factors other than the length of women's hemlines?

14.26 (Recommended for use with a computer.) The six-months cost of maintenance of shipping tractors seems to increase with the age of the tractor. The following data were collected.

Age (Year)	Cost (6 months)	Age (Year)	Cost (6 months)	Age (Year)	Cost (6 months)
4.5	619	0.5	163	2.0	397
4.5	1049	0.5	182	2.0	496
4.5	1033	6.0	764	2.5	473
4.0	495	6.0	1373	3.0	716
4.0	723	1.0	978	3.5	821
4.0	681	1.0	466	3.5	697
5.0	890	1.0	549	3.0	452
5.0	1522	1.5	623	6.0	1203
5.5	987	1.5	478		
5.0	1194	2.5	516		

(*a*) Do a complete simple linear regression analysis using $\alpha = 0.05$ for all tests.

(*b*) Construct a 95% confidence interval for the average six-months cost of all seven-year-old tractors.

(*c*) Construct a 95% confidence interval for the average six-months cost of an arbitrary seven-year-old tractor.

Level Two

14.27 Show that $\Sigma\,(x_i - \bar{x})(Y_i - \bar{Y}) = \Sigma\,(x_i - \bar{x})Y_i$.

14.28 Show that $\Sigma\,(Y_i - \hat{Y}_i)(\hat{Y}_i - \bar{Y}) = 0$.

14.29 Show that $\Sigma\,\Sigma\,(Y_{ij} - \bar{Y}_j)(\bar{Y}_j - \hat{Y}_j) = 0$.

14.30 Refer to Exercise 14.26 and determine a 90% confidence interval for the change in average six-months maintenance costs for a tractor from one year to the next.

14.31 Refer to Exercise 14.19. Occult Power wishes to build a generating plant in an area whose population is expected to stabilize at 1,600,000 residential customers. The plant will service only these customers, but OP's president, Steve Seance, is worried about blackouts. What capacity plant should OP build to ensure that the probability of a future blackout is no more than 0.001?

14.32 Derive the formulas for calculating the bounds for a $1 - \alpha$ confidence interval for the sample mean of a set of k new observations on Y at $x = x_p$, that is, for $\overline{Y}_{x_p}$.

14.33 Refer to Exercises 14.26 and 14.32. Use your formula from Exercise 14.32 to construct a 95% confidence interval for the average six-months maintenance cost of the current six-year-old tractors if they are used for one more year, that is, for $\overline{Y}_7$.

In many situations we will be interested in predicting values for a response variable over time. That is, we would like to use time as the independent variable. To do so, we will have to code the actual time periods to obtain a quantitative variable for use in the equations. One way of assigning numerical values to the time periods is to assign $x = 0$ to the first period in the sequence and increase x by 1 for each succeeding period. It is mandatory that the time period corresponding to $x = 0$ and the units for x (the time change corresponding to a one-unit change in x) be identified with the equation.

14.34 Savings and loan institutions are a major source of home mortgage loans. Their chief source of funds for loans is deposits. To aid in managing its investments and financial planning, a local savings and loan is attempting to forecast monthly deposits. To this end they have presented you with the following information.

Month/ Year	Deposits (Million \$)	Month/ Year	Deposits (Million \$)	Month/ Year	Deposits (Million \$)
1–76	3.5	1–77	4.5	1–78	5.3
2–76	3.1	2–77	3.8	2–78	4.6
3–76	3.4	3–77	3.9	3–78	4.6
4–76	3.9	4–77	4.6	4–78	5.2
5–76	3.2	5–77	4.1	5–78	4.6
6–76	3.6	6–77	4.5	6–78	5.1
7–76	3.8	7–77	4.6	7–78	5.5
8–76	3.8	8–77	4.3	8–78	5.0
9–76	4.2	9–77	5.0	9–78	5.7
10–76	4.3	10–77	4.7	10–78	5.1
11–76	4.7	11–77	5.1	11–78	5.6
12–76	5.1	12–77	5.5	12–78	6.1

Use these data to develop a simple linear trend equation and test it for usefulness at $\alpha = 0.05$. (This exercise is recommended for use with a computer.)

14.35 The Loving Massage Co. invented and developed a water-pulse shower head in the mid-1960s. Since then they have grown rapidly as their sales expanded to a national market. Like many new companies, they really haven't taken time to analyze their past operations, but have responded more to the pressures of the moment. Now they have reached a level calling for more careful planning and analysis.

Here's where you come in. You get a summer job in their newly formed analysis department, and they present you with the following record of their sales (in units) for the past 10 years (by quarters).

	Quarter			
Year	I	II	III	IV
1969	1,110	1,952	2,578	3,249
1970	1,429	2,082	3,480	4,203
1971	2,237	2,871	4,811	6,593
1972	2,585	3,613	4,363	5,908
1973	2,124	3,000	3,906	4,702
1974	1,886	2,260	4,712	5,359
1975	2,938	4,724	7,256	8,738
1976	4,555	6,437	9,355	11,942
1977	4,824	6,565	10,568	10,220
1978	5,297	6,240	8,226	10,034

Your assignment, if you accept, is to forecast their sales for each quarter of 1979. Do so by obtaining and using a simple linear trend equation. Test the usefulness of your equation using $\alpha = 0.01$. (Recommended for use with a computer.)

14.36 The freight manager of an airline has determined, using a sample of 25 flights, a relationship between the total weight Y (in pounds) of luggage stored in an aircraft's baggage compartment and the number of passengers (x) on the flight manifest. The relationship is $\hat{Y}_x = 250 + 27x$. This will be used by superintendents at airports to determine how much additional freight can safely be stored on a flight, after taking into consideration fuel load and the weight of the passengers themselves. The manager has computed $s_{Y|x} = 100$ lb and $s_Y = 234.83$ lb, $\Sigma x^2 = 64{,}000$, $\bar{x} = 50$. Assuming that the model is correct, is the equation useful at $\alpha = 0.05$?

Level Three

14.37 Refer to Exercise 14.36 and construct a 95% confidence interval for the average weight difference in stored luggage between flights of 100 and flights of 75 passengers.

14.38 Derive the mean and variance of b_0.

14.39 In many situations it is logical to assume that the Y intercept β_0 equals zero. Use your results from Exercise 14.38 to construct a test for this assumption.

14.40 Refer to Exercise 14.19 and use your test from Exercise 14.39 to decide whether or not the argument that no residential customers means no usage of electricity is supported at $\alpha = 0.05$.

14.41 Refer to Exercise 14.1 and determine whether the data support the assumption that no automobile registrations implies no sales of radial tires at $\alpha = 0.10$.

14.42 Prove that b_1 is an unbiased estimate of β_1.

14.43 Prove that $\dfrac{r_{xy}\sqrt{n-2}}{\sqrt{1-r_{xy}^2}} = \dfrac{b_1}{s_{b_1}}$.

14.44 Prove that $E[\text{MS}(b_1)] = \sigma_{Y|x}^2 + \beta_1^2 \Sigma (x - \bar{x})^2$ where $\text{MS}(b_1) =$ MSREG.

14.45 Identify some experiments in which it is logical to assume that $\beta_0 = 0$. Choose one of them and collect the data for it, then perform a regression analysis and test to determine whether or not it appears necessary to include β_0 in the model.

References

1. Draper, Norman R., and Harry Smith, Jr. *Applied Regression Analysis.* New York: John Wiley & Sons, Inc., 1966.
2. Neter, John, and William Wasserman. *Applied Linear Statistical Models.* Homewood, Ill.: Richard D. Irwin, Inc., 1974.

Chapter 15
Improving the Ability to Estimate—The Multiple Case

KEY WORDS AND TERMS

Multiple Linear Model
Quadratic Effects
Pseudoreplication
Correlation Matrix
Partial Correlation Coefficient
SS (Full)
SS(Reduced)
Stepwise Regression
All Possible Regressions
Multicollinearity
Autocorrelation
Durbin-Watson Statistic
Indicator Variables

15.1 INTRODUCTION

In Chapter 14 we considered the use of a single variable as a predictor for reducing the estimate of the error variance for the response variable. Most response variables can be related to more than one independent variable; even if a single independent variable is used, it may be desirable to use higher-order terms of it as predictors also. This chapter considers the estimation of the parameters for a multiple linear model, tests for determining the usefulness of the model and for reducing the model, and a brief discussion of possible problems that can arise.

15.2 THE MULTIPLE LINEAR REGRESSION MODEL

In extending the simple linear model to include multiple predictors (independent variables), we maintain the requirement of a linear relationship. But by a linear relationship we mean that the response variable Y is a linear combination of the parameters $\beta_0, \beta_1, \beta_2, \ldots, \beta_k$, with k being the number of predictors in the model. That is, our model has the form

$$Y_{\boldsymbol{x}} = \beta_0 + \beta_1 x_1 + \cdots + \beta_k x_k + \epsilon \tag{15.1}$$

where $\boldsymbol{x}$ represents the combined set of predictors, that is, $\boldsymbol{x} = (x_1, x_2, \ldots, x_k)$.

The restrictions imposed on the model are that $x_j, j = 1, \ldots, k$, are nonrandom variables and ϵ is a random variable with a mean of zero. Under these restrictions, Y_x becomes a random variable with a mean given by

$$\mu_{Y|x} = \beta_0 + \beta_1 x_1 + \cdots + \beta_k x_k \tag{15.2}$$

Since $E[\epsilon] = 0$, Equation 15.2 is simply an alternative way of expressing the model. The x_j may be completely different predictors, different-order terms of the same predictor, or even products (interactions) of two or more different predictors. To see this, we note that if x_j and x_i are nonrandom variables, then so are x_j^m $(m = 2, 3, \ldots)$ and $x_i x_j$.

If we express the model as a polynomial function of x, it becomes

$$\mu_{Y|x} = \beta_0 + \beta_1 x + \beta_2 x^2 + \cdots + \beta_k x^k$$

Setting $x_j = x^j$ transforms the polynomial model (which was also linear in the $\beta_j, j = 0, 1, \ldots, k$) into the form of Equation 15.2. For three different independent variables with interactions, the model is

$$\mu_{Y|x} = \beta_0 + \beta_1 x_1 + \beta_2 x_2 + \beta_3 x_3 + \beta_{12} x_1 x_2 + \beta_{13} x_1 x_3 + \beta_{23} x_2 x_3 + \beta_{123} x_1 x_2 x_3$$

If we let $\beta_{12} = \beta_4$, $x_1 x_2 = x_4$, $\beta_{13} = \beta_5$, $x_1 x_3 = x_5$, etc., this model also is transformed to the form of Equation 15.2.

These models are all linear in nature initially. In many situations, the logical relationship will be nonlinear. Some nonlinear models still can be estimated using linear regression procedures. For example, we can transform the model

$$\mu_{Y|x} = \beta_0 x_1^{\beta_1} \times x_2^{\beta_2} \times \cdots \times x_k^{\beta_k}$$

by taking logarithms to obtain

$$\log_{10} \mu_{Y|x} = \log_{10} \beta_0 + \beta_1 \log_{10} x_1 + \beta_2 \log_{10} x_2 + \cdots + \beta_k \log_{10} x_k$$

or

$$\mu'_{Y|x'} = \beta'_0 + \beta_1 x'_1 + \beta_2 x'_2 + \cdots + \beta_k x'_k$$

This is now in the form of Equation 15.2. The fitted regression equation is used to predict the logarithm of the response, then we get the prediction of the actual response by finding the antilogarithm. A second nonlinear model that can be transformed into a linear model is

$$\mu_{Y|x} = \beta_0 \exp(\beta_1 x_1 + \beta_2 x_2 + \cdots + \beta_k x_k)$$

Taking the natural logarithm of the equation gives

$$\ln \mu_{Y|x} = \ln \beta_0 + \beta_1 x_1 + \beta_2 x_2 + \cdots + \beta_k x_k$$

which is once more in the form of Equation 15.2.

We shall consider only models in the form of Equation 15.2. That is, we assume that the model is initially linear or else has been transformed to a linear one.

15.3 ESTIMATING THE PARAMETERS

Our approach to estimating the parameters for a multiple linear model is identical to the approach we took in the simple linear case. That is, we wish to determine estimates of $\beta_0, \beta_1, \ldots, \beta_k$ that will minimize the sum of the squared deviations of the observed responses from the predicted responses. This is the least-squares estimation (LSE) procedure and we determine $b_0, b_1, \ldots, b_k$ as estimates such that we minimize

$$\sum_{i=1}^{n} (Y_i - \hat{Y}_i)^2$$

where

$$\hat{Y}_i = b_0 + b_1 x_1 + b_2 x_2 + \cdots + b_k x_k$$

The difficulty that arises in the multiple linear case is that minimizing the sum of squares results in one normal equation for each b. These equations are of the form

$$\left.\begin{array}{l} nb_0 + b_1 \Sigma x_{i1} + b_2 \Sigma x_{i2} + \cdots + b_k \Sigma x_{ik} = \Sigma Y_i \\ b_0 \Sigma x_{i1} + b_1 \Sigma x_{i1}^2 + b_2 \Sigma x_{i1}x_{i2} + \cdots + b_k \Sigma x_{i1}x_{ik} = \Sigma x_{i1} Y_i \\ b_0 \Sigma x_{i2} + b_1 \Sigma x_{i1}x_{i2} + b_2 \Sigma x_{i2}^2 + \cdots + b_k \Sigma x_{i2}x_{ik} = \Sigma x_{i2} Y_i \\ \vdots \\ b_0 \Sigma x_{ik} + b_1 \Sigma x_{i1}x_{ik} + b_2 \Sigma x_{i2}x_{ik} + \cdots + b_k \Sigma x_{ik}^2 = \Sigma x_{ik} Y_i \end{array}\right\} \quad (15.3)$$

Since the summations all represent numerical values calculated from the sample data, we are faced with simultaneously solving $k + 1$ equations in $k + 1$ unknowns. As you will recall from basic algebra, this is a very tedious and time-consuming task whenever $k > 2$. Fortunately, the computer has eliminated the need to solve the normal equations manually.

Computer programs for performing the computations required in estimating a multiple linear regression model are universally available. Therefore, although we shall present a simple example using two predictors, we shall emphasize the interpretation of results given in a computer printout. We present the following example as a practical illustration of the normal equations and to refresh your memory of solving simultaneous equations.

EXAMPLE 15.1 We have believed for some time that the best preparation for a student coming into an introductory statistics class is a solid background in basic algebra and arithmetic as well as in basic logic. To evaluate this belief, we have given our students a simple exam on the first day of class. The exam consists of two parts, the first covering basic algebra and arithmetic and the second covering basic logic. The following data represent the scores of ten randomly selected students from past classes on the mathematics section (x_1) and the logic section (x_2), and the students' final class averages in statistics (Y). The final average is based on 100 points, whereas the mathematics and logic sections of the exam each consist of ten questions worth 1 point each and scored as either right or wrong.

Student:	1	2	3	4	5	6	7	8	9	10
Average Y:	72	64	88	93	77	75	68	52	81	74
Math score (X_1):	5	6	8	10	7	6	8	3	9	9
Logic score (X_2):	6	3	7	9	5	5	5	2	8	4

Calculate the estimated equation for the multiple linear model relating Y to x_1 and x_2 using LSE.

SOLUTION Our model is

$$Y = \beta_0 + \beta_1 x_1 + \beta_2 x_2 + \epsilon$$

We find the least-squares estimates by solving Equation 15.3 with $k = 2$ and $n = 10$. To write the equations, we need to calculate $\Sigma\, Y_i$, $\Sigma\, x_{i1}$, $\Sigma\, x_{i2}$, $\Sigma\, x_{i1}^2$, $\Sigma\, x_{i2}^2$, $\Sigma\, x_{i1} Y_i$, $\Sigma\, x_{i1} x_{i2}$, and $\Sigma\, x_{i2} Y_i$. These are, respectively, 744, 71, 54, 545, 334, 5462, 413, and 4225. Thus the normal equations for these data are

$$10b_0 + 71b_1 + 54b_2 = 744$$
$$71b_0 + 545b_1 + 413b_2 = 5462$$
$$54b_0 + 413b_1 + 334b_2 = 4225$$

To solve these equations simultaneously, we solve the first equation for b_0, substitute the result into each of the remaining two equations, then solve the resulting two equations simultaneously for b_1 and b_2. Doing the arithmetic gives $b_2 = 3.6906$, $b_1 = 1.7202$, and $b_0 = 42.2573$. The resulting prediction equation is

$$\hat{Y}_x = 42.2573 + 1.7202x_1 + 3.6906x_2$$

where Y is a student's final class average in statistics, x_1 is the score on the mathematics section, and x_2 is the score on the logic section.

An estimate of the final class average in statistics for a student scoring 7 on both sections is

$$\hat{Y}_{7,7} = 42.2573 + 1.7202(7) + 3.6906(7) = 80.1329$$

Note that 80.1329 is also an estimate of the mean class average of all students scoring 7 on both sections.

The interpretation of the b_j's corresponds to the interpretation of b_0 and b_1 in the simple case. That is, b_0 is an estimate of the average response when all x_j are equal to zero. The regression coefficient b_j is an estimate of the change in the response corresponding to a one-unit change in the predictor x_j. As usual, we should be careful when we apply these interpretations to x_j values outside the original range of experimentation. For Example 15.1, the value of 42.2573 for b_0 implies that students scoring 0 on both sections would have a mean class average of approximately 42. The problem with this interpretation is that a student who would score 0 on both sections probably would be out of school before ever making it to the statistics class. The value of 1.7202 for b_1 is an estimate of the change in average for a one-point change in the mathematics score, and 3.6906 is an estimate of the change in average for a one-point change in the logic score.

As with simple linear regression, the fact that we can determine an equation does not imply that the equation is useful. We determine whether or not the equation is useful by testing to see whether the reduction in the sum of squares attributed to error is significant. That is the topic of the next section.

EXERCISES

15.3.1 Identify three variables that might be useful in predicting sales of an industrial product.

15.3.2 Identify three variables that might be useful in predicting sales of a consumer product.

15.3.3 The yield of a polymerization process is believed to be linearly related to the temperature of the water bath and the amount of catalyst used in the reaction. In order to estimate the actual relationship, six experiments were run at various temperature-catalyst combinations. Use the following results to estimate the linear relationship of the yield to the temperature and the amount of catalyst.

Yield (kg):	21.2	15.7	16.9	19.4	24.5	17.6
Temperature (°C):	70	50	60	70	80	60
Catalyst (mg):	6	5	3	4	6	5

15.3.4 Most universities require applicants to their MBA programs to take the Graduate Management Admissions Test (GMAT) and to attain a particular score. They do this because the GMAT score is assumed to be an indicator of the student's ability to succeed in an MBA program. Another factor that is considered in addition to the GMAT score is the student's grade point average (GPA) over the last two years of undergraduate work. The following data represent the GMAT scores for randomly selected students, their GPAs over the last two years of undergraduate work, and their GPAs at the end of one year in an MBA program.

Student:	1	2	3	4	5	6	7	8
GMAT score (X_1):	492	515	564	471	459	572	486	528
GPA (two years) (X_2):	2.94	3.26	3.41	2.82	3.08	3.26	3.37	3.69
One-year GPA (Y):	3.12	3.48	3.26	2.90	2.86	3.51	3.22	3.74

Estimate the linear equation for predicting a student's GPA after one year in an MBA program from that student's GMAT score and GPA for the last two years of undergraduate work.

15.4 TESTING THE MODEL FOR USEFULNESS

As we indicated earlier, it is always possible to determine the estimating equation, but the equation will not always be an improvement over the estimator $\overline{Y}$. To make this determination we construct an hypothesis test for the usefulness of the model, following the same line of reasoning that we used in the simple case. If the model is useful, we can use the set of predictors to remove a significant amount of variation from the total variation of Y. That is, the variation in the values of the independent variables will explain a significant proportion of the total variation. As a result of chance alone, one variable in a sample will always explain some of the variation in another variable. To determine whether the amount explained is more than can reasonably be attributed to chance, we make the following assumptions about the distribution of the error term ϵ:

1. ϵ is $N(0, \sigma^2)$ for all values of x_j.
2. The errors are independent of one another and of the predictors.

In the single-predictor situation of Chapter 14, we saw that the equation was useful only if β_1 was nonzero. For the multiple linear model, it follows directly that the proposed estimator is an improvement over $\overline{Y}$ (or $Y = \beta_0 + \epsilon$) only if at least one of the β_j's is nonzero. Thus our hypotheses take the form

$$H_0\colon\ \beta_j = 0 \text{ for all } j \qquad H_a\colon\ \beta_j \neq 0 \text{ for some } j \tag{15.4}$$

or, equivalently,

$$H_0\colon\ Y = \beta_0 + \epsilon \qquad H_a\colon\ Y = \beta_0 + \beta_1 x_1 + \cdots + \beta_k x_k + \epsilon \tag{15.5}$$

In other words, the null hypotheses of Equations 15.4 and 15.5 state that the true model is a "plane" parallel to the "plane" of the x_j axes; that is, none of the independent variables contain any information about Y, and thus they do not improve upon $\overline{Y}$ as an estimator for $\mu_{Y|x}$. Figure 15.1 on page 596 illustrates this for $k = 2$. The alternative hypotheses, on the other hand, indicate that at least some of the x_j's contain useful information about the variable Y. Put simply, the hypotheses are

Figure 15.1 Graphical Representations of Two Predictor Models with (a) No Relation, and (b) A Linear Relation

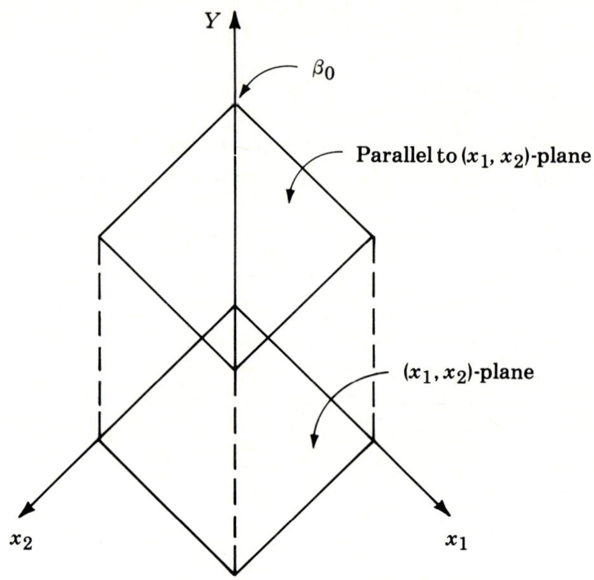

(a) No Relation Between Y and (x_1, x_2)

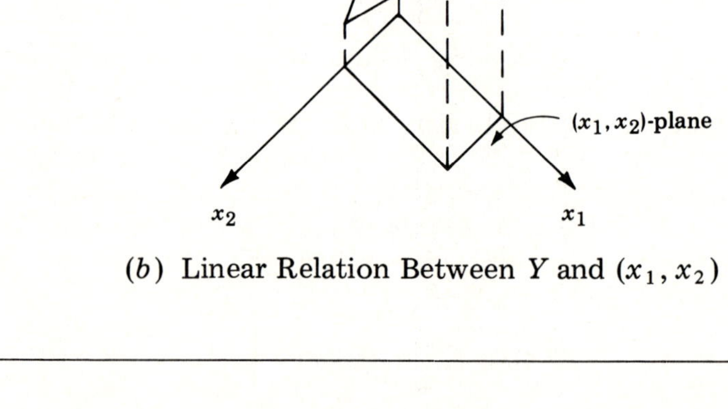

(b) Linear Relation Between Y and (x_1, x_2)

$$H_0: \text{ model is not useful} \qquad H_a: \text{ model is useful} \tag{15.6}$$

To construct the test, we note that if H_0 is true, then

$$\text{SSTO} = \Sigma\,(Y_i - \bar{Y})^2$$

is a sum of squared deviations for a random sample from a normal population and has a $\sigma^2\chi^2_{n-1}$ distribution. As usual, we partition the sum of squares into smaller sums of squares. That is,

$$\Sigma\,(Y_i - \bar{Y})^2 = \Sigma\,(Y_i - \hat{Y}_i + \hat{Y}_i - \bar{Y})^2 = \Sigma\,(Y_i - \hat{Y}_i)^2 + \Sigma\,(\hat{Y}_i - \bar{Y})^2$$

$$\text{SSTO} = \text{SSE} + \text{SSREG}$$

The difference between this partition and the corresponding partition in Chapter 14 is in the degrees of freedom. SSTO still has $n - 1$ degrees of freedom, but for SSE one degree of freedom is lost for each normal equation, resulting in $n - (k + 1)$ degrees of freedom for SSE. The regression sum of squares contains $k + 1$ variables (b_0 through b_k) with a single restriction imposed by $\bar{Y}$. Thus the degrees of freedom for SSREG are $(k + 1) - 1 = k$. Since

$$(n - 1) = n - (k + 1) + k$$

we can apply Cochran's theorem to obtain

$$\text{SSE} \sim \sigma^2\chi^2_{n-k-1} \qquad \text{if the model is correct}$$

and

$$\text{SSREG} \sim \sigma^2\chi^2_k \qquad \text{if } H_0 \text{ is true}$$

and the two are independent.

If the model is correct, then the MSE is an unbiased estimate of σ^2 independent of H_0. If H_0 is also true, then MSREG will also provide an unbiased estimate of σ^2. Therefore, if the model is correct and H_0 is true, MSE and MSREG are unbiased estimates of the same thing and should be close to each other in value. That is, their ratio should deviate from 1 only as a result of chance or sampling variation. Since the ratio of two independent χ^2 variables divided by their respective degrees of freedom has an F distribution, we have that, if H_0 is true,

$$\frac{\text{MSREG}}{\text{MSE}} \sim F_{k,\,n-k-1}$$

If H_0 is false, $\hat{Y}_x$ deviates from $\overline{Y}$ as a result of the model as well as chance, and hence we can expect MSREG to be larger than MSE. To see this, note that if x_{ij} deviates from $\overline{x}_j$ and the remaining predictors are set equal to their means, then $\hat{Y}_x$ deviates from $\overline{Y}$ as a result of chance plus the differential effect of $b_j(x_{ij} - \overline{x}_j)$. The expected value of this deviation when squared would be $\sigma^2 + \beta_j^2 (x_{ij} - \overline{x}_j)^2$. In general, the expected value of MSREG can be shown to equal

$$\sigma^2 + \frac{1}{k} \sum_j \beta_j^2 \sum_i (x_{ij} - \overline{x}_j)^2 + 2 \sum_{m \neq j} \beta_m \beta_j \sum_i (x_{im} - \overline{x}_m)(x_{ij} - \overline{x}_j)$$

Thus, if all $\beta_j = 0$, $E(\text{MSREG}) = \sigma^2$. Otherwise $E(\text{MSREG}) > \sigma^2$. Therefore our test is to reject H_0 (in any of Equations 15.4 through 15.6) if the probability of a ratio as large or larger than the observed MSREG/MSE is smaller than α, or equivalently if

$$\frac{\text{MSREG}}{\text{MSE}} > f_{k,\, n-k-1,\, \alpha} \tag{15.7}$$

To calculate the necessary sums of squares, we use the computational formulas. As always, the computational formula for SSTO is

$$\text{SSTO} = \Sigma\, Y_i^2 - \frac{(\Sigma\, Y_i)^2}{n} \tag{15.8}$$

For the SSE we start with the formula given for the simple case and subtract the contribution of each b_j as given by $b_j \,\Sigma\, x_{ij} Y_i$. That is,

Table 15.1 ANOVA Table for a Multiple Linear Regression with k Predictors

Source	SS	df	MS	F Calc
Regression $(b_1, b_2, \ldots, b_k \mid b_0)$	SSREG	k	MSREG	$\frac{\text{MSREG}}{\text{MSE}}$
Error	SSE	$n - k - 1$	MSE	
Total	SSTO	$n - 1$		

$$SSE = \Sigma\, Y_i^2 - b_0\, \Sigma\, Y_i - b_1\, \Sigma\, x_{i1} Y_i - \cdots - b_k\, \Sigma\, x_{ik} Y_i \tag{15.9}$$

The SSREG is most easily obtained by subtracting the SSE from SSTO. That is,

$$SSREG = SSTO - SSE \tag{15.10}$$

To provide an arithmetic check, the SSREG may be calculated directly using

$$SSREG = b_0\, \Sigma\, Y_i + b_1\, \Sigma\, x_{i1} Y_i + \cdots + b_k\, \Sigma\, x_{ik} Y_i - \frac{(\Sigma\, Y_i)^2}{n} \tag{15.11}$$

All calculations are summarized in an ANOVA table as shown in Table 15.1.

The notation $(b_1, b_2, \ldots, b_k | b_0)$ in Table 15.1 indicates that the effect of the intercept b_0 has been removed from the SSREG. Another way of stating this is that the b_j's are determined with b_0 already in the model.

EXAMPLE 15.2 In Example 15.1 we calculated an equation for predicting a student's average total points in an introductory statistics course from scores based on the logic (x_2) and mathematics (x_1) sections of an exam. Test that equation for usefulness as a predictor at $\alpha = 0.05$.

SOLUTION The prediction equation is

$$\hat{Y}_x = 42.2573 + 1.7202 x_1 + 3.6906 x_2$$

In addition, we obtained the following totals in Example 15.1: $\Sigma\, Y_i = 744$, $\Sigma\, x_{i1} Y_i = 5462$, and $\Sigma\, x_{i2} Y_i = 4225$. Calculating $\Sigma\, Y_i^2$ gives 56,592. Thus, using Equations 15.8, 15.9, and 15.10,

$$SSTO = 56{,}592 - \frac{(744)^2}{10} = 1238.400$$

$$\begin{aligned} SSE &= 56{,}592 - 42.2573(744) - 1.7202(5462) - 3.6906(4225) \\ &= 164.052 \end{aligned}$$

$$SSREG = 1238.4 - 164.052 = 1074.348$$

Constructing the ANOVA table, we have

Source	SS	df	MS	F Calc	F Table
Regression	1074.348	2	537.174	22.921	4.74
Error	164.052	7	23.436		
Total	1238.400	9			

Our hypotheses are

$$H_0: \beta_1 = \beta_2 = 0 \qquad H_a: \text{ some } \beta_j \neq 0$$

with H_0 being rejected, since $22.921 > 4.74$. We conclude that there appears to be a useful linear relation between a student's final class average in introductory statistics and that student's scores on the mathematics and logic sections of the exam.

Another way of measuring the model's usefulness is to determine the reduction in the variation due to the model. This reduction is called the *multiple coefficient of determination* and is calculated using the same basic formula that applies in the simple case. That is,

$$R^2_{Y|x} = \frac{\text{SSREG}}{\text{SSTO}} \tag{15.12}$$

The *multiple correlation coefficient* $R_{Y|x}$ is the square root of $R^2_{Y|x}$. In effect, $R_{Y|x}$ measures the strength of the linear relationship between Y and the *set* of predictors, whereas $R^2_{Y|x}$ is the proportion of the variation in Y that is explained by the model. Usually, but not always, the closer $R^2_{Y|x}$ is to 1.00, the better the model is for estimating $\mu_{Y|x}$ (or for predicting Y_x).

An $R^2_{Y|x}$ near 1.00 is not always an indicator of the goodness of the model because if enough predictors are added to the model, $R^2_{Y|x}$ will be forced to equal 1. This is because of the dimensionality of the problem, or, equivalently, because all the degrees of freedom for error will be used up. To see this, consider the simple linear case and recall that only two points are required to determine a straight line. Thus, if we try to fit a simple linear model with only two observations, we will always obtain a perfect (and usually meaningless) fit, and $R^2_{Y|x}$ will equal 1.00. To illustrate this point, let's reconsider Example 15.1 with some additional predictors.

EXAMPLE 15.3 Refer to Example 15.1 and add the following independent variables as predictors: x_3 = sex of student ($M = 0$, $F = 1$);

x_4 = student's number of sisters; x_5 = bubble gum chewer (chews = 1, doesn't chew = 0); x_6 = student's weight; x_7 = student's height; x_8 = golf handicap of student's father; x_9 = mother's bowling average. The data for the revised problem are given in Table 15.2. Fit the model

$$Y = \beta_0 + \beta_1 x_1 + \beta_2 x_2 + \cdots + \beta_9 x_9 + \epsilon$$

to these data, determine $R^2_{Y|x}$ for both the original model and the enlarged model, and comment on the results.

SOLUTION For the original model we had (from Example 15.2) SSREG = 1074.348 and SSTO = 1238.400. Thus

$$R^2_{Y|x_1x_2} = \frac{1074.348}{1238.400} = 0.8675$$

which implies that x_1 and x_2 together explain 86.75% of the variation in Y. Table 15.3 on page 602 shows the results of computer runs using (*a*) x_1 through x_4 as predictors and (*b*) the full model or x_1 through x_9 as predictors. Looking at the printout in Table 15.3*a*, we see that R^2 increased from 0.8675 to 0.8782 when we included the sex and number of sisters variables. Note that even though R^2 increased, the standard error increased also, from 4.841 to 5.493, which is not desirable.

For the printout of the full model in Table 15.3*b*, we see that including the additional five nonsense variables has explained 100% of the variation in Y. That is, $R^2_{Y|x} = 1.00$, which is to be expected

Table 15.2 Data for Example 15.3

x_1	x_2	x_3	x_4	x_5	x_6	x_7	x_8	x_9	Y
5	6	0	3	1	154	69	18	187	72
6	3	0	1	0	173	73	7	134	64
8	7	1	0	0	122	65	24	156	88
10	9	0	0	0	143	68	13	162	93
7	5	1	4	1	118	66	4	143	77
6	5	1	2	1	138	63	15	118	75
8	5	0	2	0	187	72	9	167	68
3	2	1	3	0	105	62	12	171	52
9	8	1	1	1	112	65	16	130	81
9	4	0	5	0	207	71	27	125	74

Table 15.3 Printouts for Example 15.3 Using (a) x_1 through x_4 as Predictors and (b) the Full Model

```
REGRESSION COEFFICIENTS

VAR COEFFICIENT(B)  STD ERROR       T
1   1.993756558     1.328043267     1.501273797
2   3.559863589     1.458711018     2.440417289
3   2.421511018     3.711113991     0.6525024625
4   0.138772298     1.363918947     0.1017452675
    39.51888772     (CONSTANT)

STANDARD ERROR OF ESTIMATE: 5.493271664
R SQUARED:                  0.8781652391

ANALYSIS OF VARIANCE

SOURCE      SUM OF SQUARES  DF   MEAN SQUARE     F
REGRESSION  1087.519832     4    271.879958      9.009797703
RESIDUAL    150.8801679     5    30.17603358
```

```
REGRESSION COEFFICIENTS

VAR COEFFICIENT(B)  STD ERROR        T
1    ¯26.05951994   2.369949467      ¯10.99581249
2   37.59072182     2.903428568      12.94701107
3   51.77013752     4.751408028      10.89574653
4   15.57068084     1.35746639       11.47039879
5    ¯65.05356364   5.576448855      ¯11.6657689
6   0.4389288908    0.05305321209    8.273370707
7   5.730872371     0.5511737568     10.39757844
8    ¯1.202945141   0.1317805979     ¯9.12839341
9   ¯0.9891311036   0.08370544206    ¯11.8168076
    ¯261.309924     (CONSTANT)

STANDARD ERROR OF ESTIMATE: 1
R SQUARED:                  1

ANALYSIS OF VARIANCE

SOURCE      SUM OF SQUARES   DF   MEAN SQUARE    F
REGRESSION  1238.4           9    137.6          137.6
RESIDUAL    0                0    1
```

since SSE and df_E both equal 0. (MSE $= 1.00$ because the program has a built-in override that equates the undefined operation 0/0 to 1.0.) The actual value of the standard error, $s_{Y|x}$, is indeterminate, since division by 0 is an illegal operation.

As you can see from Example 15.3, $R^2_{Y|x}$ can be a misleading measure of the usefulness of the model. A rule of thumb some authors recommend is to try to obtain at least three times as many distinct observations on Y as there are predictors in the model. By *distinct* we mean different combinations of levels of the predictor variables. If we obtain multiple observations by resetting the independent variables to a previously used combination, then we simply have replication of the original response. For example, if we take 42 observations on Y by using six combinations of 5 independent variables and replicating each combination 7 times, we are likely to have a very high value for $R^2_{Y|\mathbf{x}}$. This is because the regression procedure essentially fits the model to the means of the Y values; that is, to $\overline{Y}_{x_j}$, of which there are only six. Hence a five-predictor (six-parameter) model should give a nearly perfect fit, since all degrees of freedom for error correspond to the pure error sum of squares. That is, the model *will* provide a perfect fit for the six Y means.

These same problems can occur even without replication. Suppose that certain combinations of the independent variables are quite close to each other (for example, one predictor is changed by a small amount and the others remain fixed). Then we have what might be called pseudoreplication and the same phenomena can occur. The difficulty in this case is that it may be very hard to detect. One way to check for pseudoreplication is to construct residual plots and to look for clustering of the points.

EXERCISES

15.4.1 Refer to Exercise 15.3.3 and evaluate the usefulness of the model at $\alpha = 0.05$.

15.4.2 Refer to Exercise 15.3.4 and decide whether or not the GMAT score and GPA are useful in predicting one's GPA in an MBA program. Use $\alpha = 0.01$.

15.4.3 The following data represent the starting salaries (Y), cumulative GPA (x_1), number of years of work experience (x_2), and undergraduate majors for 12 recent graduates from the undergraduate program in the College of Business. The variables representing major are to be coded as $x_3 = 1$ if finance, 0 otherwise; $x_4 = 1$ if accounting, 0 otherwise; and $x_5 = 1$ if marketing, 0 otherwise. Thus, if x_3, x_4, and x_5 all equal 0, the student's major was management. Use these data to construct a prediction equation for starting salaries and test its usefulness at $\alpha = 0.05$.

Graduate:	1	2	3	4	5	6
GPA (x_1):	2.71	3.58	2.24	2.37	2.35	2.92
Experience (x_2):	1.0	0.5	2.0	0.0	1.5	2.5
Majors:	Mgt.	Mkt.	Acc.	Fin.	Fin.	Mgt.
Salary (Y):	13,500	14,200	14,000	12,500	13,000	14,500

(data continued on page 604)

Graduate:	7	8	9	10	11	12
GPA (x_1):	3.08	2.16	2.59	2.69	3.21	2.78
Experience (x_2):	3.0	1.5	0.0	3.5	0.0	5.0
Majors:	Acc.	Mkt.	Mkt.	Acc.	Mgt.	Fin.
Salary (Y):	15,500	11,500	12,400	14,600	13,300	14,800

15.5 CHECKING THE APPROPRIATENESS OF THE MODEL

Just as in the simple linear case, merely specifying and fitting a model does not mean that the model is appropriate. We may want to add higher-order terms for some of the predictors, or possibly to include interaction terms for certain combinations of independent variables, such as x_3x_1 or $x_2x_3x_6$. We will restrict our considerations to the same approaches taken in Chapter 14, namely, testing for lack of fit when possible and using residual plots.

15.5.1 Testing for Lack of Fit

To be able to construct a test for lack of fit in the model, we must have replication at some combinations of the independent variable. By replication we do not mean "setting" the x_j values to be used and then taking multiple readings without resetting them. This eliminates any variation that may occur as a result of slight errors in the "settings." To include this variation as part of the pure error term, we use the term *replication* to mean changing the predictor values and then resetting them to obtain a repeat observation on Y.

Replication can be difficult and costly in a multiple linear model, especially when we are using a large number of predictors. It requires that specific combinations of values for the independent variables be used more than once in obtaining a response for Y. For example, a value for Y when $x_1 = 1, x_2 = 13, x_3 = -4$ is *not* a replicate of the response for $x_1 = 0, x_2 = 13, x_3 = -4$, since x_1 differs even though x_2 and x_3 are unchanged. (It might be a pseudoreplicate, though.) The major difficulty with replication in a business environment is that time is often a factor in the model, and this makes obtaining a true replicate impossible. In addition, replication requires an increase in the number of observations, and this can become prohibitive from a cost standpoint.

Assuming that replicates are available, our approach is identical to that of Chapter 14. That is, we partition the SSE into two parts as follows:

$$\sum_{c=1}^{m} \sum_{i=1}^{n_c} (Y_{ic} - \hat{Y}_c)^2 = \sum_{c=1}^{m} \sum_{i=1}^{n_c} (Y_{ic} - \overline{Y}_c)^2 + \sum_{c=1}^{m} \sum_{i=1}^{n_c} (\overline{Y}_c - \hat{Y}_c)^2 \qquad (15.13)$$

or

$$\text{SSE} = \text{SSPE} + \text{SSLOF}$$

with degrees of freedom given by

$$n - k - 1 = \sum_{c=1}^{m} (n_c - 1) + (m - k - 1)$$

where

$n = \sum_{c=1}^{m} n_c$ is the total number of observations

m is the number of different combinations of the x_j

n_c is the number of replicates at combination c

The hypotheses to be evaluated are

$$H_0\colon \mu_{Y|x} = \beta_0 + \sum_{j=1}^{k} \beta_j x_j \qquad H_a\colon \mu_{Y|x} \neq \beta_0 + \sum_{j=1}^{k} \beta_j x_j$$

If H_0 is true and our assumptions are correct, then applying Cochran's theorem and making the appropriate adjustments gives an F distribution with

$$m - k - 1 \quad \text{and} \quad \sum_{c=1}^{m} (n_c - 1) = n - m$$

degrees of freedom for the statistic MSLOF/MSPE. As before, both MSPE and MSLOF are estimates of σ^2 if H_0 is true, with MSLOF overestimating it if H_a is true. Thus our test is to reject H_0 at α if

$$\frac{\text{MSLOF}}{\text{MSPE}} > f_{m-k-1,\, n-m,\, \alpha} \qquad (15.14)$$

EXAMPLE 15.4 Refer to Example 14.13, in which the Police Department of Akersville was attempting to construct a model relating an automobile's stopping distance to the speed at brake application. We found that a simple linear model was useful but that we could improve upon it by possibly adding higher-order terms in x. Use the data there and fit the model

$$Y = \beta_0 + \beta_1 x + \beta_2 x^2 + \epsilon$$

to them. Test the model for appropriateness at $\alpha = 0.01$ and for usefulness at $\alpha = 0.05$.

SOLUTION The data with the term in x^2 included are given in Table 15.4 and the results of the computer run are presented in Table 15.5.

Since our second predictor, x^2, is a function of x, replicates at values of x are automatically replicates of any combination that includes that value of x. Thus to calculate SSPE we have $n = 18$, $m = 6$, $n_c = 3$ for all $c = 1, 2, \ldots, 6$. Therefore, the computational form for SSPE is

$$\text{SSPE} = \sum_{c=1}^{m} \sum_{i=1}^{n_c} (Y_{ic} - \overline{Y}_c)^2 = \sum_{c=1}^{m} \sum_{i=1}^{n_c} Y_{ic}^2 - \sum_{c=1}^{m} \frac{\left(\sum_{i=1}^{n_c} Y_{ic}\right)^2}{n_c} \tag{15.15}$$

Since $n_c = 3$ for all combinations, the last term (in this case) reduces to

$$\sum_{c=1}^{m} \frac{\left(\sum_{i=1}^{n_c} Y_{ic}\right)^2}{n_c} = \frac{1}{3} \sum_{c=1}^{m} \left(\sum_{i=1}^{n_c} Y_{ic}\right)^2$$

For illustration, the data are reorganized in Table 15.6.

Table 15.4 Data for Example 15.4

x (mph):	20	20	20	30	30	30	40	40	40
x^2 (mph)^2:	400	400	400	900	900	900	1600	1600	1600
Y (distance):	45	55	50	95	90	85	150	140	160
x (mph):	50	50	50	60	60	60	70	70	70
x^2 (mph)^2:	2500	2500	2500	3600	3600	3600	4900	4900	4900
Y (distance):	230	250	240	360	375	345	515	535	525

Table 15.5 Printout of Regression Results for the Data of Example 15.4 for a Second-Order Model

```
REGRESSION COEFFICIENTS

VAR COEFFICIENT(B)    STD ERROR           T
1    -4.544642857     0.8384149907      -5.420517175
2    0.1544642857     0.009195417189    16.79796387
     82.5             (CONSTANT)

STANDARD ERROR OF ESTIMATE: 9.731514833
R SQUARED:                  0.9970879473

ANALYSIS OF VARIANCE

SOURCE      SUM OF SQUARES   DF    MEAN SQUARE     F
REGRESSION  486391.9643      2     243195.9821     2568.002828
RESIDUAL    1420.535714      15    94.70238095
```

The $\Sigma\ \Sigma\ Y_{ic}^2$ is 1,488,925 and

$$\frac{1}{3}\sum_c \left(\sum_i Y_{ic}\right)^2 = \frac{1}{3}(4{,}463{,}325) = 1{,}487{,}775$$

Thus,

$$\text{SSPE} = 1{,}488{,}925 - 1{,}487{,}775 = 1150$$

The degrees of freedom for pure error are $n - m$ or $18 - 6 = 12$, giving MSPE $= (1/12)(1150) = 95.833$. Subtracting SSPE from the SSE given on the printout in Table 15.5 gives

$$\text{SSLOF} = 1420.536 - 1150 = 270.536$$

The degrees of freedom for lack of fit are $m - k - 1 = 6 - 2 - 1 = 3$ and MSLOF $= (1/3)(270.536) = 90.179$.

Table 15.6 Reorganization of Data in Example 15.4 to Facilitate Calculating SSPE

	COMBINATION (c)					
	1 (20, 400)	2 (30, 900)	3 (40, 1600)	4 (50, 2500)	5 (60, 3600)	6 (70, 4900)
	45	95	150	230	360	515
	55	90	140	250	375	535
	50	85	160	240	345	525
$\sum_{i=1}^{n_c} Y_{ic}$:	150	270	450	720	1080	1575

For the hypotheses

$$H_0\colon \mu_{Y|x} = \beta_0 + \beta_1 x + \beta_2 x^2 \qquad H_a\colon \mu_{Y|x} \neq \beta_0 + \beta_1 x_1 + \beta_2 x^2$$

we reject H_0 at 0.01 if

$$\frac{\text{MSLOF}}{\text{MSPE}} > f_{3,\,12,\,0.01} = 5.95$$

For our data

$$\frac{\text{MSLOF}}{\text{MSPE}} = 0.941 < 5.95$$

which implies nonrejection of H_0. We conclude that the quadratic model in x appears to be appropriate.

The hypotheses for testing the model for usefulness are

$$H_0\colon \beta_1 = \beta_2 = 0 \qquad H_a\colon \text{at least one } \beta_j \neq 0$$

Since α is 0.05, H_0 is to be rejected if

$$\frac{\text{MSREG}}{\text{MSE}} > F_{2,\,15,\,0.05} = 3.68$$

Our calculated value from the printout is 2568.003 which definitely exceeds 3.68. Thus we can reject H_0 and conclude that the quadratic model is a very useful predictor.

Note that the SSPE in Example 15.4 is identical to the SSPE for the simple linear model of Example 14.12. This is to be expected, since a replicate of any value for x is also a replicate for x^2 or any other higher-order term in x. As you can see, polynomial models are much easier to replicate than ordinary multiple linear models.

15.5.2 Residual Plots

In those situations in which we cannot obtain replication, we can check the appropriateness of the model by constructing residual plots. These plots will also provide a rough check on the basic assumptions underlying the test procedures. As a check on the normality assumption, the residuals are plotted about 0 as before. This plot should exhibit approximate symmetry about 0 and have approximately 68% of the points within ± MSE and 95% within ± 2 MSE of 0, respectively.

To check for a need to include higher-order terms in x_j, $j = 1, 2, \ldots, k$, and homogeneity of variance for all values of x_j, we plot the residuals versus each of the independent variables. If no higher-order terms are required for a given x_j, then the residuals should be dispersed randomly about the x_j axis. Any type of curvilinear pattern in the residuals indicates that we should consider higher-order terms for the predictor being evaluated. If the homogeneity of variance assumption is satisfied, then bands drawn about the residuals should be equidistant from each other for all values of x_j (refer to Figure 14.10*a*). If a higher-order term is indicated, the bands still may be equidistant but will not be straight lines.

EXAMPLE 15.5 Construct residual plots as a check on the underlying assumptions and on the necessity of incorporating third- or higher-order terms in the Akersville model of Example 15.4.

SOLUTION The residuals for the model of Example 15.4 are given in Table 15.7. The residuals are plotted by themselves in Figure 15.2 on page 610. There is no obvious lack of symmetry, and the percentages within ± MSE and ±2 MSE (±10.7 and ±21.4) are 77.8% and 94.4%. There is no reason to doubt the normality assumption in this case.

To check for inclusion of additional terms in x and for homogeneity of variance, we have plotted the residuals versus x in Figure 15.2*b* and versus x^2 in Figure 15.2*c*. These plots exhibit no apparent curvature effect, but there is possible indication of a slight violation of the constant variance assumption. The variance may be increasing

Table 15.7 Printout of the Residuals for Example 15.5

```
DO YOU WANT TO SEE THE RESIDUALS?
Y
THIS MATRIX SHOWS THE CASE NUMBER, ACTUAL VALUE
OF THE DEPENDENT VARIABLE, PREDICTED VALUE, AND RESIDUAL.
   1          45          53.39285714     ¯8.392857143
   2          55          53.39285714      1.607142857
   3          50          53.39285714     ¯3.392857143
   4          95          85.17857143      9.821428571
   5          90          85.17857143      4.821428571
   6          85          85.17857143     ¯0.1785714286
   7         150         147.8571429       2.142857143
   8         140         147.8571429      ¯7.857142857
   9         160         147.8571429      12.14285714
  10         230         241.4285714     ¯11.42857143
  11         250         241.4285714       8.571428571
  12         240         241.4285714      ¯1.428571429
  13         360         365.8928571      ¯5.892857143
  14         375         365.8928571       9.107142857
  15         345         365.8928571     ¯20.89285714
  16         515         521.25           ¯6.25
  17         535         521.25           13.75
  18         525         521.25            3.75
```

Figure 15.2 Residual Plots for Example 15.5

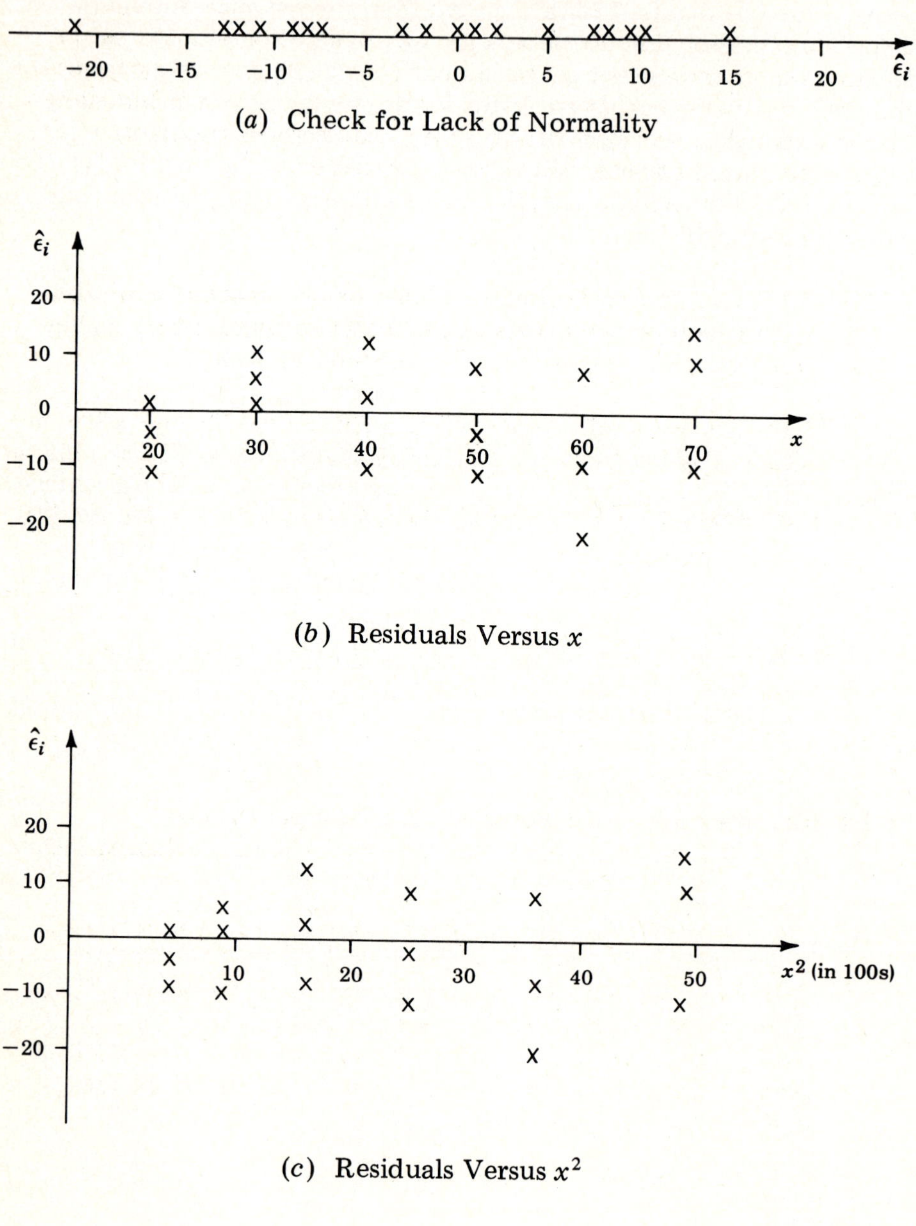

with x, but this appearance may be due only to the two largest residuals. Since the F test is robust, the slight departure from the assumptions, if it exists, should have very little effect.

As we saw in Example 15.5, residual plots can be very inconclusive. Without a large number of observations they seldom will detect violations of assumptions unless the violations are quite marked. But since the F test is robust, or insensitive to minor violations, it is only imperative that the more extreme violations be detected.

EXERCISES

15.5.1 Refer to Exercise 15.3.3 and assume that the experiment was replicated to provide a check on the linearity assumption. The new yields and predictor combinations are:

Yield (kg):	19.8	16.3	17.1	19.0	23.1	17.9
Temperature (°C):	70	50	60	70	80	60
Catalyst (mg):	6	5	5	4	6	5

Test the assumption of a linear relationship between yield, temperature, and catalyst at $\alpha = 0.05$.

15.5.2 Refer to Exercise 15.3.4 and construct residual plots. Do there appear to be any gross violations of assumptions?

15.5.3 For Exercise 15.4.3, construct residual plots to check the assumptions and to check for the possibility of including higher-order terms in some of the predictors.

15.5.4 Mr. Salz N. Pfeffer, the Service Director for Iceberg, Michigan, is trying to construct a model that will predict the average amount of salt (in tons) required to be available per day from forecasts of the median temperature and inches of snow for the day. To do so, he has recorded the following data (p. 612) from his records. Temperature and snowfall are values forecasted by the weather service.

Median temperature (°F):	29	36	21	15	18	4	22	10
Snowfall (inches):	5	2	6	4	4	1	0	0
Salt used (tons):	50	21	63	78	74	51	16	30

Temperature:	29	−12	6	−22	21	10	12	15	−6	4
Snowfall:	2	2	0	1	6	0	2	4	2	1
Salt used:	37	67	0	65	73	25	46	81	77	38

Temperature:	20	22	31	7	6	−5	10	−8	10	12
Snowfall:	7	10	3	0	0	3	3	2	0	2
Salt used:	89	112	46	18	20	85	72	69	0	56

Construct the least-squares estimate of a linear model for Salz, test it for lack of fit at $\alpha = 0.05$, and test it for usefulness at $\alpha = 0.05$. Also construct residual plots to check both the assumptions of your tests and the need to include higher-order terms.

15.6 REDUCING THE MODEL

It is not always advantageous to maintain the full model. Even if the test for usefulness has indicated that the model is an improvement over $\overline{Y}$ as an estimator, this does not imply that all the independent variables are useful. Instead we can only conclude that at least one of the β_j's is nonzero. In many cases we can do just as well with a reduced model, that is, with some of the predictors, as with the full model. In fact, we can often improve our predictive accuracy by reducing the model, that is, by deleting some of the predictors from the model.

There are several approaches that we can take in building or choosing the final prediction equation. One approach is to test sets of the predictors simultaneously; for example, we could test all quadratic terms at once. We can also test each variable individually to determine whether it should be included in the model. All approaches have one point in common. They all look at the correlation coefficients to determine in what order the variables are to be considered.

15.6.1 Correlations

If you recall from Chapter 14, the correlation coefficient is a measure of the strength of the linear relationship between two variables. In the single-predictor model its usefulness is limited because it contains no information that isn't available in the ANOVA table. That is not the case in the multiple linear model, however. The ANOVA table and $R^2_{Y|\boldsymbol{x}}$ pertain to the total model and give no information about the predictors individually. Thus the correlation coefficients become an important part of the analysis of a multiple linear model.

When there is more than one independent variable in the model, two types of correlation coefficients can be obtained. The simple linear correlation coefficient was discussed in Chapter 14. Its drawback in the multiple-predictor situation lies in the fact that its calculation ignores the presence of the other independent variables. Therefore, unless the independent variables are totally uncorrelated with each other, we can't determine from r_{Yx} the percentage of variation in Y that is explained by x alone. For example, suppose that $r_{Yx_1} = 0.90$ and $r_{Yx_2} = 0.80$. Then x_1, by itself, explains 81% and x_2, by itself, explains 64% of the variation in Y. The implication, which is obviously incorrect, is that together they would explain 145% of the variation in Y. The seeming contradiction here arises because part of the 81% being explained by x_1 is also being explained by x_2. To avoid this problem, we define a correlation coefficient that corrects for the effects of the other variables in the model.

DEFINITION 15.1 The partial correlation coefficient, denoted $r_{Yx_j|\boldsymbol{x}^*}$, where $\boldsymbol{x}^*$ is a set of variables already in the model, is a measure of the strength of the linear relationship between Y and x_j after both have been adjusted for their linear relations to $\boldsymbol{x}^*$. That is,

$$r_{Yx_j|\boldsymbol{x}^*} = (\text{sign of } b_j)\sqrt{\frac{\text{SS}(b_j|\boldsymbol{b}^*)}{\text{SSE}(\boldsymbol{b}^*)}} \tag{15.16}$$

where $\text{SS}(b_j|\boldsymbol{b}^*)$ is the increase in the regression SS due to b_j with the other variables already in the model and $\text{SSE}(\boldsymbol{b}^*)$ is the error SS for the model based only on $\boldsymbol{x}^*$.

The square of the partial correlation coefficient measures the percent of reduction in the sum of squares error that we can attain by adding x_j to the model when the other variables are already in. In effect, $r_{Yx_j|\boldsymbol{x}^*}$ is the simple correlation coefficient between the sets of residuals obtained by regressing both Y and x_j on $\boldsymbol{x}^*$.

The usefulness of the partial correlation coefficient lies in either deciding which variable is to enter the model next or which variable should leave the model next. For a more complete discussion of the partial correlation coefficient and its uses, see either Draper and Smith or Neter and Wasserman.

For our puposes we assume that the full model has been fitted and we are checking to see whether it can be reduced. Thus, it will suffice to say that we will consider deleting variables whose partial correlations are near zero in absolute value. That is, we start by first deleting that variable whose absolute partial correlation with Y is smallest, then we drop the one with the next smallest, and so on until our predictability (precision) starts to deteriorate. Precision is lost whenever the standard error, $s_{y|x}$, increases.

EXAMPLE 15.6 Refer to Example 15.2 and verify that the partial correlation coefficients between the student's averages and math and logic scores are 0.5170 and 0.7970, respectively. Use them to indicate which variable should be considered first for being dropped from the model.

SOLUTION For the full model we calculated

$$Y_x = 42.2573 + 1.7202x_1 + 3.6906x_2$$

with

$$\text{SSREG} = \text{SS}(b_1, b_2 \mid b_0) = 1074.348 \qquad \text{SSE}(b_1, b_2) = 164.052$$

and

$$\text{SSTO} = 1238.4$$

We can easily verify that the simple linear equations corresponding to x_1 and x_2 are

$$\hat{Y}_{x_1} = 43.2225 + 4.3912x_1 \qquad \text{and} \qquad \hat{Y}_{x_2} = 47.9858 + 4.8915x_2$$

respectively. To determine $r_{Yx_1|x_2}$ we calculate $\text{SS}(b_2 \mid b_0)$, which is

$$\text{SS}(b_2 \mid b_0) = b_2\left[\Sigma\, x_2 Y - \frac{(\Sigma\, x_2)(\Sigma\, Y)}{n}\right]$$
$$= 4.8915(207.4) = 1014.4971$$

Thus,

$$\text{SSE}(b_2) = 1238.4 - 1014.4971 = 223.9029$$

and

$$SS(b_1 | b_0, b_2) = 1074.348 - 1014.4971 = 59.8509$$

Therefore,

$$r_{Yx_1|x_2} = (\text{sign } b_1) \sqrt{\frac{59.8509}{223.9029}} = +0.5170$$

For $r_{Yx_2|x_1}$, we have that

$$SS(b_1 | b_0) = b_1 \left[\Sigma x_1 Y - \frac{(\Sigma x_1)(\Sigma Y)}{n}\right]$$
$$= 4.3912(179.6) = 788.6595$$

Thus,

$$SSE(b_1) = 1238.4 - 788.6595 = 449.7405$$

and

$$SS(b_2 | b_0, b_1) = 1074.348 - 788.6595 = 285.6885$$

The partial correlation between Y and x_2 given x_1 is then

$$r_{Yx_2|x_1} = (\text{sign } b_2) \sqrt{\frac{285.6885}{449.7405}} = +0.7970$$

Therefore, if we wish to consider dropping one of the predictors from the model, x_1 should be the first to go. But note that the standard error for the full model is

$$s_{Y|x} = \sqrt{\frac{SSE(b_1, b_2)}{n - 3}} = \sqrt{\frac{164.052}{7}} = 4.8411$$

whereas

$$s_{Y|x_2} = \sqrt{\frac{SSE(b_2)}{n - 2}} = \sqrt{\frac{223.9029}{8}} = 5.2904$$

Since the standard error increases with the deletion of x_1, it would not be desirable to drop the math score from the model.

Note that the values for b_0, b_1, and b_2 each change with changes in the model. This occurs because we introduce bias into the model whenever we omit terms. That is, both of the single-variable models are somewhat biased because of the exclusion of the other variable. Our primary interest in Example 15.6 was the partial correlation coefficient, but simply looking at the correlation coefficients, both simple and partial, does not constitute a test. In the next two sections we formally construct test procedures to determine whether a variable's contribution is significant.

15.6.2 Testing the β_j Individually

To determine whether or not a specific independent variable is making a significant contribution to the model requires that we test the hypotheses

$$H_0\colon\ \beta_j = 0 \qquad H_a\colon\ \beta_j \neq 0 \tag{15.17}$$

In this case the hypotheses actually refer to testing whether or not x_j is making a significant contribution given that the other independent variables are already in the model. Thus, the hypotheses in Equation 15.17 theoretically should be stated as

$$H_0\colon\ \beta_j = 0|\beta^* \qquad H_a\colon\ \beta_j \neq 0|\beta^* \tag{15.18}$$

where β^* represents the regression coefficients for the predictors already in the model.

Though the hypotheses in Equation 15.18 are theoretically more precise, they are also more cumbersome and it is common practice to state them as in Equation 15.17. We will adhere to the standard form but stress that it is important for you to remember that the interpretation is that of Equation 15.18.

To construct the test, we refer to the original partition of the sum of squares. This is

$$\text{SSTO} = \text{SSE} + \text{SSREG}$$

where SSREG can also we written as

$$\text{SSREG} = \text{SS}(b_1, \ldots, b_k | b_0) = \text{SS}(b | b_0)$$

It can be shown that $\text{SS}(b_1, \ldots, b_k | b_0)$ can be partitioned further into

$$\text{SS}(b|b_0) = \text{SS}(b_1|b_1^*) + \text{SS}(b_2|b_2^*) + \cdots + \text{SS}(b_k|b_k^*) \tag{15.19}$$

where $b_j^* = (b_0, b_1, b_2, \ldots, b_{j-1})$. Each of the sums of squares on the right of Equation 15.19 has one degree of freedom and, under our assumptions, will have a $\sigma^2\chi_1^2$ distribution if the null hypothesis of Equation 15.17 is true. Thus, if H_0 is true (we assume x_j is the last variable to enter the model)

$$\text{MS}(b_j|\boldsymbol{b}_j^*) = \frac{\text{SS}(b_j|\boldsymbol{b}_j^*)}{1}$$

is an estimate of σ^2. If H_0 is not true, then $\text{MS}(b_j|\boldsymbol{b}_j^*)$ estimates σ^2 plus a positive constant. Regardless of H_0, the SSE for the full model, $\text{SSE}(\boldsymbol{b})$, estimates σ^2 and has a $\sigma^2\chi^2_{n-k-1}$ distribution. Thus, if H_0 is true, the ratio

$$\frac{\text{MS}(b_j|\boldsymbol{b}_j^*)}{\text{MSE}(\boldsymbol{b})} \sim F_{1,\,n-k-1}$$

is expected to be near 1. If H_0 is false, then its expected value is greater than 1.

It follows from these considerations that the test is to reject H_0 if

$$\frac{\text{MS}(b_j|\boldsymbol{b}_j^*)}{\text{MSE}(\boldsymbol{b})} > f_{1,\,n-k-1,\,\alpha} \tag{15.20}$$

Since an F random variable with one degree of freedom in the numerator and v degrees of freedom in the denominator is equivalent to the square of a t random variable with v degrees of freedom, an equivalent test is to reject H_0 if

$$|t_j| = +\sqrt{\frac{\text{MS}(b_j|\boldsymbol{b}_j^*)}{\text{MSE}(\boldsymbol{b})}} > t_{n-k-1,\,\alpha/2} \tag{15.21}$$

We present both Equations 15.20 and 15.21 because some computer programs give the F statistic and others calculate the t statistic.

EXAMPLE 15.7 We refer once more to the examples pertaining to a student's success in an introductory statistics class. In Example 15.6 we calculated $\text{SS}(b_1|b_0, b_2)$ and $\text{SS}(b_2|b_0, b_1)$. Use them and the fact that $\text{SSE}(\boldsymbol{b}) = 164.052$ with seven degrees of freedom to test each β_j individually at $\alpha = 0.10$ by (a) an F test and (b) a t test.

SOLUTION The hypotheses of interest are

1. $H_0\colon \beta_1 = 0 \qquad H_a\colon \beta_1 \neq 0$
2. $H_0\colon \beta_2 = 0 \qquad H_a\colon \beta_2 \neq 0$

(a) Using Equation 15.20 the decision rule for both hypothesis sets is to reject H_0 if

$$\frac{\text{MS}(b_j|\boldsymbol{b}_j^*)}{\text{MSE}(\boldsymbol{b})} > f_{1,\,7,\,0.10}$$

The value for $f_{1,\,7,\,0.10}$ cannot be found in Table A.9, but it can be calculated by squaring $t_{7,\,0.05} = 1.895$ to obtain $f_{1,\,7,\,0.10} = 3.59$.

From Example 15.6 we have $\text{SS}(b_1 | b_1^*) = 59.8509$ and $\text{SS}(b_2 | b_2^*) = 285.6885$. Thus, for (1) the calculated F is

$$\frac{59.8509}{23.436} = 2.552$$

Since $2.552 < 3.59$, H_0 cannot be rejected, and we should *consider* dropping x_1 from the model.

For (2) the calculated F is

$$\frac{285.6885}{23.436} = 12.190$$

This value exceeds 3.59 and we reject H_0, concluding that x_2 should be retained as a predictor.

(*b*) For a t test our decision rule is to reject if

$$|t_j| > t_{7,\,0.05} = 1.895$$

The calculated t's are found by taking the square roots of the corresponding F values and are

$$|t_1| = \sqrt{2.552} = 1.597 \qquad \text{and} \qquad |t_2| = \sqrt{12.190} = 3.491$$

For (1), H_0 cannot be rejected, since $1.597 < 1.895$, and for (2), $3.491 > 1.895$, implying rejection of H_0. As before, we conclude that x_2 should be retained in the model and that we might *consider* dropping x_1.

Note that in Example 15.7 rejection of H_0 implies retention of x_j in the model, but nonrejection of H_0 does not automatically lead to the elimination of x_j. Instead we have concluded only that we should consider deleting x_j. There are two reasons for this apparent inconsistency. First, our primary goal is to obtain the best equation for predicting Y, and the word *best* means different things to different people. Our interpretation of best relates to how close the estimate of $\mu_{Y|\mathbf{x}}$ is expected to be to the true value. The primary measure of closeness for an estimator is its standard error, which is a function of $s_{Y|\mathbf{x}}$ in this case. Thus we recommend only that you consider dropping x_j from the model, since if doing so results in an increase in $s_{Y|\mathbf{x}}$, it implies a loss of precision.

The second reason arises from the fact that the independent variables may be highly correlated. This is known as *multicollinearity* and will be discussed in more detail in Section 15.8.1. The problem that can arise with correlated x_j's is that the t values (and F values) for the individual tests may be incorrect because of rounding error. In fact, when this happens, the whole analysis may be incorrect, since the b_j's will be in error and can even have the wrong sign. In this case we recommend that you check the model both with x_j present and with x_j deleted. The decision then should take into account the effect on $s_{Y|x}$. If $s_{Y|x}$ increases when the variable is deleted, then the variable should be retained.

NOTE: We issue a word of caution pertaining to $s_{Y|x}$. As you saw in Section 15.4, R^2 can be a misleading measure of the usefulness of a model, since it can be forced to equal 1.00. The same comment applies to $s_{Y|x}$ when replication is available. If enough variables are put into the model, $s_{Y|x}$ can be forced to decrease to $\sqrt{\text{MSPE}}$ through a loss of degrees of freedom. Thus, if a model having replication is saturated, that is, if $k + 1$ equals the number of different combinations of the x_j's, then dropping a variable will usually result in an increase in $s_{Y|x}$. This would not imply that the deleted variable is necessarily useless.

In Example 15.7 we used a higher α for testing

$$H_0\colon\ \beta_j = 0 \qquad H_a\colon\ \beta_j \neq 0$$

than we used for testing the full model. The reason for this is that we prefer not to delete a variable if it might possibly be making a contribution. That is, we are attempting to decrease the chance of making a Type II error. A common practice is to double or even quadruple the α used for the overall test of the model.

We make one final comment with respect to Examples 15.6 and 15.7. Another justification for retaining x_1 in the model is the value of its partial correlation with Y. We saw in Example 15.6 that this value is 0.517. When we square this, we see that x_1 is explaining 26.7% of the variation in Y that is not being explained by the logic score. In this case that 26.7% results in an appreciable reduction in $s_{Y|x}$.

In this section we developed procedures for analyzing the contribution of a single predictor. In many situations it is desirable to shortcut the testing procedure by evaluating more than one predictor at a time. We address this point in the next section.

15.6.3 Testing a Subset of the β_j

In the last section we considered testing a subset consisting of a single β_j. In this case we turn our attention to testing p ($\geqslant 1$) of the β_j simultaneously. We are assuming, as we did in Section 15.6.2, that the full model has already been deemed useful. In addition, we assume, without loss of generality, since the numbering of the variables is irrelevant, that the p coefficients to be tested correspond to the last p variables in the model. That is, for our model,

$$\mu_{Y|x} = \beta_0 + \beta_1 x_1 + \beta_2 x_2 + \cdots + \beta_{k-p} x_{k-p} + \beta_{k-p+1} x_{k-p+1} + \cdots + \beta_k x_k$$

or

$$\mu_{Y|x} = \beta_0 + \sum_{j=1}^{k-p} \beta_j x_j + \underbrace{\sum_{q=1}^{p} \beta_{k-p+q} x_{k-p+q}}_{\text{(last } p \text{ variables)}} \tag{15.22}$$

We wish to test the coefficients contained in the second summation of Equation 15.22.

The hypotheses to be tested are

$$H_0\colon\ \beta_j = 0 \text{ for } j = k - p + 1, \ldots, k \qquad H_a\colon\ \text{some } \beta_j \neq 0 \tag{15.23}$$

To construct the test we refer to the partition in Equation 15.19. For ease of reference it is repeated here.

$$\text{SSREG} = \text{SS}(\boldsymbol{b}|b_0) = \text{SS}(b_1|\boldsymbol{b}_1^*) + \cdots + \text{SS}(b_k|\boldsymbol{b}_k^*)$$

where $\boldsymbol{b}_j^* = (b_0, b_1, \ldots, b_{j-1})$, that is, those coefficients that remain after x_j is deleted from the model. Note that, as was the case with the test of an individual β_j, to be theoretically precise the hypotheses of Equation 15.23 should be stated as

$$H_0\colon\ \boldsymbol{\beta}_p = \boldsymbol{0}|\boldsymbol{\beta}_p^* \qquad H_a\colon\ \boldsymbol{\beta}_p \neq \boldsymbol{0}|\boldsymbol{\beta}_p^*$$

where

$$\boldsymbol{\beta}_p = (\beta_{k-p+1}, \beta_{k-p+2}, \ldots, \beta_k)$$
$$\boldsymbol{0} = (0, 0, \ldots, 0)$$
$$\boldsymbol{\beta}_p^* = (\beta_0, \beta_1, \ldots, \beta_{k-p})$$

That is, we are testing for the necessity of including x_{k-p+1} through x_k in the model, given that x_1 through x_{k-p} are already present.

The construction of the test follows the same line of reasoning used in Section 15.6.2. Namely, each of the $SS(b_j|\boldsymbol{b}_j^*)$ in Equation 15.19 has a $\sigma^2\chi_1^2$ distribution if H_0 is true. Hence, by Cochran's theorem,

$$SS(\boldsymbol{b}_p|\boldsymbol{b}_p^*) = SS(b_{k-p+1}|\boldsymbol{b}_{k-p+1}^*) + \cdots + SS(b_k|\boldsymbol{b}_k^*) \tag{15.24}$$

has a $\sigma^2\chi_p^2$ distribution under H_0, since there are p terms on the right. As usual, the mean square,

$$MS(\boldsymbol{b}_p|\boldsymbol{b}_p^*) = \frac{SS(\boldsymbol{b}_p|\boldsymbol{b}_p^*)}{p} \tag{15.25}$$

is an estimate of σ^2 if H_0 is true and overestimates σ^2 if H_a is true. On the other hand, the MSE for the full model estimates σ^2 in either case. Thus the test for hypotheses in Equation 15.23 is to reject H_0 if

$$\frac{MS(\boldsymbol{b}_p|\boldsymbol{b}_p^*)}{MSE} > f_{p,\,n-k-1,\,\alpha} \tag{15.26}$$

Rejection of H_0 implies only that at least one of the β_j is nonzero, whereas failure to reject indicates that all the $\beta_j, j = k - p + 1, \ldots, k$, appear to be zero.

Actually calculating $SS(\boldsymbol{b}_p|\boldsymbol{b}_p^*)$ is much simpler than it would appear from Equation 15.24. Note that if we have x_1 through x_{k-p} in the model, we have their contribution as given by $SS(b_1, \ldots, b_{k-p}|b_0)$. We shall call this the sum of squares for the reduced model, or SS(Reduced). Adding the remaining p variables to the model gives the full model with the corresponding $SS(b_1, b_2, \ldots, b_k|b_0)$, or sum of squares for the full model. We call this SS(Full). The increase in the sum of squares regression in going from the reduced model to the full model (or the decrease in going from the full to the reduced) represents the net contribution of the set of predictors being considered. That is,

$$SS(\boldsymbol{b}_p|\boldsymbol{b}_p^*) = SS(\text{Full}) - SS(\text{Reduced}) \tag{15.27}$$

and has p degrees of freedom. Hence the mean square is given by

$$MS(\boldsymbol{b}_p|\boldsymbol{b}_p^*) = \frac{SS(\text{Full}) - SS(\text{Reduced})}{p} \tag{15.28}$$

You must remember that this mean square is to be compared with the MSE for the full model. This is because the MSE for the reduced model is not independent of $MS(\boldsymbol{b}_p|\boldsymbol{b}_p^*)$, since SSE(Reduced) includes $SS(\boldsymbol{b}_p|\boldsymbol{b}_p^*)$. Thus the MSE(Reduced) is an estimate of σ^2 only if H_0 is true also.

EXAMPLE 15.8 Pete Pusher, the manager of Tree Lawn Realty, is trying to find a model that he can give his salespersons for estimating the expected selling price of a home. He would like to use the model to obtain an initial estimate, then permit the salesperson to adjust the estimate according to the condition of the property. The final estimate is then to be recommended to the seller as an asking price.

Pete believes that the following variables have a significant linear effect on the selling price: living area in square feet; lot size in acres; number of bathrooms; and number of bedrooms. In addition, he feels that the number of bathrooms and number of bedrooms may have quadratic effects and may interact. To this end Pete has collected the data given in Table 15.8. Since he is a good salesman but somewhat weak in the area of statistics, he has asked you to determine an equation for him based on these data.

First determine the least-squares estimate of the full model that Pete has indicated and test it for usefulness at $\alpha = 0.01$. Second, determine whether his feeling about the quadratic effects and interaction are supported at $\alpha = 0.05$. Third, select a final equation for him, using $\alpha = 0.05$ for any additional tests.

Table 15.8 Data for Example 15.8

(x_1) Area (ft^2)	(x_2) Lot Size (Acres)	(x_3) Number Baths	(x_4) Number Bedrooms	x_3^2	x_4^2	x_3x_4	(Y) Selling Price ($1000)
2000	0.50	2.5	4	6.25	16	10	83
1250	0.20	1.0	2	1.00	4	2	39.5
1350	1.25	1.5	2	2.25	4	3	43.5
1900	0.60	3.0	3	9.00	9	9	62.0
1450	0.40	2.0	2	4.00	4	4	40.0
2200	0.50	2.0	4	4.00	16	8	76.5
3000	1.50	4.0	5	16.00	25	20	115.0
2450	0.75	2.5	4	6.25	16	10	92.0
1750	3.00	2.5	3	6.25	9	7.5	94.0
1600	1.00	1.5	3	2.25	9	4.5	55.5
2500	2.00	2.5	5	6.25	25	12.5	136.0
2250	1.50	2.5	4	6.25	16	10	104.0
1800	0.15	1.0	3	1.00	9	3	37.5
1700	0.20	2.0	3	4.00	9	6	48.5
1650	0.25	1.5	3	2.25	9	4.5	57.5
2150	0.50	2.5	4	6.25	16	10	51.5
2700	4.25	3.5	5	12.25	25	17.5	155.0
2000	1.10	1.5	3	2.25	9	4.5	83.5
1550	0.18	2.0	2	4.00	4	4	33.5

SOLUTION Table 15.9 gives the printout for the full model. For testing the full model, the hypotheses are

$$H_0\colon\ \beta_j = 0 \text{ all } j \qquad H_a\colon\ \text{some } \beta_j \neq 0$$

with H_0 being rejected if

$$\frac{\text{MSREG}}{\text{MSE}} > f_{7,\,11,\,0.01} = 4.89$$

From the printout we see that MSREG/MSE = 21.352 exceeds 4.89, and so we can reject H_0. We conclude that at least one of the predictors is useful. Thus, for the model

$$\mu_{Y|x} = \beta_0 + \beta_1 x_1 + \beta_2 x_2 + \beta_3 x_3 + \beta_4 x_4 + \beta_5 x_3^2 + \beta_6 x_4^2 + \beta_7 x_3 x_4$$

our estimator is

$$\hat{Y}_x = -14.5949 + 0.0277x_1 + 16.3650x_2 + 16.0454x_3 - 8.9757x_4$$
$$-4.3903x_3^2 + 3.0870x_4^2 - 0.2544x_3x_4$$

To test for the need to include the nonlinear terms for x_3 and x_4, our hypotheses are

$$H_0\colon\ \beta_5 = \beta_6 = \beta_7 = 0 \qquad H_a\colon\ \text{some } \beta_j \neq 0, j = 5, 6, 7$$

Table 15.9 Printout for the Full Regression Model of Example 15.8

REGRESSION COEFFICIENTS

VAR	COEFFICIENT(B)	STD ERROR	T-VALUE
X1	0.02770	0.02279	1.21526
X2	16.36496	3.37251	4.85246
X3	16.04539	18.86824	0.85039
X4	-8.97574	23.58191	-0.38062
X5	-4.39033	9.48770	-0.46274
X6	3.08700	5.32047	0 58021
X7	-0.25440	12.3247	-0.02064
	-14.59492	(CONSTANT)	

STANDARD ERROR OF ESTIMATE: 11.72282
R SQUARED: 0.93145

ANALYSIS OF VARIANCE

SOURCE	SUM OF SQUARES	DF	MEAN SQUARE	F-VALUE
REGRESSION	20539.61995	7	2934.23142	21.35159
ERROR=SS(FULL)	1511.66952	11	137.42450	

That is, our null hypothesis specifies that the model is

$$\mu_{Y|x} = \beta_0 + \beta_1 x_1 + \beta_2 x_2 + \beta_3 x_3 + \beta_4 x_4$$

whereas the alternative indicates that at least one of the nonlinear terms should be retained. A printout for the reduced model is given in Table 15.10.

To test the nonlinear terms in x_3 and x_4, we reject H_0 if

$$\frac{\text{MS}(b_3 | b_3^*)}{\text{MSE(Full)}} > f_{3,\,11,\,0.05} = 3.59$$

Comparing the printouts in Tables 15.9 and 15.10, we have

$$\text{MS}(b_3 | b_3^*) = \frac{\text{SS(Full)} - \text{SS(Reduced)}}{3}$$

$$= \frac{20539.6200 - 20332.2594}{3} = 69.1203$$

with MSE(Full) = 137.4245. Thus,

$$\frac{\text{MS}(b_5, b_6, b_7 | b_0, b_1, b_2, b_3, b_4)}{\text{MSE(Full)}} = \frac{69.1203}{137.4245} = 0.503$$

Since $0.503 < 3.59$, we cannot reject H_0, and we must conclude that the contribution of the nonlinear terms appears to be due to chance

Table 15.10 Printout for the Reduced Model of Example 15.8

```
REGRESSION COEFFICIENTS

VAR COEFFICIENT(B)   STD ERROR        T-VALUE
X1       0.02085      0.01992          1.04673
X2      16.83537      3.00057          5.61072
X3      -3.03573      5.86068         -0.51798
X4      12.94240      8.21334          1.57578
       -21.30823        (CONSTANT)

STANDARD ERROR OF ESTIMATE:   11.08097
R SQUARED                 :    0.92204

ANALYSIS OF VARIANCE

SOURCE        SUM OF SQUARES   DF   MEAN SQUARE      F-VALUE
REGRESSION      20332.25942     4    5083.06486      41.39713
ERROR            1719.03005    14     122.78786
```

alone when x_1 through x_4 are already in the model. Thus we may drop them from the model. Note that $s_{Y|x}$ decreased with the deletion.

Finally, to determine whether the model can be reduced further, we check the t values given in Table 15.10 for the four remaining predictors. The hypotheses for the individual tests are

$$H_0: \beta_j = 0 \qquad H_a: \beta_j \neq 0$$

for $j = 1, 2, 3, 4$, and we reject H_0 for each if

$$|t_j| > t_{14,\,0.025} = 2.145$$

The $|t_j|$ are 1.0467, 5.6107, 0.5180, and 1.5758, respectively. Only $t_2 = 5.6107$ exceeds 2.145, and so we reject $H_0: \beta_j = 0$ for $j = 2$ but cannot reject it for $j = 1$, 3, and 4. Therefore we consider dropping x_1, x_3, and x_4 from the model.

When we fit the model with only x_1 as a predictor, the standard error increased dramatically. Therefore, we decided to try the model

$$\mu_{Y|x} = \beta_0 + \beta_2 x_2 + \beta_4 x_4$$

The printout for this model is given in Table 15.11. Note that for $\alpha = 0.05$, $t_{16,\,0.025} = 2.120$, implying that both variables are making a significant contribution to reducing the SSE. Also, the standard error has dropped from 11.0810 in Table 15.10 to 10.7637 in Table 15.11 with R^2 decreasing by only 0.0061. Thus, we recommend the prediction equation given by

$$\hat{Y} = -12.6764 + 16.4448x_2 + 20.6680x_4$$

Table 15.11 Printout for the Model Using x_2 and x_4 for the Data of Example 15.8

```
REGRESSION COEFFICIENTS

VAR COEFFICIENT(B)  STD ERROR        T
2    16.44475045    2.748856397      5.982397071
4    20.66800673    2.911728003      7.098192795
     ¯12.67641224   (CONSTANT)

STANDARD ERROR OF ESTIMATE: 10.76373473
R SQUARED:                  0.9159356296

ANALYSIS OF VARIANCE

SOURCE      SUM OF SQUARES  DF   MEAN SQUARE     F
REGRESSION  20197.56171     2    10098.78085     87.16516886
RESIDUAL    1853.727765     16   115.8579853
```

In Example 15.8 we did not examine the correlation matrix (a table listing the simple correlations between all variables in the model). Had we done so, our model would have consisted of x_2 and x_4^2, lot size and number of bedrooms squared, resulting in a reduction of the standard error of estimate from 10.7637 to 10.6515. The reason for ignoring the correlations was that we were interested in the need for the quadratic terms once the other variables were already in the model. Remembering who is to use the model and the fact that they must do so by hand in this case would indicate that we should attempt to keep it as simple as possible so long as the precision is not affected appreciably. We will revisit this example in the Supplementary Examples.

In choosing a final model from a given set of predictors, there are several approaches that we can take. In addition to the two just considered, we might look at all possible regressions, that is, all possible combinations of the x_j's (of which there will be $2^k - 1$), and examine the results of each to make the choice. For large models this can be a time-consuming and expensive approach. Most computer regression packages have alternative approaches available. These approaches are decidedly more efficient than looking at all possible regressions. We will consider two methods of model building, entering the variables one at a time from scratch or starting with the full model and then deleting variables one at a time. The first way is known as *forward inclusion* and the second as *backward elimination.* The basic idea is to examine each time the partial correlation coefficient between Y and all variables, holding constant those that are currently in the model. The variable with the most significant partial correlation (of those not in the model) is entered or the variable in the model with the smallest absolute partial correlation (of those that are insignificant) is deleted. The procedure stops when all variables not in the model are *insignificant* in forward selection and when all variables in the model are *significant* in backward elimination. We will use the forward selection procedure in the Supplementary Examples.

EXERCISES

15.6.1 Refer to Exercise 15.5.1 and determine the partial correlation coefficients between the temperature and the yield and between the amount of catalyst and the yield.

15.6.2 Refer to Exercise 15.3.4 and determine which variable should be considered first for deletion. Test the one you have chosen at $\alpha = 0.10$.

15.6.3 Refer to Exercise 15.6.1 and add the quadratic terms x_1^2, x_2^2, and $x_1 x_2$ to the model. Test the quadratic terms as a set to see whether any of them are necessary. Use $\alpha = 0.05$.

15.6.4 Refer to Exercise 15.4.3. It is believed that an interaction between major and GPA may have an effect on salary, and the major also may have a quadratic effect. Test both of these simultaneously, then test all variables individually. Use $\alpha = 0.05$ for the F test and $\alpha = 0.10$ for t tests.

15.7 THE EQUATION AS AN ESTIMATOR

The reason for obtaining our equation is to use it as an estimator and predictor. Once we have determined the equation, it is simple to generate point estimates or point predictions. All that we need to do is to substitute the specified x_j values into the prediction equation.

EXAMPLE 15.9 Sally Silvertongue, a salesperson for Tree Lawn Realty, has received a call from a seller who is considering listing with Tree Lawn. From the telephone conversation Sally learned that the prospective property is a four-bedroom home situated on a 3/4-acre lot. What possible selling price should Sally indicate to the owner without having seen the property?

SOLUTION The prediction equation that was selected in Example 15.8 is

$$\hat{Y}_x = -12.6764 + 16.4448x_2 + 20.6680x_4$$

where Y is in \$1000, x_2 is the lot size in acres, and x_4 is the number of bedrooms in the house. In Sally's case $x_2 = 0.75$ and $x_4 = 4$. Substituting these into the regression equation gives

$$\begin{aligned}\hat{Y}_{(0.75,\,4)} &= -12.6764 + 16.4448(0.75) + 20.6680(4)\\ &= -12.6764 + 12.3336 + 82.6720 = 82.3292\end{aligned}$$

or \$82,329.

Now, Sally knows, as do you and I, that the house in Example 15.9 is not going to sell for exactly \$82,329. It might sell for \$70,000, \$95,000 or any amount in between, but it is not going to sell for \$82,329. If Sally were to say that the property will sell for that, the owner would think she's crazy and list elsewhere. Thus, Sally would like to be able to tell the owner \$83,000, then indicate reasonable maximum and minimum prices. That is, she would like to state a confidence interval.*

*The following material requires a basic knowledge of matrix algebra. For the student who does not have this background, the material may be omitted without a loss of continuity.

Exact interval estimates can be quite difficult to calculate in a multiple linear regression, since the computations usually involve matrix algebra. Some computer programs include an option for determining exact confidence intervals, but many of the standard programs do not. We present the basic formulas for these intervals for those students who are familiar with matrix algebra.

Recall that in the single predictor model, the formula for a $1 - \alpha$ confidence interval for $E(Y_{x_p})$ is given by

$$\hat{Y}_{x_p} \pm t_{n-2,\,\alpha/2} \sqrt{\text{MSE}\left[\frac{1}{n} + (x_p - \bar{x})^2/\Sigma\,(x - \bar{x})^2\right]}$$

while the interval for Y_{x_p} (a particular value) is

$$\hat{Y}_{x_p} \pm t_{n-2,\,\alpha/2} \sqrt{\text{MSE}\left[1 + \frac{1}{n} + (x_p - \bar{x})^2/\Sigma\,(x - \bar{x})^2\right]}$$

In the multiple linear situation, the degrees of freedom for t become $n - k - 1$ and the term

$$\frac{1}{n} + \frac{(x_p - \bar{x})^2}{\Sigma\,(x - \bar{x})^2}$$

must be replaced by its corresponding form in matrix notation since $\boldsymbol{x}_p$ and $\bar{\boldsymbol{x}}$ are now vectors. Making this substitution, the indicated term becomes

$$\boldsymbol{x}_p'\,(\boldsymbol{x}'\boldsymbol{x})^{-1}\boldsymbol{x}_p$$

where

$$\boldsymbol{x}_p' = (1, x_1, x_2, x_3, \ldots, x_k)_p$$

denotes the vector of values for the x_j's that is used to calculate $\hat{Y}_{\boldsymbol{x}_p}$ with the 1 included for b_0. The term $\boldsymbol{x}$ represents the matrix of values for the x_j's that was used in obtaining the equation for $\hat{Y}_{\boldsymbol{x}}$, augmented with an initial column of 1's for the intercept, b_0. That is,

$$\boldsymbol{x} = \begin{bmatrix} 1 & x_{11} & x_{12} & \cdots & x_{1k} \\ 1 & x_{21} & x_{22} & \cdots & x_{2k} \\ 1 & x_{31} & x_{32} & \cdots & x_{3k} \\ \cdot & \cdot & \cdot & & \cdot \\ \cdot & \cdot & \cdot & & \cdot \\ 1 & x_{n1} & x_{n2} & \cdots & x_{nk} \end{bmatrix}$$

and $\boldsymbol{x}'$ is the transpose of $\boldsymbol{x}$.

Substituting these into our previous confidence intervals gives

$$\hat{Y}_{x_p} \pm t_{n-k-1,\, \alpha/2} \sqrt{\text{MSE}[x_p'(x'x)^{-1}x_p]} \tag{15.29}$$

for $E(Y_{x_p})$ and

$$\hat{Y}_{x_p} \pm t_{n-k-1,\, \alpha/2} \sqrt{\text{MSE}[1 + x_p'(x'x)^{-1}x_p]} \tag{15.30}$$

for Y_{x_p}. The use of these equations is illustrated in the next two examples.

EXAMPLE 15.10 Sally comes to you and asks you to give her a feel for the limits that she can give the owner. Provide her with an interval having a 90% chance of containing the actual selling price.

SOLUTION From Example 15.8, we obtain $n - k - 1 = 16$ and MSE $= 115.858$. Sally's prediction was \$82,329, or $\hat{Y}_x = 82.329$, and $t_{16,\,0.05} = 1.746$.

The matrix, x, is constructed with the first column being 1's and the second and third columns being the entries for x_2 (lot size) and x_4 (number of bedrooms) from Table 15.8. That is,

$$x = \begin{bmatrix} 1 & .5 & 4 \\ 1 & .2 & 2 \\ \cdot & \cdot & \cdot \\ \cdot & \cdot & \cdot \\ \cdot & \cdot & \cdot \\ 1 & .18 & 2 \end{bmatrix}$$

Thus,

$$x'x = \begin{bmatrix} 19.0000 & 19.8300 & 64.0000 \\ 19.8300 & 41.3649 & 76.7100 \\ 64.0000 & 76.7100 & 234.0000 \end{bmatrix}$$

and

$$(x'x)^{-1} = \begin{bmatrix} .7072 & .0502 & -.2099 \\ .0502 & .0652 & -.0351 \\ -.2099 & -.0351 & .0732 \end{bmatrix}$$

For Sally, $x_p' = (1, 0.75, 4)$ and $x_p'(x'x)^{-1}x_p = 0.1006$. Substituting these results into Equation 15.30 gives

$$82.329 \pm 1.746\sqrt{115.858(1 + 0.1006)} = 82.329 \pm 19.716$$

or \$62,613 to \$102,045. From a practical standpoint, Sally probably would tell the owner \$62,500 to \$102,000.

Note that the limits in Example 15.10 are practically useless because they are so wide. Rather than give the owner limits for a particular house (his) on the basis of the information available, she probably would prefer to indicate the range for an average four-bedroom home on a ¾-acre lot.

EXAMPLE 15.11 Calculate 90% limits for Sally for the selling price of an average property with the stated characteristics.

SOLUTION The only change required is the dropping of the value 1 from under the radical in Equation 15.30. Thus, we have

$$82.329 \pm 1.746\sqrt{115.858(0.1006)} = 82.329 \pm 5.961$$

or \$76,368 to \$88,290. Therefore, Sally would probably tell the owner that an average four-bedroom house on ¾ acre would sell for between \$76,400 and \$88,300.

We stress that the intervals given by Equation 15.29 and 15.30 usually involve laborious calculations and are generally calculated using a computer. If the computer program available to you calculates exact $1 - \alpha$ confidence sets, then we recommend that it be used in doing the following exercises.

EXERCISES

15.7.1 Refer to Exercise 15.3.3 and find a 90% confidence interval for the average yield of all reactions run at 75°C with 5 mg of catalyst.

15.7.2 Refer to Exercise 15.3.4 and calculate a 95% confidence interval for the average one-year GPA of all MBA students who had an undergraduate GPA of 3.00 and scored 550 on the GMAT.

15.7.3 Suppose you are a degree candidate in marketing with a GPA of 2.90 and 4 years experience. Obtain a 90% confidence interval for your starting salary. (Refer to Exercise 15.4.3 and use the information contained therein.)

15.8 OTHER CONSIDERATIONS

As indicated previously, the calculated values for F (or t) and the individual coefficients must be used with caution. One reason for this is the possibility of intercorrelations between the independent variables. We address this problem in the next section.

15.8.1 Multicollinearity

From a theoretical viewpoint it is not necessary to assume that the independent variables are uncorrelated with each other. But, in practice, intercorrelations between the x_j's can create a problem. The problem that arises is known as *multicollinearity* and actually relates to computational accuracy. If two or more predictors are highly correlated, there may be large rounding errors in the calculation of the individual F (or t values). The effect of that rounding is to cause the calculated values to be much smaller in absolute value than they should be.

The reason is that the computational process requires division by what is called the *determinant* of a matrix, obtained from the x values. If two or more x_j's are perfectly correlated, this matrix will have two or more columns that are identical except for multiplication by a constant, resulting in what is known as a *singular* matrix. The determinant of a singular matrix is zero, which implies that for perfectly correlated x_j's we would attempt to divide by zero. Since this is an undefined operation, the computer should return an error message and be unable to run the problem.

In practice we would not expect perfectly correlated x_j's, but high absolute correlations between predictors are common. When they occur, the determinant does not equal zero but may be very close to it. Thus when the computer calculates s_{b_j}, it divides by a number that is very close to zero, and significant rounding errors may result. This, in turn, introduces error into the value for s_{b_j} and can even cause b_j to have the wrong sign. When this happens, the corresponding F (or t) value is incorrect, and the result may be an inability to reject H_0 when b_j may actually be significant.

One immediate indication of a multicollinearity problem is the rejection of H_0 for the full model but nonrejection of H_0 for all individual tests when the same α is used for all tests. If this happens we have an apparent contradiction, since rejecting H_0 in the overall test implies at least one $\beta_j \neq 0$, whereas nonrejection for all individual tests implies that all $\beta_j = 0$. This occurs with highly correlated independent variables, since the individual test statistics are calculated assuming that all other variables are already in the model. Thus, if two variables are explaining essentially the same thing and one is already in the model, there is very little left for the other to explain. As an illustration we revisit Example 15.8.

EXAMPLE 15.12 The correlation matrix for the data of Example 15.8 is presented in Table 15.12 and the results of regressing Y on x_2, x_4, and x_4^2 are presented in Table 15.13. Recall that with x_2 and x_4 in the model, both individual tests were highly significant. With the inclusion of x_4^2, we see from Table 15.13 that neither x_4 nor x_4^2 is even close to being significant. Explain this apparent discrepancy.

SOLUTION Note that x_4 is highly correlated with variables x_1, x_4^2, and x_3x_4 (the circled elements in Table 15.12). Thus, any model containing x_4 and any of the other three simultaneously may have problems with multicollinearity.

With only x_2 and x_4 in the model, the t value for x_4 was 7.098 (from Table 15.11). When we include x_4^2 in Table 15.13, the t values for both x_4 and x_4^2 are less than 1.00 and insignificant. The reason for this is that 98% ($0.99^2 = 0.9801$) of the variation in one is explained by the other. Thus if one, say x_4^2, is in the model, there is very little variation left in Y that can be explained by the other, in this case x_4. We would find the exact amount by calculating the partial correlation between x_4 and Y with x_4^2 held constant and squaring the result. The partial correlation between x_4 and Y with x_2 and x_4^2 already in the model is actually 0.104. Thus, with both x_2 and x_4^2 in the model, x_4 can explain only 1.08% of the remaining variation in Y.

Table 15.12 Matrix of Simple Correlation Coefficients for All Variables of Example 15.8

	Y	x_1	x_2	x_3	x_4	x_3^2	x_4^2	x_3x_4
Y	1.00	.84	.81	.70	.85	.69	.87	.83
x_1	.84	1.00	.49	.79	.95	.80	.95	.93
x_2	.81	.49	1.00	.55	.51	.55	.54	.59
x_3	.70	.79	.55	1.00	.71	.98	.74	.93
x_4	.85	(.95)	.51	.71	1.00	.71	(.99)	(.90)
x_3^2	.69	.80	.55	.98	.71	1.00	.74	.94
x_4^2	.87	.95	.54	.74	.99	.74	1.00	.92
x_3x_4	.83	.93	.59	.93	.90	.94	.92	1.00

If we refer to Table 15.9, which contains the printout for the full model, we see that for the individual tests only b_2 is significant. This is to be expected, since x_2 is the only variable not highly correlated with any of the other independent variables. To illustrate the rounding problems, in this case the determinant of the matrix mentioned in our earlier discussion is 0 to at least five decimal places. Obviously, division by this value may cause significant rounding errors.

If multicollinearity exists in a model, the main point to remember is that it is the model and the standard error that are important, not the individual t or F values. Often dropping predictors that are highly correlated with other independent variables will decrease the standard error (or $\sqrt{\text{MSE}}$). Thus we advise you to check various reduced models before selecting the final prediction equation.

15.8.2 Autocorrelation

A second problem that may arise when data are observed over time (for example, in forecasting sales from historical data) is correlation between the residuals. One of the assumptions for our tests was uncorrelated error terms. If the residuals are dependent on one another across time, they are said to be *autocorrelated* or *serially correlated*. Thus the assumption is violated. This usually happens with time series data when an important independent variable has been omitted.

The difficulty that results from autocorrelation is that *all* test statistics will be larger than they should be because the SSE will be under-

Table 15.13 Y Regressed on x_2, x_4, and x_4^2 for the Tree Lawn Realty Data of Example 15.8

REGRESSION COEFFICIENTS

VAR	COEFFICIENT(B)	STD ERROR	T-VALUE
X2	15.74344	2.97013	5.30060
X4	7.65696	18.91613	0.40478
X4SQ=X6	1.92997	2.77133	0.69641
	8.11308 (CONSTANT)		

ANALYSIS OF VARIANCE

SOURCE	SUM OF SQUARES	DF	MEAN SQUARE	F-VALUE
REGRESSION	20255.61938	3	6751.87313	56.40128
ERROR	1795.67009	15	119.71134	

estimated. This does not affect the calculation of the b_j's, but it does affect both the individual test statistics and the overall F ratio. The net result is that insignificant results may appear to be significant, leading to the conclusion that the model is useful when in fact it is *not.*

We can make a simple check for autocorrelation by using the runs test of Section 10.6.3. The rationale for the test is that if the residuals are uncorrelated, their occurrences above and below zero should be random over time. If the randomness hypothesis is rejected, there is an indication of autocorrelation.

A more common test for autocorrelation, though it is sometimes inconclusive, was developed by Durbin and Watson. This statistic is calculated in the residual analysis presented by most computer packages. For a complete discussion of the Durbin-Watson test and possible approaches to compensate for autocorrelation, see pages 352–366 of Neter and Wasserman.

15.8.3 Indicator Variables

Most of the predictors that we have used so far have been quantitative in nature: math score, logic score, temperature forecast, lot size, etc. The regression procedure requires that all predictors be specified numerically, but they need not necessarily be quantitative. It is possible to use *indicator* (dummy) variables to identify qualitative factors. We did this in Example 15.3, where x_3 represented the sex of the student and x_5 represented whether or not the student chewed gum. In that example x_3 and x_5 were truly dummy, but there are many situations in business in which the inclusion of qualitative variables in the model is desirable. For example, they may be used to identify seasonal components in a forecasting model, the differential effects in production due to shifts, or the differential effect on sales of differing package designs.

Indicator variables can be defined in many ways, but it is generally advisable to keep the coding as simple as possible. If a qualitative variable has only two levels, the easiest coding is to identify one level as 0 and the other as 1. For more than two levels we advise using a number of (0, 1) indicator variables equal to the number of levels minus one. For seasonal effects (assuming four seasons) we would define

$$x_1 = \begin{matrix} 1 & \text{if spring} \\ 0 & \text{otherwise} \end{matrix}$$

$$x_2 = \begin{matrix} 1 & \text{if summer} \\ 0 & \text{otherwise} \end{matrix}$$

$$x_3 = \begin{cases} 1 & \text{if fall} \\ 0 & \text{otherwise} \end{cases}$$

and omit a variable for winter from the model (actually winter simply equals 0). Thus each β_j measures the differential effect between winter and the season represented by $x_j = 1$.

After the indicator variables have been defined, the regression analysis proceeds as before. If all the predictors in a model are qualitative, we are merely performing an analysis of variance as presented in Chapter 12.

EXERCISES

15.8.1 Identify four qualitative variables that might be useful predictors of sales and define indicator variables for each.

15.8.2 Using a computer, determine the correlation matrix for Exercise 15.5.4 and check it for indications of multicollinearity.

15.8.3 The GMAT of Exercise 15.3.4 is designed to measure a student's preparedness for graduate study in business colleges. Define indicator variables that take into account whether a student's undergraduate major was in business, science and engineering, or other. Then, using the data of Exercise 15.3.4, regress the GMAT scores on the student's two-year undergraduate GPA and major. For your information: student 3 was a mathematics major; student 8 was an engineering major; students 1, 2, 5, and 6 were business majors; and students 4 and 7 majored in psychology and languages, respectively. Check for multicollinearity and test to see whether any variables can be omitted.

15.8.4 Use the runs test at $\alpha = 0.05$ to test for autocorrelation in the residuals of Exercise 15.5.4. Assume that the data are sequenced over time.

15.9 SUPPLEMENTARY EXAMPLES

In Section 15.6 we discussed alternative approaches to using all possible regressions to select our model or to relying solely on individual tests. One of these approaches was to use a forward selection procedure. This procedure is available in most of the standard statistical computer packages. An excellent package that is ideal for teaching purposes because of its interactive nature is the A.D.E.P.T. package developed by the University of Akron. Not only is it interactive, but its computational accuracy surpasses most of the other packages currently available. The following example illustrates the stepwise approach. All printouts are from the A.D.E.P.T. package using the forward selection regression subroutine.

EXAMPLE 15.13 We saw in Example 15.9 that the correlation matrix for the Tree Lawn Realty data indicated that x_4^2 (number of bedrooms squared) had the highest absolute correlation with Y. It is unlikely that we would have detected this using individual tests or tests for the quadratic term. Using forward selection regression, determine the best equation, based on the data of Examples 15.8 and 15.9, for predicting selling price. Use $\alpha = 0.05$ for all overall tests and $\alpha = 0.20$ to test for inclusion of variables.

SOLUTION The first step in the procedure is to check the simple correlations between the response variable Y and each of the predictors. From Table 15.12, these are:

Predictor (x_j):	x_1	x_2	x_3	x_4	x_3^2	x_4^2	x_3x_4
r_{yx_j}:	.84	.81	.70	.85	.69	.87	.83

Upon scanning these correlations, the computer selects x_4^2 as the first variable to enter the equation. The results of step 1 are presented in Table 15.14.

After calculating the equation with x_4^2 in the model and testing it for significance, the program calculates the partial correlations between Y and the remaining variables with x_4^2 held constant (that is, the simple correlations between the residuals and the other predictors). The computer then selects the variable with the largest absolute partial correlation and tests it for significance (at $\alpha = 0.20$ in this example); if that variable is significant, the computer enters it into the model in step 2. [The beta coefficients correspond to the standardized x_j, that is, to $(x_j - \bar{x}_j)/s_{x_j}$.] Looking at the partial correlations in Table 15.14, we see that if it is significant at 0.20, x_2 will be the next variable to enter the model. The printout for step 2 is given in Table 15.15.

After entering x_2 into the model, the program calculates the partial correlations of Y with the remaining independent variables with x_2 and x_4^2 both held constant. We see that when the new partial correlations are tested for significance, the program terminates. This is because the probability of the partial correlation differing from 0

Table 15.14 Stepwise Regression Printout for Step 1 in Example 15.13

```
UNIVARIATE REGRESSION

DEPENDENT VARIABLE:          Y;SELLING PRICE
INDEPENDENT VARIABLE(S):     X1;LIVING AREA
                             X2;LOT SIZE
                             X3;NUMBER OF BATHROOMS
                             X4;NUMBER OF BEDROOMS
                             X3SQ;NUMBER OF BATHROOMS SQUARED
                             X4SQ;NUMBER OF BEDROOMS SQUARED
                             X3X4;BEDROOM BATHROOM INTERACTION

STEPWISE REGRESSION



ITERATION 1

REGRESSION EQUATION: Y  DEPENDENT
                        'B'              STD. ERR      BETA
                       COEFFS             OF 'B'       COEFFS    F-VALUES   PROB.
X4SQ                4.30687507          .58616432      .87208    53.98658   .00000
CONSTANT           21.06269647

STATISTICS FOR THE ENTIRE EQUATION

ANALYSIS OF VARIANCE
  SOURCE             SS          DF         MS                F           PROB.
REGRESSION     16770.40484        1    16770.40484        53.98658       .00000
ERROR           5280.88463       17      310.64027
TOTAL          22051.28947       18

      MULTIPLE R      MULT. R-SQR          DETER XTRSX
        .87208           .76052            1.0000000000

PARTIALS FOR VARIABLES NOT IN THE EQUATION
X1                 .05991
X2                 .81009
X3                 .18419
X4               -.15185
X3SQ               .14457
X3X4               .16457
```

Table 15.15 Stepwise Regression Printout for Step 2 in Example 15.13

```
UNIVARIATE REGRESSION

DEPENDENT VARIABLE:          Y;SELLING PRICE
INDEPENDENT VARIABLE(S):     X1;LIVING AREA
                             X2;LOT SIZE
                             X3;NUMBER OF BATHROOMS
                             X4;NUMBER OF BEDROOMS
                             X3SQ;NUMBER OF BATHROOMS SQUARED
                             X4SQ;NUMBER OF BEDROOMS SQUARED
                             X3X4;BEDROOM BATHROOM INTERACTION


ITERATION 2

REGRESSION EQUATION: Y ,DEPENDENT
                       'B'                  STD. ERR       BETA
                      COEFFS                 OF 'B'        COEFFS     F-VALUES    PROB.
X2                 15.43074692             2.79196618      .47242     30.54595   .00005
X4SQ                3.03794811              .42214012      .61514     51.79018   .00000
CONSTANT           20.58570692
STATISTICS FOR THE ENTIRE EQUATION

ANALYSIS OF VARIANCE
  SOURCE              SS          DF          MS                F          PROB.
REGRESSION      20236.00458        2    10118.00229        89.18051     .00000
ERROR            1815.28490       16      113.45531
TOTAL           22051.28947       18

     MULTIPLE R     MULT. R-SQR          DETER XTRSX
       .95796           .91768           .7041943887

PARTIALS FOR VARIABLES NOT IN THE EQUATION
X1                .21315
X3               -.04466
X4                .10395
X3SQ             -.11288
X3X4             -.11277
ASSIGNED  0.2 ALPHA LEVEL EXCEEDED
```

(in either direction) by 0.21315 or more is greater than 0.2, the assigned value for α. Thus our final equation is

$$\hat{Y}_{x_2, x_4^2} = 20.58571 + 15.43075x_2 + 3.03795x_4^2$$

which explains 91.768% of the total variation in selling price for these data.

In our discussion of multicollinearity in Section 15.8.1 we mentioned the determinant of a particular matrix. This matrix is called X transpose X and should be familiar to the student who has taken matrix algebra. Note that one of the statistics given in the printout is DETER XTRSX. This is the determinant of that matrix; in this case it equals 0.70419. If this value becomes 0 or very close to it, there is likely to be a problem with multicollinearity. (*Note*: The remaining variables were forced to enter the equation, and with the entry of each succeeding variable the standard error $\sqrt{\text{MSE}}$ deteriorated, that is, increased.)

Another problem area discussed earlier is pseudoreplication, or combinations of x_j values that are not true replicates but are nearly so. We illustrate this possibility and the use of an indicator variable in our final example.

EXAMPLE 15.14 Alice Avon, the vice president of marketing for Bard Products, a company that produces cosmetics for sale through direct contact in the customer's home, is trying to build a sales forecasting model to be used in predicting dollar sales for each sales district. She has selected the type of area (rural or urban) in which the district is located, the number of active salespersons, the level of local advertising (in \$1000), and the level of national advertising (in \$1,000,000) as predictors. She believes that a linear model will be quite adequate and has specified the following model:

$$Y_x = \beta_0 + \beta_1 x_1 + \beta_2 x_2 + \beta_3 x_3 + \beta_4 x_4 + \epsilon$$

where

Y is district sales, in \$10,000

x_1 = 0 if the district is a rural area
= 1 if the district is an urban area

x_2 = number of active salespersons during the month

x_3 = amount spent on local advertising, in \$1000

x_4 = amount spent on national advertising, in \$1,000,000

To obtain her data, Alice randomly selected six districts (three urban and three rural), then selected four months at random from the past five years and recorded the district sales and corresponding x_j values for each of the four months. Her data are presented in Table 15.16 along with the average sales per district for the four months.

Using Alice's data, perform a regression analysis for her for the full model, then use the stepwise procedure. Use $\alpha = 0.05$ for tests of models and 0.25 for inclusion of variables. In addition, assume that the four observations within a district constitute four replicates and find the SSPE. Comment on the results.

Table 15.16 Data for Example 15.14 ($\overline{Y}_c$ Represents the District Average for the Four Months Used)

Sales (Y) (\$10,000)	Area Type (x_1) (Rural = 0/ Urban = 1)	Number of Salespersons (x_2)	Local Advertising Expense (x_3) (\$1000)	National Advertising Expense (x_4) (\$1,000,000)	$\overline{Y}_c$
7.5	1	15	22	2.5	8.9
10.2	1	15	22	2.7	8.9
9.1	1	16	24	2.5	8.9
8.8	1	15	21	2.4	8.9
14.2	1	21	28	2.5	13.225
16.8	1	20	29	2.7	13.225
12.6	1	20	28	2.5	13.225
9.3	1	21	27	2.4	13.225
15.4	1	28	37	2.5	16.7
18.7	1	28	36	2.7	16.7
19.8	1	28	38	2.5	16.7
12.9	1	27	38	2.4	16.7
4.8	0	8	10	2.5	5.375
6.4	0	8	10	2.7	5.375
5.3	0	8	12	2.5	5.375
5.0	0	8	11	2.4	5.375
4.6	0	5	9	2.5	3.8
4.7	0	5	8	2.7	3.8
3.2	0	4	8	2.5	3.8
2.7	0	5	8	2.4	3.8
6.1	0	10	12	2.5	7.175
8.4	0	12	13	2.7	7.175
6.9	0	11	14	2.5	7.175
7.3	0	11	12	2.4	7.175

SOLUTION The printout for the full model and the correlation matrix are given in Table 15.17 on page 642. For

$$H_0: \beta_j = 0 \text{ all } j \qquad H_a: \beta_j \neq 0 \text{ some } j$$

H_0 is rejected, since the calculated value for F of $58.72 > f_{4,\,19,\,0.05} = 2.90$. Thus the full model appears to be useful. For the individual tests, we see that only x_2 and x_4 have probabilities less than 0.25. Thus for each β_j,

$$H_0: \beta_j = 0 \qquad H_a: \beta_j \neq 0$$

H_0 is rejected for $j = 2$ and 4 but cannot be rejected for 1 and 3. The correlations in the matrix and the value of DETER XTRSX indicate a possible problem with multicollinearity.

Going to a stepwise procedure, we have the results, after two iterations, given in Table 15.18 on page 643. Variables x_1 and x_3 remain insignificant (also, all models including either of them have larger MSEs). Alice's equation at this point is

$$\hat{Y}_{x_2, x_4} = -24.15701 + 0.57427x_2 + 9.90176x_4$$

with

$$x_2 = \text{number of active salespersons}$$
$$x_4 = \text{national advertising in \$1,000,000}$$

Since there is little variation between the combinations of the predictors within a district, we are somewhat suspicious of the high R^2 for this model. Thus we ran a stepwise regression using YBAR as the dependent variable and also a one-way ANOVA letting each district represent a treatment level. Table 15.19 on page 644 gives these results.

Note that the regression stops after x_2 is entered and that SSPE (equals SSE in the ANOVA table) is approximately 70.26. For the original model, x_2 entered first in the stepwise procedure and the SSE at that point was 71.16. Thus, it would appear that almost all of SSE is really pure error. Alice should exercise extreme caution in using the equation that was selected.

Plots of the residuals versus x_2 and x_4 showed that the residuals tended to be grouped in sets of four, also illustrating pseudoreplication. In actuality Ms. Avon should redesign her experiment, possibly selecting 24 different sales districts with one month randomly selected for each district from the time span of interest.

Table 15.17 Correlation Matrix and Regression Printout for the Full Model of Example 15.14

```
THE CORRELATION MATRIX IS:

                     Y          X1          X2          X3          X4
Y              1.00000      .78107      .93332      .92383      .23301
X1              .78107     1.00000      .85175      .89502      .00000
X2              .93332      .85175     1.00000      .98796      .00860
X3              .92383      .89502      .98796     1.00000     -.00092
X4              .23301      .00000      .00860     -.00092     1.00000

N OF CASES = 24


UNIVARIATE REGRESSION

DEPENDENT VARIABLE:          Y;SALES IN 10,000 DOLLARS
INDEPENDENT VARIABLE(S):     X1;AREA TYPE, RURAL=0, URBAN=1
                             X2;NUMBER OF SALESMEN
                             X3;LOCAL ADVERTISING IN 1000 DOLLARS
                             X4;NATIONAL ADVERTISING IN MILLION DOLLARS

MULTIPLE REGRESSION

REGRESSION EQUATION: Y  DEPENDENT
                      'B'           STD. ERR        BETA
                    COEFFS          OF 'B'          COEFFS     F-VALUES     PROB.
X1               -1.29671035      1.53075701      -.13519       .71759     .40743
X2                 .36194371       .28424523       .58702      1.62142     .21823
X3                 .21484966       .25013086       .46509       .73779     .40103
X4               10.05098629      2.76893765       .22838     13.17623     .00179
CONSTANT        -25.06795366

STATISTICS FOR THE ENTIRE EQUATION

ANALYSIS OF VARIANCE
  SOURCE             SS       DF         MS              F        PROB.
REGRESSION       510.68031     4     127.67008       58.72123    .00000
ERROR             41.30928    19       2.17417
TOTAL            551.98958    23

     MULTIPLE R      MULT. R-SQR        DETER XTRSX
       .96185          .92516           .0036870060
```

Table 15.18 Second (Final) Iteration for Stepwise Regression for Example 15.14

```
DEPENDENT VARIABLE:          Y;SALES IN 10,000 DOLLARS
INDEPENDENT VARIABLE(S):     X1;AREA TYPE, RURAL=0, URBAN=1
                             X2;NUMBER OF SALESMEN
                             X3;LOCAL ADVERTISING IN 1000 DOLLARS
                             X4;NATIONAL ADVERTISING IN MILLION DOLLARS

STEPWISE REGRESSION


ITERATION 2

REGRESSION EQUATION: Y  DEPENDENT
                         'B'          STD. ERR      BETA
                        COEFFS         OF 'B'      COEFFS    F-VALUES    PROB.
X2                   .57427361      .03764840      .93139   232.67235   .00000
X4                  9.90176182     2.68720700      .22499    13.57759   .00138
CONSTANT          -24.15701071

STATISTICS FOR THE ENTIRE EQUATION

ANALYSIS OF VARIANCE
  SOURCE              SS        DF          MS             F           PROB.
REGRESSION        508.77461      2     254.38730      123.61765      .00000
ERROR              43.21497     21       2.05786
TOTAL             551.98958     23

     MULTIPLE R      MULT. R-SQR          DETER XTRSX
       .96006          .92171             .9999259923

PARTIALS FOR VARIABLES NOT IN THE EQUATION
X1            -.08354
X3             .08942
ASSIGNED  0.25 ALPHA LEVEL EXCEEDED
```

Table 15.19 (a) One-Way ANOVA by District and (b) Stepwise Regression for $\overline{Y}_c$ for Example 15.14

```
                                  TABLE I
SOURCE          SUM SQ       D OF F     MEAN SQ         F RATIO
MEANS         481.7271          5       96.3454         24.6820
WITHIN         70.2625         18        3.9035
TOTAL         551.9896         23
PROBABILITY OF F≥            24.6820 WITH 5 AND 18 D.F.= .0023


DEPENDENT VARIABLE:          YBAR;THE DISTRICT AVERAGE FOR THE 4 MONTH USED
INDEPENDENT VARIABLE(S):     X1;AREA TYPE, RURAL=0, URBAN=1
                             X2;NUMBER OF SALESMEN
                             X3;LOCAL ADVERTISING IN 1000 DOLLARS
                             X4;NATIONAL ADVERTISING IN MILLION DOLLARS

STEPWISE REGRESSION



ITERATION 1

REGRESSION EQUATION: YBAR  DEPENDENT
                          'B'              STD. ERR       BETA
                        COEFFS             OF 'B'        COEFFS    F-VALUES     PROB.
X2                     .57274658          .01303639      .99435 1930.23676   .00000
CONSTANT               .86714351

STATISTICS FOR THE ENTIRE EQUATION

ANALYSIS OF VARIANCE
  SOURCE              SS          DF         MS              F             PROB.
REGRESSION         476.29844       1      476.29844     1930.23676       .00000
ERROR                5.42864      22         .24676
TOTAL              481.72708      23

     MULTIPLE R     MULT. R-SQR          DETER XTRSX
       .99435          .98873            1.0000000000

PARTIALS FOR VARIABLES NOT IN THE EQUATION
X1             -.19509
X3              .05687
X4             -.08058
ASSIGNED  0.25 ALPHA LEVEL EXCEEDED
```

15.10 SUMMARY

In this chapter we have investigated the concept of a multiple linear model, the estimation of its parameters, and various tests that are associated with the model. By linear we mean linear in the parameters rather than the predictors. The basic model is given by

$$Y_x = \beta_0 + \sum_{j=1}^{k} \beta_j x_j + \epsilon$$

where the x_j are nonrandom variables and ϵ is a random error term that is distributed as $N(0, \sigma^2)$ for all combinations of the x_j. To estimate the parameters β_0 through β_k, we used the least-squares procedure, which makes the solving of $k + 1$ simultaneous equations in the $k + 1$ parameters necessary. With computers readily available, the estimates are easily and rapidly obtained.

After the equation has been calculated, it must be tested for usefulness; if possible, it should also be checked for the necessity of including higher-order terms. To test for usefulness, we compare the average variation of the predicted values from $\overline{Y}$ with the average variation of the observed values about the equation, that is, with the MSE. If this ratio is too large, we reject H_0 and conclude that the equation is useful. For the test for inclusion of higher-order terms, some of the combinations of the predictors must be replicated so that an estimate of the error variance can be calculated independent of the model. Then we compare the average variation of $\overline{Y}_c$ from $\hat{Y}_c$ ($\overline{Y}_c$ and $\hat{Y}_c$ represent the mean and predicted values, respectively, for combination c) with the MSPE. A large value for this ratio indicates rejection of H_0 and implies that a higher-order model should be investigated.

If replication is not available, we can make rough checks on the assumptions and the need for higher-order terms using residual plots. In this case we construct residual plots for each predictor (other than those that are simply indicator variables). Once we have decided on the order of the model, we generally test various combinations of the variables to see whether we can improve the model by reducing the number of predictors. To do so we partition the SSREG into two parts, $SS(\boldsymbol{b}_{k-p}|b_0)$ and $SS(\boldsymbol{b}_p|b_0, \boldsymbol{b}_{k-p})$, where p is the number of variables to be tested and $k - p$ represents the remaining variables that are assumed to be already in the model. For $p = 1$ either a t test or an F test may be used; if $p > 1$, then we must use an F test. In any case, the estimate used for σ^2 is the MSE for the full model. Nonrejection of H_0 indicates that we should consider deleting the corresponding variables.

To obtain exact confidence intervals or prediction intervals in the multiple-predictor model requires mathematical techniques known as matrix algebra. A few computer programs will calculate these intervals for you, but most don't.

In concluding our discussion of multiple linear regression, we mentioned some problems that can arise in the analysis. One is multicollinearity, or linear relationships between the independent variables. The effect of this is that the individual tests may fail to reject H_0 as a result of excessive rounding errors. To guard against incorrect decisions

Table 15.20 Summary of Formulas for Chapter 15

Basic Model:

$$Y_x = \beta_0 + \sum_{j=1}^{k} \beta_j x_j + \epsilon$$

where $x_j, j = 1, \ldots, k$, is a nonrandom variable, and ϵ is a random variable with a mean of zero.

Prediction Equation:

$$\hat{Y}_x = b_0 + \sum_{j=1}^{k} b_j x_j$$

where b_0 represents the Y intercept (all $x_j = 0$), and b_j represents the change in Y for a one-unit change in x_j with all other predictors held constant.

All b's are least-squares estimates determined by solving the set of normal equations.

Sums of Squares

$$\text{SSTO} = \sum_i Y_i^2 - \frac{\left(\sum_i Y_i\right)^2}{n}$$

$$\text{SSE} = \sum_i Y_i^2 - b_0 \sum_i Y_i - \sum_j \left(b_j \sum_i Y_i x_{ij}\right)$$

$$\text{SSREG} = \text{SS}(\boldsymbol{b}|b_0) = b_0 \sum_i Y_i + \sum_j \left(b_j \sum_i Y_i x_{ij}\right) - \frac{\left(\sum_i Y_i\right)^2}{n}$$

$$\text{SSPE} = \sum_c \sum_i Y_{ic}^2 - \sum_c \frac{\left(\sum_i Y_{ic}\right)^2}{n_c}$$

where c represents a distinct combination of the x_j's.

$$\text{SSLOF} = \text{SSE} - \text{SSPE}$$

$$\text{SS}(b_p|b_0, \boldsymbol{b}_p^*) = \text{SS(Full)} - \text{SS}(\boldsymbol{b}_p^*|b_0) \qquad p = 1, \ldots, k$$

on this basis, the correlation matrix should be checked. If any independent variables appear to be highly intercorrelated with one another, then the test statistics corresponding to the individual b_j's should be used cautiously. A second possible problem, known as autocorrelation, arises with time-sequenced data. In this case the residuals may be correlated (which violates one of our test assumptions), causing the mean square error to underestimate σ^2 and the test statistics to be larger than they should be.

Most regression models use quantitative variables as predictors, but qualitative variables can be used also. To do so we must define an indicator variable (or variables). For ease in interpretation we recommend, if possible, the use of zero/one variables, where a one indicates that the level is present and a zero indicates that it is not.

The formulas and tests of this chapter are summarized in Tables 15.20 and 15.21, respectively.

Table 15.21 Summary of Inferences for Chapter 15

Tests of Hypotheses	
Hypotheses	Reject H_0 if
H_0: $\beta_j = 0$ for all j $\quad H_a$: some $\beta_j \neq 0$	$\dfrac{\text{MSREG}}{\text{MSE}} > f_{k,\, n-k-1,\, \alpha}$
H_0: $\mu_{Y\mid x} = \beta_0 + \sum_j \beta_j x_j$ H_a: $\mu_{Y\mid x} \neq \beta_0 + \sum_j \beta_j x_j$	$\dfrac{\text{MSLOF}}{\text{MSPE}} > f_{m-k-1,\, n-m,\, \alpha}$
H_0: $\beta_j = 0$ $\quad H_a$: $\beta_j \neq 0$	$\dfrac{\text{MS}(b_j \mid b_j^*)}{\text{MSE}} > f_{1,\, n-k-1,\, \alpha}$ (or if $\lvert t_j \rvert > t_{n-k-1,\, \alpha/2}$)
H_0: $\boldsymbol{\beta}_p = \mathbf{0}$ $\quad H_a$: $\boldsymbol{\beta}_p \neq \mathbf{0}$	$\dfrac{\text{MS}(\boldsymbol{b}_p \mid \boldsymbol{b}_p^*)}{\text{MSE}} > f_{p,\, n-k-1,\, \alpha}$
where $\boldsymbol{\beta}_p = (\beta_{k-p+1}, \ldots, \beta_k)$ and $\boldsymbol{\beta}_p^* = (\beta_0, \beta_1, \ldots, \beta_{k-p})$	
$1 - \alpha$ Confidence Intervals	
For $\mu_{Y\mid \boldsymbol{x}_p}$:	$\hat{Y}_{\boldsymbol{x}} \pm t_{n-k-1,\, \alpha/2} \sqrt{\text{MSE}[\boldsymbol{x}_p'(\boldsymbol{x}'\boldsymbol{x})^{-1}\boldsymbol{x}_p]}$
For $Y_{\boldsymbol{x}_p}$:	$\hat{Y}_{\boldsymbol{x}} \pm t_{n-k-1,\, \alpha/2} \sqrt{\text{MSE}[1 + \boldsymbol{x}_p'(\boldsymbol{x}'\boldsymbol{x})^{-1}\boldsymbol{x}_p]}$

REVIEW EXERCISES*

Level One

15.1 Explain why the simple correlation between Y and x_3 does not equal the partial correlation between Y and x_3 with x_1, x_2, and x_4 held constant.

15.2 Using the computer printout provided,
(*a*) Write the equation.
(*b*) Test the equation for significance at $\alpha = 0.05$.
(*c*) Assuming that no multicollinearity exists, determine which, if any, terms should be thrown out at $\alpha = 0.2$.

```
REGRESSICN CCEFFICIENTS

VAR   COEFFICIENT(B)    STD ERRCR       T-VALUE
X2       ¯0.83742        0.13676       ¯6.12322
X3       ¯0.48610        0.54393       ¯0.89368
X4       ¯0.24902        0.18364       ¯1.35599
         16.62648       (CCNSTANT)

STANDARD ERRCR CF ESTIMATE: 0.61788
R SQUARED:                  0.98909

ANALYSIS CF VARIANCE

SOURCE         SUM CF SQUARES   DF   MEAN SQUARE    F-VALUE

REGRESSICN       69.23645        3     23.07882     60.45146
ERRCR             0.76355        2      0.38177
```

15.3 Refer to Exercise 15.2 and give one reason why such a large R^2 might be somewhat misleading in this case.

15.4 In an attempt to forecast production requirements for the next two years, a multiple regression equation was determined using four independent variables. The partial analysis-of-variance table is as follows. Complete it and test the model for usefulness at $\alpha = 0.05$.

Source	SS	df	MS
Regression			1000
Error	2000		
Total		29	

*Exercises with an asterisk may require matrix algebra.

15.5 Consider the following partial computer printout for a multiple regression problem:

Regression Coefficients

Variable	Coefficient (B)	Standard Error
x_1	2.6015	0.5203
x_2	10.8312	3.6104
x_3	4.2154	0.3011
x_4	12.8000	25.6000
	5.3112	(constant)

(*a*) Write the estimate of the model.
(*b*) Predict Y if $x_1 = 13$, $x_2 = 1.8$, $x_3 = -0.5$, and $x_4 = 3.6$.
(*c*) Test the individual coefficients for significance at $\alpha = 0.10$.

15.6 Additional information from the printout of Exercise 15.5 is as follows:

Analysis of Variance

Source	Sum of Squares	df	Mean Square
Regression	298.2432	4	74.5608
Residual	31.0670	5	6.2134

Correlation Matrix

1.0000	.5200	.6132	.5200	.0625
.5200	1.0000	.2700	.6500	.0050
.6132	.2700	1.0000	.4000	.0600
.5200	.6500	.4000	1.0000	.0800
.0625	.0050	.0600	.0800	1.0000

(*a*) How many observations were there?
(*b*) What is the standard error of the estimate $s_{Y|x}$?
(*c*) Is the full model useful at $\alpha = 0.05$?
(*d*) With which variables might there be a problem due to multicollinearity?

*15.7 Refer to Exercises 15.5 and 15.6. Determine exact 90% confidence limits for

(*a*) $\mu_{Y|\boldsymbol{x}}$ when $\boldsymbol{x} = (13, 1.8, -0.5, 3.6) = \bar{\boldsymbol{x}}$.

(*b*) $\hat{Y}_{\boldsymbol{x}}$ when $\boldsymbol{x} = (13, 1.8, -0.5, 3.6) = \bar{\boldsymbol{x}}$.

15.8 You have been hired by Manpower, Ltd., of Halifax to design a multiple regression equation that will help to select applicants for the job of ditch digger. The response variable of interest is the number of linear meters of ditch, 1 meter deep, the applicant will be able to dig in one day after being on the job for one month. The variables being considered as predictors are: age, weight, time required to run 800 meters, and number of years of formal education. Twenty current employees, whose ditch-digging ability is known, are randomly selected and their attributes of interest recorded. These data are given here.

Employee	Meters/ Day	Age	Weight (Stones)	Minutes for 800 Meters	Years of Education
1	4.2	18	9.6	3.2	10
2	7.3	26	13.5	2.9	12
3	6.1	23	8.8	3.1	11
4	8.2	24	16.3	4.6	17
5	8.1	19	13.6	2.7	9
6	5.7	37	11.9	3.5	20
7	6.4	28	13.1	4.1	13
8	7.5	20	12.6	3.8	14
9	5.9	17	8.5	2.8	11
10	6.0	33	9.9	3.4	14
11	6.3	25	10.6	2.4	15
12	5.7	49	14.1	5.2	10
13	6.4	35	11.8	3.9	21
14	5.6	24	10.1	3.4	11
15	6.3	42	12.6	2.9	12
16	4.3	18	17.5	6.3	10
17	4.6	19	8.7	3.2	13
18	6.9	21	12.1	3.8	12
19	5.8	20	11.3	4.0	11
20	7.1	32	13.1	3.3	8

(*a*) Determine the prediction equation for Manpower.
(*b*) Test it for usefulness at $\alpha = 0.05$.

(c) An applicant just walked into your office. You have the applicant fill out an application form and run 800 meters, and you find that the person is 20 years old, weighs 8.2 stones, has completed two years of college, and can run 800 meters in 2.5 minutes. What is your prediction of the applicant's ditch-digging ability?

*15.9 Determine 90% bounds on the ditch-digging ability of the applicant in Exercise 15.8*c*.

15.10 Refer to Exercise 15.8 and test the predictors individually at $\alpha = 0.20$ to see whether any can be dropped from the model.

15.11 Determine the correlation matrix for Exercise 15.8 and evaluate your decisions in Exercise 15.10.

15.12 The following model was obtained to predict prices per share for a certain stock by quarters:

$$Y = 10 + 2X_1 + 3X_2 - X_3 + 0.5X_4$$

where X_1 is a leverage factor, X_2 is millions of dollars of sales in the previous quarter, X_3 is an inventory factor, and X_4 is the adjusted price per share of the previous quarter. Given the following additional information:

$$n = 20 \qquad \Sigma\, Y = 250 \qquad \Sigma\, Y^2 = 5125 \qquad \Sigma\, X_1 Y = 350$$

$$\Sigma\, X_2 Y = 200 \qquad \Sigma\, X_3 Y = 175 \qquad \Sigma\, X_4 Y = 600$$

(a) Determine whether the model is useful (use $\alpha = 0.10$).
(b) Given the values $X_1 = 2$, $X_2 = 1$, $X_3 = 3$, and $X_4 = 6$, predict the stock's price for this quarter.
(c) Determine the multiple correlation coefficient and explain its usefulness.

15.13 There has been much publicity in recent years about declining college enrollments. Identify four variables that might be useful for predicting the number of entering freshmen for a college five years from now. Indicate a model for your four variables.

15.14 WAM, Inc., recently developed a new product and is currently trying to determine how price sensitive the market is. To do so they test marketed the product in ten different cities of varying size at ten different price levels for one month. All advertising was via network television during that time. At the end of the period the unit sales (in 1000s) were recorded for each city along with the price ($) and population (in 10,000 residents). These data are given on page 652.

City:	1	2	3	4	5
Price ($):	24.95	19.95	29.95	26.95	22.95
Population (10,000s):	5	24	18	142	87
Sales (1000s):	7	51	14	162	169

City:	6	7	8	9	10
Price ($):	21.00	22.50	30.00	27.50	25.49
Population (10,000s):	8	22	69	10	31
Sales (1000s):	14	34	42	11	40

Determine the prediction equation for WAM based on these data and test it for usefulness at $\alpha = 0.10$.

15.15 Refer to Exercise 15.14. What is the sensitivity of unit sales to a $1 change in price controlling for population size? Test to determine if there is a linear effect due to price at $\alpha = 0.20$.

Level Two

15.16 Refer to Exercise 15.14 and construct residual plots to check for the necessity of including higher-order terms.

15.17 WAM has decided that it would like to formally test for the possibility of higher-order terms and a possible interaction effect between price and city size. To do so, they selected ten more cities, similar in size and type of area to the first ten. The results of the second market test are:

City:	1′	2′	3′	4′	5′
Price ($):	24.95	19.95	29.95	26.95	22.95
Population (10,000s):	6	26	17	156	85
Sales (1000s):	8	57	22	179	160

City:	6′	7′	8′	9′	10′
Price ($):	21.00	22.50	30.00	27.50	25.49
Population (10,000s):	7	21	73	11	34
Sales (1000s):	11	38	46	11	42

Assuming that these data are reasonable replicates of the data in Exercise 15.14, construct an approximate test for lack of fit in the linear model at $\alpha = 0.01$ for the combined data.

15.18 WAM still feels that there should be a quadratic term in price and an interaction between price and the city's size. State the model implied by WAM and estimate it using the combined data of Exercise 15.14 and 15.17. Simultaneously test the quadratic and interaction terms at $\alpha = 0.20$.

15.19 Recall that β_j represents the change in $\mu_{Y|x}$ for an incremental change in x_j in the presence of the other predictors in the model at that time. Derive the equations for a $1 - \alpha$ confidence interval for the effect on $\mu_{Y|x}$ for a one-unit change in x_j.

15.20 Use your results from Exercise 15.19 to determine a 95% confidence interval for the change in mean sales for WAM, Inc., when they change their price from 26.95 to 24.95 within a given city. Assume the linear equation of Exercise 15.17.

15.21 WAM has decided to market the product in a city of 732,000 people and would like to choose the price that will maximize their total contribution for this product in that area. If variable costs per unit for manufacturing and shipping this product are $16.32, what would you recommend assuming the equation of Exercise 15.17? Construct a 90% confidence interval for average monthly contribution for this product in this city at your price. (Contribution = total sales − total variable costs.) (*Requires calculus.*)

15.22 Refer to Example 15.8. Pete Pusher, the manager of Tree Lawn Realty, has decided that he overlooked some important variables in his original model, namely, the size of the garage (in cars) and the school district in which the residence is located. He provides you with this information for the 19 houses in your sample and asks you to reevaluate the present model with the additional information. Tree Lawn operates in four school districts, Hudtown, Stroh, Ghent, and Westhampton. The present model is

$\mu_{Y|x} = \beta_0 + \beta_2 x_2 + \beta_6 x_4^2$, where x_2 is lot size in acres and x_4^2 is the number of bedrooms squared. Pete's additional information is (corresponding to the order in Example 15.8):

School District:	G	S	W	S	G	S	H	W	H	S
Garage Size:	2	1	1	1.5	0	2	3	2	2.5	2
Residence:	1	2	3	4	5	6	7	8	9	10

School District:	H	H	W	G	S	G	H	W	S
Garage Size:	3.5	2.5	0	2	2	3	4.0	1.5	0
Residence:	11	12	13	14	15	16	17	18	19

Define appropriate variables to detect any differential effects due to school district. Estimate the new model containing the additional variables and test it for usefulness at $\alpha = 0.01$. Test to determine if the additional variables, as a group, make a significant contribution to the model at 0.10.

15.23 Refer to Exercise 15.22; test the additional variables individually at $\alpha = 0.10$, then obtain the correlation matrix and evaluate your decisions.

15.24 Refer to Exercise 15.22 and obtain a 90% confidence interval for the differential effect between a given property's being located in Hudtown versus Stroh.

15.25 Two models have been proposed for forecasting enlistments in the armed forces. The first model has an R^2 of 0.87, a significant F value of 68.37, and an $s_{Y|x}$ of 1.58. The second model, using the same Y values, enlistments in 100,000's, has an R^2 of 0.81, a significant F of 57.91 and an $s_{Y|x}$ of 1.24. Which model would you choose and why?

Level Three

15.26 An investor is considering investing in a particular stock and is interested primarily in its growth potential, that is, the increase in the stock price over the long run. He has collected data on the stock's average quarterly price over the last ten years, its PE ratio, total assets, R and D expenditures, and quarterly dividend per share. Three years ago the stock had a three-for-one split. Construct a model which he might consider that will utilize all the information above.

15.27 Refer to Exercise 15.8 and obtain all possible regressions (regression equations corresponding to all possible combinations of the x_j's). Choose the model that you think best predicts ditch-digging ability and explain your criteria for choosing the model you did.

15.28 Refer to Exercises 15.27 and 15.11. Use the information therein to do your own stepwise regression, using the forward selection approach. Use $\alpha = 0.05$ for tests of the model and 0.20 for variable inclusion.

15.29 Refer to Exercise 15.22 and do a forward selection using $\alpha = 0.01$ for tests of the models and $\alpha = 0.20$ for inclusion of variables.

15.30 Deposits are the fuel that keeps commercial banks running. They provide the majority of bank funds for making loans and other financial transactions of the bank. The Cattlemen and Miners' Bank of Prineville is trying to improve its forecasts of deposits so that it can better plan its investment activities. To this end the bank has presented you, their financial consultant, with the following data for the past six years.

	Quarter	Deposits	Prime Rate	Unemployment Rate
YEAR 1	1	39.93	8.20	0.082
	2	40.67	8.25	0.075
	3	41.74	8.25	0.062
	4	44.01	8.35	0.057
YEAR 2	1	42.24	8.25	0.068
	2	43.13	8.55	0.065
	3	43.88	8.70	0.065
	4	45.29	9.00	0.058
YEAR 3	1	44.49	9.40	0.078
	2	44.77	9.50	0.076
	3	46.19	9.55	0.059
	4	46.75	9.75	0.062
YEAR 4	1	46.16	10.30	0.082
	2	45.78	10.60	0.082
	3	46.87	11.00	0.066
	4	47.90	11.50	0.062

(data continued on page 656)

YEAR 5	1	48.10	11.30	0.094
	2	45.07	11.45	0.089
	3	49.23	11.75	0.073
	4	50.16	11.90	0.069
YEAR 6	1	52.28	12.00	0.081
	2	47.69	12.15	0.081
	3	53.81	12.20	0.066
	4	54.57	12.40	0.063

For these data, deposits are in millions of dollars, the prime rate is in percent, and the unemployment rate is the proportion of the work force unemployed in the bank's area. The prime rate and unemployment rate are the previous quarter government forecasts. Define an additional variable (or variables) to account for any seasonality in the deposits, then do a complete regression analysis. After you have selected a model, obtain the residuals and test them for autocorrelation. Use $\alpha = 0.20$ for inclusion of variables and 0.05 for all other tests.

References

1. Draper, Norman R., and Harry Smith, Jr. *Applied Regression Analysis.* New York: John Wiley & Sons, Inc., 1966.
2. Mendenhall, William, and James T. McClave. *A Second Course in Business Statistics: Regression Analysis.* San Francisco, CA: Dellen Pub. Co., 1981.
3. Neter, John, and William Wasserman. *Applied Linear Statistical Models.* Homewood, Ill.: Richard D. Irwin, Inc., 1974.

Chapter 16
Prediction Over Time

KEY WORDS AND TERMS

Time Series
Prediction
Forecast
Trend Component
Cyclical Component
Irregular Component
Curvilinear Regression
Weighted Moving Averages
Exponential Smoothing

16.1 INTRODUCTION

Imagine that you are responsible for ordering stock in a department store and that you are currently planning your order for the Christmas season. You will lose sales if you under order, but inventory evaluations will be excessively high in February if you over order. In this imagined situation, you would be well advised to look at past sales volume amounts and use these figures to estimate the demand for the coming season. In this case you would be using a time-oriented series of figures (the past sales volumes) to produce an estimate for a future figure (the Christmas season demand). Technically, you would be engaging in a mild form of *time series analysis*, which is a key element in many business decision environments. Perhaps you can appreciate the role time series analysis plays in production scheduling when drastic changes in assembly sequencing and machining are necessary in order to change from the production of one type of product to another. This chapter begins a short journey into the realm of time-oriented decision making that will help us understand the complexity of some of these problems.

16.2 TIME-SERIES MODELS

Our initial comments on time series models begin with the definition of a time series.

DEFINITION 16.1 Any sequence of events that occurs over time is called a *time series.*

Common examples of time series are sales demand per week (or month), rainfall per year, tons of ground fish processed per season, national debt per quarter, and so forth. Independent of the particular time series being analyzed, the usual objective will be a prediction or forecast for values we anticipate are likely to occur in the future. At this point, it is useful to make a distinction between what is meant by a prediction and what is meant by a forecast.

DEFINITION 16.2 The choice of an event from among many possible events is called a *prediction*.

DEFINITION 16.3 A set of possible events together with their corresponding probabilities is called a *forecast*.

We can illustrate the difference between a prediction and a forecast using a random variable. In this case, to predict means to choose one of the random variable range values, whereas to forecast is to provide a statement of the suspected distribution of the random variable.

The process of predicting a time-series value that corresponds to a time in the future is a form of *extrapolation*, which can be thought of as the opposite of interpolation. Thus, in this chapter we will present some techniques for time-series extrapolation. The first step in the extrapolation process is to choose a basic time-series model. The "classical" method of analysis assumes that a time-series value for time period t, denoted by S_t, is a function of four basic components, which are

1. The trend component T_t
2. The seasonal component SE_t
3. The cyclical component C_t
4. The irregular component, I_t

The basic idea is to produce effective techniques for determining the components, then to use the results of the analysis to formulate a time-series prediction.

The *trend component* describes any long-term consistency or tendency over time. The tendency may be either long-term growth or decline that is essentially linear in character (although nonlinear trends are common).

The *seasonal component* describes any periodic fluctuations over time that occur regularly according to the year, season, month, week, or day. The role of the seasonal component is apparent for retail outlets that experience a surge of demands during the Christmas season and then a drastic decline during January and February.

The *cyclical component* describes any fluctuation that occurs as a result of the prevailing economic climate, changes in the marketplace, or other sources that affect the time-series values over longer time periods and with less predictable intensity than that produced by the seasonal effect.

The *irregular component* explains the variation not covered by the other three. It includes the effect due to random fluctuation that can be expected as a part of any natural phenomenon.

Obviously, we could form a function of these four variables in many ways, but the two simplest are

$$S_t = T_t \times SE_t \times C_t \times I_t \tag{16.1}$$

and

$$S_t = T_t + SE_t + C_t + I_t \tag{16.2}$$

The model in Equation 16.1 is called the *multiplicative model.* In this model S_t and T_t are expressed in the observed units and the other components are expressed as ratios. Equation 16.2 is called the *additive model.* In this model S_t and T_t are expressed in the observed units and the other components are expressed in terms of positive or negative deviations from some norm.

We shall keep this discussion at an elementary level; that is, we will center our discussion on the multiplicative model and, through independent analysis of the components, will ignore the interaction effects between the components. Readers interested in more detailed coverage should consult a good text on econometrics or a standard reference on time-series analysis.

EXERCISES

16.2.1 Francine lives in Quebec, where temperatures vary between −35 and +30 degrees Celsius. Francine is trying to analyze her oil fuel bills using classical models for time-series analysis. She needs your help to understand how each of the component effects would apply to her particular time series. Be a good neighbor and offer her some assistance.

16.2.2 The Dow Jones and other composite stock price indices provide a good example of a time series. Explain each of the component effects as they would apply to such a time series.

16.2.3 Super Handy Open Hours, Inc., buys coffee beans in gross amounts for packaging and distribution (under its own "white label") to its numerous retail outlets. The company president's daughter is enrolled in a business program at a local university. She offers to provide planning assistance in the coffee purchasing area using the time-series analysis component approach that she is currently studying. Explain each of the component effects as they would apply to such a time series.

16.2.4 Succulent Sea Delicacies ships lobster throughout North America as well as doing a huge volume of lobster sales at a local level. The company accountant keeps track of the pounds of lobster sold on a weekly basis. Explain how each of the component effects would apply to such a time series.

16.2.5 "Mendall the Monk" (an endearment meant to preserve anonymity) grows beans for Beans, Beans the Magical Fruit, Ltd. Mendall takes delight in tending his special patch of beans, in which he has produced many advances of considerable report. Among his current projects is the documentation of the effects of a growth formula that is supposed to measure yield and thereby increase company profits. Assume that Mendall documents growth (using a complicated formula that employs a sequence of measurements on each plant) on a weekly basis. Explain how each of the component effects would apply to such a time series.

16.3 THE TREND COMPONENT

Our objective in this section is to show some of the fundamental techniques for determining, or isolating, the trend component. Basic to all approaches in this area is an assumption about the character of the trend (for example, it is linear) on which the analysis builds. The following dichotomy is standard in this respect: The trend is linear or it is nonlinear. If we assume that the trend component is essentially linear in character, we can use the method of semiaverages, linear regression, or some other method of linear extrapolation to predict or forecast a trend component for the next time period. If we assume that the trend component is essentially nonlinear in character, we can use curvilinear regression, weighted moving averages, exponential smoothing, or some other form of nonlinear extrapolation. This section shows how to isolate according to the linear trend assumption.

The character of linear trend components can be estimated using the actual time-series values. We can plot these values in ordered pairs of the type $\{(t, S_t) | t = 1, 2, \ldots, n\}$, where n denotes the number of time periods over which the linear trend is to be estimated. Once these points are plotted, we can use a ruler to draw what looks like the best-fitting line through the points. In this case the trend component estimate would be given by

$$\hat{T}_t = b_0 + b_1 t \tag{16.3}$$

where b_0 is the point at which the drawn line intersects the S_t axis and b_1 is the slope of the line. Such a subjective procedure is entirely acceptable and could produce results that are accurate enough for decision-making purposes. However, we will now present two other methods that can be classified as objective techniques for the determination of linear trend estimates.

The simpler of the two objective techniques is called the *method of semiaverages*. In this method the trend line estimate is the line passing through the two points

$$\left(\frac{n+2}{4}, S'\right) \quad \text{and} \quad \left(\frac{3n+2}{4}, S''\right) \tag{16.4}$$

$$S' = \frac{\sum_{t=1}^{n/2} S_t}{n/2}$$

$$S'' = \frac{\sum_{t=(n+2)/2}^{n} S_t}{n/2}$$

and we assume n is an even number. Note that this approach assumes uniform time intervals between the time-series observations. For an odd number of observations, the semiaverage trend-line estimate is the line passing through the two points

$$\left(\frac{n+1}{4}, S'\right) \quad \text{and} \quad \left(\frac{3(n+1)}{4}, S''\right) \tag{16.5}$$

in which

$$S' = \frac{\sum_{t=1}^{(n-1)/2} S_t}{(n-1)/2}$$

and

$$S'' = \frac{\sum_{t=(n+3)/2}^{n} S_t}{(n-1)/2}$$

The following two examples illustrate the use of the semiaverage approach.

EXAMPLE 16.1 Sales volumes over a ten-month period beginning in January and ending in October of the same year are 16, 14, 18, 20, 24, 22, 28, 32, 26, and 34 units, respectively. Use the method of semiaverages to estimate a linear trend component.

SOLUTION The ordered pairs of Equation 16.4 apply here, since there are an even number (10) of time-series observations. The calculations are as follows:

1. $\dfrac{n+2}{4} = \dfrac{10+2}{4} = 3$

2. $S' = \dfrac{\sum_{t=1}^{5} S_t}{5} = \dfrac{16 + 14 + 18 + 20 + 24}{5} = 18.4$

3. $\dfrac{3n+2}{4} = \dfrac{30+2}{4} = 8$

4. $S'' = \dfrac{\sum_{t=6}^{10} S_t}{5} = \dfrac{22 + 28 + 32 + 26 + 34}{5} = 28.4$

This means that the estimated trend line is determined by the two points

$$(3, 18.4) \qquad \text{and} \qquad (8, 28.4)$$

Applying traditional methods to these two points, we get the slope of the line:

$$b_1 = \frac{28.4 - 18.4}{8 - 3} = 2$$

and so

$$\frac{\hat{T}_t - 18.4}{t - 3} = 2$$

or

$$\hat{T}_t = 12.4 + 2t$$

EXAMPLE 16.2 Electricity consumption figures for North American Light and Power over the last seven weeks are 100.62, 89.51, 93.25, 86.37, 82.48, 84.73, and 83.61 units, respectively. Use the method of semiaverages to estimate a linear trend component.

SOLUTION The ordered pairs of Equation 16.5 apply here, since there are an odd number (7) of time-series observations. The calculations are as follows:

1. $\dfrac{n+1}{4} = \dfrac{7+1}{4} = 2$

2. $S' = \dfrac{\sum_{t=1}^{3} S_t}{3} = \dfrac{100.62 + 89.51 + 93.25}{3} = 94.46$

3. $\dfrac{3(n+1)}{4} = \dfrac{3(7+1)}{4} = 6$

4. $S'' = \dfrac{\sum_{t=5}^{7} S_t}{3} = \dfrac{82.48 + 84.73 + 83.61}{3} = 83.61$

This means that the estimated trend line is determined by the two points

$$(2, 94.46) \qquad \text{and} \qquad (6, 83.61)$$

Thus

$$b_1 = \frac{83.61 - 94.46}{6 - 2} = -2.7125$$

and so

$$\frac{\hat{T}_t - 94.46}{t - 2} = -2.7125$$

or

$$\hat{T}_t = 99.89 - 2.7125t$$

Notice that for an odd number of time-series values, the effect of the semiaverages formulas is to omit the middle time-series value from consideration.

Another popular method for estimating a linear trend component uses the techniques of simple linear regression of Section 14.2. Recall that this method chooses b_0 and b_1 in a manner that minimizes the total given by

$$\sum_{t=1}^{n} [S_t - (b_0 + b_1 t)]^2$$

In our case, assuming equal time intervals between the time-series observations, we have

$$b_1 = \left[\frac{6}{n-1}\right]\left[\frac{2\,\Sigma\, tS_t}{n(n+1)} - \overline{S}\right] \tag{16.6}$$

and

$$b_0 = \overline{S} - \frac{b_1(n+1)}{2} \tag{16.7}$$

in which

$$\overline{S} = \frac{\sum_{t=1}^{n} S_t}{n}$$

EXAMPLE 16.3 Use Equations 16.6 and 16.7 to find a linear trend estimate for the sales volume figures in Example 16.1.

SOLUTION We need $\sum_{t=1}^{n} tS_t$ and $\overline{S}$ in order to determine b_0 and b_1. First,

$$\sum_{t=1}^{10} tS_t = 1(16) + 2(14) + 3(18) + 4(20) + 5(24) + 6(22) + 7(28) + 8(32) + 9(26) + 10(34) = 1456$$

Next

$$\overline{S} = \frac{16 + 14 + 18 + 20 + 24 + 22 + 28 + 32 + 26 + 34}{10} = 23.4$$

Thus,

$$b_1 = \left[\frac{6}{10-1}\right]\left[\frac{2(1456)}{10(10+1)} - 23.4\right] = \left(\frac{6}{9}\right)(26.473 - 23.4) = 2.048$$

and

$$b_0 = 23.4 - 2.048\left(\frac{11}{2}\right) = 12.136$$

Thus, the least-squares linear trend estimate is given by

$$\hat{T}_t = 12.136 + 2.048t$$

EXAMPLE 16.4 Use Equations 16.6 and 16.7 to find a linear trend estimate for the electricity consumption figures in Example 16.2.

SOLUTION First,

$$\sum_{t=1}^{7} tS_t = 1(100.62) + 2(89.51) + 3(93.25) + 4(86.37) + 5(82.48) + 6(84.73) + 7(83.61) = 2410.92$$

Next,

$$\overline{S} = \frac{100.62 + 89.51 + 93.25 + 86.37 + 82.48 + 84.73 + 83.61}{7} = 88.65$$

Thus,

$$b_1 = \left[\frac{6}{7-1}\right]\left[\frac{2(2410.92)}{7(7+1)} - 88.65\right] = -2.55$$

and

$$b_0 = 88.65 - (-2.55)\left(\frac{8}{2}\right) = 98.85$$

Thus, the least-squares linear trend estimate is given by

$$\hat{T}_t = 98.85 - 2.55t$$

We can now introduce some flexibility into the approach by showing how to alter the linear trend equation estimate for either a shift in the location of the origin or a change in unit of measurement. To transform an equation to a system in which the origin is at (x, y) (assume that the new axes are parallel to the old), simply substitute t^* for t and $\hat{T}_{t*}$ for $\hat{T}_t$ in the equation, where

$$t^* = t - x$$

and

$$\hat{T}_{t*} = \hat{T}_t - y$$

Thus,

$$\hat{T}_t = b_0 + b_1 t \qquad \text{becomes} \qquad \hat{T}_{t*} = (b_1 x + b_0 - y) + b_1 t^* \tag{16.8}$$

We can extend this transformation by including the effects of a change in unit measurement. If r_1 is the scale change ratio for measurement of S_t and r_2 is the scale change ratio for measurement of time, then the new equation adjusted for scale changes is given by

$$\hat{T}_{t*} = r_1 b_0 + \left(\frac{r_1}{r_2}\right) b_1 t* \tag{16.9}$$

The changes given by Equations 16.8 and 16.9 can be combined to give the total picture as in Equation 16.10.

$$\hat{T}_{t*} = r_1(b_1 x + b_0 - y) + \left(\frac{r_1}{r_2}\right) b_1 t* \tag{16.10}$$

We should emphasize that both r_1 and r_2 are expressed as the reciprocal of the number of old units per new unit.

The next example illustrates the use of Equation 16.10.

EXAMPLE 16.5 The linear trend equation estimate for sales volume,

$$\hat{T}_t = 12.136 + 2.048t$$

is developed using the least-squares technique from the following information:

Date	Assigned Sales Volume in Units
January 15, 1984	16
February 15, 1984	14
March 15, 1984	18
April 15, 1984	20
May 15, 1984	24
June 15, 1984	22
July 15, 1984	28
August 15, 1984	32
September 15, 1984	26
October 15, 1984	34

The estimate is constructed with the origin at December 15, 1983, since Equations 16.6 and 16.7 assume that the first time-series observation is at $t = 1$. Also, the time unit is one month.

(*a*) Change the origin to the beginning of 1984.

(*b*) Starting with $\hat{T}_t = 12.136 + 2.048t$, change the origin to (0, 23.4), so that, since $\overline{S} = 23.4$, values are negative for volumes below the average and positive for volumes above the average.

(c) Starting with $\hat{T}_t = 12.136 + 2.048t$, change the time unit to three months.

(d) Assume that the original volumes represent the number of packages of product sold and that we want to change the values to represent the number of crates (20 packages per crate) sold. Starting with $\hat{T}_t = 12.136 + 2.048t$, change the sales volume unit so that $\hat{T}_{t*}$ is the trend estimate in crates sold.

(e) Use Equation 16.10 to do parts (a) through (d) simultaneously.

SOLUTION

(a) To change the origin to the beginning of 1984 would be to shift its location from (0, 0) to (1/2, 0), so that $x = 1/2$ and $y = 0$ in Equation 16.8. Thus, the new equation is

$$\begin{aligned}\hat{T}_{t*} &= (b_1x + b_0 - y) + b_1t^* \\ &= \left[2.048\left(\frac{1}{2}\right) + 12.136 - 0\right] + 2.048t^* = 13.16 + 2.048t^*\end{aligned}$$

(b) To change the origin from (0, 0) to (0, 23.4), we get $x = 0$ and $y = 23.4$, or

$$\begin{aligned}\hat{T}_t &= [2.048(0) + 12.136 - 23.4] + 2.048t^* \\ &= -11.264 + 2.048t^*\end{aligned}$$

(c) To change the time unit from one month to three months, we get $r_2 = 1/3$ and (since there is no change in sales volume units) $r_1 = 1$. Thus, Equation 16.9 gives

$$\hat{T}_{t*} = (1)(12.136) + \left(\frac{1}{1/3}\right)(2.048)t^* = 12.136 + 6.144t^*$$

(d) To change the sales volume unit to one 20 times as large as the current unit, we have $r_1 = 1/20$ and $r_2 = 1$. Using Equation 16.9, we get

$$\hat{T}_{t*} = \frac{1}{20} \times (12.136) + \left(\frac{1/20}{1}\right)(2.048)t^* = 0.6068 + 0.1024t^*$$

(e) If we execute all four of the changes simultaneously, then

$$r_1 = \frac{1}{20} \qquad r_2 = \frac{1}{3} \qquad x = \frac{1}{2} \qquad y = 23.4$$

so that Equation 16.10 gives

$$\begin{aligned}\hat{T}_{t*} &= (r_1)[b_1(x) + (b_0 - y)] + \left(\frac{r_1}{r_2}\right)b_1 t^* \\ &= \left(\frac{1}{20}\right)\left[2.048\left(\frac{1}{2}\right) + (12.136 - 23.4)\right] + \left(\frac{1/20}{1/3}\right)(2.048)t^* \\ &= -0.512 + 0.307t^*\end{aligned}$$

Part (d) of Example 16.5 gives a linear trend equation estimate that uses a quarterly sales volume figure. This sales volume figure is "normalized" as a function of the average sales volume over the ten-month period. Note that since analytical procedures vary considerably for this type of problem, we must report both the assumed origin and the units of measurement. If we do this, the trend estimates delivered are independent of the equation used (that is, independent of the original or a transformation of the original). The next example shows this independence for the results of Example 16.5.

EXAMPLE 16.6 Use the original and final trend equation estimates of Example 16.5 to predict a trend component value for November 15, 1984.

SOLUTION The equation

$$\hat{T}_t = 12.136 + 2.048t$$

has origin at December 15, 1983, and a time unit of one month, so that November 15, 1984 corresponds to $t = 11$. Thus, the trend component prediction for November 15, 1984, is

$$\hat{T}_t = 12.136 + 2.048(11) = 34.664$$

The equation

$$\hat{T}_{t*} = -0.512 + 0.307t^*$$

has origin at the beginning of 1984 with a quarterly time unit. Thus, November 15, 1984, corresponds to $t^* = 3.5$. So

$$\hat{T}_{t*} = -0.512 + 0.307(3.5) = 0.563$$

Note that the value 0.563 represents 0.563×20 of the original units above the original unit average of 23.4. That is,

$$0.563(20) + 23.4 = 34.664$$

The next step would be to show how nonlinear trend component estimates can be established. Those who wish to explore trends to this depth should proceed to the chapter appendix.

EXERCISES

16.3.1 A company accountant notices that overhead costs for the last nine weeks are 1213.43, 1246.72, 1257.69, 1296.35, 1311.21, 1423.92, 1442.12, 1463.51, 1517.01, respectively.

(*a*) Use the method of semiaverages to estimate a linear trend component.

(*b*) Use simple linear regression to estimate a linear trend component.

(*c*) Use your answers to parts (*a*) and (*b*) to get two different predictions for the trend component of overhead costs for the next week.

16.3.2 A homeowner keeps a strict accounting of oil fuel costs per gallon throughout the year and notices the following pattern.

Billing Date	Cost per Gallon in Dollars
January 15, 1983	0.56
February 15, 1983	0.58
March 15, 1983	0.58
April 15, 1983	0.59
May 15, 1983	0.59
June 15, 1983	0.59
July 15, 1983	0.60
August 15, 1983	0.63
September 15, 1983	0.65
October 15, 1983	0.65
November 15, 1983	0.66
December 15, 1983	0.66

(*a*) Use the method of semiaverages to estimate a linear trend component.

(*b*) Use simple linear regression to estimate a linear trend component.

(*c*) Use your answer to parts (*a*) and (*b*) to get two different predictions for the trend component of the cost per gallon in dollars for January 15, 1984.

16.3.3 The oil company of Exercise 16.3.2 announces plans to change the billing date to the first day of each month. Use Equation 16.8 to alter your answer to Exercise 16.3.2*b* so that the new origin is January 1, 1984.

16.3.4 Refer to Exercise 16.3.2. Use Equation 16.10 to alter your answer to Exercise 16.3.2*b* so that the new origin is July 1, 1983, and equate the cost per gallon in dollars of January 15, 1983, to zero. Further, change the cost measurement from dollars to cents. Use the converted linear trend component estimate to predict the trend component value (in number of cents over the base cost) for January 15, 1984. Show that this calculation is equivalent to your answer to Exercise 16.3.2*c*.

16.4 THE SEASONAL COMPONENT

There are many practical examples that illustrate how seasonal component effects as well as trend component effects determine time-series observations. Power companies are well aware of how electrical consumption provides such an example. In this case, there may be a general increasing trend in overall consumption, and each day there are surges of consumption during the morning and then later during the late afternoon and early evening hours. This latter phenomenon is an expression of what we call the *seasonal* component of the time series of consumption values. In general, the seasonal component accounts for variations due to repeating patterns, whether the patterns are repeated on a yearly, monthly, weekly, daily, or hourly basis. The method we give for isolating the seasonal component effect can be applied regardless of the time frame over which the pattern is repeated.

One of the most popular techniques for estimating the seasonal component is called the *ratio-to-moving-average method*. The method provides two formulas, and the formula chosen depends on the number of time-series observations on the time interval over which the pattern is established. If the pattern is weekly and S_t is tabulated each day, then an odd number of observations (7) are used in the calculation of the moving average. In such a case (when n is *odd*), the seasonal component estimate is given by

$$\hat{SE}_t = \frac{nS_t}{\sum_{k=-(n-1)/2}^{(n-1)/2} S_{t+k}} \qquad (16.11)$$

for each value of $t \geqslant (n + 1)/2$.

If the pattern is yearly and S_t is tabulated each month, then an even number of observations (12) are used in the calculation of the moving average. In such a case (when n is *even*), the seasonal component is given by

$$\hat{SE}_t = \frac{S_t}{(m_1 + m_2)/2} \tag{16.12}$$

in which

$$m_1 = \frac{\sum_{k=-n/2}^{(n-2)/2} S_{t+k}}{n} \quad \text{and} \quad m_2 = \frac{\sum_{k=-n/2}^{(n-2)/2} S_{(t+1)+k}}{n}$$

for each value of $t \geqslant (n + 2)/2$.

Before we proceed with the next two examples, it should be understood that Equations 16.11 and 16.12 are used in connection with the multiplicative model given in Equation 16.1. Further, the seasonal component estimate is usually *reported* after multiplication by 100.

EXAMPLE 16.7 A route popular among commuters in a metropolitan area is monitored between the hours of 7 and 8 A.M., with the number of vehicles passing a certain point tabulated on a daily basis. The following table gives the results of this tabulation over a 21-day period. Establish the seasonal component estimate for each day of the week using Equation 16.11.

Day	Number of Vehicles	Day	Number of Vehicles
T	543	Sa	87
W	562	S	15
Th	461	M	857
F	382	T	572
Sa	75	W	578
S	12	Th	501
M	867	F	396
T	593	Sa	104
W	566	S	24
Th	473	M	891
F	401		

SOLUTION Since $n = 7$, Equation 16.11 becomes

$$\hat{SE}_t = \frac{7S_t}{\sum_{k=-3}^{3} S_{t+k}}$$

for $t \geqslant 4$. Thus,

$$\hat{SE}_4 = \frac{7(382)}{543 + 562 + 461 + 382 + 75 + 12 + 867} = 0.9214$$

Notice that we can calculate $\hat{SE}_t$ for $4 \leqslant t \leqslant 18$; we will do so using a tabular format as given here.

Day	t	S_t	$\sum_{k=-3}^{3} S_{t+k}$	$\hat{SE}_t = \frac{7S_t}{\sum_{k=3}^{3} S_{t+k}}$
T	1	543	—	—
W	2	562	—	—
Th	3	461	—	—
F	4	382	2902	0.9214
Sa	5	75	2952	0.1778
S	6	12	2956	0.0284
M	7	867	2968	2.0448
T	8	593	2987	1.3897
W	9	566	2999	1.3211
Th	10	473	3002	1.1029
F	11	401	2992	0.9382
Sa	12	87	2971	0.2050
S	13	15	2983	0.0352
M	14	857	3011	1.9924
T	15	572	3006	1.3320
W	16	578	3023	1.3384
Th	17	501	3032	1.1567
F	18	396	3066	0.9041
Sa	19	104	—	—
S	20	24	—	—
M	21	891	—	—

Notice that each day of the week has at least two seasonal component estimates in the fourth column of the table. We can derive a better estimate of the seasonal component for any given day of the week by averaging its estimates, which appear in the fourth column. This is done in tabular format here.

Day	SE_t Estimates	Average of SE_t Estimates
M	2.0448, 1.9924	2.0186
T	1.3897, 1.3320	1.3609
W	1.3211, 1.3384	1.3298
Th	1.1029, 1.1567	1.1298
F	0.9214, 0.9382, 0.9041	0.9212
Sa	0.1778, 0.2050	0.1914
S	0.0284, 0.0352	0.0318

Before we proceed with the next example, we should note that the final SE_t estimates, *when reported to management*, are multiplied by a factor of 100. We also may need to adjust the estimates so that the "Average of SE_t Estimates" column totals to n. Sometimes this adjustment is necessary because of roundoff error. However, it *should* be made so that the data conform to our expectation that the seasonal effects cancel each other out over the pattern period. The following table shows how to make this adjustment to get the final result.

Day	Unadjusted Average of SE_t Estimates	Adjusted Average of SE_t Estimates
M	2.0186	2.0234
T	1.3609	1.3641
W	1.3298	1.3329
Th	1.1298	1.1325
F	0.9212	0.9234
Sa	0.1914	0.1918
S	0.0318	0.0319
Totals	6.9835	7.0000

We derive the values in the third column by multiplying each value in the second column by the ratio 7.0000/6.9835. In general, the procedure is to multiply by the ratio

$$\frac{n}{\text{UAT}} \tag{16.13}$$

in which UAT represents the total of the "Unadjusted Average of SE_t Estimates" column.

EXAMPLE 16.8 Electricity consumption rates for homeowners in a small North American city are monitored and recorded over a two-day period, with the results shown. Calculatc a scasonal componcnt estimate for each hour of the day, using the 24 consumption rate values.

SOLUTION In this example $n = 24$, so

$$\hat{SE}_t = \frac{S_t}{(m_1 + m_2)/2}$$

in which

$$m_1 = \frac{\sum_{k=-12}^{11} S_{t+k}}{24} \qquad m_2 = \frac{\sum_{k=-12}^{11} S_{(t+1)+k}}{24} \qquad t \geqslant 13$$

Thus, we have the table on page 675. We compute

$$\hat{SE}_{15} = \frac{S_{15}}{(m_1 + m_2)/2} = \frac{66}{[(2296/24) + (2300/24)]/2}$$
$$= \frac{66}{95.75} = 0.6893$$

and for the seventh-hour estimate we compute

$$\hat{SE}_{31} = \frac{132}{[(2287/24) + (2291/24)]/2} = \frac{132}{95.38} = 1.3839$$

As this example illustrates, the moving average will be "off center" (centered midway *between* observation time values) for an even number of observations per period, so that we must average the two closest moving averages at each point before we can compute the ratio. The last column is derived by multiplying the SE_t column by Equation 16.13 (which in this case is 24.0000/24.0850) so that the seasonal effects balance out to give a total of 24.0000 over a 24-hour period.

Hour	Consumption Rate in Standard Units, Day 1	Consumption Rate in Standard Units, Day 2	Moving Totals, Day 1	Moving Totals, Day 2	$(m_1 + m_2)/2$, Day 1	$(m_1 + m_2)/2$, Day 2	$\hat{SE}_t$	Adjusted $\hat{SE}_t$
1	41	43				96.08	0.4475	0.4459
				2304				
2	39	36				96.04	0.3748	0.3735
				2306				
3	35	39				95.98	0.4063	0.4049
				2301				
4	33	30				95.88	0.3129	0.3118
				2301				
5	37	35				95.73	0.3656	0.3643
				2294				
6	84	89				95.44	0.9325	0.9292
				2287				
7	125	132				95.38	1.3839	1.3790
				2291				
8	91	96				95.40	1.0063	1.0027
				2288				
9	114	115				95.33	1.2063	1.2020
				2288				
10	90	97				95.33	1.0175	1.0139
				2288				
11	110	109				95.23	1.1446	1.1406
				2283				
12	112	101				95.17	1.0613	1.0576
			2297	2285				
13	86	82			95.75		0.8982	0.8950
			2299					
14	77	79			95.73		0.8043	0.8015
			2296					
15	66	61			95.75		0.6893	0.6869
			2300					
16	114	114			95.77		1.1904	1.1862
			2297					
17	137	130			95.67		1.4320	1.4269
			2295					
18	199	192			95.73		2.0788	2.0715
			2300					
19	181	185			95.98		1.8754	1.8688
			2307					
20	173	170			96.23		1.7978	1.7915
			2312					
21	132	132			96.35		1.3700	1.3651
			2313					
22	91	91			96.52		0.9428	0.9395
			2320					
23	76	71			96.65		0.7863	0.7835
			2319					
24	54	56			96.40		0.5602	0.5582
			2308					
						Totals	24.0850	24.0000

We would like to emphasize some fundamental characteristics of this method that may not be apparent. First, the computation of the moving average eliminates the seasonal effect but retains the trend effect. Thus, when we calculate the ratio in Equation 16.11 (assuming a multiplicative model), we get an estimate of the seasonal effect. This estimate will include some cyclical and irregular effects, but we try to minimize them by averaging a sequence of SE_t estimates for each time period t. Thus, because of the averaging process, the seasonal effect estimates in Example 16.7 are less likely to contain cyclical and irregular effects that those in Example 16.8.

EXERCISES

16.4.1 A fast food restaurant in a popular shopping center has seen the following number of orders on a daily basis over a four-week period. Establish the seasonal component estimate for each day of the week using Equation 16.11.

	Number of Orders			
	Week			
Day	1	2	3	4
M	103	190	130	128
T	131	172	179	119
W	362	236	337	299
Th	437	382	496	404
F	649	675	692	654
Sa	870	876	840	861
S	596	544	561	546

16.4.2 A financial analyst compiles a composite index representing the prime rate offered across six major financial institutions in the immediate area. The results of this tabulation on a quarterly basis over an eight-year period follow. Establish a seasonal component estimate for each quarter using Equation 16.12.

Year	Composite Index Value Quarter 1	2	3	4
1	4.8	4.2	4.6	4.6
2	5.1	4.8	5.4	5.5
3	6.0	5.8	5.9	6.1
4	6.5	6.4	6.5	6.4
5	7.2	6.8	7.1	7.1
6	7.6	7.1	7.4	7.7
7	8.0	7.9	8.1	8.2
8	8.6	8.3	8.5	8.4

16.4.3 A marketing group for a radio station is trying to decide on an advertising package to be offered to local businesses. They plan to inspect their own average rating index over three time intervals per day on an 18-day period and incorporate the information as a part of their decision making. The results of their data collection follow. Establish a seasonal component estimate for each time period using Equation 16.11.

Day	Average Rating Time Interval: 6 A.M. to 8 A.M.	11 A.M. to 1 P.M.	5 P.M. to 7 P.M.
1	90	24	50
2	99	39	69
3	83	29	67
4	83	37	60
5	81	31	52
6	81	31	68
7	80	20	65
8	97	26	66
9	89	34	65
10	82	36	60
11	85	36	62
12	90	38	64
13	98	39	61
14	88	31	58
15	85	34	51
16	83	27	52
17	109	24	68
18	96	20	50

16.5 THE CYCLICAL AND IRREGULAR COMPONENTS

The basic assumptions used in the last section may not apply when we are considering techniques for isolating the cyclical component. For instance, in the seasonal analysis we assume that we are able to determine the period. Such an assumption is not unreasonable, since there are many weekly, daily, monthly, or hourly patterns that are universally assumed as a part of normal business planning. Also, the statement "the period" carries an implication of stability (the period remains the same throughout the planning horizon). With cyclical effects we often have periods longer than the seasonal period that are much less stable in length and "less predictable." This last phrase, "less predictable," can also be explained in terms of the seasonal material of the last section. In the seasonal isolation, we made the assumption that the seasonal effect remains stable over time. That is, when we see a variation in the same problem (that is, when two estimates for SE_t are not the same), we attribute the variation to the cyclical and irregular component effects and average the results for each time period to help minimize the extraneous variation. This is an assumption of amplitude stability, and it means that the effect for any given period remains the same throughout the planning period. We cannot depend on this assumption for the cyclical component, and thus we are faced with no regularity in either period or amplitude when we try to isolate their effects. This is not to say that we cannot effectively discover the character of past cyclical component effects (especially when we use the multiplicative model), but using these results to forecast or predict values is often futile.

The irregular component introduces the variability that makes it impossible to predict the exact time-series value for a future time period and so acknowledges that a forecast is the least we can do for planning purposes. In the previous chapter, the prediction can never be precisely correct, since it is actually a point estimate for a future value. In the same sense, the forecast allows the counterpart of an interval estimate for a future time-series value. The best we can do in isolating the irregular component is to try to establish a probability distribution that governs its effect on the time series and use the distribution to produce a forecast in the original time series.

We will not discuss typical methods of isolating either the cyclical or irregular components. The methods must often employ advanced techniques to produce valuable results (like using a sinusoidal model for the cyclical effect), and we consider them beyond the intended scope of this book.

16.6 PREDICTING USING TIME-SERIES ANALYSIS

The objective of this section is to show how the multiplicative model,

$$S_t = T_t \times SE_t \times C_t \times T_t$$

can be used to predict a future time-series value. In this process (since it is not a forecast) we let $I_t = 1$ and for simplification (an understatement if one refers to the last section) let $C_t = 1$. Thus, our formula for prediction is given by the equation

$$\hat{S}_t = \hat{T}_t \times \hat{SE}_t \tag{16.14}$$

So we predict a value for the future by isolating the trend and seasonal components in the time series and then multiplying the results. Note that this is a simplification, since we are ignoring the cyclical effect. Further, we are predicting a value for a time period that lies *outside* the original time span over which the components were isolated. This extrapolation process contains inherent dangers, as was explained in Chapter 14. With these warnings in hand, we can now show how to use Equation 16.14 to predict a time-series value.

EXAMPLE 16.9 Use the information in Example 16.7 to predict a time-series value for Monday of the next week.

SOLUTION To use Equation 16.14 we need estimates of both trend and seasonal components. From Example 16.7, the seasonal component estimate for a Monday is 202.34, which becomes 2.0234 for use in the prediction equation. We also need to calculate the trend component, and we refer to Equations 16.6 and 16.7 for this purpose. First,

$$\sum_{t=1}^{21} tS_t = 97{,}950 \qquad \text{and} \qquad \overline{S} = \frac{8960}{21} = 426.67$$

So

$$b_1 = \left(\frac{6}{21-1}\right)\left[\frac{2(97{,}950)}{21(22)} - 426.67\right] = -0.7932$$

and

$$b_0 = 426.67 + 0.7932\left(\frac{21+1}{2}\right) = 435.39$$

Then the trend component estimate is given by

$$\hat{T}_t = b_0 + b_1 t = 435.39 - 0.7932t$$

and Equation 16.14 becomes

$$\hat{S}_{28} = \hat{T}_{28} \times \hat{SE}_{28} = [435.39 - 0.7932(28)] \times (2.0234)$$
$$= (413.1804)(2.0234) = 836.03$$

EXERCISES

16.6.1 Sales volumes over a ten-month period beginning in January and ending in October of the same year are 16, 14, 18, 20, 24, 22, 28, 32, 26, and 34 units, respectively. Use this information to predict the sales volume for March of the next year (refer to Example 16.3).

16.6.2 Electricity consumption figures for North American Light and Power over the last seven days (beginning on a Saturday) are 100.62, 89.51, 93.25, 86.37, 82.48, 84.73, and 83.61. Predict the electric consumption for Monday of the next week.

16.6.3 Refer to Example 16.8. Predict the consumption rate for the seventh hour of the following day.

16.6.4 Discuss the quality of the prediction in Exercise 16.6.3 compared with the quality of the prediction in Exercise 16.6.2.

16.6.5 Refer to Exercise 16.4.3. Predict the average rating for the period between 11 A.M. and 1 P.M. on day 20.

16.7 SUPPLEMENTARY EXAMPLES

EXAMPLE 16.10 The management of a retail outlet is trying to iron out cash flow problems. They decide to use some time-series analysis to help determine the current financial stability for the outlet, then use the analysis to predict sales for the next quarter. A review of the past 20 quarters by company accountants gave the sales volumes as shown in the table on the following page. Use this information to predict the sales volume for the second quarter of the sixth year.

Year	Sales Volume in Units of \$10,000 Quarter 1	2	3	4
1	—	17	25	74
2	15	24	35	80
3	18	25	37	99
4	26	36	44	114
5	22	35	45	111
6	36	—	—	—

SOLUTION The following table will be useful in determining the trend and seasonal component estimates.

Quarter	t	S_t	Moving Totals	$\frac{m_1 + m_2}{2}$	$\hat{SE}_t$
2	1	17			
3	2	25			
			131		
4	3	74		33.625	2.2007
			138		
1	4	15		35.750	0.4196
			148		
2	5	24		37.750	0.6358
			154		
3	6	35		38.875	0.9003
			157		
4	7	80		39.375	2.0317
			158		
1	8	18		39.750	0.4528
			160		
2	9	25		42.375	0.5900
			179		
3	10	37		45.750	0.8087
			187		
4	11	99		48.125	2.0571
			198		
1	12	26		50.375	0.5161
			205		
2	13	36		53.125	0.6776
			220		
3	14	44		54.500	0.8073
			216		
4	15	114		53.875	2.1160
			215		
1	16	22		53.875	0.4083
			216		
2	17	35		53.625	0.6527
			213		
3	18	45		55.000	0.8182
			227		
4	19	111			
1	20	36			

In this problem $n = 4$, so

$$m_1 = \sum_{k=-2}^{1} \frac{S_{t+k}}{4} = \frac{S_{t-2} + S_{t-1} + S_t + S_{t+1}}{4}$$

and

$$m_2 = \frac{S_{t-1} + S_t + S_{t+1} + S_{t+2}}{4}$$

for $t \geqslant 3$. For instance,

$$\hat{SE}_5 = \frac{24}{[(148/4) + (154/4)]/2} = \frac{24}{37.750} = 0.6358$$

We now average the estimate for $\hat{SE}_t$ across quarters to minimize the effect of cyclical and irregular variation.

Quarter	$\hat{SE}_t$ Values	Average of $\hat{SE}_t$ Values	Adjusted SE_t Estimate
1	0.4196, 0.4528, 0.5161, 0.4083	0.4492	0.4466
2	0.6358, 0.5900, 0.6776, 0.6527	0.6390	0.6353
3	0.9003, 0.8087, 0.8073, 0.8182	0.8336	0.8288
4	2.2007, 2.0317, 2.0571, 2.1160	2.1014	2.0893
	Totals	4.0232	4.0000

The last column in the table uses Equation 16.13 (which in this case is 4.0000/4.0232) as a multiplier on each value in the next to last column. Thus the seasonal component estimate appropriate for the second quarter of the sixth year would be 0.6353, or $\hat{SE}_{21} = 0.6353$.

Now we need to use Equations 16.6 and 16.7 to find $\hat{T}_{21}$. First,

$$\Sigma\, ts_t = 10{,}759 \qquad \text{and} \qquad \overline{S} = 45.9$$

So

$$b_1 = \left(\frac{6}{20-1}\right)\left[\frac{2(10{,}759)}{20(21)} - 45.9\right] = 1.6842$$

and

$$b_2 = 45.9 - 1.6842\left(\frac{20+1}{2}\right) = 28.2158$$

Thus

$$\hat{T}_{21} = b_0 + b_1(21) = 28.2158 + 1.6842(21) = 63.5840$$

This means that the prediction for the second quarter of the sixth year is given by

$$\hat{S}_{21} = \hat{T}_{21} \times \hat{SE}_{21} = 63.5840 \times 0.6353 = 40.395$$

EXAMPLE 16.11 The manager of a fast food restaurant is trying to decide on a good scheduling routine for his employees that will hold customer service times at a reasonable level. He also is interested in the trend effects so that he can adjust his cash flow to prepare for either hiring or firing of workers in the future. You decide to look at the number of orders for each meal type (breakfast, lunch, and dinner) over the past 20 days, beginning January 10 with breakfast and ending January 29 with dinner. Your idea is to construct a formula for predicting the number of orders to expect on any given day and meal type. Since you know the manager has a calendar with the days of the year numbered from 1 to 365, you decide to construct the formula so that he can insert the number for the day of the year as an argument. Construct the formula appropriate for the following data.

Number of Orders

	Meal Type				Meal Type		
Day	B	L	D	Day	B	L	D
1	13	79	40	11	18	79	62
2	11	76	47	12	20	91	52
3	17	72	38	13	29	88	57
4	16	80	49	14	26	88	61
5	12	89	47	15	28	92	62
6	18	85	40	16	25	97	61
7	20	87	39	17	29	99	57
8	19	82	49	18	24	92	57
9	17	82	48	19	22	99	61
10	14	80	54	20	26	96	64

SOLUTION In this example $n = 3$, and so we use Equation 16.11 to obtain the seasonal estimates. Thus

$$\hat{SE}_t = \frac{3S_t}{\sum_{k=-1}^{1} S_{t+k}}$$

and we form the following table.

t			S_t			$S_{t-1} + S_t + S_{t+1}$		
1	2	3	13	79	40	—	132	130
4	5	6	11	76	47	127	134	140
7	8	9	17	72	38	136	127	126
10	11	12	16	80	49	134	145	141
13	14	15	12	89	47	150	148	154
16	17	18	18	85	40	150	143	145
19	20	21	20	87	39	147	146	145
22	23	24	19	82	49	140	150	148
25	26	27	17	82	48	148	147	144
28	29	30	14	80	54	142	148	152
31	32	33	18	79	62	151	159	161
34	35	36	20	91	52	173	163	172
37	38	39	29	88	57	169	174	171
40	41	42	26	88	61	171	175	177
43	44	45	28	92	62	181	182	179
46	47	48	25	97	61	184	183	187
49	50	51	29	99	57	189	185	180
52	53	54	24	92	57	173	173	171
55	56	57	22	99	61	178	182	186
58	59	60	26	96	64	183	186	—

The averages of the SE_t estimates for breakfast, lunch, and dinner are 0.3826, 1.6421 and 0.9755, respectively. Since their total is 3.0002, we multiply by the ratio 3.0000/3.0002 to get 0.3826, 1.6420, and 0.9754 for the adjusted seasonal component estimates.

Next we use Equations 16.6 and 16.7 to find the trend component estimate and find

$$\Sigma\, tS_t = 103{,}947 \qquad \text{and} \qquad \bar{S} = 53.0\bar{3}$$

Thus

$$b_1 = \left(\frac{6}{60 - 1}\right)\left[\frac{2(103{,}947)}{60(61)} - 53.0\overline{3}\right] = 0.3832$$

and

$$b_0 = 53.0\overline{3} - \frac{0.3832(60 + 1)}{2} = 41.3452$$

This means the trend component estimate is given by

$$\hat{T}_t = 41.3452 + 0.3832t$$

We now need to adjust this trend component estimate so that the day of the year (1 to 365) can be used as an argument. The data now begin on January 10 and we have three observations per day. Thus, we need to change the origin to the beginning of the year and the time scale to days from its current rate of three units per day. Looking at Equation 16.10 we have

$$r_1 = 1 \qquad r_2 = \frac{1}{3} \qquad x = -26 \qquad y = 0$$

so

$$\begin{aligned}\hat{T}_{t*} &= [(0.3832)(-26) + 41.3452] + 3(0.3832)t^* \\ &= 31.3820 + 1.1496t^*\end{aligned}$$

This means that the prediction for the number of orders for breakfast on day k of the year is given by

$$\hat{S}_k = (31.3820 + 1.1496k)(0.3826)$$

that for lunch on day k of the year is given by

$$\hat{S}_{k+(1/3)} = \left[31.3820 + 1.1496\left(k + \frac{1}{3}\right)\right](1.6420)$$

and that for dinner on day k of the year is given by

$$\hat{S}_{k+(2/3)} = \left[31.3820 + 1.1496\left(k + \frac{2}{3}\right)\right](0.9754)$$

In simplified form these predictions are

Breakfast: $0.4398k + 12.0068$
Lunch: $1.8876k + 51.5292$
Dinner: $1.1213k + 30.6100$

16.8 SUMMARY

Basic time-series analysis involves the assumption of one of the two classical formulations for a time series. These are called the multiplicative and additive models. This chapter uses the multiplicative model because it is easy to manipulate. The two models characterize a time series as a function of four variables: trend, seasonal, cyclical, and irregular. Techniques for isolating the contribution of a linear trend include graphical, semiaverage and least-squares. The chapter appendix discusses nonlinear trend assumptions. The method of "ratio to moving average" is effective for isolating the seasonal contribution, especially if a number of seasonal estimates for the same season are averaged to get the final estimate for each season. The averaging works to minimize the influence of the cyclical and irregular variation contribution, so that the final estimate for the seasonal effect is as pure as possible. Techniques for isolating the contribution of the cyclical and irregular components have been omitted because of their complex character and the scope of this text. This omission forces us to predict (producing a point estimate for) a future value rather than forecasting a future value (a forecast is a list of the possible time-series values together with their likelihood of occurrence). Finally, the process of isolating the seasonal effect for a set of time-series values reminds us that many time series are published in what is called a *seasonally adjusted* form, which simply means that the seasonal effect has been removed. In the multiplicative model this is accomplished by simple division by the appropriate seasonal index. Refer to the exercises for some practice in this method.

Time series often are expressed as a sequence of composite indices that are meant to reflect the basic characteristics of a marketplace, financial climate, or any collection of heterogeneous composition that can be measured in some fashion. Since an objective of such an analysis is a prediction for the future, one wants the underlying data to be of "good quality." In many cases, the decision maker can evaluate the quality if the "type" of index being used in the time series can be established. Thus, it is important to examine the different methods of constructing such composite indices. The next chapter will provide some background in this area.

REVIEW EXERCISES

Level One

16.1 The marketing manager for a restaurant in a resort area tabulates a composite index that estimates the occupancy rate for hotels and motels in the area. The index values over the last three years are given here. Use the graphical method to determine a linear trend component estimate.

Month	Occupancy Rate (in %) Year		
	1981	1982	1983
January	53	50	58
February	74	74	79
March	39	42	37
April	12	15	18
May	14	18	16
June	31	34	38
July	69	66	75
August	86	83	94
September	21	24	27
October	51	50	53
November	46	45	49
December	67	68	65

16.2 Use the method of semiaverages to determine a trend component estimate for the data in Exercise 16.1.

16.3 Use the least-squares method to determine a linear trend component estimate for the time series given in Exercise 16.1.

16.4 Revise your answer to Exercise 16.3 so that time is measured in years instead of months and the origin is April 1982.

16.5 Find the seasonal component estimates for the data in Exercise 16.1.

16.6 Refer to Exercise 16.1. Predict an occupancy rate for August of 1984 using your answers to Exercises 16.3 and 16.5.

16.7 An accountant is working closely with the comptroller of a stock exchange firm to analyze the trend in company phone bills over the last five years. The company is billed quarterly, and the time series is given by the following data. Use the graphical method to determine a linear trend component estimate.

Year	Phone Bills in Dollars Quarter			
	1	2	3	4
1979	298.20	345.37	379.49	610.23
1980	344.05	361.52	391.17	664.99
1981	374.49	383.29	428.65	709.59
1982	389.35	419.61	439.72	744.26
1983	417.44	446.57	469.49	819.73

16.8 Use the method of semiaverages to determine a trend component estimate for the data in Exercise 16.7.

16.9 Use the least-squares method to determine a linear trend component estimate for the time series given in Exercise 16.7.

16.10 Find the seasonal component estimates for the data in Exercise 16.7.

16.11 Refer to Exercise 16.7. Predict the phone bill for the second quarter of 1985.

16.12 The unemployment rate for a major metropolitan area for July is reported as 4.72%. The report also states that the figure is seasonally adjusted, meaning that the seasonal effect has been removed from the actual unemployment figure. If the seasonal index for July is 1.237, what was the actual unemployment rate for the area (assume the multiplicative time-series model)?

16.13 A ferry system operating in the twin-city area carries many commuters and shoppers from one city across the harbor to the other city during three operation time intervals each day. The following figures represent the number of users over the last 18 days. Determine a trend component estimate for this information using the method of least squares.

	Number of Users Operation Time Interval		
Day	6 A.M. to 8 A.M.	11 A.M. to 1 P.M.	4 P.M. to 6 P.M.
1	610	145	607
2	666	110	674
3	665	165	664
4	644	159	646
5	664	155	656
6	695	132	700
7	724	160	723
8	783	173	784
9	748	182	755
10	745	283	750
11	798	276	798
12	656	276	647
13	717	215	723
14	714	289	713
15	756	225	753
16	729	244	723
17	763	250	763
18	784	286	785

16.14 Calculate the seasonal component estimates for the data in Exercise 16.13.

16.15 Use your answers to Exercises 16.13 and 16.14 to predict the number of users between 11 A.M. and 1 P.M. of the next day.

Level Two

Another method for isolating the seasonal component effect is called the *ratio to trend method.* In this method the estimate is given by

$$\hat{SE}_t = \frac{S_t}{\hat{T}_t}$$

and we average the seasonal estimates for each season in order to minimize the contribution of the cyclical and irregular components.

16.16 Use the ratio-to-trend method to estimate the seasonal components for the time series given in Exercise 16.1.

16.17 Use the ratio-to-trend method to estimate the seasonal components for the time series given in Exercise 16.7.

16.18 Use the ratio-to-trend method to estimate the seasonal components for the time series given in Exercise 16.13.

16.19 A plot of the time-series data may indicate nonlinearity in the trend component. One way to handle this is to use the techniques of multiple linear regression given in the last chapter. Use this technique to fit a parabolic trend to the following data: 4.00, 4.37, 5.86, 3.21, 4.24, 5.79, 3.14, 4.10, 5.57, 3.09, 5.34, 3.05, 3.83, 5.02, 3.01, 3.71, 4.94, 2.98, 3.60, 4.79, 2.96, 3.50, 4.65, 3.40, 4.51.

Level Three

16.20 Refer to Exercise 16.1. Use the *additive* time-series model to predict an occupancy rate for August of 1984. (Note that your answer should be the same as that for Exercise 16.6.)

16.21 Refer to Exercise 16.7. Use the *additive* time-series model to predict the phone bill for the second quarter of 1985. (Note that your answer should be the same as that for Exercise 16.11.)

16.22 Refer to Exercise 16.13. Use the *additive* time series model to predict the number of users between 11 A.M. and 1 P.M. of the next day. (Note that your answer should be the same as that for Exercise 16.15.)

16.23 Refer to the data given in Exercise 16.13. Assume that the combined effect of the cyclical and irregular components can be adequately approximated using a normal distribution with expected value zero. Use an additive model to produce a forecast for the number of users between 11 A.M. and 1 P.M. of the next day, then use this forecast to produce a 95% confidence interval for the number of users between 11 A.M. and 1 P.M. of the next day.

16.9 APPENDIX: NONLINEAR TRENDS

Time-series values examined over short time intervals can lend themselves nicely to linear trend component estimates. However, more detailed analysis over larger time intervals often requires additional methods that can free us from the assumption of a linear trend character. The following three subsections present some common methods for handling situations in which the trend component is assumed to be nonlinear.

16.9.1 Curvilinear Regression

Curvilinear regression techniques can be used whenever an order-preserving transformation of the time-series values will give a linear function whose coefficients can be estimated using standard linear regression techniques. The assumption of geometric growth in the trend component is a good example. In this case the assumption is

$$\hat{T}_t = c(d^t) \tag{16.15}$$

which, under the order-preserving natural logarithm transformation, becomes

$$Y = \ln \hat{T}_t = \ln c + (\ln d)t$$

or

$$Y = \ln c + (\ln d)t \tag{16.16}$$

EXAMPLE 16.12 The number of paperback books published in a nameless country over the last 20 years, in four-year intervals, is given by the time series 5123, 8986, 21,616, 38,125, and 76,437. Use curvilinear regression techniques to get a geometric trend component estimate as given by Equation 16.15.

SOLUTION The solution is easy to follow if we use a tabular representation.

t	S_t	$S_t^* = \ln S_t$
1	5,123	8.541
2	8,986	9.103
3	21,616	9.981
4	38,125	10.549
5	76,437	11.244

Using the linear regression formulas 16.6 and 16.7, we have

$$\ln d = \left(\frac{6}{4}\right)\left(\frac{2\ \Sigma\ tS_t^*}{30} - 9.8836\right) = (1.5)\left[\frac{2(155.106)}{30} - 9.8836\right]$$
$$= 0.6852$$

and

$$\ln c = 9.8836 - 0.6852\left(\frac{6}{2}\right) = 7.828$$

thus,

$$d = e^{\ln d} = e^{0.6852} = 1.984$$

and

$$c = e^{\ln c} = e^{7.828} = 2509.903$$

This means the trend estimate is given by

$$\hat{T}_t = 2509.903(1.984)^t$$

The advanced exercise section of this chapter gives additional situations in which the curvilinear technique could be used to advantage.

16.9.2 Weighted Moving Averages

Weighted moving averages provide a wide variety of techniques for situations in which the linear assumption on the trend component is unacceptable. The general model for the method is given by

$$\hat{T}_t = w_1 S_{t-1} + w_2 S_{t-2} + \cdots + w_m S_{t-m} \qquad (16.17)$$

in which we assume

1. $m \geqslant 2$
2. $0 \leqslant w_t \leqslant 1 \qquad t = 1, 2, \ldots, m$
3. $\sum_{t=1}^{m} w_t = 1$

The simplest such model is called the moving average for which $w_t = 1/m, t = 1, 2, \ldots, m$. Thus, one predicts the trend component for the next period by using the average of the last m time-series values.

EXAMPLE 16.13 Use the moving average method, with $m = 3$, to predict a trend component value for the next period for the time series 16, 14, 18, 20, 24, 22, 28, 32, 26, and 34 (previously given in Example 16.1).

SOLUTION In this case $m = 3$, so that $w_1 = w_2 = w_3 = 1/3$ and

$$\hat{T}_{11} = w_1 S_{10} + w_2 S_9 + w_3 S_8 = \left(\frac{1}{3}\right)(34) + \left(\frac{1}{3}\right)(26) + \left(\frac{1}{3}\right)(32)$$
$$= 30.6$$

The moving average gives equal weight to the last m time series of observations. Thus it is slow to react to quick changes in the values of the trend component. If we classify this as a conservative characteristic, then perhaps it is desirable to stay "conservative" and continually lag the true component value. However, we can produce a more responsive technique by giving more weight to the most recent events than to the others. We can produce this weight differential by subjective means or by more sophisticated analytical procedures. One example of a weight sequence is the sum-of-the-years-digits method, which gets its name from the familiar depreciation routine used in accounting. In this method,

$$w_i = \frac{(m+1) - i}{m(m+1)/2} \qquad i = 1, 2, \ldots, m \tag{16.18}$$

EXAMPLE 16.14 Use the sum-of-the-years digits method to determine the weights in the weighted moving average technique (use $m = 3$), and use these weights to predict a trend component value for the next period for the time series given in Example 16.13.

SOLUTION First, $m(m + 1)/2 = 3(4)/2 = 6$ (this is actually the sum $\sum_{i=1}^{m} i = 1 + 2 + \cdots + m$, or in our case $1 + 2 + 3 = 6$). Thus

$$w_1 = \frac{4-1}{6} = \frac{3}{6}$$

$$w_2 = \frac{4-2}{6} = \frac{2}{6}$$

and

$$w_3 = \frac{4-3}{6} = \frac{1}{6}$$

The trend component prediction is

$$\begin{aligned}\hat{T}_{11} &= w_1 S_{11-1} + w_2 S_{11-2} + w_3 S_{11-3} \\ &= \left(\frac{3}{6}\right)(34) + \left(\frac{2}{6}\right)(26) + \left(\frac{1}{6}\right)(32) = 31\end{aligned}$$

This slightly larger value of 31 (compared with the previous prediction of $30.\overline{6}$) is due to the increased weight given to the most recent time-series observation. Notice that the weights in the example sum to one and that each of the weights is between zero and one.

A variation on the weighted moving average that is often used in forecasting studies is called the *exponential smoothing* method. This interesting technique is presented in the next subsection.

16.9.3 Exponential Smoothing

The basic model for predicting the trend component for the next period using the exponential smoothing method is given by

$$\hat{T}_t = \alpha S_{t-1} + (1 - \alpha)\hat{T}_{t-1} \tag{16.19}$$

in which $0 < \alpha < 1$. Although this model looks like it uses only the last observed time-series value and the last trend component prediction, it actually uses all past time-series values. Explicitly, if we set $\hat{T}_1 = S_1$, then

$$\hat{T}_t = \left[\alpha \sum_{k=0}^{t-3} (1 - \alpha)^k S_{t-1-k}\right] + (1 - \alpha)^{t-2} S_1 \tag{16.20}$$

Notice that the coefficients in Equation 16.20 sum to one, so that the exponential smoothing method finds a weighted average of all previous time-series observations.

EXAMPLE 16.15 Use exponential smoothing with $\alpha = 0.8$ to predict a trend component value for the next period using the same series 16, 14, 18, 20, 24, 22, 28, 32, 26, 34.

SOLUTION First, we set $\hat{T}_1 = S_1 = 16$ and then apply the formula given by Equation 16.16. Thus

$$\hat{T}_2 = 0.8S_1 + 0.2\hat{T}_1 = 0.8(16) + 0.2(16) = 16$$
$$\hat{T}_3 = 0.8S_2 + 0.2\hat{T}_2 = 0.4(14) + 0.2(16) = 14.4$$
$$\hat{T}_4 = 0.8(18) + 0.2(14.4) = 17.28$$
$$\hat{T}_5 = 0.8(20) + 0.2(17.28) = 19.456$$
$$\hat{T}_6 = 0.8(24) + 0.2(19.456) = 23.0912$$
$$\hat{T}_7 = 0.8(22) + 0.2(23.0912) = 22.2182$$
$$\hat{T}_8 = 0.8(28) + 0.2(22.2182) = 26.8436$$
$$\hat{T}_9 = 0.8(32) + 0.2(26.8436) = 30.9687$$
$$\hat{T}_{10} = 0.8(26) + 0.2(30.9687) = 26.9937$$

so that

$$\hat{T}_{11} = 0.8(34) + 0.2(26.9937) = 32.5887$$

One advantage of the exponential smoothing method is ease of use, since we need only two pieces of information in order to predict the trend component value. One disadvantage is the need to specify a value for alpha in the model. Values of alpha close to one make the model sensitive to the most recent observed values, whereas the opposite is true for values of alpha close to zero. Of course, the other models also have their disadvantages. For instance, the weighted moving average requires that m *and* w_i, $i = 1, 2, \ldots, m$, be specified, and the curvilinear techniques assume a basic model that should be tested for validity, with periodic changes in the parameter values (the constants in the model) made when an update is deemed appropriate.

EXERCISES

16.9.1 The following five time-series observations represent the number of albums sold for a particular cut over a 15-week period with observations tabulated every three weeks: 176,427, 111,020, 73,012, 48,011, 32,007. Assume a geometric trend component and predict the trend component for the next time-series observation.

16.9.2 Use Equation 16.18 with $m = 5$ to predict the trend component for the next time-series observation for the data in Exercise 16.9.1.

16.9.3 Use Equation 16.19 with $\alpha = 0.9$ to predict the trend component for the next time-series observation for the data in Exercise 16.9.1.

16.9.4 Use Equation 16.20 with $\alpha = 0.9$ to predict the trend component for the next time-series observation for the data in Exercise 16.9.1. Compare your results with those in Exercise 16.9.3.

16.9.5 Revise your answer to Exercise 16.9.1 so that time is measured in weeks.

16.9.6 Refer to Exercise 16.9.1. If the seasonal component for the next three weeks is 1.342, predict the number of albums to be sold during the next three weeks.

16.9.7 Find a value of α for use in Equation 16.19 that gives the best approximation to the semiaverage trend component estimates for the following data: 5, 11, 25, 20, 32, 47. (*Hint*: Use the method of least squares to choose the best α value.)

Chapter 17
Index Numbers

KEY WORDS AND TERMS

Index Number	Laspeyres Index
Price Relative	Paasche Index
Composite Index	Value Weights
Aggregative Index	Mean of Price Relatives

17.1 INTRODUCTION

Anyone who follows the news or keeps a finger on the pulse of business will be familiar with the important concept of an index number. We have the Consumer Price Index, the Unemployment Index, a number of stock exchange indices, and a multitude of general business indices of various types and emphases. One very important question we can all ask is, "How good is the index?" This means that we would like to know how the index values are determined. We have already said how important different sampling plans can be in delivering good estimates, and we have devoted an entire chapter to the discussion of how to manipulate the sample information to give us a "good" statistic that properly measures the population characteristics of interest. This chapter introduces some of the most common methods of constructing an index, or, at least, the fundamental components of index construction and their use. We begin our presentation with a definition of what we mean by an index number.

DEFINITION 17.1 An *index number* is any measure used for the purpose of expressing comparisons or changes over time.

Index numbers are extremely important to business people as key measures of the prevailing climate in which they must operate. We were first exposed to index numbers in the last chapter, when the seasonal index was introduced. As we mentioned then, the seasonal index is reported as a percentage of the average time-series value over the pattern period. That is, the seasonal index compares the value for a given period with the average value across all periods in the cycle. It does this by means of a simple ratio, and the resulting index is reported as the percentage the ratio represents. For instance, if the seasonal index is 1.0325, the reported seasonal index will be 103.25. In this case the interpretation would be that the particular period under study gives, on the average, a time-series value that is *3.25% of the average in excess of the average*. Thus, if yearly sales are averaging 56,732 units per month and the current month has this seasonal index, then we would expect the number of units sold in the current month to increase by $0.0325 \times 56{,}732 = 1843.79$. Calculations of this type are very important in that they can give us an indication of whether or not the sales in a given time period are "reasonable" when seasonal effects are present. Of course, one question we have not answered is, "What is the time interval over which the basic seasonal pattern is repeated?" Sometimes this time interval is easy to determine by a quick look at records, sometimes it is dictated by generally accepted accounting practices, and at other times it can be determined only by trial and error. However, there are many more important questions to be asked besides this one, and we will mention some of the more pertinent ones in the chapter summary. In general, we should understand the implications of using a particular type of index as a measure of the character of a given collection of objects at a given point in time. In order to simplify matters, we will be presenting methods for constructing *price* indexes, which are used to compare prices in one time period with prices in another time period.

17.2 THE PRICE RELATIVE

A price index for a single commodity is called a *price relative*. Typically, a base time period is stated along with the price relative value so that the direction of the change is apparent to the individual who is going to use the information.

DEFINITION 17.2 A price relative is given by the ratio $100 \times [p(t)/p(0)]$, in which $p(t)$ represents the price of a commodity in time period t and $p(0)$ represents the price of the same commodity in the base time period.

Thus, the price relative compares values for a single commodity; it is the simplest type of index number.

EXAMPLE 17.1 The price of a head of lettuce in December was 89 cents; then in one month's time the price jumped to $1.29. Compute the price relative for lettuce with December as the base time period.

SOLUTION In this example, as in others, we can use either cents or dollars in the price relative ratio. Using cents, we get

$$100 \times \left(\frac{129}{89}\right) = 144.94$$

as the current price relative for lettuce.

One benefit of price relatives can be quickly seen in Example 17.1, which shows that the price of lettuce has increased by 44.94% in one month. Although this information is valuable, it may not be a reliable measure of the overall tendency in the price of produce during the one-month time period. That is, something else may have decreased in price and helped to balance out the effect of the huge increase in the commodity being measured. Thus, we are interested in computing price relatives for a number of different commodities so that we can compute an "overall" price relative. Common names given to such indices are *composite* and *aggregate*. The next two sections are devoted to discussions of techniques for computing such combined measures of price changes.

EXERCISES

17.2.1 Soybean prices in the commodities exchange over a 12-month period were:

January	6.800
March	6.920
May	7.000
July	7.045
September	6.800
November	6.670

Compute a time series of price relatives using these prices, with January as the base time period.

17.2.2 The prices of Maine potatoes over a three-month time period were: March, 5.98; April, 6.31; and May, 7.10. Compute price relatives for April and May with March as the base time period.

17.2.3 Refer to Exercise 17.2.1. Compute a time series of price relatives using November as the base time period.

17.3 COMBINING COMMODITIES

As you can imagine, there are many ways of combining the prices of commodities to get a value that is intended to measure the overall characteristics of a population of prices. One of the most basic of these is called the *simple aggregative index.*

17.3.1 The Simple Aggregative Index

We use the word *aggregative* because the index is determined by adding prices for an aggregate (a group of different items considered as a whole). The word *simple* has more connotations than the obvious one, as the examples of this subsection will illustrate.

DEFINITION 17.3 The simple aggregative index is given by the ratio

$$100 \times \left[\frac{\sum_{k=1}^{n} p_k(t)}{\sum_{k=1}^{n} p_k(0)} \right] \tag{17.1}$$

in which $p_k(t)$ represents the price of commodity k in time period t, $p_k(0)$ represents the price of the same commodity k in the base time period, and n represents the number of commodities used in computing the index.

EXAMPLE 17.2 The following table lists prices in dollars for three grain products in March and December of the same year. Using March as the base period, compute the simple aggregative index using all three commodities.

Commodity	March	December
Wheat	3.515	3.340
Corn	2.310	2.495
Oats	1.360	1.580

SOLUTION Referring to Ratio 17.1, we have that $n = 3$ and the following correspondence:

Commodity	k	$p_k(0)$	$p_k(t)$
Wheat	1	3.515	3.340
Corn	2	2.310	2.495
Oats	3	1.360	1.580

The sum of the last column gives the numerator in Ratio 17.1, and the sum of the third column gives the denominator. Thus the desired simple aggregative index is given by

$$100 \times \left[\frac{7.415}{7.185}\right] = 103.20$$

EXAMPLE 17.3 A cooperative of artisans computes an index for their business that gives them a measure of the change in energy costs. The specifics are shown in the following table, which compares prices for two successive months and includes the units of measurement.

Energy Source	Unit of Measurement	Prices	
		Jan.	Feb.
Wood	Dollars/truckload	210	240
Oil	Dollars/kiloliter	316	322

Using January as the base period, compute the simple aggregative index using the two energy types.

SOLUTION Referring to Ratio 17.1, we have that $n = 2$ and that

Energy Type	k	$p_k(0)$	$p_k(t)$
Wood	1	210	240
Oil	2	316	322

Thus, the simple aggregative index is given by

$$100 \times \left(\frac{240 + 322}{210 + 316}\right) = 106.84$$

EXAMPLE 17.4 Refer to Example 17.3. Suppose that the cooperative sees a change in management and decides to change the units of measurement as follows: Measure wood prices in dollars per cord and oil in dollars per 100 gallons. They assume that this change will make no difference in the index value but will improve their record-keeping practices. Just to make sure, they use the same data and the following equivalence:

$$1 \text{ truckload} = 2 \text{ cords}$$

$$1 \text{ kiloliter} = 264.18 \text{ gallons}$$

On this basis they compute the following *equivalent* table:

Energy Source	Units of Measurement	Prices Jan.	Prices Feb.
Wood	\$/cord	105	120
Oil	\$/100 gallons	119.62	121.89

Use January as the base period and compute the simple aggregative index using the two energy types.

SOLUTION Using Ratio 17.1, we get

$$100 \times \left(\frac{120 + 121.89}{105 + 119.62}\right) = 107.69$$

These two examples illustrate why the simple aggregative index is not used extensively in business. That is, its value is not independent of the units of measurement. The arithmetic mean of price relatives eliminates this irritation.

17.3.2 The Arithmetic Mean of Price Relatives

Any mean of price relatives will eliminate the problem illustrated in Examples 17.3 and 17.4. The most common type of mean used for this purpose is the *arithmetic mean.*

DEFINITION 17.4 The arithmetic mean of a price relatives index is given by the ratio

$$100 \times \frac{1}{n} \sum_{k=1}^{n} \left[\frac{p_k(t)}{p_k(0)}\right] \tag{17.2}$$

in which n, $p_k(t)$, and $p_k(0)$ are as defined in Definition 17.3.

We now show that the arithmetic mean of price relatives avoids the problem posed by changes in units.

EXAMPLE 17.5 Compute Ratio 17.2 for the price sets given in Examples 17.3 and 17.4.

SOLUTION For Example 17.3 we have

$$100 \times \frac{1}{2} \times \left(\frac{240}{210} + \frac{322}{316}\right) = 108.09$$

and for Example 17.4 we have

$$100 \times \frac{1}{2} \times \left(\frac{120}{105} + \frac{121.89}{119.62}\right) = 108.09$$

Note that Example 17.5 shows that the arithmetic mean of price relatives is invariant under changes in units (any small difference is due to round-off error).

EXAMPLE 17.6 Compute the arithmetic mean of price relatives for the data given in Example 17.2.

SOLUTION The value of n is 3 and, referring to Ratio 17.2, we get

$$100 \times \frac{1}{3} \times \left(\frac{3.340}{3.515} + \frac{2.495}{2.310} + \frac{1.580}{1.360}\right) = 106.40$$

Perhaps you have noticed that neither of the last two indices used weights that indicate the relative "value" of the prices or relatives used in their construction. To emphasize this point, we can refer to Example 17.3. Should all commodities be valued equally if wheat accounts for 90% of the market action between March and December? Perhaps in this type of situation we should assign weights that are proportional to the share of the market each commodity commands. The possibilities in this area lead us to indices that employ weighted averages of different types. Some of these are discussed in the next section.

EXERCISES

17.3.1 Compute the arithmetic mean of price relatives for the following information using April as the base time period.

Commodity	April	October
Live beef	63.60	62.80
Live hogs	48.00	44.80
Shell eggs	50.45	57.15

17.3.2 Compute the arithmetic mean of price relatives for the following information using May as the base time period.

Commodity	May	July
Soybeans	7.00	7.04
Soybean oil	25.10	25.03
Soybean meal	190.00	190.20
Coffee	125.50	126.00

17.3.3 Compute the arithmetic mean of price relatives for the following information, using March as the base time period.

Commodity	March	September
Gold	219.70	232.80
Silver	590.00	623.00
Copper	72.50	76.35

17.3.4 Compute a time series of price relatives for the following information, using March as the base time period.

Commodity	March	May	July	Sept.	Dec.
Wheat	3.515	3.35	3.21	3.24	3.34
Corn	2.310	2.39	2.43	2.46	2.49
Oats	1.360	1.43	1.48	1.53	1.58

17.4 WEIGHTING COMMODITIES

We can usually solve the problem of assigning weights by using the relative frequency concept introduced in Chapter 7. We will illustrate how this is done after we provide formal definitions for the most common types of weighted indices.

17.4.1 The Weighted Aggregative Index

The *weighted aggregative index* is the simplest of the weighted indices. Its definition is given as follows:

DEFINITION 17.5 The weighted aggregative index is given by the ratio

$$100 \times \left\{ \frac{\sum_{k=1}^{n} [p_k(t) \times w_k(t)]}{\sum_{k=1}^{n} [p_k(0) \times w_k(0)]} \right\} \tag{17.3}$$

in which $w_k(t)$ is the weight assigned to commodity k in time period t, $w_k(0)$ is the weight assigned to the same commodity k in the base time period, and n, $p_k(t)$, and $p_k(0)$ are defined as given in Definition 17.2.

In most cases $w_k(t) = w_k(0)$, $k = 1, 2, \ldots, n$, although this is not a requirement.

EXAMPLE 17.7 Consider the following commodity information:

Commodity	March	September
Gold	219.70	232.80
Silver	590.00	623.00
Copper	72.50	76.35

Using March as the base period, calculate

(*a*) A weighted aggregative index with weights 0.2, 0.1, and 0.7 for gold, silver, and copper, respectively.
(*b*) A weighted aggregative index with weights 0.3, 0.6, and 0.1 for gold, silver, and copper, respectively.
(*c*) A weighted aggregative index with weights 1/3, 1/3, and 1/3.

SOLUTION

(*a*) Looking at definition 17.5 gives us the following table:

Commodity	k	$p_k(0)$	$w_k(0)$	$p_k(t)$	$w_k(t)$
Gold	1	219.70	0.2	232.80	0.2
Silver	2	590.00	0.1	623.00	0.1
Copper	3	72.50	0.7	76.35	0.7

Thus,

$$100 \times \frac{(232.80 \times 0.2) + (623.00 \times 0.1) + (76.35 \times 0.7)}{(219.70 \times 0.2) + (590.00 \times 0.1) + (72.50 \times 0.7)}$$
$$= 105.61$$

is the weighted aggregative index using the weights of part (*a*).

(*b*) The form is as given above, but with new weights, so that we get

$$100 \times \frac{(232.80 \times 0.3) + (623.00 \times 0.6) + (76.35 \times 0.1)}{(219.70 \times 0.3) + (590.00 \times 0.6) + (72.50 \times 0.1)}$$
$$= 105.65$$

as the weighted aggregative index using the weights of part (*b*).

(*c*) When the weights are all equal, Ratio 17.3 simplifies to become the simple aggregative index of Definition 17.2. Thus,

$$100 \times \frac{232.80 + 623.00 + 76.35}{219.70 + 590.00 + 72.50} = 105.66$$

is the weighted aggregative index using the equal weights of 1/3, 1/3, and 1/3 given in part (*c*).

Example 17.7 uses weights that sum to one but this is *not* a general requirement for computing a weighted aggregative index. In fact, many examples of this type of index determine the weights by the amount of each commodity traded or produced or consumed during one of the two periods under comparison. If the base period quantities are used as weights, then the resulting weighted aggregative index is called a *Laspeyres* index. If the current period's quantities are used as the weights, then the resulting weighted aggregative index is called a *Paasche* index. The next example illustrates these special cases.

EXAMPLE 17.8 The following information represents price and volume figures for four types of stock on a major exchange:

Company	Monday Volume	Monday Price	Tuesday Volume	Tuesday Price
A	100	149	130	158
B	4,000	44	3000	41
C	4,351	8	5562	9
D	12,500	50	3700	48

Assume Monday as the base period and

(*a*) Calculate a Laspeyres index (use the Monday volume figures as weights;

(*b*) Calculate a Paasche index (use the Tuesday volume figures as weights).

SOLUTION

(*a*) The Laspeyres index is given by

$$100 \times \frac{(158 \times 100) + (41 \times 4000) + (9 \times 4351) + (48 \times 12{,}500)}{(149 \times 100) + (44 \times 4000) + (8 \times 4351) + (50 \times 12{,}500)}$$
$$= 96.27$$

(*b*) The Paasche index is given by

$$100 \times \frac{(158 \times 130) + (41 \times 3000) + (9 \times 5562) + (48 \times 3700)}{(149 \times 130) + (44 \times 3000) + (8 \times 5562) + (50 \times 3700)}$$
$$= 97.46$$

Although the two answers in Example 17.7 are almost the same, the two approaches can yield results that are substantially different. We should also note that the Laspeyres index is easier to calculate when establishing a time series (using a fixed base year), since the denominator of the weighted aggregative index would remain the same for each time period in the series.

We now provide one more example of the weighted aggregative index. This time we will use relative frequency weights and use different weights for each time period.

EXAMPLE 17.9 Refer to the price and volume figures given in Example 17.8. Calculate a weighted aggregative index using relative frequency weights based on volume amounts, with Monday volumes used in the base period weight calculations and Tuesday volumes used in the current period weight calculations.

SOLUTION To compute relative frequency weights we need the total volume across all four stocks. Since Tuesday's total volume is 12,392 and Monday's total volume is 20,951, we get

k	$w_k(0)$	$w_k(t)$
1	100/20,951 (0.0048)	130/12,392 (0.0105)
2	4000/20,951 (0.1909)	3000/12,392 (0.2421)
3	4351/20,951 (0.2077)	5562/12,392 (0.4488)
4	12,500/20,951 (0.5966)	3700/12,392 (0.2986)

Using the decimal equivalents of these weights we get the weighted aggregative index as given by

$$100 \times \frac{(158 \times 0.0105) + (41 \times 0.2421) + (9 \times 0.4488) + (48 \times 0.2986)}{(149 \times 0.0048) + (44 \times 0.1909) + (8 \times 0.2077) + (50 \times 0.5966)}$$
$$= 73.77$$

We can see the drastic change in the index value that results from changing weights by comparing the results of the last two examples. The choice of appropriate weights is highly judgmental and sometimes depends on what the resulting index will be used for after its determination. The weights used in Example 17.9 give a measure of the change in the level of total money transfer, whereas the weights used in Example 17.8 give a measure of the change in the level of the average stock price.

17.4.2 The Weighted Mean of Price Relatives

The *weighted mean of price relatives* is a simple extension of the arithmetic mean of price relatives. The general form for this measure is given in the following definition.

DEFINITION 17.6 The weighted mean of price relatives is given by the ratio

$$100 \times \sum_{k=1}^{n} \left[\frac{p_k(t)}{p_k(0)} \times w_k \right] \tag{17.4}$$

where n, $p_k(t)$, and $p_k(0)$ are as given in Definition 17.3 and we require that

1. $0 \leqslant w_k \leqslant 1$
2. $\sum_{k=1}^{n} w_k = 1$

Although we can set the weights by subjective means or by relative frequency procedures, it is typical to use value weights. *Value weights* are relative frequency weights based on the multiplication of price with quantity. Their construction is illustrated in the next example.

EXAMPLE 17.10 Part of the information given in Example 17.8 is reproduced here. Use value weights based on Tuesday's volume to compute a weighted mean of price relatives for the four stocks.

Company	Monday Price	Tuesday Price	Tuesday Volume
A	149	158	130
B	44	41	3000
C	8	9	5562
D	50	48	3700

SOLUTION First, we calculate the total Tuesday price times Tuesday volume figure:

$$(158 \times 130) + (41 \times 3000) + (9 \times 5562) + (48 \times 3700) = 371{,}198$$

Next, the value weights are given by

k	w_k
1	$158 \times 130/371{,}198$ (0.0553)
2	$41 \times 3000/371{,}198$ (0.3314)

k	w_k
3	$9 \times 5562/371{,}198$ (0.1349)
4	$48 \times 3700/371{,}198$ (0.4784)

Thus, the weighted mean of price relatives is given by

$$100 \times \left(\frac{158}{149} \times 0.0533\right) + \left(\frac{41}{44} \times 0.3314\right) + \left(\frac{9}{8} \times 0.1349\right) + \left(\frac{48}{50} \times 0.4784\right) = 97.85$$

Before doing the exercises, note that if $w_k = 1/n$ for $k = 1, 2, \ldots, n$, then the weighted mean of price relatives becomes the arithmetic mean of price relatives. Also note that if we use value weights for the base year, the weighted mean of price relatives becomes identical to the Laspeyres index given in the last subsection.

EXERCISES

17.4.1 Hank the handyman does a quick price check periodically to find out how the prices of plywood are changing over time. Since quality measures vary from store to store, Hank uses his own quality grades of A (high), B, C, and D. The average prices across the eight stores in his survey at each grade are tabulated

Plywood Grade	Summer Survey	Winter Survey
A	38.50	40.95
B	27.95	31.50
C	22.50	23.50
D	14.95	15.50

Hank also uses his own subjective weights for the four grades according to his estimates of their popularity. These weights are 0.05 for grade A, 0.15 for grade B, 0.45 for grade C, and 0.35

for grade D. Calculate the weighted aggregative index and the weighted mean of price relatives using these weights.

17.4.2 The following data represent the volume and price figures for three stocks over a two-day period. Using Tuesday as the base period, calculate a Laspeyres and a Paasche index for this information.

Stock	Tuesday Volume	Tuesday Price	Wednesday Volume	Wednesday Price
A	200	47.50	900	48.25
B	1500	43.75	4000	44.50
C	800	70.00	200	67.75

17.4.3 Refer to Exercise 17.4.2. Use Tuesday as a base period and calculate a weighted mean of price relatives with
(*a*) Wednesday price-volume value weights
(*b*) Tuesday price-volume value weights (you should get the Laspeyres index value for the previous exercise as your answer)

17.4.4 A wood craftsman decides to establish a time series of price indices for the items he sells at a local farmer's market. His business partner keeps a strict accounting of sales volume and prices for income tax purposes. She provides the following information:

	Season 1		Season 2		Season 3		Season 4	
Item	Volume	Price	Volume	Price	Volume	Price	Volume	Price
Truck A	7	18.00	8	18.00	7	19.00	10	22.00
Truck B	5	13.50	3	14.00	3	14.00	6	16.00
Truck C	4	8.50	5	8.50	4	9.00	8	9.00
Car A	10	8.50	6	9.00	6	9.00	10	10.00
Car B	6	7.50	8	7.50	6	8.00	12	8.00
Train	8	35.00	8	37.50	8	39.00	10	40.00
Wrecker	2	28.00	1	28.00	2	29.00	3	29.00
Doll carriage	3	15.00	3	16.00	4	16.00	3	17.50
Baby rocker	3	12.00	3	12.00	4	13.50	3	14.00
Candle holders A	5	5.00	5	5.00	6	6.00	6	6.00
Candle holders B	6	6.00	6	6.00	6	7.00	6	7.00
Candle holders C	5	7.00	5	7.00	6	8.00	6	8.00

Calculate a time series of price indices using season 1 as the base time period and use a Paasche index.

17.5 USING INDEX NUMBERS

Now that we have spent some time showing how to calculate various types of indices, we can indicate how they can be used to provide some valuable interpretations of other time-series data. Consider a point close to home for all of us, the comparison of the two time series of wages per year and the Consumer Price Index each year. Thus, we are going to take another look at the time series of wages per year in light of what the time series of Consumer Price Indices looks like over the same time period. This would be a valuable comparison, since the index is supposed to measure the change in the average price a consumer must pay in order to maintain a fixed standard of living.

EXAMPLE 17.11 A dock worker's annual salary increases from \$21,375 to \$24,000 in one year. The Consumer Price Index increases from 95 to 110 in the same year. Determine whether the worker's purchasing power has actually increased.

SOLUTION One way to make the desired comparison is to compute how much the old salary would need to increase in order to maintain the same purchasing power. Since the Consumer Price Index increases by a factor of 110/95, we would expect that the old wages would need to increase by this same factor in order to keep the same purchasing power. Since $21{,}375 \times (110/95) = 24{,}750$, we see that "real" wages, or purchasing power, have actually decreased.

EXAMPLE 17.12 An employer has just announced that workers will receive a wage increase of 8% in the coming year. If the Consumer Price Index is currently 93, what value must it reach in order to cancel out the benefit of the wage increase?

SOLUTION If a worker's salary is represented by A, then the new salary is represented by $A + 0.08A$, or $1.08A$. An increase in the index value I would mean that the old wages would need to increase by a factor of $I/93$ to maintain the same purchasing power. Thus, we need to solve the following equation:

$$A\left(\frac{I}{93}\right) = 1.08A$$

or

$$\frac{I}{93} = 1.08$$

to get

$$I = 100.44$$

Examples 17.10 and 17.11 illustrate the importance of some of the common indices that abound in the business world. In fact, some unions demand that their salary amounts be tied to the Consumer Price Index to ensure that the actual purchasing power of their checks does not decrease. Whether or not the index was meant to be used in this manner is irrelevant, since it will continue to be used in a wage context for the purpose of comparing old and new salaries. That is, the use of the index becomes an established pattern, just as the Dow Jones index has established itself as a measure of stock exchange performance. Just as these indices are "accepted" by the general public, there are many other indices in almost every type of business that are accepted as valid standards of measure.

EXERCISES

17.5.1 The average net income of physicians by geographic area of major United States cities is given for the years 1970 and 1972. The Consumer Price Index increased from 116.3 to 125.3 over the same period. Convert the 1972 income values to their equivalent in 1970 dollars to see whether income increased in "real" dollars.

Geographic Division	1970	1972
North	37.5	46.2
Northeast	38.4	45.3
East	41.5	51.5
Southeast	46.6	49.3
South	41.0	50.3
Southwest	44.5	56.4
West	45.1	50.7
Northwest	45.5	46.5

Source. *Statistical Abstract of the United States*, 1977.

17.5.2 The value of farm products sold (in thousands of dollars) by geographic area of a country in 1964 and 1974 is given here. The country's Consumer Price Index changed from 92.9 to 147.7 during the same period. Convert the 1964 values to their equivalents in 1974 dollars and compare your results with the 1974 values in the table.

Geographic Division	1964	1974
Northeast	2,755	4,921
Southeast	14,388	36,537
Southwest	10,856	23,634
Northwest	7,923	17,159

Source. *Statistical Abstract of the United States*, 1977.

17.6 SUPPLEMENTARY EXAMPLES

EXAMPLE 17.13 The concept of a price relative can be applied in situations that do not involve prices. Show that this is the case by applying the price relative principle to the following time series of life expectancy for females of a nameless country using 1950 as the base time period.

Year	1950	1960	1965	1970
Life expectancy	70.83	74.17	75.18	76.36

SOLUTION With time measured in years, we have

t	0	10	15	20
$p(t)$	70.83	74.17	75.18	76.36

so that the time series of "price" relatives as applied to female life expectancy is defined by

t	0	10	15	20
$100 \times [p(t)/p(0)]$	100.0	104.72	106.14	107.81

EXAMPLE 17.14 Compute a simple aggregative index for the data given in Exercise 17.5.1, using 1970 as the base time period.

SOLUTION The simple aggregative index using 1970 as the base time period is

$$100 \times \frac{46.2 + 45.3 + 51.5 + 49.3 + 50.3 + 56.4 + 50.7 + 46.5}{37.5 + 38.4 + 41.5 + 46.6 + 41.0 + 44.5 + 45.1 + 45.5}$$

$$= 100 \times \frac{396.2}{340.1} = 116.50$$

EXAMPLE 17.15 Compute the arithmetic mean of price relatives for the data given in Exercise 17.5.1, using 1970 as the base time period.

SOLUTION The arithmetic mean of price relatives using 1970 as the base time period is

$$100 \times \left(\frac{1}{8}\right)\left(\frac{46.2}{37.5} + \frac{45.3}{38.4} + \frac{51.5}{41.5} + \frac{49.3}{46.6} + \frac{50.3}{41.0} + \frac{56.4}{44.5} + \frac{50.7}{45.1} + \frac{46.5}{45.5}\right)$$

$$= 100 \times \frac{1}{8} \times 9.3510 = 116.89$$

EXAMPLE 17.16 The weighted aggregative index and the weighted mean of price relatives index can be applied in situations in which no prices are involved. To see this, consider the following data. The harvested area in 100-acre units and the production of two types of seed in 1000-pound amounts per 1000-acre unit in 1974 and 1975 are:

	Sunflower		Mustard	
	1974	1975	1974	1975
Harvested area	21	62	350	163
Production	866	1064	742	677

Using 1974 as the base time period, calculate

(*a*) The Laspeyres index.

(*b*) The Paasche index.

(*c*) A weighted aggregative index using relative frequency weights based on harvested area–production amounts, with 1974 amounts used in the base period weight calculations and 1975 amounts used in the current period weight calculations.

(*d*) A weighted mean of price relatives using average acreage relative frequency weights.

SOLUTION

(*a*) The Laspeyres index is given by

$$100 \times \left[\frac{(1064.00 \times 21) + (677.00 \times 350)}{(866.00 \times 21) + (742.00 \times 350)}\right] = 93.31$$

(*b*) The Paasche index is given by

$$100 \times \left[\frac{(1064.00 \times 62) + (677.00 \times 163)}{(866.00 \times 62) + (742.00 \times 163)}\right] = 100.96$$

(*c*) First note that

$$w_1(t) = \frac{62 \times 1064.00}{(62 \times 1064.00) + (163 \times 677.00)} = 0.3741$$

$$w_2(t) = 1 - w_1(t) = 0.6259$$

$$w_1(0) = \frac{21 \times 866.00}{(21 \times 866.00) + (350 \times 742.00)} = 0.0654$$

$$w_2(0) = 1 - w_1(0) = 0.9346$$

Thus, the weighted aggregative index is given by

$$100 \times \left[\frac{1064.00(0.3741) + 677.00(0.6259)}{866.00(0.0654) + 742.00(0.9346)}\right] = 109.55$$

(*d*) The average acreage over the two years for the sunflower and mustard seed crops are 41.5 and 256.5, respectively, and so

$$w_1 = \frac{41.5}{41.5 + 256.5} = 0.1393$$

$$w_2 = 1 - w_1 = 0.8607$$

Thus, the weighted mean of price relatives using average acreage relative frequency weights is given by

$$100 \times \left[\frac{1064.00}{866.00}(0.1393) + \frac{677.00}{742.00}(0.8607)\right] = 95.65$$

EXAMPLE 17.17 Since the data values in Example 17.15 could just as well have been price and quantity amounts, explain how there can be such a wide range for the first three answers.

SOLUTION The Laspeyres index compares the 1975 production rates with the 1974 rates as though the 1974 acreage had been planted in 1975. Thus, applying the 1975 production rates to the 1974 acreage amounts indicates a decline of 6.69% in productivity on the basis of planting a crop of 21,000 acres of sunflower and 350,000 acres of mustard seed. The Paasche index compares the 1975 production rates with the 1974 rates as though the 1975 acreage had been planted in 1974. Thus, applying the 1974 rates to the 1975 acreage amounts indicates a stable productivity on the basis of planting a crop of 62,000 acres of sunflower and 163,000 acres of mustard seed crop. The relative frequency index compares the productivity for the two time periods on a per acre basis. It does this by calculating the average productivity rate for 1975 and dividing the result by the average productivity rate for 1974. This indicates an increase in production per acre of 9.55% *without* regard to the number of acres of each seed planted. Thus, the difference in the three index values is essentially a result of fixing the acreage composition in the first two and relaxing this condition in the third.

17.7 SUMMARY

The simplest type of index is one that compares values of a single commodity in two periods. In a price frame of reference, the index is called a price relative and is given by multiplying a ratio of the two prices by 100. When a measure for a single commodity is not satisfactory as a characterization of the change for a group of commodities, the index should assume a "composite" character. A composite index is a function of the two period prices for more than one commodity. Some of the questions that arise when choosing the type of composite index most appropriate for the situation at hand are:

1. How many commodity prices should be monitored?
2. Which commodity prices should be monitored?
3. What type of prices should be used?
4. How will the prices be monitored?
5. What base time period should be used?
6. What special problems are involved (for example, for a seasonal index, the pattern period must be determined)?
7. What are the units of measurement?
8. How will the index values be used?
9. What will the index actually measure?
10. How will lay people interpret the index values, and what kind of behavioral changes might they induce?

There are two basic types of composite indices used today. These are called the weighted aggregative index and the weighted mean of price relatives. Many different kinds of weighting scheme are possible with each type of index, but the most popular is called value weights. The weighted mean of price relatives is invariant under simple transformations in the units of measurement (of the type $y_k = a_k x_k$, in which y_k represents the new units of measurement, a_k is a constant, x_k represents the old units of measurement, and k represents the commodity being measured). Other variants of these indices are given in the advanced exercises.

Although our presentation has been oriented toward price indices, we have shown in Section 17.6 that these indices can be applied in situations in which prices are not involved. The nonprice applications shown involved quantity comparisons, but the same principles can be applied to measure changes in value.

REVIEW EXERCISES

Level One

17.1 Suppose that the average net income (in thousands) of physicians by geographic division for metropolitan areas with under 1 million inhabitants over a four-year period is:

	Year			
Geographic Division	1978	1980	1981	1982
North	76	92	94	98
East	74	90	86	100
South	94	104	102	112
West	92	112	110	124

Calculate a time series of price relatives for each division using 1978 as the base time period.

17.2 Refer to Exercise 17.1. Calculate a time series of simple aggregative indices using 1978 as the base time period.

17.3 Refer to Exercise 17.1. Suppose that the Consumer Price Index levels for 1978 and 1982 are, respectively, 249.3 and 316.6. Calculate the 1978 equivalent of the 1982 incomes, then compare these with the 1978 incomes using an arithmetic mean of price relatives and 1978 as the base time period.

17.4 Use the following information to calculate a time series of Laspeyres indices with 1960 as the base time period. The fictitious information refers to scheduled passenger/cargo air carriers.

	Domestic		International	
Year	Number	Revenue	Number	Revenue*
1960	40	48.57	12	52.69
1965	38	85.50	10	119.90
1970	33	192.34	4	478.50
1975	28	370.53	3	1021.33
1980	30	407.58	7	2620.57

*Average revenue per carrier in millions of dollars.

17.5 Refer to Exercise 17.4. Using 1960 as a base time period, calculate a time series of weighted aggregative indices using relative frequency weights based on number–revenue amounts, with 1960 amounts used in the base period weight calculations and the current year amounts used in the numerator weight calculations.

17.6 Refer to Exercise 17.4. Suppose that the Consumer Price Index levels for 1965 amd 1975 are, respectively, 142.8 and 187.3. Using 1965 as a base time period, calculate the arithmetic mean of price relatives for the two years with 1975 revenues expressed in equivalent 1965 dollars.

17.7 Suppose the quantity (in 1000 cubic meters) and value (per 1000 cubic meters) of veneer and plywood shipments in 1980, 1981, and 1982 by type were:

	1980		1981		1982	
Type	Quantity	Value	Quantity	Value	Quantity	Value
Veneer						
Softwoods	596.0	104.78	607.5	122.02	488.8	132.74
Hardwoods	151.1	515.54	143.6	632.58	118.5	742.96
Plywood						
Softwoods	1996.0	241.38	2148.9	269.60	1842.1	284.72
Hardwoods	225.0	387.24	249.6	475.52	218.2	528.10

Using 1980 as the base time period, calculate a time series of Paasche indices for the four commodities.

17.8 Refer to Exercise 17.7. Using 1980 as a base time period, calculate a time series of weighted aggregative indices using relative frequency weights based on quantity–value amounts, with 1980 amounts used in the base period weight calculations and the current year amounts used in the numerator weight calculations.

17.9 Suppose that the area (in 1000 acres) and production per 1000 acres (in 1000 bushels) for nine field crops in 1981 and 1982 were as follows:

	1981		1982	
Crop	Area	Production	Area	Production
Peas	80	21.27	76	23.91
Beans	175	20.75	150	22.80
Soybeans	410	24.80	385	34.56
Buckwheat	66	19.00	50	19.17
Waxed grains	1800	44.59	1825	48.94
Flax seed	1440	9.52	1380	12.50
Rapseed	3150	16.23	4300	17.85
Shelled corn	1470	69.48	1570	91.27
Potatoes	280	195.55	255	178.58

Repeat parts (*a*) through (*d*) of Example 17.16 using this information.

17.10 The life expectancy at birth for males and females in a certain country from 1930 through 1980 are given. Calculate a time series of arithmetic mean of "price" relative indices for this information, using 1930 as the base time period.

	Sex			Sex	
Year	Male	Female	Year	Male	Female
1930	53.6	54.6	1965	66.6	73.1
1940	58.1	61.6	1970	66.8	73.7
1950	60.8	65.2	1975	67.1	74.8
1955	65.6	71.1	1980	68.7	76.5
1960	66.7	72.8			

17.11 Refer to Exercise 17.10. Calculate a time series of simple aggregative indices for the information, using 1930 as the base time period.

17.12 The commercial nuclear power generation (in billion kilowatt hours per reactor) and the number of reactors in Exland, Whyland, and Zeeland for the years 1976 and 1982 were as follows:

	Number of Reactors		Gross Electricity Generated	
Country	1976	1982	1976	1982
Exland	5	12	2.900	3.360
Whyland	30	32	1.547	1.875
Zeeland	62	70	3.475	3.981

Calculate a Laspeyres index and a Paasche index for this information, using 1976 as the base time period.

17.13 Refer to Exercise 17.12. Calculate a weighted mean of "price" relatives for the information, using weights of your choice.

17.14 Interpret your answers for Exercises 17.12 and 17.13.

17.15 Collect your own price-volume information for two different time periods for at least three commodities. Calculate each of the types of composite index presented in this chapter for the information you collect, and interpret your results.

Level Two

17.16 Suppose we define a time series of simple aggregative indices by the equation

$$I(t) = \frac{\sum_{k=1}^{n} p_k(t)}{\sum_{k=1}^{n} p_k(t-1)}$$

Apply this formula to get a time series for the information given in Exercise 17.10 and interpret your results.

17.17 Suppose there is a proposal to establish an index that represents the cost, to the customer, of food goods available in the supermarkets of your area. How would you argue the commodities should be selected for determination of the index if you were:

1. A representative of the consumers
2. A representative of the grocers
3. An unbiased researcher

17.18 The geometric mean of price relatives is given by

$$G(t) = \left\{ \prod_{k=1}^{n} \left[\frac{p_k(t)}{p_k(0)} \right] \right\}^{1/n}$$

Calculate the geometric mean of price relatives for the information given in Examples 17.3 and 17.4.

17.19 The harmonic mean of price relatives is given by

$$H(t) = \left\{ \sum_{k=1}^{n} \left[\frac{p_k(0)}{p_k(t)} \right] \right\}^{-1} \times n$$

Calculate the harmonic mean of price relatives for the information given in Examples 17.3 and 17.4.

Level Three

17.20 Compare your answers for Exercises 17.18 and 17.19 with that for Example 17.5. In general,

$$H(t) \leqslant G(t) \leqslant A(t)$$

where $A(t)$ represents the arithmetic mean of price relatives. Prove that this inequality system is always valid for positive prices.

17.21 The generalized formula for the average of a set of positive numbers $x_1, x_2, x_3, \ldots, x_n$ is given by

$$\text{AV}(y) = \left[\frac{\sum_{i=1}^{n} w_i x_i^y}{\sum_{i=1}^{n} w_i} \right]^{1/y}$$

where $w_i > 0$, $y \neq 0$, and n is a positive integer. Show that the arithmetic mean and harmonic mean are special cases of the generalized formula.

17.22 Refer to Exercise 17.21. Apply the generalized formula for the average (with a value for y of other than 1 or -1) to the information of Example 17.8 and interpret your results.

Appendix A
Review of Basic Set Theory and Set Notation

A set can be thought of as any collection of objects, and *set theory* is a phrase that describes our attempts to quantify the basic relationships that can exist between different sets. Typically, capital letters are used to denote the different collections of objects of interest ($A, B, C, \ldots$), and corresponding small letters are used to denote those objects accounted for by the chosen sets ($a, b, c, \ldots$). The terms *set* and *membership* are accepted without formal definition (that is, they are primitive terms).

1. If the object denoted by a is a member of the set denoted by A, then we write

$$a \in A$$

   which is read "a is an element of A," so that the notation $\in$ implies membership. Not surprisingly, $a \notin A$ indicates lack of membership and is read "a is not an element of A."

2. A set A can be referenced by listing its members. The accepted notation is to place the elements in brackets and separated by commas, as in

$$\{a_1, a_2, a_3\}$$

   If a_1, a_2, and a_3 are elements of the set A, we say that A contains three elements, and we have the equivalence

$$A = \{a_1, a_2, a_3\}$$

3. Set A is said to be a *subset* of set B if $x \in A$ implies $x \in B$. In this case we write

$$A \subset B$$

   and say "A is contained in (or is a subset of) B." It is useful at this point to introduce geometrical interpretations of the concepts we are presenting. Thinking of A and B as *disks* (not circles alone but circles with all interior points included), we would interpret $A \subset B$ as

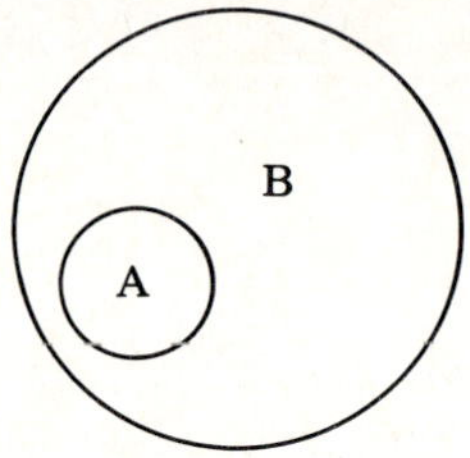

with B indicated by the larger *disk*. Notice that the graphic

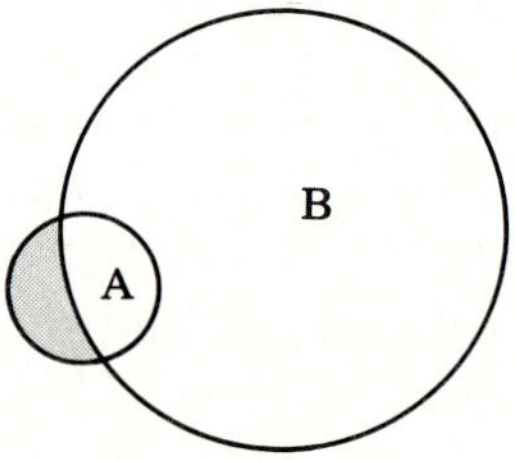

would indicate that $A \not\subset B$ (read "A is not a subset of B"), since the points in the shaded region of disk A are not within disk B.

4. Every set has a subset called the *null* set that contains no elements. The null set is denoted by $\emptyset$ and is required to make the theory consistent and complete.

5. Two sets are *equal* if and only if $A \subset B$ and $B \subset A$. When equality is uncertain, we need only show that for any $x \in A$ we have $x \in B$ and vice versa in order to say A and B are equal. Geometrically, the perimeters of the A and B disks would coincide, and we write

$$A = B$$

6. If *all* sets under consideration are to be formed from the elements of set I, then I is called the *universal* set. The universal set is depicted geometrically as a rectangle *together with all interior points*. Throughout the remainder of this appendix, I will denote a universal set.

7. The *complement* of set A is the subset of elements in I that are *not* in A. We use A' to denote the complement of set A. Geometrically, we have

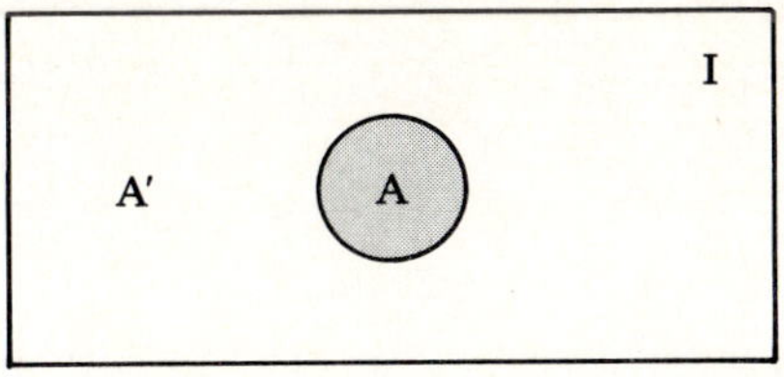

so that A' (read "A complement") contains all points within the universal set rectangle that are not in the shaded A disk. Before we go to the next point, we should note that

$$\emptyset' = I \qquad \text{and} \qquad I' = \emptyset$$

8. The set of elements that are either elements of A, elements of B, or elements of both A and B is called the *union* of sets A and B. Such a union of sets is denoted by

$$A \cup B$$

read "A union B." Set union can be illustrated geometrically by a *merging* of the A and B disks (assuming that there are elements in I that are common to both A and B). This is shown here by the shaded region.

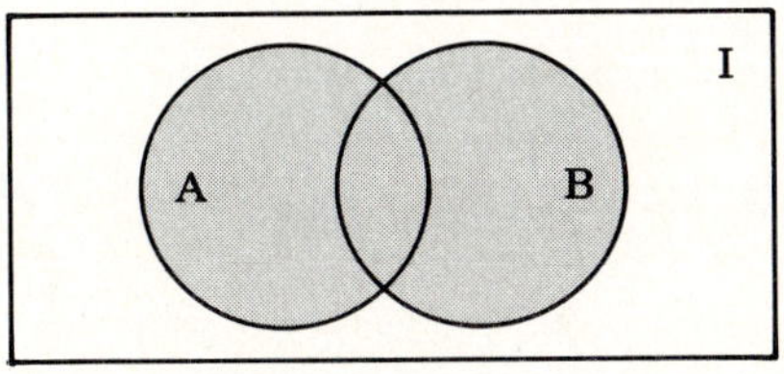

9. The set of elements that are both in A and in B concurrently (that is the set of elements common to both A and B) is called the *intersection* of sets A and B. The intersection of sets A and B is denoted by

$$A \cap B$$

Geometrically, the intersection of sets A and B is the region of overlap of the two A and B disks. This idea, illustrated here by the shaded region,

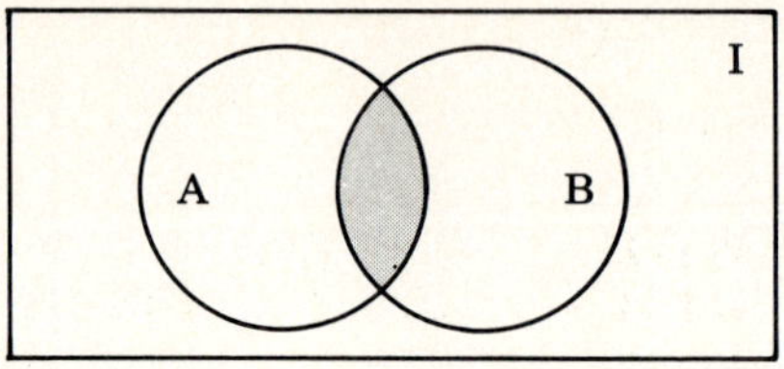

includes all perimeter points touching the shaded region.

10. The following relationships (theorems) are sometimes useful for manipulating the universal set elements. Letting A, B, and C be subsets of I, we have

(a) $A \cup A' = I$

(b) $A \cup B = B \cup A$ and $A \cap B = B \cap A$

(c) $A \cap (B \cup C) = (A \cap B) \cup (A \cap C)$ and
$A \cup (B \cap C) = (A \cup B) \cap (A \cup C)$

(d) $(A \cup B)' = A' \cap B'$ and $(A \cap B)' = A' \cup B'$

(e) $A \cup A' = I$ and $A \cap A' = \emptyset$

11. A partition of the universal set I is any collection of subsets $B_1, B_2, \ldots, B_m$ such that

(a) $I = B_1 \cup B_2 \cup B_3 \cup \cdots \cup B_m$ and

(b) $B_i \cap B_j = \emptyset$ whenever $i \neq j$

In this case, for any subset A of I we have

$$A = (A \cap B_1) \cup (A \cap B_2) \cup (A \cap B_3) \cup \cdots \cup (A \cap B_m)$$

EXERCISES

A.1 If $I = \{2, 3, 4, 5, 6, 7, 8\}$; $A = \{2, 3, 4, 5\}$; and $B = \{3, 6, 8\}$, determine

(a) A' (b) B' (c) $A \cup B$ (d) $A \cap B$

(e) $(A \cup B)'$ (f) $A' \cap B'$ (g) $(A \cap B)'$ (h) $A' \cup B'$

A.2 If $I = \{2, 3, 4, 5, 6, 7, 8, 9, 10\}$; $A = \{2, 3, 5, 6, 8, 9\}$; $B = \{3, 5, 7, 9\}$; and $C = \{6, 7, 8, 9, 10\}$, determine

(a) $B \cup C$ (b) $A \cap (B \cup C)$ (c) $A \cap B$

(d) $A \cap C$ (e) $(A \cap B) \cup (A \cap C)$ (f) $B \cap C$

(*g*) $A \cup (B \cap C)$ (*h*) $A \cup B$ (*i*) $A \cup C$
(*j*) $(A \cup B) \cap (A \cup C)$

A.3 If $I = \{a, b, c, d, e, f, g, h\}$; $B_1 = \{a, c, d\}$; $B_2 = \{f, g\}$; $B_3 = \{b, e, h\}$; and $A = \{c, d, e, f\}$, determine

(*a*) $A \cap B_1$ (*b*) $A \cap B_2$ (*c*) $A \cap B_3$
(*d*) $B_1 \cup B_2 \cup B_3$ (*e*) $B_1 \cap B_2$ (*f*) $B_1 \cap B_3$
(*g*) $B_2 \cap B_3$ (*h*) $(A \cap B_1) \cup (A \cap B_2) \cup (A \cap B_3)$

Answers

A.1 (*a*) $\{6, 7, 8\}$ (*b*) $\{2, 4, 5, 7\}$ (*c*) $\{2, 3, 4, 5, 6, 8\}$
(*d*) $\{3\}$ (*e*) $\{7\}$ (*f*) $\{7\}$
(*g*) $\{2, 4, 5, 6, 7, 8\}$ (*h*) same as (*g*)

A.2 (*a*) $\{3, 5, 6, 7, 8, 9, 10\}$ (*b*) $\{3, 5, 6, 8, 9\}$
(*c*) $\{3, 5, 9\}$ (*d*) $\{6, 8, 9\}$
(*e*) $\{3, 5, 6, 8, 9\}$ (*f*) $\{7, 9\}$
(*g*) $\{2, 3, 5, 6, 7, 8, 9\}$ (*h*) $\{2, 3, 5, 6, 7, 8, 9\}$
(*i*) $\{2, 3, 5, 6, 7, 8, 9, 10\}$ (*j*) $\{2, 3, 5, 6, 7, 8, 9\}$

A.3 (*a*) $\{c, d\}$ (*b*) $\{f\}$ (*c*) $\{e\}$ (*d*) I
(*e*) $\emptyset$ (*f*) $\emptyset$ (*g*) $\emptyset$ (*h*) A

Appendix B
Review of Differential and Integral Calculus

B.1 DIFFERENTIAL CALCULUS

The derivative of a function is essentially an expression of the function's *rate of change* at a given point in its domain. Such expressions of the rate of change are common in the business world and are of interest to all consumers. Perhaps you have heard statements like, "If the rate of inflation continues at its current level, the yearly change will be . . . ." When such changes are linear in nature (that is, when they can be described by a straight line), we can easily compute the rate of change by choosing two points on the line and calculating a slope using the values associated with the two points. This idea is shown graphically in Figure B.1.

Figure B.1

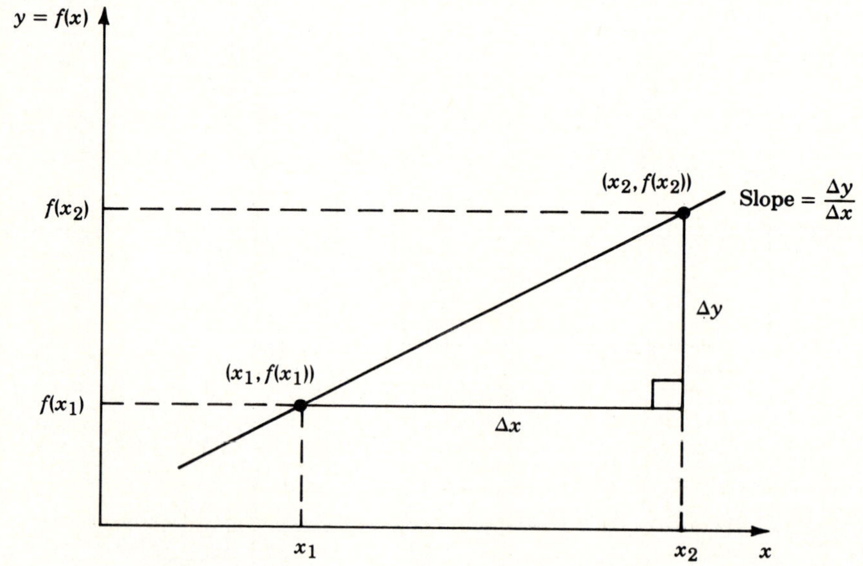

In Figure B.1,

$$\Delta y = f(x_2) - f(x_1)$$

and

$$\Delta x = x_2 - x_1$$

We say, "The slope is given by the change in y over the change in x." We find the slope *at a given point* on the line by allowing any other point on the line to slide toward the point in question. At any position of the sliding point, we can calculate a slope by referring to the stationary point in question. The derivative of the function at the stationary point is the final slope that could be calculated in this intuitive matter. For straight lines the derivative at any point is given by the slope of the line. For nonlinear functions the end product of this intuitive process is more difficult to determine. Figure B.2 shows the complication. The rate of change in this case is given by

$$\frac{\Delta y}{\Delta x} = \frac{f(x_0 + \Delta x) - f(x_0)}{\Delta x}$$

Figure B.2

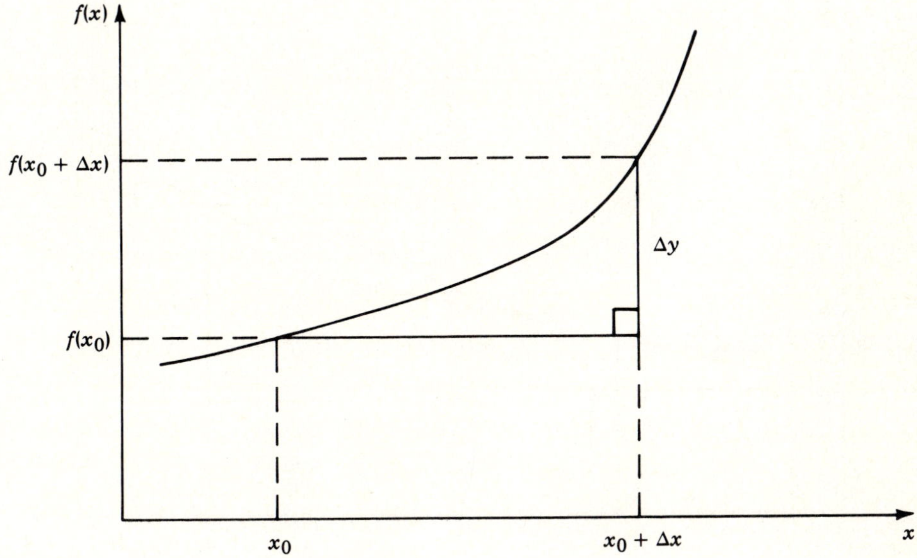

and the derivative at x_0 is the limiting value of $\Delta y/\Delta x$ as Δx approaches zero. Geometrically, this limit is the slope of the line tangent to $f(x)$ at the point $[x_0, f(x_0)]$. This relationship is shown in Figure B.3, with T denoting the tangent line.

EXAMPLE B.1 Find the derivative at x_0 for the quadratic function $f(x) = x^2 + 2x + 1$.

SOLUTION First,

$$\begin{aligned} f(x_0 + \Delta x) &= (x_0 + \Delta x)^2 + 2(x_0 + \Delta x) + 1 \\ &= x^2 + 2(\Delta x)x_0 + \Delta x^2 + 2x_0 + 2\Delta x + 1 \end{aligned}$$

Thus,

$$f(x_0 + \Delta x) - f(x_0) = 2(\Delta x)x_0 + \Delta x^2 + 2\Delta x$$

Finally,

$$\frac{\Delta y}{\Delta x} = 2x_0 + \Delta x + 2$$

Figure B.3

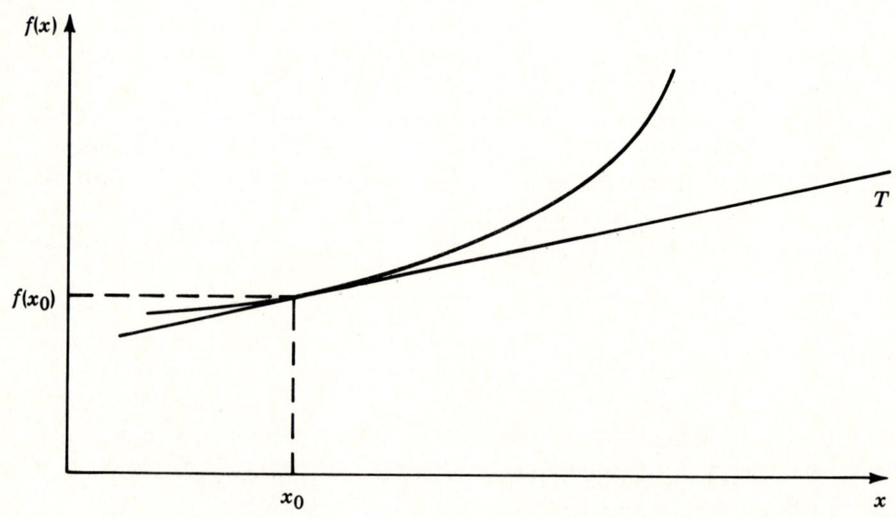

Clearly, as Δx approaches zero, $\Delta y/\Delta x$ approaches $2x_0 + 2$. This means that the derivative at x_0 must be given by $2x_0 + 2$, so that the slope of the tangent line changes according to the value of x_0.

Example B.1 illustrates both the proper computation of the derivative for a nonlinear function and the typical end product of the derivative being *another* function of x. Notationally, we write the derivative of $f(x)$ as

$$f'(x)$$

so that in the example we would indicate the derivative by

$$f'(x) = 2x + 2$$

This calculation of what can be thought of as the *instantaneous rate of change* is complicated for some functions, and the following rules can help to eliminate confusion.

1. If $f(x) = c$, where c is any constant, then $f'(x) = 0$.
2. If $f(x) = x$, then $f'(x) = 1$.
3. If $f(x) = x^m$, then $f'(x) = mx^{m-1}$.

Further, if $f(x)$ and $g(x)$ are two functions, then

1. The derivative of $f(x) + g(x)$ is $f'(x) + g'(x)$.
2. The derivative of $f(x)g(x)$ is $g(x)f'(x) + f(x)g'(x)$.
3. The derivative of $f(x)/g(x)$ is $[g(x)f'(x) - f(x)g'(x)]/[g(x)]^2$.

The table of derivatives on page 732 contains some useful results.

Local minima and maxima of functions can be found by use of the derivative. Figure B.4 on page 733 illustrates how this is intuitively reasonable from inspection of the tangent lines for selected points on the curve. Notice that at all three points (a, b, and c) the slope of the tangent (and hence the derivative) is zero, so that solving the equation

$$f'(x) = 0$$

for x gives us only those points at which local minima and maxima *may* exist. Here a is a local maximum and c is a local minimum, but b is neither a local maximum nor a local minimum. The curve causes us a complication at point b because it is concave upward over the interval $[s, b]$ but concave downward over the interval $[b, t]$.

Table B.1 Derivatives of Some Common Functions*

Function	Derivative
e^{ax}	ae^{ax}
a^x	$(\ln a)a^x$
$\ln x$	$\frac{1}{x}$
$\log_a x$	$\frac{1}{x(\ln a)}$
$f(x)^{g(x)}$	$\{g(x)[f(x)]^{g(x)-1}\}f'(x) + \{\ln(f(x))[f(x)]^{g(x)}\}g'(x)$
$\sin x$	$\cos x$
$\cos x$	$-\sin x$
$\tan x$	$\sec^2 x$
$\sec x$	$(\sec x)(\tan x)$

*a is a constant.

If we let $f''(x)$ denote "the derivative of the derivative," the following procedure suffices for categorization.

1. Let x_0 be a value such that $f'(x_0) = 0$.
2. If $f''(x_0) < 0$, then $f(x)$ has a local maximum at x_0.
3. If $f''(x_0) > 0$, then $f(x)$ has a local minimum at x_0.
4. If $f''(x_0) = 0$, compute $f'(x_0 - d)$ and $f'(x_0 + d)$ for a small value of d. If the two values thus computed are identical in sign, then x_0 is neither a local minimum nor a local maximum.

EXAMPLE B.2 Given $f(x) = 2x^3 + 3x^2 - 36x + 24$, find all local minima and maxima.

SOLUTION First, $f'(x) = 6x^2 + 6x - 36$. Setting $f'(x) = 0$ and solving for x gives

$$x = 2 \quad \text{and} \quad x = -3$$

Next, $f''(x) = 12x + 6$, so that

$$f''(2) = 12(2) + 6 = 30 \quad \text{and} \quad f''(-3) = 12(-3) + 6 = -30$$

Figure B.4

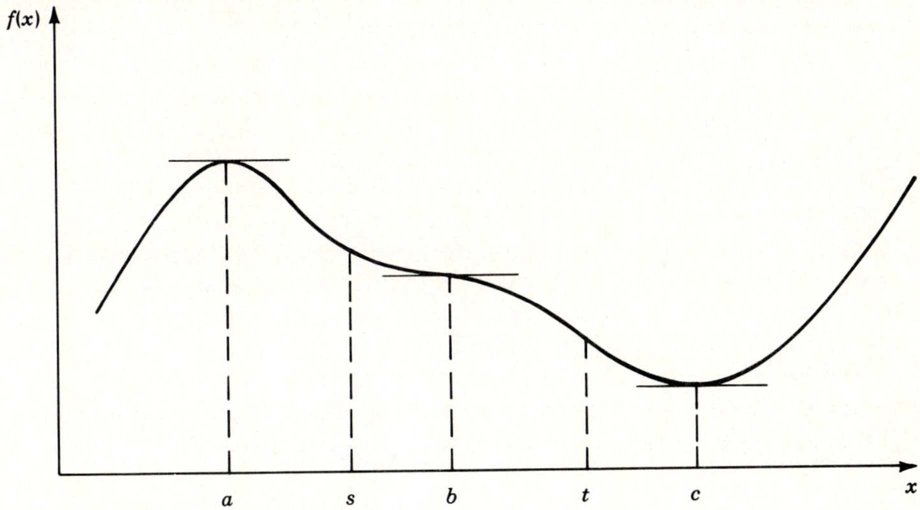

Since $f''(2) > 0$, $f(x)$ has a local minimum [equal to $f(2)$ or -20] at $x = 2$. Since $f''(-3) < 0$, $f(x)$ has a local maximum [equal to $f(-3)$ or 105] at $x = -3$.

EXAMPLE B.3 Given $f(x) = x^3 - 6x^2 + 12x - 4$, find all local minima and maxima.

SOLUTION First, $f'(x) = 0$ gives

$$3x^2 - 12x + 12 = 0$$

whose solution is $x = 2$. Now, $f''(x) = 6x - 12$, so that $f''(2) = 0$. In this case $x_0 = 2$ and we let $d = 1$, so that

$$f'(x_0 + d) = f'(3) = 3$$
$$f'(x_0 - d) = f'(1) = 3$$

Since each of these is positive, 2 is neither a local minimum nor a local maximum (that is, $x = 2$ gives the situation depicted at point b in Figure B.4), and so the function has no minimum or maximum.

B.2 INTEGRAL CALCULUS

Integral calculus allows us to understand many of the concepts associated with the calculation of probabilities for continuous phenomena. In these cases we compute probabilities by calculating areas under curves that, most often, are nonlinear in composition. For the sake of simplicity, let us assume that we have a function $f(x)$ that is continuous and nonnegative on the interval $[a, b]$. The problem at hand is to provide a method for computing the shaded area shown in Figure B.5.

We could use simple geometrical methods to approximate this area in a very direct fashion. We could do this by dividing the interval $[a, b]$ into n intervals in order to form n rectangles whose base is of width $(b - a)/n$ and whose height is $f(m_i)$, where m_i is the midpoint of the ith cell. This idea is illustrated in Figure B.6.

In Figure B.6 the width of the ith interval as shown is $(b - a)/n$, so that $w = v + (b - a)/n$ and $m_i = (v + w)/2$. Thus, the area approximation is given by

$$S_n = \sum_{i=1}^{n} f(m_i)\left(\frac{b - a}{n}\right)$$

Notice that we can improve the accuracy by increasing the number of rectangles and that the area we are after is actually given by the limit

Figure B.5

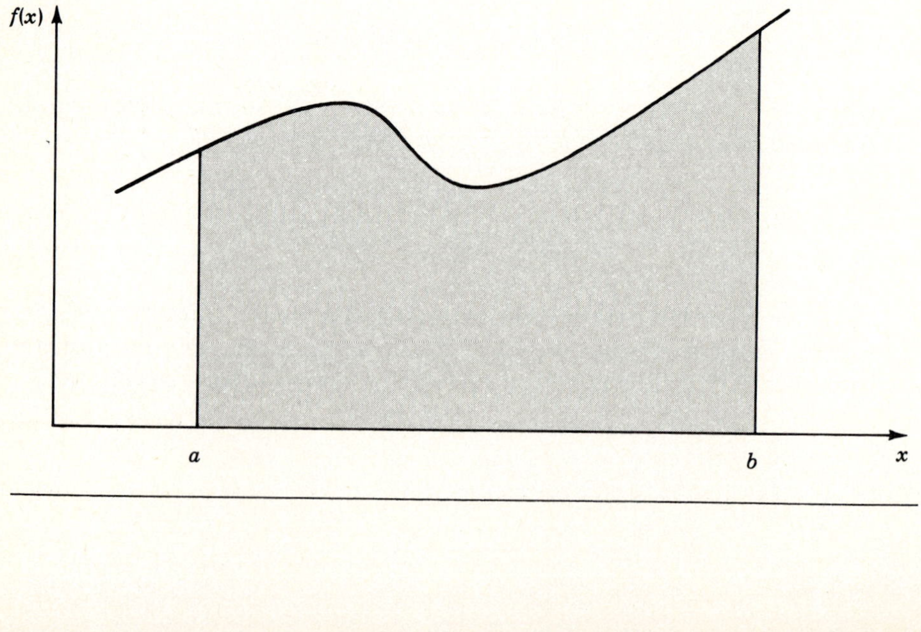

$$\lim_{n \to \infty} S_n$$

In fact, we can obtain the desired area (for a continous function) *independent* of how the rectangles are constructed so long as

1. The rectangle heights are given by $f(x_i)$, where x_i is any point in the ith subinterval of $[a, b]$ and $a < x_1 < x_2 < x_3 < \cdots < x_n < b$.
2. The maximum rectangle width (that is, the maximum subinterval width) approaches zero as $n \to \infty$.

This limit is written

$$_a\int^b f(x)\,dx$$

and is read "the definite integral of $f(x)$, with respect to x, from $x = a$ to $x = b$." Further, $f(x)$ is called the *integrand* and a and b are called the *lower* and *upper limits of integration*, respectively. Perhaps the most astounding result associated with this procedure can be found in the following statement known as the *fundamental theorem of integral calculus.*

If $f(x)$ is continuous on the interval $a \leqslant x \leqslant b$ and if $F'(x) = f(x)$ over this same interval, then

Figure B.6

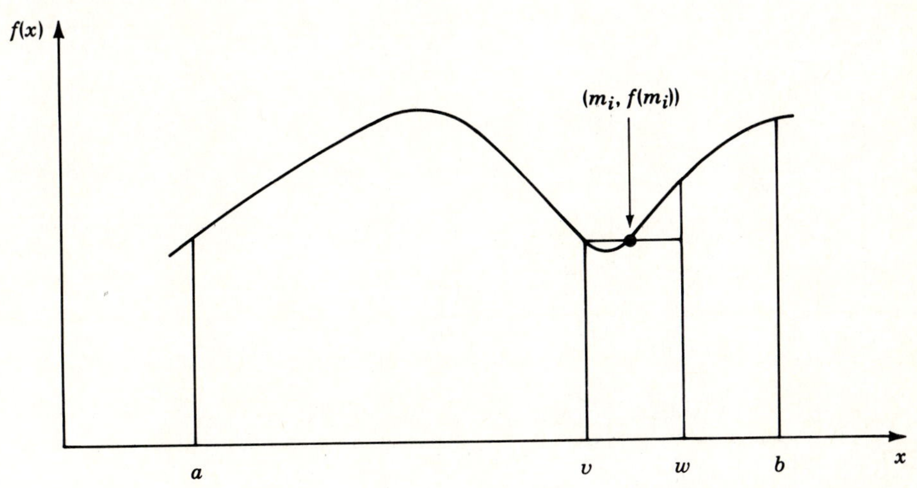

$$\int_a^b f(x)\,dx = F(b) - F(a)$$

Thus, we can find the exact area by using an extremely simple procedure:

1. Finding a function $F(x)$ whose derivative is $f(x)$
2. Evaluating $F(b)$ and $F(a)$, then subtracting the latter from the former

The simplicity of this procedure is apparent if we refer to Table B.1. That is, to find the functions $F(x)$ of interest, we simply locate $f(x)$ in the right-hand column, then read $F(x)$ as the corresponding entry in the left-hand column. For your convenience, refer to Table B.2, which contains some common expressions for $f(x)$.

The following properties of definite integrals are also useful in the computation of areas: Assuming that $f(x)$ and $g(x)$ are continuous over the interval $[a, b]$, we have

1. $\int_a^a f(x)\,dx = 0$

Table B.2 $F(x)$ Expressions for Common Choices of $f(x)$*

$f(x)$	$F(x)$
a	ax
x^a	$\dfrac{x^{a+1}}{a+1} \quad a \neq -1$
$\dfrac{1}{x}$	$\ln \lvert x \rvert$
a^x	$\dfrac{a^x}{\ln a}$
e^{ax}	$\dfrac{e^{ax}}{a}$
$\sin x$	$-\cos x$
$\cos x$	$\sin x$
$\tan x$	$\ln \lvert \sec x \rvert$

*a is a constant.

2. $\int_a^b cf(x)\,dx = c \int_a^b f(x)\,dx$ c being any constant

3. $\int_a^b [f(x) \pm g(x)]\,dx = \int_a^b f(x)\,dx \pm \int_a^b g(x)\,dx$

If, in addition we assume that

$$F'(x) = f(x) \qquad \text{and} \qquad G'(x) = g(x)$$

then

4. $\int_a^b F(x)g(x)\,dx = [F(b)G(b) - F(a)G(a)] - \int_a^b G(x)f(x)\,dx$

The execution of the last rule is called *integration by parts*; it is useful in a surprising number of elementary situations.

EXAMPLE B.4 Evaluate $\int_1^3 x^2\,dx$.

SOLUTION In this case $f(x) = x^2$ and $F(x) = \frac{1}{3}x^3$. We obtain this $F(x)$ function by referring to the second row of Table B.2, with $a = 2$. Thus,

$$\int_1^3 x^2\,dx = \frac{1}{3}(3)^3 - \frac{1}{3}(1)^3 = 9 - \frac{1}{3} = 8\frac{2}{3}$$

EXAMPLE B.5 Evaluate $\int_0^2 2e^{-2x}\,dx$

SOLUTION Referring to rule 2 for integrals, we get

$$\int_0^2 2e^{-2x}\,dx = 2\int_0^2 e^{-2x}\,dx$$

Next we need to find a function $F(x)$ such that $F'(x) = e^{-2x}$. From line 5 of Table B.2, if $F(x) = e^{-2x}$, then $F'(x) = -2e^{-2x}$. This means that if we choose $F(x) = -\frac{1}{2}e^{-2x}$, then $F'(x) = e^{-2x}$. Therefore,

$$2\int_0^2 e^{-2x}\,dx = 2\left[-\frac{1}{2}e^{-2(2)} - \left(-\frac{1}{2}\right)e^{-2(0)}\right] = 1 - e^{-4} = 0.9817$$

EXAMPLE B.6 Evaluate $\int_0^3 (3 - 2x + x^2)\,dx$.

SOLUTION Using Rule 3, we get

$$\int_0^3 (3 - 2x + x^2)\,dx = \int_0^3 3\,dx - \int_0^3 2x\,dx + \int_0^3 x^2\,dx$$

Next, we apply Rule 2 to get

$$\int_0^3 3\,dx - \int_0^3 2x\,dx + \int_0^3 x^2\,dx = \int_0^3 3\,dx - 2\int_0^3 x\,dx + \int_0^3 x^2\,dx$$

Finally, we need to determine $F_1(x)$, $F_2(x)$, and $F_3(x)$ such that $F_1'(x) = 3$; $F_2'(x) = x$; and $F_3'(x) = x^2$. Table B.2 gives us the correct expressions:

$F_1(x) = 3x$ (line 1 of Table B.2 with $a = 3$)

$F_2(x) = \frac{1}{2}x^2$ (line 2 of Table B.2 with $a = 1$)

$F_3(x) = \frac{1}{3}x^3$ (line 2 of Table B.2 with $a = 2$)

This means that

$$\int_0^3 (3 - 2x + x^2)\,dx = [3(3) - 3(0)] - 2\left[\frac{1}{2}(3)^2 - \frac{1}{2}(0)^2\right] + \frac{1}{3}[(3)^3 - (0)^3]$$

$$= 9 - 2\left(\frac{1}{2}\right)(9) + \left(\frac{1}{3}\right)(27) = 9$$

EXAMPLE B.7 Evaluate $\int_0^\infty xe^{-x}\,dx$

SOLUTION Referring to Rule 4, we choose

$$F(x) = x \quad \text{and} \quad g(x) = e^{-x}$$

From Rule 3 for derivatives, $F'(x) = 1(x)^0 = 1$, so that $f(x) = 1$, and from line 5 of Table B.2, $G(x) = -e^{-x}$. Thus, using integration by parts, we get

$$
\begin{aligned}
{}_0\int^{\infty} xe^{-x}\,dx &= {}_0\int^{\infty} F(x)g(x)\,dx \\
&= [F(\infty)G(\infty) - F(0)G(0)] - {}_0\int^{\infty} G(x)f(x)\,dx \\
&= \left[\lim_{x\to\infty} -x(e^{-x}) - 0(1)\right] - {}_0\int^{\infty} (-e^{-x})(1)\,dx \\
&= [0 - 0] + {}_0\int^{\infty} e^{-x}\,dx = {}_0\int^{\infty} e^{-x}\,dx \\
&= G(\infty) - G(0) = [-e^{-(\infty)}] - [-e^{-0}] = 0 - (-1) = 1
\end{aligned}
$$

EXERCISES

B.1 Evaluate ${}_0\int^{3} (2 + x)\,dx$.

B.2 Evaluate ${}_0\int^{3} (x + 1)^2\,dx$.

B.3 Evaluate ${}_0\int^{2} x^3(x^2 + 1)\,dx$.

B.4 Evaluate ${}_1\int^{2.5} \ln x\,dx$.

B.5 Evaluate ${}_1\int^{e} \ln x\,dx$.

B.6 Evaluate ${}_0\int^{\infty} \left(x - \frac{1}{4}\right)^2 (4e^{-4x})\,dx$. (*Hint*: Use integration by parts for this calculation of the variance of a negative exponential distribution, with $\beta = 4$.)

Answers

1. 10.5000
2. 21.0000
3. 14.6667
4. 0.7907
5. 1.0000
6. 0.0625

Appendix C
Review of Summation Notation and Basic Convergence Properties of Series

Series occur with surprising regularity in common business problems, especially in the areas of accounting, finance, econometrics, and management science. Therefore, the following fundamental principles are assembled for your convenience and review.

A *sequence* is a set of numbers, and a *series* is nothing more than the sum of the numbers in a sequence. Thus,

$$\{2, 3, 5, 7\}$$

is a sequence as well as a set, whereas the addition $2 + 3 + 5 + 7$ is considered a series. In the former case the numbers are called *elements*, and in the latter they are called *terms.* Any series that contains an infinite number of terms is called an *infinite series*; otherwise it is called a *finite series.*

We use summation notation to help us categorize and convey the information that may be contained in a series. The next few conventions are universal in their application.

First, the terms in the series are denoted by a *subscripted variable.* Common choices for subscripts are the letters i, j, k, l, m, and n.

Second, a series notation has been developed that utilizes a capital sigma. Using this notation we would write the series

$$a_1 + a_2 + a_3 + a_4 + a_5$$

as

$$\sum_{j=1}^{5} a_j \quad \text{or as} \quad \sum_{k=1}^{5} a_k \quad \text{or as} \quad \sum_{m=1}^{5} a_m$$

The differences in the three approaches lie only in the choice of the letter used to denote the subscripts.

In general, the number at the bottom of the capital sigma is called the *lower limit of summation*; it tells us which sequence element starts the series. The number at the top of the sigma is called the *upper limit of summation*; it tells us which sequence element ends the series. Thus

$$a_5 + a_6 + a_7 + a_8 + \cdots + a_{21}$$

would be written

$$\sum_{k=5}^{21} a_k$$

Notice that the implicit assumption is that we move from the lower limit of 5 to the upper limit of 21 in *increments of one unit.*

The following rules for summation can be useful. Assume that $\Sigma\, a_k$ and $\Sigma\, b_k$ result in a finite sum and that they have the same limits of summation. If c is a constant, then

1. $\Sigma\, ca_k = c\, \Sigma\, a_k$
2. $\Sigma\, (a_k \pm b_k) = (\Sigma\, a_k) \pm (\Sigma\, b_k)$

EXAMPLE C.1 Express the series $2 + 4 + 6 + 8 + 10$ using summation notation.

SOLUTION The series is of the form

$$2(1) + 2(2) + 2(3) + 2(4) + 2(5) = 2(1 + 2 + 3 + 4 + 5) = 2 \sum_{i=1}^{5} i$$

EXAMPLE C.2 Evaluate the series $\sum_{m=1}^{4} (2m + 3)$.

SOLUTION $[2(1) + 3] + [2(2) + 3] + [2(3) + 3] + [2(4) + 3]$
$= (2 + 3) + (4 + 3) + (6 + 3) + (8 + 3) = 20 + 3(4)$
$= 32$

Alternatively, we could use Rules 1 and 2 to get

$$\begin{aligned}\sum_{m=1}^{4} (2m + 3) &= \left(\sum_{m=1}^{4} 2m\right) + \left(\sum_{m=1}^{4} 3\right) \\ &= 2\left(\sum_{m=1}^{4} m\right) + (3 + 3 + 3 + 3) \\ &= 2(1 + 2 + 3 + 4) + 12 \\ &= 2(10) + 12 = 20 + 12 = 32\end{aligned}$$

Example C.2 illustrates two useful results in the area of series calculations, namely

3. $\sum_{k=1}^{n} k = \dfrac{n(n+1)}{2}$ (a formula for the sum of the first n integers)

4. $\sum_{k=1}^{n} c = c(n)$ where c is a constant

The following example emphasizes another convention.

EXAMPLE C.3 Evaluate the series $\sum_{m=1}^{4} 2m + 3$.

SOLUTION

$$\sum_{m=1}^{4} 2m + 3 = \left(\sum_{m=1}^{4} 2m\right) + 3 = 2\left(\sum_{m=1}^{4} m\right) + 3$$

$$= 2\left[\frac{4(5)}{2}\right] + 3 \qquad \text{Rule 3 with } n = 4$$

$$= 23$$

Comparing Examples C.2 and C.3, we see that the expression to be evaluated for each value of the subscript will be enclosed in parentheses if it contains more than one term. This accounts for the difference between

$$\sum_{m=1}^{4} 2m + 3 \qquad \text{and} \qquad \sum_{m=1}^{4} (2m + 3)$$

A very important series used in business analysis is the geometric series. The series is of the form

$$a + ar + ar^2 + ar^3 + \cdots + ar^{n-1} + \cdots$$

in which a and r are constants. Notice that the ratio of any term to its immediate predecessor is constant at value r. If $r = 1$, the sum of the first n terms is given by

$$\sum_{k=0}^{n-1} ar^k = \sum_{k=0}^{n-1} a = a + \sum_{k=1}^{n-1} a$$

$$= a + a(n - 1) \qquad \text{by Rule 4}$$

$$= an$$

If $r \neq 1$, the sum of the first n terms is given by

$$\sum_{k=0}^{n-1} ar^k = a\left(\frac{1 - r^n}{1 - r}\right)$$

EXAMPLE C.4 Evaluate the series $\sum_{k=0}^{5} \left(\frac{1}{2}\right)^k$.

SOLUTION In this case $a = 1$, $r = \frac{1}{2}$, and $n - 1 = 5$. Thus,

$$\sum_{k=0}^{5} \left(\frac{1}{2}\right)^k = 1\left[\frac{1 - (1/2)^6}{1 - (1/2)}\right] = 2\left[1 - \left(\frac{1}{2}\right)^6\right] = 1.96875$$

EXAMPLE C.5 Evaluate the series $\sum_{k=1}^{m} p(1 - p)^{k-1}$.

SOLUTION This is a geometric series with $a = p$, $r = 1 - p$, and $n = m$. Thus, its sum is given by

$$p\left[\frac{1 - (1 - p)^m}{1 - (1 - p)}\right] = 1 - (1 - p)^m$$

This last result is the expression for the cumulative distribution function for a geometric distribution with parameter p.

EXERCISES

C.1 Express $2^0 + 2^1 + 2^2 + 2^3 + 2^4$ using summation notation.

C.2 Express $[3(1) + 0] + [3(2) + 1] + [3(3) + 2] + [3(4) + 3]$ using summation notation.

C.3 Evaluate $\sum_{k=1}^{6} 5k + 2$.

C.4 Evaluate $\sum_{m=1}^{5} (3^m + 2)$.

C.5 Evaluate $\sum_{n=7}^{30} 5$.

C.6 Evaluate $\sum_{j=0}^{10} 30\left(\frac{1}{2}\right)^j$.

C.7 Evaluate $\sum_{i=5}^{10} (3i + i^2)$.

C.8 Evaluate $\sum_{x=0}^{n} \binom{n}{x} p^x (1-p)^{n-x}$, where $0 < p < 1$.

Answers

1. $\sum_{i=0}^{4} 2^i$

2. $\sum_{k=0}^{3} [3(k+1)+k]$ or $\sum_{k=1}^{4} [3k+(k-1)]$ or $\sum_{k=0}^{3} (4k+3)$

3. 107

4. 373

5. 120

6. $30\left[\dfrac{1-(1/2)^{11}}{1-(1/2)}\right] \approx 60[1-0.0005] = 59.97 \approx 60$

7. 490

8. 1

LIST OF TABLES

Table A.1 Individual Probabilities for Selected Binomial Distributions

n=2

	p									
	0.010	0.050	0.100	0.150	0.200	0.250	0.300	0.400	0.500	
0	0.9801	0.9025	0.8100	0.7225	0.6400	0.5625	0.4900	0.3600	0.2500	2
1	0.0198	0.0950	0.1800	0.2550	0.3200	0.3750	0.4200	0.4800	0.5000	1
2	0.0001	0.0025	0.0100	0.0225	0.0400	0.0625	0.0900	0.1600	0.2500	0

n=3

	p									
	0.010	0.050	0.100	0.150	0.200	0.250	0.300	0.400	0.500	
0	0.9703	0.8574	0.7290	0.6141	0.5120	0.4219	0.3430	0.2160	0.1250	3
1	0.0294	0.1354	0.2430	0.3251	0.3840	0.4219	0.4410	0.4320	0.3750	2
2	0.0003	0.0071	0.0270	0.0574	0.0960	0.1406	0.1890	0.2880	0.3750	1
3	0.0000	0.0001	0.0010	0.0034	0.0080	0.0156	0.0270	0.0640	0.1250	0

n=4

	p									
	0.010	0.050	0.100	0.150	0.200	0.250	0.300	0.400	0.500	
0	0.9606	0.8145	0.6561	0.5220	0.4096	0.3164	0.2401	0.1296	0.0625	4
1	0.0388	0.1715	0.2916	0.3685	0.4096	0.4219	0.4116	0.3456	0.2500	3
2	0.0006	0.0135	0.0486	0.0975	0.1536	0.2109	0.2646	0.3456	0.3750	2
3	0.0000	0.0005	0.0036	0.0115	0.0256	0.0469	0.0756	0.1536	0.2500	1
4	0.0000	0.0000	0.0001	0.0005	0.0016	0.0039	0.0081	0.0256	0.0625	0

n=5

	p									
	0.010	0.050	0.100	0.150	0.200	0.250	0.300	0.400	0.500	
0	0.9510	0.7738	0.5905	0.4437	0.3277	0.2373	0.1681	0.0778	0.0312	5
1	0.0480	0.2036	0.3280	0.3915	0.4096	0.3955	0.3601	0.2592	0.1562	4
2	0.0010	0.0214	0.0729	0.1382	0.2048	0.2637	0.3087	0.3456	0.3125	3
3	0.0000	0.0011	0.0081	0.0244	0.0512	0.0879	0.1323	0.2304	0.3125	2
4	0.0000	0.0000	0.0005	0.0022	0.0064	0.0146	0.0283	0.0768	0.1562	1
5	0.0000	0.0000	0.0000	0.0001	0.0003	0.0010	0.0024	0.0102	0.0313	0

n=10

	p									
	0.010	0.050	0.100	0.150	0.200	0.250	0.300	0.400	0.500	
0	0.9044	0.5987	0.3487	0.1969	0.1074	0.0563	0.0282	0.0060	0.0010	10
1	0.0914	0.3151	0.3874	0.3474	0.2684	0.1877	0.1211	0.0403	0.0098	9
2	0.0042	0.0746	0.1937	0.2759	0.3020	0.2816	0.2335	0.1209	0.0439	8
3	0.0001	0.0105	0.0574	0.1298	0.2013	0.2503	0.2668	0.2150	0.1172	7
4	0.0000	0.0010	0.0112	0.0401	0.0881	0.1460	0.2001	0.2508	0.2051	6
5	0.0000	0.0001	0.0015	0.0085	0.0264	0.0584	0.1029	0.2007	0.2461	5
6	0.0000	0.0000	0.0001	0.0012	0.0055	0.0162	0.0368	0.1115	0.2051	4
7	0.0000	0.0000	0.0000	0.0001	0.0008	0.0031	0.0090	0.0425	0.1172	3
8	0.0000	0.0000	0.0000	0.0000	0.0001	0.0004	0.0014	0.0106	0.0439	2
9	0.0000	0.0000	0.0000	0.0000	0.0000	0.0000	0.0001	0.0016	0.0098	1
10	0.0000	0.0000	0.0000	0.0000	0.0000	0.0000	0.0000	0.0001	0.0010	0

(Table A.1 continued)

n = 15

	p									
	0.010	0.050	0.100	0.150	0.200	0.250	0.300	0.400	0.500	
0	0.8601	0.4633	0.2059	0.0874	0.0352	0.0134	0.0047	0.0005	0.0000	15
1	0.1303	0.3658	0.3432	0.2312	0.1319	0.0668	0.0305	0.0047	0.0005	14
2	0.0092	0.1348	0.2669	0.2856	0.2309	0.1559	0.0916	0.0219	0.0032	13
3	0.0004	0.0307	0.1285	0.2184	0.2501	0.2252	0.1700	0.0634	0.0139	12
4	0.0000	0.0049	0.0428	0.1156	0.1876	0.2252	0.2186	0.1268	0.0417	11
5	0.0000	0.0006	0.0105	0.0449	0.1032	0.1651	0.2061	0.1859	0.0916	10
6	0.0000	0.0000	0.0019	0.0132	0.0430	0.0917	0.1472	0.2066	0.1527	9
7	0.0000	0.0000	0.0003	0.0030	0.0138	0.0393	0.0811	0.1771	0.1964	8
8	0.0000	0.0000	0.0000	0.0005	0.0035	0.0131	0.0348	0.1181	0.1964	7
9	0.0000	0.0000	0.0000	0.0001	0.0007	0.0034	0.0116	0.0612	0.1527	6
10	0.0000	0.0000	0.0000	0.0000	0.0001	0.0007	0.0030	0.0245	0.0916	5
11	0.0000	0.0000	0.0000	0.0000	0.0000	0.0001	0.0006	0.0074	0.0417	4
12	0.0000	0.0000	0.0000	0.0000	0.0000	0.0000	0.0001	0.0016	0.0139	3
13	0.0000	0.0000	0.0000	0.0000	0.0000	0.0000	0.0000	0.0003	0.0032	2
14	0.0000	0.0000	0.0000	0.0000	0.0000	0.0000	0.0000	0.0000	0.0005	1
15	0.0000	0.0000	0.0000	0.0000	0.0000	0.0000	0.0000	0.0000	0.0000	0

n = 20

	p									
	0.010	0.050	0.100	0.150	0.200	0.250	0.300	0.400	0.500	
0	0.8179	0.3585	0.1216	0.0388	0.0115	0.0032	0.0008	0.0000	0.0000	10
1	0.1652	0.3774	0.2702	0.1368	0.0576	0.0211	0.0068	0.0005	0.0000	9
2	0.0159	0.1887	0.2852	0.2293	0.1369	0.0669	0.0278	0.0031	0.0002	8
3	0.0010	0.0596	0.1901	0.2428	0.2054	0.1339	0.0716	0.0123	0.0011	7
4	0.0000	0.0133	0.0898	0.1821	0.2182	0.1897	0.1304	0.0350	0.0046	6
5	0.0000	0.0022	0.0319	0.1028	0.1746	0.2023	0.1789	0.0746	0.0148	5
6	0.0000	0.0003	0.0089	0.0454	0.1091	0.1686	0.1916	0.1244	0.0370	4
7	0.0000	0.0000	0.0020	0.0160	0.0545	0.1124	0.1643	0.1659	0.0739	3
8	0.0000	0.0000	0.0004	0.0046	0.0222	0.0609	0.1144	0.1797	0.1201	2
9	0.0000	0.0000	0.0001	0.0011	0.0074	0.0271	0.0654	0.1597	0.1602	1
10	0.0000	0.0000	0.0000	0.0002	0.0020	0.0099	0.0308	0.1171	0.1762	20
11	0.0000	0.0000	0.0000	0.0000	0.0005	0.0030	0.0120	0.0710	0.1602	19
12	0.0000	0.0000	0.0000	0.0000	0.0001	0.0008	0.0039	0.0355	0.1201	18
13	0.0000	0.0000	0.0000	0.0000	0.0000	0.0002	0.0010	0.0146	0.0739	17
14	0.0000	0.0000	0.0000	0.0000	0.0000	0.0000	0.0002	0.0049	0.0370	16
15	0.0000	0.0000	0.0000	0.0000	0.0000	0.0000	0.0000	0.0013	0.0148	15
16	0.0000	0.0000	0.0000	0.0000	0.0000	0.0000	0.0000	0.0003	0.0046	14
17	0.0000	0.0000	0.0000	0.0000	0.0000	0.0000	0.0000	0.0000	0.0011	13
18	0.0000	0.0000	0.0000	0.0000	0.0000	0.0000	0.0000	0.0000	0.0002	12
19	0.0000	0.0000	0.0000	0.0000	0.0000	0.0000	0.0000	0.0000	0.0000	11

n = 25

	p									
	0.010	0.050	0.100	0.150	0.200	0.250	0.300	0.400	0.500	
0	0.7778	0.2774	0.0718	0.0172	0.0038	0.0008	0.0001	0.0000	0.0000	25
1	0.1964	0.3650	0.1994	0.0759	0.0236	0.0063	0.0014	0.0000	0.0000	24
2	0.0238	0.2305	0.2659	0.1607	0.0708	0.0251	0.0074	0.0004	0.0000	23
3	0.0018	0.0930	0.2265	0.2174	0.1358	0.0641	0.0243	0.0019	0.0001	22
4	0.0001	0.0269	0.1384	0.2110	0.1867	0.1175	0.0572	0.0071	0.0004	21
5	0.0000	0.0060	0.0646	0.1564	0.1960	0.1645	0.1030	0.0199	0.0016	20
6	0.0000	0.0010	0.0239	0.0920	0.1633	0.1828	0.1472	0.0442	0.0053	19
7	0.0000	0.0001	0.0072	0.0441	0.1108	0.1654	0.1712	0.0800	0.0143	18
8	0.0000	0.0000	0.0018	0.0175	0.0623	0.1241	0.1651	0.1200	0.0322	17
9	0.0000	0.0000	0.0004	0.0058	0.0294	0.0781	0.1336	0.1511	0.0609	16
10	0.0000	0.0000	0.0001	0.0016	0.0118	0.0417	0.0916	0.1612	0.0974	15
11	0.0000	0.0000	0.0000	0.0004	0.0040	0.0189	0.0536	0.1465	0.1328	14
12	0.0000	0.0000	0.0000	0.0001	0.0012	0.0074	0.0268	0.1140	0.1550	13
13	0.0000	0.0000	0.0000	0.0000	0.0003	0.0025	0.0115	0.0760	0.1550	12
14	0.0000	0.0000	0.0000	0.0000	0.0001	0.0007	0.0042	0.0434	0.1328	11
15	0.0000	0.0000	0.0000	0.0000	0.0000	0.0002	0.0013	0.0212	0.0974	10
16	0.0000	0.0000	0.0000	0.0000	0.0000	0.0000	0.0004	0.0088	0.0609	9
17	0.0000	0.0000	0.0000	0.0000	0.0000	0.0000	0.0001	0.0031	0.0322	8
18	0.0000	0.0000	0.0000	0.0000	0.0000	0.0000	0.0000	0.0009	0.0143	7
19	0.0000	0.0000	0.0000	0.0000	0.0000	0.0000	0.0000	0.0002	0.0053	6
20	0.0000	0.0000	0.0000	0.0000	0.0000	0.0000	0.0000	0.0000	0.0016	5
21	0.0000	0.0000	0.0000	0.0000	0.0000	0.0000	0.0000	0.0000	0.0004	4
22	0.0000	0.0000	0.0000	0.0000	0.0000	0.0000	0.0000	0.0000	0.0001	3
23	0.0000	0.0000	0.0000	0.0000	0.0000	0.0000	0.0000	0.0000	0.0000	2

(Table A.1 continued)

n=50

	p									
	0.010	0.050	0.100	0.150	0.200	0.250	0.300	0.400	0.500	
0	0.6050	0.0769	0.0052	0.0003	0.0000	0.0000	0.0000	0.0000	0.0000	50
1	0.3056	0.2025	0.0286	0.0026	0.0002	0.0000	0.0000	0.0000	0.0000	49
2	0.0756	0.2611	0.0779	0.0113	0.0011	0.0001	0.0000	0.0000	0.0000	48
3	0.0122	0.2199	0.1386	0.0319	0.0044	0.0004	0.0000	0.0000	0.0000	47
4	0.0015	0.1360	0.1809	0.0661	0.0128	0.0016	0.0001	0.0000	0.0000	46
5	0.0001	0.0658	0.1849	0.1072	0.0295	0.0049	0.0006	0.0000	0.0000	45
6	0.0000	0.0260	0.1541	0.1419	0.0554	0.0123	0.0018	0.0000	0.0000	44
7	0.0000	0.0086	0.1076	0.1575	0.0870	0.0259	0.0048	0.0000	0.0000	43
8	0.0000	0.0024	0.0643	0.1493	0.1169	0.0463	0.0110	0.0002	0.0000	42
9	0.0000	0.0006	0.0333	0.1230	0.1364	0.0721	0.0220	0.0005	0.0000	41
10	0.0000	0.0001	0.0152	0.0890	0.1398	0.0985	0.0386	0.0014	0.0000	40
11	0.0000	0.0000	0.0061	0.0571	0.1271	0.1194	0.0602	0.0035	0.0000	39
12	0.0000	0.0000	0.0022	0.0328	0.1033	0.1294	0.0838	0.0076	0.0001	38
13	0.0000	0.0000	0.0007	0.0169	0.0755	0.1261	0.1050	0.0147	0.0003	37
14	0.0000	0.0000	0.0002	0.0079	0.0499	0.1110	0.1189	0.0260	0.0008	36
15	0.0000	0.0000	0.0001	0.0033	0.0299	0.0888	0.1223	0.0415	0.0020	35
16	0.0000	0.0000	0.0000	0.0013	0.0164	0.0648	0.1147	0.0606	0.0044	34
17	0.0000	0.0000	0.0000	0.0005	0.0082	0.0432	0.0983	0.0808	0.0087	33
18	0.0000	0.0000	0.0000	0.0001	0.0037	0.0264	0.0772	0.0987	0.0160	32
19	0.0000	0.0000	0.0000	0.0000	0.0016	0.0148	0.0558	0.1109	0.0270	31
20	0.0000	0.0000	0.0000	0.0000	0.0006	0.0077	0.0370	0.1146	0.0419	30
21	0.0000	0.0000	0.0000	0.0000	0.0002	0.0036	0.0227	0.1091	0.0598	29
22	0.0000	0.0000	0.0000	0.0000	0.0001	0.0016	0.0128	0.0959	0.0788	28
23	0.0000	0.0000	0.0000	0.0000	0.0000	0.0006	0.0067	0.0778	0.0960	27
24	0.0000	0.0000	0.0000	0.0000	0.0000	0.0002	0.0032	0.0584	0.1080	26
25	0.0000	0.0000	0.0000	0.0000	0.0000	0.0001	0.0014	0.0405	0.1123	25
26	0.0000	0.0000	0.0000	0.0000	0.0000	0.0000	0.0006	0.0259	0.1080	24
27	0.0000	0.0000	0.0000	0.0000	0.0000	0.0000	0.0002	0.0154	0.0960	23
28	0.0000	0.0000	0.0000	0.0000	0.0000	0.0000	0.0001	0.0084	0.0788	22
29	0.0000	0.0000	0.0000	0.0000	0.0000	0.0000	0.0000	0.0043	0.0598	21
30	0.0000	0.0000	0.0000	0.0000	0.0000	0.0000	0.0000	0.0020	0.0419	20
31	0.0000	0.0000	0.0000	0.0000	0.0000	0.0000	0.0000	0.0009	0.0270	19
32	0.0000	0.0000	0.0000	0.0000	0.0000	0.0000	0.0000	0.0003	0.0160	18
33	0.0000	0.0000	0.0000	0.0000	0.0000	0.0000	0.0000	0.0001	0.0087	17
34	0.0000	0.0000	0.0000	0.0000	0.0000	0.0000	0.0000	0.0000	0.0044	16
35	0.0000	0.0000	0.0000	0.0000	0.0000	0.0000	0.0000	0.0000	0.0020	15
36	0.0000	0.0000	0.0000	0.0000	0.0000	0.0000	0.0000	0.0000	0.0008	14
37	0.0000	0.0000	0.0000	0.0000	0.0000	0.0000	0.0000	0.0000	0.0003	13
38	0.0000	0.0000	0.0000	0.0000	0.0000	0.0000	0.0000	0.0000	0.0001	12
39	0.0000	0.0000	0.0000	0.0000	0.0000	0.0000	0.0000	0.0000	0.0000	11

Table A.2 Cumulative Probabilities for Selected Binomial Distributions

n=2

p

	0.010	0.050	0.100	0.150	0.200	0.250	0.300	0.400	0.500	
0	0.9801	0.9025	0.8100	0.7225	0.6400	0.5625	0.4900	0.3600	0.2500	2
1	0.9999	0.9975	0.9900	0.9775	0.9600	0.9375	0.9100	0.8400	0.7500	1
2	1.0000	1.0000	1.0000	1.0000	1.0000	1.0000	1.0000	1.0000	1.0000	0

n=3

p

	0.10	0.050	0.100	0.150	0.200	0.250	0.300	0.400	0.500	
0	0.9703	0.8574	0.7290	0.6141	0.5120	0.4219	0.3430	0.2160	0.1250	3
1	0.9997	0.9927	0.9720	0.9392	0.8960	0.8437	0.7840	0.6480	0.5000	2
2	1.0000	0.9999	0.9990	0.9966	0.9920	0.9844	0.9730	0.9360	0.8750	1
3	1.0000	1.0000	1.0000	1.0000	1.0000	1.0000	1.0000	1.0000	1.0000	0

n=4

p

	0.010	0.050	0.100	0.150	0.200	0.250	0.300	0.400	0.500	
0	0.9606	0.8145	0.6561	0.5220	0.4096	0.3164	0.2401	0.1296	0.0625	4
1	0.9994	0.9860	0.9477	0.8905	0.8192	0.7383	0.6517	0.4752	0.3125	3
2	1.0000	0.9995	0.9963	0.9880	0.9728	0.9492	0.9163	0.8208	0.6875	2
3	1.0000	1.0000	0.9999	0.9995	0.9984	0.9961	0.9919	0.9744	0.9375	1
4	1.0000	1.0000	1.0000	1.0000	1.0000	1.0000	1.0000	1.0000	1.0000	0

n=5

p

	0.010	0.050	0.100	0.150	0.200	0.250	0.300	0.400	0.500	
0	0.9510	0.7738	0.5905	0.4437	0.3277	0.2373	0.1681	0.0778	0.0312	5
1	0.9990	0.9774	0.9185	0.8352	0.7373	0.6328	0.5282	0.3370	0.1875	4
2	1.0000	0.9988	0.9914	0.9734	0.9421	0.8965	0.8369	0.6826	0.5000	3
3	1.0000	1.0000	0.9995	0.9978	0.9933	0.9844	0.9692	0.9130	0.8125	2
4	1.0000	1.0000	1.0000	0.9999	0.9997	0.9990	0.9976	0.9898	0.9687	1
5	1.0000	1.0000	1.0000	1.0000	1.0000	1.0000	1.0000	1.0000	1.0000	0

n=10

p

	0.010	0.050	0.100	0.150	0.200	0.250	0.300	0.400	0.500	
0	0.9044	0.5987	0.3487	0.1969	0.1074	0.0563	0.0282	0.0060	0.0010	10
1	0.9957	0.9139	0.7361	0.5443	0.3758	0.2440	0.1493	0.0464	0.0107	9
2	0.9999	0.9885	0.9298	0.8202	0.6778	0.5256	0.3828	0.1673	0.0547	8
3	1.0000	0.9990	0.9872	0.9500	0.8791	0.7759	0.6496	0.3823	0.1719	7
4	1.0000	0.9999	0.9984	0.9901	0.9672	0.9219	0.8497	0.6331	0.3770	6
5	1.0000	1.0000	0.9999	0.9986	0.9936	0.9803	0.9527	0.8338	0.6230	5
6	1.0000	1.0000	1.0000	0.9999	0.9991	0.9965	0.9894	0.9452	0.8281	4
7	1.0000	1.0000	1.0000	1.0000	0.9999	0.9996	0.9984	0.9877	0.9453	3
8	1.0000	1.0000	1.0000	1.0000	1.0000	1.0000	0.9999	0.9983	0.9893	2
9	1.0000	1.0000	1.0000	1.0000	1.0000	1.0000	1.0000	0.9999	0.9990	1
10	1.0000	1.0000	1.0000	1.0000	1.0000	1.0000	1.0000	1.0000	1.0000	0

(Table A.2 continued)

n = 15

	p									
	0.010	0.050	0.100	0.150	0.200	0.250	0.300	0.400	0.500	
0	0.8601	0.4633	0.2059	0.0874	0.0352	0.0134	0.0047	0.0005	0.0000	15
1	0.9904	0.8290	0.5490	0.3186	0.1671	0.0802	0.0353	0.0052	0.0005	14
2	0.9996	0.9638	0.8159	0.6042	0.3980	0.2361	0.1268	0.0271	0.0037	13
3	1.0000	0.9945	0.9444	0.8227	0.6482	0.4613	0.2969	0.0905	0.0176	12
4	1.0000	0.9994	0.9873	0.9383	0.8358	0.6865	0.5155	0.2173	0.0592	11
5	1.0000	0.9999	0.9978	0.9832	0.9389	0.8516	0.7216	0.4032	0.1509	10
6	1.0000	1.0000	0.9997	0.9964	0.9819	0.9434	0.8689	0.6098	0.3036	9
7	1.0000	1.0000	1.0000	0.9994	0.9958	0.9827	0.9500	0.7869	0.5000	8
8	1.0000	1.0000	1.0000	0.9999	0.9992	0.9958	0.9848	0.9050	0.6964	7
9	1.0000	1.0000	1.0000	1.0000	0.9999	0.9992	0.9963	0.9662	0.8491	6
10	1.0000	1.0000	1.0000	1.0000	1.0000	0.9999	0.9993	0.9907	0.9408	5
11	1.0000	1.0000	1.0000	1.0000	1.0000	1.0000	0.9999	0.9981	0.9824	4
12	1.0000	1.0000	1.0000	1.0000	1.0000	1.0000	1.0000	0.9997	0.9963	3
13	1.0000	1.0000	1.0000	1.0000	1.0000	1.0000	1.0000	1.0000	0.9995	2
14	1.0000	1.0000	1.0000	1.0000	1.0000	1.0000	1.0000	1.0000	1.0000	1

n = 20

	p									
	0.010	0.050	0.100	0.150	0.200	0.250	0.300	0.400	0.500	
0	0.8179	0.3585	0.1216	0.0388	0.0115	0.0032	0.0008	0.0000	0.0000	20
1	0.9831	0.7358	0.3917	0.1756	0.0692	0.0243	0.0076	0.0005	0.0000	19
2	0.9990	0.9245	0.6769	0.4049	0.2061	0.0913	0.0355	0.0036	0.0002	18
3	1.0000	0.9841	0.8670	0.6477	0.4114	0.2252	0.1071	0.0160	0.0013	17
4	1.0000	0.9974	0.9568	0.8298	0.6296	0.4148	0.2375	0.0510	0.0059	16
5	1.0000	0.9997	0.9887	0.9327	0.8042	0.6172	0.4164	0.1256	0.0207	15
6	1.0000	1.0000	0.9976	0.9781	0.9133	0.7858	0.6080	0.2500	0.0577	14
7	1.0000	1.0000	0.9996	0.9941	0.9679	0.8982	0.7723	0.4159	0.1316	13
8	1.0000	1.0000	0.9999	0.9987	0.9900	0.9591	0.8867	0.5956	0.2517	12
9	1.0000	1.0000	1.0000	0.9998	0.9974	0.9861	0.9520	0.7553	0.4119	11
10	1.0000	1.0000	1.0000	1.0000	0.9994	0.9961	0.9829	0.8725	0.5881	10
11	1.0000	1.0000	1.0000	1.0000	0.9999	0.9991	0.9949	0.9435	0.7483	9
12	1.0000	1.0000	1.0000	1.0000	1.0000	0.9998	0.9987	0.9790	0.8684	8
13	1.0000	1.0000	1.0000	1.0000	1.0000	1.0000	0.9997	0.9935	0.9423	7
14	1.0000	1.0000	1.0000	1.0000	1.0000	1.0000	1.0000	0.9984	0.9793	6
15	1.0000	1.0000	1.0000	1.0000	1.0000	1.0000	1.0000	0.9997	0.9941	5
16	1.0000	1.0000	1.0000	1.0000	1.0000	1.0000	1.0000	1.0000	0.9987	4
17	1.0000	1.0000	1.0000	1.0000	1.0000	1.0000	1.0000	1.0000	0.9998	3
18	1.0000	1.0000	1.0000	1.0000	1.0000	1.0000	1.0000	1.0000	1.0000	2

n = 25

	p									
	0.010	0.050	0.100	0.150	0.200	0.250	0.300	0.400	0.500	
0	0.7778	0.2774	0.0718	0.0172	0.0038	0.0008	0.0001	0.0000	0.0000	25
1	0.9742	0.6424	0.2712	0.0931	0.0274	0.0070	0.0016	0.0001	0.0000	24
2	0.9980	0.8729	0.5371	0.2537	0.0982	0.0321	0.0090	0.0004	0.0000	23
3	0.9999	0.9659	0.7636	0.4711	0.2340	0.0962	0.0332	0.0024	0.0001	22
4	1.0000	0.9928	0.9020	0.6821	0.4207	0.2137	0.0905	0.0095	0.0005	21
5	1.0000	0.9988	0.9666	0.8385	0.6167	0.3783	0.1935	0.0294	0.0020	20
6	1.0000	0.9998	0.9905	0.9305	0.7800	0.5611	0.3407	0.0736	0.0073	19
7	1.0000	1.0000	0.9977	0.9745	0.8909	0.7265	0.5118	0.1536	0.0216	18
8	1.0000	1.0000	0.9995	0.9920	0.9532	0.8506	0.6769	0.2735	0.0539	17
9	1.0000	1.0000	0.9999	0.9979	0.9827	0.9287	0.8106	0.4246	0.1148	16
10	1.0000	1.0000	1.0000	0.9995	0.9944	0.9703	0.9022	0.5858	0.2122	15
11	1.0000	1.0000	1.0000	0.9999	0.9985	0.9893	0.9558	0.7323	0.3450	14
12	1.0000	1.0000	1.0000	1.0000	0.9996	0.9966	0.9825	0.8462	0.5000	13
13	1.0000	1.0000	1.0000	1.0000	0.9999	0.9991	0.9940	0.9222	0.6550	12
14	1.0000	1.0000	1.0000	1.0000	1.0000	0.9998	0.9982	0.9656	0.7878	11
15	1.0000	1.0000	1.0000	1.0000	1.0000	1.0000	0.9995	0.9868	0.8852	10
16	1.0000	1.0000	1.0000	1.0000	1.0000	1.0000	0.9999	0.9957	0.9461	9
17	1.0000	1.0000	1.0000	1.0000	1.0000	1.0000	1.0000	0.9988	0.9784	8
18	1.0000	1.0000	1.0000	1.0000	1.0000	1.0000	1.0000	0.9997	0.9927	7
19	1.0000	1.0000	1.0000	1.0000	1.0000	1.0000	1.0000	0.9999	0.9980	6
20	1.0000	1.0000	1.0000	1.0000	1.0000	1.0000	1.0000	1.0000	0.9995	5
21	1.0000	1.0000	1.0000	1.0000	1.0000	1.0000	1.0000	1.0000	0.9999	4
22	1.0000	1.0000	1.0000	1.0000	1.0000	1.0000	1.0000	1.0000	1.0000	3

(Table A.2 continued)

n = 50	p									
	0.010	0.050	0.100	0.150	0.200	0.250	0.300	0.400	0.500	
0	0.6050	0.0769	0.0052	0.0003	0.0000	0.0000	0.0000	0.0000	0.0000	50
1	0.9106	0.2794	0.0338	0.0029	0.0002	0.0000	0.0000	0.0000	0.0000	49
2	0.9862	0.5405	0.1117	0.0142	0.0013	0.0001	0.0000	0.0000	0.0000	48
3	0.9984	0.7604	0.2503	0.0460	0.0057	0.0005	0.0000	0.0000	0.0000	47
4	0.9999	0.8964	0.4312	0.1121	0.0185	0.0021	0.0002	0.0000	0.0000	46
5	1.0000	0.9622	0.6161	0.2194	0.0480	0.0070	0.0007	0.0000	0.0000	45
6	1.0000	0.9882	0.7702	0.3613	0.1034	0.0194	0.0025	0.0000	0.0000	44
7	1.0000	0.9968	0.8779	0.5188	0.1904	0.0453	0.0073	0.0001	0.0000	43
8	1.0000	0.9992	0.9421	0.6681	0.3073	0.0916	0.0183	0.0002	0.0000	42
9	1.0000	0.9998	0.9755	0.7911	0.4437	0.1637	0.0402	0.0008	0.0000	41
10	1.0000	1.0000	0.9906	0.8801	0.5836	0.2622	0.0789	0.0022	0.0000	40
11	1.0000	1.0000	0.9968	0.9372	0.7107	0.3816	0.1390	0.0057	0.0000	39
12	1.0000	1.0000	0.9990	0.9699	0.8139	0.5110	0.2229	0.0133	0.0002	38
13	1.0000	1.0000	0.9997	0.9868	0.8894	0.6370	0.3279	0.0280	0.0005	37
14	1.0000	1.0000	0.9999	0.9947	0.9393	0.7481	0.4468	0.0540	0.0013	36
15	1.0000	1.0000	1.0000	0.9981	0.9692	0.8369	0.5692	0.0955	0.0033	35
16	1.0000	1.0000	1.0000	0.9993	0.9856	0.9017	0.6839	0.1561	0.0077	34
17	1.0000	1.0000	1.0000	0.9998	0.9937	0.9449	0.7822	0.2369	0.0164	33
18	1.0000	1.0000	1.0000	0.9999	0.9975	0.9713	0.8594	0.3356	0.0325	32
19	1.0000	1.0000	1.0000	1.0000	0.9991	0.9861	0.9152	0.4465	0.0595	31
20	1.0000	1.0000	1.0000	1.0000	0.9997	0.9937	0.9522	0.5610	0.1013	30
21	1.0000	1.0000	1.0000	1.0000	0.9999	0.9974	0.9749	0.6701	0.1611	29
22	1.0000	1.0000	1.0000	1.0000	1.0000	0.9990	0.9877	0.7660	0.2399	28
23	1.0000	1.0000	1.0000	1.0000	1.0000	0.9996	0.9944	0.8438	0.3359	27
24	1.0000	1.0000	1.0000	1.0000	1.0000	0.9999	0.9976	0.9022	0.4439	26
25	1.0000	1.0000	1.0000	1.0000	1.0000	1.0000	0.9991	0.9427	0.5561	25
26	1.0000	1.0000	1.0000	1.0000	1.0000	1.0000	0.9997	0.9686	0.6641	24
27	1.0000	1.0000	1.0000	1.0000	1.0000	1.0000	0.9999	0.9840	0.7601	23
28	1.0000	1.0000	1.0000	1.0000	1.0000	1.0000	1.0000	0.9924	0.8389	22
29	1.0000	1.0000	1.0000	1.0000	1.0000	1.0000	1.0000	0.9966	0.8987	21
30	1.0000	1.0000	1.0000	1.0000	1.0000	1.0000	1.0000	0.9986	0.9405	20
31	1.0000	1.0000	1.0000	1.0000	1.0000	1.0000	1.0000	0.9995	0.9675	19
32	1.0000	1.0000	1.0000	1.0000	1.0000	1.0000	1.0000	0.9998	0.9836	18
33	1.0000	1.0000	1.0000	1.0000	1.0000	1.0000	1.0000	0.9999	0.9923	17
34	1.0000	1.0000	1.0000	1.0000	1.0000	1.0000	1.0000	1.0000	0.9967	16
35	1.0000	1.0000	1.0000	1.0000	1.0000	1.0000	1.0000	1.0000	0.9987	15
36	1.0000	1.0000	1.0000	1.0000	1.0000	1.0000	1.0000	1.0000	0.9995	14
37	1.0000	1.0000	1.0000	1.0000	1.0000	1.0000	1.0000	1.0000	0.9998	13
38	1.0000	1.0000	1.0000	1.0000	1.0000	1.0000	1.0000	1.0000	1.0000	12

Table A.3 Individual Probabilities for Selected Poisson Distributions

X	μR .1	.2	.3	.4	.5	.6	.7	.8	.9	1.0
0	0.9048	0.8187	0.7408	0.6703	0.6065	0.5488	0.4966	0.4493	0.4066	0.3679
1	0.0905	0.1637	0.2222	0.2681	0.3033	0.3293	0.3476	0.3595	0.3659	0.3679
2	0.0045	0.0164	0.0333	0.0536	0.0758	0.0988	0.1217	0.1438	0.1647	0.1839
3	0.0002	0.0011	0.0033	0.0072	0.0126	0.0198	0.0284	0.0383	0.0494	0.0613
4	0.0000	0.0001	0.0003	0.0007	0.0016	0.0030	0.0050	0.0077	0.0111	0.0153
5	0.0000	0.0000	0.0000	0.0001	0.0002	0.0004	0.0007	0.0012	0.0020	0.0031
6	0.0000	0.0000	0.0000	0.0000	0.0000	0.0000	0.0001	0.0002	0.0003	0.0005
7	0.0000	0.0000	0.0000	0.0000	0.0000	0.0000	0.0000	0.0000	0.0000	0.0001
8	0.0000	0.0000	0.0000	0.0000	0.0000	0.0000	0.0000	0.0000	0.0000	0.0000

X	μR 1.1	1.2	1.3	1.4	1.5	1.6	1.7	1.8	1.9	2.0
0	0.3329	0.3012	0.2725	0.2466	0.2231	0.2019	0.1827	0.1653	0.1496	0.1353
1	0.3662	0.3614	0.3543	0.3452	0.3347	0.3230	0.3106	0.2975	0.2842	0.2707
2	0.2014	0.2169	0.2303	0.2417	0.2510	0.2584	0.2640	0.2678	0.2700	0.2707
3	0.0738	0.0867	0.0998	0.1128	0.1255	0.1378	0.1496	0.1607	0.1710	0.1804
4	0.0203	0.0260	0.0324	0.0395	0.0471	0.0551	0.0636	0.0723	0.0812	0.0902
5	0.0045	0.0062	0.0084	0.0111	0.0141	0.0176	0.0216	0.0260	0.0309	0.0361
6	0.0008	0.0012	0.0018	0.0026	0.0035	0.0047	0.0061	0.0078	0.0098	0.0120
7	0.0001	0.0002	0.0003	0.0005	0.0008	0.0011	0.0015	0.0020	0.0027	0.0034
8	0.0000	0.0000	0.0001	0.0001	0.0001	0.0002	0.0003	0.0005	0.0006	0.0009
9	0.0000	0.0000	0.0000	0.0000	0.0000	0.0000	0.0001	0.0001	0.0001	0.0002
10	0.0000	0.0000	0.0000	0.0000	0.0000	0.0000	0.0000	0.0000	0.0000	0.0000

X	μR 2.1	2.2	2.3	2.4	2.5	2.6	2.7	2.8	2.9	3.0
0	0.1225	0.1108	0.1003	0.0907	0.0821	0.0743	0.0672	0.0608	0.0550	0.0498
1	0.2572	0.2438	0.2306	0.2177	0.2052	0.1931	0.1815	0.1703	0.1596	0.1494
2	0.2700	0.2681	0.2652	0.2613	0.2565	0.2510	0.2450	0.2384	0.2314	0.2240
3	0.1890	0.1966	0.2033	0.2090	0.2138	0.2176	0.2205	0.2225	0.2237	0.2240
4	0.0992	0.1082	0.1169	0.1254	0.1336	0.1414	0.1488	0.1557	0.1622	0.1680
5	0.0417	0.0476	0.0538	0.0602	0.0668	0.0735	0.0804	0.0872	0.0940	0.1008
6	0.0146	0.0174	0.0206	0.0241	0.0278	0.0319	0.0362	0.0407	0.0455	0.0504
7	0.0044	0.0055	0.0068	0.0083	0.0099	0.0118	0.0139	0.0163	0.0188	0.0216
8	0.0011	0.0015	0.0019	0.0025	0.0031	0.0038	0.0047	0.0057	0.0068	0.0081
9	0.0003	0.0004	0.0005	0.0007	0.0009	0.0011	0.0014	0.0018	0.0022	0.0027
10	0.0001	0.0001	0.0001	0.0002	0.0002	0.0003	0.0004	0.0005	0.0006	0.0008
11	0.0000	0.0000	0.0000	0.0000	0.0000	0.0001	0.0001	0.0001	0.0002	0.0002
12	0.0000	0.0000	0.0000	0.0000	0.0000	0.0000	0.0000	0.0000	0.0000	0.0001
13	0.0000	0.0000	0.0000	0.0000	0.0000	0.0000	0.0000	0.0000	0.0000	0.0000

X	μR 3.1	3.2	3.3	3.4	3.5	3.6	3.7	3.8	3.9	4.0
0	0.0450	0.0408	0.0369	0.0334	0.0302	0.0273	0.0247	0.0224	0.0202	0.0183
1	0.1397	0.1304	0.1217	0.1135	0.1057	0.0984	0.0915	0.0850	0.0789	0.0733
2	0.2165	0.2087	0.2008	0.1929	0.1850	0.1771	0.1692	0.1615	0.1539	0.1465
3	0.2237	0.2226	0.2209	0.2186	0.2158	0.2125	0.2087	0.2046	0.2001	0.1954
4	0.1733	0.1781	0.1823	0.1858	0.1888	0.1912	0.1931	0.1944	0.1951	0.1954
5	0.1075	0.1140	0.1203	0.1264	0.1322	0.1377	0.1429	0.1477	0.1522	0.1563
6	0.0555	0.0608	0.0662	0.0716	0.0771	0.0826	0.0881	0.0936	0.0989	0.1042
7	0.0246	0.0278	0.0312	0.0348	0.0385	0.0425	0.0466	0.0508	0.0551	0.0595
8	0.0095	0.0111	0.0129	0.0148	0.0169	0.0191	0.0215	0.0241	0.0269	0.0298
9	0.0033	0.0040	0.0047	0.0056	0.0066	0.0076	0.0089	0.0102	0.0116	0.0132
10	0.0010	0.0013	0.0016	0.0019	0.0023	0.0028	0.0033	0.0039	0.0045	0.0053
11	0.0003	0.0004	0.0005	0.0006	0.0007	0.0009	0.0011	0.0013	0.0016	0.0019
12	0.0001	0.0001	0.0001	0.0002	0.0002	0.0003	0.0003	0.0004	0.0005	0.0006
13	0.0000	0.0000	0.0000	0.0000	0.0001	0.0001	0.0001	0.0001	0.0002	0.0002
14	0.0000	0.0000	0.0000	0.0000	0.0000	0.0000	0.0000	0.0000	0.0000	0.0001
15	0.0000	0.0000	0.0000	0.0000	0.0000	0.0000	0.0000	0.0000	0.0000	0.0000

(Table A.3 continued)

X	μR 4.1	4.2	4.3	4.4	4.5	4.6	4.7	4.8	4.9	5.0
0	0.0166	0.0150	0.0136	0.0123	0.0111	0.0101	0.0091	0.0082	0.0074	0.0067
1	0.0679	0.0630	0.0583	0.0540	0.0500	0.0462	0.0427	0.0395	0.0365	0.0337
2	0.1393	0.1323	0.1254	0.1188	0.1125	0.1063	0.1005	0.0948	0.0894	0.0842
3	0.1904	0.1852	0.1798	0.1743	0.1687	0.1631	0.1574	0.1517	0.1460	0.1404
4	0.1951	0.1944	0.1933	0.1917	0.1898	0.1875	0.1849	0.1820	0.1789	0.1755
5	0.1600	0.1633	0.1662	0.1687	0.1708	0.1725	0.1738	0.1747	0.1753	0.1755
6	0.1093	0.1143	0.1191	0.1237	0.1281	0.1323	0.1362	0.1398	0.1432	0.1462
7	0.0640	0.0686	0.0732	0.0778	0.0824	0.0869	0.0914	0.0959	0.1002	0.1044
8	0.0328	0.0360	0.0393	0.0428	0.0463	0.0500	0.0537	0.0575	0.0614	0.0653
9	0.0150	0.0168	0.0188	0.0209	0.0232	0.0255	0.0281	0.0307	0.0334	0.0363
10	0.0061	0.0071	0.0081	0.0092	0.0104	0.0118	0.0132	0.0147	0.0164	0.0181
11	0.0023	0.0027	0.0032	0.0037	0.0043	0.0049	0.0056	0.0064	0.0073	0.0082
12	0.0008	0.0009	0.0011	0.0013	0.0016	0.0019	0.0022	0.0026	0.0030	0.0034
13	0.0002	0.0003	0.0004	0.0005	0.0006	0.0007	0.0008	0.0009	0.0011	0.0013
14	0.0001	0.0001	0.0001	0.0001	0.0002	0.0002	0.0003	0.0003	0.0004	0.0005
15	0.0000	0.0000	0.0000	0.0000	0.0001	0.0001	0.0001	0.0001	0.0001	0.0002
16	0.0000	0.0000	0.0000	0.0000	0.0000	0.0000	0.0000	0.0000	0.0000	0.0000

X	μR 5.1	5.2	5.3	5.4	5.5	5.6	5.7	5.8	5.9	6.0
0	0.0061	0.0055	0.0050	0.0045	0.0041	0.0037	0.0033	0.0030	0.0027	0.0025
1	0.0311	0.0287	0.0265	0.0244	0.0225	0.0207	0.0191	0.0176	0.0162	0.0149
2	0.0793	0.0746	0.0701	0.0659	0.0618	0.0580	0.0544	0.0509	0.0477	0.0446
3	0.1348	0.1293	0.1239	0.1185	0.1133	0.1082	0.1033	0.0985	0.0938	0.0892
4	0.1719	0.1681	0.1641	0.1600	0.1558	0.1515	0.1472	0.1428	0.1383	0.1339
5	0.1753	0.1748	0.1740	0.1728	0.1714	0.1697	0.1678	0.1656	0.1632	0.1606
6	0.1490	0.1515	0.1537	0.1555	0.1571	0.1584	0.1594	0.1601	0.1605	0.1606
7	0.1086	0.1125	0.1163	0.1200	0.1234	0.1267	0.1298	0.1326	0.1353	0.1377
8	0.0692	0.0731	0.0771	0.0810	0.0849	0.0887	0.0925	0.0962	0.0998	0.1033
9	0.0392	0.0423	0.0454	0.0486	0.0519	0.0552	0.0586	0.0620	0.0654	0.0688
10	0.0200	0.0220	0.0241	0.0262	0.0285	0.0309	0.0334	0.0359	0.0386	0.0413
11	0.0093	0.0104	0.0116	0.0129	0.0143	0.0157	0.0173	0.0190	0.0207	0.0225
12	0.0039	0.0045	0.0051	0.0058	0.0065	0.0073	0.0082	0.0092	0.0102	0.0113
13	0.0015	0.0018	0.0021	0.0024	0.0028	0.0032	0.0036	0.0041	0.0046	0.0052
14	0.0006	0.0007	0.0008	0.0009	0.0011	0.0013	0.0015	0.0017	0.0019	0.0022
15	0.0002	0.0002	0.0003	0.0003	0.0004	0.0005	0.0006	0.0007	0.0008	0.0009
16	0.0001	0.0001	0.0001	0.0001	0.0001	0.0002	0.0002	0.0002	0.0003	0.0003
17	0.0000	0.0000	0.0000	0.0000	0.0000	0.0001	0.0001	0.0001	0.0001	0.0001
18	0.0000	0.0000	0.0000	0.0000	0.0000	0.0000	0.0000	0.0000	0.0000	0.0000

X	μR 6.1	6.2	6.3	6.4	6.5	6.6	6.7	6.8	6.9	7.0
0	0.0022	0.0020	0.0018	0.0017	0.0015	0.0014	0.0012	0.0011	0.0010	0.0009
1	0.0137	0.0126	0.0116	0.0106	0.0098	0.0090	0.0082	0.0076	0.0070	0.0064
2	0.0417	0.0390	0.0364	0.0340	0.0318	0.0296	0.0276	0.0258	0.0240	0.0223
3	0.0848	0.0806	0.0765	0.0726	0.0688	0.0652	0.0617	0.0584	0.0552	0.0521
4	0.1294	0.1249	0.1205	0.1162	0.1118	0.1076	0.1034	0.0992	0.0952	0.0912
5	0.1579	0.1549	0.1519	0.1487	0.1454	0.1420	0.1385	0.1349	0.1314	0.1277
6	0.1605	0.1601	0.1595	0.1586	0.1575	0.1562	0.1546	0.1529	0.1511	0.1490
7	0.1399	0.1418	0.1435	0.1450	0.1462	0.1472	0.1480	0.1486	0.1489	0.1490
8	0.1066	0.1099	0.1130	0.1160	0.1188	0.1215	0.1240	0.1263	0.1284	0.1304
9	0.0723	0.0757	0.0791	0.0825	0.0858	0.0891	0.0923	0.0954	0.0985	0.1014
10	0.0441	0.0469	0.0498	0.0528	0.0558	0.0588	0.0618	0.0649	0.0679	0.0710
11	0.0244	0.0265	0.0285	0.0307	0.0330	0.0353	0.0377	0.0401	0.0426	0.0452
12	0.0124	0.0137	0.0150	0.0164	0.0179	0.0194	0.0210	0.0227	0.0245	0.0263
13	0.0058	0.0065	0.0073	0.0081	0.0089	0.0099	0.0108	0.0119	0.0130	0.0142
14	0.0025	0.0029	0.0033	0.0037	0.0041	0.0046	0.0052	0.0058	0.0064	0.0071
15	0.0010	0.0012	0.0014	0.0016	0.0018	0.0020	0.0023	0.0026	0.0029	0.0033
16	0.0004	0.0005	0.0005	0.0006	0.0007	0.0008	0.0010	0.0011	0.0013	0.0014
17	0.0001	0.0002	0.0002	0.0002	0.0003	0.0003	0.0004	0.0004	0.0005	0.0006
18	0.0000	0.0001	0.0001	0.0001	0.0001	0.0001	0.0001	0.0002	0.0002	0.0002
19	0.0000	0.0000	0.0000	0.0000	0.0000	0.0000	0.0001	0.0001	0.0001	0.0001
20	0.0000	0.0000	0.0000	0.0000	0.0000	0.0000	0.0000	0.0000	0.0000	0.0000

(Table A.3 continued)

X	μR 7.1	7.2	7.3	7.4	7.5	7.6	7.7	7.8	7.9	8.0
0	0.0008	0.0007	0.0007	0.0006	0.0006	0.0005	0.0005	0.0004	0.0004	0.0003
1	0.0059	0.0054	0.0049	0.0045	0.0041	0.0038	0.0035	0.0032	0.0029	0.0027
2	0.0208	0.0194	0.0180	0.0167	0.0156	0.0145	0.0134	0.0125	0.0116	0.0107
3	0.0492	0.0464	0.0438	0.0413	0.0389	0.0366	0.0345	0.0324	0.0305	0.0286
4	0.0874	0.0836	0.0799	0.0764	0.0729	0.0696	0.0663	0.0632	0.0602	0.0573
5	0.1241	0.1204	0.1167	0.1130	0.1094	0.1057	0.1021	0.0986	0.0951	0.0916
6	0.1468	0.1445	0.1420	0.1394	0.1367	0.1339	0.1311	0.1282	0.1252	0.1221
7	0.1489	0.1486	0.1481	0.1474	0.1465	0.1454	0.1442	0.1428	0.1413	0.1396
8	0.1321	0.1337	0.1351	0.1363	0.1373	0.1381	0.1388	0.1392	0.1395	0.1396
9	0.1042	0.1070	0.1096	0.1121	0.1144	0.1167	0.1187	0.1207	0.1224	0.1241
10	0.0740	0.0770	0.0800	0.0829	0.0858	0.0887	0.0914	0.0941	0.0967	0.0993
11	0.0478	0.0504	0.0531	0.0558	0.0585	0.0613	0.0640	0.0667	0.0695	0.0722
12	0.0283	0.0303	0.0323	0.0344	0.0366	0.0388	0.0411	0.0434	0.0457	0.0481
13	0.0154	0.0168	0.0181	0.0196	0.0211	0.0227	0.0243	0.0260	0.0278	0.0296
14	0.0078	0.0086	0.0095	0.0104	0.0113	0.0123	0.0134	0.0145	0.0157	0.0169
15	0.0037	0.0041	0.0046	0.0051	0.0057	0.0062	0.0069	0.0075	0.0083	0.0090
16	0.0016	0.0019	0.0021	0.0024	0.0026	0.0030	0.0033	0.0037	0.0041	0.0045
17	0.0007	0.0008	0.0009	0.0010	0.0012	0.0013	0.0015	0.0017	0.0019	0.0021
18	0.0003	0.0003	0.0004	0.0004	0.0005	0.0006	0.0006	0.0007	0.0008	0.0009
19	0.0001	0.0001	0.0001	0.0002	0.0002	0.0002	0.0003	0.0003	0.0003	0.0004
20	0.0000	0.0000	0.0001	0.0001	0.0001	0.0001	0.0001	0.0001	0.0001	0.0002
21	0.0000	0.0000	0.0000	0.0000	0.0000	0.0000	0.0000	0.0000	0.0001	0.0001
22	0.0000	0.0000	0.0000	0.0000	0.0000	0.0000	0.0000	0.0000	0.0000	0.0000

X	μR 8.1	8.2	8.3	8.4	8.5	8.6	8.7	8.8	8.9	9.0
0	0.0003	0.0003	0.0002	0.0002	0.0002	0.0002	0.0002	0.0002	0.0001	0.0001
1	0.0025	0.0023	0.0021	0.0019	0.0017	0.0016	0.0014	0.0013	0.0012	0.0011
2	0.0100	0.0092	0.0086	0.0079	0.0074	0.0068	0.0063	0.0058	0.0054	0.0050
3	0.0269	0.0252	0.0237	0.0222	0.0208	0.0195	0.0183	0.0171	0.0160	0.0150
4	0.0544	0.0517	0.0491	0.0466	0.0443	0.0420	0.0398	0.0377	0.0357	0.0337
5	0.0882	0.0849	0.0816	0.0784	0.0752	0.0722	0.0692	0.0663	0.0635	0.0607
6	0.1191	0.1160	0.1128	0.1097	0.1066	0.1034	0.1003	0.0972	0.0941	0.0911
7	0.1378	0.1358	0.1338	0.1317	0.1294	0.1271	0.1247	0.1222	0.1197	0.1171
8	0.1395	0.1392	0.1388	0.1382	0.1375	0.1366	0.1356	0.1344	0.1332	0.1318
9	0.1256	0.1269	0.1280	0.1290	0.1299	0.1306	0.1311	0.1315	0.1317	0.1318
10	0.1017	0.1040	0.1063	0.1084	0.1104	0.1123	0.1140	0.1157	0.1172	0.1186
11	0.0749	0.0776	0.0802	0.0828	0.0853	0.0878	0.0902	0.0925	0.0948	0.0970
12	0.0505	0.0530	0.0555	0.0579	0.0604	0.0629	0.0654	0.0679	0.0703	0.0728
13	0.0315	0.0334	0.0354	0.0374	0.0395	0.0416	0.0438	0.0459	0.0481	0.0504
14	0.0182	0.0196	0.0210	0.0225	0.0240	0.0256	0.0272	0.0289	0.0306	0.0324
15	0.0098	0.0107	0.0116	0.0126	0.0136	0.0147	0.0158	0.0169	0.0182	0.0194
16	0.0050	0.0055	0.0060	0.0066	0.0072	0.0079	0.0086	0.0093	0.0101	0.0109
17	0.0024	0.0026	0.0029	0.0033	0.0036	0.0040	0.0044	0.0048	0.0053	0.0058
18	0.0011	0.0012	0.0014	0.0015	0.0017	0.0019	0.0021	0.0024	0.0026	0.0029
19	0.0005	0.0005	0.0006	0.0007	0.0008	0.0009	0.0010	0.0011	0.0012	0.0014
20	0.0002	0.0002	0.0002	0.0003	0.0003	0.0004	0.0004	0.0005	0.0005	0.0006
21	0.0001	0.0001	0.0001	0.0001	0.0001	0.0002	0.0002	0.0002	0.0002	0.0003
22	0.0000	0.0000	0.0000	0.0000	0.0001	0.0001	0.0001	0.0001	0.0001	0.0001
23	0.0000	0.0000	0.0000	0.0000	0.0000	0.0000	0.0000	0.0000	0.0000	0.0000

(Table A.3 continued)

	μR									
X	9.1	9.2	9.3	9.4	9.5	9.6	9.7	9.8	9.9	10.0
0	0.0001	0.0001	0.0001	0.0001	0.0001	0.0001	0.0001	0.0001	0.0001	0.0000
1	0.0010	0.0009	0.0009	0.0008	0.0007	0.0007	0.0006	0.0005	0.0005	0.0005
2	0.0046	0.0043	0.0040	0.0037	0.0034	0.0031	0.0029	0.0027	0.0025	0.0023
3	0.0140	0.0131	0.0123	0.0115	0.0107	0.0100	0.0093	0.0087	0.0081	0.0076
4	0.0319	0.0302	0.0285	0.0269	0.0254	0.0240	0.0226	0.0213	0.0201	0.0189
5	0.0581	0.0555	0.0530	0.0506	0.0483	0.0460	0.0439	0.0418	0.0398	0.0378
6	0.0881	0.0851	0.0822	0.0793	0.0764	0.0736	0.0709	0.0682	0.0656	0.0631
7	0.1145	0.1118	0.1091	0.1064	0.1037	0.1010	0.0982	0.0955	0.0928	0.0901
8	0.1302	0.1286	0.1269	0.1251	0.1232	0.1212	0.1191	0.1170	0.1148	0.1126
9	0.1317	0.1315	0.1311	0.1306	0.1300	0.1293	0.1284	0.1274	0.1263	0.1251
10	0.1198	0.1210	0.1219	0.1228	0.1235	0.1241	0.1245	0.1249	0.1250	0.1251
11	0.0991	0.1012	0.1031	0.1049	0.1067	0.1083	0.1098	0.1112	0.1125	0.1137
12	0.0752	0.0776	0.0799	0.0822	0.0844	0.0866	0.0888	0.0908	0.0928	0.0948
13	0.0526	0.0549	0.0572	0.0594	0.0617	0.0640	0.0662	0.0685	0.0707	0.0729
14	0.0342	0.0361	0.0380	0.0399	0.0419	0.0439	0.0459	0.0479	0.0500	0.0521
15	0.0208	0.0221	0.0235	0.0250	0.0265	0.0281	0.0297	0.0313	0.0330	0.0347
16	0.0118	0.0127	0.0137	0.0147	0.0157	0.0168	0.0180	0.0192	0.0204	0.0217
17	0.0063	0.0069	0.0075	0.0081	0.0088	0.0095	0.0103	0.0111	0.0119	0.0128
18	0.0032	0.0035	0.0039	0.0042	0.0046	0.0051	0.0055	0.0060	0.0065	0.0071
19	0.0015	0.0017	0.0019	0.0021	0.0023	0.0026	0.0028	0.0031	0.0034	0.0037
20	0.0007	0.0008	0.0009	0.0010	0.0011	0.0012	0.0014	0.0015	0.0017	0.0019
21	0.0003	0.0003	0.0004	0.0004	0.0005	0.0006	0.0006	0.0007	0.0008	0.0009
22	0.0001	0.0001	0.0002	0.0002	0.0002	0.0002	0.0003	0.0003	0.0004	0.0004
23	0.0000	0.0001	0.0001	0.0001	0.0001	0.0001	0.0001	0.0001	0.0002	0.0002
24	0.0000	0.0000	0.0000	0.0000	0.0000	0.0000	0.0000	0.0001	0.0001	0.0001
25	0.0000	0.0000	0.0000	0.0000	0.0000	0.0000	0.0000	0.0000	0.0000	0.0000

	μR									
X	11	12	13	14	15	16	17	18	19	20
0	0.0000	0.0000	0.0000	0.0000	0.0000	0.0000	0.0000	0.0000	0.0000	0.0000
1	0.0002	0.0001	0.0000	0.0000	0.0000	0.0000	0.0000	0.0000	0.0000	0.0000
2	0.0010	0.0004	0.0002	0.0001	0.0000	0.0000	0.0000	0.0000	0.0000	0.0000
3	0.0037	0.0018	0.0008	0.0004	0.0002	0.0001	0.0000	0.0000	0.0000	0.0000
4	0.0102	0.0053	0.0027	0.0013	0.0006	0.0003	0.0001	0.0001	0.0000	0.0000
5	0.0224	0.0127	0.0070	0.0037	0.0019	0.0010	0.0005	0.0002	0.0001	0.0001
6	0.0411	0.0255	0.0152	0.0087	0.0048	0.0026	0.0014	0.0007	0.0004	0.0002
7	0.0646	0.0437	0.0281	0.0174	0.0104	0.0060	0.0034	0.0019	0.0010	0.0005
8	0.0888	0.0655	0.0457	0.0304	0.0194	0.0120	0.0072	0.0042	0.0024	0.0013
9	0.1085	0.0874	0.0661	0.0473	0.0324	0.0213	0.0135	0.0083	0.0050	0.0029
10	0.1194	0.1048	0.0859	0.0663	0.0486	0.0341	0.0230	0.0150	0.0095	0.0058
11	0.1194	0.1144	0.1015	0.0844	0.0663	0.0496	0.0355	0.0245	0.0164	0.0106
12	0.1094	0.1144	0.1099	0.0984	0.0829	0.0661	0.0504	0.0368	0.0259	0.0176
13	0.0926	0.1056	0.1099	0.1060	0.0956	0.0814	0.0658	0.0509	0.0378	0.0271
14	0.0728	0.0905	0.1021	0.1060	0.1024	0.0930	0.0800	0.0655	0.0514	0.0387
15	0.0534	0.0724	0.0885	0.0989	0.1024	0.0992	0.0906	0.0786	0.0650	0.0516
16	0.0367	0.0543	0.0719	0.0866	0.0960	0.0992	0.0963	0.0884	0.0772	0.0646
17	0.0237	0.0383	0.0550	0.0713	0.0847	0.0934	0.0963	0.0936	0.0863	0.0760
18	0.0145	0.0255	0.0397	0.0554	0.0706	0.0830	0.0909	0.0936	0.0911	0.0844
19	0.0084	0.0161	0.0272	0.0409	0.0557	0.0699	0.0814	0.0887	0.0911	0.0888
20	0.0046	0.0097	0.0177	0.0286	0.0418	0.0559	0.0692	0.0798	0.0866	0.0888
21	0.0024	0.0055	0.0109	0.0191	0.0299	0.0426	0.0560	0.0684	0.0783	0.0846
22	0.0012	0.0030	0.0065	0.0121	0.0204	0.0310	0.0433	0.0560	0.0676	0.0769
23	0.0006	0.0016	0.0037	0.0074	0.0133	0.0216	0.0320	0.0438	0.0559	0.0669
24	0.0003	0.0008	0.0020	0.0043	0.0083	0.0144	0.0226	0.0328	0.0442	0.0557
25	0.0001	0.0004	0.0010	0.0024	0.0050	0.0092	0.0154	0.0237	0.0336	0.0446
26	0.0000	0.0002	0.0005	0.0013	0.0029	0.0057	0.0101	0.0164	0.0246	0.0343
27	0.0000	0.0001	0.0002	0.0007	0.0016	0.0034	0.0063	0.0109	0.0173	0.0254
28	0.0000	0.0000	0.0001	0.0003	0.0009	0.0019	0.0038	0.0070	0.0117	0.0181
29	0.0000	0.0000	0.0001	0.0002	0.0004	0.0011	0.0023	0.0044	0.0077	0.0125
30	0.0000	0.0000	0.0000	0.0001	0.0002	0.0006	0.0013	0.0026	0.0049	0.0083
31	0.0000	0.0000	0.0000	0.0000	0.0001	0.0003	0.0007	0.0015	0.0030	0.0054
32	0.0000	0.0000	0.0000	0.0000	0.0001	0.0001	0.0004	0.0009	0.0018	0.0034
33	0.0000	0.0000	0.0000	0.0000	0.0000	0.0001	0.0002	0.0005	0.0010	0.0020
34	0.0000	0.0000	0.0000	0.0000	0.0000	0.0000	0.0001	0.0002	0.0006	0.0012
35	0.0000	0.0000	0.0000	0.0000	0.0000	0.0000	0.0000	0.0001	0.0003	0.0007
36	0.0000	0.0000	0.0000	0.0000	0.0000	0.0000	0.0000	0.0001	0.0002	0.0004
37	0.0000	0.0000	0.0000	0.0000	0.0000	0.0000	0.0000	0.0000	0.0001	0.0002
38	0.0000	0.0000	0.0000	0.0000	0.0000	0.0000	0.0000	0.0000	0.0000	0.0001
39	0.0000	0.0000	0.0000	0.0000	0.0000	0.0000	0.0000	0.0000	0.0000	0.0001
40	0.0000	0.0000	0.0000	0.0000	0.0000	0.0000	0.0000	0.0000	0.0000	0.0000

Table A.4 Cumulative Probabilities for Selected Poisson Distributions

X	μR .1	.2	.3	.4	.5	.6	.7	.8	.9	1.0
0	0.9048	0.8187	0.7408	0.6703	0.6065	0.5488	0.4966	0.4493	0.4066	0.3679
1	0.9953	0.9825	0.9631	0.9384	0.9098	0.8781	0.8442	0.8088	0.7725	0.7358
2	0.9998	0.9989	0.9964	0.9921	0.9856	0.9769	0.9659	0.9526	0.9371	0.9197
3	1.0000	0.9999	0.9997	0.9992	0.9982	0.9966	0.9942	0.9909	0.9865	0.9810
4	1.0000	1.0000	1.0000	0.9999	0.9998	0.9996	0.9992	0.9986	0.9977	0.9963
5	1.0000	1.0000	1.0000	1.0000	1.0000	1.0000	0.9999	0.9998	0.9997	0.9994
6	1.0000	1.0000	1.0000	1.0000	1.0000	1.0000	1.0000	1.0000	1.0000	0.9999
7	1.0000	1.0000	1.0000	1.0000	1.0000	1.0000	1.0000	1.0000	1.0000	1.0000
8	1.0000	1.0000	1.0000	1.0000	1.0000	1.0000	1.0000	1.0000	1.0000	1.0000

X	μR 1.1	1.2	1.3	1.4	1.5	1.6	1.7	1.8	1.9	2.0
0	0.3329	0.3012	0.2725	0.2466	0.2231	0.2019	0.1827	0.1653	0.1496	0.1353
1	0.6990	0.6626	0.6268	0.5918	0.5578	0.5249	0.4932	0.4628	0.4337	0.4060
2	0.9004	0.8795	0.8571	0.8335	0.8088	0.7834	0.7572	0.7306	0.7037	0.6767
3	0.9743	0.9662	0.9569	0.9463	0.9344	0.9212	0.9068	0.8913	0.8747	0.8571
4	0.9946	0.9923	0.9893	0.9857	0.9814	0.9763	0.9704	0.9636	0.9559	0.9473
5	0.9990	0.9985	0.9978	0.9968	0.9955	0.9940	0.9920	0.9896	0.9868	0.9834
6	0.9999	0.9997	0.9996	0.9994	0.9991	0.9987	0.9981	0.9974	0.9966	0.9955
7	1.0000	1.0000	0.9999	0.9999	0.9998	0.9997	0.9996	0.9994	0.9992	0.9989
8	1.0000	1.0000	1.0000	1.0000	1.0000	1.0000	0.9999	0.9999	0.9998	0.9998
9	1.0000	1.0000	1.0000	1.0000	1.0000	1.0000	1.0000	1.0000	1.0000	1.0000
10	1.0000	1.0000	1.0000	1.0000	1.0000	1.0000	1.0000	1.0000	1.0000	1.0000

X	μR 2.1	2.2	2.3	2.4	2.5	2.6	2.7	2.8	2.9	3.0
0	0.1225	0.1108	0.1003	0.0907	0.0821	0.0743	0.0672	0.0608	0.0550	0.0498
1	0.3796	0.3546	0.3309	0.3084	0.2873	0.2674	0.2487	0.2311	0.2146	0.1991
2	0.6496	0.6227	0.5960	0.5697	0.5438	0.5184	0.4936	0.4695	0.4460	0.4232
3	0.8386	0.8194	0.7993	0.7787	0.7576	0.7360	0.7141	0.6919	0.6696	0.6472
4	0.9379	0.9275	0.9162	0.9041	0.8912	0.8774	0.8629	0.8477	0.8318	0.8153
5	0.9796	0.9751	0.9700	0.9643	0.9580	0.9510	0.9433	0.9349	0.9258	0.9161
6	0.9941	0.9925	0.9906	0.9884	0.9858	0.9828	0.9794	0.9756	0.9713	0.9665
7	0.9985	0.9980	0.9974	0.9967	0.9958	0.9947	0.9934	0.9919	0.9901	0.9881
8	0.9997	0.9995	0.9994	0.9991	0.9989	0.9985	0.9981	0.9976	0.9969	0.9962
9	0.9999	0.9999	0.9999	0.9998	0.9997	0.9996	0.9995	0.9993	0.9991	0.9989
10	1.0000	1.0000	1.0000	1.0000	0.9999	0.9999	0.9999	0.9998	0.9998	0.9997
11	1.0000	1.0000	1.0000	1.0000	1.0000	1.0000	1.0000	1.0000	0.9999	0.9999
12	1.0000	1.0000	1.0000	1.0000	1.0000	1.0000	1.0000	1.0000	1.0000	1.0000
13	1.0000	1.0000	1.0000	1.0000	1.0000	1.0000	1.0000	1.0000	1.0000	1.0000

X	μR 3.1	3.2	3.3	3.4	3.5	3.6	3.7	3.8	3.9	4.0
0	0.0450	0.0408	0.0369	0.0334	0.0302	0.0273	0.0247	0.0224	0.0202	0.0183
1	0.1847	0.1712	0.1586	0.1468	0.1359	0.1257	0.1162	0.1074	0.0992	0.0916
2	0.4012	0.3799	0.3594	0.3397	0.3208	0.3027	0.2854	0.2689	0.2531	0.2381
3	0.6248	0.6025	0.5803	0.5584	0.5366	0.5152	0.4942	0.4735	0.4532	0.4335
4	0.7982	0.7806	0.7626	0.7442	0.7254	0.7064	0.6872	0.6678	0.6484	0.6288
5	0.9057	0.8946	0.8829	0.8705	0.8576	0.8441	0.8301	0.8156	0.8006	0.7851
6	0.9612	0.9554	0.9490	0.9421	0.9347	0.9267	0.9182	0.9091	0.8995	0.8893
7	0.9858	0.9832	0.9802	0.9769	0.9733	0.9692	0.9648	0.9599	0.9546	0.9489
8	0.9953	0.9943	0.9931	0.9917	0.9901	0.9883	0.9863	0.9840	0.9815	0.9786
9	0.9986	0.9982	0.9978	0.9973	0.9967	0.9960	0.9952	0.9942	0.9931	0.9919
10	0.9996	0.9995	0.9994	0.9992	0.9990	0.9987	0.9984	0.9981	0.9977	0.9972
11	0.9999	0.9999	0.9998	0.9998	0.9997	0.9996	0.9995	0.9994	0.9993	0.9991
12	1.0000	1.0000	1.0000	0.9999	0.9999	0.9999	0.9999	0.9998	0.9998	0.9997
13	1.0000	1.0000	1.0000	1.0000	1.0000	1.0000	1.0000	1.0000	0.9999	0.9999
14	1.0000	1.0000	1.0000	1.0000	1.0000	1.0000	1.0000	1.0000	1.0000	1.0000
15	1.0000	1.0000	1.0000	1.0000	1.0000	1.0000	1.0000	1.0000	1.0000	1.0000

(Table A.4 continued)

X	μR 4.1	4.2	4.3	4.4	4.5	4.6	4.7	4.8	4.9	5.0
0	0.0166	0.0150	0.0136	0.0123	0.0111	0.0101	0.0091	0.0082	0.0074	0.0067
1	0.0845	0.0780	0.0719	0.0663	0.0611	0.0563	0.0518	0.0477	0.0439	0.0404
2	0.2238	0.2102	0.1974	0.1851	0.1736	0.1626	0.1523	0.1425	0.1333	0.1247
3	0.4142	0.3954	0.3772	0.3594	0.3423	0.3257	0.3097	0.2942	0.2793	0.2650
4	0.6093	0.5898	0.5704	0.5512	0.5321	0.5132	0.4946	0.4763	0.4582	0.4405
5	0.7693	0.7531	0.7367	0.7199	0.7029	0.6858	0.6684	0.6510	0.6335	0.6160
6	0.8786	0.8675	0.8558	0.8436	0.8311	0.8180	0.8046	0.7908	0.7767	0.7622
7	0.9427	0.9361	0.9290	0.9214	0.9134	0.9049	0.8960	0.8867	0.8769	0.8666
8	0.9755	0.9721	0.9683	0.9642	0.9597	0.9549	0.9497	0.9442	0.9382	0.9319
9	0.9905	0.9889	0.9871	0.9851	0.9829	0.9805	0.9778	0.9749	0.9717	0.9682
10	0.9966	0.9959	0.9952	0.9943	0.9933	0.9922	0.9910	0.9896	0.9880	0.9863
11	0.9989	0.9986	0.9983	0.9980	0.9976	0.9971	0.9966	0.9960	0.9953	0.9945
12	0.9997	0.9996	0.9995	0.9993	0.9992	0.9990	0.9988	0.9986	0.9983	0.9980
13	0.9999	0.9999	0.9998	0.9998	0.9997	0.9997	0.9996	0.9995	0.9994	0.9993
14	1.0000	1.0000	1.0000	0.9999	0.9999	0.9999	0.9999	0.9999	0.9998	0.9998
15	1.0000	1.0000	1.0000	1.0000	1.0000	1.0000	1.0000	1.0000	0.9999	0.9999
16	1.0000	1.0000	1.0000	1.0000	1.0000	1.0000	1.0000	1.0000	1.0000	1.0000

X	μR 5.1	5.2	5.3	5.4	5.5	5.6	5.7	5.8	5.9	6.0
0	0.0061	0.0055	0.0050	0.0045	0.0041	0.0037	0.0033	0.0030	0.0027	0.0025
1	0.0372	0.0342	0.0314	0.0289	0.0266	0.0244	0.0224	0.0206	0.0189	0.0174
2	0.1165	0.1088	0.1016	0.0948	0.0884	0.0824	0.0768	0.0715	0.0666	0.0620
3	0.2513	0.2381	0.2254	0.2133	0.2017	0.1906	0.1800	0.1700	0.1604	0.1512
4	0.4231	0.4061	0.3895	0.3733	0.3575	0.3422	0.3272	0.3127	0.2987	0.2851
5	0.5984	0.5809	0.5635	0.5461	0.5289	0.5119	0.4950	0.4783	0.4619	0.4457
6	0.7474	0.7324	0.7171	0.7017	0.6860	0.6703	0.6544	0.6384	0.6224	0.6063
7	0.8560	0.8449	0.8335	0.8217	0.8095	0.7970	0.7841	0.7710	0.7576	0.7440
8	0.9252	0.9181	0.9106	0.9027	0.8944	0.8857	0.8766	0.8672	0.8574	0.8472
9	0.9644	0.9603	0.9559	0.9512	0.9462	0.9409	0.9352	0.9292	0.9228	0.9161
10	0.9844	0.9823	0.9800	0.9775	0.9747	0.9718	0.9686	0.9651	0.9614	0.9574
11	0.9937	0.9927	0.9916	0.9904	0.9890	0.9875	0.9859	0.9841	0.9821	0.9799
12	0.9976	0.9972	0.9967	0.9962	0.9955	0.9949	0.9941	0.9932	0.9922	0.9912
13	0.9992	0.9990	0.9988	0.9986	0.9983	0.9980	0.9977	0.9973	0.9969	0.9964
14	0.9997	0.9997	0.9996	0.9995	0.9994	0.9993	0.9991	0.9990	0.9988	0.9986
15	0.9999	0.9999	0.9999	0.9998	0.9998	0.9998	0.9997	0.9996	0.9996	0.9995
16	1.0000	1.0000	1.0000	0.9999	0.9999	0.9999	0.9999	0.9999	0.9999	0.9998
17	1.0000	1.0000	1.0000	1.0000	1.0000	1.0000	1.0000	1.0000	1.0000	0.9999
18	1.0000	1.0000	1.0000	1.0000	1.0000	1.0000	1.0000	1.0000	1.0000	1.0000

X	μR 6.1	6.2	6.3	6.4	6.5	6.6	6.7	6.8	6.9	7.0
0	0.0022	0.0020	0.0018	0.0017	0.0015	0.0014	0.0012	0.0011	0.0010	0.0009
1	0.0159	0.0146	0.0134	0.0123	0.0113	0.0103	0.0095	0.0087	0.0080	0.0073
2	0.0577	0.0536	0.0498	0.0463	0.0430	0.0400	0.0371	0.0344	0.0320	0.0296
3	0.1425	0.1342	0.1264	0.1189	0.1118	0.1052	0.0988	0.0928	0.0871	0.0818
4	0.2719	0.2592	0.2469	0.2351	0.2237	0.2127	0.2022	0.1920	0.1823	0.1730
5	0.4298	0.4141	0.3988	0.3837	0.3690	0.3547	0.3406	0.3270	0.3137	0.3007
6	0.5902	0.5742	0.5582	0.5423	0.5265	0.5108	0.4953	0.4799	0.4647	0.4497
7	0.7301	0.7160	0.7017	0.6873	0.6728	0.6581	0.6433	0.6285	0.6136	0.5987
8	0.8367	0.8259	0.8148	0.8033	0.7916	0.7796	0.7673	0.7548	0.7420	0.7291
9	0.9090	0.9016	0.8939	0.8858	0.8774	0.8686	0.8596	0.8502	0.8405	0.8305
10	0.9531	0.9486	0.9437	0.9386	0.9332	0.9274	0.9214	0.9151	0.9084	0.9015
11	0.9776	0.9750	0.9723	0.9693	0.9661	0.9627	0.9591	0.9552	0.9510	0.9467
12	0.9900	0.9887	0.9873	0.9857	0.9840	0.9821	0.9801	0.9779	0.9755	0.9730
13	0.9958	0.9952	0.9945	0.9937	0.9929	0.9920	0.9909	0.9838	0.9885	0.9872
14	0.9984	0.9981	0.9978	0.9974	0.9970	0.9966	0.9961	0.9956	0.9950	0.9943
15	0.9994	0.9993	0.9992	0.9990	0.9988	0.9986	0.9984	0.9982	0.9979	0.9976
16	0.9998	0.9997	0.9997	0.9996	0.9996	0.9995	0.9994	0.9993	0.9992	0.9990
17	0.9999	0.9999	0.9999	0.9999	0.9998	0.9998	0.9998	0.9997	0.9997	0.9996
18	1.0000	1.0000	1.0000	1.0000	0.9999	0.9999	0.9999	0.9999	0.9999	0.9999
19	1.0000	1.0000	1.0000	1.0000	1.0000	1.0000	1.0000	1.0000	1.0000	1.0000
20	1.0000	1.0000	1.0000	1.0000	1.0000	1.0000	1.0000	1.0000	1.0000	1.0000

(Table A.4 continued)

	μR									
X	7.1	7.2	7.3	7.4	7.5	7.6	7.7	7.8	7.9	8.0
0	0.0008	0.0007	0.0007	0.0006	0.0006	0.0005	0.0005	0.0004	0.0004	0.0003
1	0.0067	0.0061	0.0056	0.0051	0.0047	0.0043	0.0039	0.0036	0.0033	0.0030
2	0.0275	0.0255	0.0236	0.0219	0.0203	0.0188	0.0174	0.0161	0.0149	0.0138
3	0.0767	0.0719	0.0674	0.0632	0.0591	0.0554	0.0518	0.0485	0.0453	0.0424
4	0.1641	0.1555	0.1473	0.1395	0.1321	0.1249	0.1181	0.1117	0.1055	0.0996
5	0.2881	0.2759	0.2640	0.2526	0.2414	0.2307	0.2203	0.2103	0.2006	0.1912
6	0.4349	0.4204	0.4060	0.3920	0.3782	0.3646	0.3514	0.3384	0.3257	0.3134
7	0.5838	0.5689	0.5541	0.5393	0.5246	0.5100	0.4956	0.4812	0.4670	0.4530
8	0.7160	0.7027	0.6892	0.6757	0.6620	0.6482	0.6343	0.6204	0.6065	0.5925
9	0.8202	0.8096	0.7988	0.7877	0.7764	0.7649	0.7531	0.7411	0.7290	0.7166
10	0.8942	0.8867	0.8788	0.8707	0.8622	0.8535	0.8445	0.8352	0.8257	0.8159
11	0.9420	0.9371	0.9319	0.9265	0.9208	0.9148	0.9085	0.9020	0.8952	0.8881
12	0.9703	0.9673	0.9642	0.9609	0.9573	0.9536	0.9496	0.9454	0.9409	0.9362
13	0.9857	0.9841	0.9824	0.9805	0.9784	0.9762	0.9739	0.9714	0.9687	0.9658
14	0.9935	0.9927	0.9918	0.9908	0.9897	0.9886	0.9873	0.9859	0.9844	0.9827
15	0.9972	0.9969	0.9964	0.9959	0.9954	0.9948	0.9941	0.9934	0.9926	0.9918
16	0.9989	0.9987	0.9985	0.9983	0.9980	0.9978	0.9974	0.9971	0.9967	0.9963
17	0.9996	0.9995	0.9994	0.9993	0.9992	0.9991	0.9989	0.9988	0.9986	0.9984
18	0.9998	0.9998	0.9998	0.9997	0.9997	0.9996	0.9996	0.9995	0.9994	0.9993
19	0.9999	0.9999	0.9999	0.9999	0.9999	0.9999	0.9998	0.9998	0.9998	0.9997
20	1.0000	1.0000	1.0000	1.0000	1.0000	1.0000	0.9999	0.9999	0.9999	0.9999
21	1.0000	1.0000	1.0000	1.0000	1.0000	1.0000	1.0000	1.0000	1.0000	1.0000
22	1.0000	1.0000	1.0000	1.0000	1.0000	1.0000	1.0000	1.0000	1.0000	1.0000

	μR									
X	8.1	8.2	8.3	8.4	8.5	8.6	8.7	8.8	8.9	9.0
0	0.0003	0.0003	0.0002	0.0002	0.0002	0.0002	0.0002	0.0002	0.0001	0.0001
1	0.0028	0.0025	0.0023	0.0021	0.0019	0.0018	0.0016	0.0015	0.0014	0.0012
2	0.0127	0.0118	0.0109	0.0100	0.0093	0.0086	0.0079	0.0073	0.0068	0.0062
3	0.0396	0.0370	0.0346	0.0323	0.0301	0.0281	0.0262	0.0244	0.0228	0.0212
4	0.0940	0.0887	0.0837	0.0789	0.0744	0.0701	0.0660	0.0621	0.0584	0.0550
5	0.1822	0.1736	0.1653	0.1573	0.1496	0.1422	0.1352	0.1284	0.1219	0.1157
6	0.3013	0.2896	0.2781	0.2670	0.2562	0.2457	0.2355	0.2256	0.2160	0.2068
7	0.4391	0.4254	0.4119	0.3987	0.3856	0.3728	0.3602	0.3478	0.3357	0.3239
8	0.5786	0.5647	0.5507	0.5369	0.5231	0.5094	0.4958	0.4823	0.4689	0.4557
9	0.7041	0.6915	0.6788	0.6659	0.6530	0.6400	0.6269	0.6137	0.6006	0.5874
10	0.8058	0.7955	0.7850	0.7743	0.7634	0.7522	0.7409	0.7294	0.7178	0.7060
11	0.8807	0.8731	0.8652	0.8571	0.8487	0.8400	0.8311	0.8220	0.8126	0.8030
12	0.9313	0.9261	0.9207	0.9150	0.9091	0.9029	0.8965	0.8898	0.8829	0.8758
13	0.9628	0.9595	0.9561	0.9524	0.9486	0.9445	0.9403	0.9358	0.9311	0.9261
14	0.9810	0.9791	0.9771	0.9749	0.9726	0.9701	0.9675	0.9647	0.9617	0.9585
15	0.9908	0.9898	0.9887	0.9875	0.9862	0.9848	0.9832	0.9816	0.9798	0.9780
16	0.9958	0.9953	0.9947	0.9941	0.9934	0.9926	0.9918	0.9909	0.9899	0.9889
17	0.9982	0.9979	0.9977	0.9973	0.9970	0.9966	0.9962	0.9957	0.9952	0.9947
18	0.9992	0.9991	0.9990	0.9989	0.9987	0.9985	0.9983	0.9981	0.9978	0.9976
19	0.9997	0.9997	0.9996	0.9995	0.9995	0.9994	0.9993	0.9992	0.9991	0.9989
20	0.9999	0.9999	0.9998	0.9998	0.9998	0.9998	0.9997	0.9997	0.9996	0.9996
21	1.0000	1.0000	0.9999	0.9999	0.9999	0.9999	0.9999	0.9999	0.9998	0.9998
22	1.0000	1.0000	1.0000	1.0000	1.0000	1.0000	1.0000	1.0000	0.9999	0.9999
23	1.0000	1.0000	1.0000	1.0000	1.0000	1.0000	1.0000	1.0000	1.0000	1.0000

(Table A.4 continued)

X	μR 9.1	9.2	9.3	9.4	9.5	9.6	9.7	9.8	9.9	10.0
0	0.0001	0.0001	0.0001	0.0001	0.0001	0.0001	0.0001	0.0001	0.0001	0.0000
1	0.0011	0.0010	0.0009	0.0009	0.0008	0.0007	0.0007	0.0006	0.0005	0.0005
2	0.0058	0.0053	0.0049	0.0045	0.0042	0.0038	0.0035	0.0033	0.0030	0.0028
3	0.0198	0.0184	0.0172	0.0160	0.0149	0.0138	0.0129	0.0120	0.0111	0.0103
4	0.0517	0.0486	0.0456	0.0429	0.0403	0.0378	0.0355	0.0333	0.0312	0.0293
5	0.1098	0.1041	0.0986	0.0935	0.0885	0.0838	0.0793	0.0750	0.0710	0.0671
6	0.1978	0.1892	0.1808	0.1727	0.1649	0.1574	0.1502	0.1433	0.1366	0.1301
7	0.3123	0.3010	0.2900	0.2792	0.2687	0.2584	0.2485	0.2388	0.2294	0.2202
8	0.4426	0.4296	0.4168	0.4042	0.3918	0.3796	0.3676	0.3558	0.3442	0.3328
9	0.5742	0.5611	0.5479	0.5349	0.5218	0.5089	0.4960	0.4832	0.4705	0.4579
10	0.6941	0.6820	0.6699	0.6576	0.6453	0.6329	0.6205	0.6080	0.5955	0.5830
11	0.7932	0.7832	0.7730	0.7626	0.7520	0.7412	0.7303	0.7193	0.7081	0.6968
12	0.8684	0.8607	0.8529	0.8448	0.8364	0.8279	0.8191	0.8101	0.8009	0.7916
13	0.9210	0.9156	0.9100	0.9042	0.8981	0.8919	0.8853	0.8786	0.8716	0.8645
14	0.9552	0.9517	0.9480	0.9441	0.9400	0.9357	0.9312	0.9265	0.9216	0.9165
15	0.9760	0.9738	0.9715	0.9691	0.9665	0.9638	0.9609	0.9579	0.9546	0.9513
16	0.9878	0.9865	0.9852	0.9838	0.9823	0.9806	0.9789	0.9770	0.9751	0.9730
17	0.9941	0.9934	0.9927	0.9919	0.9911	0.9902	0.9892	0.9881	0.9870	0.9857
18	0.9973	0.9969	0.9966	0.9962	0.9957	0.9952	0.9947	0.9941	0.9935	0.9928
19	0.9988	0.9986	0.9985	0.9983	0.9980	0.9978	0.9975	0.9972	0.9969	0.9965
20	0.9995	0.9994	0.9993	0.9992	0.9991	0.9990	0.9989	0.9987	0.9986	0.9984
21	0.9998	0.9998	0.9997	0.9997	0.9996	0.9996	0.9995	0.9995	0.9994	0.9993
22	0.9999	0.9999	0.9999	0.9999	0.9999	0.9998	0.9998	0.9998	0.9997	0.9997
23	1.0000	1.0000	1.0000	1.0000	0.9999	0.9999	0.9999	0.9999	0.9999	0.9999
24	1.0000	1.0000	1.0000	1.0000	1.0000	1.0000	1.0000	1.0000	1.0000	1.0000
25	1.0000	1.0000	1.0000	1.0000	1.0000	1.0000	1.0000	1.0000	1.0000	1.0000

X	μR 11	12	13	14	15	16	17	18	19	20
0	0.0000	0.0000	0.0000	0.0000	0.0000	0.0000	0.0000	0.0000	0.0000	0.0000
1	0.0002	0.0001	0.0000	0.0000	0.0000	0.0000	0.0000	0.0000	0.0000	0.0000
2	0.0012	0.0005	0.0002	0.0001	0.0000	0.0000	0.0000	0.0000	0.0000	0.0000
3	0.0049	0.0023	0.0011	0.0005	0.0002	0.0001	0.0000	0.0000	0.0000	0.0000
4	0.0151	0.0076	0.0037	0.0018	0.0009	0.0004	0.0002	0.0001	0.0000	0.0000
5	0.0375	0.0203	0.0107	0.0055	0.0028	0.0014	0.0007	0.0003	0.0002	0.0001
6	0.0786	0.0458	0.0259	0.0142	0.0076	0.0040	0.0021	0.0010	0.0005	0.0003
7	0.1432	0.0895	0.0540	0.0316	0.0180	0.0100	0.0054	0.0029	0.0015	0.0008
8	0.2320	0.1550	0.0998	0.0621	0.0374	0.0220	0.0126	0.0071	0.0039	0.0021
9	0.3405	0.2424	0.1658	0.1094	0.0699	0.0433	0.0261	0.0154	0.0089	0.0050
10	0.4599	0.3472	0.2517	0.1757	0.1185	0.0774	0.0491	0.0304	0.0183	0.0108
11	0.5793	0.4616	0.3532	0.2600	0.1848	0.1270	0.0847	0.0549	0.0347	0.0214
12	0.6887	0.5760	0.4631	0.3585	0.2676	0.1931	0.1350	0.0917	0.0606	0.0390
13	0.7813	0.6815	0.5730	0.4644	0.3632	0.2745	0.2009	0.1426	0.0984	0.0661
14	0.8540	0.7720	0.6751	0.5704	0.4657	0.3675	0.2808	0.2081	0.1497	0.1049
15	0.9074	0.8444	0.7636	0.6694	0.5681	0.4667	0.3715	0.2867	0.2148	0.1565
16	0.9441	0.8987	0.8355	0.7559	0.6641	0.5660	0.4677	0.3751	0.2920	0.2211
17	0.9678	0.9370	0.8905	0.8272	0.7489	0.6593	0.5640	0.4686	0.3784	0.2970
18	0.9823	0.9626	0.9302	0.8826	0.8195	0.7423	0.6550	0.5622	0.4695	0.3814
19	0.9907	0.9787	0.9573	0.9235	0.8752	0.8122	0.7363	0.6509	0.5606	0.4703
20	0.9953	0.9884	0.9750	0.9521	0.9170	0.8682	0.8055	0.7307	0.6472	0.5591
21	0.9977	0.9939	0.9859	0.9712	0.9469	0.9108	0.8615	0.7991	0.7255	0.6437
22	0.9990	0.9970	0.9924	0.9833	0.9673	0.9418	0.9047	0.8551	0.7931	0.7206
23	0.9995	0.9985	0.9960	0.9907	0.9805	0.9633	0.9367	0.8989	0.8490	0.7875
24	0.9998	0.9993	0.9980	0.9950	0.9888	0.9777	0.9594	0.9317	0.8933	0.8432
25	0.9999	0.9997	0.9990	0.9974	0.9938	0.9869	0.9748	0.9554	0.9269	0.8878
26	1.0000	0.9999	0.9995	0.9987	0.9967	0.9925	0.9848	0.9718	0.9514	0.9221
27	1.0000	0.9999	0.9998	0.9994	0.9983	0.9959	0.9912	0.9827	0.9687	0.9475
28	1.0000	1.0000	0.9999	0.9997	0.9991	0.9978	0.9950	0.9897	0.9805	0.9657
29	1.0000	1.0000	1.0000	0.9999	0.9996	0.9989	0.9973	0.9941	0.9882	0.9782
30	1.0000	1.0000	1.0000	0.9999	0.9998	0.9994	0.9986	0.9967	0.9930	0.9865
31	1.0000	1.0000	1.0000	1.0000	0.9999	0.9997	0.9993	0.9982	0.9960	0.9919
32	1.0000	1.0000	1.0000	1.0000	1.0000	0.9999	0.9996	0.9990	0.9978	0.9953
33	1.0000	1.0000	1.0000	1.0000	1.0000	0.9999	0.9998	0.9995	0.9988	0.9973
34	1.0000	1.0000	1.0000	1.0000	1.0000	1.0000	0.9999	0.9998	0.9994	0.9985
35	1.0000	1.0000	1.0000	1.0000	1.0000	1.0000	1.0000	0.9999	0.9997	0.9992
36	1.0000	1.0000	1.0000	1.0000	1.0000	1.0000	1.0000	0.9999	0.9998	0.9996
37	1.0000	1.0000	1.0000	1.0000	1.0000	1.0000	1.0000	1.0000	0.9999	0.9998
38	1.0000	1.0000	1.0000	1.0000	1.0000	1.0000	1.0000	1.0000	1.0000	0.9999
39	1.0000	1.0000	1.0000	1.0000	1.0000	1.0000	1.0000	1.0000	1.0000	0.9999
40	1.0000	1.0000	1.0000	1.0000	1.0000	1.0000	1.0000	1.0000	1.0000	1.0000

Table A.5 EXP($-x$) for Selected Values of x where $x = \beta t$

x	e^{-x}	x	e^{-x}
0.0	1.0000	5.0	0.00674
0.1	0.9048	5.1	0.00610
0.2	0.8187	5.2	0.00552
0.3	0.7408	5.3	0.00499
0.4	0.6703	5.4	0.00452
0.5	0.6065	5.5	0.00409
0.6	0.5488	5.6	0.00370
0.7	0.4966	5.7	0.00335
0.8	0.4493	5.8	0.00303
0.9	0.4066	5.9	0.00274
1.0	0.3679	6.0	0.00248
1.1	0.3329	6.1	0.00224
1.2	0.3012	6.2	0.00203
1.3	0.2725	6.3	0.00184
1.4	0.2466	6.4	0.00166
1.5	0.2231	6.5	0.00150
1.6	0.2019	6.6	0.00136
1.7	0.1827	6.7	0.00123
1.8	0.1653	6.8	0.00111
1.9	0.1496	6.9	0.00101
2.0	0.1353	7.0	0.00091
2.1	0.1225	7.1	0.00083
2.2	0.1108	7.2	0.00075
2.3	0.1003	7.3	0.00068
2.4	0.0907	7.4	0.00061
2.5	0.0821	7.5	0.00055
2.6	0.0743	7.6	0.00050
2.7	0.0672	7.7	0.00045
2.8	0.0608	7.8	0.00041
2.9	0.0550	7.9	0.00037
3.0	0.0498	8.0	0.00034
3.1	0.0450	8.1	0.00030
3.2	0.0408	8.2	0.00027
3.3	0.0369	8.3	0.00025
3.4	0.0334	8.4	0.00022
3.5	0.0302	8.5	0.00020
3.6	0.0273	8.6	0.00018
3.7	0.0247	8.7	0.00017
3.8	0.0224	8.8	0.00015
3.9	0.0202	8.9	0.00014
4.0	0.0183	9.0	0.00012
4.1	0.0166	9.1	0.00011
4.2	0.0150	9.2	0.00010
4.3	0.0136	9.3	0.00009
4.4	0.0123	9.4	0.00008
4.5	0.0111	9.5	0.00007
4.6	0.0101	9.6	0.00007
4.7	0.0091	9.7	0.00006
4.8	0.0082	9.8	0.00006
4.9	0.0074	9.9	0.00005

Table A.6 Cumulative Probabilities for the Standard Normal Distribution

Entry = $\text{pr}(Z \leqslant z)$

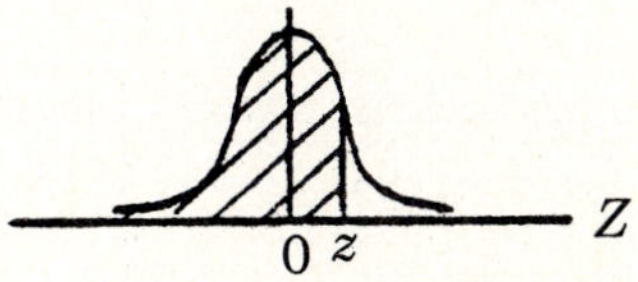

z	.00	.01	.02	.03	.04	.05	.06	.07	.08	.09
-3.9	0.0000	0.0000	0.0000	0.0000	0.0000	0.0000	0.0000	0.0000	0.0000	0.0000
-3.8	0.0001	0.0001	0.0001	0.0001	0.0001	0.0001	0.0001	0.0001	0.0001	0.0001
-3.7	0.0001	0.0001	0.0001	0.0001	0.0001	0.0001	0.0001	0.0001	0.0001	0.0001
-3.6	0.0002	0.0002	0.0001	0.0001	0.0001	0.0001	0.0001	0.0001	0.0001	0.0001
-3.5	0.0002	0.0002	0.0002	0.0002	0.0002	0.0002	0.0002	0.0002	0.0002	0.0002
-3.4	0.0003	0.0003	0.0003	0.0003	0.0003	0.0003	0.0003	0.0003	0.0003	0.0002
-3.3	0.0005	0.0005	0.0005	0.0004	0.0004	0.0004	0.0004	0.0004	0.0004	0.0003
-3.2	0.0007	0.0007	0.0006	0.0006	0.0006	0.0006	0.0006	0.0005	0.0005	0.0005
-3.1	0.0010	0.0009	0.0009	0.0009	0.0008	0.0008	0.0008	0.0008	0.0007	0.0007
-3.0	0.0013	0.0013	0.0013	0.0012	0.0012	0.0011	0.0011	0.0011	0.0010	0.0010
-2.9	0.0019	0.0018	0.0018	0.0017	0.0016	0.0016	0.0015	0.0015	0.0014	0.0014
-2.8	0.0026	0.0025	0.0024	0.0023	0.0023	0.0022	0.0021	0.0021	0.0020	0.0019
-2.7	0.0035	0.0034	0.0033	0.0032	0.0031	0.0030	0.0029	0.0028	0.0027	0.0026
-2.6	0.0047	0.0045	0.0044	0.0043	0.0041	0.0040	0.0039	0.0038	0.0037	0.0036
-2.5	0.0062	0.0060	0.0059	0.0057	0.0055	0.0054	0.0052	0.0051	0.0049	0.0048
-2.4	0.0082	0.0080	0.0078	0.0075	0.0073	0.0071	0.0069	0.0068	0.0066	0.0064
-2.3	0.0107	0.0104	0.0102	0.0099	0.0096	0.0094	0.0091	0.0089	0.0087	0.0084
-2.2	0.0139	0.0136	0.0132	0.0129	0.0125	0.0122	0.0119	0.0116	0.0113	0.0110
-2.1	0.0179	0.0174	0.0170	0.0166	0.0162	0.0158	0.0154	0.0150	0.0146	0.0143
-2.0	0.0228	0.0222	0.0217	0.0212	0.0207	0.0202	0.0197	0.0192	0.0188	0.0183
-1.9	0.0287	0.0281	0.0274	0.0268	0.0262	0.0256	0.0250	0.0244	0.0239	0.0233
-1.8	0.0359	0.0351	0.0344	0.0336	0.0329	0.0322	0.0314	0.0307	0.0301	0.0294
-1.7	0.0446	0.0436	0.0427	0.0418	0.0409	0.0401	0.0392	0.0384	0.0375	0.0367
-1.6	0.0548	0.0537	0.0526	0.0516	0.0505	0.0495	0.0485	0.0475	0.0465	0.0455
-1.5	0.0668	0.0655	0.0643	0.0630	0.0618	0.0606	0.0594	0.0582	0.0571	0.0559
-1.4	0.0808	0.0793	0.0778	0.0764	0.0749	0.0735	0.0721	0.0708	0.0694	0.0681
-1.3	0.0968	0.0951	0.0934	0.0918	0.0901	0.0885	0.0869	0.0853	0.0838	0.0823
-1.2	0.1151	0.1131	0.1112	0.1093	0.1075	0.1056	0.1038	0.1020	0.1003	0.0985
-1.1	0.1357	0.1335	0.1314	0.1292	0.1271	0.1251	0.1230	0.1210	0.1190	0.1170
-1.0	0.1587	0.1562	0.1539	0.1515	0.1492	0.1469	0.1446	0.1423	0.1401	0.1379
-.9	0.1841	0.1814	0.1788	0.1762	0.1736	0.1711	0.1685	0.1660	0.1635	0.1611
-.8	0.2119	0.2090	0.2061	0.2033	0.2005	0.1977	0.1949	0.1922	0.1894	0.1867
-.7	0.2420	0.2389	0.2358	0.2327	0.2297	0.2266	0.2236	0.2207	0.2177	0.2148
-.6	0.2743	0.2709	0.2676	0.2643	0.2611	0.2578	0.2546	0.2514	0.2483	0.2451
-.5	0.3085	0.3050	0.3015	0.2981	0.2946	0.2912	0.2877	0.2843	0.2810	0.2776
-.4	0.3446	0.3409	0.3372	0.3336	0.3300	0.3264	0.3228	0.3192	0.3156	0.3121
-.3	0.3821	0.3783	0.3745	0.3707	0.3669	0.3632	0.3594	0.3557	0.3520	0.3483
-.2	0.4207	0.4168	0.4129	0.4090	0.4052	0.4013	0.3974	0.3936	0.3897	0.3859
-.1	0.4602	0.4562	0.4522	0.4483	0.4443	0.4404	0.4364	0.4325	0.4286	0.4247
0	0.5000	0.4960	0.4920	0.4880	0.4840	0.4801	0.4761	0.4721	0.4681	0.4641

(Table A.6 continued)

z	.00	.01	.02	.03	.04	.05	.06	.07	.08	.09
0.0	0.5000	0.5040	0.5080	0.5120	0.5160	0.5199	0.5239	0.5279	0.5319	0.5359
0.1	0.5398	0.5438	0.5478	0.5517	0.5557	0.5596	0.5636	0.5675	0.5714	0.5753
0.2	0.5793	0.5832	0.5871	0.5910	0.5948	0.5987	0.6026	0.6064	0.6103	0.6141
0.3	0.6179	0.6217	0.6255	0.6293	0.6331	0.6368	0.6406	0.6443	0.6480	0.6517
0.4	0.6554	0.6591	0.6628	0.6664	0.6700	0.6736	0.6772	0.6808	0.6844	0.6879
0.5	0.6915	0.6950	0.6985	0.7019	0.7054	0.7088	0.7123	0.7157	0.7190	0.7224
0.6	0.7257	0.7291	0.7324	0.7357	0.7389	0.7422	0.7454	0.7486	0.7517	0.7549
0.7	0.7580	0.7611	0.7642	0.7673	0.7703	0.7734	0.7764	0.7793	0.7823	0.7852
0.8	0.7881	0.7910	0.7939	0.7967	0.7995	0.8023	0.8051	0.8078	0.8106	0.8133
0.9	0.8159	0.8186	0.8212	0.8238	0.8264	0.8289	0.8315	0.8340	0.8365	0.8389
1.0	0.8413	0.8438	0.8461	0.8485	0.8508	0.8531	0.8554	0.8577	0.8599	0.8621
1.1	0.8643	0.8665	0.8686	0.8708	0.8729	0.8749	0.8770	0.8790	0.8810	0.8830
1.2	0.8849	0.8869	0.8888	0.8907	0.8925	0.8944	0.8962	0.8980	0.8997	0.9015
1.3	0.9032	0.9049	0.9066	0.9082	0.9099	0.9115	0.9131	0.9147	0.9162	0.9177
1.4	0.9192	0.9207	0.9222	0.9236	0.9251	0.9265	0.9279	0.9292	0.9306	0.9319
1.5	0.9332	0.9345	0.9357	0.9370	0.9382	0.9394	0.9406	0.9418	0.9429	0.9441
1.6	0.9452	0.9463	0.9474	0.9484	0.9495	0.9505	0.9515	0.9525	0.9535	0.9545
1.7	0.9554	0.9564	0.9573	0.9582	0.9591	0.9599	0.9608	0.9616	0.9625	0.9633
1.8	0.9641	0.9649	0.9656	0.9664	0.9671	0.9678	0.9686	0.9693	0.9699	0.9706
1.9	0.9713	0.9719	0.9726	0.9732	0.9738	0.9744	0.9750	0.9756	0.9761	0.9767
2.0	0.9772	0.9778	0.9783	0.9788	0.9793	0.9798	0.9803	0.9808	0.9812	0.9817
2.1	0.9821	0.9826	0.9830	0.9834	0.9838	0.9842	0.9846	0.9850	0.9854	0.9857
2.2	0.9861	0.9864	0.9868	0.9871	0.9875	0.9878	0.9881	0.9884	0.9887	0.9890
2.3	0.9893	0.9896	0.9898	0.9901	0.9904	0.9906	0.9909	0.9911	0.9913	0.9916
2.4	0.9918	0.9920	0.9922	0.9925	0.9927	0.9929	0.9931	0.9932	0.9934	0.9936
2.5	0.9938	0.9940	0.9941	0.9943	0.9945	0.9946	0.9948	0.9949	0.9951	0.9952
2.6	0.9953	0.9955	0.9956	0.9957	0.9959	0.9960	0.9961	0.9962	0.9963	0.9964
2.7	0.9965	0.9966	0.9967	0.9968	0.9969	0.9970	0.9971	0.9972	0.9973	0.9974
2.8	0.9974	0.9975	0.9976	0.9977	0.9977	0.9978	0.9979	0.9979	0.9980	0.9981
2.9	0.9981	0.9982	0.9982	0.9983	0.9984	0.9984	0.9985	0.9985	0.9986	0.9986
3.0	0.9987	0.9987	0.9987	0.9988	0.9988	0.9989	0.9989	0.9989	0.9990	0.9990
3.1	0.9990	0.9991	0.9991	0.9991	0.9992	0.9992	0.9992	0.9992	0.9993	0.9993
3.2	0.9993	0.9993	0.9994	0.9994	0.9994	0.9994	0.9994	0.9995	0.9995	0.9995
3.3	0.9995	0.9995	0.9995	0.9996	0.9996	0.9996	0.9996	0.9996	0.9996	0.9997
3.4	0.9997	0.9997	0.9997	0.9997	0.9997	0.9997	0.9997	0.9997	0.9997	0.9998
3.5	0.9998	0.9998	0.9998	0.9998	0.9998	0.9998	0.9998	0.9998	0.9998	0.9998
3.6	0.9998	0.9998	0.9999	0.9999	0.9999	0.9999	0.9999	0.9999	0.9999	0.9999
3.7	0.9999	0.9999	0.9999	0.9999	0.9999	0.9999	0.9999	0.9999	0.9999	0.9999
3.8	0.9999	0.9999	0.9999	0.9999	0.9999	0.9999	0.9999	0.9999	0.9999	0.9999
3.9	1.0000	1.0000	1.0000	1.0000	1.0000	1.0000	1.0000	1.0000	1.0000	1.0000

Commonly used z-values for selected α's

α:	0.4000	0.3000	0.2000	0.1000	0.0500	0.0250	0.0100	0.0050	0.0010	0.0005
z:	0.253	0.524	0.842	1.282	1.645	1.960	2.326	2.576	3.090	3.291

Table A.7 Critical Values for Chi-Square Random Variables for Selected Values of υ, α

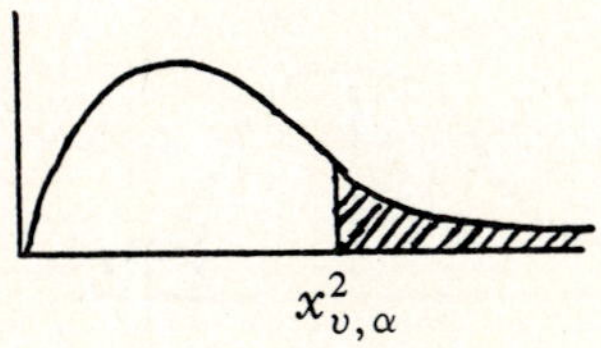

υ	α .995	.990	.975	.950	.900	.100	.050	.025	.010	.005
1				.004	.016	2.71	3.84	5.02	6.63	7.88
2	.01	.02	.05	.10	.211	4.61	5.99	7.38	9.21	10.60
3	.07	.11	.22	.35	.584	6.25	7.81	9.35	11.34	12.84
4	.21	.30	.48	.71	1.064	7.78	9.49	11.14	13.28	14.86
5	.41	.55	.83	1.15	1.610	9.24	11.07	12.83	15.09	16.75
6	.68	.87	1.24	1.64	2.20	10.64	12.59	14.45	16.81	18.55
7	.99	1.24	1.69	2.17	2.83	12.02	14.07	16.01	18.48	20.28
8	1.34	1.65	2.18	2.73	3.49	13.36	15.51	17.53	20.09	21.96
9	1.73	2.09	2.70	3.33	4.17	14.68	16.92	19.02	21.67	23.59
10	2.16	2.56	3.25	3.94	4.87	15.99	18.31	20.48	23.21	25.19
11	2.60	3.05	3.82	4.57	5.58	17.28	19.68	21.92	24.72	26.76
12	3.07	3.57	4.40	5.23	6.30	18.55	21.03	23.34	26.22	28.30
13	3.57	4.11	5.01	5.89	7.04	19.81	22.36	24.74	27.69	29.82
14	4.07	4.66	5.63	6.57	7.79	21.06	23.68	26.12	29.14	31.32
15	4.60	5.23	6.26	7.26	8.55	22.31	25.00	27.49	30.58	32.80
16	5.14	5.81	6.91	7.96	9.31	23.54	26.30	28.85	32.00	34.27
17	5.70	6.41	7.56	8.67	10.09	24.77	27.59	30.19	33.41	35.72
18	6.26	7.01	8.23	9.39	10.87	25.99	28.87	31.53	34.81	37.16
19	6.84	7.63	8.91	10.12	11.65	27.20	30.14	32.85	36.19	38.58
20	7.43	8.26	9.59	10.85	12.44	28.41	31.41	34.17	37.57	40.00
21	8.03	8.90	10.28	11.59	13.24	29.62	32.67	35.48	38.93	41.40
22	8.64	9.54	10.98	12.34	14.04	30.81	33.92	36.78	40.29	42.80
23	9.26	10.20	11.69	13.09	14.85	32.01	35.17	38.08	41.64	44.18
24	9.89	10.86	12.40	13.85	15.66	33.20	36.42	39.36	42.98	45.56
25	10.52	11.52	13.12	14.61	16.47	34.82	37.65	40.65	44.31	46.93
26	11.16	12.20	13.84	15.38	17.29	35.56	38.89	41.92	45.64	48.29
27	11.81	12.88	14.57	16.15	18.11	36.74	40.11	43.19	46.96	49.64
28	12.46	13.56	15.31	16.93	18.94	37.92	41.34	44.46	48.28	50.99
29	13.12	14.26	16.05	17.71	19.77	39.09	42.56	45.72	49.59	52.34
30	13.79	14.95	16.79	18.49	20.60	40.26	43.77	46.98	50.89	53.67
40	20.71	22.16	24.43	26.51	29.05	51.81	55.76	59.34	63.69	66.77
50	27.99	29.71	32.36	34.76	37.69	63.17	67.50	71.42	76.15	79.49
60	35.53	37.48	40.48	43.19	46.46	74.40	79.08	83.30	88.38	91.95
70	43.28	45.44	48.76	51.74	55.33	85.53	90.53	95.02	100.43	104.22
80	51.17	53.54	57.15	60.39	64.28	96.58	101.88	106.63	112.33	116.32
90	59.20	61.75	65.65	69.13	73.29	107.57	113.14	118.14	124.12	128.30
100	67.33	70.06	74.22	77.93	82.36	118.50	124.34	129.56	135.81	140.17

Table A.8 Critical Values for Student's t-distributions for Selected Values of υ, α

$$[t_{\upsilon,\alpha} \text{ such that } \mathrm{pr}(t_\upsilon > t_{\upsilon,\alpha}) = \alpha]$$

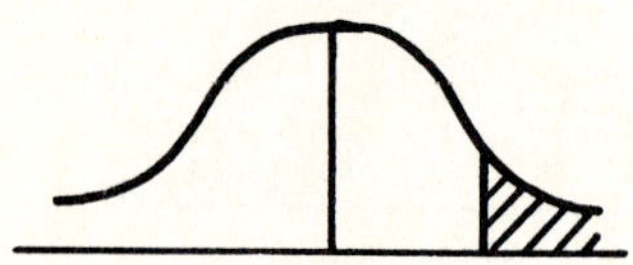

$t_{\upsilon,\alpha}$

υ	α .40	.30	.20	.10	.05	.025	.01	.005	.001	.0005
1	0.325	0.727	1.376	3.078	6.314	12.710	31.820	63.660	318.300	636.600
2	0.289	0.617	1.061	1.886	2.920	4.303	6.965	9.925	22.330	31.600
3	0.277	0.584	0.978	1.638	2.353	3.182	4.541	5.841	10.220	12.940
4	0.271	0.569	0.941	1.533	2.132	2.776	3.747	4.604	7.173	8.610
5	0.267	0.559	0.920	1.476	2.015	2.571	3.365	4.032	5.893	6.859
6	0.265	0.553	0.906	1.440	1.943	2.447	3.143	3.707	5.208	5.959
7	0.263	0.549	0.896	1.415	1.895	2.365	2.998	3.499	4.785	5.405
8	0.262	0.546	0.889	1.397	1.860	2.306	2.896	3.355	4.501	5.041
9	0.261	0.543	0.883	1.383	1.833	2.262	2.821	3.250	4.297	4.781
10	0.260	0.542	0.879	1.372	1.812	2.228	2.764	3.169	4.144	4.587
11	0.260	0.540	0.876	1.363	1.796	2.201	2.718	3.106	4.025	4.437
12	0.259	0.539	0.873	1.356	1.782	2.179	2.681	3.055	3.930	4.318
13	0.259	0.538	0.870	1.350	1.771	2.160	2.650	3.012	3.852	4.221
14	0.258	0.537	0.868	1.345	1.761	2.145	2.624	2.977	3.787	4.140
15	0.258	0.536	0.866	1.341	1.753	2.131	2.602	2.947	3.733	4.073
16	0.258	0.535	0.865	1.337	1.746	2.120	2.583	2.921	3.686	4.015
17	0.257	0.534	0.863	1.333	1.740	2.110	2.567	2.898	3.646	3.965
18	0.257	0.534	0.862	1.330	1.734	2.101	2.552	2.878	3.611	3.922
19	0.257	0.533	0.861	1.328	1.729	2.093	2.539	2.861	3.579	3.883
20	0.257	0.533	0.860	1.325	1.725	2.086	2.528	2.845	3.552	3.850
21	0.257	0.532	0.859	1.323	1.721	2.080	2.518	2.831	3.527	3.819
22	0.256	0.532	0.858	1.321	1.717	2.074	2.508	2.819	3.505	3.792
23	0.256	0.532	0.858	1.319	1.714	2.069	2.500	2.807	3.485	3.767
24	0.256	0.531	0.857	1.318	1.711	2.064	2.492	2.797	3.467	3.745
25	0.256	0.531	0.856	1.316	1.708	2.060	2.485	2.787	3.450	3.725
26	0.256	0.531	0.856	1.315	1.706	2.056	2.479	2.779	3.435	3.707
27	0.256	0.531	0.855	1.314	1.703	2.052	2.473	2.771	3.421	3.690
28	0.256	0.530	0.855	1.313	1.701	2.048	2.467	2.763	3.408	3.674
29	0.256	0.530	0.854	1.311	1.699	2.045	2.462	2.756	3.396	3.659
30	0.256	0.530	0.854	1.310	1.697	2.042	2.457	2.750	3.385	3.646
40	0.255	0.529	0.851	1.303	1.684	2.021	2.423	2.704	3.307	3.551
50	0.255	0.528	0.849	1.298	1.676	2.009	2.403	2.678	3.262	3.495
60	0.254	0.527	0.848	1.296	1.671	2.000	2.390	2.660	3.232	3.460
80	0.254	0.527	0.846	1.292	1.664	1.990	2.374	2.639	3.195	3.415
100	0.254	0.526	0.845	1.290	1.660	1.984	2.365	2.626	3.174	3.389
120	0.254	0.526	0.845	1.289	1.658	1.980	2.358	2.617	3.165	3.373
200	0.254	0.525	0.843	1.286	1.653	1.972	2.345	2.601	3.131	3.339
500	0.253	0.525	0.842	1.283	1.648	1.965	2.334	2.586	3.106	3.310
∞	0.253	0.524	0.842	1.282	1.645	1.960	2.326	2.576	3.090	3.291

Table A.9(a) Critical Values for Fisher's F Distributions for Selected Values of υ_1 and υ_2 at $\alpha = 0.05$

υ_1 = Degrees of Freedom for Numerator MS; υ_2 = Degrees of Freedom for Denominator MS

υ_2	1	2	3	4	5	6	7	8	9	10	12	15	20	24	30	40	60	120	∞
1	161·4	199·5	215·7	224·6	230·2	234·0	236·8	238·9	240·5	241·9	243·9	245·9	248·0	249·1	250·1	251·1	252·2	253·3	254·3
2	18·51	19·00	19·16	19·25	19·30	19·33	19·35	19·37	19·38	19·40	19·41	19·43	19·45	19·45	19·46	19·47	19·48	19·49	19·50
3	10·13	9·55	9·28	9·12	9·01	8·94	8·89	8·85	8·81	8·79	8·74	8·70	8·66	8·64	8·62	8·59	8·57	8·55	8·53
4	7·71	6·94	6·59	6·39	6·26	6·16	6·09	6·04	6·00	5·96	5·91	5·86	5·80	5·77	5·75	5·72	5·69	5·66	5·63
5	6·61	5·79	5·41	5·19	5·05	4·95	4·88	4·82	4·77	4·74	4·68	4·62	4·56	4·53	4·50	4·46	4·43	4·40	4·36
6	5·99	5·14	4·76	4·53	4·39	4·28	4·21	4·15	4·10	4·06	4·00	3·94	3·87	3·84	3·81	3·77	3·74	3·70	3·67
7	5·59	4·74	4·35	4·12	3·97	3·87	3·79	3·73	3·68	3·64	3·57	3·51	3·44	3·41	3·38	3·34	3·30	3·27	3·23
8	5·32	4·46	4·07	3·84	3·69	3·58	3·50	3·44	3·39	3·35	3·28	3·22	3·15	3·12	3·08	3·04	3·01	2·97	2·93
9	5·12	4·26	3·86	3·63	3·48	3·37	3·29	3·23	3·18	3·14	3·07	3·01	2·94	2·90	2·86	2·83	2·79	2·75	2·71
10	4·96	4·10	3·71	3·48	3·33	3·22	3·14	3·07	3·02	2·98	2·91	2·85	2·77	2·74	2·70	2·66	2·62	2·58	2·54
11	4·84	3·98	3·59	3·36	3·20	3·09	3·01	2·95	2·90	2·85	2·79	2·72	2·65	2·61	2·57	2·53	2·49	2·45	2·40
12	4·75	3·89	3·49	3·26	3·11	3·00	2·91	2·85	2·80	2·75	2·69	2·62	2·54	2·51	2·47	2·43	2·38	2·34	2·30
13	4·67	3·81	3·41	3·18	3·03	2·92	2·83	2·77	2·71	2·67	2·60	2·53	2·46	2·42	2·38	2·34	2·30	2·25	2·21
14	4·60	3·74	3·34	3·11	2·96	2·85	2·76	2·70	2·65	2·60	2·53	2·46	2·39	2·35	2·31	2·27	2·22	2·18	2·13
15	4·54	3·68	3·29	3·06	2·90	2·79	2·71	2·64	2·59	2·54	2·48	2·40	2·33	2·29	2·25	2·20	2·16	2·11	2·07
16	4·49	3·63	3·24	3·01	2·85	2·74	2·66	2·59	2·54	2·49	2·42	2·35	2·28	2·24	2·19	2·15	2·11	2·06	2·01
17	4·45	3·59	3·20	2·96	2·81	2·70	2·61	2·55	2·49	2·45	2·38	2·31	2·23	2·19	2·15	2·10	2·06	2·01	1·96
18	4·41	3·55	3·16	2·93	2·77	2·66	2·58	2·51	2·46	2·41	2·34	2·27	2·19	2·15	2·11	2·06	2·02	1·97	1·92
19	4·38	3·52	3·13	2·90	2·74	2·63	2·54	2·48	2·42	2·38	2·31	2·23	2·16	2·11	2·07	2·03	1·98	1·93	1·88
20	4·35	3·49	3·10	2·87	2·71	2·60	2·51	2·45	2·39	2·35	2·28	2·20	2·12	2·08	2·04	1·99	1·95	1·90	1·84
21	4·32	3·47	3·07	2·84	2·68	2·57	2·49	2·42	2·37	2·32	2·25	2·18	2·10	2·05	2·01	1·96	1·92	1·87	1·81
22	4·30	3·44	3·05	2·82	2·66	2·55	2·46	2·40	2·34	2·30	2·23	2·15	2·07	2·03	1·98	1·94	1·89	1·84	1·78
23	4·28	3·42	3·03	2·80	2·64	2·53	2·44	2·37	2·32	2·27	2·20	2·13	2·05	2·01	1·96	1·91	1·86	1·81	1·76
24	4·26	3·40	3·01	2·78	2·62	2·51	2·42	2·36	2·30	2·25	2·18	2·11	2·03	1·98	1·94	1·89	1·84	1·79	1·73
25	4·24	3·39	2·99	2·76	2·60	2·49	2·40	2·34	2·28	2·24	2·16	2·09	2·01	1·96	1·92	1·87	1·82	1·77	1·71
26	4·23	3·37	2·98	2·74	2·59	2·47	2·39	2·32	2·27	2·22	2·15	2·07	1·99	1·95	1·90	1·85	1·80	1·75	1·69
27	4·21	3·35	2·96	2·73	2·57	2·46	2·37	2·31	2·25	2·20	2·13	2·06	1·97	1·93	1·88	1·84	1·79	1·73	1·67
28	4·20	3·34	2·95	2·71	2·56	2·45	2·36	2·29	2·24	2·19	2·12	2·04	1·96	1·91	1·87	1·82	1·77	1·71	1·65
29	4·18	3·33	2·93	2·70	2·55	2·43	2·35	2·28	2·22	2·18	2·10	2·03	1·94	1·90	1·85	1·81	1·75	1·70	1·64
30	4·17	3·32	2·92	2·69	2·53	2·42	2·33	2·27	2·21	2·16	2·09	2·01	1·93	1·89	1·84	1·79	1·74	1·68	1·62
40	4·08	3·23	2·84	2·61	2·45	2·34	2·25	2·18	2·12	2·08	2·00	1·92	1·84	1·79	1·74	1·69	1·64	1·58	1·51
60	4·00	3·15	2·76	2·53	2·37	2·25	2·17	2·10	2·04	1·99	1·92	1·84	1·75	1·70	1·65	1·59	1·53	1·47	1·39
120	3·92	3·07	2·68	2·45	2·29	2·17	2·09	2·02	1·96	1·91	1·83	1·75	1·66	1·61	1·55	1·50	1·43	1·35	1·25
∞	3·84	3·00	2·60	2·37	2·21	2·10	2·01	1·94	1·88	1·83	1·75	1·67	1·57	1·52	1·46	1·39	1·32	1·22	1·00

Table A.9(b) Critical Values for Fisher's F Distributions for Selected Values of υ_1 and υ_2 at $\alpha = 0.01$

υ_1 = Degrees of Freedom for Numerator MS; υ_2 = Degrees of Freedom for Denominator MS

υ_2	1	2	3	4	5	6	7	8	9	10	12	15	20	24	30	40	60	120	∞
1	4052	4999.5	5403	5625	5764	5859	5928	5981	6022	6056	6106	6157	6209	6235	6261	6287	6313	6339	6366
2	98.50	99.00	99.17	99.25	99.30	99.33	99.36	99.37	99.39	99.40	99.42	99.43	99.45	99.46	99.47	99.47	99.48	99.49	99.50
3	34.12	30.82	29.46	28.71	28.24	27.91	27.67	27.49	27.35	27.23	27.05	26.87	26.69	26.60	26.50	26.41	26.32	26.22	26.13
4	21.20	18.00	16.69	15.98	15.52	15.21	14.98	14.80	14.66	14.55	14.37	14.20	14.02	13.93	13.84	13.75	13.65	13.56	13.46
5	16.26	13.27	12.06	11.39	10.97	10.67	10.46	10.29	10.16	10.05	9.89	9.72	9.55	9.47	9.38	9.29	9.20	9.11	9.02
6	13.75	10.92	9.78	9.15	8.75	8.47	8.26	8.10	7.98	7.87	7.72	7.56	7.40	7.31	7.23	7.14	7.06	6.97	6.88
7	12.25	9.55	8.45	7.85	7.46	7.19	6.99	6.84	6.72	6.62	6.47	6.31	6.16	6.07	5.99	5.91	5.82	5.74	5.65
8	11.26	8.65	7.59	7.01	6.63	6.37	6.18	6.03	5.91	5.81	5.67	5.52	5.36	5.28	5.20	5.12	5.03	4.95	4.86
9	10.56	8.02	6.99	6.42	6.06	5.80	5.61	5.47	5.35	5.26	5.11	4.96	4.81	4.73	4.65	4.57	4.48	4.40	4.31
10	10.04	7.56	6.55	5.99	5.64	5.39	5.20	5.06	4.94	4.85	4.71	4.56	4.41	4.33	4.25	4.17	4.08	4.00	3.91
11	9.65	7.21	6.22	5.67	5.32	5.07	4.89	4.74	4.63	4.54	4.40	4.25	4.10	4.02	3.94	3.86	3.78	3.69	3.60
12	9.33	6.93	5.95	5.41	5.06	4.82	4.64	4.50	4.39	4.30	4.16	4.01	3.86	3.78	3.70	3.62	3.54	3.45	3.36
13	9.07	6.70	5.74	5.21	4.86	4.62	4.44	4.30	4.19	4.10	3.96	3.82	3.66	3.59	3.51	3.43	3.34	3.25	3.17
14	8.86	6.51	5.56	5.04	4.69	4.46	4.28	4.14	4.03	3.94	3.80	3.66	3.51	3.43	3.35	3.27	3.18	3.09	3.00
15	8.68	6.36	5.42	4.89	4.56	4.32	4.14	4.00	3.89	3.80	3.67	3.52	3.37	3.29	3.21	3.13	3.05	2.96	2.87
16	8.53	6.23	5.29	4.77	4.44	4.20	4.03	3.89	3.78	3.69	3.55	3.41	3.26	3.18	3.10	3.02	2.93	2.84	2.75
17	8.40	6.11	5.18	4.67	4.34	4.10	3.93	3.79	3.68	3.59	3.46	3.31	3.16	3.08	3.00	2.92	2.83	2.75	2.65
18	8.29	6.01	5.09	4.58	4.25	4.01	3.84	3.71	3.60	3.51	3.37	3.23	3.08	3.00	2.92	2.84	2.75	2.66	2.57
19	8.18	5.93	5.01	4.50	4.17	3.94	3.77	3.63	3.52	3.43	3.30	3.15	3.00	2.92	2.84	2.76	2.67	2.58	2.49
20	8.10	5.85	4.94	4.43	4.10	3.87	3.70	3.56	3.46	3.37	3.23	3.09	2.94	2.86	2.78	2.69	2.61	2.52	2.42
21	8.02	5.78	4.87	4.37	4.04	3.81	3.64	3.51	3.40	3.31	3.17	3.03	2.88	2.80	2.72	2.64	2.55	2.46	2.36
22	7.95	5.72	4.82	4.31	3.99	3.76	3.59	3.45	3.35	3.26	3.12	2.98	2.83	2.75	2.67	2.58	2.50	2.40	2.31
23	7.88	5.66	4.76	4.26	3.94	3.71	3.54	3.41	3.30	3.21	3.07	2.93	2.78	2.70	2.62	2.54	2.45	2.35	2.26
24	7.82	5.61	4.72	4.22	3.90	3.67	3.50	3.36	3.26	3.17	3.03	2.89	2.74	2.66	2.58	2.49	2.40	2.31	2.21
25	7.77	5.57	4.68	4.18	3.85	3.63	3.46	3.32	3.22	3.13	2.99	2.85	2.70	2.62	2.54	2.45	2.36	2.27	2.17
26	7.72	5.53	4.64	4.14	3.82	3.59	3.42	3.29	3.18	3.09	2.96	2.81	2.66	2.58	2.50	2.42	2.33	2.23	2.13
27	7.68	5.49	4.60	4.11	3.78	3.56	3.39	3.26	3.15	3.06	2.93	2.78	2.63	2.55	2.47	2.38	2.29	2.20	2.10
28	7.64	5.45	4.57	4.07	3.75	3.53	3.36	3.23	3.12	3.03	2.90	2.75	2.60	2.52	2.44	2.35	2.26	2.17	2.06
29	7.60	5.42	4.54	4.04	3.73	3.50	3.33	3.20	3.09	3.00	2.87	2.73	2.57	2.49	2.41	2.33	2.23	2.14	2.03
30	7.56	5.39	4.51	4.02	3.70	3.47	3.30	3.17	3.07	2.98	2.84	2.70	2.55	2.47	2.39	2.30	2.21	2.11	2.01
40	7.31	5.18	4.31	3.83	3.51	3.29	3.12	2.99	2.89	2.80	2.66	2.52	2.37	2.29	2.20	2.11	2.02	1.92	1.80
60	7.08	4.98	4.13	3.65	3.34	3.12	2.95	2.82	2.72	2.63	2.50	2.35	2.20	2.12	2.03	1.94	1.84	1.73	1.60
120	6.85	4.79	3.95	3.48	3.17	2.96	2.79	2.66	2.56	2.47	2.34	2.19	2.03	1.95	1.86	1.76	1.66	1.53	1.38
∞	6.63	4.61	3.78	3.32	3.02	2.80	2.64	2.51	2.41	2.32	2.18	2.04	1.88	1.79	1.70	1.59	1.47	1.32	1.00

Table A.10 Lower Tail Critical Values for Wilcoxon's Signed-Rank Statistic

$T_{n,\alpha}$ such that $\text{pr}(T \leqslant T_{n,\alpha}) \leqslant \alpha \qquad n = n_+ + n_-$

	Nominal α			
n	0.05	0.025	0.01	0.005
5	0	-	-	-
6	2	0	-	-
7	3	2	0	-
8	5	3	1	0
9	8	5	3	1
10	10	8	5	3
11	13	10	7	5
12	17	13	9	7
13	21	17	12	9
14	25	21	15	12
15	30	25	19	15
16	35	29	23	19
17	41	34	27	23
18	47	40	32	27
19	53	46	37	32
20	60	52	43	37
21	67	58	49	42
22	75	65	55	48
23	83	73	62	54
24	91	81	69	61
25	100	89	76	68
26	110	98	84	75
27	119	107	92	83
28	130	116	101	91
29	140	126	110	100
30	151	137	120	109
31	163	147	130	118
32	175	159	140	128
33	187	170	151	138
34	200	182	162	148
35	213	195	173	159
36	227	208	185	171
37	241	221	198	182
38	256	235	211	194
39	271	249	224	207
40	286	264	238	220

Source: Reprinted with permission from James V. Bradley, *Distribution-Free Statistical Tests* (Englewood Cliffs, N.J.: Prentice-Hall, Inc., 1968), pp. 315–317.

Table A.11 Lower Tail Critical Values for the Mann-Whitney U-Statistic

$U_{n_1, n_2, \alpha}$ such that $\mathrm{pr}(U \leqslant U_{n_1, n_2, \alpha}) \leqslant \alpha$

$(n_1 \leqslant n_2)$

(n_1, n_2)	$\alpha = 0.10$	$\alpha = 0.05$	$\alpha = 0.025$	$\alpha = 0.01$	$\alpha = 0.005$
(2,3)	0	-	-	-	-
(3,3)	1	0	-	-	-
(2,4)	0	-	-	-	-
(3,4)	1	0	-	-	-
(4,4)	3	1	0	-	-
(2,5)	1	0	-	-	-
(3,5)	2	1	0	-	-
(4,5)	4	2	1	0	-
(5,5)	5	4	2	1	0
(2,6)	1	0	-	-	-
(3,6)	3	2	1	-	-
(4,6)	5	3	2	1	0
(5,6)	7	5	3	2	1
(6,6)	9	7	5	3	2
(2,7)	1	0	-	-	-
(3,7)	4	2	1	0	-
(4,7)	6	4	3	1	0
(5,7)	8	6	5	3	1
(6,7)	11	8	6	4	3
(7,7)	13	11	8	6	4
(2,8)	2	1	0	-	-
(3,8)	5	3	2	0	-
(4,8)	7	5	4	2	1
(5,8)	10	8	6	4	2
(6,8)	13	10	8	6	4
(7,8)	16	13	10	7	6
(8,8)	19	15	13	9	7
(1,9)	0	-	-	-	-
(2,9)	2	1	0	-	-
(3,9)	5	4	2	1	0
(4,9)	9	6	4	3	1
(5,9)	12	9	7	5	3
(6,9)	15	12	10	7	5
(7,9)	18	15	12	9	7
(8,9)	22	18	15	11	9
(9,9)	25	20	17	14	11
(1,10)	0	-	-	-	-
(2,10)	3	1	0	-	-
(3,10)	6	4	3	1	0
(4,10)	10	7	5	3	2
(5,10)	13	11	8	6	4
(6,10)	17	14	11	8	6
(7,10)	21	17	14	11	9
(8,10)	24	20	17	13	11
(9,10)	28	24	20	16	13
(10,10)	32	27	23	19	16

Source: Reprinted with permission from W. H. Beyer, *Handbook of Tables for Probability and Statistics* (Boca Raton, Fl.: The Chemical Rubber Co., CRC Press, Inc., 1966), pp. 167–172.

Table A.12 Critical Values for the Total Number of Runs

$R_{n_1, n_2, \alpha}$ such that $\text{pr}(R \leqslant R_{n_1, n_2, \alpha}) \leqslant \alpha$ and
$R_{n_1, n_2, 1-\alpha}$ such that $\text{pr}(R \geqslant R_{n_1, n_2, 1-\alpha}) \leqslant \alpha$

(n_1, n_2)	$\alpha = 0.025$	$\alpha = 0.05$	$\alpha = 0.10$	$\alpha = 0.90$	$\alpha = 0.95$	$\alpha = 0.975$
(2,3)	-	-	-	5	-	-
(2,4)	-	-	-	-	-	-
(2,5)	-	-	2	-	-	-
(2,6)	-	-	2	-	-	-
(2,7)	-	-	2	-	-	-
(2,8)	-	2	2	-	-	-
(2,9)	-	2	2	-	-	-
(2,10)	-	2	2	-	-	-
(3,3)	-	-	2	6	-	-
(3,4)	-	-	2	7	7	-
(3,5)	-	2	2	7	-	-
(3,6)	2	2	2	-	-	-
(3,7)	2	2	3	-	-	-
(3,8)	2	2	3	-	-	-
(3,9)	2	2	3	-	-	-
(3,10)	2	3	3	-	-	-
(4,4)	-	2	2	8	8	-
(4,5)	2	2	3	8	9	9
(4,6)	2	3	3	9	9	9
(4,7)	2	3	3	9	9	-
(4,8)	3	3	3	9	-	-
(4,9)	3	3	4	9	-	-
(4,10)	3	3	4	-	-	-
(5,5)	2	3	3	9	9	10
(5,6)	3	3	3	9	10	10
(5,7)	3	3	4	10	11	11
(5,8)	3	3	4	10	11	11
(5,9)	3	4	4	10	11	-
(5,10)	3	4	5	11	11	-
(6,6)	3	3	4	10	11	11
(6,7)	3	4	4	11	11	12
(6,8)	3	4	5	11	12	12
(6,9)	4	4	5	11	12	13
(6,10)	4	5	5	12	12	13
(7,7)	4	4	5	11	12	12
(7,8)	4	4	5	12	13	13
(7,9)	4	5	5	12	13	13
(7,10)	5	5	6	13	13	14
(8,8)	4	5	6	12	13	14
(8,9)	5	5	6	13	14	14
(8,10)	5	6	6	13	14	15
(9,9)	5	6	6	14	14	15
(9,10)	5	6	7	14	15	16
(10,10)	6	6	7	15	16	16

Source: Reprinted with permission from W. H. Beyer, *Handbook of Tables for Probability and Statistics* (Boca Raton, Fl.: The Chemical Rubber Co., CRC Press, Inc., 1966), pp. 174–175.

Table A.13 Critical Values for the Kolmogorov-Smirnov One-Sample Statistic

Sample Size (n)	Level of significance for $D =$ maximum $\lvert F_0(X) - S_n(X)\rvert$				
	0.20	0.15	0.10	0.05	0.01
1	.900	.925	.950	.975	.995
2	.684	.726	.776	.842	.929
3	.565	.597	.642	.708	.828
4	.494	.525	.564	.624	.733
5	.446	.474	.510	.565	.669
6	.410	.436	.470	.521	.618
7	.381	.405	.438	.486	.577
8	.358	.381	.411	.457	.543
9	.339	.360	.388	.432	.514
10	.322	.342	.368	.410	.490
11	.307	.326	.352	.391	.468
12	.295	.313	.338	.375	.450
13	.284	.302	.325	.361	.433
14	.274	.292	.314	.349	.418
15	.266	.283	.304	.338	.404
16	.258	.274	.295	.328	.392
17	.250	.266	.286	.318	.381
18	.244	.259	.278	.309	.371
19	.237	.252	.272	.301	.363
20	.231	.246	.264	.294	.356
25	.21	.22	.24	.27	.32
30	.19	.20	.22	.24	.29
35	.18	.19	.21	.23	.27
Over 35	$\frac{1.07}{\sqrt{N}}$	$\frac{1.14}{\sqrt{N}}$	$\frac{1.22}{\sqrt{N}}$	$\frac{1.36}{\sqrt{N}}$	$\frac{1.63}{\sqrt{N}}$

Source: Reprinted with permission from W. H. Beyer, *Handbook of Tables for Probability and Statistics* (Boca Raton, Fl.: The Chemical Rubber Co., CRC Press, Inc., 1966), p. 178.

Table A.14 Critical Values for Friedman's Statistic

$S_{c,n,\alpha}$ such that $\mathrm{pr}(S \geqslant S_{c,n,\alpha}) \leqslant \alpha$

(c, n)	$\alpha = 0.10$	$\alpha = 0.05$	$\alpha = 0.01$	$\alpha = 0.005$
(3,3)	18	18	-	-
(3,4)	24	26	32	-
(3,5)	26	32	42	50
(3,6)	32	42	54	72
(3,7)	38	50	62	86
(3,8)	42	50	72	98
(3,9)	50	56	78	114
(3,10)	50	62	96	122
(3,11)	54	72	104	146
(3,12)	62	74	114	150
(3,13)	62	78	122	168
(3,14)	72	86	126	186
(3,15)	74	96	134	194
(4,2)	20	20	-	-
(4,3)	33	37	45	-
(4,4)	42	52	64	74
(4,5)	53	65	83	105
(4,6)	64	76	102	128
(4,7)	75	91	121	161
(4,8)	84	102	138	184

Source: Reprinted with permission from D. B. Owen, *Handbook of Statistical Tables* (Reading, MA.: Addison-Wesley Press, Inc., 1962), pp. 407–419.

Table A.15 Upper Tail Critical Values for Spearman's Rank Correlation Coefficient

$r_{n,\alpha}$ such that $\text{pr}(r \geqslant r_{n,\alpha}) \leqslant \alpha$

n	$\alpha = 0.05$	$\alpha = 0.025$	$\alpha = 0.01$	$\alpha = 0.005$
5	.900	-	-	-
6	.829	.886	.943	-
7	.714	.786	.893	-
8	.643	.738	.833	.881
9	.600	.683	.783	.833
10	.564	.648	.745	.794
11	.523	.623	.736	.818
12	.497	.591	.703	.780
13	.475	.566	.673	.745
14	.457	.545	.646	.716
15	.441	.525	.623	.689
16	.425	.507	.601	.666
17	.412	.490	.582	.645
18	.399	.476	.564	.625
19	.388	.462	.549	.608
20	.377	.450	.534	.591
21	.368	.438	.521	.576
22	.359	.428	.508	.562
23	.351	.418	.496	.549
24	.343	.409	.485	.537
25	.336	.400	.475	.526
26	.329	.392	.465	.515
27	.323	.385	.456	.505
28	.317	.377	.448	.496
29	.311	.370	.440	.487
30	.305	.364	.432	.478

Source: Reprinted with permission from W. H. Beyer, *Handbook of Tables for Probability and Statistics* (Boca Raton, Fl.: The Chemical Rubber Co., CRC Press, Inc., 1966), p. 186.

Table A.16 Percentage Points for Duncan's Multiple Range Statistic

$\alpha = 0.01$

n_2	Ordered Pairs										
	2	3	4	5	6	7	8	9	10	20	50
1	90.00	90.00	90.00	90.00	90.00	90.00	90.00	90.00	90.00	90.00	90.00
2	14.00	14.00	14.00	14.00	14.00	14.00	14.00	14.00	14.00	14.00	14.00
3	8.26	8.50	8.60	8.70	8.80	8.90	8.90	9.00	9.00	9.30	9.30
4	6.51	6.80	6.90	7.00	7.10	7.10	7.20	7.20	7.30	7.50	7.50
5	5.70	5.96	6.11	6.18	6.26	6.33	6.40	6.44	6.50	6.80	6.80
6	5.24	5.51	5.65	5.73	5.81	5.88	5.95	6.00	6.00	6.30	6.30
7	4.95	5.22	5.37	5.45	5.53	5.61	5.69	5.73	5.80	6.00	6.00
8	4.74	5.00	5.14	5.23	5.32	5.40	5.47	5.51	5.50	5.80	5.80
9	4.60	4.86	4.99	5.08	5.17	5.25	5.32	5.36	5.40	5.70	5.70
10	4.48	4.73	4.88	4.96	5.06	5.13	5.20	5.24	5.28	5.55	5.55
11	4.39	4.63	4.77	4.86	4.94	5.01	5.06	5.12	5.15	5.39	5.39
12	4.32	4.55	4.68	4.76	4.84	4.92	4.96	5.02	5.07	5.26	5.26
13	4.26	4.48	4.62	4.69	4.74	4.84	4.88	4.94	4.98	5.15	5.15
14	4.21	4.42	4.55	4.63	4.70	4.78	4.83	4.87	4.91	5.07	5.07
15	4.17	4.37	4.50	4.58	4.64	4.72	4.77	4.81	4.84	5.00	5.00
16	4.13	4.34	4.45	4.54	4.60	4.67	4.72	4.76	4.79	4.94	4.94
17	4.10	4.30	4.41	4.50	4.56	4.63	4.68	4.73	4.75	4.89	4.89
18	4.07	4.27	4.38	4.46	4.53	4.59	4.64	4.68	4.71	4.85	4.85
19	4.05	4.24	4.35	4.43	4.50	4.56	4.61	4.64	4.67	4.82	4.82
20	4.02	4.22	4.33	4.40	4.47	4.53	4.58	4.61	4.65	4.79	4.79
30	3.89	4.06	4.16	4.22	4.32	4.36	4.41	4.45	4.48	4.65	4.71
40	3.82	3.99	4.10	4.17	4.24	4.30	4.34	4.37	4.41	4.59	4.69
60	3.76	3.92	4.03	4.12	4.17	4.23	4.27	4.31	4.34	4.53	4.66
100	3.71	3.86	3.98	4.06	4.11	4.17	4.21	4.25	4.29	4.48	4.64
∞	3.64	3.80	3.90	3.98	4.04	4.09	4.14	4.17	4.20	4.41	4.60

$\alpha = 0.05$

n_2	Ordered Pairs										
	2	3	4	5	6	7	8	9	10	20	50
1	18.00	18.00	18.00	18.00	18.00	18.00	18.00	18.00	18.00	18.00	18.00
2	6.09	6.09	6.09	6.09	6.09	6.09	6.09	6.09	6.09	6.09	6.09
3	4.50	4.50	4.50	4.50	4.50	4.50	4.50	4.50	4.50	4.50	4.50
4	3.93	4.01	4.02	4.02	4.02	4.02	4.02	4.02	4.02	4.02	4.02
5	3.64	3.74	3.79	3.83	3.83	3.83	3.83	3.83	3.83	3.83	3.83
6	3.46	3.58	3.64	3.68	3.68	3.68	3.68	3.68	3.68	3.68	3.68
7	3.35	3.47	3.54	3.58	3.60	3.61	3.61	3.61	3.61	3.61	3.61
8	3.26	3.39	3.47	3.52	3.55	3.56	3.56	3.56	3.56	3.56	3.56
9	3.20	3.34	3.41	3.47	3.50	3.52	3.52	3.52	3.52	3.52	3.52
10	3.15	3.30	3.37	3.43	3.46	3.47	3.47	3.47	3.47	3.48	3.48
11	3.11	3.27	3.35	3.39	3.43	3.44	3.45	3.46	3.46	3.48	3.48
12	3.08	3.23	3.33	3.36	3.40	3.42	3.44	3.44	3.46	3.48	3.48
13	3.06	3.21	3.30	3.35	3.38	3.41	3.42	3.44	3.45	3.47	3.47
14	3.03	3.18	3.27	3.33	3.37	3.39	3.41	3.42	3.44	3.47	3.47
15	3.01	3.16	3.25	3.31	3.36	3.38	3.40	3.42	3.43	3.47	3.47
16	3.00	3.15	3.23	3.30	3.34	3.37	3.39	3.41	3.43	3.47	3.47
17	2.98	3.13	3.22	3.28	3.33	3.36	3.38	3.40	3.42	3.47	3.47
18	2.97	3.12	3.21	3.27	3.32	3.35	3.37	3.39	3.41	3.47	3.47
19	2.96	3.11	3.19	3.26	3.31	3.35	3.37	3.39	3.41	3.47	3.47
20	2.95	3.10	3.18	3.25	3.30	3.34	3.36	3.38	3.40	3.47	3.47
30	2.89	3.04	3.12	3.20	3.25	3.29	3.32	3.35	3.37	3.47	3.47
40	2.86	3.01	3.10	3.17	3.22	3.27	3.30	3.33	3.35	3.47	3.47
60	2.83	2.98	3.08	3.14	3.20	3.24	3.28	3.31	3.33	3.47	3.47
100	2.80	2.95	3.05	3.12	3.18	3.22	3.26	3.29	3.32	3.47	3.47
∞	2.77	2.92	3.02	3.09	3.15	3.19	3.23	3.26	3.29	3.47	3.47

Source: Reproduced from D. B. Duncan, "Multiple Range and Multiple F Tests." *Biometrics*, 2:1-42. 1955. With permission from The Biometric Society.

Table A.17(a) Percentage Points for the Studentized Range Statistic

$\alpha = 0.05$

n_1	r = 2	3	4	5	6	7	8	9	10	11	12	13	14	15	16	17	18	19	20
1	18.0	27.0	32.8	37.1	40.4	43.1	45.4	47.4	49.1	50.6	52.0	53.2	54.3	55.4	56.3	57.2	58.0	58.8	59.6
2	6.08	8.33	9.80	10.9	11.7	12.4	13.0	13.5	14.0	14.4	14.7	15.1	15.4	15.7	15.9	16.1	16.4	16.6	16.8
3	4.50	5.91	6.82	7.50	8.04	8.48	8.85	9.18	9.46	9.72	9.95	10.2	10.3	10.5	10.7	10.8	11.0	11.1	11.2
4	3.93	5.04	5.76	6.29	6.71	7.05	7.35	7.60	7.83	8.03	8.21	8.37	8.52	8.66	8.79	8.91	9.03	9.13	9.23
5	3.64	4.60	5.22	5.67	6.03	6.33	6.58	6.80	6.99	7.17	7.32	7.47	7.60	7.72	7.83	7.93	8.03	8.12	8.21
6	3.46	4.34	4.90	5.30	5.63	5.90	6.12	6.32	6.49	6.65	6.79	6.92	7.03	7.14	7.24	7.34	7.43	7.51	7.59
7	3.34	4.16	4.68	5.06	5.36	5.61	5.82	6.00	6.16	6.30	6.43	6.55	6.66	6.76	6.85	6.94	7.02	7.10	7.17
8	3.26	4.04	4.53	4.89	5.17	5.40	5.60	5.77	5.92	6.05	6.18	6.29	6.39	6.48	6.57	6.65	6.73	6.80	6.87
9	3.20	3.95	4.41	4.76	5.02	5.24	5.43	5.59	5.74	5.87	5.98	6.09	6.19	6.28	6.36	6.44	6.51	6.58	6.64
10	3.15	3.88	4.33	4.65	4.91	5.12	5.30	5.46	5.60	5.72	5.83	5.93	6.03	6.11	6.19	6.27	6.34	6.40	6.47
11	3.11	3.82	4.26	4.57	4.82	5.03	5.20	5.35	5.49	5.61	5.71	5.81	5.90	5.98	6.06	6.13	6.20	6.27	6.33
12	3.08	3.77	4.20	4.51	4.75	4.95	5.12	5.27	5.39	5.51	5.61	5.71	5.80	5.88	5.95	6.02	6.09	6.15	6.21
13	3.06	3.73	4.15	4.45	4.69	4.88	5.05	5.19	5.32	5.43	5.53	5.63	5.71	5.79	5.86	5.93	5.99	6.05	6.11
14	3.03	3.70	4.11	4.41	4.64	4.83	4.99	5.13	5.25	5.36	5.46	5.55	5.64	5.71	5.79	5.85	5.91	5.97	6.03
15	3.01	3.67	4.08	4.37	4.59	4.78	4.94	5.08	5.20	5.31	5.40	5.49	5.57	5.65	5.72	5.78	5.85	5.90	5.96
16	3.00	3.65	4.05	4.33	4.56	4.74	4.90	5.03	5.15	5.26	5.35	5.44	5.52	5.59	5.66	5.73	5.79	5.84	5.90
17	2.98	3.63	4.02	4.30	4.52	4.70	4.86	4.99	5.11	5.21	5.31	5.39	5.47	5.54	5.61	5.67	5.73	5.79	5.84
18	2.97	3.61	4.00	4.28	4.49	4.67	4.82	4.96	5.07	5.17	5.27	5.35	5.43	5.50	5.57	5.63	5.69	5.74	5.79
19	2.96	3.59	3.98	4.25	4.47	4.65	4.79	4.92	5.04	5.14	5.23	5.31	5.39	5.46	5.53	5.59	5.65	5.70	5.75
20	2.95	3.58	3.96	4.23	4.45	4.62	4.77	4.90	5.01	5.11	5.20	5.28	5.36	5.43	5.49	5.55	5.61	5.66	5.71
24	2.92	3.53	3.90	4.17	4.37	4.54	4.68	4.81	4.92	5.01	5.10	5.18	5.25	5.32	5.38	5.44	5.49	5.55	5.59
30	2.89	3.49	3.85	4.10	4.30	4.46	4.60	4.72	4.82	4.92	5.00	5.08	5.15	5.21	5.27	5.33	5.38	5.43	5.47
40	2.86	3.44	3.79	4.04	4.23	4.39	4.52	4.63	4.73	4.82	4.90	4.98	5.04	5.11	5.16	5.22	5.27	5.31	5.36
60	2.83	3.40	3.74	3.98	4.16	4.31	4.44	4.55	4.65	4.73	4.81	4.88	4.94	5.00	5.06	5.11	5.15	5.20	5.24
120	2.80	3.36	3.68	3.92	4.10	4.24	4.36	4.47	4.56	4.64	4.71	4.78	4.84	4.90	4.95	5.00	5.04	5.09	5.13
∞	2.77	3.31	3.63	3.86	4.03	4.17	4.29	4.39	4.47	4.55	4.62	4.68	4.74	4.80	4.85	4.89	4.93	4.97	5.01

Source: Reprinted with permission from Henry Scheffe, *The Analysis of Variance* (New York: John Wiley & Sons, Inc., 1959), pp. 434–436.

Table A.17(b) Percentage Points for the Studentized Range Statistic

$\alpha = 0.01$

n_2 \ r	2	3	4	5	6	7	8	9	10	11	12	13	14	15	16	17	18	19	20
1	90.0	135	164	186	202	216	227	237	246	253	260	266	272	277	282	286	290	294	298
2	14.0	19.0	22.3	24.7	26.6	28.2	29.5	30.7	31.7	32.6	33.4	34.1	34.8	35.4	36.0	36.5	37.0	37.5	37.9
3	8.26	10.6	12.2	13.3	14.2	15.0	15.6	16.2	16.7	17.1	17.5	17.9	18.2	18.5	18.8	19.1	19.3	19.5	19.8
4	6.51	8.12	9.17	9.96	10.6	11.1	11.5	11.9	12.3	12.6	12.8	13.1	13.3	13.5	13.7	13.9	14.1	14.2	14.4
5	5.70	6.97	7.80	8.42	8.91	9.32	9.67	9.97	10.2	10.5	10.7	10.9	11.1	11.2	11.4	11.6	11.7	11.8	11.9
6	5.24	6.33	7.03	7.56	7.97	8.32	8.61	8.87	9.10	9.30	9.49	9.65	9.81	9.95	10.1	10.2	10.3	10.4	10.5
7	4.95	5.92	6.54	7.01	7.37	7.68	7.94	8.17	8.37	8.55	8.71	8.86	9.00	9.12	9.24	9.35	9.46	9.55	9.65
8	4.74	5.63	6.20	6.63	6.96	7.24	7.47	7.68	7.87	8.03	8.18	8.31	8.44	8.55	8.66	8.76	8.85	8.94	9.03
9	4.60	5.43	5.96	6.35	6.66	6.91	7.13	7.32	7.49	7.65	7.78	7.91	8.03	8.13	8.23	8.32	8.41	8.49	8.57
10	4.48	5.27	5.77	6.14	6.43	6.67	6.87	7.05	7.21	7.36	7.48	7.60	7.71	7.81	7.91	7.99	8.07	8.15	8.22
11	4.39	5.14	5.62	5.97	6.25	6.48	6.67	6.84	6.99	7.13	7.25	7.36	7.46	7.56	7.65	7.73	7.81	7.88	7.95
12	4.32	5.04	5.50	5.84	6.10	6.32	6.51	6.67	6.81	6.94	7.06	7.17	7.26	7.36	7.44	7.52	7.59	7.66	7.73
13	4.26	4.96	5.40	5.73	5.98	6.19	6.37	6.53	6.67	6.79	6.90	7.01	7.10	7.19	7.27	7.34	7.42	7.48	7.55
14	4.21	4.89	5.32	5.63	5.88	6.08	6.26	6.41	6.54	6.66	6.77	6.87	6.96	7.05	7.12	7.20	7.27	7.33	7.39
15	4.17	4.83	5.25	5.56	5.80	5.99	6.16	6.31	6.44	6.55	6.66	6.76	6.84	6.93	7.00	7.07	7.14	7.20	7.26
16	4.13	4.78	5.19	5.49	5.72	5.92	6.08	6.22	6.35	6.46	6.56	6.66	6.74	6.82	6.90	6.97	7.03	7.09	7.15
17	4.10	4.74	5.14	5.43	5.66	5.85	6.01	6.15	6.27	6.38	6.48	6.57	6.66	6.73	6.80	6.87	6.94	7.00	7.05
18	4.07	4.70	5.09	5.38	5.60	5.79	5.94	6.08	6.20	6.31	6.41	6.50	6.58	6.65	6.72	6.79	6.85	6.91	6.96
19	4.05	4.67	5.05	5.33	5.55	5.73	5.89	6.02	6.14	6.25	6.34	6.43	6.51	6.58	6.65	6.72	6.78	6.84	6.89
20	4.02	4.64	5.02	5.29	5.51	5.69	5.84	5.97	6.09	6.19	6.29	6.37	6.45	6.52	6.59	6.65	6.71	6.76	6.82
24	3.96	4.54	4.91	5.17	5.37	5.54	5.69	5.81	5.92	6.02	6.11	6.19	6.26	6.33	6.39	6.45	6.51	6.56	6.61
30	3.89	4.45	4.80	5.05	5.24	5.40	5.54	5.65	5.76	5.85	5.93	6.01	6.08	6.14	6.20	6.26	6.31	6.36	6.41
40	3.82	4.37	4.70	4.93	5.11	5.27	5.39	5.50	5.60	5.69	5.77	5.84	5.90	5.96	6.02	6.07	6.12	6.17	6.21
60	3.76	4.28	4.60	4.82	4.99	5.13	5.25	5.36	5.45	5.53	5.60	5.67	5.73	5.79	5.84	5.89	5.93	5.98	6.02
120	3.70	4.20	4.50	4.71	4.87	5.01	5.12	5.21	5.30	5.38	5.44	5.51	5.56	5.61	5.66	5.71	5.75	5.79	5.83
∞	3.64	4.12	4.40	4.60	4.76	4.88	4.99	5.08	5.16	5.23	5.29	5.35	5.40	5.45	5.49	5.54	5.57	5.61	5.65

Source: Reprinted with permission from Henry Scheffe, *The Analysis of Variance* (New York: John Wiley & Sons, Inc., 1959), pp. 434–436.

Index

ABCDEFGHIJ-BP-89876543